P9-DIG-081

No Longer the Property of
Hayner Public Library District

MAIN

REFERENCE

Grzimek's
Animal Life Encyclopedia

Second Edition

••••

Grzimek's
Animal Life Encyclopedia

Second Edition

●●●●

Volume 8
Birds I

Jerome A. Jackson, Advisory Editor
Walter J. Bock, Taxonomic Editor
Donna Olendorf, Project Editor

Joseph E. Trumpey, Chief Scientific Illustrator

Michael Hutchins, Series Editor
In association with the American Zoo and Aquarium Association

GALE®

THOMSON
★
GALE

Detroit • New York • San Diego • San Francisco • Cleveland • New Haven, Conn. • Waterville, Maine • London • Munich

HAYNER PUBLIC LIBRARY DISTRICT
ALTON, ILLINOIS

THOMSON

GALE

Grzimek's Animal Life Encyclopedia, Second Edition
Volume 8: Birds I

Project Editor
Donna Olendorf

Editorial
Deirdre Blanchfield, Madeline Harris, Christine Jeryan, Kristine M. Krapp, Kate Kretschmann, Melissa C. McDade, Mark Springer

Permissions
Kim Davis

Imaging and Multimedia
Mary K. Grimes, Lezlie Light, Christine O'Bryan, Barbara Yarrow, Robyn V. Young

Product Design
Tracey Rowens, Jennifer Wahi

Manufacturing
Dorothy Maki, Evi Seoud, Mary Beth Trimper

© 2003 by Gale. Gale is an imprint of The Gale Group, Inc., a division of Thomson Learning Inc.

Gale and Design™ and Thomson Learning™ are trademarks used herein under license.

For more information, contact
The Gale Group, Inc.
27500 Drake Rd.
Farmington Hills, MI 48331–3535
Or you can visit our Internet site at
http://www.gale.com

ALL RIGHTS RESERVED
No part of this work covered by the copyright hereon may be reproduced or used in any form or by any means—graphic, electronic, or mechanical, including photocopying, recording, taping, Web distribution, or information storage retrieval systems—without the written permission of the publisher.

For permission to use material from this product, submit your request via Web at http://www.gale-edit.com/permissions, or you may download our Permissions Request form and submit your request by fax or mail to: The Gale Group, Inc., Permissions Department, 27500 Drake Road, Farmington Hills, MI, 48331-3535, Permissions hotline: 248-699-8074 or 800-877-4253, ext. 8006, Fax: 248-699-8074 or 800-762-4058.

Cover photo of great egret (*Casmerodius albus*) by M.H. Sharp, Photo Researchers, Inc. Back cover photos of sea anemone by AP/Wide World Photos/University of Wisconsin-Superior; land snail, lionfish, golden frog, and green python by JLM Visuals; red-legged locust © 2001 Susan Sam; hornbill by Margaret F. Kinnaird; and tiger by Jeff Lepore/Photo Researchers. All reproduced by permission.

While every effort has been made to ensure the reliability of the information presented in this publication, The Gale Group, Inc. does not guarantee the accuracy of the data contained herein. The Gale Group, Inc. accepts no payment for listing; and inclusion in the publication of any organization, agency, institution, publication, service, or individual does not imply endorsement of the editors and publisher. Errors brought to the attention of the publisher and verified to the satisfaction of the publisher will be corrected in future editions.

ISBN 0-7876-5362-4 (vols. 1-17 set)
0-7876-6571-1 (vols. 8-11 set)
0-7876-5784-0 (vol. 8)
0-7876-5785-9 (vol. 9)
0-7876-5786-7 (vol. 10)
0-7876-5787-5 (vol. 11)

LIBRARY OF CONGRESS CATALOGING-IN-PUBLICATION DATA

Grzimek, Bernhard.
   [Tierleben. English]
   Grzimek's animal life encyclopedia.— 2nd ed.
      v. cm.
Includes bibliographical references.
Contents: v. 1. Lower metazoans and lesser deuterosomes / Neil Schlager, editor — v. 2. Protostomes / Neil Schlager, editor — v. 3. Insects / Neil Schlager, editor — v. 4-5. Fishes I-II / Neil Schlager, editor — v. 6. Amphibians / Neil Schlager, editor — v. 7. Reptiles / Neil Schlager, editor — v. 8-11. Birds I-IV / Donna Olendorf, editor — v. 12-16. Mammals I-V / Melissa C. McDade, editor — v. 17. Cumulative index / Melissa C. McDade, editor.
ISBN 0-7876-5362-4 (set hardcover : alk. paper)
   1. Zoology—Encyclopedias. I. Title: Animal life encyclopedia. II. Schlager, Neil, 1966- III. Olendorf, Donna IV. McDade, Melissa C. V. American Zoo and Aquarium Association. VI. Title.
QL7 .G7813 2004

590'.3—dc21
2002003351

Printed in Canada
10 9 8 7 6 5 4 3 2 1

Recommended citation: *Grzimek's Animal Life Encyclopedia*, 2nd edition. Volumes 8–11, *Birds I–IV*, edited by Michael Hutchins, Jerome A. Jackson, Walter J. Bock, and Donna Olendorf. Farmington Hills, MI: Gale Group, 2002.

Contents

Contents

. . . . .

Foreword

Earth is teeming with life. No one knows exactly how many distinct organisms inhabit our planet, but more than 5 million different species of animals and plants could exist, ranging from microscopic algae and bacteria to gigantic elephants, redwood trees and blue whales. Yet, throughout this wonderful tapestry of living creatures, there runs a single thread: Deoxyribonucleic acid or DNA. The existence of DNA, an elegant, twisted organic molecule that is the building block of all life, is perhaps the best evidence that all living organisms on this planet share a common ancestry. Our ancient connection to the living world may drive our curiosity, and perhaps also explain our seemingly insatiable desire for information about animals and nature. Noted zoologist, E.O. Wilson, recently coined the term "biophilia" to describe this phenomenon. The term is derived from the Greek *bios* meaning "life" and *philos* meaning "love." Wilson argues that we are human because of our innate affinity to and interest in the other organisms with which we share our planet. They are, as he says, "the matrix in which the human mind originated and is permanently rooted." To put it simply and metaphorically, our love for nature flows in our blood and is deeply engrained in both our psyche and cultural traditions.

Our own personal awakenings to the natural world are as diverse as humanity itself. I spent my early childhood in rural Iowa where nature was an integral part of my life. My father and I spent many hours collecting, identifying and studying local insects, amphibians and reptiles. These experiences had a significant impact on my early intellectual and even spiritual development. One event I can recall most vividly. I had collected a cocoon in a field near my home in early spring. The large, silky capsule was attached to a stick. I brought the cocoon back to my room and placed it in a jar on top of my dresser. I remember waking one morning and, there, perched on the tip of the stick was a large moth, slowly moving its delicate, light green wings in the early morning sunlight. It took my breath away. To my inexperienced eyes, it was one of the most beautiful things I had ever seen. I knew it was a moth, but did not know which species. Upon closer examination, I noticed two moon-like markings on the wings and also noted that the wings had long "tails", much like the ubiquitous tiger swallow-tail butterflies that visited the lilac bush in our backyard. Not wanting to suffer my ignorance any longer, I reached immediately for my *Golden Guide to North*

American Insects and searched through the section on moths and butterflies. It was a luna moth! My heart was pounding with the excitement of new knowledge as I ran to share the discovery with my parents.

I consider myself very fortunate to have made a living as a professional biologist and conservationist for the past 20 years. I've traveled to over 30 countries and six continents to study and photograph wildlife or to attend related conferences and meetings. Yet, each time I encounter a new and unusual animal or habitat my heart still races with the same excitement of my youth. If this is biophilia, then I certainly possess it, and it is my hope that others will experience it too. I am therefore extremely proud to have served as the series editor for the Gale Group's rewrite of *Grzimek's Animal Life Encyclopedia*, one of the best known and widely used reference works on the animal world. *Grzimek's* is a celebration of animals, a snapshot of our current knowledge of the Earth's incredible range of biological diversity. Although many other animal encyclopedias exist, *Grzimek's Animal Life Encyclopedia* remains unparalleled in its size and in the breadth of topics and organisms it covers.

The revision of these volumes could not come at a more opportune time. In fact, there is a desperate need for a deeper understanding and appreciation of our natural world. Many species are classified as threatened or endangered, and the situation is expected to get much worse before it gets better. Species extinction has always been part of the evolutionary history of life; some organisms adapt to changing circumstances and some do not. However, the current rate of species loss is now estimated to be 1,000–10,000 times the normal "background" rate of extinction since life began on Earth some 4 billion years ago. The primary factor responsible for this decline in biological diversity is the exponential growth of human populations, combined with peoples' unsustainable appetite for natural resources, such as land, water, minerals, oil, and timber. The world's human population now exceeds 6 billion, and even though the average birth rate has begun to decline, most demographers believe that the global human population will reach 8–10 billion in the next 50 years. Much of this projected growth will occur in developing countries in Central and South America, Asia and Africa—regions that are rich in unique biological diversity.

Finding solutions to conservation challenges will not be easy in today's human-dominated world. A growing number of people live in urban settings and are becoming increasingly isolated from nature. They "hunt" in super markets and malls, live in apartments and houses, spend their time watching television and searching the World Wide Web. Children and adults must be taught to value biological diversity and the habitats that support it. Education is of prime importance now while we still have time to respond to the impending crisis. There still exist in many parts of the world large numbers of biological "hotspots"—places that are relatively unaffected by humans and which still contain a rich store of their original animal and plant life. These living repositories, along with selected populations of animals and plants held in professionally managed zoos, aquariums and botanical gardens, could provide the basis for restoring the planet's biological wealth and ecological health. This encyclopedia and the collective knowledge it represents can assist in educating people about animals and their ecological and cultural significance. Perhaps it will also assist others in making deeper connections to nature and spreading biophilia. Information on the conservation status, threats and efforts to preserve various species have been integrated into this revision. We have also included information on the cultural significance of animals, including their roles in art and religion.

It was over 30 years ago that Dr. Bernhard Grzimek, then director of the Frankfurt Zoo in Frankfurt, Germany, edited the first edition of *Grzimek's Animal Life Encyclopedia*. Dr. Grzimek was among the world's best known zoo directors and conservationists. He was a prolific author, publishing nine books. Among his contributions were: *Serengeti Shall Not Die*, *Rhinos Belong to Everybody* and *He and I and the Elephants*. Dr. Grzimek's career was remarkable. He was one of the first modern zoo or aquarium directors to understand the importance of zoo involvement in *in situ* conservation, that is, of their role in preserving wildlife in nature. During his tenure, Frankfurt Zoo became one of the leading western advocates and supporters of wildlife conservation in East Africa. Dr. Grzimek served as a Trustee of the National Parks Board of Uganda and Tanzania and assisted in the development of several protected areas. The film he made with his son Michael, *Serengeti Shall Not Die*, won the 1959 Oscar for best documentary.

Professor Grzimek has recently been criticized by some for his failure to consider the human element in wildlife conservation. He once wrote: "A national park must remain a primordial wilderness to be effective. No men, not even native ones, should live inside its borders." Such ideas, although considered politically incorrect by many, may in retrospect actually prove to be true. Human populations throughout Africa continue to grow exponentially, forcing wildlife into small islands of natural habitat surrounded by a sea of humanity. The illegal commercial bushmeat trade—the hunting of endangered wild animals for large scale human consumption—is pushing many species, including our closest relatives, the gorillas, bonobos, and chimpanzees, to the brink of extinction. The trade is driven by widespread poverty and lack of economic alternatives. In order for some species to survive it will be necessary, as Grzimek suggested, to establish and enforce a system of protected areas where wildlife can roam free from exploitation of any kind.

While it is clear that modern conservation must take the needs of both wildlife and people into consideration, what will the quality of human life be if the collective impact of short-term economic decisions is allowed to drive wildlife populations into irreversible extinction? Many rural populations living in areas of high biodiversity are dependent on wild animals as their major source of protein. In addition, wildlife tourism is the primary source of foreign currency in many developing countries and is critical to their financial and social stability. When this source of protein and income is gone, what will become of the local people? The loss of species is not only a conservation disaster; it also has the potential to be a human tragedy of immense proportions. Protected areas, such as national parks, and regulated hunting in areas outside of parks are the only solutions. What critics do not realize is that the fate of wildlife and people in developing countries is closely intertwined. Forests and savannas emptied of wildlife will result in hungry, desperate people, and will, in the long-term lead to extreme poverty and social instability. Dr. Grzimek's early contributions to conservation should be recognized, not only as benefiting wildlife, but as benefiting local people as well.

Dr. Grzimek's hope in publishing his *Animal Life Encyclopedia* was that it would "...disseminate knowledge of the animals and love for them", so that future generations would "...have an opportunity to live together with the great diversity of these magnificent creatures." As stated above, our goals in producing this updated and revised edition are similar. However, our challenges in producing this encyclopedia were more formidable. The volume of knowledge to be summarized is certainly much greater in the twenty-first century than it was in the 1970's and 80's. Scientists, both professional and amateur, have learned and published a great deal about the animal kingdom in the past three decades, and our understanding of biological and ecological theory has also progressed. Perhaps our greatest hurdle in producing this revision was to include the new information, while at the same time retaining some of the characteristics that have made *Grzimek's Animal Life Encyclopedia* so popular. We have therefore strived to retain the series' narrative style, while giving the information more organizational structure. Unlike the original *Grzimek's*, this updated version organizes information under specific topic areas, such as reproduction, behavior, ecology and so forth. In addition, the basic organizational structure is generally consistent from one volume to the next, regardless of the animal groups covered. This should make it easier for users to locate information more quickly and efficiently. Like the original Grzimek's, we have done our best to avoid any overly technical language that would make the work difficult to understand by non-biologists. When certain technical expressions were necessary, we have included explanations or clarifications.

Considering the vast array of knowledge that such a work represents, it would be impossible for any one zoologist to have completed these volumes. We have therefore sought specialists from various disciplines to write the sections with which they are most familiar. As with the original *Grzimek's*, we have engaged the best scholars available to serve as topic editors, writers, and consultants. There were some complaints about inaccuracies in the original English version that may have been due to mistakes or misinterpretation during the complicated translation process. However, unlike the original *Grzimek's*, which was translated from German, this revision has been completely re-written by English-speaking scientists. This work was truly a cooperative endeavor, and I thank all of those dedicated individuals who have written, edited, consulted, drawn, photographed, or contributed to its production in any way. The names of the topic editors, authors, and illustrators are presented in the list of contributors in each individual volume.

The overall structure of this reference work is based on the classification of animals into naturally related groups, a discipline known as taxonomy or biosystematics. Taxonomy is the science through which various organisms are discovered, identified, described, named, classified and catalogued. It should be noted that in preparing this volume we adopted what might be termed a conservative approach, relying primarily on traditional animal classification schemes. Taxonomy has always been a volatile field, with frequent arguments over the naming of or evolutionary relationships between various organisms. The advent of DNA fingerprinting and other advanced biochemical techniques has revolutionized the field and, not unexpectedly, has produced both advances and confusion. In producing these volumes, we have consulted with specialists to obtain the most up-to-date information possible, but knowing that new findings may result in changes at any time. When scientific controversy over the classification of a particular animal or group of animals existed, we did our best to point this out in the text.

Readers should note that it was impossible to include as much detail on some animal groups as was provided on others. For example, the marine and freshwater fish, with vast numbers of orders, families, and species, did not receive as

detailed a treatment as did the birds and mammals. Due to practical and financial considerations, the publishers could provide only so much space for each animal group. In such cases, it was impossible to provide more than a broad overview and to feature a few selected examples for the purposes of illustration. To help compensate, we have provided a few key bibliographic references in each section to aid those interested in learning more. This is a common limitation in all reference works, but *Grzimek's Encyclopedia of Animal Life* is still the most comprehensive work of its kind.

I am indebted to the Gale Group, Inc. and Senior Editor Donna Olendorf for selecting me as Series Editor for this project. It was an honor to follow in the footsteps of Dr. Grzimek and to play a key role in the revision that still bears his name. *Grzimek's Animal Life Encyclopedia* is being published by the Gale Group, Inc. in affiliation with my employer, the American Zoo and Aquarium Association (AZA), and I would like to thank AZA Executive Director, Sydney J. Butler; AZA Past-President Ted Beattie (John G. Shedd Aquarium, Chicago, IL); and current AZA President, John Lewis (John Ball Zoological Garden, Grand Rapids, MI), for approving my participation. I would also like to thank AZA Conservation and Science Department Program Assistant, Michael Souza, for his assistance during the project. The AZA is a professional membership association, representing 205 accredited zoological parks and aquariums in North America. As Director/William Conway Chair, AZA Department of Conservation and Science, I feel that I am a philosophical descendant of Dr. Grzimek, whose many works I have collected and read. The zoo and aquarium profession has come a long way since the 1970s, due, in part, to innovative thinkers such as Dr. Grzimek. I hope this latest revision of his work will continue his extraordinary legacy.

Silver Spring, Maryland, 2001
Michael Hutchins
Series Editor

How to use this book

Grzimek's Animal Life Encyclopedia is an internationally prominent scientific reference compilation, first published in German in the late 1960s, under the editorship of zoologist Bernhard Grzimek (1909–1987). In a cooperative effort between Gale and the American Zoo and Aquarium Association, the series is being completely revised and updated for the first time in over 30 years. Gale is expanding the series from 13 to 17 volumes, commissioning new color images, and updating the information while also making the set easier to use. The order of revisions is:

Vol 8–11: Birds I–IV
Vol 6: Amphibians
Vol 7: Reptiles
Vol 4–5: Fishes I–II
Vol 12–16: Mammals I–V
Vol 1: Lower Metazoans and Lesser Deuterostomes
Vol 2: Protostomes
Vol 3: Insects
Vol 17: Cumulative Index

Organized by order and family

The overall structure of this reference work is based on the classification of animals into naturally related groups, a discipline known as taxonomy—the science through which various organisms are discovered, identified, described, named, classified, and catalogued. Starting with the simplest life forms, the protostomes, in Vol. 1, the series progresses through the more complex animal classes, culminating with the mammals in Vols. 12–16. Volume 17 is a stand-alone cumulative index.

Organization of chapters within each volume reinforces the taxonomic hierarchy. Opening chapters introduce the class of animal, followed by chapters dedicated to order and family. Species accounts appear at the end of family chapters. To help the reader grasp the scientific arrangement, each type of chapter has a distinctive color and symbol:

▲ = Family Chapter (yellow background)

● = Order Chapter (blue background)

◖ = Monotypic Order Chapter (green background)

As chapters narrow in focus, they become more tightly formatted. General chapters have a loose structure, reminiscent of the first edition. While not strictly formatted, order chapters are carefully structured to cover basic information about member families. Monotypic orders, comprised of a single family, utilize family chapter organization. Family chapters are most tightly structured, following a prescribed format of standard rubrics that make information easy to find and understand. Family chapters typically include:

Thumbnail introduction
    Common name
    Scientific name
    Class
    Order
    Suborder
    Family
    Thumbnail description
    Size
    Number of genera, species
    Habitat
    Conservation status
Main essay
    Evolution and systematics
    Physical characteristics
    Distribution
    Habitat
    Behavior
    Feeding ecology and diet
    Reproductive biology
    Conservation status
    Significance to humans
Species accounts
    Common name
    Scientific name
    Subfamily
    Taxonomy
    Other common names
    Physical characteristics
    Distribution
    Habitat
    Behavior
    Feeding ecology and diet
    Reproductive biology

Conservation status
Significance to humans
Resources
 Books
 Periodicals
 Organizations
 Other

Color graphics enhance understanding

Grzimek's features approximately 3,500 color photos, including approximately 480 in four Birds volumes; 3,500 total color maps, including almost 1,500 in the four Birds volumes; and approximately 5,500 total color illustrations, including 1,385 in four Birds volumes. Each featured species of animal is accompanied by both a distribution map and an illustration.

All maps in *Grzimek's* were created specifically for the project by XNR Productions. Distribution information was provided by expert contributors and, if necessary, further researched at the University of Michigan Zoological Museum library. Maps are intended to show broad distribution, not definitive ranges, and are color coded to show resident, breeding, and nonbreeding locations (where appropriate).

All the color illustrations in *Grzimek's* were created specifically for the project by Michigan Science Art. Expert contributors recommended the species to be illustrated and provided feedback to the artists, who supplemented this information with authoritative references and animal skins from University of Michigan Zoological Museum library. In addition to species illustrations, *Grzimek's* features conceptual drawings that illustrate characteristic traits and behaviors.

About the contributors

The essays were written by expert contributors, including ornithologists, curators, professors, zookeepers, and other reputable professionals. *Grzimek's* subject advisors reviewed the completed essays to insure that they are appropriate, accurate, and up-to-date.

Standards employed

In preparing these volumes, the editors adopted a conservative approach to taxonomy, relying primarily on Peters Checklist (1934–1986)—a traditional classification scheme. Taxonomy has always been a volatile field, with frequent arguments over the naming of or evolutionary relationships between various organisms. The advent of DNA fingerprinting and other advanced biochemical techniques has revolutionized the field and, not unexpectedly, has produced both advances and confusion. In producing these volumes, Gale consulted with noted taxonomist Professor Walter J. Bock as well as other specialists to obtain the most up-to-date information possible. When scientific controversy over the classification of a particular animal or group of animals existed, the text makes this clear.

Grzimek's has been designed with ready reference in mind and the editors have standardized information wherever fea-

sible. For ***Conservation status,*** *Grzimek's* follows the IUCN Red List system, developed by its Species Survival Commission. The Red List provides the world's most comprehensive inventory of the global conservation status of plants and animals. Using a set of criteria to evaluate extinction risk, the IUCN recognizes the following categories: Extinct, Extinct in the Wild, Critically Endangered, Endangered, Vulnerable, Conservation Dependent, Near Threatened, Least Concern, and Data Deficient. For a complete explanation of each category, visit the IUCN web page at http://www.iucn.org/themes/ssc/redlists/categor.htm

In addition to IUCN ratings, essays may contain other conservation information, such as a species' inclusion on one of three Convention on International Trade in Endangered Species (CITES) appendices. Adopted in 1975, CITES is a global treaty whose focus is the protection of plant and animal species from unregulated international trade.

Grzimek's provides the following standard information on avian lineage in ***Taxonomy*** rubric of each Species account: [First described as] *Muscicapa rufifrons* [by] Latham, [in] 1801, [based on a specimen from] Sydney, New South Wales, Australia. The person's name and date refer to earliest identification of a species, although the species name may have changed since first identification. However, the organism described is the same.

Other common names in English, French, German, and Spanish are given when an accepted common name is available.

Appendices and index

For further reading directs readers to additional sources of information about birds. Valuable contact information for ***Organizations*** is also included in an appendix. While the encyclopedia minimizes scientific jargon, it also provides a ***Glossary*** at the back of the book to define unfamiliar terms. An exhaustive ***Aves species list*** records all known species of birds, categorized according to Peters Checklist (1934–1986). And a full-color ***Geologic time scale*** helps readers understand prehistoric time periods. Additionally, each of the four volumes contains a full ***Subject index*** for the Birds subset.

Acknowledgements

Gale would like to thank several individuals for their important contributions to the series. Michael Souza, Program Assistant, Department of Conservation and Science, American Zoo and Aquarium Association, provided valuable behind-the-scenes research and reliable support at every juncture of the project. Also deserving of recognition are Christine Sheppard, Curator of Ornithology at Bronx Zoo, and Barry Taylor, professor at the University of Natal, in Pietermaritzburg, South Africa, who assisted subject advisors in reviewing manuscripts for accuracy and currency. And, last but not least, Janet Hinshaw, Bird Division Collection Manager at the University of Michigan Museum of Zoology, who opened her collections to *Grzimek's* artists and staff and also compiled the "For Further Reading" bibliography at the back of the book.

. . . . .

Advisory boards

Series advisor

Michael Hutchins, PhD
Director of Conservation and William Conway Chair
American Zoo and Aquarium Association
Silver Spring, Maryland

Subject advisors

Volume 1: Lower Metazoans and Lesser Deuterostomes
Dennis Thoney, PhD
Director, Marine Laboratory & Facilities
Humboldt State University
Arcata, California

Volume 2: Protostomes
Dennis Thoney, PhD
Director, Marine Laboratory & Facilities
Humboldt State University
Arcata, California

Sean F. Craig, PhD
Assistant Professor, Department of Biological Sciences
Humboldt State University
Arcata, California

Volume 3: Insects
Art Evans, PhD
Entomologist
Richmond, Virginia

Rosser W. Garrison, PhD
Systematic Entomologist, Los Angeles County
Los Angeles, California

Volumes 4–5: Fishes I–II
Paul Loiselle, PhD
Curator, Freshwater Fishes
New York Aquarium
Brooklyn, New York

Dennis Thoney, PhD

Director, Marine Laboratory & Facilities
Humboldt State University
Arcata, California

Volume 6: Amphibians
William E. Duellman, PhD
Curator of Herpetology Emeritus
Natural History Museum and Biodiversity Research Center
University of Kansas
Lawrence, Kansas

Volume 7: Reptiles
James B. Murphy, PhD
Smithsonian Research Associate
Department of Herpetology
National Zoological Park
Washington, DC

Volumes 8–11: Birds I–IV
Walter J. Bock, PhD
Permanent secretary, International Ornithological Congress
Professor of Evolutionary Biology
Department of Biological Sciences,
Columbia University
New York, New York

Jerome A. Jackson, PhD
Program Director, Whitaker Center for Science,
Mathematics, and Technology Education
Florida Gulf Coast University
Ft. Myers, Florida

Volumes 12–16: Mammals I–V
Valerius Geist, PhD
Professor Emeritus of Environmental Science
University of Calgary
Calgary, Alberta
Canada

Devra Gail Kleiman, PhD
Smithsonian Research Associate
National Zoological Park
Washington, DC

Advisory boards

Library advisors

James Bobick
Head, Science & Technology Department
Carnegie Library of Pittsburgh
Pittsburgh, Pennsylvania

Linda L. Coates
Associate Director of Libraries
Zoological Society of San Diego Library
San Diego, California

Lloyd Davidson, PhD
Life Sciences bibliographer and head, Access Services
Seeley G. Mudd Library for Science and Engineering
Evanston, Illinois

Thane Johnson
Librarian
Oaklahoma City Zoo
Oaklahoma City, Oklahoma

Charles Jones
Library Media Specialist
Plymouth Salem High School
Plymouth, Michigan

Ken Kister
Reviewer/General Reference teacher
Tampa, Florida

Richard Nagler
Reference Librarian
Oakland Community College
Southfield Campus
Southfield, Michigan

Roland Person
Librarian, Science Division
Morris Library
Southern Illinois University
Carbondale, Illinois

Contributing writers

Birds I–IV

Michael Abs, Dr. rer. nat.
Berlin, Germany

George William Archibald, PhD
International Crane Foundation
Baraboo, Wisconsin

Helen Baker, PhD
Joint Nature Conservation Committee
Peterborough, Cambridgeshire
United Kingdom

Cynthia Ann Berger, MS
Pennsylvania State University
State College, Pennsylvania

Matthew A. Bille, MSc
Colorado Springs, Colorado

Walter E. Boles, PhD
Australian Museum
Sydney, New South Wales
Australia

Carlos Bosque, PhD
Universidad Simón Bolívar
Caracas, Venezuela

David Brewer, PhD
Research Associate
Royal Ontario Museum
Toronto, Ontario
Canada

Daniel M. Brooks, PhD
Houston Museum of Natural Science
Houston, Texas

Donald F. Bruning, PhD
Wildlife Conservation Society
Bronx, New York

Joanna Burger, PhD
Rutgers University
Piscataway, New Jersey

Carles Carboneras
SEO/BirdLife
Barcelona, Spain

John Patrick Carroll, PhD
University of Georgia
Athens, Georgia

Robert Alexander Cheke, PhD
Natural Resources Institute
University of Greenwich
Chatham, Kent
United Kingdom

Jay Robert Christie, MBA
Racine Zoological Gardens
Racine, Wisconsin

Charles T. Collins, PhD
California State University
Long Beach, California

Malcolm C. Coulter, PhD
IUCN Specialist Group on Storks,
Ibises and Spoonbills
Chocorua, New Hampshire

Adrian Craig, PhD
Rhodes University
Grahamstown, South Africa

Francis Hugh John Crome, BSc
Consultant
Atheron, Queensland
Australia

Timothy Michael Crowe, PhD
University of Cape Town
Rondebosch, South Africa

H. Sydney Curtis, BSc
Queensland National Parks &
Wildlife Service (Retired)
Brisbane, Queensland
Australia

S. J. J. F. Davies, ScD
Curtin University of Technology
Department of Environmental Biology
Perth, Western Australia
Australia

Gregory J. Davis, PhD
University of Wisconsin-Green Bay
Green Bay, Wisconsin

William E. Davis, Jr., PhD
Boston University
Boston, Massachusetts

Stephen Debus, MSc
University of New England
Armidale, New South Wales
Australia

Michael Colin Double, PhD
Australian National University
Canberra, A.C.T.
Australia

Rachel Ehrenberg, MS
University of Michigan
Ann Arbor, Michigan

Contributing writers

Eladio M. Fernandez
Santo Domingo
Dominican Republic

Simon Ferrier, PhD
New South Wales National Parks and
Wildlife Service
Armidale, New South Wales
Australia

Kevin F. Fitzgerald, BS
South Windsor, Connecticut

Hugh Alastair Ford, PhD
University of New England
Armidale, New South Wales
Australia

Joseph M. Forshaw
Australian Museum
Sydney, New South Wales
Australia

Bill Freedman, PhD
Department of Biology
Dalhousie University
Halifax, Nova Scotia
Canada

Clifford B. Frith, PhD
Honorary research fellow
Queensland Museum
Brisbane, Australia

Dawn W. Frith, PhD
Honorary research fellow
Queensland Museum
Brisbane, Australia

Peter Jeffery Garson, DPhil
University of Newcastle
Newcastle upon Tyne
United Kingdom

Michael Gochfeld, PhD, MD
UMDNJ-Robert Wood Johnson
Medical School
Piscataway, New Jersey

Michelle L. Hall, PhD
Australian National University
School of Botany and Zoology
Canberra, A.C.T.
Australia

Frank Hawkins, PhD
Conservation International
Antananarivo, Madagascar

David G. Hoccom, BSc
Royal Society for the Protection of
Birds
Sandy, Bedfordshire
United Kingdom

Peter Andrew Hosner
Cornell University
Ithaca, New York

Brian Douglas Hoyle PhD
Bedford, Nova Scotia
Canada

Julian Hughes
Royal Society for the Protection of
Birds
Sandy, Bedfordshire
United Kingdom

Robert Arthur Hume, BA
Royal Society for the Protection of
Birds
Sandy, Bedfordshire
United Kingdom

Gavin Raymond Hunt, PhD
University of Auckland
Auckland, New Zealand

Jerome A. Jackson, PhD
Florida Gulf Coast University
Ft. Myers, Florida

Bette J. S. Jackson, PhD
Florida Gulf Coast University
Ft. Myers, Florida

Darryl N. Jones, PhD
Griffith University
Queensland, Australia

Alan C. Kemp, PhD
Naturalists & Nomads
Pretoria, South Africa

Angela Kay Kepler, PhD
Pan-Pacific Ecological Consulting
Maui, Hawaii

Jiro Kikkawa, DSc
Professor Emeritus
University of Queensland,
Brisbane, Queensland
Australia

Margaret Field Kinnaird, PhD
Wildlife Conservation Society
Bronx, New York

Guy M. Kirwan, BA
Ornithological Society of the Middle
East
Sandy, Bedfordshire
United Kingdom

Melissa Knopper, MS
Denver Colorado

Niels K. Krabbe, PhD
University of Copenhagen
Copenhagen, Denmark

James A. Kushlan, PhD
U.S. Geological Survey
Smithsonian Environmental Research
Center
Edgewater, Maryland

Norbert Lefranc, PhD
Ministère de l'Environnement,
Direction Régionale
Metz, France

P. D. Lewis, BS
Jacksonville Zoological Gardens
Jacksonville, Florida

Josef H. Lindholm III, BA
Cameron Park Zoo
Waco, Texas

Peter E. Lowther, PhD
Field Museum
Chicago, Illinois

Gordon Lindsay Maclean, PhD, DSc
Rosetta, South Africa

Steve Madge
Downderry, Torpoint
Cornwall
United Kingdom

Albrecht Manegold
Institut für Biologie/Zoologie
Berlin, Germany

Jeffrey S. Marks, PhD
University of Montana
Missoula, Montana

Juan Gabriel Martínez, PhD
Universidad de Granada
Departamento de Biologia
Animal y Ecologia
Granada, Spain

Barbara Jean Maynard, PhD
Laporte, Colorado

Cherie A. McCollough, MS
PhD candidate, University of Texas
Austin, Texas

Leslie Ann Mertz, PhD
Fish Lake Biological Program
Wayne State University
Biological Station
Lapeer, Michigan

Derek William Niemann, BA
Royal Society for the Protection of
Birds
Sandy, Bedfordshire
United Kingdom

Malcolm Ogilvie, PhD
Glencairn, Bruichladdich
Isle of Islay
United Kingdom

Penny Olsen, PhD
Australian National University
Canberra, A.C.T.
Australia

Jemima Parry-Jones, MBE
National Birds of Prey Centre
Newent, Gloucestershire
United Kingdom

Colin Pennycuick, PhD, FRS
University of Bristol
Bristol, United Kingdom

James David Rising, PhD
University of Toronto
Department of Zoology
Toronto, Ontario
Canada

Christopher John Rutherford Robertson
Wellington, New Zealand

Peter Martin Sanzenbacher, MS
USGS Forest & Rangeland Ecosystem
Science Center
Corvallis, Oregon

Matthew J. Sarver, BS
Ithaca, New York

Herbert K. Schifter, PhD
Naturhistorisches Museum
Vienna, Austria

Richard Schodde PhD, CFAOU
Australian National Wildlife
Collection, CSIRO
Canberra, A.C.T.
Australia

Karl-L. Schuchmann, PhD
Alexander Koenig Zoological Research
Institute and Zoological Museum
Bonn, Germany

Tamara Schuyler, MA
Santa Cruz, California

Nathaniel E. Seavy, MS
Department of Zoology
University of Florida
Gainesville, Florida

Charles E. Siegel, MS
Dallas Zoo
Dallas, Texas

Julian Smith, MS
Katonah, New York

Joseph Allen Smith
Baton Rouge, Louisiana

Walter Sudhaus, PhD
Institut für Zoologie
Berlin, Germany

J. Denis Summers-Smith, PhD
Cleveland, North England
United Kingdom

Barry Taylor, PhD
University of Natal
Pietermaritzburg, South Africa

Markus Patricio Tellkamp, MS
University of Florida
Gainesville, Florida

Joseph Andrew Tobias, PhD
BirdLife International
Cambridge
United Kingdom

Susan L. Tomlinson, PhD
Texas Tech University
Lubbock, Texas

Donald Arthur Turner, PhD
East African Natural History Society
Nairobi, Kenya

Michael Phillip Wallace, PhD
Zoological Society of San Diego
San Diego, California

John Warham, PhD, DSc
University of Canterbury
Christchurch, New Zealand

Tony Whitehead, BSc
Ipplepen, Devon
United Kingdom

Peter H. Wrege, PhD
Cornell University
Ithaca, New York

Contributing illustrators

**Drawings by Michigan
Science Art**

Joseph E. Trumpey, Director, AB, MFA
Science Illustration, School of Art and Design, University of Michigan

Wendy Baker, ADN, BFA

Brian Cressman, BFA, MFA

Emily S. Damstra, BFA, MFA

Maggie Dongvillo, BFA

Barbara Duperron, BFA, MFA

Dan Erickson, BA, MS

Patricia Ferrer, AB, BFA, MFA

Gillian Harris, BA

Jonathan Higgins, BFA, MFA

Amanda Humphrey, BFA

Jacqueline Mahannah, BFA, MFA

John Megahan, BA, BS, MS

Michelle L. Meneghini, BFA, MFA

Bruce D. Worden, BFA

Thanks are due to the University of Michigan, Museum of Zoology, which provided specimens that served as models for the images.

Maps by XNR Productions

Paul Exner, Chief cartographer
XNR Productions, Madison, WI

Tanya Buckingham

Jon Daugherity

Laura Exner

Andy Grosvold

Cory Johnson

Paula Robbins

• • • • •

Topic overviews

What is a bird?

Birds and humans

Avian migration and navigation

Avian song

Avian flight

What is a bird?

Birds

Everyone recognizes birds. They have feathers, wings, two legs, and a bill. Less uniquely, they have a backbone, are warm-blooded, and lay eggs. All but a few birds can fly. Birds have much in common with reptiles, from which they have evolved. They share several skeletal characteristics, nucleated red blood cells, and their young develop in cleidoic eggs. The main difference is feathers, which are modified scales. Not only do feathers allow flight, they are insulated, more so than mammalian hair, enabling birds to maintain steady internal temperatures and stay active even in extreme climates. The acquisition of flight and homeothermia has influenced the evolution of other anatomical and physiological changes in birds and led to increased cerebral and sensory development. It has freed them to travel the globe, colonizing most environments and diversifying to fill many ecological niches. Consequently, it is not surprising that birds are the most successful of the vertebrates, outnumbering the number of mammal species twofold.

Evolution and systematics

The fossil record of birds is patchy and their evolutionary history is poorly known. The first feathered animal, *Archaeopteryx*, has been identified in Upper Jurassic deposits, from 150 million years ago (mya). However, while it does appear intermediate between early reptiles and birds, there is some disagreement over whether it is a direct ancestor of present day birds. Fossils unequivocally of birds do not appear until the Cretaceous period, 80–120 mya, although the number of species suggests that they radiated earlier. The earliest remains are of large flightless diving birds, *Hesperornis* spp., with primitive teeth. Other toothed sea birds also lived during the Cretaceous, including the flighted ichthyosaurs. Also appearing in the Early Cretaceous were the *Enantiornithes*, a little understood group of seemingly primitive birds. At the end of the period, the toothed birds disappeared with the dinosaurs. Since then, only toothless birds have been found in the record and it is not clear how or when they arose, though it is thought that it was during the Cretaceous. By the Eocene (c. 50 mya), many modern forms were recognizable. These are non-passerines, including ostriches, penguins, storks, ducks, hawks, cuckoos, and kingfishers. The passerines (small songbirds) appear to have diversified 36–45 mya, along with

flowering plants and insects. Several other forms, mostly large birds, were also present in the Eocene but died out. Other giant birds such as the larger moas of New Zealand and the elephant birds of Africa and Madagascar survived until about 10,000 years ago when they were exterminated by humans.

The evolutionary success of birds is evidenced by the wide variety of present-day forms. They have long been popular subjects of study for taxonomists. Traditional classifications are based mainly on morphological and anatomical differences in structure, plumage, and so forth. More recently, behavioral traits, song, and biochemical techniques (including DNA) have been employed. Yet, while there is general agreement as to the families to which the 9,000 or so extant bird species belong, a variety of opinions exists on the relationships within and between families.

Structure and function

General structure

Birds have adapted to a multitude of situations. For this reason, they occur in a wide diversity of shapes, sizes, and colors. Weighing up to 285 lbs (130 kg), and reaching 9 ft (2.75 m), the flightless ostrich is the largest of the living birds. At a mere 0.7 oz (2 g) and around 2.4 in (6 cm), the tiny bee hummingbird is the smallest. Even closely related forms can look very different (adaptive radiation). A famous example is the enormous range of bill shapes and sizes in Charles Darwin's Galápagos finches; a single over-water colonizing species is thought to have undergone repeated evolutionary divergence to produce the 14 or so contemporary species on the different islands. Conversely, unrelated species can closely resemble each other (convergent evolution) because they have evolved for the same lifestyle. Examples of this are the Old World and New World vultures, which belong to the diurnal birds of prey and storks, respectively.

Body shapes vary enormously, from the flexible, long-necked form of the cranes and ibises to short-necked, stiff-backed falcons and penguins. These latter species, the speedy, predatory hunters of the air and the seas, have torpedo-shaped bodies to minimize drag. Bills and beaks take a variety of forms that generally reflect their major function in feeding: from the sturdy, seed-cracking bills of finches to the long, soil-probing

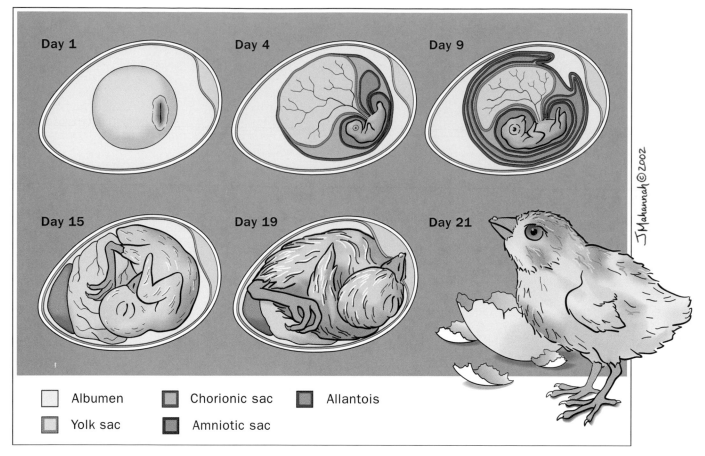

Embryonic development in birds. (Illustration by Jacqueline Mahannah)

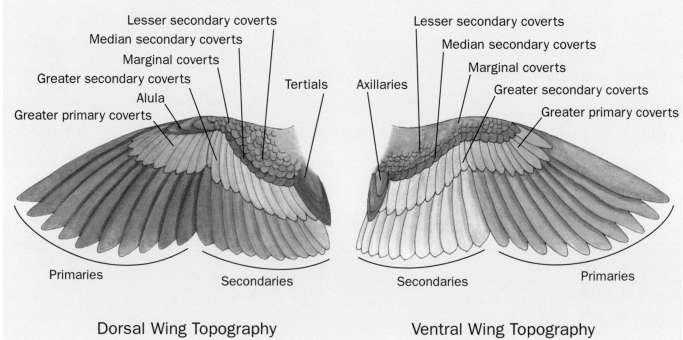

Dorsal and ventral views of a birds wing, showing the different feather groups. (Illustration by Marguette Dongvillo)

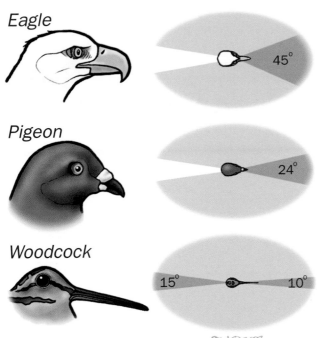

Eagle

Pigeon

Woodcock

Fields of view: an eagle has about 45° of frontal binocular vision, with 147.5° of monocular vision on both sides and 20° behind out of its visual range. A pigeon has about 24° of frontal binocular, 158° of monocular on each side, and 20° behind out of its visual range. The woodcock has 10° of frontal binocular vision, 15° of binocular behind it, and 167.5° of monocular vision to either side. (Illustration by Bruce Worden)

tween species, variation can be quite marked within species, either geographically or between the sexes.

Species often vary in size clinally (with environmental or geographic change), usually increasing in size between from hotter to cooler parts of their range (Bergman's rule); races at either end of the cline can be remarkably different. A few species even have different forms, for example, the large- and small-beaked snail kites. Males are often larger (sexual dimorphism), but in some species, including birds of prey, some seabirds, and game birds, the female is larger.

The senses

Birds' active lifestyles require highly developed senses. For the vast majority of species, sight is the dominant sense and the eyes are relatively large. The eyes are generally set to the sides of the head, allowing a wide field of view, (about 300°), presumably useful for detecting approaching predators. For predatory birds (insectivores and raptors), the eyes are set

bill of a kiwi, the delicate curve of a nectar feeder's bill and the massive bone-shearing beak of a large vulture. In a few species, bills also serve as signals of breeding condition and sexual ornaments to attract the opposite sex. For example, the bills of cattle egrets turn from yellow to orange-yellow in the breeding season and the huge gorgeously rainbow-hued bills of the sulphur-breasted toucans may separate species. Similarly, birds' feet and legs suit their lifestyle: webbed for swimming; short and flat for ground dwellers; longer and grasping for perching species; powerful and heavily taloned for raptorial species. Stilt-like legs and spider-like toes with a span the length of the bird's body are a feature of the lily-pad walking jacanas. The legs are almost nonexistent in swifts and other birds that spend much of their lives on the wing, and long and muscular in ostriches and emus that stride and run across the plains. The ostrich has two toes, and a few species such as those that run on hard surfaces have lost the first (hind) toe or it is very small. Most species have four toes but their arrangement differs: in most perching birds, toes two, three, and four point forward and the hind toe opposes them; some species have two toes pointing forward, two back; others can move a toe to have either arrangement; swifts have all four toes pointing forward.

Wings are less variable than lower limbs, although their different forms can be extreme: much reduced in the flightless ratites, put to good use as fins in penguins, and at their most extended in gliding species that spend much of their lives riding air currents. Not only are there differences be-

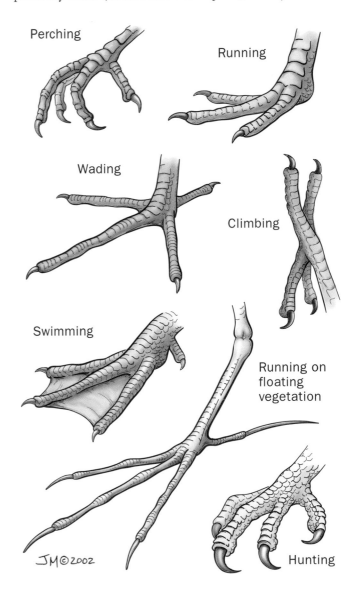

Perching

Running

Wading

Climbing

Swimming

Running on floating vegetation

Hunting

Birds' feet have different shapes and uses. (Illustration by Jacqueline Mahannah)

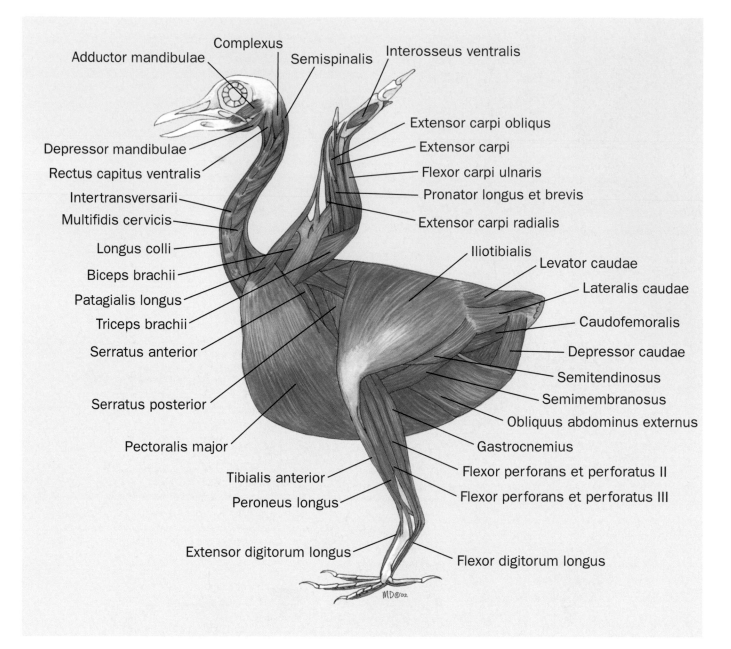

Bird musculature. (Illustration by Marguette Dongvillo)

more forward to give a greater overlap in the field of vision of the two eyes. This increase in binocular vision is important for depth perception. Compared with mammals' eyes, birds' eyes are relatively immobile. They compensate by being able to rotate the head by as much as 270° in species such as owls, which have the most forward facing eyes. Their eyes are protected by a nictitating membrane, which closes from the inside to the outside corner, and a top and bottom eyelid. Birds can focus their eyes rapidly, which is important in flight and when diving underwater. In general, they may not have exceptional visual acuity compared with humans. However, birds have a larger field of sharp vision, good color perception, and can also discriminate in the ultraviolet part of spectrum and in polarized light. Nocturnal species have more

rods than cones in their retinas to enhance their vision in dim light.

The ear of birds is simpler than that of mammals, but their sense of hearing appears to be at least as sensitive. Some species, such as some of the owls, have a disc of stiff feathers around the face that directs sound to the ears, and asymmetrically placed ear openings and enlarged inner ears to enhance discrimination of direction and distance of the source of the sound. Oilbirds and some swiftlets that live in caves use echolocation. They emit audible clicks to help them navigate and locate prey in the dark.

The great number of sensory receptors and nerve endings distributed about the body indicate that birds' sense of touch,

A collection of eggs from British birds. (Photo by E. & D. Hosking. Photo Researchers, Inc. Reproduced by permission.)

A bird cares for its feathers by grooming and preening. (Illustration by Jonathan Higgins)

What *Archaeopteryx* may have looked like. (Illustration by Brian Cressman)

pain, and temperature is keen. By contrast, the olfactory system is poorly developed and few birds seem to make great use of smell. Exceptions include the New World vultures and the kiwi, which can detect prey by its scent.

Plumage

Feathers distinguish birds from all other living animals (there is recent evidence that some dinosaurs were feathered but this remains controversial). Light, strong, and colorful, feathers are extraordinarily multifunctional. They provide warmth, protection from the elements, decoration and camouflage, and are specialized for aerodynamics and flight (most birds), hydrodynamics and diving (e.g., penguins), or to cope with both elements (e.g., cormorants). A few species use them to make sound (e.g., snipe) or carry water to their young (e.g., sandgrouse). Not least, they identify species and subspecies, may vary with age, sex and breeding condition, and signal emotion.

Feathers are made of keratin and, once grown, are entirely dead tissue. They are of six main types. The most obvious are the long, stiff feathers of the wings and tail that provide the flight surfaces; more flexible, contour feathers make the

sculpted outer covering for the body; and down makes a soft insulative underlayer. Semi-plumes, which are between down and contour feathers in form, help to provide insulation and fill out body contours so that air (or water) flows easily over the body. Two types of feathers are mainly sensory in function: stiff bristles that are usually found around the face (around the feet in the *Tyto* owls) like the whiskers of a cat or a net around the gape of some insect-eating species, and filoplumes, which are fine, hair-like feathers with a tuft of barbs at the tip that lie beside contour feathers and monitor whether the plumage is in place. In some species, modified feathers form features such as crests, ornamental bristles, cheek tufts, plumes, and tail flags and trains.

The contour, flight feathers, and semi-plumes have a central shaft, and a vane made up of barbs and barbules that interlock with each other and sometimes with neighboring feathers. The bird carefully maintains these links by nibbling and pulling the feathers through its bill. Many birds bathe regularly, most in water, but a few in dust. Some such as herons and elanine kites have powder down that grows continuously and crumbles into a fine powder that is spread through the feathers for cleaning and water resistance. Other

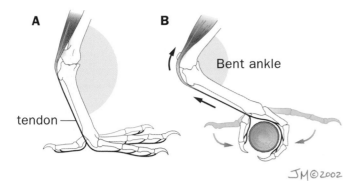

A

B

Bent ankle

tendon

JM©2002

When a bird perches, its ankle bends and contracts the tendons in its foot, forcing its foot to close around the perch (B). (Illustration by Jacqueline Mahannah)

birds have oil glands at the base of the tail for the same purpose. Sunbathing also helps to maintain the health and curvature of feathers. Some birds, including many passerines, appear to use biting ants in feather maintenance, perhaps to control ectoparasites, either by wallowing among the swarm or by wiping individual ants through their plumage.

Over time, feathers become worn and bleached, and damaged by parasites. They are completely replaced annually in most except very large birds such as eagles and albatrosses, which spread the molt over two or so years. Some species, notably those that change from dull winter plumage into bright breeding colors (e.g., American goldfinch), have two molts a year: a full molt after breeding, and a partial body molt into breeding plumage. Most species shed their feathers sequentially to maintain their powers of flight. However, a few, particularly waterbirds that can find food in the relative safety of open water, replace all their flight feathers at once and are grounded for about five weeks.

Spectacular colors are a feature of birds, from the soft, mottled leaf patterns of nightjars (cryptic) to the gaudy, ornate plumes of a peacock (conspicuous). Cryptic colors conceal the bird from predators or rivals; conspicuous colors are used in courtship or threat. The colors themselves are produced by pigments in the feathers themselves or by structural features that interact with the pigment and the light to produce iridescent color, which can only be seen from certain angles, or non-iridescent color, that can be seen from any angle. In many species, the sexes are similar in color. In others, the sexes differ, and usually the male is showier and the female resembles a juvenile bird. In these species, sexual selection is thought to have favored dichromatism (and dimorphism) through female preference for partners with bright colors (and extravagant ornamentation). In the few polyandrous species, the reverse is that case and the females are more vibrant. Plumage may also vary geographically, with races from warm, humid climates tending to be more heavily pigmented than those from cool, dry regions (Gloger's rule).

Anatomy and physiology

The skeleto-muscular system of birds combines light weight with high power for flight. Muscle mass is concentrated near the center of gravity—around the breast and bases of the wings and legs—which gives a compact, aerodynamic form. Long tendons control movements at the ends of the limbs. Flighted birds have more massive breasts and wing muscles; in terrestrial birds, much of the muscle mass is in the upper legs. In perching birds, the tendon from the flexor muscle loops behind the ankle; when the ankle bends on landing, the toes automatically close around the perch and maintain the grip without effort, anchoring the bird even in sleep. In many species, the toe tendons have ridges, which also help to lock the feet around the perch.

In contrast to the mammalian skeleton, birds' bones are hollow and less massive and several have fused to form a strong, light frame. A bird's skeleton constitutes only about 5% of its mass. Another distinctive feature is that the bones, including the skull, are pneumatized: their core is filled with air via a system of interconnecting passages that connect with the air sacs of the respiratory system and nasal/tympanic cavities. Flighted species tend to have extensive pneumatization but it is reduced or lacking in diving birds, which would be hindered by such buoyancy. Even birds' bills are light—the horny equivalent of the heavy, toothed muzzle of mammals.

Respiration, circulation, and body temperature

Unlike mammals, birds lack a diaphragm. Instead, air is drawn into the rigid lungs by bellow-like expansion and contraction of the air sacs surrounding the lungs and another group in the head, which is driven by muscles that move the ribs and sternum up and out and back again. Features of birds' circulatory and respiratory systems make their respiration more efficient than that of most mammals, allowing them to use 25% more oxygen from each breath. This enables them to sustain a high metabolic rate and, among other benefits, as-

An ostrich (*Struthio camelus*) shades its chicks in Etosha National Park, Namibia. (Photo by Nigel Dennis. Photo Researchers, Inc. Reproduced by permission.)

Bill morphology differs with use: 1. Greater flamingo filters microorganisms from water; 2. Peregrine falcon tears prey; 3. Roseate spoonbill sifts water for fish; 4. Dalmation pelican scoops fish in pouch; 5. Anna's hummingbird sips nectar; 6. Brown kiwi probes soil for invertebrates; 7. Green woodhoopoe probes bark crevices for insects; 8. Rufous flycatcher captures insects in flight; 9. Java sparrow eats seeds; 10. Papuan frogmouth trawls for insects; 11. Bicornis hornbill eats fruit; 12. American anhinga spears fish; 13. Rainbow lorikeet cracks nuts. (Illustration by Jacqueline Mahannah)

sists those high-flying migrants that cross the world's tallest mountains and reach altitudes up to 29,500 ft (9,000 m), where the oxygen content of the air is low.

Birds and mammals both generate their own body heat, but birds' high metabolic rate helps to maintain theirs at around 100°F (38–42°C), depending on species; 5–7°F (3–4°C) hotter than most mammals. When it becomes difficult to obtain enough energy to stay warm and active, a few species become torpid (lower their body temperature and become inactive) on the coldest days or during bad weather, usually for a few days

or overnight. Other birds cope with cold by increasing their metabolic rate slightly, growing denser feathers, having a layer of fat, or by behavioral means such as huddling with others, tucking up a leg to decrease the heat loss surface, and fluffing out the feathers to trap more air. Lacking sweat glands to shed body heat, they may pant, lower their metabolic rate, seek shade, or raise their feathers to catch the breeze in hot weather.

Digestion and excretion

To provide the energy needed for flight, and so that they are not weighed down for a long time by the food they have

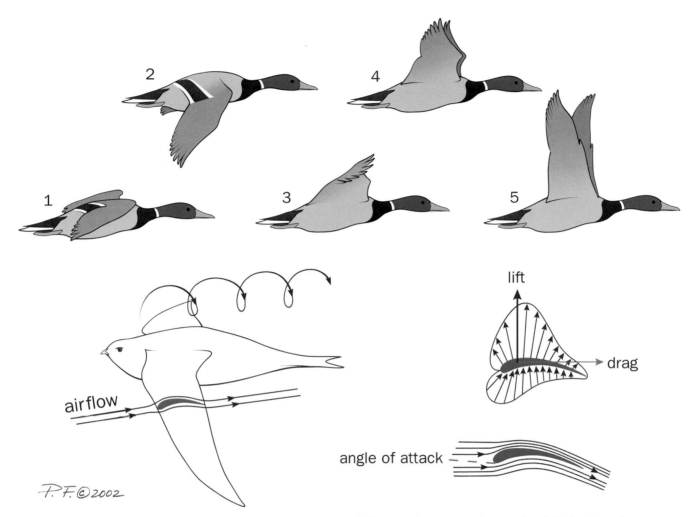

1. Downstroke; 2. Wings sweep forward at the end of the downstroke; 3. Feathers fan open on the upstroke; 4. Each primary becomes a small propeller in the upstroke; 5. Primaries swoop back, ready for a new downstroke. (Illustration by Patricia Ferrer)

consumed, birds have a high metabolic rate and digest food rapidly and efficiently. Their digestive tract is modified accordingly, and many species have the second part of their two-part stomach modified into a muscular gizzard where hard food is physically ground down so that gastric juices can penetrate easily. Some species swallow pebbles to assist with this breakdown. The digestive tract tends to be long in grazers, fish-eaters, and seed-eaters, and short in meat- and insect-eaters.

Birds have three ways to rid themselves of excess water, salts, and waste products: through breathing and the skin; the renal system; and salt glands. The salt glands are located in the orbit of each eye. They secrete sodium chloride and, therefore, are well developed in seabirds that have a salty diet, and non-functional in some other groups. Birds' kidneys are more complex than those of mammals: they produce concentrated urine with nitrogen waste in the form of insoluble uric acid, rather than urea. Such a water-efficient system does not require a (heavy) bladder, with obvious advantages for flight. It also enables many species, especially those with a moist diet (carnivores, insectivores, and frugivores), to drink seldom or not at all. Both the white urine and dark fecal mat-

ter are voided through the anus, and bird droppings often contain both.

Life history and reproduction

Life history features

To a large extent, body size determines life history. Compared with smaller species, larger species tend to live longer, breed at a later age, have a longer breeding cycle, and, at each breeding attempt, produce fewer young with a greater chance of survival. There are always exceptions, and climate and risk of predation and other factors impinge on this overlying pattern. For example, some small temperate zone Australian passerines live up to 18 years and have small clutches, whereas ground-nesting grouse may live a few years and have large clutches. For convenience, species may be classed as fast-breeders (r-selected), which have many large clutches and short periods of nestling care, and slow-breeders (K-selected) that have a few small clutches and extended periods of offspring care. In reality, there is a continuum between the two extremes.

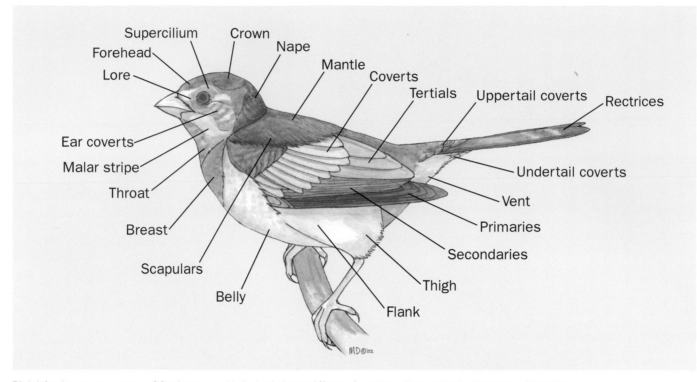

Birds' feather tracts—types of feathers on a bird's body have different functions. (Illustration by Marguette Dongvillo)

Mating systems

The majority of bird species are at least socially monogamous, that is, a pair of birds cooperates to raise young. They may stay together for the breeding attempt or mate for life. However, many other arrangements exist. Some birds have a polygynous mating system, particularly species that use rich resources that are clumped, so that a male can support more than one female (e.g., New World blackbirds, some harriers). Successive polygyny is less common, mainly practiced by species in which the female alone raises the chicks and visits a lek where males display. The female may mate with several (e.g., black grouse and some birds of paradise) males. Less frequent again is polyandry, where the females mate with various males and leave them to care for the eggs and chicks (e.g., emus, buttonquail, and jacanas). Within these systems, there is also scope for cheating, and the advent of DNA fingerprinting has revealed that in many monogamous species, there are broods of mixed paternity. At its extreme are species such as the superb fairy-wren in which about three-quarters of the chicks are raised by males that are not the biological father.

Most birds nest as solitary pairs, but some 13% of species are colonial, particularly seabirds. Colonies may be a few pairs (e.g., king penguin) or millions (e.g., queleas). In between are species that nest in loose colonies, either regularly or when conditions are favorable. Nest spacing varies enormously from a few inches/centimeters in some colonies to several miles/kilometers in species that have large territories. Spacing is linked to food and nest sites; nests are closer together where resources are plentiful.

Breeding seasons and nests

Most birds have a regular breeding season, timed to coincide with the most abundant season of the year, but some, particularly those adapted to unpredictable climates, are opportunistic, only breeding when conditions allow. The largest species may breed every two years, but most attempt to breed at least annually, some raising several broods over the breeding season.

Birds build their nests of several substrates; the most important issues are protection from predators and the elements. Therefore, species that nest on predator-free islands often build close to the ground but, on the mainland, species build in higher locations. The nest itself must hold and shelter the eggs; in form, they vary enormously from simple scrapes in the dirt to large complex stick nests and hanging structures, woven together or glued with cobwebs or mud, and lined with soft material. Tree holes and holes in banks or cliffs also make good nests but competition for them may be fierce. The megapodes construct a mound in which they bury their eggs and maintain the temperature at about 90–95°F (32–35°C) by scratching soil on or off. Many species use the same nest area, nest site, or actual nest year after year.

Eggs and incubation

Birds' eggs are beautiful in their variety. They may be plain or colored, marked or unmarked, oval or round. All are slightly more pointed at one end so that the egg tends to roll in a circle. Species that lay in the open where eggs may roll tend to have long-oval, cryptically colored eggs, and those that lay in holes tend to have rounder, unmarked eggs. Egg laying can be energetically costly, especially for small birds: for a hum-

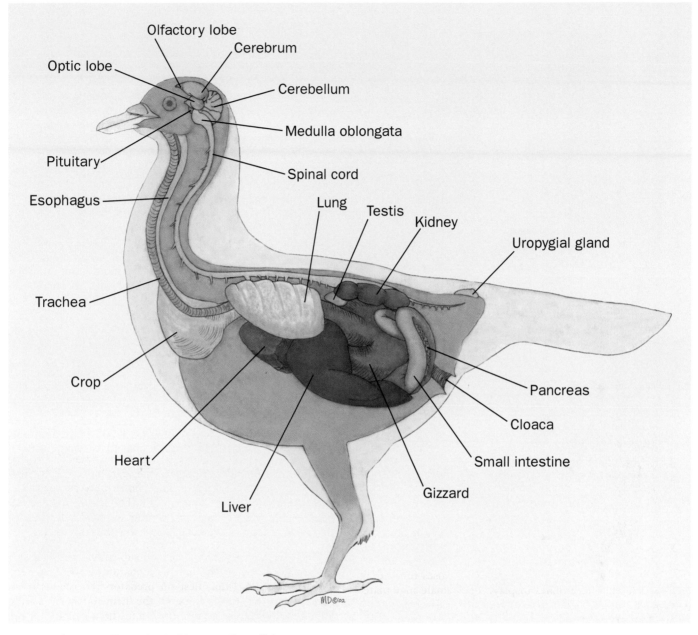

Bird internal organs. (Illustration by Marguette Dongvillo)

mingbird, each egg (0.01 oz/0.3g) represents 25% of the bird's mass; for an ostrich, the 50 oz (1,500 g) egg is 1% of the hen's weight. Birds that have precocial chicks tend to have large eggs (about 35% yolk compared with 20% in altricial [more helpless] species) because the chick must be advanced and well-developed when it hatches. Clutch size varies enormously both within and between species. Nevertheless, most species have a typical number of eggs; one in the kiwi, perhaps 20 in some ducks. Larger species tend to have fewer eggs. In some species, two are laid but only one ever hatches. Across species, there tends to be a trade-off between egg-size and clutch-size, some species lay a few large eggs, others many smaller eggs. In some species, the clutch size is fixed (determinate layers), in others, if the eggs are lost or removed within the breeding season, the

bird will go on laying eggs (indeterminate layers). The majority of species lay every second day until the clutch is complete; in a few of the largest species, the interval is four days. In the vast majority of species incubation is carried out by one or both of the parents, but a few species, such as some ducks, nest-dump (lay some of their eggs in a neighbor's nest), and some such as the cuckoos are parasites and lay their eggs in the nest of another species. Incubation varies little within species. Among species, it ranges in length from 10 days for some woodpeckers to about 80 days for albatross and the very large-egged kiwi.

Egg formation and embryo development

To most people, eggs and birds go together. Certainly all bird species lay shelled eggs, but so too do some reptiles. In

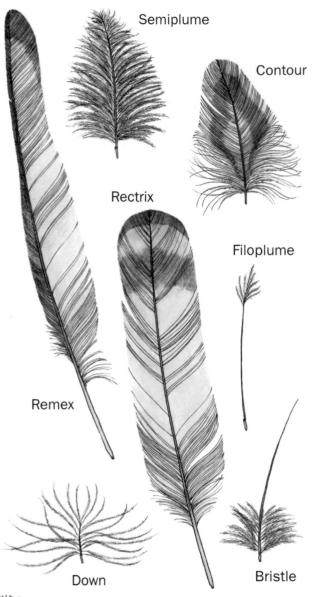

Types of bird feathers. (Illustration by Marguette Dongvillo)

The egg contains nutrients for the developing embryo. There is some exchange of gases and water across the shell, and waste from the embryo is stored in a sac that develops outside of the embryo, but within the shell (the allantois). Another sac develops into an air sac for ventilation. Once incubation begins, development is rapid, triggered by heat either from a brooding parent, as in most birds, or the environment, as in mound builders, which bury their eggs to capture heat from the sun. Within days, the embryo has large eyes and rudimentary organs. As it develops, the yolk sac is absorbed and the embryo fills more of the shell. By hatching, the yolk is fully absorbed and the embryo has moved its bill into the air sac and begins to breathe air. By this time, mainly through water loss, the egg is about 12–15% lighter than at the start of incubation. The embryo uses an egg tooth on the tip of its bill to chip away a ring around one end of the shell (already thinned by loss of calcium to the embryo) and hatch. If it is altricial (e.g., songbirds and seabirds), the chick will be naked or downy, and helpless; the chick of a precocial species (e.g., ostriches, ducks, and game birds) will be feathered, able to regulate its own body temperature, and can follow its parent and feed itself almost immediately. The chicks of precocial species hatch synchronously (at the same time), those of other species are sometimes asynchronous, resulting in a mixed-age brood.

Growth and care of young

Most birds grow remarkably fast and, in many altricial species, are more or less fully grown by the time they leave the nest. At the extremes, chicks stay in the nest about 10–20 days

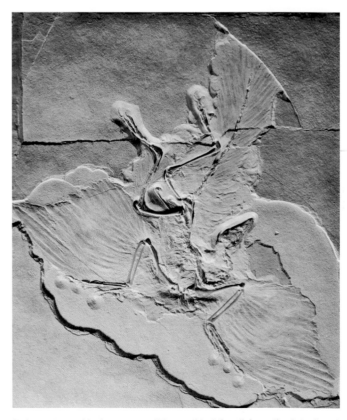

Archaeopteryx (*Archaeopteryx lithographica*) fossil from the late Jurassic period, found in Bavaria, Germany limestone in 1860. (Photo by François Gohier. Photo Researchers, Inc. Reproduced by permission.)

most bird species, only the left ovary develops. The ovary holds a large number of oocytes. During the breeding season, a few oocytes (immature eggs) start to develop. They are covered in a follicle that lays down yolk, which is manufactured in the liver and carried in the blood. The one with the most yolk is shed first; the follicle ruptures and the oocyte moves into the first part of the oviduct where it may be fertilized from sperm stored in the sperm storage glands. During about 20 hours it passes down the oviduct where it is covered in albumin and, towards the end, the outer layer calcifies to form the shell and any markings are laid down from blood (red-brown) or bile (blue-green) pigments. About this time, if another egg is to be laid, another oocyte is released. In a few large species, the process may take longer.

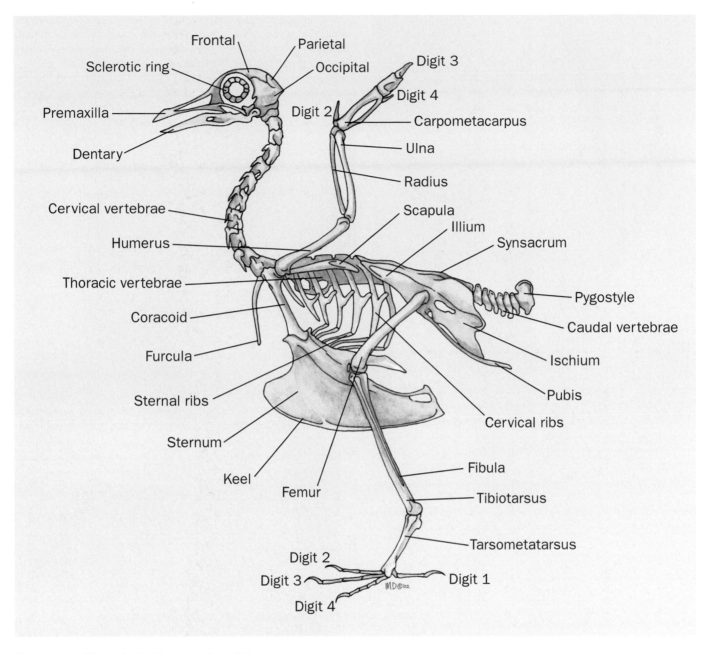

Bird skeleton. (Illustration by Marguette Dongvillo)

in passerines and 150–250 days in albatrosses. Precocial species, which are free of the nest, are slower growing—their rate of growth is approximately one-third that of altricial species.

The amount and type of care given is diverse. Nidicolous species, in which the chicks stay in the nest, put in considerable effort bringing food to the nest, either bringing many small items (e.g., insectivores) or a few large ones (raptors). Some species regurgitate food for the chicks (e.g., seabirds). These species must also keep the chicks warm and dry and protect them from predators, and may also keep the nest free of droppings. Nudifugous species, in which the chicks follow adult(s), are either fed by the adult (some seabirds) or, more commonly, feed themselves on plant material (e.g., many waterfowl and game birds).

Parental care varies accordingly. As protection against predators, some waterfowl carry their young on their backs and, in colonial species, several neighbors unite to try to drive off an intruder. In some species, the breeding pair is accompanied by "helpers," which may be offspring from earlier breeding attempts or unrelated individuals, males or females or both sexes, juveniles or adults. Helpers help particularly in those species in which prey is difficult to find or catch.

Ecology

All species have a niche—the range of environmental conditions under which they can survive and reproduce. The

Emperor penguin (*Aptenodytes forsteri*) adult and chick, Antarctica. (Photo by Art Wolfe. Photo Researchers, Inc. Reproduced by permission.)

few birds have the ability to digest cellulose. Not surprisingly, birds are pollinators and seed dispersers for a great variety of plants. Some species specialize, others are more varied in their diet. Methods of collecting food are numerous, but most involve sight.

Birds are infected by a variety of diseases, both internal and external, caused by bacteria, viruses, and parasites. Some of these can also affect humans and their livestock. In general, healthy birds can carry both internal and external parasites without obvious harm. Nevertheless, disease outbreaks occur; for example, following floods, outbreaks of insect-borne pox can occur in wild birds. Birds can also suffer from exposure to toxic substances of both natural (e.g., toxic algal blooms) and human-made origin (e.g., oil spills, pesticides, lead-shot, and other pollutants). Predation is a fact of life for the majority of bird species, particularly for their eggs and young. Their life history has evolved to allow for losses, which are naturally high. For example, in passerines, perhaps 5% of eggs result in adult birds and annual mortality of adults may be as high as 60%. For larger species, the rate of mortality tends to be lower. Provided that they are not too prolonged or severe, starvation, predation, disease, and other causes of

Down feathers form part of a bird's insulating underplumage. The central axis, or rachis, bears the flat vane of the feather made up of many slender barbs. Feathers allow a bird to fly, and provide waterproofing, camouflage, and color and shape to attract mates. (Photo by Gusto Productions/Science Photo Library. Photo Researchers, Inc. Reproduced by permission.)

niche may be broad and unspecialized or narrow and highly specialized. Coexisting species tend not to overlap much in their niche requirements. This is particularly obvious with food: the type and where, when, and how it is collected. For example, three hawks coexist in some Australian woodlands and they roughly segregate by species and sex: the small male sparrow hawk takes passerines in the canopy; the medium-sized brown goshawk hunts birds and small mammals in the air and on the ground in more open spaces; and the largest, the female gray goshawk, captures medium-sized birds and mammals on or near the ground.

To provide energy for flight, birds need highly nutritive food. For this reason, most birds eat at least some arthropods, especially insects. They may do this incidentally, for example, mixed in with nectar, but mostly they are actively captured. In addition to this, there are four basic diet groups: the carnivores (those that eat other vertebrates, including the fish-eaters); berry- and seedeaters; eaters of non-flowering plants (fungi, mosses, algae, and so forth); eaters of flowering plants (roots, tubers, leaves, nectar). Grasses and herbage are low in nutritive value and must be consumed in too-large quantities to be of major importance to many species, and

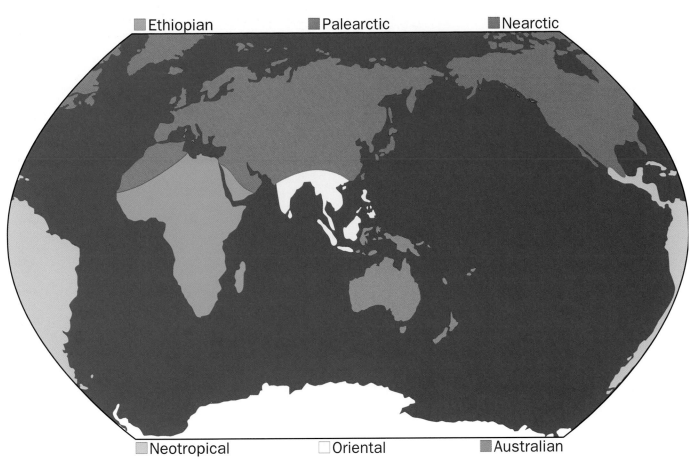

Regions of bird distribution. (Illustration by Emily Damstra)

Distribution and biogeography

Birds are distributed across all continents and on most islands, and in all major habitats from caves to mountaintops,
deserts to rainforests. Many factors limit the distribution of
species, all relating to their ecology: climate, habitat availability, and the presence of predators, competitors, and food.
There may also be physical barriers such as mountain ranges,
oceans, and impassable expanses of unsuitable habitat. The
breeding range often differs from the nonbreeding range because of seasonal or other movements to remain in the most
favorable conditions. There are broad patterns to general distribution; for example, woodpeckers are found in many regions of the world but are absent from Australia, New
Zealand, and Madagascar. The ratites are southern in distribution, pointing to their early evolution in Gondwana (the
huge southern super-continent that split into the southern
continents). These patterns reveal six major biogeographical
realms: Neotropical, Nearctic, Ethiopian, Palearctic, Oriental, and Australian. The greatest number of species is found
in the Neotropics (South and Central America, the West Indies, and southern Mexico) with roughly 3,000 species, and

31 endemic families. The Nearctic (North America, Greenland,
Iceland), with about 1,000 breeding species and no endemic
bird families, is among those areas with the fewest species.

Bird populations vary enormously in their abundance and
density. Some species are widely scattered across the landscape (e.g., the solitary eagles), others live in crowded colonies
of millions (queleas). There are several general patterns: population density is related to body size (smaller species tend to
be more numerous); the number of species and of individuals tends to be greatest in complex habitats (e.g., forest compared with grassland) and where several habitats meet
(ecotones); the number of species increases with the size of
the habitat patch; and, the number of species and overall densities tend to increase from the poles to the equator.

Behavior

Bird species can be social or solitary; they may nest, roost,
and feed in small or large groups for part of their lives (e.g.,
when young or in the nonbreeding season) or their entire
lives. Some form feeding flocks with other bird species or nest
near more powerful species for protection. Some associate
loosely with other vertebrates, such as following monkey
troops to catch the insects they flush. A few live more or less

mortality are compensatory and, in general, bird populations
recover quickly.

commensally, for example, by feeding on ticks from ungulates. Some species spend much of their lives on the ground (terrestrial species), others in the air, in water, or in various combinations. The majority of species are active by day (diurnal), but many are crepuscular (active in twilight) or truly nocturnal. During inactive periods, most retire to a safe roost where they socialize, preen, relax, or sleep. They appear to need to sleep several hours a day.

Birds have developed complex communication systems, including various calls and song and visual signals involving posture, movements, facial expressions, display of certain characteristics of plumage, and, in some species, flushing of the skin (e.g., vultures) or popping of the eyes (Australian choughs). Some behaviors, like the dance of brolgas and sychronized swimming of swans, is ritualized, others appear more spontaneous. Courtship is a highly ritualized sequence of displays, on the ground or in the air, and can involve court-ship feeding, which may be a way for females to judge the quality of their partner.

As a group, birds are cerebrally advanced. Much of their behavior is innate but they also learn by experience throughout life. They have good memories, such as retrieving cached food items from several stores several weeks after they were hidden. Some species appear to be particularly adaptable if not intelligent. Certainly, their behavior can be complex and interesting. Individual rooks wait to cache nuts if a non-hoarding rook is passing by, yet are unconcerned by rooks that are also hoarding. Jays that steal food are more likely to move their caches to prevent theft than are jays that are not thieves. Parrots have a complex social system and enjoy playing with each other and with found objects. Black kites and green-backed herons have learned to place bread on water to attract fish. One of the Galápagos finches uses a cactus thorn to probe bark crevices for insects that it cannot reach with its short tongue.

Resources

Books

Brooke, M., and T. Birkhead. *The Cambridge Encyclopedia of Ornithology.* Cambridge: Cambridge University Press, 1991.

Campbell, B., and E. Lack, eds. *A Dictionary of Birds.* Calton: Poyser, 1985.

de Juana, E. *Handbook of the Birds of the World.* Vol. 1. *Ostrich to Ducks,* edited by J. A. del Hoyo, A. Elliot, and J. Sargatal. Barcelona: Lynx Edicions, 1992.

Farner, D. S., J. R. King, and K. C. Parkes, eds. *Avian Biology.* Vols. 6, 7, and 8. London: Academic Press, 1982 through 1985.

Farner, D. S., and J. R. King, eds. *Avian Biology.* Vols. 1, 2, 3, 4, and 5. London: Academic Press, 1971 through 1975.

Perrins, C. M. *The Illustrated Encyclopedia of Birds.* London: Headline, 1990.

Perrins, C. M., and T. R. Birkhead. *Avian Ecology.* Glasgow: Blackie, 1983.

Welty, J. C., and L. Baptista. *The Life of Birds.* 4th ed. London: Poyser, 1988.

Organizations

American Ornithologists' Union. Suite 402, 1313 Dolley Madison Blvd, McLean, VA 22101 USA. E-mail: AOU@BurkInc.com Web site: <www.aou.org>

BirdLife International. Wellbrook Court, Girton Road, Cambridge, Cambridgeshire CB3 ONA United Kingdom. Phone: +44 1 223 277 318. Fax: +44-1-223-277-200. E-mail: birdlife@birdlife.org.uk Web site: <http://www.birdlife.net>

Birds Australia. 415 Riversdale Road, Hawthorn East, Victoria 3123 Australia. Phone: +61 3 9882 2622. Fax: +61 3 9882 2677. E-mail: mail@birdsaustralia.com.au Web site: <http://www.birdsaustralia.com.au>

British Trust for Ornithology. The Nunnery, Thetford, Norfolk IP24 2PU United Kingdom. Phone: +44 0 1842 750050. Fax: +44 0 1842 750030. E-mail: info@bto.org Web site: <www.bto.org>

National Audubon Society. 700 Broadway, New York, NY 10003 USA. Phone: (212) 979-3000. Fax: (212) 978-3188. Web site: <www.Audubon.org>

Royal Society for the Protection of Birds. Admail 975 Freepost ANG 6335, The Lodge, Sandy, Bedfordshire SG19 2TN United Kingdom. Web site: <www.rspb.org.uk>

Penny Olsen, PhD

• • • • •

Birds and humans

Introduction

Why are we so fascinated with birds? From the earliest cave paintings and ceramic effigies of prehistoric humans to the present, we find close links between birds and ourselves. Those links are related to several things that draw us together: (1) our fascination with and envy of the ability of birds to fly; (2) the meat and eggs they provide us; (3) their colorful plumage that we admire and often use to decorate our own attire; (4) their down feathers that we use for insulation, and other feathers that we've used as writing instruments, to fletch arrows, to fan royalty, and to dust our homes; (5) their beautiful voices; (6) their hollow bones which we have at times used to produce tools and even flutes with which to emulate their songs; (7) their ritualistic courtship behavior; (8) their attentive parental care; (9) the vigor of their territorial defense; (10) their seasonal migrations; and (11) their diversity.

Certainly the utilitarian aspects of our association with birds would remain even if birds were less conspicuous. Our fascination with birds might be found in many seemingly simple things we share with them: perception of color, song, concern for the decor of our homes, parental care, the awkwardness of adolescence, territorial defense—the list could go on. Our fascination with birds might also be found in things deeper within our psyche—things we don't really share with birds, but that we can imagine as sharing: emotional things such as love, hate, fear, pride, sadness, and hope. Yes, as one recent book title has it, "Hope is the thing with feathers."

Cultural significance of birds

We, like most birds, are most active in the light of day—a time when the light-sensitive cone cells of our respective eyes allow our brains to interpret the world in color. Color has a common basic value to primitive humans and to birds. It allows both to identify ripe fruit from that which is not ripe, poisonous fruits from those that are edible. With the ability to see color come a number of gifts—lagniappe—something a little extra. From the human perspective we might think of them as aesthetic, but from the birds' perspective they seem utilitarian as well. The diversity of patterns and colors in the plumages of birds facilitate recognition of members of their own species, just as colors of uniforms allow us to recognize players for the home team. We use color we find pleasing as

decoration for ourselves and our surroundings—to please ourselves and those we wish to please. Birds do the same thing. The bowerbirds of Australia have drab plumage, but decorate their courtship bowers with objects of specific hues. Natural selection has favored the development of colors and patterns that help fulfill needs among birds for the attraction of a mate and defense of a territory, and camouflage for protection from enemies. In many ways we share such needs and such benefits of color. We differ from most mammals in being able to see color, a trait we share with most birds. The beauty we recognize in the colors of birds might be looked upon as a celebration of our uniqueness.

Birds and music

The purity of tones, diversity of melody, and the predictability of the rhythms of bird songs and mechanical sounds are music to our ears. For birds the sounds are messages: "This territory is occupied," "I'm an available and desirable suitor." We have intercepted—no, merely eavesdropped—on their conversations. We have borrowed them for our own uses and embellishment. For us the songs are messages as well, reflections of our well-being and desires. Bird songs have influenced the works of great composers. The common cuckoo (*Cuculus canorus*), nightingale (*Luscinia megarhynchus*), and common quail (*Coturnix coturnix*) can all be heard in Beethoven's Sixth Symphony, the "Pastoral Symphony." Béla Bartók recorded bird songs in musical notation and included them in his compositions. His final work, "Piano Concerto no. 3" includes bird songs he heard during his stay in North Carolina. Antonín Dvořák also used bird songs and the red-winged blackbird's (*Agelaius phoeniceus*) territorial "oak and leo" call can be heard in his Opus 96 from his days in Spillville, Iowa.

Birds imitate us as we imitate them; they have borrowed sounds from our musical repertoire. My African gray parrot (*Psittacus erithacus*), whose name is "Smoky" dutifully sings "On Top of Old Smoky"—but his rendition is no better than mine. In a possible turn about of truly musical influence, Mozart was passing a pet store near his home in May of 1784 when he heard the strains of the allegretto theme from his G major concerto which he had written just five weeks earlier. He immediately went into the store and purchased the European starling (*Sturnus vulgaris*) that was singing it!

Hundreds of turkeys are raised for food at the Givat Hayim kibbutz near Natanya, Israel. (Photo by Rick Browne. Photo Researchers, Inc. Reproduced by permission.)

We have not only borrowed from the music of birds, but around the world various cultures have also incorporated elements of bird courtship displays in their own dances. The Blackfoot Indians of the northwestern United States mimicked the foot-stomping and strutting of displaying male sage grouse (*Centrocercus urophasianus*) and even wore feathered costumes that mimicked the birds' plumage. The Jivaros of Amazonia, perhaps best known for shrinking human heads, mimic the courtship displays of the brilliant orange cock-of-the-rock (*Rupicola rupicola*). The people of Monumbo of Papua New Guinea mimic the courtship of the cassowary (*Casuarius* sp.).

The behavior of birds has always provided the basis for metaphors, symbolism, mythology, and lessons to be learned. The courtship cooing of doves has long been symbolic of human love and courtship, the powerful beak and talons of eagles have similarly served as a metaphor for strength and a symbol of our armies. The silence, stillness, attentiveness to our presence, and human-like expression created by the forward-facing eyes and facial disc of feathers on an owl have led to owls becoming symbolic of wisdom. Yet owls are not wise, but highly instinctive in their behavior, the symbolism resulting from the accident of their diurnal inactivity and the convergence of their binocular vision with our own—an adaptation that gives both of us the ability to judge distance and size—critical abilities for an animal to possess if leaping from limb to limb as our ancestors did or if diving suddenly to pounce on a mouse in near darkness, as an owl does.

Many expressions in human languages draw symbolism from the language and actions of birds. At times visual metaphors are obvious, such as our association of the peacock (*Pavo cristatus*) with vanity—or with the kaleidoscope of colors on a television network. At other times the links are fanciful, such as the notion that white storks (*Ciconia ciconia*) bring babies or that ravens (*Corvus corax*) foretell death. The link between magpies (*Pica pica*) and talkative humans is a clear reference to the chatter of the birds, as is the use of the word "parrot" to refer to an individual who repeats what has been said. "Eagle-eye" is a reasonable descriptor for a human who picks up on details. To have one's "ducks in a row" is a verbal expression of the visual impression of organization provided by a view of a duck being followed by a brood of ducklings in single file behind her. The impact of "bird words" on our language and culture is immense, though at times the link to birds has been lost and only the expression remains. Have you ever been "goosed?" Really? If you've walked across a pasture that was patrolled by a vigilant gander you might understand. Yes, they do come up from behind in a surprise attack with a sharp "goose."

Saker falcon (*Falco cherrug*), a bird used in falconry. (Photo by Eric & David Hosking. Photo Researchers, Inc. Reproduced by permission.)

Birds used to hunt and fish

Falconry, using a falcon or other bird of prey to hunt, was the sport of kings in the Middle Ages. It was practiced from the steppes of Asia to the sands of the Middle East and the British Isles. Social position not only allowed time for the sport, but also dictated which species of bird could be used. The techniques and vocabulary of falconry were ritualized. Falconry persisted through the centuries and in the twentieth century experienced a popularity that, without regulation, contributed to the decline of such prized species as the peregrine (*Falco peregrinus*) and gyrfalcon (*F. rusticolus*). By the mid-twentieth century laws were in place to protect falcons, but raiding of traditional nest sites continued. Then came the pesticide years when birds of prey suffered as a result of biomagnification of organochlorines in their tissues and populations of birds like the peregrine plummeted. In the last decades of the twentieth century, following bans on the pesticides, the tools of falconry were used to save the birds.

Other birds have also played the role of hunter for humans. For more than 1,000 years, Japanese and Chinese fishermen have used trained cormorants to capture fish for them. A ring is often placed around a cormorant's neck to prevent it from swallowing large fish, but well-trained birds need no such ring.

Domestication of birds/artificial selection

Domestication of birds has occurred in several human cultures, typically involving species kept for meat or eggs,

but at times birds were domesticated for other reasons first. Chickens were likely the earliest domesticated birds; archaeological evidence suggests domestication in the Indus Valley more than 5,000 years ago. There is no historic record of the domestication of the red jungle fowl (*Gallus gallus*), a bird native to Southeast Asia, but by the early days of the Roman Empire it was already a commonly kept bird throughout the civilized world. No bird has been so important to humans as the chicken—nor so selectively bred. Varieties have been developed not only for meat and egg production and for fighting, but also for eggs of specific shell color, ability to tolerate crowding, and for a diversity of fancy plumages.

As urban human populations have grown, so have needs for mass production of foods. Factory farming of poultry now provides considerable protein to populations around the world. During the last two decades of the 20th century the production of chicken meat increased by an average of 6% per year and by the year 2000, factory farming of chickens was producing more than 20 billion broiler chickens per year.

While we have developed chickens that produce more meat and eggs, there are consequences to such efforts. Animal rights advocates question conditions under which the chickens are kept, charging that we maximize production in minimal space, processing chickens as if they were widgets at a manufacturing plant. Confinement of birds in high-density populations makes them more susceptible to disease, a problem we have addressed by dosing industrial flocks with antibiotics. These are often the same or similar antibiotics used by humans, and widespread use for poultry production has led to development of bacterial resistance to antibiotics. Now these resistant strains of bacteria are making humans sick and treatment difficult. Industrial poultry farming involving millions of chickens produced per farm has also resulted in high nitrogen runoff into rivers and streams, contaminating water supplies and contributing to other environmental problems.

Turkeys (*Meleagris gallopavo*) were domesticated in Mexico about 2,000 years ago and were first taken to Europe in the sixteenth century. They spread rapidly through Europe as a domesticated bird and were taken back to the New World early in the seventeenth century. The turkeys served by the Pilgrims on that first Thanksgiving were likely descendants of the Mexican birds rather than the native wild turkeys. Modern commercial turkeys are the result of hybridization of the domesticated Mexican form with the wild turkey of the eastern United States.

Ducks (the mallard, *Anas platyrhynchos*) and the greylag goose (*Anser anser*) were domesticated centuries ago in temperate Eurasia for meat, eggs, and down. Mute swans (*Cygnus olor*) were domesticated in Britain for their meat and, had they not been domesticated, might have become extinct as a result of overhunting. Two species of guineafowl (*Numida* spp.) were domesticated in Africa for meat and eggs.

The only South American bird to be truly domesticated is the muscovy duck (*Cairina moschata*). The ostrich (*Struthio camelus*), emu (*Dromaius novaehollandiae*), and rhea (*Rhea*

Temple of Horus, a god represented by a falcon, in Egypt. (Photo by C.J. Collins. Photo Researchers, Inc. Reproduced by permission.)

americana) are kept for meat, feathers, skins, and eggs. Ostrich eggs were made into elaborately carved and decorated cups by Egyptians centuries ago, and in the late twentieth century as ostrich farming grew in popularity, there has been a rebirth of artistic uses of ostrich eggs.

Birds as companions

Although the first association of birds and humans almost certainly included birds in the role of lunch, they have also been kept as messengers, for the sport of racing, for fighting, hunting, and as pets. The origins of these associations are lost in prehistory. The early Greeks and Romans kept birds for meat and for mail service. Early armies—and even armies of the twentieth century—used homing pigeons to send word of the tide of battle home from the front lines.

One 2002 estimate suggests that there are over 31 million pet birds in the United States alone. Companion animals such as birds long have been recognized as making us "feel good" and as helping to relieve our stress. Medical specialists now recognize the therapeutic role that pet birds can have.

Birds in entertainment

Our recognition of the behaviors and voices of birds have become so much a part of our culture that some have become stylized as icons of entertainment and sports that have been enjoyed now by several generations of humans. Donald Duck's waddle may not quite be that of a real bird, but the caricature fits, as does Donald's human-duck hybrid voice. Woody Woodpecker's appearance and voice are movie icons borrowed in stylized form from the North American pileated woodpecker. Big Bird is much more generic, but a lovable character recognized by young and old. The role of birds as mascots for athletic teams is really big business—and sometimes contentious. When the University of Central Florida opened its doors in the 1970s, the student body twice voted for "Vincent the Vulture" as their mascot, conjuring up images of a trained turkey vulture (*Cathartes aura*) circling the opponents' bench. But the administration overruled the choice, settling instead for Pegasus, a mythical winged horse.

Real birds have had roles in human entertainment as well, ranging from both staged and real cock fights, to the many birds in Alfred Hitchcock's "The Birds," to pets that have helped define a star's screen persona, such as the television private investigator, Beretta, who had a cockatoo named Fred, and the African gray parrot in the movie *Being John Malkovich*.

Modern birding

The interest that humans have shown in birds has evolved from seeing them as a source of food, feathers, or other products and as a source of awe, to keeping birds for sport or as pets, to collecting the skins and eggs of birds, to attracting wild birds to feeders, to observing birds as a pastime, to tallying observations of birds as a sport—sometimes even under competitive circumstances. Like the collectors of old, birders today find thrill in seeking birds. Unlike them, the culmination of the hunt is not measured in skins and eggs, but in lists of species seen or heard. Among the earliest of organized birding efforts are Christmas Bird Counts, initiated by the National Association of Audubon Societies in 1900 to replace the traditional "after Christmas dinner" shooting of birds. Birders keep life lists, year lists, state lists, and yard lists. Many birders do "Big Days" in which they strive to see as many species as possible within a 24-hour period. A "World Series of Birding" in New Jersey pits teams against one another in a grand 24-hour hunt through the state.

Some modern birders pursue their sport from an easy chair in front of a television set, carefully listing those caught, sometimes deliberately, sometimes by accident, on film or sound tracks in the background of movie sets. As an ornithologist, I have often smiled at old westerns that always seem to include a view of a turkey vulture circling high over a dead or dying desperado. Directors of those films always seemed compelled to include the "skreee" cry of a red-tailed hawk dubbed in as if produced by the naturally mute vulture. Equally amusing are the Star Trek episodes with scenes on far off planets that have Carolina wrens (*Thryothorus ludovicianus*) calling in the background. Identifying those televsion

Aborigine emu dancers in Northern Territory, Australia. (Photo by Gordon Gaman. Photo Researchers, Inc. Reproduced by permission.)

and movie birds is often a challenge and can reveal just how intimately some birders come to know birds. The challenge contributes to learning more, and it's all part of the lure of the list.

Old Tarzan movies with the story set in Africa often had the loud call of a pileated woodpecker in the background. These were not all chance recordings. Some were likely the result of deliberate dubbing in of a call that had been recorded in the 1930s by Arthur A. Allen, an ornithologist at Cornell University. Allen collaborated with Hollywood filmmakers in developing techniques for recording natural sounds. The results of those efforts not only added "realism" to movies, but also contributed much to the scientific study of bird sounds.

The contributions of birders to the science of ornithology

Amateurs have often made important contributions to science through their avocations, but there are few sciences where amateurs have made as many contributions as in ornithology. The bird-finding and bird-identification expertise of modern birders has contributed greatly to our knowledge and understanding of birds through both independent research and through organized research programs that enlist the aid of am-

ateurs. The Christmas Bird Count was not initiated as a research program, but over the years data from it have been used in efforts to monitor bird populations. The Breeding Bird Survey of the United States Fish and Wildlife Service makes use of this large cadre of skilled amateurs. So too does Project Feeder Watch and other programs through the Cornell Laboratory of Ornithology. While watching birds has grown from a simple pastime to a competitive sport for some, we are now seeing an emphasis on "birding with a purpose"—furthering our knowledge and abilities to conserve birds and their habitats through collaborative scientific studies.

Feeding wild birds

Feeding backyard birds became popular at the close of the nineteenth century and at the close of the twentieth century it was enjoying further resurgence, especially in temperate areas of developed nations, developing into a multi-billion-dollar a year industry. In 1980–81, about 20% of North American adults purchased seed to feed wild birds; by 1997, about 30% of North Americans over age 16 were involved in feeding wild birds. Bird feeding was long ignored by scientists, viewed with apathy, or even viewed as harmful to bird populations, but in the late twentieth century it became the subject of several scientific studies. Clearly, feeding backyard birds has tremendous educational potential, bringing birds to within easy

A fancy domestic pigeon—Old Dutch Capuchine breed. (Photo by Kenneth W. Fink. Photo Researchers, Inc. Reproduced by permission.)

viewing distance for young and old. Range expansion of many North American bird species has likely been facilitated by bird feeding. Such expansions may include the northward movements of the tufted titmouse (*Parus bicolor*) and red-bellied woodpecker (*Melanerpes carolinus*), the southward expansion of the evening grosbeak, and the eastward expansion of wintering rufous hummingbirds (*Selasphorus rufus*). The extent to which such range expansions can be attributed to bird feeding, versus the extent to which it is simply easier to document the expansions as a result of bird feeding, is not clear. Many factors are usually involved in range expansions, including such things as habitat modifications, changing climate, and elimination or reduction of predators or competitors, as well as increased or more dependable food supplies.

Introduced species

North America has been the "melting pot" not only for immigrant humans from many nations, but also for more than 120 bird species that have been deliberately or accidentally released in sufficient numbers to establish breeding populations. About 40 of those have been successful; some, such as the European starling, rock dove (*Columba livia*), and house sparrow (*Passer domesticus*), have been incredibly successful. Each has its own story. Rock doves were brought with early immigrants as semi-domesticated birds that were kept for food and sport. House sparrows were brought to control insect pests in the gardens of immigrants, and though not prone to long distance dispersal, succeeded in conquering North America and many other areas as a result of deliberate introductions. For example, many wagon trains heading west from St. Louis took along a cage of house sparrows to assure pest control in gardens at the end of the journey. Early attempts to introduce the European starling apparently failed, but an attempt in 1890 succeeded beyond all expectations. That effort was for the sole purpose of introducing a bird that had been mentioned in

Shakespeare's Henry IV! Today the European starling is touted as the most abundant bird in North America.

The greatest diversity of successfully introduced exotic birds can be found in warmer areas. Florida, southern California, and Hawaii all have significant exotic bird populations. The most successful of these birds are ones that can live in association with humans. Indeed, many have been inadvertent introductions as a result of escape of pet birds and birds destined for the pet trade. Some, such as the Eurasian tree sparrow (*Passer montanus*) and crested myna (*Acridotheres cristatellus*) in North America have established small, relatively stable breeding populations and such small populations are rarely linked to problems. As numbers increase, however, competition with native species and other problems have often become evident. Many of the introduced exotics are secondary cavity nesters that compete with native cavity nesters—birds that nest in cavities such as a hole in a tree. Populations of many native North American birds have suffered as a result of competition with house sparrows and European starlings. The budgerigar (*Melopsittacus undulatus*) has been reported competing with purple martins (*Progne subis*) for nest sites in Florida. Aside from competition with native species, European starlings, house sparrows, and rock doves are often a nuisance in making messy nests on buildings, have the potential for dispersing diseases and parasites to poultry, other birds, and humans. Their droppings have damaged buildings and monuments and have created health hazards for humans. While the monk parakeet (*Myiopsitta monachus*) is considered a serious agricultural pest in Argentina, it has for many years been imported into the United States for the pet trade. In 1972, a crate of monk parakeets destined for the pet trade was dropped as it was being unloaded during a snowstorm in New York. The birds survived and began nesting in the area. Other escaped monk parakeets have likely augmented their population and the species now nests along the Atlantic coast from New York to Florida and west to Louisiana. In Florida the species has become a serious problem as a result of building its large stick nests on electric transformers.

In the last years of the twentieth century the Eurasian collared dove (*Streptopelia decaocto*) made it to south Florida after escaping captivity in the Bahamas. It succeeded in establishing breeding populations there and by 2002 was nesting in much of eastern North America. Other exotic doves have become locally established, such as the ringed turtle doves (*Streptopelia risoria*) in several North American cities, but have shown no tendency for wide dispersal.

Some introduced birds have been moved about the world not only with government approval, but also by various governments. Exotic game bird programs of the United States Fish and Wildlife Service and various state and provincial governments have resulted in the introduction of many species, few of which have been successful and most of which have left us with continuing questions. Even for the ring-necked pheasant (*Phasianus colchicus*), often touted as a highly successful introduction, questions have been raised about competition with native species. In at least some areas, populations have not been sustained and more pheasants are released from game farms than are shot by hunters each year. These programs have fallen out of favor because of the recognition of problems with such introductions, but the "potential" for new game birds is continually discussed and there have been past subsidence and resurgence in interest in introducing exotic game birds. Memory of the problems and failures often seems short-lived.

Among the famous failures were repeated introductions to North America of the European common quail (*Coturnix coturnix*), a migratory species that simply disappeared following introductions. Some apparently followed migration pathways long incorporated into their genetic make-up. Instinct seems to have programmed them to head a certain direction, and fly a certain distance or length of time to reach wintering areas. But when moved from Europe to North America, following that genetic roadmap at least sometimes dumped them in the Gulf of Mexico.

Birds as vectors of human disease

In the summer of 1997, a particularly lethal form of influenza appeared in Hong Kong and chickens were identified as an intermediate link in its transmission to humans. Thousands of chickens were killed and a human catastrophe was averted. One of the consequences of maintaining high numbers of birds near human population centers is the hazard posed by their role as intermediate hosts for human diseases.

In August 1999, Americans were shaken by the news of a new disease in North America—one that was killing humans, horses, and birds. West Nile virus, originally discovered in Africa, had spread to Europe, then to North America, and wild birds were part of the transmission cycle. This mosquito-borne virus in its most serious form causes encephalitis, an often fatal inflammation of the brain. As of July 2001, more than 70 species of North American birds have tested positive for the disease, though most were crows (*Corvus* spp.). Birds don't "cause" the disease, but their role as an intermediate host, and the potential for rapid spread of the disease through bird migrations, have generated some negative attitudes towards birds. As of 2002 the disease has been found along the Atlantic coast from New York to Florida and as far west as Louisiana. The threat to humans is important, and the threat to wild birds is very serious. Since members of the crow and jay family (Corvidae) seem particularly vulnerable, an increased incidence of the disease in Florida could seriously threaten the endangered Florida scrub-jay (*Aphelocoma coerulescens*). Thus far there is no evidence that West Nile virus can be transmitted directly from birds to humans, though those handling dead birds are urged to take appropriate cautions. In this case, chickens serve us in the fight to prevent the disease. Penned chickens have been placed in many areas and their blood is regularly checked for evidence of the virus, thus these sentinel chickens serve as an early warning system.

Other bird-human disease links have generated concern in the past, although most of the diseases involved are not commonly found in humans. Psittacosis also known as parrot fever, is a rare human disease caused by a chlamydia, a parasite closely related to bacteria. Humans usually get the disease by inhaling spores in dust from dried bird droppings or from handling infected birds. In humans the disease causes flu-like or pneumonia-like symptoms and it is usually not fa-

Totem poles in Vancouver, Canada, carved by Native Americans of that area. (Photo by Porterfield/Chickering. Photo Researchers, Inc. Reproduced by permission.)

tal. We know now that birds other than parrots can transmit the disease and the disease is increasingly referred to by the more appropriate names "chamydiosis" or "ornithosis."

Aside from publicity associated with the discovery of West Nile virus in North America, one of the most frequently reported "disease links" between birds and humans in the United States is with the fungal disease histoplasmosis. In this case, birds do not deserve the negative association made: they neither harbor nor carry the disease. The disease-causing organism, *Histoplasma capsulatum*, is a fungus that grows in nitrogen-rich soils in the southeastern United States. It is spread to humans when such soil is disturbed and generally when spores or bits of the fungus are blown by the wind and inhaled by humans. The link with birds is related to the millions of blackbirds that are sometimes found in individual winter roosts in the Southeast. Excrement from the roosting birds enriches the soil and provides optimum conditions for the growth of the fungus. But many other sources of soil enrichment also occur: including cattle feed lots, poultry farms, heavily fertilized agricultural fields, and roosting concentration of bats (which also get blamed for the disease). The problem comes when such enriched soil is disturbed, dries out, and the fungus becomes airborne.

Problems between birds and humans have always included bird consumption of foods we might otherwise eat or feed to livestock. Flocking birds such as crows, starlings (Sturnidae), doves (Columbidae), blackbirds (Icteridae), gulls (Laridae), and parrots (Psittacidae) are among the primary offenders. The same and other flocking birds also create problems in urban centers, around airports, and other areas. Whole industries have built up around the development, sale, and deployment of devices and chemicals to thwart depredation of crops by birds. Yet in some areas the problem remains very serious. In parts of Argentina, for example, the monk parakeet is said to consume about 50% of the annual grain crop. Such figures, however, must be viewed with a skeptical eye. There is no doubt that the birds take grain, some of which would otherwise make it to human tables. But how much of the grain taken is spilled grain coming from the ground? And

Metal bands are often used to track endangered birds, such as this Victoria crowned-pigeon (*Goura victoria*), found on the island of New Guinea. (Photo by Robert J. Huffman. Field Mark Publications. Reproduced by permission.)

how many weed seeds and harmful insect pests do these same birds take? The charges against birds are rarely presented in the form of a "cost-benefit" analysis, but they should be. Here is a challenge for economists and where greater understanding is needed.

Ornamental uses of bird feathers and bills

Native cultures around the world have employed bird feathers in art and decorative efforts. Some of these efforts have contributed to the extinction of species. For example, the first humans to reach the Hawaiian Islands made cloaks for their royalty of the yellow-feathered skins of the Kauai oo (*Moho braccatus*). It took feathers from thousands of birds to make a single cloak. Similarly, the Maoris of New Zealand made use of the skins and feathers of the huia (*Heteralocha acutirostris*), contributing to their extinction.

In the late 1800s feathers—and even whole birds—became popular ornamentation for ladies' hats in Western cultures. At one point the aigrette feathers of egret species sold for as

much as 32 dollars an ounce; they were literally worth their weight in gold. The decline in populations of species such as the snowy egret (*Egretta thula*) was recognized before it was too late and became a cause that led to the Audubon movement in North America and a new era of conservation activity around the world.

Birds as religious symbols

In most cultures, birds have always played major roles as symbols. A few of these include the sacred ibis (*Threskiornis aethiopicus*) of Egypt symbolized the moon god, Thoth, a deity of wisdom, apparently because its curved bill resembled the crescent moon. Cranes were symbolic of Apollo, the Greek god of the sun. The hoopoe (*Upupa epops*) plays a major role in the "The Conference of the Birds" in Islamic mysticism. Doves are well recognized as symbols of love and peace, and the Holy Spirit in Judeo-Christian cultures is often symbolized as a dove.

Birds in myths, literature, and art

The legend of the Roc brought back from Marco Polo's travels may well have a basis in fact in the elephant birds (Aepyornithidae) that appear to have survived into historic times on Madagascar. These birds were flightless, stood 11 feet tall (335 cm) and laid an egg that was large enough to hold the contents of seven ostrich eggs—or as one author suggested, 10,000 hummingbird eggs! Elephant bird eggs are still occasionally found buried on Madagascar and have at times been salvaged by native peoples for use as buckets. A more profitable use has been their sale to eclectic collectors.

The causes of endangerment

Certainly the single most important cause of endangerment of modern birds is habitat destruction. Deforestation and conversion of natural habitats to landscaped, irrigated, and fertilized areas dominated by exotic species not only replace native habitats, but also fragment remaining natural environments, thus limiting movement of habitat-limited birds among populations. Exotic animals often compete with or prey on native birds. Particularly on islands, the impacts of exotics can be devastating. On the island of Guam, for example, the brown tree snake was accidentally introduced in the 1950s. Without natural controls, it has increased in abundance and has decimated native forest birds. At least twelve species of birds have disappeared from the island. For a brief time from the 1940s into the early 1970s, organochlorine pesticides were an incredibly serious threat to birds high on the food chain such as herons, egrets, hawks, falcons, eagles, pelicans, and their relatives. These pesticides were entirely human-made and had been developed to be persistent—to stay around and continue to control insect pests without frequent spraying. The developers succeeded in their efforts only to learn that vertebrates could not break down the chemicals and they were stored in fatty tissue, becoming more concentrated, "biomagnified," with every meal.

Studies of the accumulation of toxins in bird tissues taught us that humans too were vulnerable to these chemicals and birds became recognized as "environmental barometers" capable of providing us an early warning of serious environmental problems. In the mid-twentieth century, we learned that heavy metals such as lead and mercury were biomagnified in living tissues. So too was another class of human-made compounds, the polychlorinated biphenyls—known as PCBs. PCBs had been developed for use in electrical transformers, but are now recognized as posing serious threats to both birds and humans.

Hope for a feathered future

By the early twenty-first century, nearly every state in the United States had a state Ornithological Society or state Audubon Society that published a journal including results of original observations and research with birds. Most developed nations of the world have at least one national ornithological journal—some have several. In North America *The Auk*, *The Wilson Bulletin*, *The Condor*, and *The Journal of Field Ornithology* are international journals focusing primarily on the results of original research with birds. The list merely begins with these because they encompass the whole field of ornithology. We could continue with specialty journals such as *Waterbirds*, *Birding*, and *The Journal of Raptor Research*. In the United Kingdom, *The Ibis*, and in Germany, the *Journal für Ornithologie*, stand among the oldest of continuously published scientific journals. In short, our knowledge and understanding of birds is increasing at a geometric rate. This knowledge is no longer remaining within the realm of science, but is now quickly passed on to a growing birding public through the pages of such popular magazines as *Birder's World*, *Bird Watcher's Digest*, and *WildBird* in North America, and literally scores of popular bird magazines published around the world.

In the first half of the twentieth century motion pictures brought birds to audiences around the world, often through organized programs such as the Audubon Screen Tours in North America. Then television brought such programming into our homes. Then satellites sent such programming around the world. In the early twenty-first century, the Internet has further revolutionized the way we disseminate knowledge about birds. This information age has, of course, also brought growth of human populations and accelerated destruction of habitats. But with new understanding of birds and their needs, and the ability for individuals around the world to band together on behalf of troubled species, conservation efforts have reached new levels and provide us with great hope. Efforts to conserve birds and their habitats and to reduce the rate of human-induced extinction of species grew immensely in the closing decades of the twentieth century. These endeavors are intimately linked not only to research and dissemination of information, but to the approval and implementation of laws such as the Endangered Species Act of 1973 in the United States. This monumental piece of legislation has a primary focus on North America's endangered species, but ramifications that are global. In concert with the Convention on International Trade in Endangered Species (CITES), and endangered species laws in other countries around the world, a framework for global protection of the diversity of birds and other living creatures has been put in place. The key is to keep the momentum going, but for many species and in many areas of the world, it is a close race with extinction.

There are many successes and signs of hope in this race. Following the banning of many organochlorine pesticides in the 1970s, there have been major increases in populations of species such as the double-crested cormorant (*Phalacrocorax auritus*) and other fish-eating birds whose numbers had declined precipitously. It was a sad day when the last free-living California condors (*Gymnogyps californicus*) were taken into captivity in a partnership involving zoos, federal and state agencies, and conservation organizations. It was a last ditch effort to save the species through captive breeding, but the effort paid off. Condor numbers grew through careful stewardship and, though still critically endangered, the species once again flies free. Along the way a great deal was learned that has been applied to conservation efforts for other species.

Other captive breeding and release programs have brought nesting peregrine falcons back to North American cities, and increased bald eagle (*Haliaeetus leucocephalus*) populations dramatically. Combinations of captive-breeding, habitat-restoration, nest-box, and educational programs in the early twenty-first century are assisting many species on the brink of extinction. In Hawaii, captive rearing of the 'alala, or Hawaiian crow (*Corvus hawaiiensis*) is underway. In Puerto Rico the Puerto Rican parrot (*Amazona vittata*) teeters on the edge of extinction, aided by captive breeding and nest boxes, but so low in numbers that hurricanes, predators, and poachers continually threaten it. In Mauritius, captive breeding and nest boxes saved the Mauritius kestrel (*Falco punctatus*) from imminent extinction, bringing a low of only six known individuals in 1974 to nearly 300 birds by 1994.

Other unique, often high-technology, tools have contributed to conservation efforts at the beginning of the twenty-first century. These include biochemical tools such as DNA-DNA hybridization to look at relationships among birds and examination of trace minerals in feathers to determine site of origin of individuals. Miniature transmitters and satellite radio-tracking have been used to monitor movements of some species. Ultralight aircraft have been used to teach whooping cranes (*Grus americana*) a new migratory pathway.

The ultimate keys to conservation of biodiversity rest neither with laws and their enforcement, nor with captive breeding and other costly and labor-intensive manipulations of the creatures. All of these are good, and at times critically necessary. But they are not inherently sustainable. The keys to conservation of biodiversity are to be found in providing the habitats needed for naturally reproducing populations and in understanding the values of biodiversity. What does each species contribute to that diversity? How does each interact with its world and with us? Sound research and education are the ultimate keys, because to know is to love—and we protect what we love. If we know the wonders of these feathered beings we share our world with, we will want to protect them.

Resources

Books

Dunn, E. H., and D. L. Tessaglia-Hymes. *Birds at your Feeder.* New York: W.W. Norton and Company. 1999.

Gibbons, F., and D. Strom. *Neighbors to the Birds.* New York: W.W. Norton and Co. 1988.

Hyams, E. 1972. *Animals in the Service of Man.* Philadelphia, PA: Lippincott. 1972.

Jackson, J. A. *Bird Watch.* Washington, D.C.: Starwood Publishing Co. 1994.

Jackson, J. A., and B. J. S. Jackson. "Avian Ecology." In *The Birds Around Us.* San Francisco: Ortho Publishing Co, 1987.

Zeuner, F. E. *A History of Domesticated Animals.* New York: Harper and Row. 1963.

Periodicals

Crawford, R. D. "Introduction to Europe and diffusion of domesticated Turkeys from the Americas." *Archivos de Zootechnia* 41 (154, extra): 307–314.

Jackson, J. A. "Century of birdwatching." *Birder's World* 13 (6): 58–63.

Other

Geiss, A. D. *Relative attractiveness of different foods at wild bird feeders.* Washington, D.C.: United States Fish and Wildlife Service, Special Scientific Report, Wildlife No. 233: 1–11.

Jerome A. Jackson, PhD

• • • • •

Avian migration and navigation

What is migration?

Ornithologists typically think of migration in terms of the dramatic round-trip journeys undertaken by species that move between high and low latitudes. Even in birds, however, migrations of many types occur that vary in regularity of occurrence, duration, and distance covered. The theme that ties the various types of migration together is that they are all evolved adaptations to fluctuating environmental conditions that render some areas uninhabitable during some portion of the year.

The adaptations to fluctuating environmental conditions that render some areas uninhabitable range from irruptive movements to true migration. Irruptive movements involve irregular dispersal from an unfavorable area to a more favorable area. In contrast, true migration characteristically involves return to the place of origin when conditions improve.

Bird migration includes a broad spectrum of movements by individuals that range from irregular eruptions of individuals to the long-distance round-trip flights that we typically think of when migration is mentioned. There are between 9,000 and 10,000 species of birds and more than half of these migrate regularly. Billions of individual birds are involved in these migrations. Depending upon species, the migration might comprise a journey on foot up and down a mountain (as in some grouse), or it might involve a flight that literally spans the globe. Some species fly by day, others almost exclusively at night; some migrate alone, others in flocks that may reach immense size; many migrations involve a return, often with uncanny precision, to localities previously occupied.

In terms of sheer magnitude, the migrations of many seabirds are the most impressive. The famous Arctic tern (*Sterna paradisea*) nests as far north as open ground exists and migrates the length of the oceans to spend the winter in Antarctic waters, a round-trip of some 25,000 mi (40,000 km) performed every year of the bird's life. Some of the great albatrosses, such as the wandering albatross (*Diomedia exulans*), circumnavigate the globe by moving west to east over the turbulent oceans within the "roaring forties" latitudes south of the tips of the southern continents. Sooty shearwaters (*Puffinus griseus*) are extremely abundant seabirds that breed on islands deep in the Southern Hemisphere, mostly around New

Zealand and the southern tip of South America. In late spring and throughout the northern summer, sooty shearwaters migrate northward and circle the basins of the northern Pacific and northern Atlantic Oceans. Flocks of many thousands of individuals may be seen along the Pacific coast of North America. By late summer they are headed back across the ocean to their distant nesting islands, having circled the ocean in the process.

Many shorebirds nest at high latitudes in the Arctic and spend the winter far into the Southern Hemisphere. Their chicks are precocial and thus require relatively little parental care. Adult birds often depart on autumn migration before juveniles, leaving the inexperienced youngsters to make their way to the wintering grounds on their own. Typically, shorebirds migrate in flocks, often at night, but also during the day when crossing large ecological barriers such as oceans, gulfs, or deserts requires extremely long flights. American golden-plovers (*Pluvialis dominica*) make a non-stop flight from the Maritime Provinces of Canada across the western Atlantic Ocean to South America, often flying at altitudes that exceed 20,000 ft (6,000 m). In spring, the species follows a different route northward, crossing the Caribbean and Gulf of Mexico and then heading north through the interior of North America. Its western cousin, the Pacific golden-plover (*P. fulva*), departs its Alaskan nesting areas and flies over the Pacific to Hawaii and beyond, performing single flights of 25,000 mi (7,500 km) or more.

Most waterfowl (swans, geese, and ducks) are shorter-distance migrants, typically nesting and overwintering on the same continent. They tend to migrate in cohesive flocks that often contain family groups and repeatedly use traditional stop-over locations to rest and refuel. Often flying both day and night, migrating waterfowl are strongly influenced by weather conditions. When the conditions are right, they can cover many hundreds of miles in a single, high-altitude flight.

Soaring birds can take advantage of a free energy-subsidy from the atmosphere. Warming of the earth's surface induces columns of rising warmer air (thermals). Hawks, eagles, vultures, storks, and cranes use this atmospheric structure by finding a thermal and then circling within the column of rising air, gaining altitude with almost no expenditure of energy. Once a substantial altitude has been reached, the birds glide

Blackpoll Warbler

South America

Migration route of the blackpoll warbler (*Dendroica striata*). (Illustration by Emily Damstra)

typically fly alone or in only very loosely organized flocks and because of their lesser powers of flight, most make shorter individual flights and overall migrations than larger, stronger fliers. A typical night migration under good flying conditions might encompass 200 mi (320 km) and be followed by two or three days on the ground during which the birds rest, feed, and deposit fat supplies that will fuel the next leg of the migration. Some small songbirds regularly cross the Gulf of Mexico and the Mediterranean Sea-Sahara Desert. Such flights, initiated at dusk, often require more than the night to complete, thus the birds continue to fly into the following day until a suitable landing place is reached. If bad weather is encountered, especially over water, many birds may become exhausted and perish.

The blackpoll warbler (*Dendroica striata*) of North America is exceptional. Weighing about 0.4 oz (11 g) at the end of its breeding season, blackpolls may accumulate enough subcutaneous fat in autumn to increase its body weight to approximately 0.7 oz (21 g) before departing from northeastern North America on a non-stop flight over the western Atlantic Ocean. The trip takes from three to four days to complete, and with an in-flight fat consumption rate of 0.6% of its body weight per hour, the blackpoll has enough added fuel for approximately 90 hours of flight. There is essentially no place to stop enroute and most individuals overfly the Antilles and make first landfall on the continent of South America. Their autumn journey is surely among the greatest feats of endurance in the bird world.

Why migrate?

Despite the obvious diversity in migration strategies among birds, they all represent adaptations to variability in resources and the predictability of that variability. These two variables determine to a large extent what kind of migratory behavior will evolve. Resources are the necessities of life: food, water, shelter, etc. Many animals respond to the seasonal disappearance of some essential resource by entering an inactive or dormant state (e.g., hibernation). Most birds, however, preadapted by the ability to fly long distances, have responded to variable resources by moving to more hospitable areas.

The more variable the resources on which a bird depends, the stronger will be the selection pressure favoring the evolution of migration. Another important consideration is the predictability with which the resources fluctuate. There is no option for an insectivorous bird such as the blackpoll warbler to spend the winter in its breeding range. Selection favoring obligatory migratory behavior will be strong because any individual that fails to migrate will not survive the winter. In more temperate areas, differences in the severity of winter from year to year might make it possible for a bird to overwinter in some years, but not in others. Selection in these less predictable environments should favor the evolution of more flexible strategies such as partial migration in which some individuals of a population migrate whereas others remain year round in the breeding area. Some fluctuations in resources do not always follow regular seasonal changes in climate. The coniferous seeds and buds eaten by various cardueline finches, such as crossbills, fluctuate not only seasonally, but also from year to year and region to region. These fluctuations may be

off, covering ground as their path slowly descends. After covering a considerable distance, often without the need to flap their wings, the birds need to locate another thermal and repeat the process. Under the right weather conditions, large distances over the ground can be covered with very little energetic cost. Because thermals are present only during the warmer portions of the day, soaring migrants are almost exclusively diurnal and selective in terms of the weather conditions under which they migrate.

Some common landbirds (including the passerines) migrate during daylight hours. These include swifts, some woodpeckers, swallows, some New World flycatchers, jays, crows, bluebirds, American robin (*Turdus migratorius*), New World blackbirds, European starling (*Sturnus vulgaris*), larks, pipits, some buntings, cardueline finches, and others. Most songbirds, however, migrate almost exclusively at night. Nocturnal migrants include many thrushes, flycatchers, sylviid and parulid warblers, vireos, orioles, tanagers, and many buntings and New World sparrows. Night migrants

Migration routes of the arctic tern (*Sterna paradisaea*), sooty shearwater (*Puffinus griseus*), and the wandering albatross (*Diomedea exulans*). (Illustration by Emily Damstra)

quite unpredictable, so migration in these species must be facultative, responding directly to local conditions. These movements are termed irruptive because large numbers of birds emigrate from the boreal forests in some years, but remain there in others. Nomadism implies that individuals are perpetually on the move. It is not clear that any bird species are truly nomadic, although birds of the desert interior of Australia are often cited as examples.

Origin and evolution of migration

Seasonal migration is found on all the continents and among species as diverse as penguins, owls, parrots, and hummingbirds. Evidence of the first origins of migration are probably lost forever, but recent phylogenetic reconstructions suggest that migratory behavior has appeared and disappeared repeatedly in avian lineages. Its first appearance may well have coincided with the acquisition of efficient long-distance flight capability. Although present patterns of migration may have been influenced by global climatic events such as glaciation, migration on a large scale probably predated these events.

There is considerable evidence that migratory behavior can appear and disappear quite rapidly in populations. House finches (*Carpodacus mexicanus*) introduced into the more seasonal climate of the northeastern United States from a sedentary population in southern California have evolved full-scale partial migration in fewer than 50 years. Experiments with a sylviid warbler, the blackcap (*Sylvia atricapilla*), have demonstrated strong genetic control over several aspects of migratory behavior. The blackcap is widespread in western Europe from Scandinavia south to areas around the Mediterranean, and island populations live on the Canary and Cape Verde Islands. This single species exhibits a wide range of migratory behavior from obligate long distance migrants in the north to non-migratory populations on the Cape Verdes. Cross-breeding of Canary Island birds with northern obligate migrants produced individuals that showed almost exactly intermediate levels of migratory activity in captivity. This activity, known as *Zugunruhe* or migratory restlessness, is exhibited twice a year in captive migratory birds. It is characterized by a daily rhythm in which these birds rest for a short

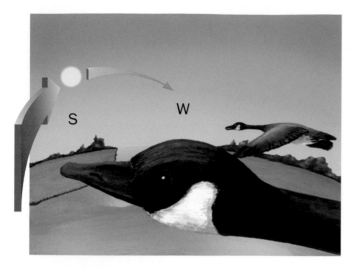

During daylight, birds can tell direction as the sun moves through the sky from east to west. (Illustration by Emily Damstra)

period in the evening and then awaken and flutter vigorously on their perches throughout much of the night.

Control of the annual cycle

Migratory behavior is an integral part of the annual cycle of those species that are obligate migrants. Like other events that occur during each year of a bird's life (e.g., molt, breeding), migration is under the control of an endogenous clock called a circannual rhythm, a clock that runs with a periodicity of approximately one year. In birds held under constant environmental conditions (constant temperature, constant dim light), the events of the annual cycle, including migratory behavior, continue to recur in the proper sequence for years. Thus, although migration in the real world is certainly influenced by ambient environmental conditions (climate, weather, food supply, changes in photoperiod), the underlying stimulus comes from within. In nature, circannual rhythms are synchronized with the real world changes in seasons through the associated systematic changes in day-length (photoperiod).

Energetics of migration

Birds are remarkably adapted to flight by virtue of a lung and air-sac system that permits maximal oxygen uptake, hollow bones and other weight-reduction adaptations, and extraordinarily efficient hemoglobin. Nonetheless, long migratory flights are extremely strenuous and energy-demanding. Gram for gram, fat is the most calorie-rich substance that animals can sequester in their bodies. It provides about twice the calories per gram as a carbohydrate or protein, and oxidation of fat is a more efficient metabolic process. Fat deposition prior to migration results from changes in diet, increases in food intake, and changes in metabolism. Migratory fat is deposited throughout a bird's body, including within internal organs such as the heart and liver, but most is laid down in subcutaneous "fat bodies." The amount of fat deposited varies greatly among migratory species. Long-distance migrants and those that must cross large ecological barriers (e.g., the blackpoll warbler) deposit larger amounts

of fat prior to initiating flight. Typical nonmigratory birds carry 3–5% of fat-free mass as fat; in passerine migrants this figure may reach 60–100% at the beginning of a long flight.

Once in flight, the migrant's fat deposits are depleted. Fuel stores must be replenished during a migratory stop-over before the bird will be able to execute another flight segment. The rate of fuel recovery will depend upon the quality of the habitat in which the bird lands and lays over. Typically, songbirds are able to deposit fat at rates of 2–3% per day and under optimal conditions the rate may reach 10% of fat-free body mass. Given the obvious importance of stored fat to the success of a continuing migration, finding stop-over habitat in which fuel supplies can be quickly replenished is critical to a migrant's success.

Altitude of migration and flight speed

Most bird migration proceeds at rather low altitudes, as has been revealed by radar studies. Larger, faster-flying species such as shorebirds and waterfowl tend to fly at the highest altitudes. It is usual to find them up to 15,000 ft (4,500 m) or higher when migrating over land. Shorebirds on a trans-oceanic flight have been detected passing over Puerto Rico at 23,000 ft (7,000 m). Bar-headed geese (*Anser indicus*), regularly migrate over the tops of the highest Himalayas. Songbirds migrate at substantially lower altitudes. At night, most typically fly below 2,000 ft (600 m) and nearly all will be below 6,500 ft (2,000 m).

The speed at which a migrating bird makes progress over the ground depends upon how fast it flies (its air speed) and the wind direction and speed where it is flying. Most passerine migrants fly at relatively slow air speeds of 20–30 mph (32–48 km/hr). Flying in a tail wind could easily double their speed over the ground; likewise a head wind could substantially retard progress and a cross-wind could cause the bird to drift from its preferred heading direction. Waterfowl and shorebirds fly faster, with air speeds in the 30–50 mph (48–80 km/hr) range. Given the potential influence of winds on the progress and success of migration, it is not surprising that migrating birds show quite refined selectivity in terms of the weather conditions under which they initiate flight.

Orientation and navigation

One of the most remarkable things about birds, and migrating birds in particular, is their ability to return to specific spots on the earth with amazing precision. This phenomenon is termed homing. Many different kinds of animals exhibit homing ability, but the behavior reaches its pinnacle in birds where the scale of homing flights may reach thousands of miles in a long-distance migrant. Not all migratory species show strong fidelity to previously occupied places, but many do, returning year after year to exactly the same place to nest and to exactly the same place to spend the winter. How they are able to navigate with such precision over great distances is a question that we still cannot answer completely.

Many young birds on their first migration travel alone, without the potential aid of more experienced parents or other older birds. Having never been to the winter range of their population before, they are not navigating toward a precise locality, but rather toward the general region in which their relatives spend the winter. But, how do they know in which direction the wintering place lies and how far it is from their birthplace? At least some of this information is genetically coded and controlled by the endogenous circannual rhythms discussed above. This migratory program provides the young bird with information about the direction it should fly on its first migration. This can be demonstrated readily by captive nocturnal migrants during the seasons of migration when the birds become restless and active at night. Displaying a behavior known as *Zugunruhe*, birds placed in a circular orientation cage will exhibit hopping that indicates the direction in which they would migrate if free-flying. Cross-breeding experiments with European blackcaps from populations with quite different migration directions have shown that the bearing taken up by inexperienced first-time migrants is strongly heritable and controlled by a small number of genes. Some species have complicated migration routes involving large changes in direction. The garden warbler (*Sylvia borin*) and blackcap are examples that have been studied. Western European populations initially migrate southwestward toward the Iberian Peninsula and then change direction to a more southward heading that takes them into Africa. Hand-raised warblers kept in Germany for the entire season and tested repeatedly in orientation cages showed this change in direction at approximately the right time during the migration season.

It is less clear how the distance of the first migration might be coded. It is known that the amount and duration of migratory restlessness exhibited by different species and populations is correlated with the distances that they migrate and the lengths of their migration seasons, suggesting that some component of the endogenous time-based program also controls the distance migrated. However, migrants are often stopped and held up by bad weather and it is not clear how the endogenous program might adjust for these changes.

Once having bred or spent the winter in a specific place, many birds will show strong site fidelity to those places. To home to a specific site requires more than a simple direction and distance program. The animal must be able to navigate to a particular goal, compensating for errors made along the way or for departures from course caused by wind. This has been demonstrated with individuals of many species that have homed after being artificially transported to places where they have never been before. In the case of various seabirds, homing distances were thousands of miles. Animals employ a variety of different mechanisms in order to home. For short distances, a number of relatively simple strategies will work: laying down a trail that is retraced; maintaining sensory contact with the goal by sight, sound, or smell; inertial navigation; referring to learned landmarks; or even random or systematic search. But for the very large distances involved in some cases of goal navigation in birds, including long-distance homing by pigeons, some other explanation is required. The best evidence suggests that birds employ a navigational mechanism based on possessing an extensive map that provides the individual with information about its spatial location relative to its goal (home), and a compass that is used to identify the course direction indicated by the map. Obviously, neither

Birds tell direction at night by using the stars as a guide. (Illustration by Emily Damstra)

component will work by itself. A compass will be useless without information about the direction toward home from the map. Likewise, the map, telling the bird that home lies to the north of its present location, will also be ineffective in the absence of a compass to indicate the direction "north."

Homing pigeons, and probably other birds as well, employ a number of strategies when attempting to return to a specific goal. There is good evidence that they use information perceived during the displacement journey to the release site to determine the direction of displacement (route-based navigation). They probably also use familiar landmarks when navigating close to home. These means will be of limited effectiveness at great distances, yet even when released at unfamiliar sites hundreds of miles from their home loft, pigeons with some homing experience return directly and rapidly. Therefore, whereas pigeons may make use of various strategies when homing, it appears that a map based solely on stimuli perceived at a distant and unfamiliar location are sufficient to provide the requisite homing information.

During the last half of the twentieth century, a great deal of experimental effort was devoted to discovering the physical basis of the compass and map employed in homing navigation by birds. Interestingly, compasses have been studied primarily in migratory birds, especially those species that migrate at night, whereas the map component of navigation has been studied almost exclusively in nonmigratory homing pigeons.

Bird compasses

A compass is involved not only in map and compass navigation such as occurs when a homing pigeon is taken away from its loft; it is also a fundamental component of orientation by migrating birds. Two main experimental approaches have been used to study bird compasses. First, various manipulations have been performed on homing pigeons and their effects monitored by observing the initial flight direction taken by pigeons when they were released at distant sites. Second, orientation in migratory birds has been studied by observing captive birds in a migratory condition. These birds will hop and flutter in the intended migratory direction when placed in a circular orientation cage (a behavior known as *Zugunruhe*). The studies on pigeons and captive migrants revealed that birds possess several different compass capabilities.

Magnetic compass

The ability to detect the earth's magnetic field is widespread among animals and microbes. The magnetic compass in birds has been demonstrated by predictably altering the migratory orientation of birds through manipulating the magnetic field within their cages, and by altering the initial homing bearings of pigeons by changing the magnetic field around a bird's head with miniature magnetic coils. By independently manipulating the horizontal and vertical components of an earth-strength magnetic field, it has been shown that the magnetic compass of birds is not based on the polarity of the field as is our technical compass. Rather, the bird compass relies on the inclination or dip angle of magnetic field lines to determine the direction toward the pole or toward the equator. Within both the Northern and Southern Hemispheres, magnetic lines of force dip downward toward the poles, so the same magnetic compass mechanism is effective for species living within either hemisphere. Birds that cross the magnetic equator will face a problem and in two species of trans-equatorial migrants it has been shown that the birds' directional response, with respect to magnetic inclination, reverses after they experience a period in a magnetic field without inclination, simulating an equatorial field.

A magnetic orientation capability develops spontaneously in young birds. The mechanism by which birds perceive the magnetic field remains uncertain. There is experimental evidence to support a receptor based on tiny particles of magnetite located in the anterior region of the head and evidence suggesting that photoreceptive pigments in the eye provide the sensor.

Sun compass

The sun compass is found in many animals. To use the sun as a compass, of course, one must take time of day into account. Animals accomplish this because the sun compass is linked to the internal circadian clock possessed by all animals. With this time-compensation mechanism in place, a bird can orient in a given compass direction at any time of day. The link between the circadian clock and sun-compass orientation can be demonstrated by changing the bird's internal clock by confining it in a room with a light:dark cycle that differs by several hours from that outside. If this is done with a homing pigeon that is then released on a sunny day, it will make a predictable error in identifying the direction in which to fly. It will take the time indicated by its internal clock to be correct and thus misinterpret the direction of the sun. With this sort of experiment it can be shown, for example, that a pigeon will mistake a midday sun, high in the sky, for a rising or setting sun that would be near the horizon. This is because only the azimuth direction of the sun is used in sun compass orientation; the sun's elevation is ignored. Birds apparently have to learn the sun's path across the sky and that path varies considerably with latitude and with season. Once the path is

learned, compass directions must be assigned to it, i.e., the pigeon must learn that the sun rises in the east, etc. There is evidence that the magnetic compass provides the compass directions that are then associated with the azimuths of the sun's path. The sun compass is the primary compass employed by homing pigeons to identify the direction that its map indicates is homeward. Its role in migratory birds is less clear, but it may be important for species that migrate at very high latitudes in the Arctic where daylight is nearly continuous and magnetic directions often unreliable.

Star-pattern compass

Night-migrating birds are the only animals known to use the stars as a compass. Once learned, orientation is based on the fixed spatial relationships among stars and groups of stars: birds can orient even under the fixed sky of a planetarium. Young birds learn the relationships among star patterns by observing the rotation of the night sky that results from the earth's rotation on its axis. In this way they are able to localize the center of rotation (Polaris) and thus true north. The innate migratory direction is coded with respect to stellar rotation in young birds.

Polarized-light compass

Most nocturnal migrants initiate flights shortly after sunset and diurnal migrants often take flight around sunrise. At these times of day, patterns of polarized skylight, resulting from the sun's light passing through the earth's atmosphere are very conspicuous across the vault of the sky. Because it is sunlight that is being polarized, the patterns of polarized light in the sky change in parallel with the movement of the sun and rotate around the celestial pole similar to stars at night. Thus observation of celestial rotation via polarized light in the sky can also reveal true north. Experiments with migrants in orientation cages in which polarized light has been manipulated and skylight polarization patterns have been eliminated show that the birds employ a polarized light compass.

Multiple compasses and interactions

It is impossible to know why evolution has endowed birds with so many compasses, but presumably it is advantageous to have back-up systems available when navigational problems such as overcast skies or magnetic anomalies are encountered enroute. The compasses appear to be related to one another hierarchically. In homing pigeons, for example, the sun compass is used preferentially whenever the sun is visible; the magnetic compass seems to provide an overcast sky back-up. In adult migrants, the relationship among compasses is somewhat less clear and there may be differences among species. Based on the information available as of 2001, the magnetic compass appears to be central to migratory orientation, providing the primary directional information. When an immediate choice of direction needs to be made, visual compasses may be used, but ultimately the birds seem to rely on their magnetic compass.

In many areas, especially at high latitudes, magnetic and geographic compass directions may differ substantially (magnetic declination). In these situations, the true or geographic compass directions will be the more reliable indicators of the directions that birds need to travel. During the development of compass capabilities in young songbirds, true compass directions identified by observing celestial rotation of stars at night and skylight polarization patterns during the day are used to calibrate the preferred magnetic migratory direction. In this way, the visual and non-visual compasses are brought into conformity.

Navigational maps

The search for the physical basis of the navigational map sense has been one of the more controversial issues in the study of animal behavior. There are currently two hypotheses to explain the map sense of birds. They may not be mutually exclusive and as is the case with the compass, birds may have more than one map to consult when faced with a navigational problem. Experimental studies of the navigational map have been confined almost exclusively to homing pigeons.

The olfactory map hypothesis is based almost entirely on work done with homing pigeons. It states that pigeons learn an odor map of the region by associating different odors with the different directions from which winds carry them past the loft. A large body of experimental evidence involving both manipulations of the pigeon's olfactory system and manipulations of the odor environment supports the odor map. Experienced pigeons seem to be able to use an odor-based map over surprisingly large distances (up to 250 mi [400 km]). How odorous substances could provide reliable spatial information while being transported in an often turbulent atmosphere has been a contentious issue. Very recently, analysis of trace gases in the atmosphere has shown that spatially stable patterns of odorants sufficient to account for the known levels of precision exhibited by homing pigeons do exist, although the odors actually used by the pigeons remain unknown.

The other hypothesis is that birds might use the earth's magnetic field as a map. Components of the magnetic field vary systematically over the earth, particularly as a function of latitude. There is good evidence that homing pigeons can detect very small differences in magnetic fields that would be required to use the information as a map (a much more challenging task than using the magnetic field as a compass). The best evidence supporting the idea of a magnetic map is indirect. Homing pigeons are often disoriented when released at magnetic anomalies—places where the earth's field is disturbed, usually as a result of large iron deposits underground. The effect occurs only when the pigeons are forced to make their navigational decisions right at the anomaly: if released outside the anomaly, they are not disturbed by flying across it on the way home. That the pigeons are disturbed even under sunny skies when their sun compass is readily available further suggests that the magnetic anomalies are somehow affecting the map step of navigation rather than the compass.

We know that migratory birds exhibit striking site accuracy and therefore must possess sophisticated homing abilities. However, very little is known about navigational maps in these species. Although many migrants return with precision to previously occupied places, we do not know whether they are really goal-orienting throughout the migratory journey or only when they get closer to the destination.

Resources

Books

Able, Kenneth P., and Mary A. Able. "Migratory Orientation: Learning Rules for a Complex Behaviour." In *Proceedings of the 22nd International Ornithological Congress, Durban*, edited by Nigel J. Adams and Robert H. Slotow. Johannesburg: Birdlife South Africa, 1999.

Berthold, Peter. *Bird Migration*. 2nd ed. New York: Oxford University Press, 2001.

Berthold, Peter. *Control of Bird Migration*. New York: Chapman and Hall, 1996.

Wiltschko, Roswitha, and Wolfgang Wiltschko. "Compass Orientation as a Basic Element in Avian Orientation and Navigation." In *Wayfinding Behavior: Cognitive Mapping and Other Spatial Processes*, ed. R. G. Golledge. Baltimore: Johns Hopkins University Press, 1999.

Wiltschko, Roswitha, and Wolfgang Wiltschko. *Magnetic Orientation in Animals*. Berlin: Springer-Verlag, 1995.

Periodicals

Able, Kenneth P. "The Concepts and Terminology of Bird Navigation." *Journal of Avian Biology* 32 (2000): 174–183.

Able, Kenneth P. "The Debate Over Olfactory Navigation by Homing Pigeons." *Journal of Experimental Biology* 199 (1999): 121–124.

Phillips, John B. "Magnetic Navigation." *Journal of Theoretical Biology* 180 (1996): 309–319.

Wallraff, Hans G. "Navigation by Homing Pigeons: Updated Perspective." *Ethology, Ecology and Evolution* 13 (2001): 1–48.

Wallraff, Hans G., and M. O. Andreae. "Spatial Gradients in Ratios of Atmospheric Trace Gases: A Study Stimulated by Experiments on Bird Navigation." *Tellus* 52B (2000): 1138–1157.

Wiltschko, Wolfgang, et al. "Interaction of Magnetic and Celestial Cues in the Migratory Orientation of Passerines." *Journal of Avian Biology* 29 (1998): 606–617.

Other

Have Wings, Will Travel: Avian Adaptations to Migration. 25 July 1997. <http://www.umd/umich.edu/dept/rouge_river/migration.html>

Bird, David M. "Migration: Your Questions Answered." *Bird Watcher's Digest* <http://www.petersononline.com/birds/bwd/backyard/watching.shtml>

Kenneth Paul Able, PhD

Avian song

Introduction

Many people have been captivated by the soaring melodies of a singing bird, or have been amazed by the diversity of sounds and the sheer noisiness of a dawn chorus in the rainforest. Keen birdwatchers soon realize that listening more closely leads to an appreciation not only of the aesthetic beauty of song, but also of differences between songs: the whistles, warbles, trills, chirrups, squeaks, and buzzes that distinguish species. Following a single bird reveals even more; it may sing different songs when it interacts with its mate compared to to those it uses when it interacts with a neighbor, or give shrieking alarm calls when danger threatens. Detailed scientific studies of bird song have revealed the intricate complexity of communication in birds, and contributed to diverse fields such as ethology (study of animal behavior), neurobiology, evolutionary biology, and bioacoustics (study of sound and living systems).

There is tremendous variation in the sounds that birds produce. Some species produce non-vocal sounds in addition to vocalizations. Woodpeckers use their bills to drum loudly and rhythmically on tree trunks. Snipes have modified tail feathers that vibrate to produce a bleating sound as the bird plummets downward. The oscines, or songbirds, which comprise nearly half of all the species of birds in the world, are known for their well-developed singing abilities. Songs vary considerably, not only between species, but also at a finer level between different populations of the same species, between different individuals in the same population, and even within a single individual that sings a repertoire of different song types. Songs are sometimes distinguished from calls, a difference that is obvious in some species but not in others. Generally, calls are short, simple, stereotyped, and innate, while songs are longer, more complex and varied, and have to be learned. However, in some species, calls are also learned; two non-oscine groups, the parrots and hummingbirds, learn their vocalizations.

Bird species vary in who does the singing. In some species, only male birds sing. This is the case in many of the long-studied birds of the north temperate regions like Europe and North America, but it is not true of many tropical and south temperate species where it is quite common for females to sing as well. In some of the species where both sexes sing, breeding partners may coordinate their songs to produce duets that can be so precisely coordinated that it sounds as if only one bird is singing. Duetting is particularly common among the *Thryothorus* wrens of South America and the shrikes of Africa. Differences in the singing behavior of males and females are reflected in their brain structure. In species where only males sing, the song control regions in the brain are much larger in males than females, but in duetting species, females also have well-developed song control regions.

There are some patterns in when and where birds sing that are common to many species, and that hint at the function of song. Many species have what is known as a dawn chorus, where they start the day with high song rates, and some also have a smaller chorus at dusk. As well as this daily variation in song rates, there are also seasonal fluctuations. Spring is heralded in many parts of the world by a profusion of birdsong. This peak in song rates coincides with the time that many birds are establishing territories and finding mates, suggesting that song has a role in these activities. Birds often sing from a song post, where they have a clear view and may be near the border of their territory. When one bird sings, its neighbor often sings immediately afterwards, also suggesting that song is used in maintaining territory boundaries.

Methods of studying bird song

Looking and listening

Much can be learned about bird song using nothing more than a keen pair of eyes and ears, and a notebook. Many different vocalizations can be distinguished from one another by careful listening, though specialized sound analysis equipment may be necessary to detect some of the more subtle variation in form. Counting the number of songs given in different contexts allows variation in song rates associated with time of day, season, and stage of the breeding season to be identified. Observing the social context of singing can indicate whether song is directed at a partner, a neighbor, or an intruder, and what kind of response it elicits.

Recording song

Tape recorders and microphones are used to record birdsong so that it can be analyzed or used for playback in more detailed studies. Anyone can record the songs of a bird,

Marsh wren (*Cistothorus palustris*) singing in Alpena County, Michigan. (Photo by Rod Planck. Photo Researchers, Inc. Reproduced by permission.)

in pitch; for example, middle C has a frequency of 261 cycles per second while the C an octave higher has a frequency of 783 Hz. The amplitude, or intensity, of the pressure in the sound waves determines how loud the sound is. Georgia State University's Hyperphysics Web site (<hyperphysics .phy-astr.gsu.edu/hbase/hframe.html>) has more about the properties of sound.

For sound analysis, sound spectrograph machines and sound analysis computer software measure the physical properties of sound and convert them into pictures. These images allow different song types to be categorized by visual inspection of patterns, or by measuring particular features of the song such as length, frequency range, or maximum amplitude. Sonograms are images that show changes in frequency over time, while "waveforms" graph amplitude against time. Looking at a sonogram can give an idea of what a song sounds like, remembering that the vertical axis shows the pitch of the sound. A whistle at a single pitch appears as a horizontal line on a sonogram, while a descending whistle appears as a descending line. A very short sound spanning a wide frequency range appears as a vertical line, and sounds like a click, while buzzing sounds consist of a series of these clicks given in rapid succession.

Experimenting

Experiments are an important component of the scientific study of birdsong. One of the techniques used most commonly is playback, in which recorded songs are broadcast through a speaker and the response of birds to the song is monitored. Playback experiments allow responses to the song itself to be measured without any influence of other effects, like the behavior of the singing bird. Many birds respond dramatically to playback, flying toward the speaker and singing as if there were an intruder. Differences in the intensity of response to different playback suggest different levels of threat. The closer the speaker is to the center of their territory, the more aggressive the response. Even on the territory boundary, if an unfamiliar song is broadcast, it will elicit a more aggressive response than songs of a familiar neighbor. Also, if a neighbor's songs are not played from the appropriate territory boundary but from the opposite boundary, then they too will elicit a more aggressive response. Playback experiments have thus been used to show that birds can recognize different neighbors solely on the basis of their songs.

Innovative technology allows more advanced, interactive playback experiments to be conducted. Experimenters link a computer to the speaker and choose which songs to broadcast, depending on the behavior of the bird being tested. For example, the experimenter might choose to play a song that is the same length or the same type as the song the bird itself has just sung, to see how that influences its response.

The approach

Niko Tinbergen, a pioneer in the study of animal behavior, highlighted four approaches to answering questions about animal behavior. First, there are causal factors that can be studied, including both internal mechanisms and external fac-

though getting a high quality recording can be challenging. Getting as close as possible to the bird without disturbing it is a good start. It is also important to choose a time when song rates are high and there is little other noise around; dawn is often best. Parabolic reflectors or highly directional microphones focused on the singing bird help to reduce background noise. Advanced recording techniques such as microphone arrays involve recording the songs of several birds simultaneously onto a series of microphones, and using differences in the time it takes a sound to reach each microphone to map the position and movements of the birds as they interact.

Analyzing song

Analyzing birdsong requires an understanding of some of the properties of sound. Sound consists of alternating waves of high and low pressure generated by a vibrating object, and the length of one complete cycle of high and low pressure is known as the wavelength of a sound. The number of cycles reaching an observation point per second depends on the wavelength and is called the frequency, measured in hertz (Hz). Frequency differences are heard as differences

tors such as the environment. Second, there is the developmental perspective, for example how birds learn to sing and who they learn from. Third, the survival value, or function, of a behavior can be studied, such as the role of birdsong in territorial defense and mate attraction. Fourth, the evolution of the behavior over time can be investigated.

Causes and mechanisms

Song production

Birds sing using an organ called the syrinx, which operates in much the same way as the mammalian larynx. Both are associated with the windpipe, or trachea; the larynx is at the top of the trachea, while the syrinx is at the bottom where it splits into two bronchi before entering the lungs. When they sing, birds push thin tympaniform membranes into the flow of air passing through the syrinx, causing the membranes to vibrate and generate sound waves. Because of the location of the syrinx at the junction of the bronchi, birds have two voices and can produce two harmonically unrelated sounds at once. Suthers implanted tiny devices for measuring air oscillations in each side of the syrinx to show that gray catbirds (*Dumetella carolinensis*) and brown thrashers (*Toxostoma rufum*) use both sides of the syrinx in producing their songs, while some birds, like the canary (*Serinus canaria*), use predominantly one side or the other.

Sounds generated in the syrinx are modified as they pass through the trachea and bill higher up in the vocal tract. The vocal tract selectively filters overtones produced by the syrinx, emphasizing some frequencies and filtering out others. Nowicki found that birds in helium-enriched air, which is less dense and allows sound to travel faster, produce songs with more harmonic overtones. So, like humans, higher pitched components of the sound that are normally filtered out by the vocal tract become audible in helium. Postural changes also affect song. When birds sing, they move their bills, puff out their throats, and stretch their necks, modifying the effective length of the trachea and influencing the resonance properties of their vocal tract and the frequencies of the sounds they produce. Trumpet manucodes (*Manucodia keraudrenii*) have greatly elongated trachea coiled in their breasts that give resonance to their loud, deep, trumpet-like vocalizations.

The production of diverse and complex sounds by the syrinx is achieved by an integrated system involving muscles, nerves, and hormones. The muscles of the syrinx control the tympaniform membranes to finely modify the physical properties of the sound. Syringeal muscles, in turn, are controlled by nerve impulses from specialized areas of the brain including the higher vocal center (HVC). Individual nerve cells in these areas are specialized and active only when particular songs are sung. Seasonal changes in song production are associated with seasonal changes in the size of the HVC, and also with changes in testosterone levels. Implanting males with testosterone in the nonbreeding season causes them to sing more, and females in species that do not normally sing can be induced to sing by injections of testosterone. The relationship between birdsong and hormones goes both ways; as well as hormones inducing song, hearing the song of male birds causes hormonal changes in females that induce them to start building nests.

Transmission of birdsong

The farther away the singer is, the fainter its song sounds. This attenuation (weakening) of sound is partly just because of a physical principle, the inverse square law. This states that sound radiates out in all directions from a source, so doubling the distance from the source causes a four-fold decrease in sound intensity, while trebling the distance decreases intensity nine-fold, and so on. The air itself also absorbs sound, thus windy, hot, or humid weather conditions cause sound to be attenuated more rapidly. Obstacles in the environment, such as vegetation, reflect and scatter sound waves, causing further attenuation. This attenuation by air and obstacles in the environment is frequency dependent; longer wavelengths do not attenuate as quickly, so lower frequency sounds travel farther than high frequency sounds. As well as becoming attenuated as it passes through the environment, sound is also degraded and the different elements of a song start to blur as sound is reflected off the ground, canopy, and tree trunks. Songs that have rapidly changing amplitudes or frequencies are especially susceptible to this degradation, known as reverberation.

These various constraints on sound transmission affect how, when, and where birds sing. Birds seem to have songs that are structured to transmit most effectively in their environment; for example, forest-dwelling birds have songs with acoustic properties that minimize the effects of reverberation due to vegetation. Song structure is also related to function, so species with larger territories have territorial songs that are louder and will travel farther. Many birds do a lot of their singing from special song posts in their territories, choosing a perch well above the ground with not much vegetation around, so that interference from these obstacles is minimized and the song will carry farther. Birds tend to sing less on windy days when their songs are less likely to be heard. They also avoid singing at the same time as other birds, and interactions between neighboring birds often take the form of countersinging where the song of one bird is followed almost immediately by a song from its neighbor, and then the first bird may sing again as soon as its neighbor finishes. In some species, overlapping a neighbor's song is a sign of aggression that can escalate into conflict.

Hearing songs

Bird ears are not very obvious because there is no outer ear funneling sound waves into the inner ear, and feathers cover the opening of the ear. Some birds have special feather structures that serve a similar purpose to an outer ear; the dish-shaped faces of owls concentrate sound waves on their ear openings. Internally, bird ears are similar to other vertebrates, with a thin tympanic membrane that detects sound waves. Vibrations of the tympanic membrane are transferred via a small bone, the columella, to the fluid-filled cochlea of the inner ear. Movements of fluid in the cochlea stimulate underlying hair cells that are sensitive to different sound frequencies. Generally, birds are most sensitive to frequencies in the range of 1–5 kHz, but there is variation between species that influences the design of signals. Prey species can produce vocalizations that take advantage of subtle differences in the hearing abilities of their predators, enabling them to communicate with members of their own species using frequencies that predators are less sensitive to, and reducing the risk of eavesdropping.

Signals from the inner ear are transferred by the nervous system to specialized areas of the brain that extract all sorts of information about the sound. Different frequencies activate different nerve cells so that certain song types only activate particular cells. The location of the sound is worked out by integrating the signals from each ear; comparing differences in arrival time and intensity of sound at each ear tells what direction it came from. Relative intensity, reverberation, and degree of attenuation of different frequencies provide information about the distance of a sound. Barn owls (*Tyto alba*) hunt in the dark, pinpointing prey by listening. A special arrangement of their ears, with one slightly higher than the other, and one pointing downward while the other points upward, increases the detectability of the differences between ears that the brain uses to locate the source of the sound with extreme precision in three dimensions. Again, mechanisms of sound perception have implications for signal design because aspects of the structure of a song, like timing and frequency range, influence how easily listeners can locate it. Song structure, whether its acoustic properties reveal or hide the location of the singer, can therefore reveal something about song function.

Learning to sing

Nature and nurture

Like human speech, birdsong in the songbirds (Oscines) is something that has to be learned. The way in which song develops is one of the most well studied processes in animal behavior, and illustrates the complex interactions between nature and nurture, or instinct and learning, that are involved in the development of behavior. Young songbirds make only simple begging noises at first to solicit food from their parents, but later they begin trying to produce adult-sounding songs that initially are wobbly and not very precise. This "subsong" gradually improves with practice to become "plastic song," which sounds more like adult song but is still quite variable, before crystallizing to the stereotyped adult song.

Experiments have been used to identify various components of the normal song-learning process. Young birds raised in isolation in captivity develop songs that are crude approximations of normal adult song. Only if they hear adult song do they develop normal adult song. Some species copy songs heard from tapes, while others only copy songs heard from live tutors. Some species will copy adult songs of other species, but only if it is similar to that of their own. It seems that young birds hatch with a genetically determined template of song that enables them to recognize species-specific songs. The innate template allows them to produce something resembling a species-specific song even if they have never heard one, but memorizing songs they hear refines the template and enables them to produce normal songs. The second component in the process of song development is lots of practice,

known as the sensorimotor phase because it involves both sound production and perception. Young birds deafened before they begin singing are unable to produce anything resembling a normal adult song. Auditory feedback is essential at this stage; in other words, they have to hear themselves sing to be able to match the sounds they produce to the songs they are copying.

When do they learn?

The timing of song learning varies considerably among species. Some have only a very short sensitive period, and if they do not hear adult songs during this period, they never develop normal songs. Other species, open-ended learners, continue learning new songs throughout their lives. Social influences are very important for facilitating learning. White-crowned sparrows (*Zonotrichia leucophrys*) exposed to taped song are sensitive only during the period from 10 to 50 days old, and do not copy songs of other species. However, when housed with live tutors, they learn songs after 50 days of age, and copy songs of other species. Changes in social circumstances at later stages in life can also lead to learning; captive budgerigars (*Melopsittacus undulatus*) moved to a new flock modify the structure of their calls to more closely match those of their new flock mates.

What do they learn?

Most birds copy the songs of their fathers or neighbors. Who birds learn from depends partly on when they learn. If they learn only after leaving the territory where they were born, then they may learn from other males in the neighborhood where they settle. Sharing song types with neighbors is important for communication in song sparrows (*Melospiza melodia*), and Beecher and his colleagues found that young males learn the songs of three or four neighboring males, selecting song types that are shared by the neighbors. In another population, where song sharing is less important, Hughes and her colleagues found that young birds copy parts of songs and recombine song segments, rather than copying whole songs.

The number of songs that each individual learns varies considerably among species, with some learning only a single song type, while others sing hundreds or even thousands of different songs. The number of song types that birds can learn seems to be related to brain space. Individuals with large repertoire sizes have larger song-control areas in the brain. Brenowitz argues that, rather than the size of song control areas being determined by the number of songs learned, it is likely that other factors such as genetic differences or early development may cause differences in the size of song-control areas that determine how many song types can be learned.

Some species are virtuosos when it comes to learning; they sing not only species-specific songs, but also the songs of other species. A survey by Baylis showed that a diverse range of birds engage in vocal mimicry. The function of vocal mimicry is not well understood, and may vary between species. Some mimicry is associated with brood parasitism, where females of one species lay their eggs in nests of another species, the host that then provides parental care. Male indigo birds (*Vidua* spp.) learn the songs of their host species. Single nestling cuckoos (*Cuculus canorus*) mimic nestling begging calls of a whole brood of host chicks to stimulate their host parents to feed at high rates, though this deceptive mimicry is innate rather than learned. Many accomplished mimics are not brood parasites. Lyrebirds (*Menura novaehollandiae*) sing the songs of many different species, and even incorporate other sounds into their repertoires. Once these sounds have been incorporated into a repertoire, they are learned along with species-specific songs by young birds. Male lyrebirds have display areas where they try to attract females with spectacular visual as well as vocal displays. It is possible that mimicry increases their repertoire sizes and makes them more attractive to females.

Functions of song

Signaling

Song is one of the most obvious and important ways that birds communicate with one another. Although birds also use visual signals like bright colors and displays, acoustic signals can often be heard from farther away than visual signals can be seen, and can be used to either avoid or mediate closer contact. Signals, of whatever variety, are used to convey information to others, and induce a change in behavior that benefits the signaler. Birds use song to convey information about themselves, such as identity or location. Birds may also convey information about the environment, for example, by using alarm calls to warn of approaching danger. Because song facilitates recognition, it plays an important role in mediating social interactions, hence what is known as the dual function of song in territorial defense and mate attraction.

Recognition

The immense variation in the structure of songs allows birds to recognize others as belonging to the same species, to identify their sex, and even to recognize particular individuals, be it a neighbor, mate, or offspring. Recognition at all these levels is vital for survival and successful reproduction. Sound analysis and playback experiments have been used to identify the features of song that permit recognition, and to show that recognition occurs. For example, in some species, males and females sing different song types, or there may be subtle differences in male and female renditions of the same song types. Birds respond differently to playback of their neighbors, their mate, and unfamiliar birds, indicating that they can distinguish between them on the basis of song alone. Vocal recognition is particularly important for parent-offspring recognition in species that breed in large colonies.

Defending a territory

Song is vital as a first line of defense for birds that hold territories. Many species defend a territory during their breeding season, and some species defend a territory throughout the year. Song serves as a long-distance proclamation of territory ownership, and as a threat to intruders that, if unheeded, may be followed up by physical aggression. The fact that birds respond to playback by singing suggests that song has a role in territorial defense. More direct evidence comes from experiments where birds have been removed from their territories and replaced by a speaker. Far fewer intruders are

seen in territories where the speaker is playing their songs than where the speaker is playing control songs of a different species. Clearly, song alone, even without the physical presence of the bird, serves to deter intruders. A different type of experiment, involving temporarily muting birds, also shows that birds that are unable to sing are slower to establish territories and suffer more intrusions by other birds. Most studies on song have focused on male song, but research on song by female birds, reviewed by Langmore, indicates that their song also has a territorial function. It seems therefore that song is used to repel birds of the same sex.

Studies on interactions between birds occupying territories adjacent to one another have revealed some interesting details about the way songs are used in territorial defense. Birds save time and energy in defense by recognizing the songs of individuals living around them. In playback experiments, birds respond less aggressively to playback of their neighbors' songs than strangers' songs. Nevertheless, birds may direct song at neighbors, approaching the common boundary and singing in reply to their neighbor's songs, or countersinging. Birds also use their repertoires to communicate with neighbors; when individuals share song types with neighbors, they can direct songs to specific birds by matching song types. Burt and his colleagues used interactive playback experiments with song sparrows to show that song-type matching (replying with the same song type) elicits a more aggressive response than repertoire matching (replying with a shared song type other than the one just sung), and can escalate conflict between neighbors.

Although birds have ways of indicating that a song is intended for a particular recipient, McGregor and Dabelsteen argue that, because sound is transmitted in all directions, communication networks are set up where individuals can eavesdrop on signals between others. By eavesdropping on neighbors and listening to their vocal interactions with intruders, birds obtain an early warning if the intruder comes their way. They may also learn something about the competitive ability of the intruder by listening to how it fares in an interaction with a neighbor whose competitive ability is known. Females may also eavesdrop on singing interactions between males to assess their relative competitive abilities.

Attracting a mate

Male song rates are highest in spring, suggesting that song has a mate-attracting function. Consistent with this, males of some species stop singing once they have a partner, and experiments removing the female of a mated pair cause an increase in male song rates until the female is returned. More direct evidence that song attracts females comes from experiments on birds that breed in nestboxes. Nestboxes fitted with a trap to catch prospecting females and with speakers broadcasting male song catch more females than nestboxes from which no song is broadcast.

Male song can be quite elaborate, and it appears that females are not only attracted by song, but may choose between males on the basis of the amount or variety of song they produce. Males that have the time and energy to sing at high rates often have a good territory with a plentiful supply of food. Females also seem to prefer males with larger repertoires of song types. Female great reed warblers (*Acrocephalus arundinaceus*) pair with males that have better territories and bigger repertoires. They also sometimes mate with a male other than their social partner, and these males have larger repertoires. Hasselquist and his colleagues found that the offspring of males with larger repertoires of song types generally have higher survival rates, so it seems as if females are getting better quality males when they choose males that sing more song types.

As well as attracting females, male song also stimulates females to reproduce. Kroodsma played male song to captive female canaries, stimulating them to build nests and lay eggs. Again, larger repertoires seem to be better, because when he played large song repertoires to some females and small song repertoires to others, those hearing the larger repertoires built nests faster and laid more eggs than those hearing small repertoires.

Evolution of song

Regional variation

Fine-scale geographic variation is most obvious in species where each individual sings only one or two song types and neighbors share song types. Boundaries between song types are often sharp, creating dialect areas where males all sing the same song. Differences between areas are less obvious in species where each individual has large repertoires. Nevertheless, because birds generally learn their songs from parents or neighbors, songs tend to differ more between populations than within populations. Differences between populations can also be due partly to adaptations for sound transmission in different environments. Social influences might also contribute to differences between areas if individuals modify their songs to be more similar to their neighbors because there are benefits from sharing songs with neighbors.

Cultural change

Because there are slight changes every time birds learn songs, the songs in a population undergo a process of cultural evolution over time that is similar to genetic evolution. Also, because copying errors occur more frequently in the song-learning process than mutations occur in genetic evolution, birdsong provides an excellent system for studying evolutionary processes. Movement of birds between populations introduces new songs to a population and increases diversity within populations. Selection can act on song if habitats limit transmission of some songs, or if females prefer some songs.

Song and speciation

Because song is important in mate choice, it can be a powerful isolating mechanism leading to the formation of new species. Gradual changes in the form of songs could accumulate to the point where individuals no longer recognize one another as being from the same species, and therefore do not reproduce. In Darwin's finches of the Galápagos Islands, natural selection caused changes in bill size and shape, which in turn influenced the acoustics of song production. Podos suggested that these changes in the temporal and frequency structure of songs may have caused reproductive isolation and facilitated the rapid speciation that occurred.

Resources

Books

Baylis, Jeffrey R. "Avian Vocal Mimicry: Its Function and Evolution." In vol. 2 of *Acoustic Communication in Birds*, edited by D. E. Kroodsma and E. H. Miller, 51–83. New York: Academic Press, 1982.

Bradbury, J. W., and S. L. Vehrencamp. *Principles of Animal Communication.* Sunderland, MA: Sinauer Associates Inc., 1998.

Catchpole, C. K., and P. J. B. Slater. *Bird Song: Biological Themes and Variations.* Cambridge: Cambridge University Press, 1995.

McGregor, P. K., and T. Dabelsteen. "Communication Networks." In *Ecology and Evolution of Acoustic Communication in Birds*, edited by D. E. Kroodsma and E. H. Miller. Ithaca: Cornell University Press, 1996.

Periodicals

Beecher, M. D., S. E. Campbell, and P. K. Stoddard. "Correlation of Song Learning and Territory Establishment Strategies in the Song Sparrow." *Proceedings of the National Academy of Sciences of the United States of America* 91 (1994): 1450–1454.

Brenowitz, E. A. "Comparative Approaches to the Avian Song System." *Journal of Neurobiology* 33 (1997): 517–531.

Burt, J. M., S. E. Campbell, and M. D. Beecher. "Song Type Matching as Threat: A Test Using Interactive Playback." *Animal Behavior* 62 (2001): 1163–1170.

Kroodsma, D. E. "Reproductive Development in a Female Songbird: Differential Stimulation by Quality of Male Song." *Science* 192 (1976): 574–575.

Hasselquist, D., S. Bensch, and T. Vonschantz. "Correlation between Male Song Repertoire, Extra-Pair Paternity and Offspring Survival in the Great Reed Warbler." *Nature* 381 (1996): 229–232.

Hughes, M., S. Nowicki, W. A. Searcy, and S. Peters. "Song-Type Sharing in Song Sparrows—Implications for Repertoire Function and Song Learning." *Behavioral Ecology & Sociobiology* 42 (1998): 437–446.

Langmore, N. E. "Functions of Duet and Solo Songs of Female Birds." *Trends in Ecology and Evolution* 13 (1998): 136–140.

Nowicki, S. "Vocal Tract Resonance in Oscine Bird Sound Production: Evidence from Birdsongs in a Helium Atmosphere." *Nature* 325 (1987): 53–55.

Podos, J. "Correlated Evolution of Morphology and Vocal Signal Structure in Darwin's Finches." *Nature* 409 (2001): 185–188.

Suthers, R. A. "Contributions to Birdsong from the Left and Right Sides of the Intact Syrinx." *Nature* 347 (1990): 473–477.

Tinbergen, N. "On Aims and Methods of Ethology." *Zeitschrift für Tierpsychologie* 20 (1963): 410–433.

Other

Nave, C. R. *Hyperphysics.* Georgia State University. 2000. <hyperphysics.phy-astr.gsu.edu/hbase/hframe.html>

Macaulay Library of Natural Sounds. Cornell Laboratory of Ornithology. <www.birds.cornell.edu/LNS/>

Michelle L. Hall, PhD

Avian flight

Advantages of flight

Birds have developed the power of flight to an extraordinary degree that sets them apart from other vertebrates, and they have done it with minimal loss of other forms of locomotion. Unlike bats, most birds can walk and run, and many can swim and dive well enough to catch fish and squid. Migrating birds fly airline distances over mountains, seas, and deserts, and thus gain access to remote habitats such as the arctic tundra, which are highly productive during a short season, but uninhabitable for several months of the year. On a shorter time scale, wading birds can exploit tidal mudflats, which are inaccessible on foot. Many food sources are accessible only to flying animals, from the "aerial plankton" of flying invertebrates to the fruits of forest trees that rely on birds and bats for dispersal of seeds. Trees, cliffs, and islands provide flying animals with nesting and roosting places where terrestrial predators cannot reach them. Although bats are supreme at catching insects in the dark, birds have the advantage in most other respects, thanks to a unique set of morphological and physiological adaptations.

The bird body plan

All birds share the same basic body plan, with relatively minor variations, despite being adapted to a tremendous range of habitats and lifestyles. Birds have inherited the bipedal body plan of the Archosaur branch of the reptiles, which also includes dinosaurs, pterosaurs, and crocodiles, but they have a number of distinctive modifications to the basic plan. The loss of the long balancing tail of bipedal dinosaurs means that the weight of the upper body no longer acts through the hip joint, but far ahead of it. Bipedal standing and walking with this "unbalanced hip" is made possible by the characteristic bird *synsacrum*. This is an expanded and elongated pelvis that is fused rather than articulated to the vertebrae. Postural muscles pull downward and forward on the rear end of the synsacrum, which acts as a lever holding the forward part of the body up. The bird ancestor's bipedal stance left the forelimb free to evolve into a wing that is structurally independent of the legs. Unlike bats and pterosaurs, in which the leg supports the inner end of a wing membrane, birds have been free to evolve the legs for perching, walking, running, and swimming, while also being able to use them for other functions altogether, especially catching and manipulating prey.

Birds have no diaphragm like that of mammals. Instead the curved synsacrum and sternum form the two halves of a bellows, which pumps air in and out of the body cavity. The lungs are small and compact, with "air capillaries" through which air is drawn into a system of air sacs beyond the lungs. The air and blood capillaries are arranged transversely to each other to form a "cross-current" arrangement, which is more effective than the dead-end pouches (alveoli) of mammal lungs for extracting oxygen from air at high altitudes.

The bird wing

The evolution of the small archosaur forelimb into a gliding wing required the wing span and surface area to be greatly increased, while retaining sufficient bending and torsional strength to support the weight of the body, suspended from the shoulder joints. The arm skeleton, consisting of the upper arm (humerus) and forearm (radius and ulna) provides the bending strength for the inner half of the wing only. Beyond the hand skeleton, the keratin shafts of the primary flight feathers provide the bending strength of the hand wing. Their bases are tightly bound by connective tissue to the reduced and rudimentary hand skeleton, with no freedom of movement in any direction. The bases of the secondary flight feather shafts are bound to bumps on the back side of the ulna, with some freedom to rotate downward and inward, but not upward or outward. The surface area of the hand wing is made up of the expanded vanes of the primary feathers, while the secondaries make up the area of the inner part of the wing. A leading-edge tendon with an elastic section in the middle joins the shoulder to the wrist, and supports a small triangular membrane ahead of the elbow joint. Smaller covert feathers smooth over the bases of the flight feather shafts, and seal the gaps between them. The free ends of each row of coverts overlap the feathers behind, in the manner of a tiled roof.

All of the aerodynamic force acting on a bird's wing is ultimately collected at the humerus, which has to support bending and twisting loads, with little or no compression or tension. The humerus shaft is a thin-walled, hollow cylinder of large diameter, adapted to carry these loads with a minimum amount of material. The central cavity is connected to the air sac system, and filled with air. Internal struts (trabeculae) prevent buckling of the load-bearing bony wall.

Bird adaptations for flight. (Illustration by Emily Damstra)

to the wing skeleton, and therefore could not transmit bending loads to the humerus. Unlike the vanes of flight feathers, the wing membrane had no bending strength, and had to be stretched between two bony supports, the wing finger and the leg. Like modern birds, later pterosaurs (pterodactyls) reduced the tail so that it lost its original function of balancing the weight of the upper body about the hip joint, but unlike birds, pterodactyls did not expand the pelvis to provide an alternative balancing mechanism for bipedal standing and walking. On the ground they must have been quadrupedal, somewhat like vampire bats, although the front "foot" was the end of the metacarpus rather than the wrist, as in bats. As judged by the wing span, pterosaurs ranged from sparrow-sized forms to 20 ft (6 m) pterodactyls. These thrived throughout the Cretaceous period, with even larger forms just at the end. However, all of them had slender bodies and large wings, somewhat like frigate birds. None were heavy-bodied like ducks or auks.

Comparison with bats

Bats are mammals, unrelated to either birds or pterosaurs, and are first known from the Eocene mammal radiation. They have a membrane wing somewhat similar to that of pterosaurs, but with all five digits contributing to the wing structure. Digit 1 (the thumb) projects forward from the wrist and is used for climbing. Digits 2 and 3 together form a stiff panel that allows the outer part of the wing to be pulled forward, creating tension that runs through the membrane to pull against the legs. Digit 3 supports the wing tip, and digits 4

Likewise, the primary and secondary feather shafts are hollow, and filled with a keratin foam (parenchyma) which maintains the shape of the load-bearing keratin walls. The tail feathers are structurally similar to the flight feathers, with their bases attached to the rudimentary tail skeleton (pygostyle). They can be spread fanwise, forming an auxiliary lifting surface, which has been likened to an expandable delta wing, behind the main wing.

Comparison with pterosaurs

Both birds and pterosaurs originated in early Mesozoic times from the ancestral archosaur stock, which also gave rise to crocodiles and dinosaurs. They represent radically different solutions to the mechanical problems of flight, each modifying the basic archosaur body plan in a different way, unrelated to the other. All of the bending loads in a pterosaur's wing were carried by the bones of the arm and hand, ending in one enormously elongated "wing finger" (digit 4). There were no load-bearing keratin structures corresponding to the flight feather shafts of birds. Even those who interpret the parallel ridge patterns on pterosaur wing impressions as "stiffening fibers" concede that these ridges were not connected

Greater flamingo (*Phoenicopterus ruber*) in flight. (Photo by Art Wolfe. Photo Researchers, Inc. Reproduced by permission.)

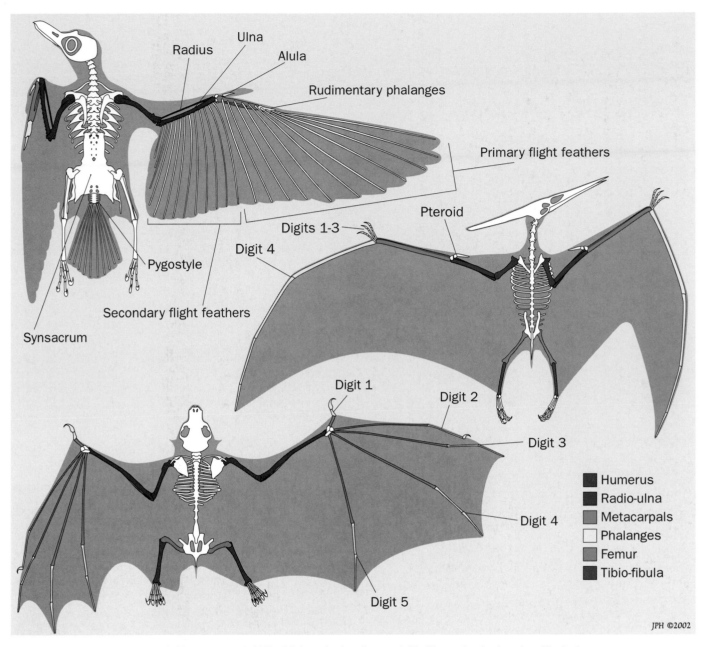

Wing comparisons of a bird (top left), pterosaur (middle right), and a bat (bottom left). (Illustration by Jonathan Higgins)

and 5 control the profile shape of the wing membrane. The ankle joints and feet are quite similar to those of pterosaurs, and are used to curl the posterior edge of the wing membrane downward. There are muscles running fore-and-aft in the membrane, which are not attached to the skeleton but are used to flatten the profile shape of the wing. Few bat species are active in full daylight, perhaps because the wing membrane is susceptible to sunburn, or because of bats' dependence on convective cooling. Many bats make seasonal migrations within continental areas, proceeding in many short stages, but bats are apparently not capable of flying as high as birds, nor of covering such long nonstop distances. Whether pterosaurs could match the performance of birds in these respects is unknown.

Flight muscles and sternum

In both birds and bats, the spherical head of the humerus articulates with the shoulder girdle, and is free to swing forward, back, up, and down, and also to rotate about its own axis, within limits set by a complicated arrangement of ligaments. The pectoralis or "breast" muscle of both birds and bats does most of the work in powered flight, by pulling the humerus downward. It inserts on the underside of a ridge that projects forward from the base of the humerus. In birds, the inner end of the pectoralis muscle is attached to the body skeleton along a prominent keel, which projects from the midline of the sternum (breastbone), and also along the outer edge of the sternum where it joins the ribs. The humerus is raised

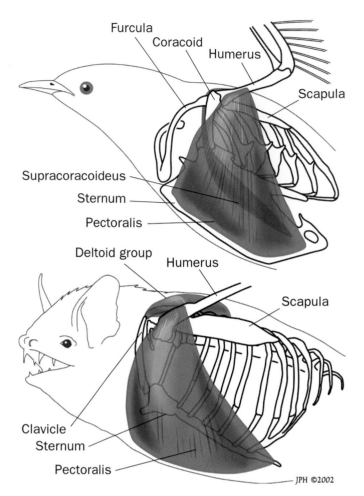

Comparison of muscles used for flight in birds (top) and bats (bottom). (Illustration by Jonathan Higgins)

by the supracoracoideus muscle, which is attached to the sternum along the base of the keel, where it is entirely surrounded and covered by the pectoralis muscle. Its tendon runs up the coracoid, through a channel between the bones of the shoulder girdle, and curves over the shoulder joint, to insert on the top side of the humerus.

This arrangement is found only in birds, not in bats, which fly very well without an expanded keel on the sternum. In their case the two pectoralis muscles pull directly against one another, with only a small keel between them, or none at all. This shows that the keel, which is such a prominent and characteristic feature of the bird skeleton, is not a requirement for flapping flight. In bats the humerus is raised by the deltoid group of muscles on the dorsal side, as in other mammals and also in reptiles.

Heat disposal in flight

The flight muscles generate a large amount of heat in flapping flight, and this heat has to be disposed of in a controlled manner to maintain the body temperature within the required limits. Provided that the air temperature is well below the blood temperature, most or all of the excess heat can be lost

by convection. Birds do this by passing a copious blood flow just below thinly insulated areas of skin that are exposed to the air flow, especially the sides of the body and the undersides of the inner parts of the wings. These areas are covered when the wings are folded, thus avoiding loss of heat when the bird is not flying. If the air temperature is too high for convective cooling, the bird opens its beak and flutters the throat pouch, thereby cooling the blood by evaporating water from the upper respiratory tract. Evaporation also takes place from the lining of the air sacs, which penetrate many organs, including the interior of the pectoralis muscles. The keel of the sternum keeps these cavities open when the muscles contract, and its function is probably to permit direct evaporative cooling of the interior of the muscles.

In bats, the huge area of skin exposed to the air flow is ready-made for convective heat disposal. The wing membrane contains a system of blood vessels in which the flow is controlled to regulate body temperature. Most bats rely mainly or wholly on convective cooling, and only resort to evaporative cooling in emergencies.

Muscle power for level flight

Because a bird's body is much denser than air, the flight muscles have to work continuously, accelerating air downward so as to produce an upward reaction that balances the weight. The rate at which work is required (power) to support the weight is highest when the air speed is zero (hovering), and decreases at medium and high speeds. However, additional power is required to overcome the drag of the body, and this increases with speed. Because of their opposite trends, these two components of power produce a characteristic U-shaped curve when added together. There is a well-defined "minimum power speed" at which the mechanical power required to fly is lower than at either slower or faster speeds.

Canadian geese (*Branta canadensis*) fly during courting. (Photo by Jack A. Barrie. Bruce Coleman Inc. Reproduced by permission.)

A bald eagle (*Haliaeetus leucocephalus*) soaring. (Photo by Joe Mc-Donald. Bruce Coleman Inc. Reproduced by permission.)

The margin of power available over power required dwindles as the size of the bird increases, eventually defining an upper limit to the mass of viable birds, which appears empirically to be in the region of 35 lb (16 kg). Modern flying birds are restricted to small sizes, when compared with the much wider size range of walking and swimming animals. Antelope-sized birds, such as ostriches and emus, exist but are necessarily flightless. Small birds have sufficient muscle power to fly over a wide range of speeds, but large birds like swans have only just enough power to fly very near the minimum power speed. Birds heavier than any living forms might have been possible under special circumstances in the past, for instance in a combination of landscape and climate that allowed reliable soaring, taking off from slopes in hang-glider fashion.

Metabolic power and aerobic capacity

Physiological experiments measure the metabolic power, which is the rate of consumption of fuel energy, as distinct from the mechanical power, which is the rate at which the muscles do work. A bird's "aerobic capacity" is the maximum sustained metabolic power of which it is capable, and this is determined by the capacity of the heart and lungs, not by the muscles. Only hummingbirds have sufficient aerobic capacity to hover continuously, although many small and medium-sized birds can hover anaerobically for short periods. Some very large birds such as condors have insufficient aerobic capacity for sustained level flight at any speed, and are forced to resort to soaring for sustained flight.

Water birds

Gulls and ducks float on the surface of the water and use their feet in a fore-and-aft rowing motion, whereas more ex-tremely adapted water birds such as loons and grebes have the legs set far back, and swing them in a more lateral motion, using the feet as hydrofoils. Auks and diving-petrels have wings of reduced size, forcing them to fly rather fast, with high wingbeat frequencies, but they also use their wings for propulsion under water. The aquatic wing motion is quite similar to flight, but at a much reduced frequency, with the wings partly folded. Gannets, petrels, and some albatrosses can also swim under water in this manner to a limited extent, diving a few meters below the surface. Penguins and the great auk (*Pinguinus impennis*) carried this line of evolution further, with wings too small to fly, but optimized as hydrofoils. The wings of penguins beat up and down (not in a rowing motion) and are convergent on the flippers of sea lions and marine turtles. Frigate birds do not swim or alight on the water at all, although their dispersal movements show that they spend weeks or months at a time over the open ocean, flying day and night.

Takeoff and landing

To take off, a bird has to acquire sufficient air speed over the wings to sustain its weight, either from forward motion, or by flapping the wings relative to the body, or (usually) by a combination of both. Birds up to the size of pigeons or small ducks can jump into the air from a standing start, and accelerate into forward or climbing flight, but larger birds have to run to get flying speed on a level surface. Swans use their large webbed feet alternately in a running motion to help them accelerate over a water surface, while cormorants and pelicans more often use both feet together. Large birds taking off from a tree or cliff drop to convert height into air speed. All birds head into the wind when taking off from the ground or water, as the wind then supplies part of the air speed that they have to acquire. If the wind is strong enough, petrels and albatrosses simply spread their wings and levitate into the air from the crest of a wave.

Landing into wind is obligatory, but newly fledged birds have to learn this, and often make spectacular errors by attempting to land across or down wind. In light winds, most birds slow down when preparing to land, by increasing the frequency and amplitude of the wing beat, tilting the wing-beat plane until the wings are beating nearly horizontally, and spreading and lowering the tail. Any residual downward and forward velocity (relative to the ground) is absorbed by the legs. Glide landings are often possible in moderate wind strengths, even for large birds. The body and wings are tilted up as the bird flares, with one final wing beat sweeping the wings forward almost horizontally, just before the wings are folded. Auks and loons land on water at rather high speeds, lowering their bellies into the water with the feet trailing behind, whereas ducks and swans swing their feet forward and use them like water skis. Gannets often enter the water in a shallow dive rather than alighting on the surface, while petrels and albatrosses slow down while gliding, and drop gently onto the surface. Guillemots (murres) nest on cliff ledges although they are not capable of flying slowly enough to land safely in such places. Their landing technique involves diving toward the cliff at high speed, then pulling up into a near-vertical climb. If this is accurately judged, the guillemot's

Mallard duck (*Anas platyrhynchos*) in flight. (Photo by Neal & MJ Mishler. Photo Reseearchers, Inc. Reproduced by permission.)

speed drops to zero just above the landing ledge, but if not, it has to dive away from the cliff, fly out to sea, and repeat the whole procedure.

Migration

The longest known nonstop migration is that of the Alaskan bar-tailed godwit (*Limosa lapponica*), which flies from the Alaskan Peninsula across the equator to the North Island of New Zealand, a distance of about 6,400 mi (10,300 km). The godwits build up fat before they depart, at the same time reducing the mass of organs such as the digestive system and liver, until about 55% of the total mass consists of fat. Like all long-distance migrants, they supplement the primary fuel (fat) by progressively consuming protein from the flight muscles in the course of the flight, "burning the engine" as the power required decreases. Although the fuel reserves are ample for the distance, this migration is remarkable because it is a formidable feat of navigation, requiring at least eight days and nights of continuous flight. The red knot (*Calidris canutus*) is another arctic breeding wader that migrates across the equator to high southern latitudes, with nonstop stages lasting several days on some routes. Many small passerine species cross the Mediterranean Sea and the Sahara Desert without stopping, while their American counterparts fly directly from the New England coast to the Caribbean Islands. Even longer distances are flown by species that are able to feed along the way, notably the Arctic tern (*Sterna paradisaea*), in which some individuals migrate from arctic to antarctic latitudes and back again each year.

Soaring over land

"Soaring" should not be confused with "gliding," which means flight without flapping the wings. The term "soaring" refers to behavior whereby the bird extracts energy from movements of the atmosphere, and uses this in place of work done by the flight muscles. Soaring birds do usually glide, but it is also possible to soar while flapping. Soaring is obligatory for many large birds, because of marginal muscle power. Slope soaring is the simplest method, in which the bird exploits rising air that is deflected upward as the wind blows against a hillside, or some smaller obstruction such as a tree or build-

ing. Thermals are vortex structures that float along balloon-like with the wind, containing an updraft in the central core and downdrafts around the outside. A gliding bird can gain height by circling in the core, but is carried along by the wind while doing so. At the top of the thermal, the bird glides off in a straight line, losing height until it finds another thermal and repeats the climb. When thermals are marked by cumulus clouds, soaring birds climb up to the cloudbase level. Away from sea coasts, this may be as high as 7,000 ft (2,000 m) above sea level in temperate latitudes, and higher in drier parts of the tropics and subtropics. Thermal soaring is the characteristic method of cross-country flight in large soaring birds such as storks, pelicans, and migratory eagles, while many raptors also use thermals to patrol in search of food. Lee waves are stationary wave systems that form downwind of hills. They can be exploited to higher altitudes than slope lift or thermals, but the technique is difficult. Canada geese (*Branta canadensis*) are known to use lee waves when migrating, and it is possible that some migratory swans may also occasionally use this method. Being slower than direct flapping flight, soaring migration is most advantageous to large birds, in which basal metabolism is only a small fraction of the power required for flapping flight. In small birds the energy cost of basal metabolism offsets any direct gains from soaring, because of the longer flight time.

Soaring over the sea

The trade wind zones of the tropical oceans cover a vast area in which the weather is predominantly fair with small, regularly spaced cumulus clouds. These are the visible signs of thermals, which are caused by the air mass being convected toward the equator over progressively warmer water. Although weaker than thermals caused by direct solar heating of a land surface, trade-wind thermals continue reliably at all hours of the day and night, and provide frigate birds with the means to disperse across the oceans without ever alighting on the surface. The middle latitudes, where stronger winds prevail, are the home of the petrels and albatrosses, especially in the southern hemisphere. The medium-sized and large members of this group skim with no apparent effort in and out of the wave troughs, sometimes very close to the surface, sometimes pulling up to 50 ft (15 m) or so, seldom flapping their wings. A petrel or albatross replenishes its air speed with a pulse of kinetic energy each time it pulls up out of the sheltered zone in the lee of a wave, into the unobstructed wind above. As the energy comes from the relative motion between the air and the waves, birds that use this technique are confined to the interface between air and water, just above the surface. Albatrosses can also slope-soar in zero wind by gliding along the leading slopes of moving waves. Pelicans, boobies, and other pelecaniform birds soar over slopes and cliffs when they come ashore to breed, but use mainly flapping flight at sea, as do gulls and auks.

Altitude of bird flight

Most birds fly near the earth's surface most of the time, except for soaring species, which climb to cloudbase or as high as convection allows. With the exception of frigate birds, most seabirds spend their entire lives within 100 ft (30 m) of the sea surface, except when they come to land, and soar in slope lift or thermals. Radar studies reveal that passerines typically fly at heights up to 10,000 ft (3,000 m) above sea level on long migration flights, while waders may fly as high as 20,000 ft (6000 m). The reduced air density at high altitudes requires birds to fly faster than at sea level, and to produce more muscle power, but there may be a small increase in range due to reduced wastage on basal metabolism, caused by the shorter flight time. The "cross-current" lungs of birds appear to give them an advantage over mammals (including bats), when it comes to extracting oxygen from low-density air. Lower air temperatures aloft reduce or eliminate the need for evaporative cooling, but reports of swans migrating at 27,000 ft (8,200 m) are apocryphal. Such large birds cannot climb to great heights by muscle power alone. It is conceivable that they could do so by exploiting lee waves, but air temperatures below −58°F (−50°C) would present physiological problems.

Origin of flight

Scientists realized in the nineteenth century that both birds and pterosaurs belong to the same branch of the reptiles as dinosaurs, and it is a matter of definition whether birds actually "are" dinosaurs, or whether they developed as one or more distinct strands of the original archosaur stock. Among living mammals, cobegos, flying squirrels, and their marsupial counterparts suggest an obvious route for the evolution of flying forms from arboreal ancestors. Pterosaurs and birds could have originated in a similar way, if we suppose that there were two parallel groups of small, arboreal archosaurs early in the Mesozoic. One such group, ancestral to the pterosaurs, would initially have resembled a flying squirrel, with a membrane stretched between the fore and hind limbs, and would then have extended the wing span by lengthening the wing finger. The other group, ancestral to birds, would have developed flight feathers from modified scales, extending the wing span and area without involving the legs in the wing support structure. Whether or not this is exactly what occurred, the two groups must have diverged in Triassic times, long before any of the known Jurassic birds or birdlike reptiles.

Flight in past times and on other planets

The two physical factors that most strongly constrain flying animals are gravity and air density. On a planet with stronger surface gravity than Earth, the maximum size of flying animals would be even more restricted than it is here, while a higher air density would ease this limit, irrespective of the atmospheric composition. It is likely that the Mesozoic atmosphere was indeed denser than the modern one, and this helps to explain the prolonged success of pterosaurs with 20 ft (6 m) wing spans. However, the largest pterodactyls, which flourished briefly in the last days of the Cretaceous period, would appear to require a reduction in gravity to make them feasible, and that is more difficult to explain.

Resources

Books

Alerstam, T. *Bird Migration*. Cambridge: Cambridge University Press, 1990.

Burton, R. *Bird Flight*. New York: Facts on File, 1990.

Norberg, U. M. *Vertebrate Flight*. Berlin: Springer, 1990.

Pennycuick, C. J. *Animal Flight*. London: Edward Arnold, 1972.

Pennycuick, C. J. *Bird Flight Performance*. Oxford: Oxford University Press, 1989.

Rüppell, G. *Bird Flight*. New York: Van Nostrand Reinhold, 1977.

Tennekes, H. *The Simple Science of Flight*. Cambridge, MA: MIT Press, 1996.

Wellnhofer, P. *The Illustrated Encyclopedia of Pterosaurs*. London: Salamander Books, 1991.

Periodicals

Pennycuick, C. J. "Mechanical Constraints on the Evolution of Flight." *Mem. California Acad. Sci.* 8 (1986): 83–98.

Colin Pennycuick, PhD, FRS

•
Struthioniformes
(Tinamous and ratites)

Class Aves
Order Struthioniformes
Number of families 6 families of living birds
Number of genera, species 15 genera; 58 species

Photo: Southern cassowaries (*Casuarius casuarius*). (Photo by Eric Crichton. Bruce Coleman Inc. Reproduced by permission.)

Evolution and systematics

While most birds fly, there are several groups of birds that do not fly and have anatomical adaptations for a life on land. Some of the largest living birds make up the group of flightless birds generally called the ratites. Historically, some taxonomists have placed most of these large birds in the order Struthioniformes. Many recent taxonomists have divided the ratite group into separate orders and others into separate suborders or families. Most recently the *Handbook of the Birds of the World* has once again placed these birds in one large order, Struthioniformes, with several families: Struthionidae, the ostriches; Rheidae, the rhea; Casuariidae, the cassowaries; Dromaiidae, the emus; and Apterygidae, the kiwis. Additionally, the extinct moas, genus *Dinornis* (Dinornithidae), from New Zealand and the elephant birds, genus *Aepyornis* (Aepyornidae), from Madagascar and Africa were probably closely related and have been placed in separate orders or families as well. The tinamous, which are included with the Struthioniformes here, may now be considered to be in the group called Tinamiformes. Unlike ratites, tinamous have a keeled sternum and can fly, although weakly. Ratites are mostly located in central and southern Africa, central and southern South America, New Guinea and surrounding achipelagos, Australia, and New Zealand.

Ratites were considered to be very ancient birds, more primitive than most other birds. Their anatomical features, once thought to be primitive, led early taxonomists to believe that ratites descended from birds prior to the development of flight. However, if this were true, many of the anatomical features of these birds would not make much sense. The current interpretation is that these birds evolved from birds that could fly, but have developed a number of special adaptations for a non-flying existence. Ratites have wing skeletons that are not fundamentally different from those of flying birds, but are used for purposes other than flying. Ostriches and rheas, for example, use their wings for both courtship and distraction

displays. Other ratites such as cassowaries, emus, and kiwis have various degrees of degeneration of the basic wing structures, but their wings are still derived from the basic wing structure of flying birds. Ratite wings still bear flight feathers and coverts in some groups, thus clearly suggesting an origin from flying birds and not directly from bipedal dinosaurs. The increase in size of most ratites has resulted in significant changes in bones, muscles, and plumage. The long, muscular legs of large ratites are well adapted for running.

Early taxonomists considered ratite birds to be a good example of convergent evolution on all the southern continents, but as the theory of continental drift emerged and evolved into plate tectonics, it became much easier to assume that ratites arose from common ancestors which became isolated as the continents drifted apart. Most families have evolved in isolation from the others. The only exception to this are the cassowaries and emus, which evolved on the same continent, Australia, but separately in different habitats, so they did not evolve in direct competition with each other. The emu, following the pattern of the ostrich and rhea, lives in more open grassland, while the cassowary lives primarily in dense rainforest. The debate on the origin and relationship of ratites continues, focusing on the exact level of relationship at the order or a higher level. Taxonomists generally agree that ratites are closer to each other and to tinamous than they are to any other bird groups. One question that has not been adequately answered is why these large flightless birds evolved in only the Southern Hemisphere. The answer to this question may well lie in the fact that major mammal predators evolved mostly in the Northern Hemisphere. Small flightless birds are very vulnerable as has been demonstrated when predators are introduced to islands with flightless birds. Ancestors of ratites had to evolve into larger and faster animals since they could not escape by flying, and this would have been much easier with a minimal number of large mammalian predators. The one exception to this is the kiwi, which evolved as a secretive forest

The emu (*Dromaius novaehollandiae*) is found throughout Australia. (Photo by Janis Burger. Bruce Coleman Inc. Reproduced by permission.)

retrogressed or have been converted to decorative plumes, and a loss of feather vanes, which means that oiling the plumage is not necessary, and as a result there is no preen gland. There is also no separation of skin bearing contour feathers, or feather tracts (pterylae), and the area of skin devoid of contour feathers (apteria). Ratites have a palaeognathous (meaning "old jaw") palate which is found in no other bird groups except the tinamous, which are considered to be the closest phylogenetically to the ratites and probably evolved from a common ancestor.

Ostriches show the greatest dimorphism with males being generally black with white plumes and the females being brown instead of black. Rheas show some dichromatism during the breeding season when the males' color grows darker black and their posture also changes. Emus have little dimorphism except the males are usually a little larger than females and their posture during the breeding season can be used to identify the sex. Cassowaries are dimorphic in size with the females being larger and more aggressive than the males. Kiwis have little dimorphism other than a small size difference.

While ratites share features such as the strong development of feather aftershafts that are often nearly as large as the main shaft, there are also many differences between families and species as well. Ostriches have their toe number reduced to two, and one is much more prominent than the other. Ostriches are also the tallest and heaviest of modern ratites. Cassowaries have developed long inner toenails that can be used defensively. While the largest cassowaries can weigh almost as much as some ostriches, they are not nearly as tall. Ostriches and rheas both have prominent wings, which, along with flight feathers, play a significant role in courtship displays. They also use their wings in distracting displays and maneuvers to evade predators or draw them away from their nests. These behaviorisms are shared with a number of ground-nesting birds that can fly, which also probably suggests their ancestry among flying birds.

Feeding ecology and diet

Ratites also share a number of other similar behavioral and ecological adaptations. Ratite eggs are very thick-shelled and difficult for most predators to break. Chicks are well developed and can walk or run within a very short time after hatching. The diets of chicks are much more insectivorous and omnivorous than are the diets of adults. While the ostrich, rhea, and emu seem to share similar ecological habitats, the digestive tract of each shows a basic difference in diet. The ostrich has the longest digestive tract, up to 46 ft (14 m) in length, suggesting an almost exclusive vegetarian diet. Ostriches are noted for eating almost anything if they have an opportunity, including stones or pebbles to help grind up the plant material in their diet. The rhea's digestive tract is the second longest, up to 25–30 ft (8–9 m) in length with the addition of large caeca (tubes branching off the junction of the small and large intestines). They are largely vegetarian as adults and eat a great variety of broad-leafed plants, including thistles and other "weeds." However, they also eat almost all varieties of agricultural crops, making them unpopular with farmers. Rhea chicks eat mostly insects in the first few days

bird in New Zealand, where the lack of large mammals may have allowed it to maintain its small size. The large number of moa species that also developed in New Zealand, and follow the pattern of ostriches, rheas, and emus, fell prey to humans when they arrived, just as the elephant birds did in Madagascar. The evolution of ratites in the absence of large mammalian predators seems to make sense. However, as with everything there is one major exception, the ostrich. Ostriches must have evolved in Africa with large mammal predators, but to compensate, they developed very large size, acute eyesight, and great speed. However, the ostrich may have evolved in very arid areas where the numbers and varieties of large predators were greatly reduced. Ostriches are also the only ratite to have spread into the Northern Hemisphere even though they have since disappeared from most of their range north of the equator. Fossil records show that ostriches once occurred from Greece to southern Russia, India, and Mongolia.

Physical characteristics

The basic characteristics of the ratite group include the following: degenerated breast muscles, lack of a keel of the sternum, an almost absent wishbone (furcula), a simplified wing skeleton and musculature, strong legs, leg bones without air chambers except in the femur, flight and tail feathers that have

Reproductive biology

The social structures of these groups have similarities and differences, but generally support the close relationship among groups. Ostriches have a social system wherein one dominant pair has a nest, but additional females may lay eggs in it and assist with incubation and chick rearing. Males incubate at night and females during the day. Both parents rear the chicks although the females dominate in the care of young. Rheas show a reversal of sex roles where the male incubates and cares for the chicks, even though the male initially gathers a harem of up to 10 or 12 females. Female rheas move from one male to the next during the breeding season. Emus are relatively close to rheas in their system where several females, usually two to four, lay eggs in one nest for a male who incubates and rears the chicks. Cassowaries are similar, with the male incubating the eggs and rearing the chicks, but usually only one or two females lay in each male's nest. Tinamous are quite similar to rheas in this reversed role of the sexes. Kiwis are monogamous and nocturnal, which differentiate them from other ratites. However, like other ratites, the males generally incubate the eggs. The larger ratites—ostriches, rheas, and emus—tend to congregate in flocks during the non-breeding season and some yearling birds remain in flocks until they become sexually mature.

Significance to humans

Ratite eggs have been used by humans for centuries. Ostrich eggs have been used as water containers by local bushmen and Sudanese. They are also used to make bracelets and necklaces and are considered to have mystical powers by some local peoples. Ratite eggs are carved and decorated by artists around the world. People also have used feathers of ratites for centuries. During the eighteenth century the soft white feathers of the male ostrich wings and tail came into fashion for ladies hats. This led to widespread hunting of the wild birds, and as a result, large declines in populations. Ostrich farms developed in southern Africa and many remain today. Ostrich farming spread to Australia, North America, and eventually around the world as use of the feathers, eggs, meat, and hides became popular. Emu farms also sprang up for similar reasons and for emu oil as well. Rhea feathers have long been used for feather dusters, and their eggs and meat are used to feed chickens and pets in South America. Hides of ratites are used to produce shoes and other leather products. Ostriches that have escaped from farms in Australia have thrived in some arid habitats there.

This elegant crested tinamou (*Eudromia elegans*) is found in Argentina. (Photo by Jeff Foott. Bruce Coleman Inc. Reproduced by permission.)

of life and over the course of several months convert to a largely vegetarian diet, but will eat insects and a variety of other things when the opportunity arises. The emu's intestine is of medium length, under 22 ft (7 m), suggesting a more varied but largely vegetarian diet. Emu chicks are also largely insectivorous when small and become more vegetarian as they grow older, but all emus eat insects and other small creatures when given the chance. The cassowary has the shortest digestive tract, under 12 ft (4 m), and is clearly much more omnivorous in its diet. As chicks, cassowaries eat largely insects and fruits but will eat nearly anything as chicks and as adults. Kiwis are adapted to feeding on earthworms, insects, and other similar creatures, and as a result they also have relatively short digestive tracts.

Resources

Books

Bertram, B. C. R. *The Ostrich Communal Nesting System.* Princeton: Princeton University Press, 1992.

Brown, L. H., E .K. Urban, and K. Newman.*The Birds of Africa.* Vol. 1. New York: Academic Press, 1982.

Bruning, D. F., and E. P. Dolenzek. "Ratites (Struthioniformes, Cassuariiformes, Rheiformes, Tinamiformes, and Apterygiformes)." In *Zoo and Wild Animal Medicine*, edited by M. E. Fowler. Philadelphia: W. B. Saunders Co., 1986.

del Hoyo, et al., eds. *Handbook of the Birds of the World.* Vol. 1. Barcelona: Lynx Edicions, 1992.

Resources

Periodicals

Bruning, D. F. "Social Structure and Reproductive Behavior in the Greater Rhea." *Living Bird* (1974): 251–294.

Other

Laufer, B. "Ostrich Egg-shell Cups of Mesopotamia and the Ostrich in Ancient and Modern Times." *Leaflet No. 23.* Chicago: Field Museum of Natural History, 1926.

Museum of Paleontology. U of California, Berkley. 1 Jan. 1996 (20 Jan. 2002). <http://www.ucmp.berkeley.edu/diapsids/birds/palaeognathae.html>

Viegas, Jennifer. "DNA Reveals Ancient Big Birds." *Discovery News* 13 Feb. 2001 (20 Jan. 2002). <http://dsc.discovery.com/news/briefs/20010212/moa.html>

Donald F. Bruning, PhD

Tinamous
(Tinamidae)

Class Aves
Order Struthioniformes
Suborder Tinami
Family Tinamidae

Thumbnail description
Small to medium-sized ground-dwelling birds with fully developed wings and capable of flight; cryptically colored in grays and browns, with three or four toes

Size
8–21 in (20–53 cm), 0.09–5 lb (43–2,300 g)

Number of genera, species
9 genera; 47 species

Habitat
Rainforest, secondary forest, savanna woodland, montane steppe, grassland, and cropland

Conservation status
Critically Endangered: 2 species; Vulnerable: 5 species; Near Threatened: 4 species

Distribution
South and Central America, as far north as Mexico

Evolution and systematics

In older natural history books, the description of birds usually began with ratites, including rheas, ostriches, cassowaries, kiwis, and the extinct moas and elephant birds of Madagascar. Because these large flightless birds have no keel on the sternum (breastbone), they are known as flat-chested birds, in contrast to all modern keel-breasted (Carinatae) birds. Today we know that the ratites descended from flying keel-breasted ancestors. Tinamous (Tinamidae) are a still-living primitive bird family that may be close to the ancestral group of ratites.

Physical characteristics

Length 8–21 in (20–53 cm), weight 0.09–5 lb (43–2,300 g). Tinamous are ground birds with a compact form, a slender neck, a small head, and a short, slender bill that curves slightly downward. The wings are short and capacity for flight is poor. The feet are strong; there are three well-developed forward toes, and the hind toe is in a high position and either retrogressed or absent. The tail is very short and in some species it is hidden under tail coverts; this and abundant rump feathering give the body a rounded shape. Powder down and

a preen gland are present. In contrast to gallinaceous birds, they do not scratch for food with their feet, but do use their feet to dig nest scrapes. A copulatory organ is present. The plumage is inconspicuous, although crown feathers of some species can be raised as a crest. Males and females have similar plumage, or females may be somewhat brighter in color and often larger than males.

Distribution

The range of tinamous has not changed significantly in recorded time. They live in tropical parts of Central and South America. In the north they live only slightly beyond the tropic of Cancer (Zimttao in northwest Mexico), but in South America they are widespread, distributed throughout the continent. One species has been successfully introduced to Easter Island in the Pacific.

Habitat

Many species live in dense forest, especially those in the genera *Tinamus*, *Nothocercus*, and *Crypturellus*. Other species live in savanna, on grassland, and in the montane or puna regions, at high altitudes. Some grassland forms have established themselves successfully in cropland; others can live on ranches among grazing cattle.

Behavior

Walking or running, tinamous move almost exclusively on the ground. When approached by people, they hide in thick ground cover or steal away unobserved. When hard-pressed in open country, they crawl into holes dug by other animals. Some species are very reluctant to fly. When surprised by a larger animal or when followed too closely, they rise suddenly with loud, frightening wing beats, and often they call. They disappear swiftly and alight in thick vegetation. Before a surprised hunter can bring up his gun, they have disappeared.

The initial burst of wing fluttering is often followed by a long glide and again by renewed wing beats. Although many tinamous cover long distances on foot rather than in flight, D. A. Lancaster observed a brushland tinamou (*Nothoprocta cinerascens*) that regularly flew 660 ft (200 m) from nest to feeding place. In other species, flights of 660–4,900 ft (200–1,500 m) have been observed. In Patagonia, the elegant crested tinamou (*Eudromia elegans*) forms strings of six to 30 or more birds. But most adult tinamous live singly except during the reproductive season. Many tinamous roost on the ground, but in some—especially those in the genus *Tinamus*—the hind surface of the tarsus is rough. These birds roost in trees, squatting with the tarsus across a branch rather than gripping the branch with their toes as songbirds do.

Hearing tinamou calls is one of the most unforgettable experiences one can have in tropical America. In contrast to most tinamous, the male highland tinamou (*Nothocercus bonapartei*) utters a rough crowing or barking call that can be heard for several miles (kilometers) through mountain forests. In some species, males and females are distinguishable by their calls. Tinamous sing mainly during the breeding season; they are noisiest in early morning and late evening. When startled, they fly off or follow one another, hoarsely shrieking or crowing.

Feeding ecology and diet

Tinamous eat mainly small fruits and seeds that they pick from the ground or from plants they can reach from the ground; they may jump up 4 in (10 cm) to reach a particularly tempting fruit. Seeds with wing-like appendages that make swallowing difficult are beaten against the ground or vigorously shaken. They also eat opening buds, tender leaves, blossoms, and even roots. For variety they catch insects and their larvae, worms, and, in moist places, mollusks. They swallow small animals whole; they first peck at larger ones, then shake them or beat them against the ground. When searching for food, they scatter fallen leaves and other ground cover with their bills, but do not scratch with their feet. When searching for worms or larvae in moist places, many species fling the earth aside with their bills, digging down 0.7–1.2 in (2–3 cm).

A puna tinamou (*Tinamotis pentlandii*) nest in the high Andes of Peru. (Photo by F. Gohier. Photo Researchers, Inc. Reproduced by permission.)

Reproductive biology

In ornate and highland tinamous, and in taos and many other species, there are about as many females as males. But in the variegated tinamou there may be four times as many males as females.

Reproductive behavior differs from that of most other birds. Only males care for eggs and young. In the few species that are adequately studied, males live in polygyny and females in polyandry. A male in breeding condition attracts two, three, or more females by continually calling. Females lay in the same nest and leave incubation to the male. Females leave to lay eggs in the nests of other males. When the male has raised his young or lost the eggs, he begins to call again and attracts new hens that supply him with another nest full of eggs. This breeding behavior has been observed in such different species as the highland tinamou, the brushland tinamou, and the slaty-breasted tinamou (*Crypturellus boucardi*). The variegated tinamou (*Crypturellus variegatus*) cares for only a single egg; the clutch of four to nine eggs incubated by the male ornate tinamou (*Nothoprocta ornata*) seems to come from a single female. In this mountain species, the

Highland tinamou (*Nothocercus bonapartei*) with hatching chicks in Costa Rica. (Photo by Michael Fogden. Bruce Coleman Inc. Reproduced by permission.)

larger and more aggressive female defends the breeding territory; in other tinamous this is done by males.

Tinamous almost always nest on the ground, often in densest herbage or between projecting root buttresses of large trees. Many tinamous lay their eggs directly on the ground or on leaves that happen to be on the chosen spot. Ornate tinamou build a proper nest base of dry earth, or a mixture of earth and moss turf. On this they erect a firm structure of grass that is worked into the base to form a circle. Shiny tinamou eggs are among the most beautiful natural products known. They may be green, turquoise blue, purple, wine red, slate gray, or a chocolate color, and often have a purple or violet luster. They are always uniformly colored, without spots or blotches. Their shape is oval or elliptical.

Incubating male tinamous sit continuously on eggs for many hours. In most species they leave the clutch once a day to look for food, usually during mornings or in the afternoon, depending on the weather, and are away from the nest for 45 minutes up to five hours.

Although the eggs are conspicuous and have no protective coloration, some tinamous do not cover the eggs when leaving the nest. A brushland tinamou covers its eggs only when they are about to hatch. The ornate tinamou regularly covers the eggs with feathers, giving them some protection from the harsh climate in the 13,000-ft (4,000-m) Peruvian puna. In the warm forests of Central America, a slaty-breasted tinamou always threw leaves toward the nest quite carelessly; often more than half the eggs were left uncovered.

Incubating tinamous sit so firmly on the nest that one can approach very closely. Although they will not let themselves

be touched by a hand, an observer can sometimes touch them with the end of a long stick. If someone approaches, some species—including the ornate tinamou and several *Crypturellus* species—press head and body close to the ground and raise the hind end, sometimes until the stumpy tail and lower tail coverts stand almost vertically. This posture resembles that assumed by certain tinamous in courtship display or when they are alarmed while walking. It seems to serve no special purpose in the incubating cocks, because raising the hind end may expose the shiny eggs. If one approaches an incubating tinamou cock too closely, it flies off the nest and out of sight in a burst of speed. An exception is the tataupa (*Crypturellus tataupa*), which flutters over the ground as if hurt and unable to fly when displaced from its eggs.

Newly hatched tinamous are densely covered with long soft down that in some species is marked in dulled colors. On the first day after hatching, the male leads the young out of the nest, moving slowly and calling them with repeated soft whistles or whining tones. Now and then he picks up an insect from the ground and moves it in his bill, then lays the insect before one of the young to be picked up. On leaving the nest, chicks of the small tinamous are very delicate, but they move so skillfully through the dense vegetation that little is known about them after they leave the nest. They probably develop rapidly and soon leave their father. A slaty-breasted tinamou at 20 days differed little in color and size from adults.

Conservation status

Eleven species are threatened, mainly because clearing and development have fragmented their habitats. The rarest are the Magdalena tinamou (*Crypturellus saltuarius*), known from one specimen, and Kalinowski's tinamou (*Nothoprocta kalinowskii*), not seen for many years. Both of these species are considered Critically Endangered by the IUCN. The dwarf tinamou (*Taoniscus*) and the lesser nothura (*Nothura minor*) are badly affected by development—IUCN considers these species Vulnerable. Three other species with very small ranges are also considered Vulnerable—the black tinamou (*Tinamus osgoodi*), the Choco tinamou (*Crypturellus kerriae*), and Taczanowski's tinamou (*Nothoprocta taczanowskii*). Four additional species are listed as Near Threatened—the yellow-legged tinamou (*Crypturellus noctivagus*), the pale-browed tinamou (*Crypturellus transfasciatus*), the Colombian tinamou (*Crypturellus columbianus*), and the solitary tinamou (*Tinamus solitarius*).

Significance to humans

Tinamous are hunted for food by humans throughout their range. A Brazilian rural family consumed about 60 nothuras a year, but this is not thought to be too severe a drain on the birds. Many attempts have been made to introduce the birds to other parts of the world, so far, except for Easter Island, without success.

1. Variegated tinamou (*Crypturellus variegatus*); 2. Female spotted nothura (*Nothura maculosa*); 3. Highland tinamou (*Nothocercus bonapartei*); 4. Male slaty-breasted tinamou (*Crypturellus boucardi*); 5. Male thicket tinamou (*Crypturellus cinnamomeus*); 6. Red-winged tinamou (*Rhynchotus rufescens*); 7. Elegant-crested tinamou (*Eudromia elegans*); 8. Female brushland tinamou (*Nothoprocta cinerascens*); 9. Black tinamou (*Tinamus osgoodi*); 10. Great tinamou (*Tinamus major*). (Illustration by Bruce Worden)

Species accounts

Black tinamou
Tinamus osgoodi

TAXONOMY
Tinamus osgoodi Conover, 1949, Curzo, Peru. Two subspecies.

OTHER COMMON NAMES
French: Tinamou noir; German: Schwartztinamu; Spanish: Tinamú Negro.

PHYSICAL CHARACTERISTICS
17 in (43 cm). Females are slightly larger. Sooty brown belly; vent is chestnut with black speckling.

DISTRIBUTION
Known only from two restricted and widely separated localities—the upper Magdalena valley in southern Colombia (subspecies *T. o. hershkovitzi*) and the Marcapata valley in southeastern Peru (*T. o. osgoodi*).

HABITAT
Humid, high-altitude tropical forest, 5,000–7,000 ft (1,500–2,100 m), where epiphytes, tree ferns, bromeliads, and moss abound.

BEHAVIOR
The call is a simple, descending whistle.

FEEDING ECOLOGY AND DIET
Not known.

REPRODUCTIVE BIOLOGY
The only nest found was on the ground and contained two glossy blue eggs.

CONSERVATION STATUS
Vulnerable. Threatened by habitat destruction.

SIGNIFICANCE TO HUMANS
None known. ◆

Great tinamou
Tinamus major

TAXONOMY
Tinamus major Gmelin, 1789, Cayenne.

OTHER COMMON NAMES
English: Mountain hen; French: Grand tinamou; German: Großtinamu; Spanish: Tinamú Oliváceo.

PHYSICAL CHARACTERISTICS
17.5 in (44 cm), 2.5 lb (1.1 kg). Female slightly larger. Overall color ranges from light to dark olive brown. Whitish on throat and center of belly.

Tinamus osgoodi
  Resident

Tinamus major
  Resident

DISTRIBUTION
Widely distributed, with seven subspecies in Belize, Bolivia, Brazil, Colombia, Costa Rica, Ecuador, Guatemala, French Guiana, Honduras, Mexico, Nicaragua, Panama, and Venezuela.

HABITAT
Dense tropical and subtropical forest, preferably with an open floor, at altitudes of 1,000–5,000 ft (300–1,500 m).

BEHAVIOR
Usually solitary, maintaining a home range. The call is a series of musical, tremulous whistles.

FEEDING ECOLOGY AND DIET
Feeds on the forest floor, taking fruits and seeds, especially of the Lauraceae, Annonaceae, Myrtaceae, and Sapotaceae.

REPRODUCTIVE BIOLOGY
The breeding season is long, extending from mid-winter to late summer. The nest, built between buttresses of a forest tree, contains 3–6 glossy turquoise or violet eggs. The male alone incubates eggs and rears the brood.

CONSERVATION STATUS
Not threatened.

SIGNIFICANCE TO HUMANS
It is hunted as a game bird, especially around towns, but has survived better than other game species. The great tinamou has various roles in native American folklore in Brazil, Colombia, and Panama. ◆

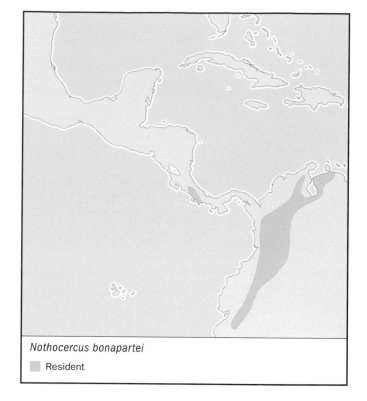

Nothocercus bonapartei
■ Resident

Highland tinamou
Nothocercus bonapartei

TAXONOMY
Nothocercus bonapartei Gray, 1867, Aragua, Venezuela. Five subspecies.

OTHER COMMON NAMES
English: Bonaparte's tinamou; French: Tinamou de Bonaparte; German: Bergtinamu; Spanish: Tinamú Serrano.

PHYSICAL CHARACTERISTICS
15 in (38.5 cm), 2 lb (925 g). Mottled or barred with black and cinnamon on back and wings. Throat is variable rufous color.

DISTRIBUTION
Colombia, Costa Rica, Panama, and Venezuela.

HABITAT
Tropical and subtropical forest, mainly above 5,000 ft (1,500 m), favoring damp areas, especially those with bamboo thickets.

BEHAVIOR
The call is loud and hollow, repeated many times, given by the male from his home range, which he occupies throughout the year.

FEEDING ECOLOGY AND DIET
Feeds on fallen fruits and small animals.

REPRODUCTIVE BIOLOGY
The male defends a small territory in his home range, attracting one or more females with calls and a display known as "follow feeding." The nest, which may contain eggs from several females in a clutch of four to 12, is concealed in ground vegetation. Incubation is by the male alone.

CONSERVATION STATUS
Not threatened.

SIGNIFICANCE TO HUMANS
Hunted as a game bird; populations in Costa Rica and Peru have declined as a result. ◆

Thicket tinamou
Crypturellus cinnamomeus

TAXONOMY
Crypturellus cinnamomeus Lesson, 1842, La Union, El Salvador.

OTHER COMMON NAMES
English: Rufescent tinamou; French: Tinamou cannelle; German: Beschtinamu; Spanish: Tinamú Canelo.

PHYSICAL CHARACTERISTICS
10.8 in (27.5 cm), 1 lb (440 g). Barred black on back and flanks; white throat and cinnamon or rufous cheeks and breast.

DISTRIBUTION
This tinamou, with nine subspecies, is widespread in Central America and has populations in Belize, Costa Rica, El Salvador, Guatemala, Honduras, Mexico, and Nicaragua. Its distribution extends farther north than that of any other tinamou.

HABITAT
Thick undergrowth, with an overstory—a foliage layer in a forest canopy including the trees in a timber stand—ranging from arid scrub to secondary forest.

Crypturellus cinnamomeus
■ Resident

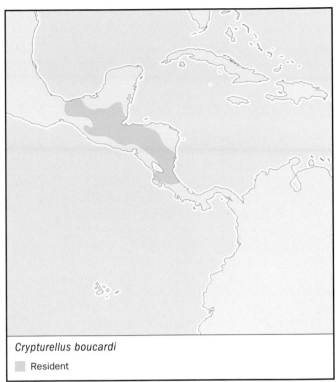

Crypturellus boucardi
■ Resident

BEHAVIOR
The monotonous call sounds like a steam whistle when heard at close quarters. The bird lives singly, in pairs, or in family parties. When disturbed it walks, rather than flies, away.

FEEDING ECOLOGY AND DIET
Feeds on fruit, seeds, and insects, searching for food in small parties that attract attention by crackling dry leaves as they feed.

REPRODUCTIVE BIOLOGY
The nest is placed on the ground at the base of a tree. The clutch is usually three, but may be up to seven glossy purplish eggs. Hybrids have been found between this species and the slaty-breasted tinamou.

CONSERVATION STATUS
Not threatened.

SIGNIFICANCE TO HUMANS
Because it is so unwilling to fly, it is not regarded as an important game bird. ◆

Slaty-breasted tinamou
Crypturellus boucardi

TAXONOMY
Crypturellus boucardi Sclater, 1859, Oaxaca, Mexico. Two subspecies.

OTHER COMMON NAMES
English: Boucard tinamu; French: Tinamu de Boucard; German: Graukehltinamu; Spanish: Tinamú Pizarroso.

PHYSICAL CHARACTERISTICS
10.8 in (27.5 cm), 1 lb (470 g). Pink to bright red legs; slaty breast, blackish head, and white throat. Back is blackish to chestnut. Female has barring on wings.

DISTRIBUTION
Belize, Costa Rica, Guatemala, Honduras, Mexico, and Nicaragua.

HABITAT
Sea level to 6,000 ft (1,800 m), sometimes favors thick undergrowth. Also humid forest with little undergrowth at ground level. Sometimes common in regenerating plantations and is often in damp areas, especially near forest edges.

BEHAVIOR
The call has three notes and is lower than the calls of many tinamous. It may be given in long bouts, up to five hours in one case. Calls of individual males are recognizable, and mellower and less variable than female calls. It is solitary, remaining in its home range throughout the year.

FEEDING ECOLOGY AND DIET
Feeds on fruits and seeds, tossing leaves aside with its bill in its search. It takes insects, including ants and termites.

REPRODUCTIVE BIOLOGY
In the breeding season it establishes a small territory in its home range, attracting two to four females to lay in a nest at the base of a tree or in thick vegetation. The male alone incubates; females leave to mate with another male.

CONSERVATION STATUS
Not threatened.

SIGNIFICANCE TO HUMANS
Hunted as a game bird and has become rare in some of its range, but is elsewhere still common. ◆

Variegated tinamou
Crypturellus variegatus

TAXONOMY
Crypturellus variegatus Gmelin, 1789, Cayenne.

OTHER COMMON NAMES
French: Tinamou varié; German: Rotbrusttinamu; Spanish: Tinamú Abigarrado.

PHYSICAL CHARACTERISTICS
11.5 in (29.5 cm), 0.8 lb (380 g). Black head; neck and breast rufous. Light barring on underparts.

DISTRIBUTION
Belize, Brazil, Colombia, Ecuador, Guiana, Peru, and Venezuela.

HABITAT
Tropical forest with dense undergrowth at moderate altitudes, 300–4,300 ft (100–1,300 m).

BEHAVIOR
The call is a series of five evenly pitched tremulous notes, often merging to a trill, with the first note descending and distinct from the rest of the trill.

FEEDING ECOLOGY AND DIET
Mainly seeds and fruits, with few insects.

REPRODUCTIVE BIOLOGY
The female establishes a territory, attracts a male, lays one egg in a rudimentary nest, and leaves the male incubating while she departs to establish another territory and repeat the process.

CONSERVATION STATUS
Not threatened.

SIGNIFICANCE TO HUMANS
The bird does not appear to be important as a game species. ◆

Red-winged tinamou
Rhynchotus rufescens

SUBFAMILY
Rhynchotinae

TAXONOMY
Rhynchotus rufescens Temminck, 1815, São Paulo, Brazil. Four subspecies.

OTHER COMMON NAMES
French: Tinamou isabelle; German: Pampahuhn; Spanish: Tinamú Alirrojo.

PHYSICAL CHARACTERISTICS
16 in (41 cm), 1.8 lb (830 g). Female slightly larger. Black patch on crown; rufous primaries. Light grayish brown to whitish underneath. May be black barring on flanks, abdomen, and vent.

Crypturellus variegatus

▢ Resident

Rhynchotus rufescens

▢ Resident

DISTRIBUTION
Argentina, Bolivia, Brazil, Paraguay, and Uruguay.

HABITAT
At low altitudes, below 3,300 ft (1,000 m), it lives in damp grassland and woodland edges; at higher altitudes it is found in semiarid scrub and cereal fields.

BEHAVIOR
The call, given only by males, is a long, ringing single whistle followed by shorter, mournful whistles. The birds live dispersed in the dense vegetation, and are most active in the heat of the day.

FEEDING ECOLOGY AND DIET
It is sedentary, feeding on the ground on seeds, tubers, and fruit. In the summer it takes more animal food, including earthworms, termites, and other insects. It digs for food with its bill, and so is unpopular on newly sown cropland.

REPRODUCTIVE BIOLOGY
The red-winged tinamou has many displays, the male attracting one or more females by follow feeding, and always accompanies the female to the nest when she is to lay. He alone incubates the eggs and broods the chicks.

CONSERVATION STATUS
Not threatened.

SIGNIFICANCE TO HUMANS
A popular game bird and hunted out in some regions, but elsewhere common. Because it will live in cropland it has extended its range alongside agricultural development. ◆

Nothoprocta cinerascens
▨ Resident

Brushland tinamou
Nothoprocta cinerascens

TAXONOMY
Nothoprocta cinerascens Burmeister, 1860, Tucumán, Argentina. Two subspecies.

OTHER COMMON NAMES
French: Tinamou sauvageon; German: Cordobasteißhuhn; Spanish: Tinamú Montaraz.

PHYSICAL CHARACTERISTICS
12.5 in (31.5 cm), 1.2 lb (540 g). Female slightly larger and darker. Black barring on back and wings.

DISTRIBUTION
Argentina, Bolivia, and Paraguay.

HABITAT
Favors dry savanna woodlands, usually below 3,300 ft (1,000 m), but will live in cropland and open thorn scrub.

BEHAVIOR
The advertising call is a series of seven to 10 clear whistled notes with considerable carrying power. Home ranges are about 50 acres (20 ha), maintained mainly by calls but often overlapping ranges of other males.

FEEDING ECOLOGY AND DIET
Feed on the ground, mostly on insects and small animals, but also take some fruit.

REPRODUCTIVE BIOLOGY
Males attract groups of two to four females, establish a nest site, and supervise females while they lay in it. When the females leave to join another male, the original male incubates the clutch and rears the brood alone.

CONSERVATION STATUS
Not threatened.

SIGNIFICANCE TO HUMANS
Subject to light hunting but remains common. ◆

Spotted nothura
Nothura maculosa

TAXONOMY
Nothura maculosa Temminck, 1815, Paraguay. Eight subspecies.

OTHER COMMON NAMES
French: Tinamou tacheté; German: Fleckensteißhuhn; Spanish: Tinamú Manchado.

PHYSICAL CHARACTERISTICS
10 in (25.5 cm), 0.6 lb (250 g). Female slightly larger. Variable appearance, sometimes very dark upperparts.

DISTRIBUTION
Argentina, Brazil, and Uruguay.

HABITAT
Most subspecies inhabit lowlands, living in open grassland, shrub steppe, and cropland. Its range is expanding as clearing takes place for agriculture.

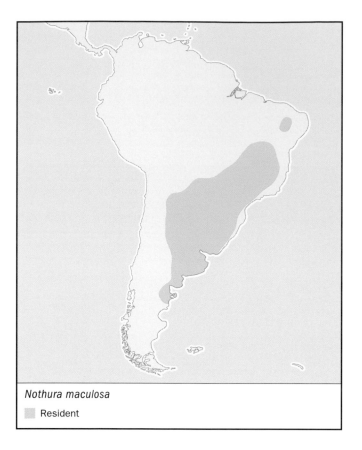

Nothura maculosa

☐ Resident

Eudromia elegans

☐ Resident

BEHAVIOR
The call is a series of brief, high-pitched piping notes, often given in response to other calling birds. Populations may be very dense in favorable country, up to a bird to every 2.5 acres (1 ha).

FEEDING ECOLOGY AND DIET
The spotted nothura feeds on vegetable and animal matter, taking more insects than plants in Argentina, but elsewhere feeding mainly on seeds, including those of pasture plants, crops, and weeds.

REPRODUCTIVE BIOLOGY
The species has a very high reproductive rate; females can mature at two months of age and have 5–6 broods in a year. Males take longer to mature, or at least to establish nests. As with other tinamous, males undertake all incubation and parenting, often attracting more than one female to lay in a single nest.

CONSERVATION STATUS
Not threatened.

SIGNIFICANCE TO HUMANS
A very popular game bird, but a high reproductive rate and early maturity ensure that it remains common. ◆

Elegant crested-tinamou
Eudromia elegans

TAXONOMY
Eudromia elegans Geoffroy St. Hillaire, 1832, South America. Eight subspecies.

OTHER COMMON NAMES
French: Tinamou élégant; German: Perlsteißhuhn; Spanish: Martineta Común.

PHYSICAL CHARACTERISTICS
15.5 in (39 cm), 1.3 lb (600 g). Leg color pale bluish to grayish brown. Lacks hind toe. Crest is long, normally carried backwards.

DISTRIBUTION
Throughout Argentina and Chile.

HABITAT
Arid and semiarid grassland and savanna, favoring open sites, ranging from sea level to 8,000 ft (2,500 m) in altitude.

BEHAVIOR
The call is a loud melancholy whistle. Unlike many tinamous, this species forms small flocks, especially in winter when it invades alfalfa crops. In spring and summer it may still be found in pairs and small groups.

FEEDING ECOLOGY AND DIET
In winter it feeds mainly on seeds and leaves of plants; in summer it takes many insects and invertebrates, including termites.

REPRODUCTIVE BIOLOGY
Breeding systems are polyandrous and polygynous, although males undertake all incubation and parenting.

CONSERVATION STATUS
Not threatened.

SIGNIFICANCE TO HUMANS
It is hunted intensely and remains common only in remote areas. ◆

Resources

Books

Davies, S. J. J. F. *Ratites and Tinamous.* Oxford: Oxford University Press, 2002.

del Hoyo, J., A. Elliot, and J. Sargatal, eds. *Handbook of the Birds of the World.* Vol. 1, *Ostrich to Ducks.* Barcelona: Lynx Editions, 1992.

Periodicals

Beebe, W. "The Variegated Tinamou *Crypturus variegatus* (Gmelin)." *Zoologica* 6 (1925): 195–227.

Lancaster, D. A. "Life History of the Boucard Tinamou in British Honduras." *Condor* 66 (1964): 165–81, 253–76.

Lancaster, D. A. "Biology of the Brushland Tinamou *Nothoprocta cinerescens*." *Bulletin of the American Museum of Natural History* 127 (1964): 271–314.

S. J. J. F. Davies, ScD

Rheas
(Rheidae)

Class Aves
Order Struthioniformes
Suborder Rheae
Family Rheidae

Thumbnail description
Very large flightless birds with long legs and three toes; plumage gray or spotted brown and white; wings used only in display, reduced with long soft feathers; no tail feathers and no casque on the head

Size
36.4–55 in (92–140 cm); 33–88 lb (15–40 kg)

Number of genera, species
2 genera; 2 species

Habitat
Savanna, grassland, and high mountain plains

Conservation status
Not threatened. Populations are now fragmented and numbers have declined, but still abundant in some areas

Distribution
Argentina, Bolivia, Brazil, Chile, Paraguay, Peru, and Uruguay

Evolution and systematics

Rheas belong to the group of large, flightless birds known as the ratites, which all lack a keel to the sternum and a distinctive palate. The origin of these birds has recently been clarified by the discovery of numerous good fossils in North America and Europe. Whereas it used to be thought that ratites had a southern origin, in the old continent of Gondwana, new fossil evidence has shown flying ratites inhabited the northern hemisphere in the Paleocene and Eocene, 40–70 million years ago. The present southern hemisphere distribution of ratites probably results from the spread of flying ancestors of the group from the north. Fossil rheas have been found in the Upper Pleistocene of Argentina. They lived there about two million years ago; it is thought that rheas are related to Tinamidae.

Physical characteristics

Rheas are smaller and more slender than ostriches: standing upright they reach 5.6 ft (1.7 m). They may weigh up to 88 lb (40 kg), the head, neck, rump, and thighs are feathered, and their plumage is soft and loose. There are three front toes, and the hind toe is absent. The tarsus has horizontal plates in front. The gut and the caeca are very long. Urine is

stored in an expansion of the cloaca and is eliminated in liquid form. The copulatory organ is extrudable. In the greater rhea (*Rhea americana*) total height reaches 5.6 ft (1.7 m); height of the back is 3.3 ft (100 cm), wingspan reaches 8.2 ft (250 cm), tarsal length is 12–14.5 in (30–37 cm), and bill length is 3.5–4.7 in (9–12 cm). Males are larger than females. The tarsus has about 22 horizontal plates in front. The lesser rhea (*Pterocnemia pennata*) is smaller than the greater with a height at the back of 3.0 ft (90 cm). The tarsus is 11.0–11.8 in (28–30 cm) and has about 18 horizontal plates.

Distribution

Rheas are confined to South America—Argentina, Bolivia, Brazil, Chile, Paraguay, Peru, and Uruguay.

Habitat

Rheas are birds of grassland, the greater rhea and one subspecies of lesser rhea, *Pterocnemia pennata pennata*, of lowland grassland or pampas. The other two subspecies of lesser rhea live in the puna zone of the Andes, inhabiting deserts, salt puna, heath, and pumice flats. Although *P.p. pennata* feeds on lowland grassland in the non-breeding season, it usually

The lesser rhea (*Pterocnemia pennata*) with chicks on their nesting ground just north of Punta Arenas, Chile. (Photo by N.H. [Dan] Cheatham. Photo Researchers, Inc. Reproduced by permission.)

breeds in upland areas where bunch grass grows, around 5–6,000 ft (2,000 m).

Behavior

Rheas are silent except as chicks, when they give a plaintive whistle, and during the breeding season when males make a deep booming call, sometimes described as like the last tone of a siren or as a di-syllabic grunt. In any case one of the renditions of the call "nandu" has become a common name for the bird. Males perform an elaborate display while giving these calls, raising the front of the body, with the neck held stiffly upwards and forward and the plumage greatly ruffled. Wings are raised and extended, and after calling the bird may run some distance, sometimes flipping the wings up and down alternately. Usually this display, the call display, is given near females and may be followed by the wing display, that is directed at a specific female. The male spreads his wings, lowers his head, and walks in this posture beside or in front of a female, holding the display for 10 minutes or more. Females seem to be attracted to a male displaying in this way. As the display becomes more intense the male waves his neck from side to side in a figure-eight pattern, often attracting females to watch him for several minutes before they move off to feed. If a female remains beside a displaying male, she may solicit and copulation follows.

During the non-breeding season the greater rhea forms flocks of 10–100 birds, while the lesser rhea lives in smaller flocks than that. As in the ostrich, birds in small flocks are more vigilant than those in large flocks. They are also more vigilant when in tall grass environments than on open plains. Rheas feed most of the day in these flocks, although males show mild aggression to each other from time to time. When fleeing in alarm, a rhea will follow a zigzag course, often raising one wing, apparently to act as a rudder and help it to turn rapidly. Dust bathing is common in captive birds. Flocks break up in the winter for the breeding season.

Feeding ecology and diet

Both species of rhea are mainly herbivores. Both take a few small animals—lizards, beetles and grasshoppers—but not in any significant quantity. Most of the present range of the greater rhea is used for cattle ranching, with the result that pastures have been seeded with fodder grasses and forbs. The greater rhea takes much alfalfa and maize. The Lesser rhea lives in less developed areas, but is mainly herbivorous, taking forbs like saltbush and fruits of cactus.

Reproductive biology

Much information has been gathered about the reproductive behavior of the greater rhea, both in the wild and in captivity. It is a polygamous bird, meaning both males and females take two or more mates. When the winter comes, the flocks break up into three types of groups—single males, flocks of

A male greater rhea (*Rhea americana*) sits on its nest. (Photo by F. Gohier. Photo Researchers, Inc. Reproduced by permission.)

two to 15 females, and large flocks of yearlings. Males soon begin posturing and challenging each other, behavior that becomes intense as the spring and summer breeding season arrives. Males attempt to attract harems of females, building a nest and leading females toward it. When a harem has chosen a male, one female will approach the nest. The male may stand but usually remains sitting or crouching, twisting his neck to follow the movements of the female as she walks around the nest. The male, at first acting aggressively and spreading his wings to cover the eggs, gradually relaxes and replaces head-forward movements with head-bobbing and neck swinging. The female crouches and lays her egg at the rim of the nest and the male rolls the egg beneath himself. In this way, with eggs from as many as 15 females, he may build up a clutch of 50 or more eggs in a week. Then he begins to incubate in earnest. As in the ostrich some eggs remain outside the nest, and these seem to act to dilute the clutch, so that it is less likely that the eggs being incubated are taken by predators, because they first take the eggs outside the nest. Recent work has shown that some males have male partners. A subordinate male may take over the clutch from the first harem of females that the dominant male attracts, incubate them, and parent the chicks. Meanwhile the dominant male attracts another harem, or the same one to another nest, incubates the second clutch and rears the second brood. Measurements show that both dominant and subordinate males are equally successful at incubating and raising broods.

The mean clutch size of the greater rhea in Argentina is 24.9 eggs and the incubation period 36–37 days. The male alone incubates, leaving the nest from time to time to feed. As in other birds, chicks in the eggs are able to communicate with each other and synchronize hatching, so that the male stays no more than 36 hours on the nest once hatching be-

gins. Breeding success is usually low, about 20%, but in some years breeding success is greater than this. Some nests are deserted when bad eggs explode during incubation, and in other cases armadillos dig under the nest and eat the eggs. Predation also accounts for the loss of many small chicks; birds of prey follow broods until one chick straggles and then snatch it. Much less is known about the breeding biology of the lesser rhea, but it appears to have a similar breeding system, clutch size, and incubation period.

Conservation status

The greater rhea is farmed in some areas for its meat and leather, but the range available for wild birds is shrinking, and many are taken by hunters. Similarly the lesser rhea suffers from development of its environment by the construction of roads that allow hunters access to country that was previously inaccessible. Both species need large, well protected reserves if they are to survive as wild populations.

Significance to humans

Feathers of rheas have always been taken for use as dusters. Skins are used as cloaks in their dried state and as fashion leather when fully tanned. Rhea meat has long been a staple food for South Americans, who are now able to hunt the birds more effectively with rifles than they were when they had only the bolas, a weapon made of three thongs of leather tied together centrally. At the outer end of each thong a small stone is attached. The bolas is whirled around by the hunter and thrown with great skill and accuracy at the running bird. The thongs wrap around the birds legs and bring it to the ground, effectively immobilizing it. The bolas is still used, even by scientists who want to catch the birds alive.

1. Lesser rhea (*Pterocnemia pennata*); 2. Greater rhea (*Rhea americana*). (Illustration by Patricia Ferrer)

Species accounts

Greater rhea
Rhea americana

TAXONOMY
Rhea americana Linnaeus, 1758, Sergipe and Rio Grande do Norte, Brazil. Five subspecies.

OTHER COMMON NAMES
English: Common rhea; French: Nandou d'Amérique; German: Nandu; Spanish: Ñandú.

PHYSICAL CHARACTERISTICS
50–55 in (127–140 cm); 44–88 lb (20–40 kg). General color gray or grayish brown above, whitish below without spotting in both sexes. The head and neck of the male are black or largely black. The female is paler. Unlike the lesser rhea, the whole length of the tarsus is bare and covered with transverse scutes.

DISTRIBUTION
Brazil, Uruguay, Paraguay, and Bolivia. The form in eastern Brazil, *Rhea americana americana* is the nominate, or first named, form. *R. a. intermedia* comes from southeastern Brazil and Uruguay, *R. a. nobilis* from Paraguay, *R. a. araneiceps* from Paraguay and Bolivia, and *R. a. albescens* from Paraguay and, possibly, Bolivia.

HABITAT
Grassland and pampas.

BEHAVIOR
Greater rheas live at densities of 0.002–0.076 birds per acre (0.05–0.19 birds/ha). In the nonbreeding season they live in flocks of 20–50 birds. Once the breeding season starts, males establish a nest site and defend its immediate vicinity, attracting groups of females to lay in the nest.

FEEDING ECOLOGY AND DIET
Herbivore, feeds on grasses and forbs.

REPRODUCTIVE BIOLOGY
Males incubate eggs laid by harem of females in a nest on the ground. The mean clutch size is 26, the eggs coming from up to seven different females. Females are attracted to the nest by male displays in which the wings are prominently displayed. The male leads the female to the nest and often sits on it while she lays outside it. He then rolls the egg into the nest. Eggs are greenish yellow color, 5 by 3.5 in (13 by 9 cm). Incubation period is 29–43 days, by the male only.

CONSERVATION STATUS
Population fragmented by agricultural development.

SIGNIFICANCE TO HUMANS
Hunted for meat, leather, and feathers; now farmed. ◆

Rhea americana

◼ Resident

Lesser rhea
Pterocnemia pennata

TAXONOMY
Pterocnemia pennata d'Orbigny, 1834, Lower Río Negro, south of Buenos Aires. Three subspecies.

OTHER COMMON NAMES
English: Darwin's rhea; French: Nandou de Darwin; German: Darwinstrauss; Spanish: Ñandú Overo.

PHYSICAL CHARACTERISTICS
36–39 in (92–100 cm); 33–55 lb (15–25 kg). The plumage is spotted brown and white. The upper part of the tarsus is partly feathered, but the rear and lower part is bare, covered with transverse scutes.

DISTRIBUTION
Argentina, Chile, Peru, and Bolivia. The nominate form, *Pterocnemia pennata pennata*, lives in southern Chile and Argentina, while *P. p. tarapacensis* lives in the Andes of Chile and *P. p. garleppi* lives in the Andes of Peru, Bolivia and northwestern Argentina.

HABITAT
Grassland, high Andes in the puna zone.

BEHAVIOR
Lives in flocks of 2–30 individuals at a mean density of 0.28 birds per mi² (0.11 birds/km²). Males defend nest sites during the breeding season. A male attracts groups of females to a

Pterocnemia pennata

☐ Resident

nest site and lay there, leaving the male to incubate the eggs alone.

FEEDING ECOLOGY AND DIET
Feeds on fruits and leaves of forbs, the items taken varying by place and season. In some places grasses are prominent, in others shrub foliage and fruits are mostly taken.

REPRODUCTIVE BIOLOGY
Males incubate eggs laid by harem of females in a nest on the ground. The greenish yellow eggs measure 4.9 by 3.4 in (12.6 by 8.7 cm) and are incubated for 30–44 days. Clutch size varies from five to 55 eggs, depending on the region. The birds mature in their third year.

CONSERVATION STATUS
Two isolated populations subject to severe hunting pressure.

SIGNIFICANCE TO HUMANS
Hunted for meat and leather. ◆

Resources

Books

Davies, S. J. J. F. *Ratites and Tinamous.* Oxford: Oxford University Press, 2002.

del Hoyo, J., A. Elliot, and J. Sargatal, eds. *Ostrich to Ducks.* Vol. 1 of *Handbook of the Birds of the World.* Barcelona: Lynx Edicions, 1992.

Periodicals

Bruning, D. F. "The Social Structure and Reproductive Behavior in the Greater Rhea." *Living Bird* 13 (1974): 251–94.

Codenotti, T. L., and F. Alvarez. "Cooperative Breeding Between Males in the Greater Rhea *Rhea americana*." *Ibis* 139 (1997): 568–71.

S. J. J. F. Davies, ScD

Cassowaries
(Casuariidae)

Class Aves
Order Struthioniformes
Suborder Casuarii
Family Casuariidae

Thumbnail description
Large flightless birds with tiny wings terminating in long spines, shiny black plumage, three toes, a casque on the head (also called a helmet or a crown), and colorful bare skin on the neck

Size
40–67 in (102–170 cm); 30–130 lb (14–59 kg)

Number of genera, species
1 genus; 3 (possibly 4) species

Habitat
Rainforest and adjacent dense vegetation

Conservation status
Potentially endangered by logging and forest clearing and by competition from feral pigs and dogs

Distribution
Cape York (Australia), New Guinea, and some surrounding islands

Evolution and systematics

Cassowaries belong to the group of large flightless birds known as the ratites that have in common a distinctive palate and the lack of a keel to the sternum. The origin of these birds has recently been clarified by the discovery of numerous good fossils in North America and Europe. Whereas it was previously thought that ratites had a southern origin, new fossil evidence has shown flying ratites inhabited the Northern Hemisphere in the Paleocene and Eocene, between 40 million and 70 million years ago. The present Southern Hemisphere distribution of the ratites probably results from the spread of flying ancestors of the group from the north. The cassowaries differ from the rheas and ostriches in their structure and way of life. All cassowary feathers consist of a shaft and loose barbules; there are no rectrices (tail feathers) nor a preen gland and only five to six large wing feathers. On the strongly retrogressed wing, the lower arm and hand are only half as long as the upper arm. The furcula (wishbone) and coracoid (shoulder blade) are degenerate. There is a special palatal structure, and the palatal bones and sphenoids touch one another. Cassowaries

are known from fossils of the Pliocene (about three million to seven million years ago) in New Guinea. Although not formally described until the nineteenth century, the first living cassowary to reach Europe was transported to Amsterdam in 1597.

Physical characteristics

Cassowaries are large, long-legged, cursorial (running) birds, with distinctive head casques of trabecular (fibrous and cordlike) bone or calcified cartilage up to 7 in (18 cm) high. The colorful skin of the neck is bare, and long neck wattles adorn two species. The birds weigh 37–130 lb (17–59 kg). Cassowary wings are small, but the shafts of five or six primary feathers remain as long curved spines. Of the three toes, the inner one is armed with a long sharp claw, an effective weapon that is capable of disemboweling an adversary—even a human. Like the emu, the aftershaft of the cassowaries' coarse, black feathers is as long as the main shaft, so that each feather appears double—almost like extremely thick hair.

Male southern cassowary (*Casuarius casuarius*) on its nest with eggs. (Photo by Frith Photo. Bruce Coleman Inc. Reproduced by permission.)

southern cassowary in north Queensland, Australia, the fruits of laurels, myrtles, and palms were most important. Opportunistically, the birds will take fungi, insects, and small vertebrates, but the basic diet consists of fruit. Disturbance of the forest can have serious consequences for cassowaries. Selective logging can remove almost all of one species of tree, so that the crop of fruit from that species is missing from the forest. If the fruit of this tree forms a significant part of the cassowary's diet, it will be left without food for weeks or months and suffer accordingly. Selective logging damages the bird's habitat more subtly than clear cutting, but equally seriously.

Reproductive biology

Cassowaries nest on a pad of vegetation on the ground. The clutch contains three to eight bright green or greenish blue eggs. Incubation lasts for 50–52 days, and is performed by the male alone. Chicks remain with the male for some months before gaining independence.

Distribution

The eastern side of Cape York in northern Australia, throughout New Guinea, New Britain, Seram, and Aru, Japen, Salawati, and Batanta islands. Humans have introduced the birds to some of these islands, and their natural distribution is uncertain.

Habitat

Cassowaries are birds of the rainforest but often stray into adjoining eucalypt forest, palm scrub, tall grassland, savanna, secondary growth, and swamp forest.

Behavior

Except during courtship and egg-laying, cassowaries are solitary birds, seldom seen in groups, and then usually at some source of abundant food such as a fruiting tree. Each bird occupies a home range, moving around within it to find food. Each species has a characteristic territorial boom call, a threatening roar, given with the head bent down under the body. The birds are able to move quietly through the rainforest until disturbed. The noise of their hasty departure as they crash through the undergrowth is often the first indication of their presence. They swim well and have been recorded reaching an island a mile and a half (2.4 km) from the coast.

Feeding ecology and diet

Cassowaries feed on the fruits of rainforest trees and shrubs. The birds collect most of these from the ground, using their bill and sometimes their casque to unearth the fallen fruit from the litter of the forest floor. As the cassowaries travel, they disperse the seeds of these fruits throughout the rainforest, thus ensuring the continuance of more than 150 species of rainforest plants. In a study of the

A male southern cassowary (*Casuarius casuarius*) on its nest with a 1-day-old chick. (Photo by Cliff Frith. Bruce Coleman Inc. Reproduced by permission.)

Conservation status

Disturbance of the forest is the main factor causing a decline in cassowary numbers. In Australia the cassowary population is estimated at 1,300 to 2,000 adults. Information on the status of New Guinea species is scant. The birds are so secretive—and the political situation so uncertain in West Irian—that any assessment is mere guesswork. It can only be said that the birds, or signs of them, can still be found whenever they are sought.

Significance to humans

Although they do not breed well in zoos, many cassowaries are kept in New Guinea villages. They are caught as chicks and raised to be killed and eaten when mature. Some of these captive birds have caused serious injury, even death, to village people tending them. They attack unexpectedly, slashing with powerful forward kicks, tearing the bodies of opponents with the long sharp claws of the inner toes with such accuracy that they are much feared.

1. Bennett's cassowary (*Casuarius bennettii*); 2. One-wattled cassowary (*Casuarius unappendiculatus*); 3. Southern cassowary (*Casuarius casuarius*). (Illustration by Marguette Dongvillo)

Species accounts

Southern cassowary
Casuarius casuarius

TAXONOMY
Casuarius casuarius Linnaeus, 1758, Seram.

OTHER COMMON NAMES
English: Double-wattled cassowary, two-wattled cassowary, Australian cassowary, kudari; French: Casoar à casque; German: Helmkasuar; Spanish: Casuario Común.

PHYSICAL CHARACTERISTICS
50–67 in (127–170 cm); female 128 lb (58 kg), male 64–75 lb (29–34 kg). Distinguished from the other cassowaries by having two wattles hanging from the neck. The bare skin of the head and neck is vividly colored in blue and red, and the legs are gray-green to gray-brown. The species has an especially long inner toenail or spike up to 5 in (12 cm) in length. Chicks are longitudinally striped with black, brown, and cream, and they have a chestnut head and neck for their first three to six months. Immatures have dark brown plumage and small casques, acquiring their colorful necks toward the end of their first year, and glossy adult plumage in about three years.

DISTRIBUTION
The Australian populations are all north of Townsville, Queensland, on the eastern side of Cape York. It is widespread in southern, eastern, and northwestern New Guinea, the Aru Islands, and Seram. The population on Seram has probably been introduced.

Casuarius casuarius

▨ Resident

HABITAT
The southern cassowary lives mainly in lowland rainforest, below 3,600 ft (about 1,100 m).

BEHAVIOR
Although usually shy, some birds will become tame enough near settlements to approach places where food is regularly put out for them. Adults are territorial, no more than two associating together, except that the chicks stay with their father for about nine months.

FEEDING ECOLOGY AND DIET
The southern cassowary feeds on the fallen fruits of rainforest trees, fungi, and a few insects and small vertebrates.

REPRODUCTIVE BIOLOGY
In Australia the southern cassowary breeds in the winter—June and July—coinciding with the abundance of forest fruit, especially laurels (of the family *Lauraceae*). The nest, on the ground, is often close to the roots of a large tree, and the clutch consists of up to four lime green eggs. Males and females hold separate territories except for a few weeks at laying time. Incubation, by the male alone, 47–61 days, with variation thought to be a response to ambient temperature. Polyandrous females may take on another male or two before mating season ends, providing a clutch of eggs for each of her partners. The chicks stay with the male for up to nine months.

CONSERVATION STATUS
The status of the southern cassowary is uncertain. It requires large areas of undisturbed rainforest to flourish. As these are logged or disturbed by roadmaking and settlement, the bird's future is put at risk. Some are killed on the roads. Feral animals such as pigs and dogs disturb the nests in their search for eggs, causing the population to shrink. In New Guinea the bird is hunted and snared for food, but while large tracts of forest remain, it is secure.

SIGNIFICANCE TO HUMANS
Both in Australia and New Guinea the southern cassowary is incorporated into the mythology of the indigenous peoples, but it is still hunted by them, and the chicks captured, to be kept in pens in the villages until they are large enough to eat. ◆

Bennett's cassowary
Casuarius bennettii

TAXONOMY
Casuarius bennettii Gould, 1857, New Britain.

OTHER COMMON NAMES
English: Dwarf cassowary, little cassowary, mountain cassowary; French: Casoar de Bennett; German: Bennettkasuar; Spanish: Casuario Menor.

Casuarius bennettii

■ Resident

PHYSICAL CHARACTERISTICS
Height 39–53 in (99–135 cm); weight 39 lb (about 18 kg). A small cassowary with a flat, low casque and a less colorful neck than the other species. A distinctive form lives on the west side of Geevink Bay, West Irian, and may merit recognition as a species, *C. papuanus*.

DISTRIBUTION
New Guinea, New Britain, and Japen Island.

HABITAT
Lives in forest and secondary growth, favoring hilly and mountainous country to 10,800 ft (3,300 m). On New Britain, where other species are absent, it lives in lowland forest as well.

BEHAVIOR
Usually solitary or in small family groups, traversing steep slopes and thick vegetation. Its call is higher pitched than that of the other species.

FEEDING ECOLOGY AND DIET
Bennett's cassowary feeds mainly on fallen fruits in the rainforest but also takes fungi, insects, and small vertebrates.

REPRODUCTIVE BIOLOGY
The clutch consists of four to six eggs. Incubation 49–52 days.

CONSERVATION STATUS
Not threatened. Although it is hunted extensively it remains widespread at low densities.

SIGNIFICANCE TO HUMANS
Widely kept as a pet and, when small, traded between localities. ◆

One-wattled cassowary
Casuarius unappendiculatus

TAXONOMY
Casuarius unappendiculatus Blyth, 1860, aviary in Calcutta.

OTHER COMMON NAMES
English: Northern cassowary; French: Casoar unicaronculé; German: Einlappenkasuar; Spanish: Casuario Unicarunculado.

PHYSICAL CHARACTERISTICS
Height 65–69 in (165–175 cm); weight females 128 lb (58 kg); males 81 lb (about 37 kg). A large cassowary with coarse black plumage, a tall casque, a colorful neck, and one central wattle.

DISTRIBUTION
Northern New Guinea, from western Vogelkop, West Irian, to Astrolabe Bay, Papua New Guinea, and on Satawati, Batanta, and Japen islands.

HABITAT
Mostly lowland areas of rainforest and swamp forest, up to 1,600 ft (490 m).

BEHAVIOR
Assumed to be similar to other cassowaries.

FEEDING ECOLOGY AND DIET
Feeds on fallen forest fruits.

REPRODUCTIVE BIOLOGY
Birds in breeding condition have been collected in May and June, but nothing else has been reported about its breeding.

Casuarius unappendiculatus

■ Resident

CONSERVATION STATUS

The status of the cassowary is uncertain. It requires large areas of undisturbed rainforest to flourish. It is hunted and snared for food, but where large tracts of forest remain, it is secure.

SIGNIFICANCE TO HUMANS

The cassowary is incorporated into the mythology of the indigenous peoples, but it is still hunted by them, and the chicks captured, to be kept in pens in the villages until they are big enough to eat. ◆

Resources

Books

Coates, B. J. *The Birds of Papua New Guinea.* Alderly, Australia: Dove, 1985.

Davies, S. J. J. F. *Ratites and Tinamous.* Oxford: Oxford University Press, 2002.

del Hoyo, J., A. Elliott, and J. Sargatal, eds. *Handbook of the Birds of the World.* Vol. 1, *Ostrich to Ducks.* Barcelona: Lynx Edicions, 1992.

Marchant, S., and P. J. Higgins. *Handbook of Australian, New Zealand and Antarctic Birds.* Vol. 1, *Ratites to Ducks.* Oxford: Oxford University Press, 1990.

Periodicals

Crome, F. H. J. "Some Observations on the Biology of the Cassowary in Northern Queensland." *Emu* 76 (1976): 8–14.

Davies, S. J. J. F. "The Natural History of the Emu in Comparison with That of Other Ratites." *Proceedings of the Sixteenth International Ornithological Congress* (1976): 109–20.

Organizations

Birds Australia. 415 Riversdale Road, Hawthorn East, Victoria 3123 Australia. Phone: +61 3 9882 2622. Fax: +61 3 9882 2677. E-mail: mail@birdsaustralia.com.au Web site: <http://www.birdsaustralia.com.au>

Other

Bredl, Rob. "Cassowaries." *Barefoot Bushman.* 5 Dec. 2001 <http://www.barefootbushman.webcentral.com.au/cassowaries.htm>

"The Cassowary." *The Living Museum: Wet Tropics.* Wet Tropics Management Authority Official Web Site. 5 Dec. 2001 <http://www.wettropics.gov.au/lm/The_Cassowary.htm>

"Double-wattled cassowary." *Zoo Discovery Kit.* Los Angeles Zoo. 5 Dec. 2001 <http://www.lazoo.org/learnfiles/ZooDiscoveryKit/Cassowaryfacts.html>

S.J.J.F. Davies, ScD

Emus
(Dromaiidae)

Class Aves
Order Struthioniformes
Suborder Casuarii
Family Dromaiidae

Thumbnail description
Large, flightless birds with tiny wings, three toes, and brown-black plumage

Size
60–75 in (150–190 cm); 75–110 lb (35–50 kg)

Number of genera, species
1 genus; 3 species

Habitat
Forest, woodlands, savanna, heath, and grasslands

Conservation status
Extinct: 2 species; Not threatened: 1 species

Distribution
Continental Australia and, formerly, Tasmania, King, and Kangaroo Islands

Evolution and systematics

Emus belong to the group of large, flightless birds known as the ratites that have in common massive, muscular legs, small wings, and a distinctive palate. Their inability to fly is due to the lack of a keel to the sternum, which in flying birds serves as the attachment point of flight muscles. The term "ratites" is derived from the Latin "ratis" for "raft", a boat without a keel. The origin of these birds has recently been clarified by the discovery of numerous good fossils in North America and Europe. Whereas it used to be thought that the ratites had a southern origin, in the old continent of Gondwana, new fossil evidence has shown flying ratites inhabited the Northern Hemisphere in the Paleocene and Eocene, 40–70 million years ago. The present Southern Hemisphere distribution of the ratites probably results from the spread of flying ancestors of the group from the north. The first evidence of ratites in Australia comes from the Eocene and the first recognizable Dromaiidae from the Miocene, 20 million years ago. The emu, *Dromaius novaehollandiae*, appears in the Pleistocene, only 2 million years ago.

Physical characteristics

Emus are large, cursorial, flightless birds, with long, scaly legs, three toes, and no preen gland. Weighing 51–120 lb (23–55 kg) and standing 6.5 ft (2 m) tall, emus are second only to ostriches in bird size. Although emus usually walk, their long, muscular legs are adapted to running, and they can run up to 30 mph (48 km/h), reaching strides of 9 ft (2.7 m) long. Their plumage is dark brown just after the annual molt, but fades during the year to pale brown. The wings are small, only one tenth the length of the body, and are hidden by the plumage. The main shaft and aftershaft of the feathers are equal in length so that every feather appears double. Pale blue skin shows clearly through the sparse feathers of the long neck. Females are slightly larger than males, weighing 90 lb (41 kg) versus male weights of 80 lb (36 kg). The females have a stronger blue coloration on the bare skin of the neck and head.

The sexes can be further distinguished during the laying period. The hen molts before laying and is a dark bird at this time, whereas the male does not molt until he is incubating,

Emu (*Dromaius novaehollandiae*). (Illustration by Patricia Ferrer)

and is paler than the female. New chicks are striped longitudinally with black, brown, and cream. They weigh 1 lb (0.5 kg) and stand approximately 5 in (12 cm) tall. Birds in their first year have black feathers covering their heads and necks, and the bare skin of the adult appears during their second year of life.

In the 15 ft (4.5 m) gut the gizzard is very muscular, capable of grinding, with the help of ingested stones, hard seeds and nuts. The emu bill is broad and soft, adaptable for grazing. The emu has a tracheal pouch, part of its air sac system, which is used for communication. The pouch is over 12 in (30 cm) long, very thin-walled, and allows the emu to produce deep guttural grunts. The pouch develops fully during breeding season and is used for courtship.

Distribution

Emus lived on Australia, Tasmania, and on King and Kangaroo islands. The only surviving species is confined to the Australian mainland. Because emus are nomadic and all members of the populations respond to the same environmental cues, large movements can occur. In Western Australia a fence has been built to protect the agricultural areas from emus moving towards the southern winter rains. Movements are often seen in Western Australia, where they can be detected on this fence, and also occur in eastern Australia, when emus swim the Murray River in large numbers.

Habitat

Emus are omnivores, at home in eucalypt forest and woodland, acacia woodland, savanna and heath, pine woodland, coastal heath, open spinifex and other tussock grassland, and in remnant vegetation along salt watercourses and on high alpine plains. They swim well and have been found on islands up to 3 mi (5 km) off the coast. Emus drink frequently, probably every day, taking 0.2–0.4 gal (600–1,500 ml) of water at a time. The chicks also need water, but usually obtain it from the leaves of succulent plants that they eat.

Behavior

Behavior has been studied only in the surviving species *Dromaius novaehollandiae*. Many accounts describe emus as flock birds but this is an artifact of observations on captive and semi-captive birds. In open country emus live as pairs in home ranges of about 12 mi² (30 km²). They will defend significant parts of these areas, but usually try to avoid meeting other emus. In confrontations, they use kicking as a defense. The members of a brood may stay together for some months after they are deserted by their father, but will eventually break up and form pairs, rather than remain as a flock.

Emus have two basic calls, each of which can vary greatly in intensity. The grunt is given mainly by the male and has an aggressive message. The drum or boom is given mainly by females and has a territorial motivation. For the drum, the bird uses the tracheal pouch as a resonating chamber and the low frequency call carries 0.6 mi (1 km). Chicks have a piping call until their voice breaks, at about five months old, when a hole opens in their trachea, and communication with the air sac in the neck is established. Emus display by fluffing out their neck while they drum and, at other times, by stretching themselves up to their full height and grunting with vigor.

An emu (*Dromaius novaehollandiae*) tends its nest. (Photo by J.P. Ferrero/Jacana. Photo Researchers, Inc. Reproduced by permission.)

Emus (*Dromaius novaehollandiae*) in a field in the Grampian Mountains area of Australia. (Photo by Joy Spurr. Bruce Coleman Inc. Reproduced by permission.)

Feeding ecology and diet

Emus feed on nutritious parts of plants, fruits, seeds, flowers, and green shoots. They also take grasshoppers, beetles, and caterpillars, often in great quantities. In order to grind up hard material, emus eat pebbles and stones. Individual stones may weigh 0.1 lb (45 g) and individual gizzards may contain 1.6 lb (745 g) of mineral material. Experiments in a zoo, where emus were fed marbles, showed that such hard materials could be retained in the gizzard for over 100 days. It was thought that this might be a useful method of marking the birds, getting them to eat marbles of different colors at different sites, but the method failed because the bird's bill was too weak to pick up the marbles. Emus also eat quantities of charcoal, but the function of this material is unknown. Emus exhibit very strong specific search images, concentrating on one food even though alternative foods are readily available. They also move great distances searching for food.

Reproductive biology

Reproductive biology has been studied only in the surviving species *D. novaehollandiae*. In southern Australia emus usually lay in autumn, on a platform 3.3 ft (1 m) in length consisting of sticks, grass, and debris on the ground. The clutch varies with the rainfall, from eight to 20 eggs, which are dark green and have a pimply texture. Each egg weighs 1.5 lb (0.7 kg), is 5–6 in (13–15 cm) high, and the shells are 0.03 in (0.8 mm) thick. Incubation is by the male alone, although in captivity hens have been observed to sit for short periods. The male chases the female from the nest when the clutch is completed; she may mate with another male, move away on migration, or remain in the territory, defending it for some weeks. Wild males do not leave the vicinity of the nest, unless disturbed, and do not eat, drink or defecate during incubation. They lose 10–20 lb (5–9 kg) during this period. The chicks hatch after about 56 days, often over four days, and the male then leaves the nest area with them. They stay with their father for 5–7 months. Emus can lay at a year old, but most do not breed until they are two. Wild birds live only six or seven years, although captive ones may live much longer. Under captive conditions some males will have two mates, but this is rare in the wild.

Conservation status

The emu is secure on the Australian mainland and has been reintroduced to Tasmania. The Tasmanian emu, a subspecies of *D. novaehollandiae*, died out in 1865, and the dwarf species *D. ater* of King Island and *D. baudinianus* of Kangaroo Island were exterminated by human activity before 1810 and 1827 respectively. Fears have long been expressed for the survival of the emu on mainland Australia. In fact the bird is surviving well and will continue to do so while two thirds of Australia is unoccupied or used only as rangeland. It seems to be one of those species that can sustain very large population fluc-

An emu (*Dromaius novaehollandiae*) drinking in western New South Wales, Australia. (Photo by Jen and Des Bartlett. Bruce Coleman Inc. Reproduced by permission.)

tuations. On an area of 1,000 mi² (2,500 km²) in Western Australia, the number of birds fluctuated between five and 970 over ten years. Estimates of the total population of emus in Australia vary from half to one million. Emus are vulnerable as eggs and hatchlings. Buzzards eat the eggs, and young emus are hunted by dingoes, eagles, non-native foxes, dogs, and cats.

Significance to humans

The emu is the national bird of Australia and appears on that country's coat of arms. For the Australian Aborigines the emu was as central to their existence as was the American buffalo to Native Americans. The emu provided food, and their fat and organs were also used for medicine. The bird was incorporated into their rituals and mythology. Emus were killed for meat by early European settlers. Later emus damaged crops in some areas, leading to organized campaigns to eradicate the birds, including the brief 1932 "emu war" in which machine guns were used. Bounties were paid on their heads in some areas, but recently fences have been used to keep emus out of areas where they might cause damage. Europeans have attempted to farm emus, because they reproduce well in captivity and grow fast, but so far have been unable to establish stable markets for any commercially viable quantity of product. Emu farming began in Australia in the 1970s. The insular species were killed for food by settlers, sealers, and whalers until they became extinct.

Species accounts

Emu
Dromaius novaehollandiae

TAXONOMY
Casuarius n. hollandiae Latham, 1790, New Holland (=Sydney, Australia).

OTHER COMMON NAMES
French: Émeu d'Australie; German: Emu; Spanish: Emú.

PHYSICAL CHARACTERISTICS
60–75 in (150–190 cm); female 57–106 lb (26–48 kg), male 39–103 lb (18–47 kg). The plumage is brown, sometimes ticked, the feathers long and shaggy.

DISTRIBUTION
Away from settled areas, the emu can be seen anywhere in Australia, but it visits the arid zone only when good rains have fallen there.

HABITAT
Emus are able to live in all types of native Australian vegetation.

BEHAVIOR
Emus live as pairs in large territories; young birds remain with their male parent for about seven months, and may then form small flocks until maturity.

FEEDING ECOLOGY AND DIET
The emu feeds on nutritious parts of plants, fruits, seeds, flowers, and green shoots.

REPRODUCTIVE BIOLOGY
Emus breed as pairs in natural conditions, the female laying up to 20 eggs that she leaves the male to incubate. The male also guards the chicks for about seven months.

CONSERVATION STATUS
Not threatened. The emu is secure in mainland Australia.

SIGNIFICANCE TO HUMANS
The emu has been seen as a pest in some areas and elsewhere as a potential farm animal from which meat, leather and fat can be harvested. ◆

King Island emu (extinct)
Dromaius ater

TAXONOMY
Dromaius ater Vieillot, 1817, King Island, Tasmania.

OTHER COMMON NAMES
English: Black emu; French: Émeu noir; German: Emu King Island; Spanish: El Emú de King Island.

PHYSICAL CHARACTERISTICS
Height 55 in (140 cm); 51 lb (c. 23 kg). A small, black emu, with grayish juveniles and striped chicks.

DISTRIBUTION
King Island, Tasmania.

HABITAT
Not known.

BEHAVIOR
Not known.

FEEDING ECOLOGY AND DIET
Ate berries, grass, and seaweed.

REPRODUCTIVE BIOLOGY
Not known.

CONSERVATION STATUS
Extinct.

SIGNIFICANCE TO HUMANS
Source of food to sealers and settlers. ◆

Kangaroo Island emu (extinct)
Dromaius baudinianus

TAXONOMY
Dromaius baudinianus Parker, 1984, Kangaroo Island, South Australia.

OTHER COMMON NAMES
English: Dwarf emu; French: Ému de Baudin; German: Emu Kangaroo Island; Spanish: El Emú de Kangaroo Island.

PHYSICAL CHARACTERISTICS
A small, black emu, slightly larger than the King Island emu.

DISTRIBUTION
Kangaroo Island, South Australia.

HABITAT
Not known.

BEHAVIOR
Not known.

FEEDING ECOLOGY AND DIET
Not known.

REPRODUCTIVE BIOLOGY
Not known.

CONSERVATION STATUS
Extinct.

SIGNIFICANCE TO HUMANS
Source of food to sealers and settlers. ◆

Resources

Books

Bird, David M. *The Bird Almanac: The Ultimate Guide to Essential Facts and Figures of the World's Birds.* Buffalo: Firefly Books, 1999.

Davies, S. J. J. F. *Ratites and Tinamous.* Oxford: Oxford University Press, 2002.

del Hoyo, J., A. Elliot, and J. Sargatal, eds. *Ostrich to Ducks.* Vol. 1, *Handbook of the Birds of the World.* Barcelona: Lynx Edicions, 1992.

Marchant, S., and P. J. Higgins. *Ratites to Ducks.* Vol 1, *Handbook of Australian, New Zealand and Antarctic Birds.* Oxford: Oxford University Press, 2001.

Whitfield, P., ed. *The MacMillan Illustrated Encyclopedia of Birds.* New York: Collier Books, 1988.

Periodicals

Davies, S. J. J. F. "The food of emus." *Australian Journal of Ecology* 3 (1978): 411–22.

Davies, S. J. J. F. "Nomadism in response to desert conditions in Australia." *Journal of Arid Environments* 7 (1984): 183–95.

Grice, D., G. Caughley, and J. Short. "Density and distribution of emus." *Australian Wildlife Research* 12 (1985): 69–73.

Organizations

Birds Australia. 415 Riversdale Road, Hawthorn East, Victoria 3123 Australia. Phone: +61 3 9882 2622. Fax: +61 3 9882 2677. E-mail: mail@birdsaustralia.com.au Web site: <http://www.birdsaustralia.com.au>

Emu Farmers Federation of Australia. c/o Secretary, Arthur Pederick, P.O Box 57, Wagin, Western Australia 6315 Australia. Phone: +61 8 9861 1136.

S. J. J. F. Davies, ScD

Kiwis

(Apterygidae)

Class Aves
Order Struthioniformes
Suborder Apteryges
Family Apterygidae

Thumbnail description
Chicken-sized birds, the smallest of the ostrich-like birds (ratites); only ratite with four toes; stout legs and feet, vestigial wings, long and curved bill with nostrils near tip; brown-black "hair-like" plumage

Size
14–22 in (35–55 cm); 2.6–8.6 lb (1.2–3.9 kg)

Number of genera, species
1 genus; 3 species

Habitat
Sub-tropical and temperate forest, woodlands, coastal heath, pasture, and tussock grasslands

Conservation status
Endangered: 2 species; Not threatened: 1 species

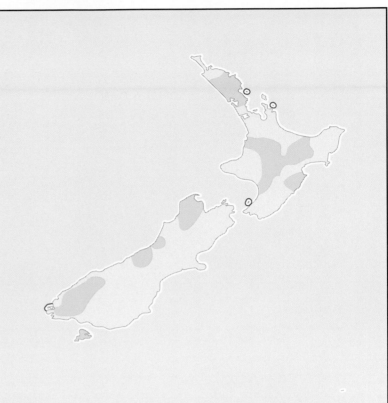

Distribution
New Zealand

Evolution and systematics

Kiwis (genus *Apteryx*) belong to the group of large, flightless birds known as the ratites that have in common the lack of a keel on the sternum and a distinctive palate. The origin of these birds has recently been clarified by the discovery of numerous good fossils in North America and Europe. Whereas it used to be thought that the ratites had a southern origin, in the old continent of Gondwana, new fossil evidence has shown flying ratites inhabited the Northern Hemisphere in the Paleocene and Eocene, 40–70 million years ago. The present Southern Hemisphere distribution of the ratites probably results from the spread of flying ancestors of the group from the north. Anatomical evidence suggests that the kiwis' closest relatives were the extinct moas of New Zealand. Biochemical evidence is conflicting.

Physical characteristics

Kiwis are medium-sized, flightless birds, with stout legs, four toes, and no preen gland. They weigh 2.6–8.6 lb (1.2–3.9 kg). The bill is long, pliable, and sensitive to touch, and the nostrils are lateral at the tip. The eye interior has a much-

reduced pectin, which normally serves to supply nutrients and oxygen to the retina and tends to be smaller in nocturnal birds. The feathers have no aftershaft and lack barbules (hooks on the barbs); therefore, the feathers are loose and project out much like coarse hair. There are large vibrissae, stiff feathers that usually have tactile function, around the gape, and there are 13 flight feathers, which are only a little stronger than the other feathers. The second finger is absent. There is no tail, only a small pygostyle (similar to a tail-bone). The legs are strong but short, and the claws are sharp. The gizzard is weak. The caeca, which aid in digestion, are long and narrow. The young are colored like the adults but with softer plumage.

Distribution

Kiwis live on the North and South Islands of New Zealand and on Stewart Island. Although formerly widespread, only the brown kiwi (*Apteryx australis*) remains common, inhabiting several areas of North Island, some parts of South Island and most of Stewart Island. The other two species are confined to a few island sanctuaries and a small area of the northwest of South Island.

Brown kiwis (*Apteryx australis*) are largely nocturnal, and roost in sheltered areas at ground level during the day. (Photo by I. Visser/VIREO. Reproduced by permission.)

Habitat

Kiwis favor subtropical and temperate podocarp and beech forest, but settlement and forest clearing has left little forest for them to use. The brown kiwi has successfully occupied plantations, even of exotic pines, as well as the fringes of farmland, sub-alpine scrub and tussock grassland. The other two species are now confined to mountainous regions and islands, but in the past they were probably widespread in podocarp forests of both lowlands and highlands.

Behavior

Most kiwis are nocturnal, but the Stewart Island form of the brown kiwi is active during the day. They form monogamous pairs, probably lasting for life, moving about their territory singly and indulging in frequent calling, sometimes as duets between males and females. Territories seem to be maintained by calling, although aggressive behavior has been observed, involving vigorous encounters and chases at territorial boundaries. Territory sizes vary with locality and species from 5 to 111 acres (2–45 ha). Only the weak, shrill "kee wee" or "kee kee" whistles of the male and the hoarse "kurr kurr" of the female betray their presence. Males call more frequently than females. Both sexes call in an upright position, with bill raised and neck and legs fully stretched. Apart from calling, few displays accompany mating, which may last 1–2 minutes. Kiwis roost alone during daylight in shallow burrows and sheltered places, mostly at ground level.

Feeding ecology and diet

Kiwis feed on invertebrates, especially earthworms, spiders, and insects from the ground and litter. They take some plant material, but the quantity is insignificant compared with their intake of animal food. The sense of smell of kiwis is very acute so that most of their food is located by scent. Sight and sound play only minor roles in food searching. While probing for hidden worms and insect larvae in the soft forest floor, they use their long bills in the same manner as snipes (family Scolopacidae). The bill is thrust deeply into the ground when feeding and the resulting characteristic holes betray the presence of the birds. Distended gizzards may contain 2 oz (50 g) of material. As in other birds, their gizzards also contain some grit that helps to grind up the food.

Reproductive biology

Kiwis are unique in the bird world in having paired functional ovaries. In most other birds, only the left ovary is functional, although some individuals of a few raptor species also have a functional right ovary. In kiwis, both ovaries function regularly, but only the left oviduct is developed, the eggs from both ovaries passing down it. The eggs are of great size, up to 1 lb (450 g), each egg a fifth to a quarter the weight of the female. Often only one egg is laid, but some two-egg clutches have been found. It may be 20–60 days between the laying of eggs in two-egg clutches. For its nest, the bird digs a burrow or selects and remodels a den in some sheltered spot. Incubation is by the male, except in the great spotted kiwi (*A. haastii*), where both sexes regularly incubate the egg(s). The incubation period lasts 63–92 days. The chick hatches in adult plumage, remains inactive in the nest burrow for some days while feeding on its yolk sac, and then emerges to feed independently.

Conservation status

New Zealand has no native mammals, but the introduction of rats, dogs, pigs, and mustelids (stoats and weasels) has caused severe predation on kiwis. Apart from the clearing of native forest, predation has been blamed for the decline of the populations of all three kiwi species. The effect has been worst on the spotted kiwis; the brown kiwi seems able to survive in spite of the presence of dogs and introduced mammals. The little spotted kiwi (*A. owenii*) is now confined to four island sanctuaries from which predators have been or are being removed. The great spotted kiwi population suffers from traps set to catch introduced possums; for example, up to half of some populations have fractured or amputated toes. Captive breeding and translocations are being undertaken by New Zealand conservation agencies.

Significance to humans

Kiwis are the national bird of New Zealand, but are of no other special significance to other people. In former times, Maoris used kiwi skins to make cloaks and they and the early European settlers hunted kiwis for food.

1. Little spotted kiwi (*Apteryx owenii*); 2. Great spotted kiwi (*Apteryx haastii*); 3. Brown kiwi (*Apteryx australis*). (Illustration by Bruce Worden)

Species accounts

Brown kiwi
Apteryx australis

TAXONOMY
Apteryx australis Shaw and Nodder, 1813, Dusky Sound, South Island, New Zealand.

OTHER COMMON NAMES
English: Common kiwi; French: Kiwi austral; German: Streifenkiwi; Spanish: Kiwi Común.

PHYSICAL CHARACTERISTICS
18–22 in (45–55 cm); female: 4.6–8.5 lb (2.1–3.9 kg), male: 3.6–6.1 lb (1.6–2.8 kg). Medium-sized, rotund, flightless bird, with no tail. Body cone-shaped, tapering to a small head with a long, slightly down-curved bill. Streaked rufous plumage, shaggy and hair-like, obscuring short wings that end in a claw. Female larger than male.

DISTRIBUTION
On North Island mainly in Northland and Taranaki, although still occurs in small pockets elsewhere. On South Island mainly in Fiordland, with small populations in Westland. Widespread on Stewart Island.

HABITAT
Subtropical and temperate forests and shrublands. Most common in dense forest but able to maintain populations in regenerating bush, pasture, and pine forest.

BEHAVIOR
Nocturnal, usually seen alone; roosts in dens or burrows by day. The name "kiwi" comes from the sound of one whistled call that has also been rendered as "ah-eel". Males call most often, with duets between partnered males and females at times.

FEEDING ECOLOGY AND DIET
The brown kiwi feeds on soil invertebrates such as earthworms, beetle larvae, snails, spiders, centipedes, and orthoptera. It uses its sense of smell to find food, probing ceaselessly into the ground, leaving characteristic cone-shaped holes in the substrate.

REPRODUCTIVE BIOLOGY
Live as monogamous pairs in territories of 12–106 acres (5–43 ha), depending on location. Nests are made in burrows, sheltered places, and beneath thick vegetation. The female lays one or two large eggs that the male incubates for up to 90 days. The young hatch in adult plumage and, after a few days in the nest, come out to feed independently. There is little evidence of parental care, but the chick may be found near its parents for up to a year.

CONSERVATION STATUS
Not threatened. Although the brown kiwi is the most common of the group, it suffers from attacks by dogs and is often caught in traps set for the introduced possum. Large populations live in Northland and on Stewart Island, but elsewhere fragmentation has reduced population sizes below sustainable levels.

SIGNIFICANCE TO HUMANS
The Maori formerly ate the birds and made cloaks from their skins. Apart from being New Zealand's national bird, the species is of no economic significance to humans now. ◆

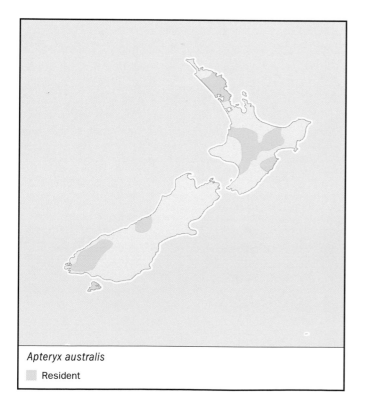

Apteryx australis
Resident

Little spotted kiwi
Apteryx owenii

TAXONOMY
Apteryx owenii Gould, 1847, New Zealand.

OTHER COMMON NAMES
English: Little gray kiwi; French: Kiwi d'Owen; German: Zwergkiwi; Spanish: Kiwi Moteado Menor.

PHYSICAL CHARACTERISTICS
Length 13.8–17.7 in (35–45 cm); males 2–2.9 lb (0.9–1.3 kg), females 2.2–4.2 lb (1.0–1.9 kg). Medium-sized, flightless, nocturnal bird with pale-mottled, gray, shaggy plumage. The body is pear-shaped with a long neck and bill.

DISTRIBUTION
Surviving on only four islands: Kapati, Red Mercury, Hen, and Long.

HABITAT
Evergreen, broadleaf forest and margins of forest up to 3,000 ft (1,000 m) with over 40 in (100 cm) annual rainfall. Favors wet forest, with rotten logs and dense undergrowth.

BEHAVIOR
Nocturnal, pairs hold territories of about 10 acres (4 ha). Pair formation in second year and maintained for life. Chases

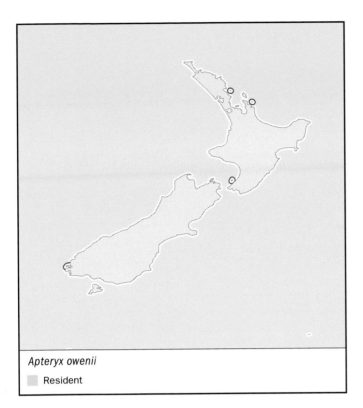

Apteryx owenii
☐ Resident

OTHER COMMON NAMES
English: Great gray kiwi; French: Kiwi roa; German: Haastkiwi; Spanish: Kiwi Moteado Mayor.

PHYSICAL CHARACTERISTICS
Length 17.7–19.7 in (45–50 cm); males 2.6–5.7 lb (1.2–2.6 kg), females 3.3–7.3 lb (1.5–13.3 kg). Medium-sized, flightless, nocturnal bird with pale, mottled-gray, shaggy plumage. The body is pear-shaped with a long neck and bill. Females larger than males, and with longer bills. Larger than the little spotted kiwi and about the size of the brown kiwi.

DISTRIBUTION
Two isolated populations in the northwest of South Island, in Nelson and Westland.

HABITAT
Densest population above 2,000 ft (700 m) in wet beech forest in mountain ranges running parallel to the coast. Also in tussock grassland, podocarp, and hardwood forests, and sometimes in coastal pasture.

BEHAVIOR
Nocturnal, pairs holding territories of about 49 acres (20 ha) or more. Most displays are vocal, using high-pitched whistles that have distinct male and female versions—the male shriller than the female. Roost in dens during the day, into which some vegetation is taken by the bird to form a small mat. A kiwi may have 100 dens in its territory, using a different one each day.

FEEDING ECOLOGY AND DIET
Omnivorous, but eats mostly soil and litter invertebrates, such as earthworms, millipedes, and larval beetles, as well as moths, crickets, and spiders, supplemented with some fruit. Crayfish

occur in defense of territory, but most displays are vocal, using high-pitched whistles that have distinct male and female versions.

FEEDING ECOLOGY AND DIET
Omnivorous, but eats mostly soil and litter invertebrates, such as earthworms, millipedes, larval beetles, as well as moths, crickets, and spiders, supplemented with some fruit.

REPRODUCTIVE BIOLOGY
Nest in burrows dug by the pair. Sometimes there is no nest material, but in other burrows some leaves and twigs have been gathered. Most clutches are composed of one egg, but about 15% have two. Only the male incubates, sitting for 63–76 days. The chick is tended (probably fed) for about four weeks after it hatches. Unlike the brown kiwi, the chick of the little spotted kiwi may stay in the nest for two to three weeks before emerging.

CONSERVATION STATUS
Endangered with a total population of about 1,000 individuals. Even on its island sanctuaries it suffers predation on its eggs from native rails (wekas) and rats.

SIGNIFICANCE TO HUMANS
None known. ◆

Great spotted kiwi
Apteryx haastii

TAXONOMY
Apteryx haastii Potts, 1872, Westland, New Zealand.

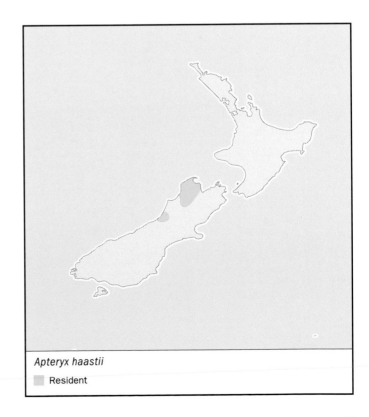

Apteryx haastii
☐ Resident

are eaten when they leave flooded streams, and some food is taken above the ground when the bird can walk out along leaning branches.

REPRODUCTIVE BIOLOGY
Pairs maintained for life with some indication of polyandry in lowland populations. Mostly nest in natural hollows and sheltered places, but a few nests are in short burrows dug by the pair. Some moss lichen, leaves, and twigs are gathered to form a thick nest. Most clutches are composed of one egg, but two

eggs have been reported. Both sexes incubate, usually the male by day and the female by night. The incubation period is not known.

CONSERVATION STATUS
Endangered, declining in lowland forests and vulnerable to traps set for possums and to attacks by dogs.

SIGNIFICANCE TO HUMANS
None known. ◆

Resources

Books

Davies, S. J. J. F. *Ratites and Tinamous*. Oxford: Oxford University Press, 2002.

Folch, A. "Apterygidae (Kiwis)." In *Handbook of the Birds of the World*. Vol. 1, *Ostrich to Ducks*, edited by Josep del Hoyo, Andrew Elliott, and Jordi Sargatal. Barcelona: Lynx Edicions, 1992.

Marchant, S., and P. J. Higgins. *Ratites to Ducks*. Vol. 1, *Handbook of Australian, New Zealand and Antarctic Birds*. Oxford: Oxford University Press, 1990.

Reid, B., and G. R. Williams. "The Kiwi." In *Biogeography and Ecology in New Zealand*, edited by G. Kuschel. The Hague, 1975.

Periodicals

Wenzel, B. M. "Olfactory Sensation in the Kiwi and Other Birds." *Annals of the New York Academy of Sciences* 188 (1971): 183–93.

Organizations

Birds Australia. 415 Riversdale Road, Hawthorn East, Victoria 3123 Australia. Phone: +61 3 9882 2622. Fax: +61 3 9882 2677. E-mail: mail@birdsaustralia.com.au Web site: <http://www.birdsaustralia.com.au>

Ornithological Society of New Zealand. c/o Secretary, P.O. Box 12397, Wellington, North Island New Zealand. E-mail: OSNZ@xtra.co.nz Web site: <http://osnz.org.nz>

S. J. J. F. Davies, ScD

Moas
(Dinornithidae)

Class Aves
Order Struthioniformes
Suborder Dinornithes
Family Dinornithidae

Thumbnail description
Large, flightless cursorial birds with no visible wings, long legs, long necks, and four toes

Size
3–12 ft (0.9–3.7 m); 48–506 lb (22–230 kg)

Number of genera, species
6 genera; 10 species

Habitat
Forest, woodland, heath, and grassland

Conservation status
Extinct

Distribution
New Zealand

Evolution and systematics

Moas belong to the group of large, flightless birds known as ratites. Ratites have a distinctive palate, and a sternum (breastbone) with no keel, so there is no anchor for the strong musculature needed for powered flight. The origin of these birds has recently been clarified by the discovery of numerous good fossils in North America and Europe. Ratites were once thought to have a southern origin in the ancient continent of Gondwana, but new fossil evidence shows that flying ratites inhabited the Northern Hemisphere in the Paleocene and Eocene, 40–70 million years ago. The present Southern Hemisphere distribution of ratites probably resulted from the spread of flying ancestors of the group from the north.

The earliest remains of moas in New Zealand are from the Upper Pliocene, about 1.5 million years ago. By then they were recognizably moas, but it is thought that the two subfamilies— the tall, graceful Dinornithinae and the short, stout-bodied Anomalopteryginae—may be descended from two different flying ratites that invaded New Zealand during the Tertiary. Many different classifications have been proposed for moas; some authors list as many as 27 species, others only 10. The

lower figure is adopted here, following a detailed review by Atholl Anderson, because many earlier classifications depended on small differences in subfossil bone size and shape that could be due to sexual dimorphism or age differences.

Physical characteristics

There are no accounts of living moas. All information is derived from subfossil material recovered from swamps, caves, and river beds. Moas were very large, flightless birds with long necks and long legs. Unlike surviving large ratites, the tibial bone of moa legs was longer than the tarsus, and they moved slowly. Except for bats, no mammals inhabited New Zealand, so moas had no predators before the Polynesians arrived and no need to flee swiftly. The height of the birds (to the back) was 3–12 ft (0.9–3.7 m) and they weighed 48–506 lb (22–230 kg). The three *Dinornis* species and *Pachyornis elephantopus* were the largest, weighing 257–506 lb (117–230 kg). Surviving feather and skin fragments indicate that, apart from the legs, the birds were fully feathered, although they had no visible wings. Each feather was double, with a well-developed aftershaft, and was brownish, sometimes with pale edging.

1. *Pachyornis elephantopus*; 2. *Euryapteryx curtus*; 3. *Euryapteryx geranoides*; 4. *Emeus crassus*; 5. *Pachyornis mappini*. (Illustration by Marguette Dongvillo)

Evidence indicates that in at least some species females were larger than males.

Distribution

Moas lived only in New Zealand. Five species (*Anomalopteryx didiformis, Euryapteryx geranoides, Dinornis struthoides, D. novaezealandiae,* and *D. giganteus*) were common to North and South Island, two have been found only on North Island (*Euryapteryx curtus* and *Pachyornis mappini*), and three only on South Island (*Megalapteryx didinus, Emeus crassus,* and *P. elephantopus*). At least one species (*Euryapteryx geranoides*), and possibly two, also lived on Stewart Island. There is debate

about the total numbers of moas, with estimates ranging from millions to thousands. Using the most recent evidence, Atholl Anderson has argued that if all 10 species were combined, there were probably only tens of thousands in total—twice as many in South Island as in North Island—with the greatest concentration on the eastern side of South Island.

Habitat

Moas apparently lived in all New Zealand environments, using coastal dunes, forest fringes, podocarp (seed-producing conifers) forests, beech forests on limestone, shrubland, and grassland up to 6,600 ft (2,000 m). *A. didiformis, D. struthoides,*

1. *Dinornis giganteus*; 2. *Megalapteryx didinus*; 3. *Anomalopteryx didiformis*; 4. *Dinornis struthoides*; 5. *Dinornis novaezealandiae*. (Illustration by Marguette Dongvillo)

and *D. novaezealandiae* lived in dense lowland conifer and broad-leaved forest, and beech forest. *M. didinus* lived only in high-altitude beech forest. *Emeus crassus*, *Euryapteryx* species, *Pachyornis* species, and *D. giganteus* lived in the lowlands—dunelands, forest fringes, and forest, shrub, and grassland mosaics.

Behavior

Most moas were probably diurnal, living in small groups rather than large flocks. There is no hard evidence about their daily lives.

Feeding ecology and diet

A number of moa gizzards have been found and analyzed, showing that the birds fed on plants, taking seeds, twigs, and leaves from different species. Nineteen gizzards from two different sites showed that the birds took plants that grew in the forest and those from open country, suggesting they often fed along the boundary between the two environments. About 80% of the material was twigs, but seeds and leaves were also abundant. It may be that the tough twigs stayed longer in the gizzard than other plants and were over-represented in the sample.

It is now clear that moas did not exist solely on ferns, as some early authors suggested, but used a number of different plant species as food. Seeds of the shrub *Comprosma* and the tree *Podocarpus* were abundant, but 29 different species were represented in 19 samples. All gizzards contained pebbles, up to 11 lb (5 kg) in a large *Dinornis*. From the nature of these pebbles, which were the kinds of rock found where the gizzard was collected, it can be deduced that moas were sedentary. Had they been migratory or nomadic, pebbles in the gizzard would have included rocks of many kinds because such pebbles remain in the gizzard for months, and would have reflected different landscapes. More information will gradually come to light about the diversity of the moa diet, because the shapes and sizes of their bills differ considerably, suggesting that various species must have selected different foods.

Reproductive biology

Moas laid small clutches, perhaps only one egg. It is thought, on very slight evidence, that males incubated, as is the case in most other ratites, but nothing is known of their routines, not even the incubation period. Some supposed nests of sticks around scrapes in the ground have been found in caves and rock shelters associated with egg shell fragments. The eggs are large—up to 9 in (23 cm) long and 7.6 in (19.5 cm) wide—and green, at least in some species.

Conservation status

Extinct. The last moas probably died out in the seventeenth century.

Significance to humans

Moa is the Maori word for a group of large, flightless birds. A great deal of thought has been given to the interaction of Maoris and moas. More than 300 sites have been identified at which moas were butchered by Maoris, some (according to carbon dating) as old as A.D. 1000. The sites provide evidence that the Maoris killed and ate moas systematically. Some sites are very large, but Atholl Anderson has analyzed their location and size to show that they fall into two classes. Sites near the coast are adjacent to large rivers and usually contain remains of many moas. This suggests that expeditions were made up river and the birds brought downstream by boat for butchering. Small sites are mostly in the mountains, or where water transport would not be available, suggesting that they represent the catch of one hunting party, the birds being butchered on the spot and probably eaten there too. Moas would provide a very valuable protein resource in a land where other game was scarce.

The evidence of such extensive hunting has been used to suggest that moas were hunted to extinction. However, the Maoris also cleared the land, mainly by burning vegetation; such habitat alteration may have contributed significantly, perhaps fatally, to the extermination of many species. There are other puzzles, too. Many moa bones are found aggregated in beds of former and existing swamps. There is much to discover about the interactions between moas and Maoris. It is known that Maoris used moa skins as cloaks and carved tools from the bones. It is surprising, considering the significance the birds must have had, how little mythology about them has survived.

Resources

Books

Davies, S. J. J. F. *Ratites and Tinamous.* Oxford: Oxford University Press, 2002.

Anderson, A. *Prodigious Birds, Moas and Moa-hunting in prehistoric New Zealand.* Cambridge: Cambridge University Press, 1989.

Periodicals

Anderson, A. "Habitat Preferences of Moas in Central Otago, A.D. 1000–1500, According to Palaeobotanical and Archaeological Evidence." *Journal of the Royal Society of New Zealand* 3 (1982): 321–36.

Rudge, M. R., ed. "Moas, Mammals and Climate in the Ecological History of New Zealand." *New Zealand Journal of Ecology* Supplement 12 (1989): 1–169.

Organizations

Ornithological Society of New Zealand. P.O. Box 12397, Wellington, North Island New Zealand. E-mail: OSNZ@xtra.co.nz Web site: <http://osnz.org.nz>

S. J. J. F. Davies, ScD

Ostriches
(Struthionidae)

Class Aves
Order Struthioniformes
Suborder Struthiones
Family Struthionidae

Thumbnail description
Very large flightless birds with large, loose-feathered wings, two toes, and black-and-white (male) or gray-brown (female) plumage

Size
5.7–9.0 ft (175–275 cm); 139–345 lb (63–157 kg)

Number of genera, species
1 genus; 1 species

Habitat
Open woodlands, savanna, arid shrubland, desert, and grasslands

Conservation status
Secure: 1 species

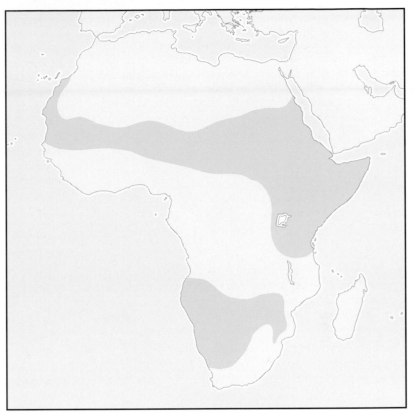

Distribution
Africa, excluding tropical forest belt adjacent to the equator

Evolution and systematics

The ostrich (*Struthio camelus*) belongs to the group of large, flightless birds known as ratites. Ratites have in common a distinctive palate and the lack of a bony keel to the sternum (breastbone), to which the powerful musculature required for flight would be attached.

Ratites were once thought to have a southern origin in the ancient continent of Gondwana, but recent fossil evidence discovered in North America and Europe shows that flying ratites inhabited the Northern Hemisphere in the Paleocene and Eocene, 40–70 million years ago. The current Southern Hemisphere distribution of ratites was likely due to the spread of flying ancestors from the north.

The ostrich is the only living representative of suborder Struthiones, family Struthionidae. Eight extinct species all belonged to the same genus. Fossil bones and egg shells show that ostrich ancestors probably originated in the Eocene (40–55 million years ago) in the Asiatic steppes as small flightless birds. In the lower Pliocene (about 12 million years ago) they developed into gigantic forms that were distributed as far as Mongolia and, later, South Africa. The present-day ostrich is somewhat smaller and originated as a new species in the Pleistocene (one to two million years ago); some of its early remains were found at home sites of prehistoric humans.

Physical characteristics

A large bird, cursorial with long legs and neck. At 5.7–9 ft (1.8–2.8 m) and 139–345 lb (63–157 kg), the ostrich is the largest living bird. Males are larger than females. The head and about two-thirds of the neck are sparsely covered with short, hair-like, degenerated feathers, making the bird appear nude. The skin is variably colored, depending on the subspecies. The legs are particularly strong and long. The tarsus in sexually mature males has red horn plates; in sexually mature females they are black. The foot has two toes: a large, strongly clawed third toe and a weaker, generally clawless fourth (outside) toe. The first and second toes are absent. The feathers have no secondary shaft or aftershaft. There are 50–60 tail feathers. The wing has 16 primaries, four alular, and 20–23 secondary feathers. Wing feathers and rectrices have changed to decorative plumes.

Ostrich (*Struthio camelus*). (Illustration by Patricia Ferrer)

toms and in desert-savanna plains, rarely above 300 ft (100 m). In eastern Africa it is in savanna; in southern Africa it is in open grassland with some shrubs. In southwest Africa its habitat is semidesert or true desert, with patches of open, stunted woodland.

Behavior

An ostrich population often becomes a mixed society of flocks, families, and individuals of all age groups whose composition changes with the season. In a series of ostrich sightings in east Africa, 49% of the sightings were of single birds, 35% were of two birds together, and 16% were of groups of three to five birds. In rainless periods, when wandering, in common grazing grounds, and at watering places, they form peaceful aggregations of hundreds of birds, but individual flocks remain recognizable. Social contacts between birds of different groups are initiated when one bird approaches another in a submissive posture, with head lowered and tail down. Often a family of one herd adopts the chicks or young of another. Single cocks may join together and form "schools" of half-grown ostriches, which then wander about for days or weeks. For communal sand baths, each flock seeks out a sandy depression.

The ostrich is diurnal but may sometimes be active on moonlit nights. It loafs and roosts by squatting on the ground, and is most active early and late in the day. Its normal walking pace is 2.5 mph (4km/hr), but when running in alarm it can reach speeds of 45 mph (70 km/hr). It is very vigilant when feeding, continually raising its head to look around. It feeds more frequently in small flocks than in large ones. The territorial call of the male ostrich is a roar, far-carrying and

The penis-like copulatory organ is retractable and can be as long as 8 in (20 cm). Food passes through three stomach segments; the gut can be as long as 46 ft (14 m). The rectum is especially expanded, and the caecae—cul-de-sac-like structures at the lower end of the gastrointestinal tract—are about 28 in (70 cm) long. Urine is concentrated in the large cloaca but, in contrast to all other living birds, is secreted separately from the feces. Unlike other birds, ostriches have pubic bones that are fused toward the rear and support the gut. The wishbone is absent, and palate formation is different from that of other ratites. Sphenoid and palatal bones are unconnected.

Distribution

The ostrich formerly occupied Africa north and south of the Sahara, east Africa, Africa south of the rainforest belt, and much of Asia Minor. Its distribution has now shrunk to Africa south of the Sahara and parts of east Africa, but most existing populations are in game parks.

Habitat

The ostrich is an open-country bird. In northern Africa it lives in the dry beds of watercourses in broad valley bot-

A male ostrich (*Struthio camelus*) displays his courting dance. (Photo by Leonard Lee Rue, III. Photo Researchers, Inc. Reproduced by permission.)

Ostrich (*Struthio camelus*) drinking at Kalahari Gemsbok National Park, South Africa. (Photo by Nigel J. Dennis. Photo Researchers, Inc. Reproduced by permission.)

During the courtship of a specific hen, the cock always drives her away from the others. Both move to a remote spot and graze, while their behavior becomes increasingly synchronized. The feeding evidently becomes secondary and evolves into a ritual that further synchronizes the partners' behavior. The smallest disruption in their movements leads promptly to a premature end of the preliminary display. When the courtship continues undisturbed, the cock excitedly and alternately flaps the right and left wings. Both birds then slow their steps and each begins to poke its bill into the ground on a sandy spot and pull out grasses. The cock then throws himself to the ground and stirs up the sand with tremendous wing beats; this seems to be a symbolic hollowing of a nest bowl. Simultaneously, he turns and winds his head in a rapid spiral motion. He continually repeats his muted courtship song while the hen circles in front of or around him in submissive posture, dragging her wings. Suddenly the cock jumps up, the hen drops to the ground and, beating his wings, the cock mounts for copulation.

The breeding season varies regionally, often correlating with the rainy season. The first female to lay in a nest becomes the major hen of the male that owns the nest; she dominates other hens, minor hens, that may also lay in it. As the time of incubation approaches the major hen discards some of the eggs of minor hens, so that the clutch is kept to a size that a bird can incubate effectively. The male incubates by night and the female by day. The clutch of wild birds averages 13 eggs and takes, on average, 42 days to hatch. Both parents care for the chicks, but as they grow, broods may amalgamate, and finally the young from one region gather into an immature flock. Breeding success is low, of the order of one chick per incubated nest, the eggs and chicks being subject to attacks from many predators, hyenas, jackals, and vultures.

Conservation status

The ostrich has declined greatly in abundance and distribution in the last 200 years. Most surviving birds are in game parks or on farms. Only in remote desert regions do truly wild birds persist, but farms and game parks ensure the preservation of the species.

Significance to humans

The ostrich has inspired human thought, religion, and art since ancient times, as indicated by 5,000-year-old records from Mesopotamia and Egypt. For today's Kalahari bushman, its egg is still a valuable vessel in which he keeps scarce water, and from the shells he makes beautiful jewelry for his wife and children. Ostriches are also farmed for feathers, meat, eggs, and leather.

resembling the roar of a lion. A soft "booh" is used as a contact call and hisses are given in threat displays.

Feeding ecology and diet

The ostrich grazes on green grass and browses on shrubs, succulents, and seeds. A few animals are taken, particularly when swarms of insects, such as the plague locust, are active.

Reproductive biology

Males are polygynous, taking more than one mate at a time. In the initial phase of courtship, the cock displays by alternating wing beats in front of the flock to attract or separate out the chosen hens. He chases yearlings away with the help of his major hen. Then the birds move together to the breeding territory, an area of 1–6 mi^2 (2–14 km^2).

A crèche of 3-ft (90-cm) tall ostrich (*Struthio camelus*) chicks is escorted by an adult across the shortgrass plains of the Serengeti in Tanzania. (Photo by Gregory G. Dimijian. Photo Researchers, Inc. Reproduced by permission.)

Resources

Books

Bertram, B. C. R. "The Ostrich Communal Nesting System." In *Monographs in Behavior and Ecology*, edited by J. R. Krebs and T. H. Clutton-Brock. New Jersey: Princeton University Press, 1990.

Davies, S. J. J. F. *Ratites and Tinamous*. Oxford: Oxford University Press, 2002.

del Hoyo, J., A. Elliot, and J. Sargatal, eds. *Handbook of the Birds of the World*. Vol. 1, *Ostrich to Ducks*. Barcelona: Lynx Edicions, 1992.

Periodicals

Bertram, B. C. R. "Ostriches Recognize Their Own Eggs and Discard Others." *Nature* 279 (1979): 233–4.

Bolwig, N. "Agonistic and Sexual Behavior of the African Ostrich (*Struthio camelus*)." *Condor* 75 (1973): 100–5.

Sauer, E. G. F., and E. M. Sauer. "The Behavior and Ecology of the South African Ostrich." *Living Bird* Supplement 5 (1966): 45–75.

Organizations

BirdLife South Africa. P. O. Box 515, Randburg, 2125 South Africa. Phone: +27-11-7895188. Web site: <http://www.birdlife.org.za>

S. J. J. F. Davies, ScD

Elephant birds

(Aepyornithidae)

Class Aves
Order Struthioniformes
Suborder Aepyornithes
Family Aepyornithidae

Thumbnail description
Extinct, large, flightless birds of massive build, known only from fragmentary fossil remains

Size
Some species probably 10 ft (3 m), 880 lb (400 kg)

Number of genera, species
2 genera; 7 species

Habitat
Thought to have inhabited woodland and forest in southwest Madagascar

Conservation status
Extinct

Distribution
Madagascar

Evolution and systematics

Elephant birds belong to the group of large, flightless birds known as ratites. Ratites had a distinctive palate, and a sternum (breastbone) with no keel, so there was no anchor for the strong musculature needed for powered flight.

The origin of these birds has recently been clarified by the discovery of numerous good fossils in North America and Europe. Ratites were once thought to have a southern origin in the ancient continent of Gondwana, but new fossil evidence shows that flying ratites inhabited the Northern Hemisphere in the Paleocene and Eocene, 40–70 million years ago. The present Southern Hemisphere distribution of ratites probably resulted from the spread of flying ancestors of the group from the north.

Another indication that ancestors of elephant birds reached Madagascar as flying birds is that no fossils of ratites or elephant birds are found in India. In the process of separation of Gondwana into multiple continents, Madagascar and India remained joined for millions of years after breaking away from Gondwana. If elephant birds had walked to Madagascar, they would surely also have reached India. On the other hand, numerous remains of birds from genera such as

Mullerornis and *Aepyornis* are known from the Quaternary period of Madagascar. They were found in rock strata that are at most two million years old.

Elephant birds seem most closely related to present-day ostriches. Two fossil birds, *Eremopezus eocaenus* and *Stromeria fajumensis*, from the lower Tertiary of Egypt are sometimes placed in the Aepyornithidae, but opinion is divided about their relationships and they are omitted from the family in this treatment.

Seven separate species of elephant bird are known to have existed: *Mullerornis betsilei, Mullerornis agilis, Mullerornis rudis, Aepyornis maximus, Aepyornis medius, Aepyornis hildebrandti,* and *Aepyornis gracilis.*

Physical characteristics

No precise estimate can be made of the size and weight of these birds. Some were very large, up to 10 ft (3 m) tall, and weighed 880 lb (400 kg). Others were probably smaller, but more fossil material is needed to give good size-range estimates. When x-rayed, some eggs reveal embryonic elephant

Elephant bird (*Aepyornis maximus*). (Illustration by Bruce Worden)

Habitat

Étienne de Flacourt, the first French governor of Madagascar, was the first to report to scientists about elephant birds. He stated that a giant bird called "vouron patra" was still frequently found in the southern half of the island in the mid-seventeenth century. It is thought that elephant birds lived in the forests and woodlands of southwestern Madagascar. When human inhabitants arrived on the island about 2,000 years ago, they fragmented and burned these environments, causing the birds to lose their livelihood and become extinct soon after Flacourt's report.

Behavior

Nothing is known of the behavior of these birds. An account by Marco Polo in which large birds seized elephants, flew into the sky, then dropped the elephants to kill them and feast on them is a delightful fairy tale that may have given elephant birds their name.

Feeding ecology and diet

Elephant birds are thought to have fed on forest fruits. They may have been important in the dispersal of some fruit-bearing plants on the island—plants that are now known only from a few very old individual trees.

Reproductive biology

It is likely that elephant birds laid small clutches, perhaps of only one egg, and therefore reproduced slowly. The first scientific data on elephant birds was a report on their eggs made when a traveler named Sganzin sent a sketch of one of the giant eggs to collector Jules Verreaux from Madagascar in 1832. The eggs would have weighed about 13 lb (6 kg) and would be some of the largest single cells ever known.

Conservation status

Extinct.

birds, giving clues about the form of the whole bird, or at least its chick. The middle bone of the leg, the tibia, is longer than the lowest bone, the tarsus, indicating the birds were not fast runners. They had no need to run because other animals on Madagascar were no larger than a cat.

Distribution

Most early reports and recent fossil material have come from southwestern Madagascar. Two intact elephant bird eggs were found on the beaches of western Australia, on the far side of the Indian Ocean from Madagascar. It was concluded that these eggs were laid near the sea, washed into the sea by rivers or brought to the coast by human inhabitants of Madagascar, and floated to western Australia. Their survival on a journey of at least 5,000 mi (8,000 km) is remarkable.

Significance to humans

Flacourt reported that the natives used remains of elephant bird eggs as vessels. The shells are several millimeters thick; they may be more than 12 in (30 cm) long, and their volume is given as more than 1.6 gal (6 l). This corresponds to more than six ostrich eggs or more than 150 chicken eggs. Even today, many broken eggshells litter the beaches of southwestern Madagascar. The eggs and the birds that laid them must have been a great food resource for local people.

Resources

Books

Davies, S. J. J. F. *Ratites and Tinamous.* Oxford: Oxford University Press, 2002.

Feduccia, Alan. *The Origin and Evolution of Birds.* New Haven and London: Yale University Press, 1996.

Heuvelmans, Bernard. *On the Track of Unknown Animals.* London: Hart-Davis, 1959.

Periodicals

Brodkorb, P. "Catalogue of Fossil Birds. Part 1." *Bulletin of the Florida State Museum (Biological Sciences)* 7 (1963): 205–7.

Wetmore, A. "Re-creating Madagascar's Giant Extinct Bird." *National Geographic* 132 (1967): 488–93.

S. J. J. F. Davies, ScD

Procellariiformes

(Tubenosed seabirds)

Class Aves

Order Procellariiformes

Number of families 4 families

Number of genera, species 23 genera; 108 species

Photo: Northern fulmar (*Fulmarus glacialis*) at Kilt Rock on the Isle of Skye, Scotland. (Photo by Art Wolfe. Photo Researchers, Inc. Reproduced by permission.)

Introduction

Procellariiformes are exclusively marine birds. Also commonly known as petrels, tubinare, or tube-noses, this order is extremely diverse: from the massive, yet majestic, wandering albatross (*Diomedea exulans*) to the tiny, wave-dancing least storm-petrel (*Halocyptena microsoma*).

Traditionally there are four families within the order Procellariiformes: the Diomedeidae (albatrosses); the Procellariidae (giant petrels, fulmars, gadfly petrels, and shearwaters); the Hydrobatidae (storm-petrels); and the Pelecanoididae (diving-petrels). These four families include 23 genera and 108 species.

Evolution and systematics

The oldest Procellariiform fossil is from the early Paleocene, some 60 million years ago. However, a DNA-based study published in 1997 suggests that the order is even older and was distinct from penguins (order Sphenisciformes) and divers (order Gaviiformes) prior to the end of the Cretaceous. Thus, like many early avian groups, the Procellariiformes survived the mass extinction event at the end of Cretaceous about 65 million years ago. The fossil record of the Procellariiformes is generally poor, but a few sixteen million-year-old fossils show that even then albatrosses and shearwaters were very similar to modern-day species.

Procellariiformes are thought to have first evolved in the Southern Hemisphere and two-thirds of extant species are still found in this region. Surprisingly, most Procellariiform fossils have been found north of the equator. Many albatross fossils from the Pliocene (2 to 5 million years ago) have been recovered from Europe, North America, and Japan. This northern bias may simply reflect relative effort and the greater amount of landmass in the Northern Hemisphere. A few fossils from Australia, South Africa, and Argentina do confirm, however, the presence of albatrosses in the southern oceans over five million years ago.

DNA-DNA hybridization and DNA sequencing have confirmed the common ancestry of all Procellariiformes, but the taxonomy within the order is complex and subject to constant revision. John Warham, in his detailed book *The Petrels*, states that "the classification and systematics of Procellariiformes have long been the subject of controversy and a general agreement on species' limits in the near future seems unlikely." His prediction proved to be accurate. For example, in 1997 it was suggested that the number of albatross species should be revised from 14 to 24.

Physical characteristics

The unifying characteristic of Procellariiformes is their tubular nostrils. In the albatrosses (Diomedeidae) the tubular nostrils protrude from each side of the bill whereas in the three other Procellariiform families the nostrils are fused and sit prominently at the base of the upper bill. Unlike most other birds, petrels are thought to have a have a highly developed sense of smell, which they use to locate food and breeding sites. The tubular nostrils may enhance this sense or the tubes could simply act to keep the salty solution produced by the nasal glands away from the face and eyes.

Another unique feature of this order is the structure of the bill. Unlike any other birds, the bills of Procellariiformes are split into seven to nine distinct horny plates. The hooked bill of petrels is formed by a plate on the upper bill called the maxillary unguis. The hooked, stout, and very sharp unguis can firmly hold slippery food items such as fish and squid. In

Kittiwakes, murres, and fulmars share this bird-nesting cliff on the Snæfellsnes Peninsula in Iceland. (Photo by Kenneth W. Fink. Bruce Coleman Inc. Reproduced by permission.)

smaller Procellariiformes the cutting edges of the plates in the lower bill (tomia) are more comb-like and form filters for feeding on plankton and other small items of food.

Procellariiformes show the greatest range in body size of any avian order. The smallest species is the least storm-petrel (*Halocyptena microsoma*), which weighs less than 1 oz (20 g) and has a wingspan of 12.5 inches (32 cm). The largest species, the wandering albatross (*Diomedea exulans*), can weigh over 24 pounds (11 kg) and has a wingspan of up to 12 feet (3.6 m).

The plumages of Procellariiformes are generally quite plain and are composed of black, brown, gray, or white feathers. The legs and feet are usually black, but some are flesh-colored or mottled. In prions, diving-petrels, and little shearwaters (*Puffinis assimilis*) the feet and legs are blue. The bills of Procellariiformes are usually dark gray or black although some have yellow, orange, or pink coloration.

Also peculiar to Procellariiformes is stomach oil. This pale oil mainly contains wax esters and triglycerides and has a dietary origin. It is stored in the large, sac-like proventriculus

that separates the esophagus and gizzard. Enzymes secreted within the proventriculus allow Procellariiformes to metabolize the wax esters. The oil is used by both chicks and adults as an energy-rich food source during the potentially long periods between meals. The oil has a strong smell and may give the Procellariiformes their characteristic musty odor. Giant petrels (Genus *Macronectes*) are particularly pungent, hence their nickname "stinkers."

The oil has a second function. If chicks or ground-nesting adults are threatened, the oil can be regurgitated from the proventriculus and sprayed over a considerable distance. When the oil cools it has a wax-like consistency and can damage the plumage of predatory birds such as skuas (Laridae; Stercorariinae).

Distribution

Procellariiformes have the widest distribution of any avian order. Antarctic petrels (*Thalassoica antarctica*) and snow petrels (*Pagodroma nivea*) breed so far south that birds have to fly over a hundred miles from their inland colony before they reach the coastline of the Antarctic continent. In the Northern Hemisphere, fulmars (*Fulmaris glacialis*) nest on the northeastern tip of Greenland, as far into the Arctic as any land reaches. Petrels occur in all oceans but are most numerous in the Southern Hemisphere and are least abundant in the tropics.

Habitat

Petrel colonies are mostly found on remote islands away from land-based predators. Those that nest on larger islands or mainland continents do so in areas with low numbers of predators, such as deserts or mountainsides.

The breeding sites of the larger petrels must be windswept. Albatrosses and other large petrels cannot take off or forage widely for food without the help of strong winds. The subantarctic islands so favored by Procellariiformes are in latitudes referred to by sailors as the 'Roaring Forties' and 'Furious Fifties' because of the powerful, westerly winds that blow throughout the year.

Outside the breeding season, Procellariiformes spend virtually all their time at sea. Their distribution is largely governed by the availability of food, which is in turn influenced by the distribution of currents, upwellings, and weather conditions. However, some Procellariiformes, such as short-tailed shearwaters (*Puffinus tenuirostris*) and Manx shearwaters (*Puffinus puffinus*), make predictable return migrations between the Northern and Southern Hemispheres. Although migrating at the same time and in the same direction, these two species do not breed in the same hemisphere. The Manx shearwater favors the northern summer whereas the short-tailed shearwater, like most petrels, breeds during the summer months of the Southern Hemisphere. Research published in 2000 has shown that wandering albatrosses (*Diomedea exulans*) also have predictable migrations. Adults that nest on the Crozet Islands near South Africa return to the same patch of ocean at the end of each breeding cycle. However, the favored area could be as far away as Australia and may be different for each adult albatross.

Behavior

Most petrels are gregarious. At sea they can occur in large multi-species flocks around natural food sources or fishing boats. Squabbles are common and the large, aggressive albatrosses and giant petrels usually displace other species.

Petrels are also gregarious during breeding and can form huge colonies. Surface-nesters usually build their nests just beyond the pecking distance of their nearest neighbor. Larger albatrosses and the northern giant petrel (*Macronectes halli*) nest on the ground but not in dense colonies. More commonly their nests are loosely scattered along hillsides, headlands, or mountain ridges. Smaller petrels nest in dense, single-species colonies but usually excavate burrows or squeeze behind rocks.

Many petrels have elaborate display rituals in order to choose a mate or maintain a pair bond. Diurnal species such as albatrosses perform terrestrial dances or synchronized "aerial-ballet" routines. The courtship display of other petrels can be equally impressive. The display of the black-winged petrel (*Pterodroma nigripennis*) consists of swooping aerial chases and loud high-pitched calls. Species that return to their breeding sites after dusk tend to have less elaborate displays but can still be extremely vocal.

Most Procellariiformes are silent at sea unless competing for food. On land, various piping calls, shrieks, croaks, and other calls are produced at the nest or burrow. Albatrosses produce a variety of calls that accompany their complex displays. Shearwaters are renowned for the eerie human-like cries they produce from within their burrows.

Feeding ecology and diet

Among seafarers, albatrosses were well known for their ability to effortlessly follow ships for thousands of miles. By gliding on long, narrow wings, albatrosses, fulmars, petrels, and shearwaters can use the ocean winds to cover vast distances in search of food. However, not all Procellariiformes fly so economically. The short, stubby wings and rounded, penguin-like body of diving-petrels is more adapted to a life under the water than above it.

Typically petrels search for their patchily distributed food either offshore or beyond the continental shelf. Their search can take them thousands of miles from their breeding colony. Studies using satellite-tracking devices have shown that some albatrosses breeding on the Crozet Islands forage up to 1,600 mi (2,600 km) away from their nest. In contrast, diving-petrels predominantly search for food in inshore waters close to their breeding sites.

Petrels usually find their food close to or on the surface of the ocean although shearwaters, diving-petrels, and even some albatrosses can dive more than 30 ft (10 m) below the surface. Squid is the principal food source for most large petrels, although they will opportunistically eat other seabirds and carrion. Carcasses of seals, whales, and cuttlefish will attract hungry albatrosses while other Procellariiformes such as gadfly petrels and storm-petrels will mop up any scraps. Only giant petrels regularly forage for food on land. Petrels also exploit the actions of whales, dolphins, sharks, and tuna.

Least storm-petrel (*Oceanodroma microsoma*). (Photo by J. Hoffman/VIREO. Reproduced by permission.)

These marine predators will push schools of fish close to the surface and within reach of the shallow-diving-petrels. Storm-petrels and prions eat zooplankton such as copepods, amphipods, and fish eggs, which they delicately pluck from the surface of the ocean.

Concentrations of sea life occur where upwellings bring nutrient-rich waters closer to the surface. But even in these regions the abundance of food is largely unpredictable and the physiology of Procellariiformes reflects the ephemeral nature of their food sources. At 100°F (38°C), the body temperature of petrels is lower than most birds (105°F; 41°C). Therefore, less energy is required to maintain body temperature and less heat is lost. When food is plentiful, layers of subdermal fat and stomach oil can store excess energy until it is needed. The digestive tract of Procellariiformes is unusual in that the esophagus passes unrestricted into the proventriculus, which fills a large proportion of the abdominal cavity. The size of the proventriculus allows very large meals to be consumed and stored.

Reproductive biology

Procellariiformes are long-lived, very slow breeders. None can breed in the first year, and the largest petrels wait over a decade before breeding for the first time. In each breeding attempt, all Procellariiformes lay a single white egg. The egg is large relative to body size and can be up to 28% of the mother's body weight. Incubation in petrels is prolonged (6 to 11 weeks): about twice as long as gull (Laridae) eggs of a similar size. A petrel chick takes between two and nine months to fledge, twice as long as gulls of the same body mass. The reasons behind such a slow growth rate are thought to be associated with breeding sites and the parents' ability to feed the young. Terrestrial predators are usually absent from the islands where petrels breed, which removes the pressure to fledge a chick quickly. Also, food is rarely abundant close to breeding sites, so a fast growing chick would be more likely to starve during the potentially long periods between meals.

All Procellariiformes form exclusive social pairings, but behavioral and DNA-based studies have shown that infidelity does occur and males are not always the genetic fathers of the chick they help to raise. Copulations occur at the nest and are often preceded by complex behaviors or mutual allopreening.

Both sexes build or excavate the nests, incubate the egg, and provision the chick. Initially, surface-nesting petrels protect the chick from potential predators. Later, both parents leave the chick while they forage for food. A healthy chick can defend itself by regurgitating the stomach oil stored in its proventriculus.

Petrels receive little or no post-fledging care. They spend their first 2–11 years at sea before returning to their natal site to breed. Rarely, a young bird will return to a different island or colony to make its first breeding attempt.

Conservation

There are 108 extant species of Procellariiformes. Of these, 23 are threatened with extinction. Only one species, the Guadalupe storm-petrel (*Oceanodroma macrodactyla*), has become extinct since 1600. The principal threats to petrels are mammalian predators introduced to breeding islands and interactions with fishing vessels.

Humans have accidentally or deliberately introduced cats, rodents, possums, pigs, mustelids, rabbits, goats, foxes, and other mammals to petrel breeding sites. They usually have a severe impact. On Marion Island in the Indian Ocean, the 2000 or so feral cats targeted diving-petrels and killed nearly half a million seabirds each year. Eradication schemes have now successfully removed mammalian predators from many petrel breeding sites. On Australia's subantarctic Macquarie Island, an intensive effort to remove cats saw immediate success when in 2000 gray petrels (*Procellaria cinerea*) bred successfully on the main island for the first time in 40 years.

Before being banned in 1991, drift-net fisheries were thought to be responsible for killing up to 500,000 seabirds each year. Currently, the greatest threat to foraging seabirds

are long-line and trawl fishing boats that annually kill thousands of petrels and have been linked to the decline of many albatross populations. One study published in 1991 estimated that 44,000 albatrosses are killed by the Japanese longline fishery each year. Longline fishers are now expected to use bird-scaring lines and other bycatch mitigation measures in an attempt to reduce the death toll of seabirds.

Significance to humans

Not surprisingly, Procellariiformes have been of most significance to fishermen, whalers, and other seafarers. They are sometimes used to pinpoint fish shoals or a surfacing whale and are the subject of many superstitions. Some thought albatrosses were good omens and to kill one would bring ill fortune. At other times, to see an albatross or touch a storm-petrel was considered to be bad luck. Procellariiformes were also seen as the embodiment of the souls of cruel captains or drowned sailors that were destined to wander the seas for all time.

Petrels have long been used as a source of food for humans and have been found among archaeological remains around the world. Petrels have also sustained many a sailor shipwrecked in the southern oceans. In Alaska, Kodiak Islanders harpooned short-tailed albatrosses (*Diomedea albatrus*) from canoes, and royal albatross chicks (*Diomedea epomophora*) were highly prized by New Zealand tribes. Until the late 1980s, the inhabitants of Tristan Island in the Indian Ocean harvested the eggs of yellow-nosed mollymawks (*Diomedea chlororhynchos*) and sooty albatrosses (*Phoebetria fusca*). In many places, humans harvested shearwater chicks, also known as "muttonbirds." Tasmanian aborigines ate short-tailed shearwaters (*Puffinus tenuirostris*) at least 2000 years ago and a highly regulated harvest continues today. In New Zealand, there is a traditional harvest of sooty shearwater chicks (*Puffinus griseus*). Meticulous records track the number of Manx shearwaters (*Puffinus puffinus*) harvested from colonies on the Isle of Man in the United Kingdom. In the mid 1600s, the annual harvest was 10,000 chicks. However, like most petrel colonies, this population was vulnerable to introduced predators. By 1789, the colony disappeared after a shipwreck introduced rats to the island.

Many places where albatrosses and other petrels breed or forage now attract humans that simply wish to marvel at their size, elegance, and beauty.

Resources

Books

del Hoyo, J., A. Elliot, and J. Sargatal, eds. "Ostrich to Ducks." Vol. 1 of *Handbook of the Birds of the World*. Barcelona: Lynx Edicions, 1992.

Marchant, S., and P.J. Higgins, eds. "Ratites to Ducks." Vol. 1 of *Handbook of Australian, New Zealand and Antarctic Birds*. Melbourne: Oxford University Press, 1990.

Tickell, W.L.N. *Albatrosses*. Sussex: Pica Press, 2000.

Robertson, G., and R. Gales, eds. *Albatross Biology and Conservation*. Chipping Norton: Surrey Beatty, 1998.

Warham, J. *The Petrels*. London: Academic Press, 1990.

Periodicals

Brothers, N. "Albatross Mortality and Associated Bait Loss in the Japanese Longline Fishery in the Southern Ocean." *Biological Conservation* 55 (1991): 255-268.

Resources

Cooper, A., and D. Penny. "Mass Survival of Birds across the Cretaceous-Tertiary Boundary: Molecular Evidence." *Science* 275 (1997): 1109-1113.

Huyvaert, K. P., D.J. Anderson, T.C. Jones, W.R. Duan, and P.G. Parker. "Extra-pair Paternity in Waved Albatrosses." *Molecular Ecology* 9 (2000): 1415-1419.

Nunn, G.B. and S.E. Stanley. "Body Size Effects and Rates of Cytochrome b Evolution in Tube-nosed Seabirds." *Molecular Biology and Evolution* 15 (1998): 1360-1371.

Roby D.D., J.R.E. Taylor, and A.R. Place. "Significance of Stomach Oil for Reproduction in Seabirds: an Interspecies Cross-fostering experiment." *The Auk* 114 (1997): 725-736.

Weimerskirch, H., N. Brothers, and P. Jouventin. "Population Dynamics of Wandering Albatross *Diomedea exulans* and Amsterdam albatross *D. amsterdamensis* in the Indian Ocean and Their Relationships with Long-line Fisheries - Conservation Implications." *Biological Conservation* 79 (1997): 257-270.

Weimerskirch, H. and R.P. Wilson. "Oceanic Respite for Wandering Albatrosses." *Nature* (2000): 955-956.

Michael Colin Double, PhD

▲ Albatrosses
(Diomedeidae)

Class Aves
Order Procellariiformes
Family Diomedeidae

Thumbnail description
The largest flying seabirds, with exceptionally long narrow wings adapted for gliding and distinctive hooked bill and plumage ranging from all dark to mostly white

Size
Wingspan 62.3–106 in (190–323 cm); 3.74–26.25 lb (1.7–11.9 kg); length: 20–80 in (50–200 cm)

Number of genera, species
4 genera; 14 species

Habitat
Oceanic, generally only approach land for breeding on remote islands

Conservation status
Critically Endangered: 1 species; Vulnerable: 8 species; Lower Risk: 4 species; Data Deficient: 1 species.

Distribution
The north and south Pacific, Indian, and south Atlantic oceans

Evolution and systematics

A range of fossil albatrosses are evidence of a wider and more cosmopolitan distribution than those extant today. The earliest identified are from the Oligocene in Germany and South Carolina. Species approaching the characteristics of modern albatrosses are from the Northern Hemisphere (Europe and both coasts of North America) in the Miocene and Pliocene, but deposits are known from Australia, South Africa and Argentina in the predominantly marine Southern Hemisphere. Albatrosses were probably widespread in the north Atlantic until the late Tertiary.

The taxonomic status of albatrosses was fragmented and confusing until long-term field studies started during the 1930s. The collation of morphological, biological, and distribution data from breeding locations, with various genetic analyses from the 1990s, suggests a division into 4 genera and 24 taxa, a term that applies to both the species and subspecies of this order (*Phoebastria* with three taxa; *Diomedea*, 7 taxa; *Thalassarche*, 12 taxa; and *Phoebetria*, two taxa). Most of the 24 recognizable taxa (by combined morphology and genetics) may warrant species status and can be considered as distinct conservation units. In this treatment, however, the more traditional count of two genera will be applied: *Diomedidae*, which encompasses the proposed genera *Phoebastria* and *Thalassarche*; and *Phoebetria*, the two species of sooty albatross. A more positive resolution of the species question awaits data from poorly studied species in remote locations.

Physical characteristics

The great albatrosses (*Diomedea*) are the largest, with wingspans that can exceed 9.8 ft (300 cm). They lack a dark back except as juveniles. All have a white underwing. The upper wing of the northern royal albatross (*D. epomophora sanfordi*) is always black, while that of the wandering albatross (*D. exulans*) and the southern royal albatross (*D. epomophora*) grow increasingly white with age, especially among males. When Antipodean (*D. antipodensis*) and Amsterdam albatrosses (*D. amsterdamensis*) breed, especially females, they are almost as dark as in as juveniles. In the wandering royal albatross (*D. exulans*), Tristan albatross (*D. dabbenena*) and Gibson's albatross (*D. gibsoni*), body plumage whitens with age, and females may retain dark markings on the chest, flank, and back. Otherwise the body is chiefly white in the royal and northern royal albatrosses. The long (5.5–7.5 in; 140–190 mm) pale, horn-colored bill has a distinctively hooked tip, and it flushes pink in adults rearing chicks.

The Northern Pacific albatrosses include four medium to small taxa with wingspans of 6.2–7.9 ft (190–240 cm), and all have short, black tails. The two largest, i.e. the short-tailed albatross (*D. albatrus*) and the waved albatross (*D. irrorata*), have distinctive yellow/golden plumage on the head and nape. Of the two smallest, the Laysan albatross (*D. immutabilis*) has a pinkish bill, white body, and dark upper wing, while the black-footed albatross (*D. nigripes*) has a black bill and is mainly dark brown, except for a white patch at the rump and a variably pale face.

Laysan albatross (*Diomedea immutabilis*) parent with chick on Midway Atoll. (Photo by Frans Lanting. Photo Researchers, Inc. Reproduced by permission.)

The mollymawks, which include 11 small to medium size taxa, are the most diverse group of albatrosses with wingspans of 5.9–8.4 ft (180–256 cm). All have black upper wings and back, variable amounts of black on the underwing, and white body. All have a variable gray eyebrow with heads and necks varying from mainly white to dark gray, some with pronounced paler cap. The shy albatross, white-capped albatross, Salvin's, and Chatham mollymawk (*D. cauta, D. cauta cauta, D. salvini* and *D. cauta eremita*) all have chiefly white underwings, while all others have variable amounts of black reaching into the underwing from the leading edge. All mollymawks have distinctive bill structures and colors which, when combined with head color, help identification. The black-browed mollymawk (*D. melanophris*) and the Campbell black-browed mollymawk (*D. impavida*) have golden yellow bills with pink tips. The gray-headed mollymawk (*D. chrysostoma*) has a dark gray head and bill, with a yellow culmen and pink nail, yellow lower mandible stripe and black intervening sides to the bill. Buller's mollymawk (*D. bulleri*) and the Pacific mollymawk (*D. platei*) have similar bills without the pink nail and have gray, with paler-capped, heads. The smallest mollymawks (*D. chlororhynchos*) have a gray washed head, and the eastern yellow-nosed mollymawk (*D. bassi*) has a chiefly white head. Both taxa have bills with a yellow culmen stripe and pink nail. All mollymawks have a colorful pink/orange fleshy facial stripe from gape to ear which is exposed during displays.

The two sooty albatrosses (*Phoebetria*), with a wing span of 6.0–7.15 ft (183–218 cm), and the longest and most pointed tails of all albatross taxa, have mainly dark bills, plumage, and legs. However, the light-mantled sooty albatross (*P. palpebrata*), normally has a paler brown mantle than the dark-mantled sooty albatross (*P. fusca*).

Distribution

Three albatrosses, namely the short-tailed, the Laysan, and the black-footed, are confined to the northern Pacific ocean.

The waved albatross is tropical, mainly found from the Galápagos Islands to the coasts of Ecuador and Peru. All other albatrosses are found in the Southern Hemisphere in a circumpolar band mainly from 65°S to 20°S, but north to 15°S on the coast of southern Africa and 5°S on the west coast of South America.

Habitat

More than 70% of an albatross' life is spent on or over the ocean, while foraging, migrating, or resting. With the exception of the waved albatross, albatrosses avoid the relatively windless tropical doldrums. Though ranging widely at sea from breeding islands and during non-breeding time, significant differences in distribution can occur within and between species or between sexes. Some species are found to forage locally over continental shelves, while others roam widely to obtain food. Significant concentrations of birds can be found in areas of ocean richness near major currents, gyres and upwellings around South America (e.g. Humbolt current), Australia, New Zealand, South Africa (Benguela current) and in the north Pacific (Bering Sea and Gulf of Alaska). The remaining time is spent ashore at the usually windswept, remote island breeding locations for courtship, nesting and chick rearing. *Diomedea* species are more commonly found on grassy slopes or plains where nests are often far apart, and rebuilt each nesting attempt, but located within sight of neighbors. The northern royal albatross uses the flat scrubby tops of small rocky islets, while *Phoebetria* species are usually widely spaced along steep grassy slopes and cliff ledges.

Behavior

Albatrosses cannot fly in calm weather, needing a good breeze to effect the soaring and tacking pattern of flight which enables large distances to be covered with little effort. Generally silent at sea, or in social resting or washing flocks. However, breeding colonies (especially close-nesting mollymawks) can be noisy with buzzing cries, clattering bills, and wailing screams accompanying a wide repertoire of body displays associated with recognition, threat, and courtship. While there are common components of display throughout the family, the dances and wing displays of the northern Pacific albatrosses have no equivalent in the Southern Hemisphere. The most intense courtship display sequences are seen among adolescent pre-breeders, often in small groups or gams. Some displays are between similar sexes within such groups. Birds that develop a pair-bond remove themselves from the group to a potential nesting site where the displays become shorter, gentler and more mutual without the flamboyance of courtship. Some (e.g. the sooty albatrosses) indulge in courtship flying interspersed with synchronous calling from both sexes both on the ground and in the air. Albatrosses generally defend small spaces associated with the nest site or territory. Fighting is not a regular occurrence, with a reliance on threat displays and charging, but the hooked bill can damage bills and eyes. Chicks at the nest site clapper their bills to discourage intruders (e.g. predatory skuas, *Catharacta*), followed by regurgitation of oily stomach contents if approached too closely. Considerable time is spent in self and mutual

A pair of black-browed albatrosses (*Diomedea melanophris*) at their nest on New Island, Falkland Islands. (Photo by Rod Planck. Photo Researchers, Inc. Reproduced by permission.)

preening of plumage by adults, and of the growing chick by the parents.

Feeding ecology and diet

Various species of squid seem to provide the main component of the albatross diet. Many of these species are bioluminescent, and can be caught during the night. Some localised feeding spots provide a regular annual supply of carrion (e.g. during the annual die-off of *Sepia* cuttlefish on the eastern Australian coast) along established migration routes for some species. The diet also inlcudes a wide range of fish including small flying-fish, lampreys (*Geotrea*), pilchards (*Sardinops*) and crustaceans such as krills (*Euphausia* sp.), amphipods, copepods and crabs. Other recorded prey items include salps, seaweeds, barnacles, and fish spawn. Other small seabirds (prions, diving-petrels, and penguins) have been found in stomach contents as have examples of carrion from dead whales and seals. Some species can feed during the breeding season within a few hundred miles (kilometers) of the breeding place, as is the case with the northern royal albatross and the shy and Chatham mollymawks. Most food is gathered at the surface, but some of the smaller mollymawks may plunge and swim a short distance (up to 16 ft [5 m]) below the surface after prey. Kleptoparisitism has been recorded among waved albatrosses chasing boobies (*Sula* spp.), black-browed albatrosses from *Phalacrocorax*, and among eastern yellow-nosed albatrosses chasing shearwaters. Food is also obtained by some species from human fisheries offal, discards, and stolen baits.

Reproductive biology

Albatrosses usually build bowl-shaped nesting mounds with grasses and small shrubs bound together with soil, peat, or even penguin feathers where no vegetation is available. The waved albatross does not build nests, and the other northern Pacific albatrosses have very rudimentary ones that are rebuilt each season. Many Buller's mollymawks and black-footed albatrosses nest under trees in open forest. Most mollymawks nest in tight colonies just out of pecking range from neighbors, and reuse previous nest mounds.

Albatrosses are generally monogamous and most are annual breeders. Albatrosses lay one large white egg with reddish brown spots at the largest end weighing 7.0–18.2 oz (200–510 g) ranging from 5 to 10% of female body weight. First eggs are narrower and lighter. Incubation lasts 65–85 days with both sexes sharing the incubation stints, which may range from one day to as long as 29 days according to species foraging methods and locations. Hatching takes 2–5 days. Immediately after hatching, during the brooding (guard) stage of chick growth (15–40 days), one parent remains with the chick at the nest. Chicks fledge at 120–180 days for all small

A waved albatross (*Diomedea irrorata*) pair courting in the Galápagos Islands. (Photo by JLM Visuals. Reproduced by permission.)

albatrosses, while *Diomedea* have a range of 220–303 days. Though breeding success can vary according to species and breeding season, fledging may be as high as 80% of eggs laid. Long-term averages can range from 25 to 67%, with the waved albatross being the lowest.

Recruitment of fledglings into the breeding population occurs at 5–15 years of age with 15–65% of those fledged surviving to breed. Biennial breeders take longer to become sexually mature. Annual mortality rates for adults range from 3 to 9%. The oldest known albatross was a northern royal albatross, still breeding at over 62 years old.

Conservation status

Albatrosses are long-lived, with delayed maturity and low reproductive output and adult mortality. This strategy ensures that only a small proportion of the population is breeding at any one time, with the remainder often in other parts of their range as adolescents or resting adult breeders. This mitigates the effects of localized disasters, but may disguise for some years any detrimental effects on populations or age groups which are more widespread. Threats which affect breeding birds have the most immediate impact, and recorded increases in adult mortality of 1–5% have significantly affected some colonies. Of the nearly 2 million breeding pairs of albatrosses worldwide in 24 recognizable taxa, 14 have populations of less than 20,000 breeding pairs. At the most populous level, the black-browed and the black-footed albatrosses have total populations that exceed 600,000 breeding pairs. However, some of these populations are fragmented, with a third of discrete groups having fewer than 100 breeding pairs. Most species have populations where there is not enough data to determine the rate of increase or decline, but have evidence of being exposed to known threats. The most vulnerable populations are those which are confined to one breeding locality. The high degree of philopatry of both juveniles and adults limits any ability to colonize new sites when facing adversity at their natal colony. Most factors adversely affecting populations involve human activities. However, climatic events have been seen to cause changes in habitat, which severely reduced breeding productivity in the northern royal albatross. Changing sea temperatures may also contribute to decline by changing food distribution and availability.

Significance to humans

The Rime of the Ancient Mariner, by S.T. Coleridge (1798), has done much to determine the popular conception of the albatross. The family name is derived from the name of the Greek hero of the Trojan War, Diomedes, whom the gods exiled to an isolated island, turning all of his deceased companions into large, white birds.

The preponderance of albatrosses breeding in locations remote from human habitation may indicate that closer populations were historically extirpated by humans. Certainly harvesting of eggs or chicks continues legally or illegally in a few locations today. During the past 250 years since the first naming of an albatross by Linnaeus, these legendary birds have been directly or indirectly exploited at sea and at their breeding colonies by those on boats—mariners, sealers, and whalers as food and artifacts, passengers as sport, and science as specimens.

1. Light-mantled albatross (*Phoebetria palpebrata*); 2. Northern royal albatross (*Diomedea epomophora sanfordi*); 3. Wandering albatross (*Diomedea exulans*); 4. Black-browed mollymawk (*Diomedea melanophris*); 5. Chatham mollymawk (*Diomedea cauta eremita*); 6. Laysan albatross (*Diomedea immutabilis*). (Illustration by Dan Erickson)

Species accounts

Royal albatross
Diomedea epomophora sanfordi

TAXONOMY
Diomedea epomophora sanfordi Murphy, 1917, Chatham Islands.

OTHER COMMON NAMES
English: Toroa; French: Albatros royal; German: Königsalbatros; Spanish: Albatros Real.

PHYSICAL CHARACTERISTICS
Wingspan 8.85–10.0 ft (270–305 cm); 13.75–18.1 lb (6.25–8.2 kg); length: c. 45 in (115 cm). Large white bodied albatross with upper wing surface black. Eyelids black, spotted white in oldest birds.

DISTRIBUTION
Breeds only on New Zealand South Island (Taiaroa Head), Chatham Islands (Sisters and Forty-Fours Islands), and Enderby Island. The only albatross to have a circumpolar range when not breeding.

HABITAT
Marine, breeding on exposed tops of small islets or headlands.

BEHAVIOR
Extensive repertoire of mutual and group displays at the breeding site, some of which are occasionally performed in the air or on the water. Once pair bond is formed the most extravagantly spread wing displays are not used.

FEEDING ECOLOGY AND DIET
Most food taken by surface seizing. Mainly cephalopods, with some fish, salps, and crustaceans. During the breeding season, feeding occurs over continental shelf breaks within 620 mi (1,000 km) of the colony. Probably an opportunistic feeder when migrating.

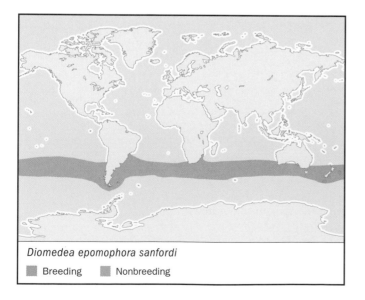

Diomedea epomophora sanfordi
■ Breeding ■ Nonbreeding

REPRODUCTIVE BIOLOGY
Lays one egg 27 October to 8 December with laying time fixed according to parentage. Nest a raised bowl of soil and vegetation rebuilt after each nesting attempt. Will also lay on bare rock with rock chips, but egg failures then are greater than 90%. On average, incubation is 79 days and fledging 240 days. Biennial breeder if successful. Monogamous, pairing usually for life. Breeding starts at 8 years and the average age of the breeding population is 20 years. Adult annual mortality is 4–5%.

CONSERVATION STATUS
Endangered. Total population c. 7,700 pairs, restricted to a tiny breeding range; the habitat supporting 99% of the population in Chatham Islands was severely degraded by storms in the 1980s. The resulting reduced productivity suggests a predicted 50% decline will occur over three generations unless the habitat improves significantly.

SIGNIFICANCE TO HUMANS
At Taiaroa Head the efforts of L.E. Richdale enabled protection of the fledgling colony by 1950. Public viewing started in 1972, and by 2001 more than 100,000 persons annually viewed the nesting colony. ◆

Wandering albatross
Diomedea exulans

TAXONOMY
Diomedea exulans Linnaeus, 1758, Cape of Good Hope. Two subspecies.

OTHER COMMON NAMES
English: Snowy albatross, white-winged albatross; French: Albatros hurleur; German: Wanderalbatros; Spanish: Albatros Viajero.

PHYSICAL CHARACTERISTICS
Wingspan 9.2–11.5 ft (280–350 cm); 13.7–25 lb (6.25–11.3 kg). One of largest albatrosses with variable plumage developing from chocolate brown. Back and belly whiten first, followed by head and rump. Wing whitens from center. Oldest males are whitest.

DISTRIBUTION
D. exulans breeds in high latitudes of the southern oceans at South Georgia, Marion and Prince Edward Islands, Crozet Island, Kerguelen Island, Heard and Macquarie Islands. Juveniles are thought to disperse northwards from these locations before developing regular downwind migrations.

HABITAT
Marine and highly pelagic over deep waters away from coastal continental shelves.

BEHAVIOR
Extensive repertoire of group and mutual displays accompanied by a wide range of screams, whistles, moans, grunts, and bill clappering.

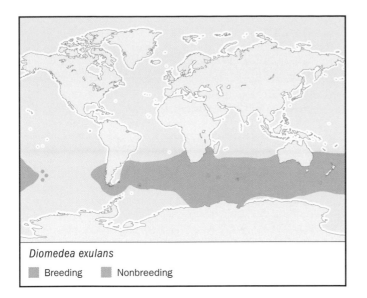

Diomedea exulans
■ Breeding ■ Nonbreeding

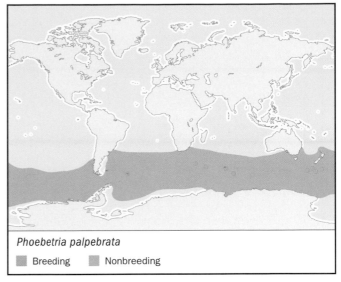

Phoebetria palpebrata
■ Breeding ■ Nonbreeding

FEEDING ECOLOGY AND DIET
Most food taken by surface seizing, but may make shallow plunge dives. Feeds primarily on cephalopods, deepwater and bioluminescent squid at night, but also fish and crustaceans. Feeds on carrion more than other albatrosses and is an extensive follower of ships for galley scraps, and fishing vessels for offal, discards, and baits. During breeding, feeding flights over deep ocean areas may be for as long as 25 days and covering some 13,000 miles (20,800 km).

REPRODUCTIVE BIOLOGY
Lays one egg between 10 December and 5 January. The nest is a raised bowl of soil peat and grassy vegetation rebuilt at each nesting. On average, incubation lasts 79 days, fledging 271 days. Usually a biennial breeder. Monogamous, pairing usually for life. Productivity c. 70%. Adolescents return by 6 years. Breeding starts at 11 years. Only c. 30% of fledglings survive. Adult annual mortality averages 5–7% with females being higher than males.

CONSERVATION STATUS
Vulnerable. The main populations are at South Georgia (2,100 pairs annually), Marion and Prince Edward Island (3,000), Crozet (1,700), and Kerguelen Island (1,400). All colonies have experienced declines in breeding populations which have been attributed to mortality associated with long-line fisheries in different parts of the southern oceans.

SIGNIFICANCE TO HUMANS
None known. ◆

Light-mantled albatross
Phoebetria palpebrata

TAXONOMY
Phoebetria palpebrata J.R. Forster, 1785, south of Cape of Good Hope. Monotypic.

OTHER COMMON NAMES
English: Light-mantled sooty, gray-mantled albatross; French: Albatros fuligineux; German: Graumantel-Rußalbatros; Spanish: Albatros Tiznado.

PHYSICAL CHARACTERISTICS
Wingspan 6.0–7.15 ft (183–218 cm); 6.1–8.1 lb (2.5–3.7 kg). Small, all dark albatross with paler mantle and a partial white eye-ring. Bill black with pale blue sulcus line.

DISTRIBUTION
Widely distributed throughout the southern oceans breeding at South Georgia, Marion, Prince Edward, Crozet, Kerguelen, Heard, Macquarie, Auckland, Campbell, and Antipodes Islands. Distributed at sea generally south of 40° latitude to the edges of Antarctica.

HABITAT
Marine. Generally breeding in isolated nests on sheltered steep slopes or cliff ledges close to a rock face.

BEHAVIOR
Aerial displays and formation flying are a distinctive feature of courtship and pair-bonding behavior. Mutual calling modulated in tone by the position of the head is an essential part of the displays. Does not have open or extended wing displays, but uses the long tail in display more than other albatrosses.

FEEDING ECOLOGY AND DIET
Mainly solitary at sea, feeding by surface seizing or surface plunging, chiefly for cephalopods and krill. Sometimes fish and carrion including remains of birds at sea. Some observed interaction with commercial fishing.

REPRODUCTIVE BIOLOGY
Lays one egg between October and November with the 2 week laying period being shorter than other albatrosses except the dark-mantled sooty albatross. Incubation lasts for 65–72 days. Have the longest incubation shifts of any albatross. Hatching takes 3–5 days. Chicks are guarded by a parent for the first three weeks. Mean fledging varies between 140 days (Macquarie) and 170 days (Marion Island). Productivity variable. Monogamous. Generally classed as a biennial breeder. Starts breeding at 8–15 years. Adult annual mortality probably about 3%.

CONSERVATION STATUS
Data Deficient, not globally threatened. Tentatively estimated to have a world population of 30,000 breeding pairs. Main causes of nesting failure seem to be starvation and desertion by parents, which along with the length of foraging stints suggests

a species with distant and restricted food sources. Incidence of fisheries bycatch not large.

SIGNIFICANCE TO HUMANS
None known. ◆

Chatham mollymawk
Diomedea cauta eremita

TAXONOMY
Diomedea cauta eremita Murphy, 1930, Pyramid Rock, Chatham Islands.

OTHER COMMON NAMES
English: Chatham Islands albatross; shy albatross; French: Albatros des Chatham; German: Chatham albatros; Spanish: Albatros de Chatham.

PHYSICAL CHARACTERISTICS
The "shy" mollymawks are the largest mollymawks. *D. eremita* is the smallest (6.8–10.4 lb; 3.1–4.7 kg) and darkest of the "shy" mollymawks. White body, dark gray head and mantle, black upper wing and tail, underwing white except for wingtip and small dark patch at base of wing leading edge. Bill chrome yellow with dark spot at tip of lower mandible. Orange cheek stripe.

DISTRIBUTION
Breeds only at The Pyramid, a small rocky cone (650 ft; 200 m high) in the Chatham Islands. Rarely recorded at sea away from breeding location. During the breeding season mainly found within 190 mi (300 km) of the colony on and along the edge of the continental shelf.

HABITAT
Marine. Small pedestal nests of soil and limited vegetation, which may collapse in periods of extended drought, on mainly bare steep rocky slopes, crevices and ledges.

Diomedea cauta eremita
■ Breeding ■ Nonbreeding

BEHAVIOR
Similar to other mollymawks with harsh buzzing bray with open mouth used in both threat and courtship. A range of displays featuring fanning of the tail, mutual jousting of bills, and tympanic grunting over the back between partly raised wings.

FEEDING ECOLOGY AND DIET
Probably surface seizing of a mix of cephalopods, krill, floating barnacles, and fish. Scavenges behind fishing vessels for baits, discards and offal.

REPRODUCTIVE BIOLOGY
Lays one egg between 20 August and 1 October. Incubation period 68–72 days shared by both parents with short stints rarely longer than 5 days. Fledging estimated at 130–140 days from hatching. Adolescents return from 4 years and first breeding recorded at 7 years. Productivity averages 60% of available nest sites. Crude estimates of annual adult mortality range between 4 and 15%. Breeds annually and seemingly monogamous, pairing for life.

CONSERVATION STATUS
One of two albatrosses classed as Critically Endangered because of tiny single breeding place, and recent evidence of deterioration of habitat. With 5,300 occupied breeding sites, the breeding population is probably c. 4,200 pairs. No evidence of population decline between 1975 and 2001. Now protected, but sporadic small harvests of tens of birds still occur.

SIGNIFICANCE TO HUMANS
None known. ◆

Black-browed mollymawk
Diomedea melanophris

TAXONOMY
Diomedea melanophris Temminck, 1828, Cape of Good Hope. Two subspecies.

OTHER COMMON NAMES
English: Black-browed albatross; French: Albatros à sourcils noirs; German: Schwarzbrauenalbatros; Spanish: Albatros Ojeroso.

PHYSICAL CHARACTERISTICS
Wingspan c. 7.9 ft (240 cm); 6.4–10.3 lb (2.9–4.7 kg). Heavily built mollymawk with mainly all white body and head, black upper wings, mantle and tail, underwing black with variable amounts of central white. Eyebrow black.

DISTRIBUTION
The most plentiful of the albatrosses with 98% of breeding concentrated in the southern Chile at Diego Ramirez, Islas Ildefonso, Diego de Almagra (Chile), south Atlantic at Falkland Islands (12 islands) forming a distinctive genetic grouping; a second genetic grouping comprising South Georgia, small numbers at Crozet, Kerguelen, Heard, and McDonald Islands in the Indian Ocean and very small numbers at Macquarie, Bishop & Clerk, Antipodes, Campbell Islands and The Snares in the southwest Pacific. Most common straggler into North Atlantic.

HABITAT
Marine, not significantly pelagic, being found close to coast more than other albatrosses. Highly colonial breeder, can be in large colonies of thousands of nests. Often on coastal tussock slopes or ledges where nests are built of packed soil and grasses

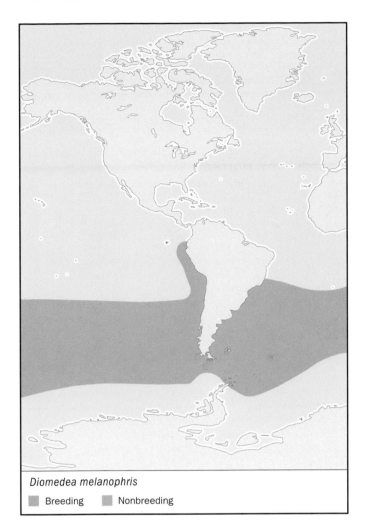

Diomedea melanophris

■ Breeding   ■ Nonbreeding

and used in successive seasons. Largest colonies in the Falkland Islands are on gently sloping rocky terrain without vegetation.

BEHAVIOR
Colonies actively noisy with strident territorial open mouthed bray with flagging of the head, and aggressive harsh cackling. Like other Mollymawks uses fanned tail extensively in display sequences.

FEEDING ECOLOGY AND DIET
Most food taken by surface seizing, with occasional shallow plunging, and swimming below surface. Mainly crustaceans, squid, fish, carrion, and fisheries discards or offal. Feeds extensively on large swarms of krill. Kleptoparisitism (prey theft) observed with stealing from surfacing shags. Sometimes follows whales. Often feeds aggressively among flocks of albatrosses and petrels.

REPRODUCTIVE BIOLOGY
Annual breeder, laying one egg with laying period covering three weeks with a mean of 10 October at South Georgia, but three weeks earlier at Falklands/Crozet and Kerguelen Islands. Incubation is 68–71 days with a guard stage of 1–4 weeks following hatching, before fledging from 120–130 days. Overall productivity averages 27% for chicks fledged from eggs laid. Pairing usually for life. Adolescents return at 2–3 years old and breed at 10 years with c. 28% or less of fledglings surviving to

breed. Adults have an annual mortality c. 8–9% with females surviving better than males.

CONSERVATION STATUS
Not globally threatened, with a population of c. 600,000 breeding pairs, but with significant declines noted in some areas. But *D. m. impavida* (26,000 pairs) is classed as vulnerable as its breeding is restricted to one island.

SIGNIFICANCE TO HUMANS
None known. ◆

Laysan albatross
Diomedea immutabilis

TAXONOMY
Diomedea immutabilis Rothschild, 1893, Laysan Island. Monotypic.

OTHER COMMON NAMES
French: Albatros de Laysan; German: Laysanalbatros; Spanish: Albatros de Laysan.

PHYSICAL CHARACTERISTICS
Wingspan 6.4–6.7 ft (195–203 cm); 5.3–9.0 lb (2.4–4.1 kg). A slender white albatross, with underwing most similar to *D. melanophris*, but with black patches at wrist and elbow. Distinctive gray patch around eye and cheeks. Bill has yellowish orange broad base, blending to pinker horn and then black at tip.

DISTRIBUTION
The most plentiful of the north Pacific albatrosses. Almost all breed in the Hawaiian chain of islands with largest colonies at Midway and Laysan Island. Tiny new populations at Mukojima in Bonin Islands (west Pacific), and in eastern Pacific off Mexican coast at Islas Guadalupe, Benedicto and Clarion. During the breeding season regularly travels to the seas separating Japan from the western Aleutians.

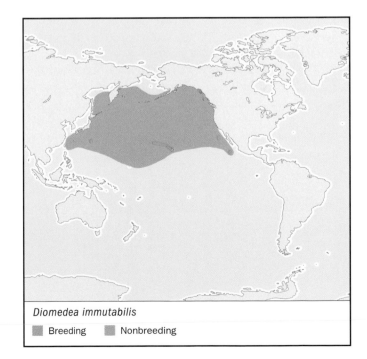

Diomedea immutabilis

■ Breeding   ■ Nonbreeding

HABITAT
Marine, combined pelagic and coastal shelves, but rarely approaches land except breeding islands.

BEHAVIOR
Colonies actively noisy during daylight with distinctive brays, whistles, groans, and calls. Along with *P. nigripes*, has a wider range of displays than other albatrosses. There are actively energetic dances with birds circling about each other, walking and standing and prancing on extreme tiptoe, swaying and jousting motions of the head, combining with sideways lifting bent-wing postures similar to all north Pacific albatrosses.

FEEDING ECOLOGY AND DIET
Mainly squid, but also fish and fish eggs, crustaceans and coelenterates. Does not often follow ships. Undertakes a mix of long and short foraging flights when chick rearing, to compensate for far distant feeding locations.

REPRODUCTIVE BIOLOGY
Nest a scrape in ground, built up around rim by debris and sand with one egg. Annual breeder laying between 20 November and 24 December. Incubation lasts an average of 64 days, with longest stints at beginning of incubation lasting more than 3 weeks. Newly hatched chicks are guarded for c. 27 days and are then left alone except for feeding visits until fledging at c. 165 days. Productivity averages 64%, though 4–24% of chicks may die before fledging through dehydration, starvation, or wandering into other territories to beg for food. Only c. 14% of fledglings may survive to breed at 9 years, and adults have annual mortality of 5%.

CONSERVATION STATUS
Not globally threatened with a world population of c. 607,000 breeding pairs, but some colonies may be decreasing. Drift gill-netting in the north Pacific has been a major source of mortality (17,500 in one year) and effects of longline fisheries not yet known. Have recorded high levels of contaminants which may affect breeding, as well as ingestion of plastic rubbish.

SIGNIFICANCE TO HUMANS
Like the oceanic "wanderer" of the Southern Hemisphere which has come to epitomize the albatross, so the "gooney" is the common image of the albatross in the countries surrounding the north Pacific. ◆

Resources

Books

BirdLife International. *Threatened Birds of the World.* Barcelona and Cambridge, UK: Lynx Edicions and BirdLife International, 2000.

Croxall, J.P. ed. *Seabirds: Feeding Ecology and Role in Marine Ecosystems.* New York: Cambridge University Press, 1987.

del Hoyo, J., A. Elliott, and J. Sargatal, eds. *Ostrich to Ducks. Vol. 1 of Handbook of the Birds of the World.* Barcelona: Lynx Edicions, 1992.

Marchant, S., Higgins, P.J., eds. *Ratites to Ducks. Vol. 1A of Handbook of Australian, New Zealand and Antarctic Birds.* Melbourne: Oxford University Press, 1990.

Robertson, G., R. Gales, eds. *Albatross Biology and Conservation.* Chipping Norton, Australia: Surrey Beatty, 1998.

Tickell, W.L.N. *Albatrosses.* Sussex: Pica Press, 2000.

Warham, J. *The Petrels: Their Ecology and Breeding Systems.* New York: Academic Press, 1990.

Warham, J. *The Behaviour, Population Biology and Physiology of the Petrels.* New York: Academic Press, 1996.

Periodicals

Cooper, J. ed. Albatross and Petrel Mortality from Longline Fishing International Workshop, Honolulu, Hawaii, USA. Report and presented papers. *Marine Ornithology* 28 (2000): 153–190.

Flint, E., K, Swift. eds. Second International Conference on the Biology and Conservation of Albatrosses and other Petrels, Honolulu, Hawaii, USA. Abstract of oral and poster presentations. *Marine Ornithology* 28 (2000): 125–152.

Nicholls, D.G., C. J. R. Robertson, P. A. Prince, M. D. Murray, K. J. Walker, G. P. Elliott. Foraging Niches of Three *Diomedea* Albatrosses. *Marine Ecology Progress Series* 231 (2002): 269–77.

Organizations

BirdLife International. Wellbrook Court, Girton Road, Cambridge, Cambridgeshire CB3 0NA United Kingdom. Phone: +44 1 223 277 318. Fax: +44-1-223-277-200. E-mail: birdlife@birdlife.org.uk Web site: <http://www.birdlife.net>

Christopher John Rutherford Robertson

Shearwaters, petrels, and fulmars
(Procellariidae)

Class Aves
Order Procellariiformes
Family Procellariidae

Thumbnail description
Medium-sized tube-nosed seabirds with hooked bill and large wingspan for gliding

Size
9.1–39 in (23–99 cm); 2.8 oz–11 lb (78 g–5 kg); wingspan 20.9–80.7 in (53–205 cm)

Number of genera, species
12 genera; 60–76 species

Habitat
Marine

Conservation status
Critically endangered: 10 species; Endangered: 6 species; Vulnerable: 20 species; Lower Risk/Near Threatened: 9 species

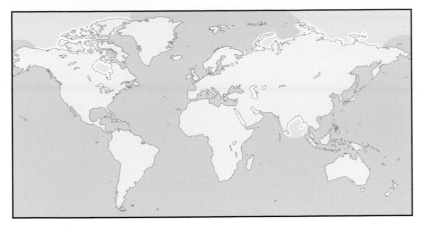

Distribution
Oceans worldwide

Evolution and systematics

The Procellariiformes is one of the most primitive bird orders. In his 1996 book, John Warham proposed that ancestral procellariiforms may have resembled *Bulweria*; that is, they may have been small in size, used natural cavities for nesting, and performed few vocalizations or visual displays. Procellariidae probably diverged from other Procellariiformes in the Eocene, about 40–50 million years ago, coinciding with a wide procellariiform radiation.

Systematics within the family are subject to debate. Because procellariids spend so much time at sea, little is known about them outside of their terrestrial breeding habits. For some species, not even that is known. In addition, the slow rate of speciation makes delineating species difficult. Therefore, debate continues regarding the taxonomy of the family. However, there are four natural groups of Procellariidae. Fulmar-petrels comprise seven species in five genera, ranging from the Arctic to Antarctica. Gadfly-petrels include two genera of 25–36 medium-sized species, some of which are the least known of the family. Prions include seven species in two genera, all of which are found in southern oceans. The shearwaters include 21–26 species in three genera.

Physical characteristics

As Procellariiformes, the procellariids have hooked bills with tubular nostrils. The sharp tip of the bill is an effective implement for handling prey; the tubular nostrils may be related to the well-developed sense of smell used for finding food at great distances and nests in the dark.

Size ranges from 9.1–11 in (23–28 cm) for fairy prions (*Pachyptila turtur*) to 31.9–39 in (81–99 cm) for giant petrels

(*Macronectes* spp.), which sport wingspans of about 2.2 yards (2 m). Plumage varies among species and consists of whites, blues, grays, browns, and blacks and does not vary by sex or season; juveniles typically resemble adults.

Most procellariidae are awkward on land because of weak legs set far back on the body. Rather than walking on their legs, they tend to shuffle along on the breast and wings. The giant petrels are the exception; they have strong legs suitable for scavenging beached carcasses.

Prions have a unique upper bill that is fringed with lamellae (thin plates) that act like baleen to filter plankton out of the water.

Distribution

Procellariidae can be found in oceans throughout the world. However, the Southern Hemisphere contains far more species than the Northern. Fulmar-petrels prefer cooler waters and are rarely found in subtropic waters. Prions are found in southern temperate subantarctic and Antarctic zones. Shearwaters, the most diverse in terms of habitats used, are found in both hemispheres, from tropical to arctic waters.

Some procellariids undergo migrations of thousands of miles each year; other species remain closer to their breeding grounds year-round. Short-tailed shearwaters (*Puffinus tenuirostris*) cover 120° of latitude in their annual trek from subantarctic nesting grounds to subarctic feeding grounds. Cory's shearwaters (*Calonectris diomedea*) tend to disperse and wander outside of the breeding season, with less specific feeding grounds.

A northern fulmar (*Fulmaris glacialis*) vomits stomach oils to defend its nest site against an intruder. (Illustration by Bruce Worden)

Habitat

Procellariids are marine; they come to land almost exclusively to breed.

Behavior

Procellariids are excellent flyers, alternating flapping and dynamic soaring. Shearwaters are named for their ability to dip and glide between waves, just above the ocean surface.

Procellariids have an amazing, but not infallible, homing ability. Records of lost procellariids are not uncommon, and in 1999, a great shearwater (*Puffinus gravis*) wandered to inland England when it should have been nesting in the South Atlantic.

Procellariids vomit smelly stomach oils on invaders. Fulmar-petrels vomit on intrusive conspecifics at breeding colonies, but other procellariids reserve this tactic for use against predators or nosy humans.

On land, prions, shearwaters, and most gadfly-petrels are nocturnal, perhaps because they are awkward on land and vulnerable to predation by gulls, raptors, and crows.

Vocal communication is most common at breeding colonies, with sounds ranging from coos to growls to shrill cries. Many species are silent at sea, but fulmar-petrels make a raucous gull-like noise when competing for food in large flocks.

Feeding ecology and diet

Almost all procellariids feed exclusively at sea on squid, fish, plankton, and discards from fishing boats. Procellariids use their keen sense of smell to locate food. Giant petrels focus on seal and penguin carcasses but switch to squid, krill, and fish when carcasses are scarce. Prions eat primarily zooplankton that they strain through their fringed upper bill. Some prions hydroplane: they submerge the bill as they fly low over the water and filter plankton as they go.

Giant petrels (*Macronectes* sp.) treading water in face-to-face feeding competition over the seal carcass floating in the water nearby (South Georgia Island, Falkland Islands). (Photo by Greg Dimijian. Photo Researchers, Inc. Reproduced by permission.)

Reproductive biology

Procellarids choose breeding grounds with ready access to the sea. Many species form huge breeding colonies: one sooty shearwater (*Puffinus griseus*) colony on the Snares Islands contains 2.5 million pairs. Other species, such as giant petrels, breed alone or in small, loose colonies.

Procellariid nests vary from mounds of grass or stones built by giant petrels to cliff ledges used by northern fulmars (*Fulmarus glacialis*) to burrows used by shearwaters, prions, and gadfly-petrels. The burrow nesters either excavate their own cavities or find abandoned rabbit dens or natural cavities. Few species are forest nesters.

Breeding is usually annual in the local spring or summer. The age of sexual maturity ranges from three to twelve years, with five to six years as the average age of first breeding. Procellariiformes are typically monogamous, mating for life, with pairs using the same nest year after year. One pair of northern fulmars reportedly has used the same nest for at least 25 years.

One white egg is laid that constitutes an average of 12–16% of the female's body weight. Both parents incubate the egg in alternating shifts of 2–14 days for an incubation period of six to nine weeks, depending on species. After hatching, both parents care for the chick, leaving it alone after 2–20 days (as soon as it can regulate its own body temperature). The parents then visit the chick only for feeding. Food consists of fat-rich stomach oils produced by partial digestion of the adults' normal diet.

Chicks put on large amounts of fat and quickly outweigh their parents, then slim down to an adult weight before fledg-ing. For example, northern fulmar chicks weigh 33.5 oz (950 g) by 40–45 days after hatching. When they fledge at 57 days they weigh 28.2 oz (800 g). The large fat deposits might be insurance against periods of starvation that could occur while parents are away foraging. However, fat stores exceed the amount necessary to survive periods of parental absence and are maintained during fledging—the weight loss prior to fledging is due largely to water loss. Another theory suggests that fat stores are for survival after fledging when chicks must learn to forage for themselves.

A northern fulmar (*Fulmarus glacialis*) and egg at Bass Rock, Scotland. (Photo by J.C. Carton. Bruce Coleman Inc. Reproduced by permission.)

Antarctic petrel (*Thalassoica antarctica*) chick camouflaged in the rocks and snow. (Photo by Joyce Photographics. Photo Researchers, Inc. Reproduced by permission.)

As chicks approach full size, they begin to flap their wings around the nest in preparation for the first flight. A week or two after the parents have abandoned the chick, it fledges, flinging itself out to sea. While most chicks survive fledging, introduced (rats and cats) and natural (gulls and raptors) predators pose a threat.

Life expectancy for procellariids is about 15–20 years, although some individuals are known to have lived much longer. Britain's oldest bird, a northern fulmar aged over 50 years, was missing and presumed dead in November 1997.

Conservation status

While some procellariids are thriving, others are among the most threatened of all birds. The breeding population of Zino's petrel (*Pterodroma madeira*) is estimated to be 20–30 pairs. Estimates put the total Chatham petrel (*Pterodroma axillaris*) population at 800–1,000 birds. It breeds only on tiny South East Island in the Chatham Islands of New Zealand, where broad-billed prions (*Pachyptila vittata*) are the primary threat because they kill chicks, eggs, and sometimes adults.

All species face the same challenges: introduced predators, habitat deterioration, and human exploitation. International conservation efforts are protecting breeding grounds. Gough Islands, a South Atlantic breeding ground for great shearwaters and others, was declared a World Heritage Site in 1995 by the United Nations science agency UNESCO. Predator extermination programs show promise. After a 100-year absence, gray (*Procellaria cinerea*) and blue (*Halobaena caerulea*) petrels returned to breed on Macquerie Island in 2001 after the complete removal of feral cats.

Significance to humans

Several cultures including Eskimos, Maoris, and Europeans have traditionally eaten procellariid eggs, chicks, and adults. Natives of New Zealand and Tasmania still harvest several thousand chicks annually for feathers, fat, flesh, oil, and down, earning the short-tailed and sooty shearwaters the local nickname of muttonbirds.

On June 20, 2001, the Agreement on the Conservation of Albatrosses and Petrels was signed by seven major fishing nations. The agreement requires member nations to manage fisheries by-catch, protect breeding sites from disturbance, promote conservation in the fishing industry, and conduct research to understand threatened species. A primary goal is to reduce seabird fatalities by long-line fishing. An estimated 300–350,000 seabirds are killed annually while attempting to get long-line bait. Long-liners present a dilemma for conservationists because this fishing technique is considered environmentally good for fish.

Flesh-footed shearwaters (*Puffinus carneipes*) in their nest in the ground in Mercury Islands, New Zealand. (Photo by Asa C. Thoresen. Photo Researchers, Inc. Reproduced by permission.)

1. Cory's shearwater (*Calonectris diomedea*); 2. Manx shearwater (*Puffinus puffinus*); 3. Bermuda petrel (*Pterodroma cahow*); 4. Southern giant petrel (*Macronectes giganteus*); 5. Northern fulmar (*Fulmarus glacialis*); 6. Broad-billed prion (*Pachyptila vittata*); 7. Bulwer's petrel (*Bulweria bulwerii*); 8. Short-tailed shearwater (*Puffinus tenuirostris*). (Illustration by Bruce Worden)

Species accounts

Cory's shearwater
Calonectris diomedea

TAXONOMY
Procellaria diomedea Scopoli, 1769, no locality: Tremiti Islands, Adriatic Sea.

OTHER COMMON NAMES
English: Mediterranean/(North) Atlantic shearwater; French: Puffin cendré; German: Gelbschnabel-Sturmtaucher; Spanish: Pardela Ceniciera.

PHYSICAL CHARACTERISTICS
17.7–18.9 in (45–48 cm); 19.8–33.7 oz (560–956 g); wingspan 39.4–49.2 in (100–125 cm). Heavy bodied. Uniformly pale underneath, with darker gray/brown plumage above; yellow bill. Has lighter cap than the similar greater shearwater.

DISTRIBUTION
Breeds on islands in the eastern North Atlantic and Mediterranean, migrates across the equator to South Atlantic and Indian oceans.

HABITAT
Marine, nesting on barren offshore islands away from mainland. Nests on rocky slopes, on cliffs, or in caves.

BEHAVIOR
Cory's shearwaters fly with a slower, more relaxed wingbeat than do other shearwaters.

FEEDING ECOLOGY AND DIET
Feeds mainly at night on fish, squid, crustaceans, and offal by plunging and surface-seizing. Follows fishing boats.

REPRODUCTIVE BIOLOGY
Breeding season starts in April. Nests in natural nooks such as burrows or rock crevices. The single white egg is incubated for 54 days. The brown chicks are brooded for approximately four days, fledging after 97 days. Sexual maturity at seven to 13 years.

CONSERVATION STATUS
Not threatened.

SIGNIFICANCE TO HUMANS
Regarding the awesome sight of migrating Cory's shearwaters, the Stuarts wrote in *Birds of Africa*: "Those that breed in the Mediterranean move in from the Atlantic at a rate of some 3,600 birds per hour through the Strait of Gibraltar. At the end of the breeding season when the adults and young birds depart for their overwintering grounds in the open ocean, they stream through the Strait in October to November at an estimated rate of 26,272 each day." ◆

Manx shearwater
Puffinus puffinus

TAXONOMY
Procellaria puffinus Brünnich, 1764, Faeroes and Norway.

OTHER COMMON NAMES
French: Puffin des Anglais; German: Schwarzschnabel-Sturmtaucher; Spanish: Pardela Pichoneta.

PHYSICAL CHARACTERISTICS
11.8–15 in (30–38 cm), 12.3–20.3 oz (350–575 g), wingspan 29.9–35 in (76–89 cm). Blackish upper body with contrasting white underneath. Upper parts are much darker than Cory's shearwater; face has more black than the little shearwater. The white undertail coverts contrast with the dark undertail coverts of the black-vented shearwater, once considered a subspecies of the Manx shearwater.

DISTRIBUTION
Breeds on islands on both sides of the North Atlantic, winters in Atlantic off Brazil, Argentina, and South Africa.

HABITAT
Marine, primarily over continental shelf.

BEHAVIOR
Gregarious, swims and dives to feed. Dives can be from the surface or from the air, and do not go deep below the water surface. To start the breeding season, males claim abandoned rabbit burrows, then call from within to attract females.

FEEDING ECOLOGY AND DIET
Feeds on small shoaling fish, squid, crustaceans, and offal. Does not normally feed in large flocks.

REPRODUCTIVE BIOLOGY
Colonial burrow nester. Breeding season begins in March. The egg, laid in mid-May, is incubated 47–55 days and fledging oc-

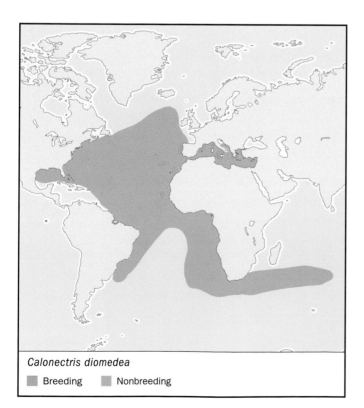

Calonectris diomedea

▮ Breeding ▮ Nonbreeding

Puffinus puffinus

■ Breeding   ■ Nonbreeding

Puffinus tenuirostris

■ Breeding   ■ Nonbreeding

curs after 62–76 days. Young fledge at night to begin a two to three week journey to wintering sites off Brazil, Argentina, and Uruguay. Sexual maturity at 5–6 years.

CONSERVATION STATUS
Not threatened.

SIGNIFICANCE TO HUMANS
Formerly hunted for food. ◆

Short-tailed shearwater
Puffinus tenuirostris

TAXONOMY
Procellaria tenuirostris Temminck, 1835, seas north of Japan and shores of Korea.

OTHER COMMON NAMES
English: Slender-billed shearwater/petrel, Tasmanian mutton-bird; French: Puffin à bec grêle; German: Kurzschwanz-Sturm-taucher; Spanish: Pardela de Tasmania.

PHYSICAL CHARACTERISTICS
15.7–17.7 in (40–45 cm), 16.9–28.2 oz (480–800 g), wingspan 37.4–39.4 in (95–100 cm). Dark brownish-gray above, lighter underneath. Pale chin, dark bill. Dark feet reach beyond short, square tail.

DISTRIBUTION
Breeds in Tasmania and southern Australia, migrates north across the equator to arctic reaches, including Alaska. Stays primarily in the Pacific.

HABITAT
Marine, found near land and in open seas. Typically breeds on grassy coastal islands.

BEHAVIOR
Forms flocks of up to 20,000.

FEEDING ECOLOGY AND DIET
Fish, crustaceans, and cephalopods. Found with whales. Social feeding is common, including flocking at dawn and dusk to feed on swarming euphausids.

REPRODUCTIVE BIOLOGY
Breeding season starts in October, forming crowded colonies of burrow nests. The single white egg is incubated for 52–55 days; the dark gray to brown chick in brooded for two to three days; fledging after 94 days. Sexual maturity at 4–6 years in males, five to seven years for females. Can live at least 30 years.

Parents on Montague Island travel up to 9,600 miles (15,450 km) round-trip on feeding voyages, the longest flights known for birds feeding young. Such long journeys may serve more to replenish the adults' reserves, than to feed the young.

CONSERVATION STATUS
Not threatened.

SIGNIFICANCE TO HUMANS
Approximately 300,000 chicks are harvested each year from Tasmania. ◆

Northern fulmar
Fulmarus glacialis

TAXONOMY
Procellaria glacialis Linnaeus, 1761, within the Arctic Circle (Spitsbergen).

OTHER COMMON NAMES
English: Arctic fulmar; French: Fulmar boréal; German: Eis-sturmvogel; Spanish: Fulmar Boreal.

PHYSICAL CHARACTERISTICS
With a wingspan of 40.2–44.1 in (102–112 cm), white head, and gray upper body, northern fulmars resemble gulls, but their wings are broader and the neck is thicker. Lighter morphs are more common in Atlantic, darker morphs in Pacific.

DISTRIBUTION
Northern Atlantic and Pacific oceans, has spread southward over much of the Atlantic Ocean. Winters farther south.

HABITAT
Marine, especially colder waters of the Northern Hemisphere.

BEHAVIOR
More aggressive in vomiting habits than other procellariids. Stiff wings held out straight from body help to distinguish them from gulls.

FEEDING ECOLOGY AND DIET
Feed on fish, squid, plankton, and fishing refuse. Feeds in flocks, frequently behind fishing boats. Will scavenge on carrion.

REPRODUCTIVE BIOLOGY
Breeding season begins in May. The single white egg is incubated 47–53 days; the white to dark gray chick is brooded for 2 weeks; fledging after 46–53 days. Nests colonially on cliff ledges and on level ground, and has expanded to buildings and rooftops.

CONSERVATION STATUS
Not threatened. One of few seabirds to increase in numbers and range since 1800. Expansion may be due to food from

fishing ship discards or to changing oceanographic conditions. The dependence on fishing ships presents an interesting conservation problem. If ships clean up their refuse, then fulmars may be pushed to feed elsewhere, potentially on smaller birds.

SIGNIFICANCE TO HUMANS
Once hunted for food. Now may be anthropophilic, following fishing ships for food. ◆

Broad-billed prion
Pachyptila vittata

TAXONOMY
Procellaria vittata G. Forster, 1777, lat. 47°.

OTHER COMMON NAMES
English: Blue/broad-billed dove-petrel, long-billed/common prion, icebird, whalebird; French: Prion de Forster; German: Großer Entensturmvogel; Spanish: Pato-petrel Piquiancho.

PHYSICAL CHARACTERISTICS
9.8–11.8 in (25–30 cm); 5.6–8.3 oz (160–235 g); wingspan 22.4–26 in (57–66 cm). The largest prion, with a wide bill. Dark patches form an "M" across back of outstretched wings.

DISTRIBUTION
Breeds in South Pacific on New Zealand's South Island and on Chatham Islands, and in the South Atlantic on Gough and Tristan da Cunha Islands.

HABITAT
Marine, stays away from land except to breed. Breeds on barren areas including lava fields, cliffs, and coastal slopes.

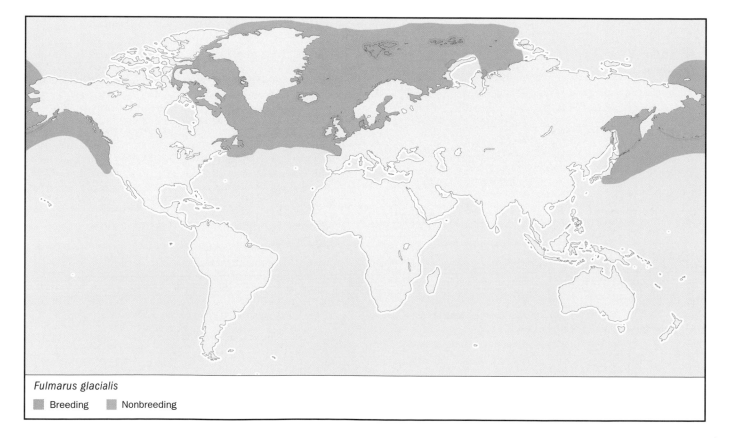

Fulmarus glacialis

Breeding   Nonbreeding

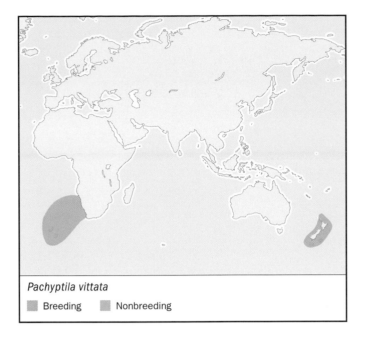

Pachyptila vittata

■ Breeding   ■ Nonbreeding

Pterodroma cahow

■ Breeding   ■ Nonbreeding

BEHAVIOR
Prions are social. Courtship displays are restricted to cover of night or burrows. Pairs defend their burrows aggressively with calls, posturing, or if the threat intensifies, with biting of each other's bill and neck.

FEEDING ECOLOGY AND DIET
Crustaceans (mostly copepods), squid, and fish. Feeds by hydroplaning and by surface-seizing. Does not tend to follow fishing boats. Feeds gregariously.

REPRODUCTIVE BIOLOGY
Breeding season starts in July or August. Forms tight colonies of burrow nests. More than one pair may occupy one nest. The egg is incubated for 50 days and fledging occurs after 50 days.

CONSERVATION STATUS
Not threatened.

SIGNIFICANCE TO HUMANS
None known. ◆

Bermuda petrel
Pterodroma cahow

TAXONOMY
Aestrelata cahow Nichols and Mowbray, 1916, Castle Island, Bermuda.

OTHER COMMON NAMES
English: Cahow; French: Pétrel des Bermudes; German: Bermudasturmvogel; Spanish: Petrel Cahow.

PHYSICAL CHARACTERISTICS
15.0 in (38 cm); wingspan 35.0 in (89 cm). Brownish-gray upper body, including a cap that covers the eye and a partial brown collar on the nape. Black bill. White underneath, except for black edges of wings. Easily confused with the larger black-capped petrel.

DISTRIBUTION
Islets in Castle Harbour, Bermuda.

HABITAT
Marine. Formerly excavated burrows in sand or soft soils, but now nests on small, rocky offshore islands and in artificial burrows.

BEHAVIOR
Little is known about the natural behavior of these birds. Their normal night-time aerial courtship has been disrupted by lights from human facilities.

FEEDING ECOLOGY AND DIET
Very little known; not known to follow ships.

REPRODUCTIVE BIOLOGY
Breeding season January to June. The single white egg is incubated for 51–54 days; the chick is brooded for one or two days; fledging after 90–100 days. Historically it bred inland in soil burrows, but rats have driven colonies to suboptimal, rocky offshore islets.

CONSERVATION STATUS
Endangered. Had been thought extinct since 1621 after colonists hunted it for food. In 1921, it was found living, and in 1951, 18 breeding pairs were found. Intensive conservation efforts began in 1961; 45 pairs were found breeding in 1994. Current threats include native species (white-tailed tropicbirds [*Phaethon lepturus*] compete for nesting sites), human disturbance (light pollution disrupts nocturnal courtship), natural disasters (flooding of nests became a problem in the 1990s, with rising sea levels), and atmospheric pollution. In 1997, the population was estimated at 180 birds.

SIGNIFICANCE TO HUMANS
Formerly hunted for food. ◆

Bulwer's petrel
Bulweria bulwerii

TAXONOMY
Procellaria bulwerii Jardine and Selby, 1828, Madeira.

OTHER COMMON NAMES
French: Petrel de Bulwer; German: Bulwersturmvogel; Spanish: Petrel de Bulwer.

PHYSICAL CHARACTERISTICS
10.2–11.0 in (26–28 cm); 2.8–4.6 oz (78–130 g); wingspan 26.8–28.7 in (68–73 cm). One of the smallest procellarids. Dark brown plumage, long wings, flies with tail narrowed to a point.

DISTRIBUTION
Throughout the tropics in eastern Atlantic and Pacific oceans. One study suggests that Bulwer's petrel prefers warm waters of intermediate salinity, while Joaquin's petrel (*Bulweria fallax*) prefers slightly higher salinity levels.

HABITAT
Marine, strongly pelaqic. Breeds on barren, remote islands.

BEHAVIOR
Nocturnal. Both sexes make a barking call known as a "woof." They do not call while flying.

FEEDING ECOLOGY AND DIET
Feeds primarily at night on fish, squid, some crustaceans and sea-striders, by seizing prey from the surface.

REPRODUCTIVE BIOLOGY
Breeding season starts in April or May. Nest in burrows, crevices, cracks, caves, or under other cover. The single white egg is incubated for 44 days; the chick is blackish when it hatches; fledging after 62 days.

CONSERVATION STATUS
Not threatened.

SIGNIFICANCE TO HUMANS
Collected for food and fish bait on Atlantic islands. ◆

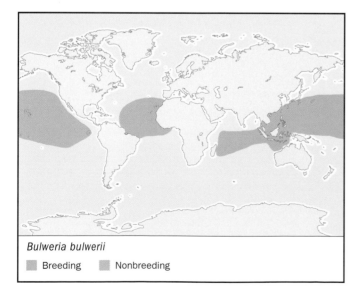

Bulweria bulwerii

■ Breeding ■ Nonbreeding

Southern giant petrel
Macronectes giganteus

TAXONOMY
Procellaria gigantea Gmelin, 1789, Staten Island, off Tierra del Fuego.

OTHER COMMON NAMES
English: Antarctic giant petrel, giant fulmar, stinker, stinkpot; French: Fulmar géant; German: Riesensturmvogel; Spanish: Abanto-marino Antártico.

PHYSICAL CHARACTERISTICS
Largest procellariid; 33.9–39 in (86–99 cm); male 11 lb (5 kg), female 6.6–17.6 lb (3–8 kg); wingspan 72.8–80.7 in (185–205 cm). Enormous yellow bill. Head, neck, and upper breast are pale; the rest of the body is mottled brown on the dark morph, with the underside lighter than the upper parts. Dark legs. The dark color morph resembles the northern giant petrel, but the less common all-white morph is distinctive.

DISTRIBUTION
Found throughout the Southern Hemisphere from Antarctica to the subtropics of Chile, Africa, and Australia.

HABITAT
Marine, feeding in coastal and pelagic southern hemisphere waters. Nests on bare or grassy, exposed ground throughout Antarctica, on the coasts of Chile and Argentina, and on subantarctic and Antarctic islands.

BEHAVIOR
The birds are so tame that researchers can walk up to brooding females and remove chicks from underneath them for study. Their strong legs, unusual for a procellariid, allow the giant petrels to scavenge beached carcasses. The larger males exclude females from carcasses, forcing the females to depend more heavily on live prey taken at sea. Both sexes visit the colony year-round, even outside of the breeding season.

FEEDING ECOLOGY AND DIET
Shunned, even by avid birders, for feeding on rotting carcasses. Parents travel up to 3,000 miles (4,830 km) to ice packs to feed on krill and squid brought to the surface by ocean upwellings. Gather in the thousands at long-line fishing boats.

REPRODUCTIVE BIOLOGY
Sexual maturity at six to seven years of age. Breeding season starts in October. Nest is made by gathering grass and moss, or small stones, into a mound, with a depression in the middle. The single white egg is incubated 55–66 days; the whitish chick is brooded for two or three weeks; fledging at 104–132 days.

CONSERVATION STATUS
Vulnerable. Total population about 36,000 breeding pairs. Many populations have inexplicably decreased by 30–35% since 1981.

SIGNIFICANCE TO HUMANS
Their long lives (the oldest known lived 50 years) make them a valuable indicator species to judge the health of the Antarctic ecosystem. ◆

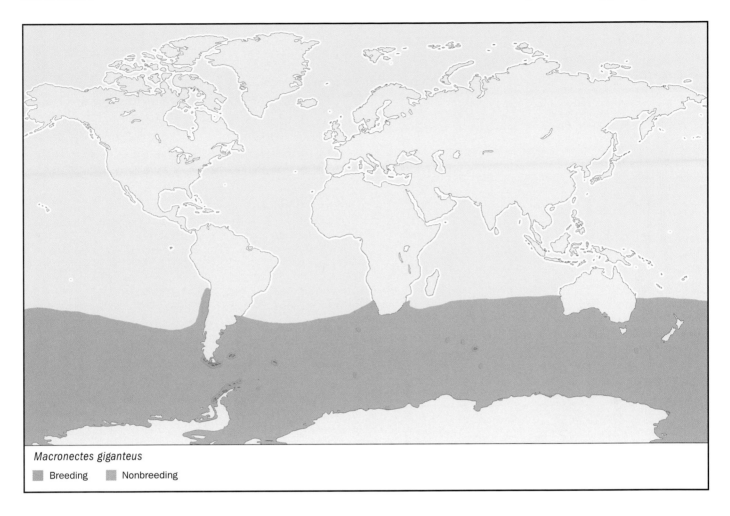

Macronectes giganteus

■ Breeding   ■ Nonbreeding

Resources

Books

BirdLife International. *Threatened Birds of the World.* Cambridge: BirdLife International, 2000.

del Hoyo, J., A. Elliot, and J. Sargatal, eds. "Ostrich to Ducks." In *Handbook of the Birds of the World.* Vol. 1. Barcelona: Lynx Edicions, 1992.

Stuart, Chris, and Tilde Stuart. "Birds of the Oceans." In *Birds of Africa from Seabirds to Seed-Eaters.* Cambridge, Massachusetts: The MIT Press, 1999.

Warham, John. *The Behavior, Population Biology and Physiology of the Petrels.* San Diego, CA: Academic Press, 1996.

Periodicals

Braasch, Gary. "Antarctic Mystery." *International Wildlife* 31 (2001): 52–57.

Phillips, R.A., and K.C. Hamer. "Postnatal Development of Northern Fulmar Chicks, *Fulmarus glacialis.*" *Physiological and Biochemical Zoology* 73 (2000): 597–604.

Schultz, Mark A., and Nicholas I. Klomp. "Does the Foraging Strategy of Adult Short-Tailed Shearwaters Cause Obesity in Their Chicks?" *Journal of Avian Biology* 31 (2000): 287–294.

Thompson, Paul M., and Janet C. Ollason. "Lagged Effects of Ocean Climate Change on Fulmar Population Dynamics." *Nature* 413 (2001): 417–420.

Other

Earth-Life Web Productions. "Shearwaters (Procellariidae)." 21 Oct. 2001 (21 Feb. 2002). <http://www.earthlife.net/birds/shearwaters.html>.

Gough, G.A., J.R. Sauer, and M. Iliff. *Patuxent Bird Identification Infocenter.* 1998. Version 97.1. Patuxent Wildlife Research Center, Laurel, MD. 28 Dec. 2000 (21 Feb. 2002). <http://www.mbr-pwrc.usgs.gov/Infocenter/infocenter.html>.

Barbara Jean Maynard, PhD

Storm-petrels
(Hydrobatidae)

Class Aves

Order Procellariiformes

Family Hydrobatidae

Thumbnail description
Small seabirds with prominent upturned nostrils and dancing flight over the waves

Size
5.5–10 in (13–26 cm)

Number of genera, species
8 genera; 21 species

Habitat
Open sea

Conservation status
Several species, whose status is unknown, are possibly rare

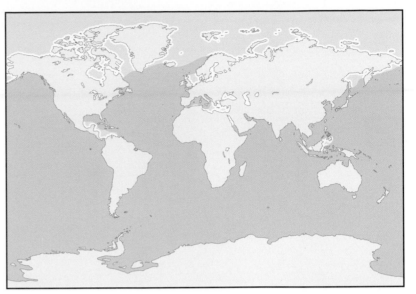

Distribution
Worldwide

Evolution and systematics

The fossil record is poor, represented only by *Oceanodroma hubbsi* from the Upper Miocene of California, *Oceanites zalascarthmus* from the South African Pliocene, and *Primodroma bournei* from the English Eocene.

A number of studies based on comparative anatomy and DNA analyses suggest that the order evolved close to the base of procellariiform radiation.

The family Hydrobatidae is divided into two subfamilies: the Oceanitinae contains seven very long-legged species found in southern seas, and the Hydrobatinae contains 14 rather shorter-legged species mainly found in northern waters.

Physical characteristics

Storm-petrels are small, delicate seabirds whose long legs are used to fend off from the water as the birds snap up food items. The wings are rounded at their tips because the tenth and outermost functional primary is shorter than that of the ninth (primaries are main flight feathers). The fused tubular nostrils are prominent and span nearly half the length of the bill. The smallest species, least storm-petrels (*Oceanodroma microsoma*), weigh only 0.7 oz (20 g) and have a wing span of about 12.6 in (32 cm); the largest, Tristram's storm-petrels (*Oceanodroma tristrami*), weigh about 2.9 oz (83 g) and have a wingspan of 22.4 in (57 cm).

Storm-petrels tend to have dark blackish or brownish plumage, but many are paler ventrally and have white rumps.

The tail may be square cut or forked. The feet of birds of the genera *Fregetta* and *Nesofregetta* bear strange spade-like claws. Female storm-petrels tend to be larger than males. Most (perhaps all) storm-petrels carry the musty body odor characteristic of tubenoses. The olfactory lobes of the brain are large and a functional sense of smell has been demonstrated in field experiments at sea and over land.

Distribution

Storm-petrels are found worldwide, but they are particularly numerous in the vast Southern Ocean. Many breed around Australasia but five species are concentrated around islands from Mexico to California. At sea they occur in all oceans but fail to penetrate Arctic seas.

Habitat

Marine distributions of storm-petrels are poorly known; being small, storm-petrels are hard to see and identify as they dart along hugging the waves. Some species prefer warm or cool waters. For example, Leach's storm-petrels (*Oceanodroma leucorhoa*) inhabit cooler water well offshore of the western United States, and wedge-rumped storm-petrels (*Oceanodroma tethys*) and Elliot's storm-petrels (*Oceanites gracilis*) seem confined to cool waters of the Humboldt Current. Some species congregate along areas of upwellings, as does the band-rumped storm-petrel, (*Oceanodroma castro*), which prefers warm water and aggregates off Florida and South Carolina along Gulf Stream eddies. Storm-petrels breed on islands that are free of mammalian predators.

Wedge-rumped storm-petrels (*Oceanodroma tethys*) feed together at the ocean's surface. (Photo by R.L. Pitman/VIREO. Reproduced by permission.)

Behavior

Most nests are in burrows, and the nest sites are retained from season to season and form the focus for the maintenance of the same pair bonds from year to year. Most visit their nests only after dark and appear to have little by way of display except for mutual preening. Aerial flight displays and chasing in species like the wedge-rumped storm-petrel and Wilson's storm-petrel (*Oceanites oceanicus*) have been described. The spectacular performances of the first species take place by day, and this species is evidently adapted to nighttime feeding.

In all species studied the sexes call differently. A variety of churring or whirring sounds is heard from burrows of northern species such as the European storm-petrel (*Hydrobates pelagicus*). On the Norwegian island of Rost, European storm-petrels flutter around the exhaust of the lighthouse engine, and the birds' calls match the rhythm of the engine. Calls may vary from bird to bird, which suggests a role in individual recognition—important for communication after dark. The voices of the southern Oceanitinae tend to be a higher pitch than those of the Hydrobatinae, and their whistles may have a ventriloqual quality.

Most storm-petrels tend to be solitary at sea but flocking also occurs. Some species, like Wilson's storm-petrels, are highly migratory. White-faced storm-petrels (*Pelagodroma marina*) from Western Australia migrate north to winter along the convergences of the northern Indian Ocean. Other species, such as Leach's and European storm-petrels, shift south after breeding, the former to the tropical Atlantic and Pacific oceans and the latter mainly to the Benguela Current off Africa. Other North Pacific species also shift south; for example, Matsudaira's and Swinhoe's storm-petrels (*Oceanodroma matsudairae* and *O. monorhis*) from Japanese seas reach the Indian Ocean via the Straits of Malacca and the Timor Sea. Most storm-petrels breeding off California just disperse into local seas, although the black storm-petrel (*Oceanodroma melania*) moves south to the Humbolt current.

Feeding ecology and diet

Crustaceans are important foods for storm-petrels. For example, euphausids are the most frequent prey item for Wilson's storm-petrels nesting around Antarctica; the euphausid species and its relative importance varies with locale and season. Farther north around the Crozet Islands, the same bird still eats crustaceans, but copepods and cirripedes are relatively more abundant in the diet. The gray-backed storm-petrel (*Garrodia nereis*) seems to specialize on barnacles that evidently are picked off floating rafts of seaweed.

Storm-petrels have a penchant for oily foods. They snip up oil droplets from the sea but seem to avoid man-made oil slicks, perhaps by using their sense of smell. Stomachs usually contain the stomach oil found in most tubenoses, and this oil, being digestible and full of energy, forms an important food for adults and chicks. It is derived directly from prey items, many of which contain heavy loads of oil droplets (especially when breeding).

Storm-petrels usually feed solitarily but will congregate around a suitable food source such as a dead seal or squid. Some species associate with pods of whales, and Wilson's storm-petrels ingest whale feces. Ship following is common; the birds eat small prey churned up by propellers. Wake followers tend to attract others, and up to 50 black storm-petrels have been seen combing the wake at one time.

Storm-petrels feed from the top few inches of the sea. They seem to have the ability to stay within bill range of a heaving sea. This ability is aided by their low wing loadings, which means that as the wave heaves, the air above moves with it and so does the bird. Both Wilson's and black storm-petrels will dive to retrieve food. The precise mode of feeding varies with the species, the length of the legs, and other factors. Some species, like Wilson's storm-petrels, hold their wings high while fending off from the surface with both feet, whereas black-bellied storm-petrels (*Fregetta tropica*) hold their wings out and skip from side to side. Feeding birds face the wind, and if a gale shifts abruptly through 90 degrees to blow straight down the wave furrows, the tiny birds may be unable to feed and be forced far down wind.

Reproductive biology

Typically monogamous, storm-petrels first visit their natal colonies as prebreeders, and they breed at 4–5 years old. Nesting occurs when surrounding seas provide plenty of food; that is in the spring or summer in middle and high latitudes. Tropical breeders often have an extended laying season.

With few exceptions (e.g., the wedge-rumped storm-petrel in the Galápagos), all activity on land occurs after dark, and for many species the behaviors culminating in mating are unclear. European storm-petrels often crash into one other, apparently deliberately, and then fall to the ground before disentangling themselves. In Alaska, fork-tailed storm-petrels (*Oceanodroma furcata*) circling overhead call in response to cries from burrows below, suggesting pair bonding. High-speed zigzagging chases have been seen among other species, but, in general, the significance of these maneuvers is obscure. Not much more is known of their behavior on the ground, but using night vision equipment, copulation of fork-tailed storm-petrels has been seen in the nest chamber and on the ground outside. There was little precopulation ceremony other than mutual preening.

Some female storm-petrels feed at sea while producing the single egg; this trek is called the prelaying exodus. In a study of Wilson's storm-petrels, the females of 31 pairs stayed away for 16–18 days, leaving their partners to visit the nests from time to time, perhaps to keep them clear of snow. In other species there seems to be no clear exodus. In Leach's storm-petrels, semen is stored in special glands in the vaginal folds of the cloaca. This arrangement allows fertilization to be delayed while the bird travels to the best feeding area while producing her egg and returning to her nest.

The single egg is ovoid, whitish, and often sprinkled with pinkish dots. It is laid within 24 hours of the female's return and is very large compared to her body size. That of the least

A white-vented storm-petrel (*Oceanites gracilis*) feeds by plucking tiny debris from the sea surface near Fernandina Island, in the Galápagos. (Photo by Tui De Roy. Bruce Coleman Inc. Reproduced by permission.)

storm-petrel is 29% of her body weight—probably the heaviest egg relative to body size of any bird. Highly migratory species lay over a relatively short period, whereas sedentary birds like the ashy storm-petrel (*Oceanodroma homochroa*) breeding off California lay over a period of about 100 days.

The nest is a chamber at the end of a tunnel bored into soft ground or in a crevice among rocks. Males seem to be the tunnel diggers and the chambers are often sparsely lined with bits of local vegetation. Gray-backed storm-petrels and white-bellied storm-petrels (*Fregetta grallaria*) burrow into the fibrous debris at the bases of tussock clumps and so are, in effect, above ground. The microclimate in a burrow is milder than the climate outside, being more humid and warmer in cold climates and cooler than ambient temperatures in tropical regions. The egg is tucked snugly into a bare, well-vascularized incubation patch. Incubation shifts vary from two to three days in cold water species but average 4.5 days in warm water species like the white-faced storm-petrel.

The female usually leaves for the sea within 24 hours of laying. The male takes the first incubation shift and the sexes take turns thereafter. The total time between laying and hatching in continuously covered eggs varies from 38 to 42 days. When eggs are temporarily abandoned, this interval is longer. Eggs are resistant to chilling, can be abandoned for several days, and yet will still produce a chick. In one extreme case, a fork-tailed storm-petrel egg left uncovered for 31 days still hatched. The embryo's resistance to chilling while still remaining viable is a valuable adaptation for a small seabird having to cope with variable weather and sea conditions which may prevent parent's return.

The chick is hatched covered in down, its eyes closed, and with incomplete control of its body temperature. It is brooded by one or other of the parents for three or more days—known as the guard stage. In their thick down the chicks look like powder puffs. They are grayish or whitish, often paler on the belly. In most, the initial down (protoptile) is replaced by a second (mesoptile) down that grows attached to the protoptile, which, in turn, is pushed out by the feathers proper (the teleoptiles). At least two of the southern birds, Wilson's storm-petrels and white-faced storm-petrels, have only one down.

Initially the chick gets small meals, perhaps mostly of stomach oil, but meal sizes increase when the guard stage is over. With the milder climate in the burrow and a layer of subdermal fat, the chick can then thermoregulate with the milder climate in the burrow. Both parents feed the chick and a fairly even division of labor seems to be the norm. Some meals can be huge, e.g., as much as the chick's own body weight (usually after both adults have fed it on the same night). Parental feeding visits are frequent at first but decline in older chicks. For instance, European storm-petrel chicks 11–20 days old received visits on 93% of the nights, those aged 31–40 days on 85% of the nights, and 51–60 day-olds on 66% of nights. Chicks can withstand fasting for 6–7 days. Starved chicks may become torpid to reduce energy needs, but their body weight falls and death may result.

A chick's growth follows a typical logistical curve, climbing steeply for about the first half of the nestling stage then stabilizing above adult weight and finally falling in the last 10–20 days before first flight. This is preceded during its last week or so by its leaving the burrow to exercise its wings and to explore the vicinity of the nest. A chick's first flight usually takes place from some nearby eminence. Parents take no part in this; the chick leaves alone and, as far as is known, once at sea is really on its own. Average nestling periods (the days between hatching and the first flight) range from 57 to 84 days. Nestling periods are much more variable than incubation periods, partly due to erratic provisioning. Breeding success (percent of eggs laid that produce flying young) varies from year to year.

Conservation status

Some species are abundant with many colonies scattered widely. These include Leach's, European, Wilson's, gray-backed, white-faced, and band-rumped storm-petrels. There are also a number of very localized species such as Matsudaira's storm-petrel, which breeds only at the Volcano Islands, and the ashy, least, and black storm-petrels, which nest off southern California and Mexico.

For some species, breeding places are virtually unknown. For example, Markham's storm-petrel (*Oceanodroma markhami*) has been found nesting on the Paracas Peninsula, Peru, but more undiscovered colonies must exist to account for the numbers of birds encountered at sea. None of these seems to be endangered, but Elliot's storm-petrel may well be as only one pair of the subspecies *Oceanites gracilis gracilis* has been found nesting off Chile. Breeding places of *O. g. galapagoensis* remain undetected.

Colonies of storm-petrels have been wiped out by predators, which are usually mammals like rats and mustelids (even mice can damage these small birds by eating their eggs). The extinction of the Guadalupe storm-petrel (*Oceanodroma macrodactyla*) has been attributed to a combination of predation by cats and erosion from grazing by goats. A particularly noteworthy conservation effort is the extermination of foxes from colonies of Leach's and forked-tailed storm-petrels in Alaska, which resulted in the reestablishment of many colonies. Another success story concerns the use of tape playbacks of Leach's storm-petrel calls plus the provision of artificial burrows on islands in Muscongus Bay, Maine. Attracted by the amplified calls, birds investigated the burrows and eventually bred there. A colony was established (on Ross Island) where there had been no previous evidence of breeding.

No storm-petrel species is threatened, although birds like Hornby's (*Oceanodroma hornbyi*) and Markham's, being quite unknown, badly need investigation. The populations of Tristram's storm-petrel on Midway Island, formerly decimated by rats, seem to be recovering since the rats were wiped out.

Significance to humans

Storm-petrels were well known to early sailors and very familiar to sealers and whalers because their hunting activities attracted birds like the cosmopolitan Wilson's storm-petrel. They lumped several species together as Mother Carey's Chickens, *Mater cara* referring to the Virgin Mary under whose protection seafarers were supposed to come. Although killing of these birds was sometimes considered fraught with danger, they were often caught for use as bait. Wilson's storm-petrels were killed at night from the stern of the ship, and many are attracted to lights. Sealers threaded wicks through the alimentary tracts of adults or fat chicks to draw out the stomach oil, which could be used as a candle.

Small though they are, several species have been eaten regularly by primitive societies: Aleuts, American Indians, and the inhabitants of the Izu Islands, Japan, all ate Leach's storm-petrels, while the Morioris of the Chatham Islands, New Zealand and aborigines of eastern Australia ate white-faced storm-petrels when available.

Species accounts

Wilson's storm-petrel
Oceanites oceanicus

SUBFAMILY
Oceanitinae

TAXONOMY
Oceanites oceanicus Kuhl, 1840, no locality. *O. o. oceanicus*: Islands off Tierra del Fuego and subantarctic islands of Atlantic and Indian Oceans, including South Georgia; *O. o. exasperatus*: South Shetland, South Sandwich, South Orkney Islands, Elephant Island, coasts and offshore islands of Antarctica.

OTHER COMMON NAMES
English: Mother Carey's chicken; French: Pétrel océanite; German: Buntfüssige Sturmschwalbe; Spanish: Paíño de Wilson.

PHYSICAL CHARACTERISTICS
7 in (18 cm); 1.3 oz (35 g). Wholly black above and below except for white rump merging into the white lower flanks and thighs and a pale band across the center of each wing. Tail square cut and black. Legs black, very long, and projecting beyond the tail when flying; webs between toes yellow. Juvenile like adult. Bill black with prominent nasal tubes reaching about halfway along the ridge of the bill. Sexes alike.

Oceanites oceanicus

DISTRIBUTION
Wholly marine except when nesting, found in all oceans particularly along coastal upwellings and fronts. It tends to be more often seen offshore compared to other storm-petrels such as Leach's and white-faced storm-petrels, which prefer deeper water.

Highly migratory, moving from April to June from the southern breeding stations to northern reaches of the oceans, but avoids Arctic seas. In the Atlantic the journey from the south is 7,000 miles (11,000 km) for some birds.

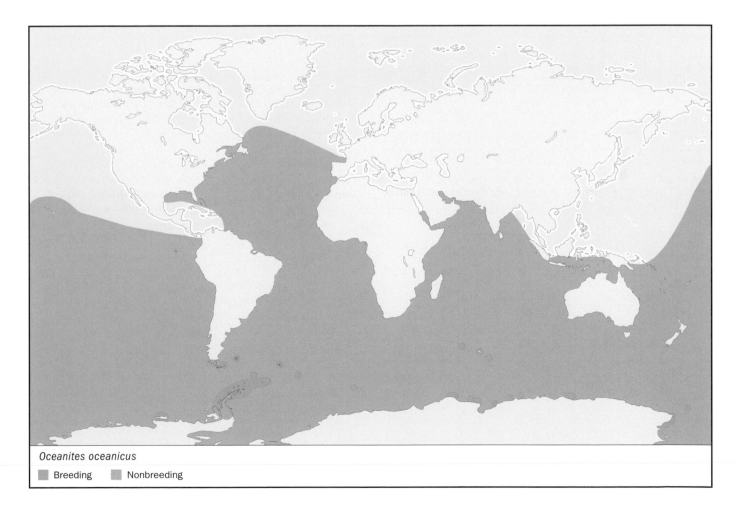

Oceanites oceanicus

■ Breeding   ■ Nonbreeding

HABITAT

These birds are concentrated along the ocean shelves during the northern summer. Although most move back to southern waters to breed during the northern winter, some remain: these are probably juveniles or birds that are taking a season off-duty—a so-called sabbatical year.

BEHAVIOR

The feeding behavior is distinctive with a flight characterized by alternate glides and wing flutterings while the long legs are drooped and often break the surface. Most food is snipped from the surface without alighting, and this is the most common ship's follower. Calls used on the breeding grounds include a grating sound used by both sexes and a chatter call used by the males to attract females.

FEEDING ECOLOGY AND DIET

Crustaceans, but fish are also eaten (these are more energy rich than crustacea) with mycophids up to 3.3 in (8.5 cm) long being fed to the chicks (quite a meal for a bird with a bill only 0.5 in [1.2 cm] long).

REPRODUCTIVE BIOLOGY

The pair-bond is held over several seasons and most pairs tend to breed annually. The nest forms their focus. There is little to suggest that they stay together during migration. Because of the short polar summers, Wilson's storm-petrels breeding around Antarctica have accelerated the development of the egg and chick: the time from laying to fledging is 91 days (the shortest period for any tubenose). Birds farther north, though slightly smaller, take longer, perhaps because the food supply is less concentrated than it is off the southern continent.

Most nests are hidden in crevices among rocks or coarse scree. The egg is laid on the bare earth in a shallow scrape, and those on the southern islands are often lined with scraps of local vegetation.

The eggs take about 40 days to hatch if continually incubated. The chick flies at 48–78 days old. A major cause of mortality is unseasonal weather that stops birds entering their nests or freezes chicks within them. Predation by skuas is not usually significant.

CONSERVATION STATUS

Not threatened. One of the most abundant seabirds. Its isolation is its major safeguard.

SIGNIFICANCE TO HUMANS

None known. ◆

Leach's storm-petrel

Oceanodroma leucorhoa

SUBFAMILY

Hydrobatinae

TAXONOMY

Oceanodroma leucorhoa Vieillot, 1818, maritime parts of Picardy. *O. l. leucorhoa*: all North Atlantic and North Pacific populations south to California. Populations in eastern Pacific vary in size and degree of white on rump; *O. l. chapmani*: breeds on San Benitos and Los Coronados islands, Mexico. *O. l. socorrensis*: breeds in summer on islets off Guadalupe Island, Mexico. *O. l.*

cheimomnestes: breeds in winter on islets off Guadalupe Island, Mexico.

OTHER COMMON NAMES

French: Pétrel culblanc; German: Wellenläufer; Spanish: Paiño de Leach.

PHYSICAL CHARACTERISTICS

7.0–8.5 in (180–220 mm); 1.3–1.9 oz (38–54 g). Medium-sized storm-petrel with white rump patch and distinctly forked tail, blackish brown above and below with paler diagonal upperwing bar from carpal joint back to trailing edge near body. Strongly downhooked bill and

Oceanodroma leucorhoa

prominent nostrils. Legs and feet black. Sexes alike. Flight less fluttering than that of Wilson's storm-petrel and wings tend to be held more horizontally than those of Wilson's storm-petrel (which raises its wings into a V). Feet not visible beyond tail and much less prone to pattering.

DISTRIBUTION

Breeds on islands in the North Atlantic and North Pacific that lack mammalian predators. Found at sea throughout these seas, migrates south into the tropics after breeding. Pairs have even been found in burrows on the Chatham Islands, New Zealand. One bird was found on St. Croix Island, South Africa, which suggests the possibility of extending the breeding range southwards. The eastern North American populations shift south to Brazilian waters but many cross to European seas like the Bay of Biscay. British breeders appear to winter mainly off tropical Africa. Japanese and Alaskan birds also winter in tropical seas.

HABITAT

Ranges widely in the open sea. California birds feed farther out in warmer, less productive seas than do ashy and fork-tailed storm-petrels with which they often share nesting islands.

BEHAVIOR

On land, overflying birds that emit calls are mainly prebreeders; nesting birds call mostly from their burrows. Little display occurs between breeding birds except persistent calling using two main types of rhythmic purrings and chatterings. Chatter calls are given from the air and the burrow, and research in Japan suggests that variation in the pitch of calls among birds of the same sex may be used for individual recognition.

FEEDING ECOLOGY AND DIET

Nekton and planktonic organisms are taken from the surface while the bird hovers facing the wind, sometimes alighting momentarily. Otherwise the flight is a mixture of gliding and rather wild dashes, with many changes of direction. Seldom follows ships but is prone to be wrecked on beaches during gales.

Diet includes a great range of fish, crustaceans, and squid as available. The birds tend to seek areas where upcurrents bring organisms to the surface. Their stomachs often contain deep-sea animals that only approach the surface at night, evidently taken after dark. They appear to find some prey using their good sense of smell.

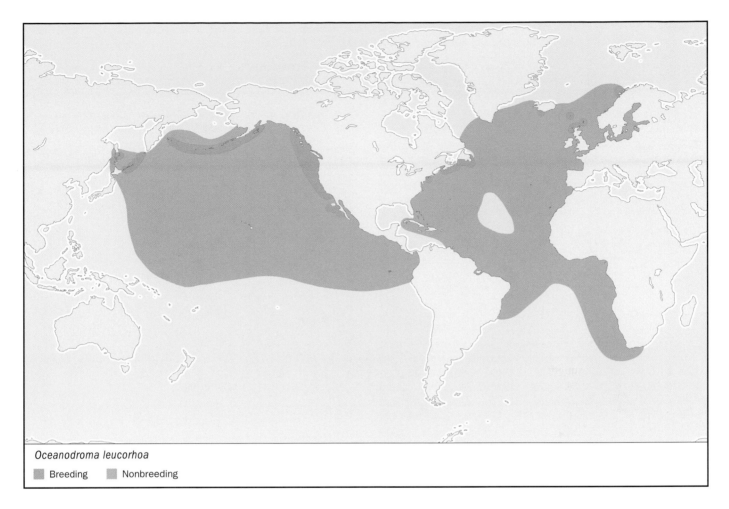

Oceanodroma leucorhoa

■ Breeding   ■ Nonbreeding

REPRODUCTIVE BIOLOGY

Most breed at around five years old. Having gained a nest site, a pair remains intact as long as they reproduce satisfactorily. Most dig burrows, but some occupy crevices among rocks or stone walls. The single egg is occasionally replaced if it is lost soon after laying. Both sexes incubate in 2–3 day shifts for about 43 days. Chick is fed almost nightly in the first few days after the brief brooding period, then less often as it grows. Unfed chicks become torpid but can recover. Growth follows the normal curve: weight climbs steadily at a constant rate, levels out at above parental weight, then falls in the last 10 or so days before fledging.

Fledging occurs when the chick is between 56 and 79 days. The number of chicks fledged per egg laid differs between seasons and places and ranges from 48 to 73%. Losses have often been due to introduced mammals like mink, cats, and others, but natural predators also include owls, eagles, corvids, and gulls.

CONSERVATION STATUS

Not threatened, but existing colonies need protection against the introduction of placental mammals and trampling of burrows.

SIGNIFICANCE TO HUMANS

None known. ◆

Resources

Books

Ainley, D.G., and R.J. Boekelheide, eds. *Seabirds of the Farallon Islands.* Stanford, CA: Stanford University Press, 1990.

Huntington, C.E., R.G. Butler, and R.A. Mauck. "Leach's Storm-petrel." In *The Birds of North America,* No. 223, edited by A. Poole and F. Gill. Philadelphia: Academy of Natural Sciences; Washington, DC: American Ornithologists' Union, 1996.

Lockley, R.M. *Flight of the Storm-petrel.* London: David and Charles, 1983.

Warham, J. *The Behaviour, Population Biology and Physiology of the Petrels.* San Diego: Academic Press, 1996.

Warham, J. *The Petrels: Their Ecology and Breeding Systems.* San Diego: Academic Press, 1990.

Resources

Periodicals

Croxall, J.P., H.J. Hill, R. Lidstone-Scott, M.J. O'Connell, and P.A. Prince. "Food and Feeding Ecology of Wilson's Storm-petrel *Oceanites oceanicus* at South Georgia." *Journal of Zoology, London* 216 (1988): 83–102.

Taoka, M., T. Sato, T. Kamada, and H. Okumura. "Sexual Dimorphism of Chatter-Calls and Vocal Sex Recognition in Leach's Storm-Petrels (*Oceanodroma leucorhoa*)." *Auk* 106 (1989): 498–501.

Other

Warham, J. *A Bibliography of the Procellariiformes or Petrels.* February 1999. (January 31 2001). <http://www.zool.canterbury.ac.nz/jwjw.htm>

John Warham, DSc

Diving-petrels
(Pelecanoididae)

Class Aves

Order Procellariiformes

Family Pelecanoididae

Thumbnail description
Small-sized, black-and-white colored, stocky, short-winged, tube-nosed seabirds with nostrils pointing upwards. Diving-petrels dive and swim for their food

Size
7–10 in (18–25 cm); 4–8 oz (120–220 g)

Number of genera, species
1 genus; 4 species

Habitat
Cool and cold oceans

Conservation status
One species is Endangered

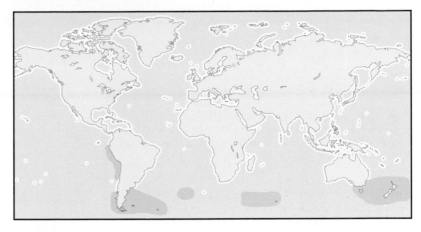

Distribution
Occurs in cool and cold oceans of the Southern Hemisphere, usually close to breeding sites

Evolution and systematics

Diving-petrels (family Pelecanoididae) are in a group of seabirds known as tubenoses (order Procellariiformes), all of which have a distinctive pair of tube-like, salt-excreting, external nostrils on the top or sides of the upper mandible. Other families in this group are the albatrosses (Diomedeidae), storm-petrels (Oceanitidae), and fulmars, petrels, shearwaters, and prions (Procellariidae).

With their rapid wing beats and stocky, short-necked appearance, diving-petrels resemble the little auks of the Northern Hemisphere (family Alcidae). This resemblance represents an example of convergent evolution between unrelated species occupying similar ecological niches in widely separated parts of the world.

Physical characteristics

Diving-petrels are small, stocky-bodied, short-winged, tube-nosed seabirds that dive and swim to catch their food. Their body length is 7–10 in (18–25 cm) and they weigh 4–8 oz (120–220 g). Their bill is small, short, broad, and slightly hooked at the tip. The nostril tubes on the upper bill are parallel, short, have a thin partition between them, and are directed upward. Diving-petrels are the only tubenoses in which the nostrils project upward rather than forward, which may be an adaptation to diving. The wings are relatively short and wide and the flight is consequently swift, direct, fluttering, and whirring. When diving and swimming, the wings are used as flippers to achieve forward propulsion. The plumage is gray, blue-gray, or black on top and whitish on the underside. The primary feathers all molt simultaneously, rendering the birds temporarily flightless.

Distribution

Diving-petrels are restricted to waters of the Southern Hemisphere, generally between latitudes 35° south and 60° south. They usually occur in coastal waters but may sometimes be found well offshore.

Peruvian and Magellan diving-petrels (*Pelecanoides garnotii* and *P. magellani*) inhabit South American waters. Common and South Georgian diving-petrels (*P. urinatrix* and *P. georgicus*) are circumpolar species.

Habitat

Diving-petrels breed on remote oceanic islands. They feed in cool and cold oceans, usually rather close to their breeding sites.

Behavior

Diving-petrels characteristically fly low, direct, and fast over water, occasionally diving and swimming to catch their prey. In rough weather, they may fly right through the crests of waves rather than around or over them. Diving-petrels are the only tubenoses that swim underwater using their wings for propulsion. Diving-petrels only come to land to breed, and they will do so only at night. This wariness is an adaptive response to predation by larger seabirds, such as skuas. Diving-petrels are not migratory, but they may wander during the nonbreeding season.

Feeding ecology and diet

Diving-petrels catch their prey of small fishes and crustaceans by flying directly into the water and then using their wings to swim underwater to pursue their food. They emerge

A common diving-petrel (*Pelecanoides urinatrix*) on Karewa Island, New Zealand. (Photo by K. Westerkov/Animals Animals. Reproduced by permission.)

from the water in a similar manner, by flying directly out into the air. Diving-petrels usually feed in flocks.

Reproductive biology

Diving-petrels breed in colonies. They nest in burrows excavated in organic turf and also in cavities among rocks and tufts of grass. Each female lays only one relatively large white egg that weighs 10–15% of the female's body weight. The incubation period is about eight weeks, and both parents tend the egg during one-day-long watches. Egg laying generally occurs between July and December. The newly hatched chick is brooded closely for its first two weeks of life. After about eight weeks the chick fledges and begins to fend for itself. After the breeding season ends, adults molt all flight feathers and are flightless until this plumage has regrown. Diving-petrels reach sexual maturity in two or three years, which is considerably faster than other tubenoses.

Conservation status

The Peruvian diving-petrel is listed as Endangered. This rare species has an extremely small breeding range on only four islands off the west coast of South America, and all of its subpopulations are declining, some quite rapidly. The declines in abundance have been caused by excessive hunting of these birds for food, disturbance of their habitat during guano collection, predation on eggs, nestlings, and adults by introduced mammals, and diminishment of their food supply by commercial overfishing of the waters around their breeding colonies.

Significance to humans

Diving-petrels are not of much importance to humans, except for the economic benefits of marine ecotourism related to birdwatching.

Species accounts

Common diving-petrel
Pelecanoides urinatrix

TAXONOMY
Pelecanoides urinatrix Gmelin, 1789, New Zealand. Six subspecies.

OTHER COMMON NAMES
English: Subantarctic diving-petrel; French: Puffinure plongeur; German: Lummensturmvolgel; Spanish: Potoyunco Común.

PHYSICAL CHARACTERISTICS
8–10 in (20–25 cm); wingspan 13–15 in (33–38 cm). Coloring similar to other *Pelecanoides* species; differentiated by configuration of its bill and nostrils.

Pelecanoides urinatrix

DISTRIBUTION
This is the most widespread of the diving-petrels, occurring in the Southern Ocean between about latitudes 35° south and 55° south. It breeds on islands off Australia, New Zealand, Chile, Argentina, and in the south Atlantic Ocean and south Indian Ocean.

HABITAT
Breeds on oceanic islands and feeds in cool and cold oceans, usually close to breeding sites.

BEHAVIOR
Flies low, direct, and fast, both through the air and in the water.

FEEDING ECOLOGY AND DIET
Dives and swims to feed on small fish and crustaceans.

REPRODUCTIVE BIOLOGY
Lays a single egg in a burrow or crevice. The egg is incubated by both parents.

CONSERVATION STATUS
Not threatened. Locally widespread and abundant.

SIGNIFICANCE TO HUMANS
None other than through economic benefits of birdwatching and ecotourism. ◆

Magellan diving-petrel
Pelecanoides magellani

TAXONOMY
Pelecanoides magellani Gray, 1871, Strait of Magellan. Monotypic.

OTHER COMMON NAMES
English: Magellanic diving-petrel; French: Puffinure de Magellan; German: Magellan-Lummensturmvogel; Spanish: Potoyunco Magallánico.

PHYSICAL CHARACTERISTICS
7.5–8 in (19–20 cm), side of neck bears a crescent-shaped half collar.

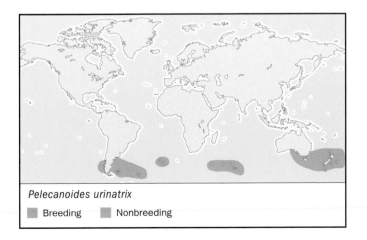

Pelecanoides urinatrix
▨ Breeding ▨ Nonbreeding

Pelecanoides magellani
▨ Breeding ▨ Nonbreeding

DISTRIBUTION
Occurs only in extreme southern South America, off southernmost Argentina and Chile.

HABITAT
Breeds on oceanic islands and feeds in cold oceans, usually close to breeding sites.

Pelecanoides magellani

BEHAVIOR
Flies low, direct, and fast, both through the air and in the water.

FEEDING ECOLOGY AND DIET
Dives and swims to feed on small fish and crustaceans.

REPRODUCTIVE BIOLOGY
Lays a single egg in a burrow or crevice. The egg is incubated by both parents.

CONSERVATION STATUS
A locally abundant species.

SIGNIFICANCE TO HUMANS
None other than through economic benefits of birdwatching and ecotourism. ◆

Resources

Books

BirdLife International. *Threatened Birds of the World*. Barcelona: Lynx Edicions and BirdLife International, 2000.

Carboneras, C. "Family Pelecanoididae (Diving-petrels)." *Handbook of the Birds of the World*. Vol. 1, edited by J. del Hoyo, A. Elliott, and J. Sargatal. Barcelona: Lynx Edicions, 1992.

Harrison, P. *Seabirds. An Identification Guide*. Beckenham, U.K.: Croom Helm Ltd., 1983.

Warham, J. *The Behaviour, Population Biology and Physiology of the Petrels*. San Diego: Academic Press, 1996.

Organizations

BirdLife International. Wellbrook Court, Girton Road, Cambridge, Cambridgeshire CB3 0NA United Kingdom. Phone: +44 1 223 277 318. Fax: +44-1-223-277-200. E-mail: birdlife@birdlife.org.uk Web site: <http://www.birdlife.net>

IUCN–The World Conservation Union. Rue Mauverney 28, Gland, 1196 Switzerland. Phone: +41-22-999-0001. Fax: +41-22-999-0025. E-mail: mail@hq.iucn.org Web site: <http://www.iucn.org>

Bill Freedman, PhD

Sphenisciformes
Penguins
(Spheniscidae)

Class Aves

Order Sphenisciformes

Family Spheniscidae

Number of families 1

Thumbnail description
Medium to large flightless seabirds with
streamlined bodies adapted for swimming and
diving underwater

Size
17.7–51.2 in (45–130 cm); 1.8–88 lb (842
g–40 kg)

Number of genera, species
6 genera; 17 species

Habitat
Marine coastal areas of the southern
hemisphere; one species found at the equator

Conservation status
Endangered: 3 species; Vulnerable: 7 species;
Lower Risk: 2 species

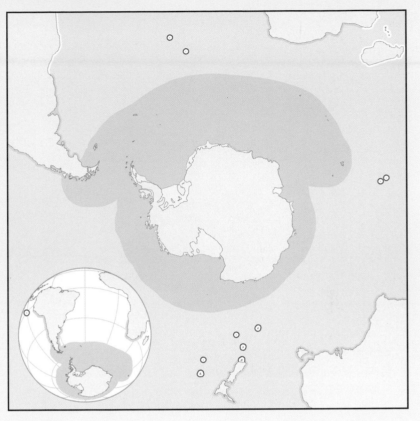

Distribution
Cool waters of the southern hemisphere, including coastal Antarctica, New Zealand,
Australia, South Africa, and South America and the Falkland Islands; one species
occurs at the equator, on the Galápagos Islands

Evolution and systematics

Ninteenth-century French explorer Dumont d'Urville described penguins as "fish-birds" when he first spotted them in Antarctic waters. Like fish, penguins have streamlined, torpedo-shaped bodies and swim easily underwater; however, they are indisputably birds, members of the order Sphenisciformes, which consists of a single family, the Spheniscidae. Their closest living relatives are thought to be petrels and albatrosses (Procellariidae) and loons and grebes (Gaviidae).

Taxonomists concur that penguins probably evolved during the Cretaceous period (140–65 million years ago) from an ancestor that could fly but also swam underwater to catch food. The ancestor might have resembled an auk (Alcidae) or a diving-petrel (Procellariidae).

More than 40 fossil penguin species have been described; a distinctive fused foot bone, the tarsometatarsus, is diagnostic. In a report published in 1990, Ewan Fordyce and C. M.

Jones suggested that mass extinctions of marine reptiles during the late Cretaceous left open ecological niches that penguins evolved to fill. In the period from 40 to 10 million years ago, penguins flourished; species diversity was higher than it is in the twenty-first century, and the penguin fauna included many species even bigger than the largest living species, the emperor penguin (*Aptenodytes forsteri*). By the Miocene period (about 15 million years ago) most of the large species were extinct, perhaps because seals and small whales had evolved and were outcompeting penguins for food.

As of 2001, scientists recognize 17 species of penguins in 6 genera. The genus *Aptenodytes* includes the two largest penguins, emperor penguins and three-foot-tall king penguins (*A. patavonicus*). The genus *Pygoscelis* includes Adelie (*P. adeliae*), chinstrap (*P. antarctica*), and gentoo (*P. papua*) penguins. Adelie penguins, with their black-and-white plumage that suggests a formal tuxedo, are what most people think of when they hear the word penguin. The genus *Eudyptes* consists of

Gentoo penguins (*Pygoscelis papua*) leave the water in the Falkland Islands. (Photo by Tim Davis. Photo Researchers, Inc. Reproduced by permission.)

macaroni (*E. chrysolophus*), rockhopper (*E. chrysocome*), Snares (*E. robustus*), erect-crested (*E. sclateri*), royal (*E. schlegeli*), and Fiordland (*E. pachyrhynchus*) penguins; these are also known as crested penguins because of the wispy yellow feathers that sprout above their eyes. African (*Spheniscus demersus*), Humboldt (*S. humboldti*), Magellanic (*S. magellanicus*), and Galapagos (*S. mendiculus*) penguins make up the genus *Spheniscus*, also referred to as black-footed penguins. Members of this group have curving white stripes on either side of their black heads and a black stripe that forms a horseshoe shape on their white chests. *Spheniscus* penguins are also nicknamed jackass penguins because of their braying calls. The final two penguin genera each consist of a single species: *Eudyptula minor*, little penguins (as its name suggests, the smallest of the penguins), and *Megadyptes antipodes*, yellow-eyed penguins.

Physical characteristics

With their large heads, elongate bodies, and upright stance, penguins look somewhat human as they waddle around on land. The most common plumage, a black back and white chest, evokes a comparison with tuxedoed waiters. All penguins share a set of anatomical features that make them uniquely adapted for life in a marine environment. They are able to dive and maneuver with agility underwa-

ter. They have solid, heavy bones; wings that are modified into stiff, flat flippers; webbed feet set well back on the body; and short, stiff feathers that repel water and provide excellent insulation.

Species vary in size. Little penguins generally weigh less than 3 lb (1100 g) and stand less than 18 in (45 cm) tall; the emperor penguin can be nearly four ft (115 cm) tall, and a male at the beginning of the breeding season may weigh as much as 88 lb (40 kg). The weight of each penguin may vary dramatically over the course of the breeding season; male emperor penguins go without eating for as long as 115 days during courtship and egg incubation and may lose 41% of their initial body weight during this period. Male penguins are slightly larger than females, especially with regard to the length of their flippers and the size of the bill, but except among the eudyptid penguins (in particular the macaroni) this difference can be hard to detect on casual observation.

All penguins have black, blue-gray, or gray feathers on the back and white feathers on the chest and belly. Many species have distinct orange or yellow plumes sprouting from the head or patches of bright yellow or orange on the face. Males and females look similar; chicks are covered with a layer of fluffy down.

Even though penguins are flightless, they have a keeled breastbone like that of flying birds (the keel anchors the pectoral muscles usually used for flight). Penguin bones differ from those of most birds in being solid and heavy instead of light and filled with air spaces; heavy bones are an adaptation for diving underwater. Penguin wings have been modified into flippers for "flying" underwater; the joints at elbow and wrist are almost fused so that the flippers do not fold up the way wings do. Penguin legs are short and stout with webbed feet; underwater, the feet trail behind, pressed against the stiff tail where they serve as a rudder.

In most birds, feathers grow only from certain sections of the skin called feather tracts, while large areas of skin between the tracts are bare. Penguins, on the other hand, have feathers over almost the entire body surface; the exception is the bare brood patch on the belly. Tropical penguins have the largest areas of bare skin to facilitate cooling. The tips of feathers overlap like scales to form a waterproof outer covering, while fluffy down at the base of each feather traps a layer of air that holds in body heat. Most species experience a complete molt annually; they stay on land and fast during the molting period of 13–34 days.

A layer of blubber provides additional insulation, and a heat-exchange system in blood vessels of the flippers and legs helps maintain body temperature while swimming in cold water. One other adaptation to life in the water is the ability to reduce blood flow to the muscles while submerged. How penguins such as emperors dive repeatedly to great depths without developing decompression sickness and nitrogen narcosis is not known.

Distribution

Penguins live almost exclusively in the southern half of the world. A single species, the Galapagos penguin, occurs just north of the equator. Popularly thought to be birds of the Antarctic, penguins are actually widely distributed, and more than half of the 17 species are never found in Antarctica. Most species live between 45 and 60° south. Seven penguin species breed on the mainland and islands of southern New Zealand; other species breed along the subtropical coasts of South America and South Africa. Only four species breed in Antarctica—the emperor, Adelie, gentoo, and chinstrap—and only the emperor and Adelie stay in the Antarctic year-round.

Habitat

Penguins spend much of their time at sea, diving underwater to catch fish, crustaceans, and squid. Like marine mammals, however, they must go ashore to rest and to breed and rear their young. Most breeding colonies are within a few hundred yards of shore, although gentoo and king penguin colonies can be almost 2 mi (3 km) inland. Breeding habitats range from the snowfields and ice sheets of Antarctica (where male emperors cradle their eggs on their feet rather than in a nest) to the famous equatorial islands off the coast of Ecuador, where Galapagos penguins breed in lava fields. Most species establish colonies in open, level terrain, often beneath coastal cliffs, although macaroni and chinstrap penguins nest on rocky slopes. Gentoo penguins nest amid mounds of tus-

An Adelie penguin (*Pygoscelis adeliae*) toboggans at Petermann Island, Antarctica. (Photo by Renee Lynn. Photo Researchers, Inc. Reproduced by permission.)

sock grass, and all of the *Spheniscus* penguins nest in protected places, either in underground burrows or beneath bushes. In Southern Chile, Magellanic penguins go ashore to lay their eggs in coastal beech forests.

Behavior

Penguins are highly social birds. They gather to breed in small groups or in large, noisy colonies, and they take to the water in flocks. Penguins interact constantly with their neighbors and they have evolved a large repertoire of complex behaviors that allow them to appease aggression, court a mate, and recognize the mate and offspring amid the throngs of birds.

To avoid aggression when entering or leaving the colony, a penguin adopts the "slender walk" behavior, lowering its head and holding its flippers forward as it threads its way past other birds. Many species use a sideways stare to signal "Keep away." Fights break out despite such submissive and defensive behaviors, and penguins will peck, bite, and hit opponents with their flippers. Some species engage in ritualized bill-jousting, using the bill like a sword to attack and parry.

A male penguin claiming a nest site puts on an "ecstatic display," standing erect with his bill pointed skyward while waving his flippers and calling loudly. When a female joins her mate, they cement the pair bond with a mutual ecstatic display and by bowing to one another. Among the smaller penguin species, courting pairs also engage in mutual preening.

Mutual displays and bowing continue after the pair have mated and one or two eggs are laid; these behaviors constitute a "nest relief ceremony" conducted when the male and female change places on the nest. Adult birds recognize one another

An adult rockhopper penguin (*Eudyptes chrysocome*) grooms a chick in the Falkland Islands. (Photo by Rod Planck. Photo Researchers, Inc. Reproduced by permission.)

by these behaviors and also by voice. Species that form the largest colonies (pygoscelids and some eudyptids) have the most elaborate mate-recognition displays. Experiments with recorded calls have shown that king penguins respond to their mate's calls but not to those of neighbors or other colony members, and chicks recognize their parents by a distinct vocal signature based on frequency modulation.

When penguins head to sea to forage, they typically stay in groups rather than hunt alone. This way each individual reduces its risk of being eaten. Foraging flocks may also be more efficient at finding food than are solitary birds.

Swimming penguins sometimes progress by porpoising, or shooting out of the water to skim above the surface for a few feet before splashing back down. While porpoising, the birds grab a breath in mid-air. On land, penguins walk upright with a waddling gait; some also progress by two-footed jumps (this method of locomotion gives the rockhopper penguin its name). Penguins can also travel by tobogganing, or sliding on their bellies over ice and snow. Some species travel hundreds of miles to inland nesting sites this way.

Feeding ecology and diet

Penguins feed at sea by diving after prey, which may be small fish, crustaceans, or squid. A bird can swallow a large number of prey items before it has to return to the surface to breathe. Different species take different prey; crested penguins (eudyptid species) eat mostly krill and other small crustaceans that occur in dense swarms. *Spheniscus* penguins and the little penguin eat small fish such as anchovies and sprats. Pygoscelid penguins eat almost nothing but krill.

Because penguins pursue their prey underwater, often in very cold water, few humans have seen a penguin capture prey. Nonetheless, using radio and satellite telemetry and miniaturized instruments that record depth, swimming speed, and duration of dives, scientists have learned much in the past decade about penguin foraging behavior. Penguins make shallow dives as they move from the shore to a foraging area, resting on the surface between dives. When they are pursuing

prey, they dive deeper and stay underwater longer. The longest documented penguin dive was 18 minutes for an emperor penguin. Emperors are also the deepest divers; one researcher documented a bird that reached a depth of 1755 feet (535 meters). Diving ability seems to be positively correlated with body size: king penguins can dive for seven to 10 minutes, most medium-sized penguins dive for three to six minutes, and little penguins rarely dive for more than a minute or deeper than 98 ft (30 m). Penguins dive deepest at midday; they rely on excellent vision to spot prey, so low light conditions at dawn and dusk probably limit them to shallower dives at these times.

Reproductive biology

Penguins do not breed until they are at least two to five years old. For example, gentoo, little, and yellow-eyed penguins attempt to breed at age two; king and emperor penguins delay breeding until they are at least three years old; and macaroni and royal penguins wait at least five years. Females are ready to breed at a younger age than males. Penguins are usually monogamous and may take the same mate year after year; however, extra-pair copulations do occur among Humboldt penguins and others.

Emperor and king penguins build no nest and simply hold the egg on their feet. Gentoo penguins build nests of stones, and the spheniscids and little penguins dig burrows and nest underground.

The two largest penguin species lay a single egg. Other species usually have a two-egg clutch but occasionally lay one or three eggs. Crested penguins lay two eggs but rarely raise two offspring; the first egg laid is usually smaller than the second and is often lost or destroyed before it hatches.

Incubation period varies among species from 33 to 64 days. Chicks in the same clutch hatch at the same time or within a

A pair of chinstrap penguins (*Pygoscelis antarctica*) in courtship at their nest made of pebbles. (Photo by George Holton. Photo Researchers, Inc. Reproduced by permission.)

day of each other. One parent broods the down-covered chicks while the other goes on a foraging trip; upon return, the parent regurgitates food for the chicks. When the chicks no longer require brooding to maintain body temperature, the parents continue to guard them from predators.

Parental care can be quite extended; for example, king penguins care for chicks for more than 12 months. Young birds are ready to leave the nest when their down is replaced by feathers. Juveniles look much like adults, though the species-specific crests or cheek patches are less bright in young birds. After young birds leave the nest, they jump in the water and pursue prey without having had any obvious training from their parents in how to forage.

Conservation status

The exact impact of human actions on penguin populations is hard to measure because these birds naturally experience dramatic population fluctuations when changing ocean conditions affect food availability. From the eighteenth century to the early twentieth century, crews of whaling and sealing ships took millions of penguins and their eggs for food and also used the birds as bait. In places, penguins were rendered to produce oil; king penguins are thought to have been eradicated on Heard Island for this reason. Human enterprise has caused other problems for penguins; for example, Humboldt penguins in South America prefer to dig breeding burrows in centuries-old mounds of accumulated guano, but most of these breeding sites have been mined for fertilizer. Among the modern problems penguins face are oil pollution from tanker spills and bilge flushing and entanglement in discarded fishing nets. Commercial fishing may also reduce food supplies for some populations. Where human populations concentrate near penguin habitat, domestic animals can be a problem: cattle and sheep trample burrows and nests; rabbits browse away the concealing vegetation around nests; and introduced species including dogs, feral cats, ferrets, and stoats prey on nestlings.

Twelve of 17 penguin species are included on the 2000 IUCN Red List of Threatened Species. Erect-crested, Galapagos, and yellow-eyed penguins are listed as Endangered. The yellow-eyed penguin, found in New Zealand, has the smallest population of any species; much of its habitat has been lost to logging and farming, and introduced predators are also a problem. Rockhopper, macaroni, Fiordland, Snares, royal, African, and Humboldt penguins are listed as Vulnerable. Two species are classified as Lower Risk: gentoo penguins and Magellanic penguins.

Significance to humans

During the era of sailing ships, penguins were taken by the hundreds of thousands, for food and for the extraction of oil. Eggs were also collected in large numbers, and penguin guano was mined for fertilizer. Though penguins are now protected in most countries, they are sometimes taken illegally for food and bait. Since the 1960s, penguin colonies in the Antarctic, Argentina, and the Galapagos have become tourist attractions, drawing thousands to tens of thousands of visitors. Penguins are also popular as advertising logos (most notably for books, coffee, and cigarettes), as hockey team mascots, and as characters in cartoons and children's books.

1. Emperor penguin (*Aptenodytes forsteri*); 2. Yellow-eyed penguin (*Megadyptes antipodes*); 3. Macaroni penguin (*Eudyptes chrysolophus*); 4. Little penguin (*Eudyptula minor*); 5. Adelie penguin (*Pygoscelis adeliae*); 6. Magellanic penguin (*Spheniscus magellanicus*). (Illustration by Patricia Ferrer)

Species accounts

Emperor penguin

Aptenodytes forsteri

TAXONOMY
Aptenodytes forsteri G. R. Gray, 1844, Antarctic Seas.

OTHER COMMON NAMES
French: Manchot empereur; German: Kaiserpinguin; Spanish: Pingüino Emperador.

PHYSICAL CHARACTERISTICS
39.4–51.2 in (100–130 cm); female weight 44.5–70.5 lb (20.2–32 kg); male 48.3–88 lb (21.9–40 kg). The largest penguin is about the same size as the smallest diving marine mammal, the Galapagos fur seal (*Arctocephalus galapagoensis*). Bright yellow ear patches contrast sharply with black head, chin, and throat. Back is dark blue-gray, underparts are white shading to pale yellow on upper breast. Bill is slender and down-curving. Eyes are brown. Upper bill is black and lower bill is pink, orange, or lilac. Feet and legs are black. Juvenile is similar to adult but smaller and duller, with white ear-patches and black bill.

DISTRIBUTION
Breed on the coast of the Antarctic continent and adjacent islands, from 66° to 78° south latitude. Rarely seen outside of the Antarctic, although migrating birds are occasionally spotted near the Falkland Islands, southern New Zealand, and southern South America.

HABITAT
Cold waters of the Antarctic zone, where pack ice forms. Usually breed on sea ice, often on level sites sheltered by ice cliffs.

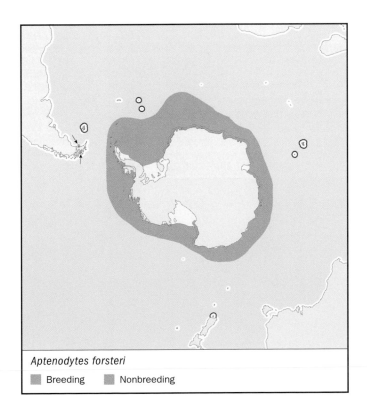

Aptenodytes forsteri

■ Breeding   ■ Nonbreeding

BEHAVIOR
Less aggressive than some penguins and behavioral repertoire is less varied, perhaps because incubating males do not defend territories but instead huddle together for warmth. Nest colonially and forage in groups. Loud vocalizations characterized as trumpeting. Horizontal head-circling signals aggression but is also common during pair formation, copulation, egg-laying, and as part of nest-relief ceremony.

FEEDING ECOLOGY AND DIET
Birds appear to coordinate their foraging at sea, diving and surfacing as a group. Main prey type varies with location; in a 1998 study, small fish made up more than 90% of the diet in three locations. Antarctic silverfish (*Pleuragramma antarcticum*) were the main prey item, and small cephalopods and crustaceans were also taken. About a third of all dives are deeper than 330 ft (100 m); birds sometimes dive as deep as 1,480 ft (450 m), and may feed near the sea bottom. Birds also feed near the surface along underside of ice where crustaceans gather to graze on algae. May travel 90–620 mi (150–1,000 km) in a single foraging trip.

REPRODUCTIVE BIOLOGY
Less mate-faithful than smaller penguins. After laying a single, large, greenish-white egg, females return to sea to feed. Males incubate alone, fasting for up to 115 days (from arrival at breeding colony to end of incubation, which lasts for 64 days). Chick has comical appearance, with black-and-white head emerging from what looks like a brown fur coat enveloping the body (actually, a layer of insulating down). Females return soon after chicks hatch and parents alternate feeding and brooding duties for 45–50 days. Chicks then form crèches (large numbers of young birds huddle together for warmth, standing close enough to touch one another). They are independent at 150 days. Adults molt after chicks leave colony.

CONSERVATION STATUS
Not threatened. Population stable or increasing; total breeding population was estimated in 1993 to be 314,000 pairs. Susceptible to human disturbance but at present face no major threats.

SIGNIFICANCE TO HUMANS
Emperor penguins are a key attraction on Antarctic ecotours, and also at Sea World in San Diego, where the Penguin Encounter exhibit is the world's only successful emperor penguin breeding colony outside of Antarctica. ◆

Adelie penguin

Pygoscelis adeliae

TAXONOMY
Catarrhactes adeliae Hombron and Jacquinot, 1841, Adelie Land.

OTHER COMMON NAMES
French: Manchot d'Adélie; German: Adeliepinguin; Spanish: Pingüino Adelia.

PHYSICAL CHARACTERISTICS
Female weight 8.6–10.5 lb (3,890–4,740 g); male 9.6–11.8 lb (4,340–5,350 g). Back, tail, and head (including face) are

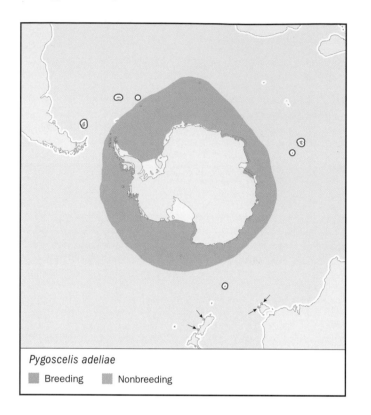

Pygoscelis adeliae
■ Breeding   ■ Nonbreeding

blue-black; underparts are white. Distinctive white eye ring. Feathers cover half of bill, which is black with orange base. Eyes are brown. Legs and feet are dull white to pink with black soles.

DISTRIBUTION
Circumpolar, associated with pack ice of Antarctic Zone. Breeds on coasts of the Antarctic continent and surrounding islands; non-breeding distribution is not well known.

HABITAT
Within home range, they breed wherever land is ice-free and access from the sea is feasible.

BEHAVIOR
Male and female defend territory vigorously; often fight with neighbors. Birds signal apprehension by raising head feathers. Common threat display is a sideways stare with crest raised and eyes rolled downward.

FEEDING ECOLOGY AND DIET
Take mostly krill but also fish and cephalopods. During incubation, the bird not tending the nest may make a very long foraging trip, traveling more than 93 mi (150 km) from the colony over the course of 9–25 days. One study of birds at Hope Bay documented a maximum dive of 558 ft (170 m); estimated prey capture rate was 1,150 krill per foraging trip (7.2 krill per minute).

REPRODUCTIVE BIOLOGY
Well studied. Monogamous, often return repeatedly to same nest site. Nest in large colonies of up to 200,000 pairs. Build nests of small stones. Considerable energy devoted to stone searching, stone stealing, and rearranging stones in nest. Two eggs laid; parents alternate incubation duties (sometimes with egg on feet) for 32–24 days. Young brooded for 22 days, then

join small crèches; fed by parents until they leave colony at 50–60 days.

CONSERVATION STATUS
Not threatened. Stable or increasing; population estimated in 1993 at 2,610,000 breeding pairs. Susceptible to disturbance from human activity.

SIGNIFICANCE TO HUMANS
Ornithologist Robert Cushman Murphy called Adelies "the type and epitome of the penguin family." Adelies are responsible for the habitual comparison of penguins to little men in evening clothes. ◆

Macaroni penguin
Eudyptes chrysolophus

TAXONOMY
Catarrhactes chrysolophus Brandt, 1837, Falkland Islands.

OTHER COMMON NAMES
English: Crested penguin, royal penguin; French: Gorfou doré; German: Goldschopfpinguin, Haubenpinguin; Spanish: Pengüino Macarrones.

PHYSICAL CHARACTERISTICS
27.9 in (71 cm); male weight 8.2–14.1 lb (3,720–6,410 g); female weight 7.0–12.6 lb (3,180–5,700 g). Comical appearance, with long, yellow and orange plumes like shaggy eyebrows growing from a patch in the center of the forehead. Males noticeably larger than females but plumage similar. Head and cheeks are black or dark gray; back is slate black with blue sheen; breast, belly, and rump patch are white. Bill is stout and dark orange-brown, often ridged in older birds. Eyes are garnet red. Juveniles are smaller than adults and have lighter plumage, smaller and more scattered crest feathers, and a more slender bill.

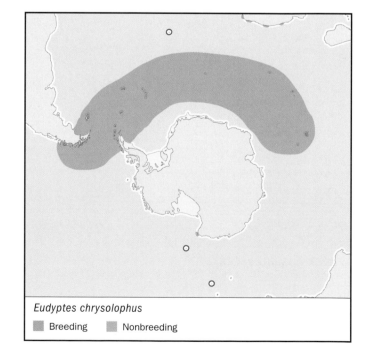

Eudyptes chrysolophus
■ Breeding   ■ Nonbreeding

DISTRIBUTION
Breeds farther south than other eudyptids, on Antarctic Peninsula and on Antarctic and subantarctic islands. In non-breeding season, probably remains in subantarctic waters.

HABITAT
Often nests on steep, rough terrain with little or no vegetation, including lava flows and scree slopes.

BEHAVIOR
Forms colonies of 100 to more than 100,000 birds. Birds on neighboring nests often fight by bill-jousting. A courting male collects pebbles and places them at female's feet. Mated pairs engage in mutual preening of feathers. Very noisy and aggressive in breeding colonies; after females begin incubating, males go to sea and females may then be attacked by unmated males. When birds return from foraging, harsh braying calls are essential for recognition between mates and between parent and chick. Parents take turns incubating and brooding young.

FEEDING ECOLOGY AND DIET
Typically dives to 66–330 ft (20–100 m) in pursuit of prey; mean dive duration 1.48 minutes. In one study, estimated prey capture rates were 4.0–16.0 krill per dive and up to 50 amphipods per dive. At first, young are fed krill exclusively; gradually, small fish and squid are added to the diet. In the last week before the chick becomes independent, parents feed it only fish and squid.

REPRODUCTIVE BIOLOGY
Female alone scrapes out depression to serve as nest; both members of the pair line nest with pebbles. Eggs rough-textured with faint blue tinge. Egg laying tightly synchronized within colonies; first-laid egg is small, weighing 61–64% of second egg, which is laid 3.2 days later. First egg almost always lost or destroyed; if second egg is destroyed, normal, healthy chick may hatch from first egg. Incubation period is 35–37 days from laying of second egg. Males guard chicks for about 20 days after hatching, when young birds form crèches. Both parents continue to feed their offspring until independence at 60–70 days.

CONSERVATION STATUS
Listed as Vulnerable because the world population appears to have decreased by at least 20% over a 36-year period. Designation was based on extrapolation from a small amount of data, so large-scale surveys will be needed to confirm this penguin's status.

SIGNIFICANCE TO HUMANS
Communities in the Falkland Islands formerly observed November 9 as a holiday on which children were excused from school to collect the eggs of macaroni and other penguins. ◆

Magellanic penguin
Spheniscus magellanicus

TAXONOMY
Aptenodytes magellanicus J. R. Forster, 1781, Strait of Magellan.

OTHER COMMON NAMES
French: Manchot de Magellan; German: Magellanpinguin; Spanish: Pingüino de Magallanes.

PHYSICAL CHARACTERISTICS
28 in (71 cm); female weight 5.9–9.0 lb (2.7–4.1 kg), male 6.4–10.6 lb (2.9–4.8 kg). Boldly striped penguin with two black

Spheniscus magellanicus
■ Breeding   ■ Nonbreeding

bands across chest. Sexes similar but female smaller than male. Cheeks and cap brownish black, divided by wide white ring. Black back and white underparts lightly splotched with black. Stubby black bill with gray band near tip. Eyes are brown. Feet are pink blotched with black. Juveniles smaller than adults and breast bands are not distinct.

DISTRIBUTION
Breeds in coastal areas and on offshore islands along the southern part of South America, occasionally in southern coastal Australia and New Zealand. Breeding distribution seems to be moving northward. Migratory outside of breeding season; birds from colonies at the tip of south America may travel as far north as Peru and southern Brazil.

HABITAT
Breeds on bare or vegetated islands, in flat areas and on cliff faces. Colonies located in areas where offshore winds cause upwelling of deep, cold, nutrient-rich waters so that primary productivity is high. Birds have best nesting success in protected sites under bushes or other vegetation. Feed inshore during breeding season and in pelagic waters during migration.

BEHAVIOR
Foraging birds may be seen porpoising in long lines, one after the other. Breed in large colonies; often return to the same nest site from year to year. Voice described as a mournful, donkey-like braying; often call in chorus at night.

FEEDING ECOLOGY AND DIET
Eat mostly small, schooling fish such as anchovies, sardines, and sprats but diet varies depending on prey availability. Most dives descend 66–164 ft (20–50 m). Underwater swimming speed measured at about 4.7 mi/hr (7.6 km/hr).

REPRODUCTIVE BIOLOGY

Where soil allows digging, they nest in burrows; otherwise they build nests on the ground. Both members of pair build nest. Two-egg clutch is laid and eggs are of similar size. Parents share incubation, brooding, and guarding duties. Chick hatched from second-laid egg less likely to survive to fledging. Chicks independent at 60–70 days.

CONSERVATION STATUS

Listed as Near Threatened. A 1994 study estimated that oil pollution kills 40,000 penguins a year on the southern coast of Argentina.

SIGNIFICANCE TO HUMANS

In the past, native peoples killed penguins for meat and for their skins. Today, Magellanic penguin rookeries at Punta Tombo in Argentina are a major ecotourism destination, attracting 50,000 visitors a year. ◆

Yellow-eyed penguin
Megadyptes antipodes

TAXONOMY

Catarrhactes antipodes Hombron and Jacquinot, 1841, Auckland Islands.

OTHER COMMON NAMES

French: Manchot antipode; German: Gelbaugenpinguin; Spanish: Pingüino de Ojos Amarillos.

PHYSICAL CHARACTERISTICS

22.0–30.7 in (56–78 cm); female weight 9.3–15.5 lb (4,200–7,500 g); male 9.70–18.7 lb (4,400–8,500 g). The only penguin with yellow eyes; band of yellow feathers extends from one

edge of mouth to the other, passing through the eyes and around the nape of the head. Head feathers are yellow with a central black streak; back and tail are slate blue; flippers are darker blue; and breast and belly are white. Long slender bill is red-brown above cream shading to red-brown below. Pale pink feet turn magenta with exertion. Juveniles lack yellow band of feathers around nape.

DISTRIBUTION

Endemic to New Zealand and nearby smaller islands. Most birds winter on or near the breeding grounds.

HABITAT

Breed in coastal areas of southern New Zealand and neighboring subantarctic islands. Birds stay near breeding sites year round, except for juveniles that move north to feeding grounds for a few months after fledging. Nest from sea level to elevation of 820 ft (250 m) on sea-facing, forested slopes and cliff tops, usually amid dense forest vegetation. Probably choose cool, shady forests to avoid overheating.

BEHAVIOR

Gregarious in winter (non-breeding season); roost communally on flat, open ground and gather in groups of 50–100 on beaches before departing for foraging areas. They forage alone at sea. Secretive and especially wary of humans. During breeding season they come ashore at night and negotiate difficult terrain to reach cliff-top breeding areas. The least colonial of penguins when breeding; nests are clustered together only because appropriate habitat is limited. Calls are less harsh than those of other penguins; Maori name, hoiho, means "noise shouter."

FEEDING ECOLOGY AND DIET

Eat mostly fish, some squid, and rarely crustaceans. While one parent guards chick, off-duty bird heads to sea to forage in afternoon and returns at dusk. Outside of the breeding season, most birds head to sea at dawn and return before dark.

REPRODUCTIVE BIOLOGY

Nest out of sight of nearest neighbors amid dense vegetation. Prefer hardwood (*Podocarpus*) forests (where nest sites are at the base of trees or alongside fallen longs) but also nest in fields of tussock grass. Nesting territory may be defended year-round. Nest is a shallow bowl of twigs and other plant matter constructed by both parents. Two eggs laid, three to five days apart. Parents alternate incubation shifts of one to seven days during 39–51 day incubation period and also take turns brooding chicks for four to six weeks.

CONSERVATION STATUS

A 1990 study indicated the population had declined at least 75% over 40 years. Changed from Vulnerable to Endangered; total breeding population estimated to be fewer than 2,000 pairs. Breeding range is very small, and habitat has been degraded, especially by clearing of hardwood forests for farming. In addition, cattle trample nests and introduced ferrets, stoats, and feral cats are significant predators. Adults also caught and killed accidentally in fishing nets. Ongoing conservation efforts may be starting to reverse population decline.

SIGNIFICANCE TO HUMANS

Yellow-eyed penguins have become a figurehead species of the New Zealand environmental movement. ◆

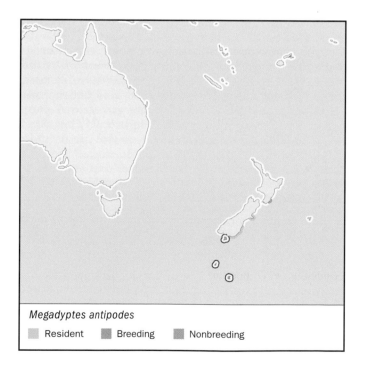

Megadyptes antipodes

■ Resident   ■ Breeding   ■ Nonbreeding

Little penguin
Eudyptula minor

TAXONOMY
Eudyptula minor J. R. Forster, 1781, Dusky Sound, South Island, New Zealand.

OTHER COMMON NAMES
English: Fairy penguin, little blue penguin, white-flippered penguin; French: Manchot pygmée; German: Weissflügelpinguin, Zwergpinguin; Spanish: Pingüino Pequeño.

PHYSICAL CHARACTERISTICS
15.7–17.7 in (40–45 cm); weight 2.2 lb (1 kg). The smallest penguin; male larger than female. Indigo-blue above, white below. Eyes are gray to hazel. Stout black bill is slightly hooked. Feet are white above with black soles. Juveniles similar to adult but smaller and with slimmer bill; plumage brighter than that of adults.

DISTRIBUTION
Southern coast of Australia; coastal New Zealand; offshore islands.

HABITAT
Temperate inshore waters; often seen in bays and estuaries. Often breeds in secluded bays, promontories, or islands, often at the base of cliffs. Prefers flat areas with protective vegetation. Nests in burrows but also under rocks, in caves, and under mounds of tussock-grass. Has adapted to nest around humans, including under houses and in culverts, and will also use artificial burrows. Require the shelter of burrows or rocks or bushes during molt.

BEHAVIOR
Colonial; adults reside at breeding sites year-round. Typically forage within 0.6 mi (1 km) of shore but may travel farther. Mated pairs stay together year-round. Roost alone or in pairs, often in burrows. The most nocturnal of all penguins. Calls include short yaps, grunts, trilling, and braying.

FEEDING ECOLOGY AND DIET
Prefer small fish or cephalopods. When swimming underwater, a bird will circle a school several times and then plunge through its middle.

REPRODUCTIVE BIOLOGY
Both parents dig the burrow and build the nest; they also share incubation and feeding duties. Nest built of grass and other

Eudyptula minor
■ Breeding ■ Nonbreeding

plant material. Two eggs laid over three to five days. Parents accept eggs other than their own and have been seen to incubate stones, golf balls, and teacups. Chicks are brooded for 10 days and are guarded for another 10–21 days.

CONSERVATION STATUS
Not threatened; however, populations described as stable or decreasing. Housing developments and farmland have replaced many breeding areas. Face predation from introduced foxes and dogs; also, livestock trample nesting sites and rabbits eat protective vegetation around nests. Erosion and run-off from agriculture affects marine water quality, which can reduce food supply and also increase rates of disease.

SIGNIFICANCE TO HUMANS
Little penguins returning from a night's fishing form a parade that is a popular tourist attraction on resort beaches. ◆

Resources

Books
Davis, L. S., and J. T. Darby, eds. *Penguin Biology.* New York: Academic Press, 1990.

Marchant, S., and P. J. Higgins, eds. *Handbook of Australian, New Zealand, and Antarctic Birds.* Vol. 1, *Ratites to Ducks.* New York: Oxford University Press, 1990.

Marion, R. *Penguins: A Worldwide Guide.* New York: Sterling Publishing Co.,1999.

Reilly, P. *Penguins of the World.* New York: Oxford University Press, 1994.

Williams, Tony D. *The Penguins: Spheniscidae.* New York: Oxford University Press, 1995.

Periodicals
Bried, J., F. Jiguet, and P. Jouventin. "Why do *Aptenodytes* Penguins Have High Divorce Rates?" *Auk* 116 (1999): 504–512.

Cherel, Y., and G. L. Kooyman. "Food of Emperor Penguins (*Aptenodytes forsteri*) in the Western Ross Sea, Antarctica." *Marine Biology Berlin* 130 (1998): 335–344.

Gandini, P., P. D. Boersma, E. Frere, M. Gandini, T. Holik, and V. Lichtschein. "Magellanic Penguins (*Spheniscus magellanicus*) Are Affected by Chronic Petroleum Pollution Along the Coast of Chubut, Argentina." *Auk* 111 (1994): 20–27.

Green, K., R. Williams, and M. G. Green. "Foraging Ecology and Diving Behavior of Macaroni Penguins *Eudyptes*

Resources

chrysolophus at Heard Island." *Marine Ornithology* 25 (1998): 27–34.

Jouventin, P., T. Aubin, and T. Lengagne. "Finding a Parent in a King Penguin Colony: The Acoustic System of Individual Recognition." *Animal Behavior* 57 (1999): 1175–1183.

Kent, S., J. Seddon, G. Robertson, and B. Wienecke. "Diet of Adelie Penguins *Pygoscelis adeliae* at Shirley Siland, East Antarctica, January 1992." *Marine Ornithology* 26 (1998): 7–10.

Lengagne, T., T. Aubin, P. Jouventin, J. Lauga. "Perceptual Salience of Individually Distinctive Features in the Calls of Adult King Penguins." *Journal of the Acoustical Society of America* 107 (2000):508–516.

Schwartz, M. K., D. J. Doness, C. M. Schaeff, P. Majluf, E. A. Perry, and R. C. Fleischer. "Female-Solicited Extra-Pair Matings in Humboldt Penguins Fail to Produce Extra-Pair Fertilizations." *Behavioral Ecology* 10 (1999): 242–250.

Scolaro, J. A., R. P. Wilson, S. Laurenti, and M. Kierspel. "Feeding Preferences of the Magellanic Penguin over its Breeding Range in Argentina." *Waterbirds* 22 (1999): 104–110.

Stokes, D. L., and P. D. Boersma. "Nest Site Characteristics and Reproductive Success in Magellanic Penguins (*Spheniscus magellanicus*)." *Auk* (1998) 115: 34–49.

Cynthia Ann Berger, MS

Gaviiformes
Loons
(Gaviidae)

Class Aves
Order Gaviiformes
Family Gaviidae
Number of families 1

Thumbnail description
Medium to large-sized, foot-propelled diving waterbirds that feed mainly on fish. Foot placement is far posterior, making walking on land difficult. Bills are sharp and dagger-like. Alternate (breeding) plumage is boldly patterned primarily with black, white, and gray; nonbreeding plumages are drab gray-brown and white. All have brilliant red iris in alternate plumage, and distinctive eerie vocalizations

Size
20.8–35.8 in (53–91 cm); 2.2–14.1 lb (1.0–6.4 kg); males slightly larger than females in all species

Number of genera, species
1 genus; 5 species

Habitat
Breed in forested and tundra lakes and ponds, winter at sea and large reservoirs

Conservation status
No species considered Endangered or Threatened

Distribution
Holarctic

Evolution and systematics

The five extant species currently recognized are descendants of an ancient bird lineage. The extinct genus *Colymboides* arose in the late Eocene to early Miocene. *Gavia* appeared in the Miocene, and radiated into three size classes by the early Pliocene. Biochemical analyses suggest that loons are most closely related to penguins (Sphenisciformes), tubenoses (Procellariiformes), frigatebirds (Fregatidae), and possibly auks and gulls (Charadriiformes). Traditionally loons have been grouped with grebes (Podicipediformes) because the two orders are convergent. Within the family, common (*Gavia immer*) and yellow-billed (*G. adamsii*) loons are very closely related, as are Arctic (*G. arctica*) and Pacific (*G. pacifica*) loons. Red-throated loons (*G. stellata*) are considered less closely related to the other four species, but phylogenetic relationships are uncertain and controversial.

Physical characteristics

Loons are medium to large, foot-propelled diving waterbirds, with anatomy specialized for pursuit and capture of fish. Overall shape is very distinctive with a short neck, pointed wings, and legs set far back on the body. Loon tarsi are flat-

tened and knife-like, cutting through the water efficiently. The feet are large and palmate with the front three toes webbed, and a free hallux. Loon feet, legs, and nails are uniquely countershaded such that when swimming white surfaces, the tops of their feet are oriented down, helping the bird blend in against light sky. Bills are medium-sized and dagger-like. All species have a distinct blood red iris in alternate plumage. Sexes are similar, with the male larger than the female in all species. Each species has four plumages after they are fully grown: juvenal, second alternate (second summer), basic (nonbreeding), and alternate (breeding). Molt times are slightly different for all species.

In alternate plumage all species have striking patterns composed of black, white, and gray. Upperparts are dark gray or black, with faint white speckling to bold white checkering. Underparts are completely white. Each species has a series of bold, thin, black-and-white parallel stripes on the neck. Head patterns are similar in Arctic/Pacific loons and common/yellow-billed loons. The head pattern of red-throated loon is unique: it is the only loon to have brick red in its plumage.

All loons have similar juvenal, second alternate, and basic plumages. Upperparts are gray-brown and underparts are

Red-throated loon (*Gavia stellata*) turning eggs during incubation in Yukon Delta National Wildlife Refuge, Alaska. (Photo by Stephen J. Krasemann. Photo Researchers, Inc. Reproduced by permission.)

white as a rule; head pattern is slightly different for all species. Juvenal and second alternate birds appear more scaly than adults due to the presence of pale fringes on many contour feathers. Identification of non-breeding birds can be difficult, and general body shape, bill shape and size, and posture can be more helpful than plumage differences for identification.

Molt is complicated and not yet fully understood. Loons have one complete pre-basic molt in fall and one partial pre-alternate molt in spring. The complete molt in fall is prolonged, and individuals may appear to be molting much of the year. Due to the high wing loading in loons, flight is impossible with the loss of a few primaries. Instead of a gradual molt, which would leave them flightless for months, loons have evolved to molt all of their primaries simultaneously, which only leaves them flightless for a couple of weeks. This molt occurs during winter, from about January to April in all species but the red-throated loon, which is small enough that it can fly during a gradual molt of flight feathers. First-year birds also molt primaries simultaneously in their first summer on salt water.

Distribution

Holarctic in distribution, loons breed from north temperate areas to the high arctic. All species migrate up and down coasts and across land to winter primarily at sea, south to coastal Baja California, Gulf of Mexico, Mediterranean Sea, and coastal China. Loons may stage on inland lakes and rivers during migration. Nonbreeding birds will often spend summer in their species' winter range.

Habitat

Loons breed in freshwater inland lakes and tundra ponds. Where sympatric, different species occupy different-sized lakes. Larger species exclude smaller species from breeding ponds, but are limited to larger ponds by minimum take-off distances. Smaller species can occupy ponds too small for larger species. Individuals usually spend winter near shore in oceans and seas, less than 62 mi (100 km) offshore. They are occasionally found in large freshwater lakes and rivers during winter.

Behavior

Loons are extremely territorial on their breeding grounds. Pairs have been observed attacking their own species, as well as other species of loons, ducks, and geese. In one study, 50% of common loons had healed fractures believed to have been caused by the bill of other common loons. Loon pairs have a series of territorial threat postures and calls to prevent fighting, which can be fatal. In winter and during migration loons may be found singly or in loose flocks.

Loons are known for their unusual vocalizations, described as yodels, wails, and tremolos. Most vocalizations are given on the breeding grounds, occasionally during winter, or on migration. Territorial yodels, most often given at night or early morning, can be heard from great distances (reports have been made of calls heard up to 16 mi [25.8 km] by humans). These male yodels are individually recognizable throughout the birds' lifetime.

Flight is powerful and direct, with wings beating constantly. Most species require running on the surface of water to gain speed for take-off. The red-throated loon is the only species that can take off from land. All species are awkward on land; the posterior positioning of the feet often forces them to push themselves along breast first on their bellies. Occasionally loons accidentally land on wet pavement after mistaking it for water, or are forced to land during storms. In this case, they are stranded and will most likely die, although adults in this situation have traveled over a kilometer to seek water at the expense of broken toes and wrists.

The agility of loons on water compensates for their inadequacies on land. Posterior placement of the feet allows for powerful swimming and diving capabilities. Adults can stay submerged for several minutes and can travel hundreds of meters underwater. Loons dive with a forward thrust using both feet simultaneously to propel them through water. Occasionally, wings are also used to supplement the feet underwater, and they can use one foot as a rudder when turning.

Feeding ecology and diet

Loons take a variety of vertebrate and invertebrate foods, but small to medium-sized fish up to about 7–8 in (18–20 cm) are the primary items. Young are also fed crustaceans, mollusks, and worms. Prey is located by sight from the surface of the water. Loons peer downward, often with their bill in the water, and dive with a thrust from both feet. Most prey is eaten underwater; fish that are too large to be handled underwater are taken to the surface. Most foraging is done close to the surface, but loons may forage as deep as 230 ft (70 m) if the water is clear enough. Very small serrations on the bill help loons hold onto prey. Adults and young consume large quantities of prey; an adult common loon consumes 1,214 kilocalories a day, and a pair may eat 2,000 lb (910 kg) of fish in a breeding season. Loons have large salt glands that remove excess salt consumed from marine environments.

Stomach contents have shown that loons consume a wide variety of fish, including sticklebacks (*Gasterosteus*), trout (*Salvelinus*), sculpin (*Leptocottus*), cods (*Gadus*), herrings (Clupeiidae), haddock (*Melanogrammus*), whitefish (*Coregonus*), capelins (*Mallotus*), minnows (Cyprinidae), and many other species. At sea, menhaden (*Brevoortia*) are extremely important. Little is known about prey selection.

Reproductive biology

Loons are monogamous, although they will quickly replace a lost mate. Extra-pair copulation has been noted with marked birds, but has not been studied extensively to determine frequency. Pair bonds, which may last for life, are first formed on breeding grounds. Pair formation is not well understood, although it includes bill-dipping and paired-swimming displays where both adults rise out of the water in various postures. Both sexes quickly build the nest, sometimes in as little as half a day to a week; returning pairs may reuse old nests. Nests are constructed of wet vegetation on land or as a floating mat, with a 15–32 in (38–82 cm) diameter. Two eggs are usually laid (rarely one or three) from May to July, depending on latitude and weather; in northernmost areas there may be only two to three months to breed. Eggs are long, subelliptical in shape, from 2.9 by 1.8 in (72.7 by 44.8 mm) to 3.5 by 2.2 in (89.4 by 55.15 mm) in size, and are olive green with brown markings. Pairs will re-lay if the first clutch is lost. Both sexes incubate, beginning with the laying of the first egg. Incubation period is 24–30 days; larger species have longer incubation periods. Chicks hatch asynchronously, are semiprecocial, and covered with dark gray natal down. Chicks leave the nest soon after they dry, but may return for brooding. Young rely on both parents for food but will begin to dive on their own in three days. Chicks will occasionally ride on a parent's back when small. Young can fly in six to eight weeks. Predators of eggs and chicks include many mammals and other birds. To avoid predators, chicks dive to the bottom of the water to stir up sediments, resurfacing in emergent vegetation for cover. Adults have few predators, and band recoveries suggest loons have a life span of 25–30 years.

Conservation status

No loons are listed on the IUCN Red List of Threatened Species of Birds. Isolation of breeding habitat protects their

Common loons (*Gavia immer*) stretch, remove water, and signal with this motion. (Photo by Gregory K. Scott. Photo Researchers, Inc. Reproduced by permission.)

numbers from human disturbance in many areas, but where overlap does occur loons have declined due to habitat encroachment and acid precipitation, which can lower pH levels enough to kill all the fish in many lakes. Loon conservation groups have formed to protect them in many areas. At sea, fishnets kill many adults, and are responsible for one-third of red-throated loon banding recoveries. Oil spills kill many loons, which are especially susceptible when adults are flightless while molting primaries. The *Exxon Valdez* spill in Alaska resulted in hundreds of loons being washed ashore. Botulism, a bacterial disease, has killed thousands of loons staging on the Great Lakes. Common loon populations, in particular, have declined in parts of northeastern United States and in Ontario, Canada, due to increase in human activities—boating, use of jet skis, and canoeing—lake acidification, and mercury poisoning.

Significance to humans

The Inuit legally hunt loons in arctic North America for subsistence purposes. Roughly 4,600 may be taken each year. They are not considered to be the best-tasting food, and may be fed to dogs. Native Americans honor loons with many stories and parables. Loons are a symbol of the north, and a symbol of tranquility. The common loon is featured in Canadian currency on the $1 coin (commonly called "loonies") and the $20 bill.

1. Red-throated loon (*Gavia stellata*); 2. Arctic loon (*Gavia arctica*); 3. Pacific loon (*Gavia pacifica*); 4. Common loon (*Gavia immer*); 5. Yellow-billed loon (*Gavia adamsii*). (Illustration by Marguette Dongvillo)

Species accounts

Red-throated loon
Gavia stellata

TAXONOMY
Colymbus stellatus Pontoppidan, 1763, Tame River, Warwick-shire, England. Monotypic.

OTHER COMMON NAMES
English: Red-throated diver; French: Plongeon catmarin; German: Sterntaucher; Spanish: Colimbo Chico.

PHYSICAL CHARACTERISTICS
20.8–27.19 in (53–69 cm); 2.2–5.9 lb (1.0–2.7 kg). The smallest and least robust in the family, with proportionally smaller, up-turned bill and smaller feet than other loons. Smaller size allows red-throated loons to take off directly from water and even from land. In alternate plumage, has grayish upperparts, white underparts, gray face, and brick red throat patch. In basic plumage, has grayish upperparts with white speckling, gray cap and nape, white underparts, throat, and face. Juvenal and second alternate plumages similar to basic plumage, with gray-brown wash on head and neck.

DISTRIBUTION
Breeding range is circumpolar, ranging farther north than other loons. Occupies coastal plain in Alaska, northern Canada, Greenland, Iceland, northern British Isles, Norway, Sweden, Finland, and across Russia. Winters on coasts on Atlantic and

Pacific Oceans north of the tropic of Cancer, occasionally found inland. Migrates mostly along coast, occasionally over land.

HABITAT
Breeds mainly on ponds in coastal tundra, occasionally inland up to 3,511 ft (1,070 m) in elevation. Where it competes with other loons, occupies smaller (sometimes fishless) ponds too small for larger loons. In the far north where it is the only loon present, will breed on larger ponds and lakes. Winters on coasts, usually within 3 mi (5 km) of shore in areas with a soft, sandy substrate. Occasionally found inland on large lakes and rivers.

BEHAVIOR
The only loon to have duet vocalizations, given by pairs on breeding ponds. May migrate singly or in loose flocks. Does not require running start from water during take-off like other loons, and is the only loon that can take off from land.

FEEDING ECOLOGY AND DIET
Feeds on variety of small freshwater and marine fish. Will feed invertebrates to small chicks, and will feed on invertebrates as adults when fish are scarce. When breeding in fishless ponds, will fly to the coast and other ponds to catch prey to bring back to the young.

REPRODUCTIVE BIOLOGY
Breeds from May to September, depending on latitude and climate. Incubation 24–27 days. Occasionally moves from breeding

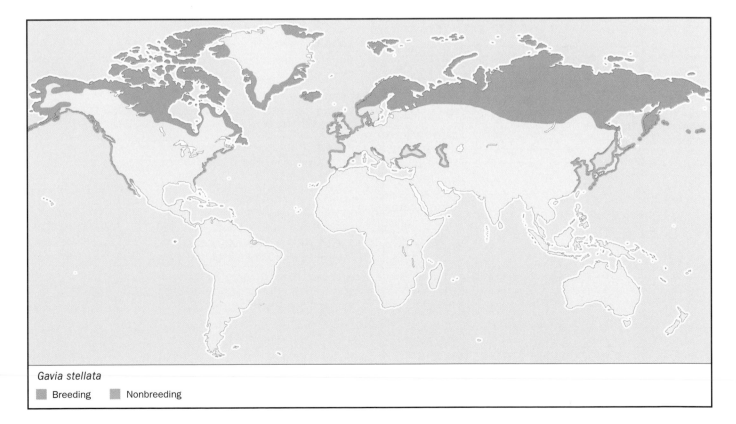

Gavia stellata
■ Breeding   ■ Nonbreeding

pond to a larger pond or the ocean. Chicks are more agile on land than are adults, and have been seen traveling over a kilometer over land. Young can fly after 38 days. Predators include Arctic fox (*Aloplex lagopus*) and other mammals, jaegers (*Stercorarius*), and gulls (*Larus*).

CONSERVATION STATUS
Declining over much of its range, although the cause is unknown. Not listed on IUCN Red List of Threatened Birds.

SIGNIFICANCE TO HUMANS
Inuit legally hunt around 4,600 loons (of all species) each year for subsistence; the proportion that are red-throated is unknown. ◆

Pacific loon
Gavia pacifica

TAXONOMY
Gavia pacifica Lawrence, 1858. Monotypic.

OTHER COMMON NAMES
English: Pacific diver; French: Plongeon du Pacifique; German: Weissnackentaucher; Spanish: Colimbo del Pacifico.

PHYSICAL CHARACTERISTICS
20–27 in (50–68 cm); 3.7 lb (1.7 kg). Very similar to the larger Arctic loon in all plumages. Bill medium-sized, straight. Black upperparts with white patches, white underparts, black throat, and gray head and neck (darker near bill). Differentiated from the Arctic loon by black flanks, paler nape, and thinner white stripes on neck. In hand, throat shows faint purple iridescence. Basic, juvenal, and second alternate plumages similar, with gray upperparts, crown, and nape and white underparts.

DISTRIBUTION
Breeds on tundra ponds from eastern Siberia to Hudson Bay; winters on Pacific Ocean from southern Japan to Siberia, Alaska to southern California. Migration chiefly occurs coastally.

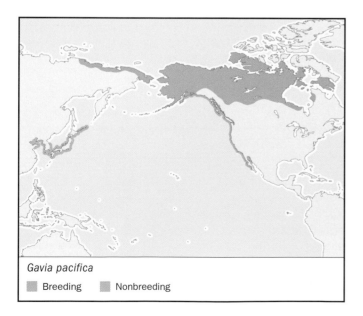

Gavia pacifica
■ Breeding   ■ Nonbreeding

HABITAT
Breeds on medium-sized lakes and ponds in northern forests and tundra. Excluded from large lakes by yellow-billed and common loons; excludes red-throated loons from medium-sized lakes. Winters in coastal areas, often farther offshore than other species.

BEHAVIOR
Males give territorial yodel call that is individually recognizable, and can be heard from miles away. Pacific loons are unafraid of humans, and allow close approach on breeding grounds. A pair was observed adopting a brood of spectacled eiders; this is the only reported case of adoption in Gaviidae.

FEEDING EC OLOGY AND DIET
Consumes a wide variety of fish; feeding and diet similar to other species.

REPRODUCTIVE BIOLOGY
Breeds on medium-sized ponds, sympatric with the Arctic loon where ranges overlap; pairs of both species have been found on the same lake. Breeds from May to September. Incubation 28–30 days, fly at 60 days. Predators include gulls (*Larus*), foxes (*Aloplex* and *Vulpes*), jaegers (*Stercorarius*), and ravens (*Corvus*).

CONSERVATION STATUS
Most populations stable. Not listed on IUCN Red List of Threatened Birds.

SIGNIFICANCE TO HUMANS
Inuit legally hunt loons for subsistence on breeding grounds; 4,600 loons (of all species) are taken yearly. ◆

Arctic loon
Gavia arctica

TAXONOMY
Colymbus arcticus Linnaeus, 1758, Sweden. Two subspecies recognized.

OTHER COMMON NAMES
English: Black-throated diver; French: Plongeon Arctique; German: Prachttaucher; Spanish; Colimbo Arctico.

PHYSICAL CHARACTERISTICS
23.6–29.6 in (60–75 cm); 5.7 lb (2.6 kg). Very similar to the smaller Pacific loon. Black upperparts with white patches, white underparts, black throat, and gray head and neck (darker near bill). Differentiated from the Pacific loon by white flanks, darker nape, more distinct white stripes on neck. In hand, throat shows faint greenish iridescence. Basic, juvenal, and second alternate plumages similar, with gray upperparts, crown, and nape and white underparts.

DISTRIBUTION
Breeds from extreme western Alaska across northern Eurasia to northern Scotland. Winters coastally from Japan to China and Europe.

HABITAT
Breeds on medium-to-large lakes and ponds in northern forests and tundra; winters on coasts.

BEHAVIOR
Similar to the Pacific loon.

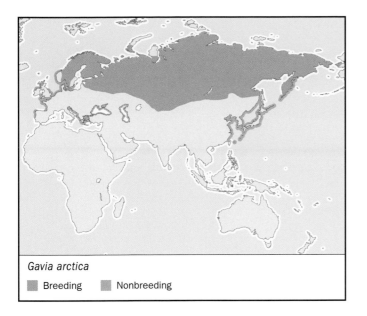

Gavia arctica
■ Breeding   ■ Nonbreeding

FEEDING ECOLOGY AND DIET
Consumes a wide variety of fish, similar to other loons. Has been observed catching frogs.

REPRODUCTIVE BIOLOGY
Incubation 28–29 days, flight 60 days. In Scotland, artificial breeding platforms have been used to increase chick production by an estimated 44%.

CONSERVATION STATUS
Most populations stable, not listed on IUCN Red List of Threatened Birds.

SIGNIFICANCE TO HUMANS
A species of interest for many people. Nest platform programs have been started in Scotland, where populations have declined. ◆

Common loon
Gavia immer

TAXONOMY
Colymbus immer Brünnich, 1764, Faeroes. Monotypic.

OTHER COMMON NAMES
English: Great northern diver; French: Plongeon huard, Plongeon Imbrin; German: Eistaucher; Spanish: Colimbo grande, Colimbo Comun.

PHYSICAL CHARACTERISTICS
26.0–35.8 in (66–91cm); 5.5–13.4 lb (2.5–6.1 kg). Very similar to the yellow-billed loon in all plumages. In alternate plumage, black upperparts with white checkering and spotting, black neck with white stripping, and black head. Underparts are white. Juvenal, second alternate, and winter plumages similar, dark gray brown upperparts, head, and nape; white underparts and throat. Bill is straight and black in alternate and dark gray with a black culmen in other plumages.

DISTRIBUTION
Breeds throughout Alaska, Canada, northern New England, northern Midwest, and parts of Greenland and Iceland. Winters in Pacific Ocean from southern Alaska to Baja California, on Atlantic Ocean from Newfoundland to Mexico, also Europe and Iceland. Migrates over land and down coasts; stages on larger lakes. Many non-breeders summer in winter range.

HABITAT
Breeds in clear, oligotrophic, forested lakes and large tundra ponds. Winters mainly on coast within 62 mi (100 km) of shore, occasionally on large inland lakes and rivers.

BEHAVIOR
Found in pairs on breeding grounds, singly or in loose flocks during migration and winter. Requires 100–650 ft (30–200 m) to take off, limiting common loons to large lakes. Extremely territorial on breeding grounds—other loons and waterbirds are chased off. Yodel call, a series of repeated two-note phrases, is recognizable to individuals and used to defend territories.

FEEDING ECOLOGY AND DIET
Feed mainly on fish and invertebrates; vegetation occasionally taken. Crayfish are a common food when fish are scarce.

REPRODUCTIVE BIOLOGY
Nests farther south than other loons, from May to October. Little is known about pair formation. Nest is made of vegetation in about a week, on land at the edge of a lake. Incubation 27–30 days. Young leave nest in one day, but may return for brooding. Young able to fly in 11 weeks. Predators include gulls (*Larus*), ravens and crows (*Corvus*), pike (*Esox*), and raccoons, weasels, and skunks (Carnivora).

CONSERVATION STATUS
Populations are stable. Not listed on IUCN Red List of Threatened Birds, but is listed as threatened or of special concern in several northeast states. Acidification of lakes, heavy metal contamination, and human encroachment threaten populations in southern range.

SIGNIFICANCE TO HUMANS
Inuit hunt 4,600 loons (of all species) per year for subsistence. Many Native American tribes have stories about common

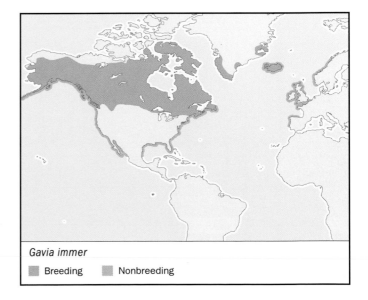

Gavia immer
■ Breeding   ■ Nonbreeding

loons. Many loon conservation groups have also been formed to protect common loons in their southern range. ◆

Yellow-billed loon

Gavia adamsii

TAXONOMY
Colymbus adamsii Gray, 1859, Alaska. Monotypic.

OTHER COMMON NAMES
English: White-billed diver; French: Plongeon a Bec Blanc; German: Gelbschnabel-Eistaucher; Spanish: Colimbo de Adams.

PHYSICAL CHARACTERISTICS
30–36 in (76–91cm); 9.0–14.1 lb (4.1–6.4 kg). Largest loon, with the largest bill. Very similar to the common loon in all plumages. In alternate plumage, black upperparts, neck, and head; white striping on the neck and white checkering on the back. Underparts white. Juvenal, second alternate, and winter plumages similar, dark gray brown upperparts, head, and nape; white underparts and throat. Bill is large and upturned, yellow to ivory.

DISTRIBUTION
Replaces common loon in high arctic, breeds in Canada and Alaska, also across Siberia. Winters along coasts in northern Pacific, and along the coast of Norway. Winters farther north than other loons.

HABITAT
Breeds in large tundra lakes; winters along coasts.

BEHAVIOR
Similar to the common loon.

FEEDING ECOLOGY AND DIET
Up-turned bill has been suggested as an adaptation for feeding along water bottom; very little data on diet, probably similar to other loons.

REPRODUCTIVE BIOLOGY
Nests from June to September on hummocks near breeding ponds. Incubation around 28 days; young leave nest within one day of hatching. Young can fly at about 11 weeks. Predators include foxes (*Aloplex*), ravens (*Corvus*), and gulls (*Larus*).

CONSERVATION STATUS
Rarest of the loons, but not threatened. Populations stable.

SIGNIFICANCE TO HUMANS
Taken occasionally for subsistence by Inuit. ◆

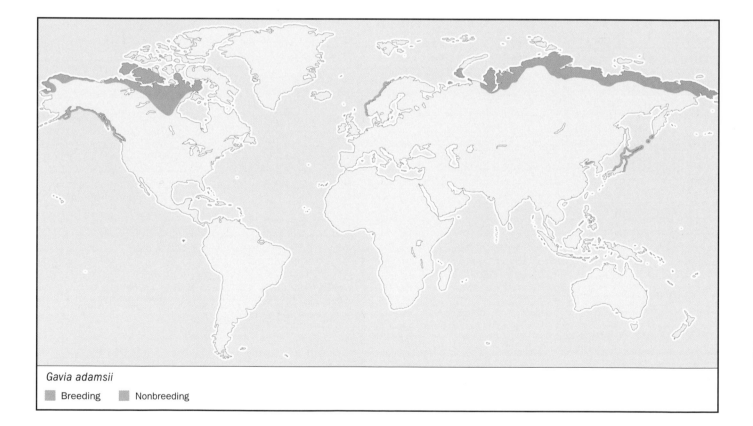

Gavia adamsii
▨ Breeding   ▨ Nonbreeding

Resources

Books

Harrison, C. J. O., and P. Castel. *Bird Nests, Eggs, and Nestlings of Britain and Europe.* Milan: New Interlew Spa, 1998.

Josson, Lars. *Birds of Europe with North Africa and the Middle East.* Princeton: Princeton University Press, 1993.

Wild Bird Society of Japan. *A Field Guide to the Birds of Japan.* Tokyo: Wild Bird Society of Japan, 1985.

Periodicals

Abraham, Kenneth F. "Adoption of Spectacled Eider Ducklings by Arctic Loons." *Condor* 80, no. 3 (1978): 339–340.

Barr, Jack F., Christine Eberl, and Judith W. McIntyre. "Red-Throated Loon (*Gavia stellata*)." *Birds of North America* no. 513 (2000): 1–28.

Hancock, Mark. "Artificial Floating Islands for Nesting Black-Throated Divers (*Gavia arctica*) in Scotland: Construction, Use and Effect on Breeding Success." *Bird Study* 47, no. 2 (2000): 165–175.

McIntyre, Judith W., and Jack F. Barr. "Common Loon (*Gavia immer*)." *Birds of North America* no. 313 (1997): 1–32.

North, Michael R. "Yellow-Billed Loon (*Gavia Adamsii*)." *Birds of North America* no. 121 (1994): 1–24.

Vallianatos, Mary, and Jon McCracken. "Loons and Type E Botulism." *Birdwatch Canada* 15 (2001): 9.

Woolfenden, Glen E. "Selection for a Delayed Simultaneous Wing Molt in Loons (Gaviidae)." *Wilson Bulletin* 79, no.4 (1967): 416–420.

Peter Andrew Hosner

Podicipediformes
Grebes
(Podicipedidae)

Class Aves
Order Podicipediformes
Family Podicipedidae
Number of families 1

Thumbnail description
Medium-sized, swimming and diving birds, recalling ducks, gallinules, or finfoots. Head small, bill pointed, neck medium to long. Rear end of body fluffy and appearing tail-less. Wings not used during diving. Most species dull-colored, a few with golden head plumes

Size
7.9–31 in (20–78 cm); 0.25–4.0 lb (112–1,826 g)

Number of genera, species
7 genera; 22 species

Habitat
Freshwater lakes, in winter also on coast

Conservation status
Two species recently Extinct, two Endangered, one Near Threatened; two with restricted range, one of them seriously declining

Distribution
Worldwide except for Antarctic and high Arctic regions

Evolution and systematics

For a long time grebes were thought to be related to loons, but it is now known that the similarity is merely owing to convergence. Despite their fairly similar lifestyle and shape, the two have very different tongues, knees, toe webbing, tail, and wing structure. DNA comparisons show that grebes have evolved independently since the early Tertiary, whereas loons branched off from the penguin-tubenose lineage much later, in mid Tertiary. The details of neck musculature, shape of sternum, and some special muscles that pull the skin of the nape forwards to raise the cheeks and crown, might link the grebes to coots (Rallidae) and finfoots (Heliornithidae). Seven genera with 22 species are currently recognized. Three species, Atitlán (*Podilymbus gigas*), Colombian (*Podiceps andinus*), and Alaotra grebes (*Tachybaptus rufolavatus*) are fairly poorly differentiated and have at some time been treated as subspecies of pied-billed (*Podilymbus podiceps*), eared (*Podiceps nigricollis*), and little grebes (*Tachybaptus ruficollis*), whereas others currently treated as subspecies might deserve status of full species. These latter include the large and nearly flightless Malvinas race of the white-tufted grebe (*Rollandia roland*), two or even three populations of silvery grebe (*Podiceps occipitalis*), and perhaps the yellow-eyed forms of the little grebe.

Physical characteristics

Grebes have rather flattened, round bodies, long necks that are nearly twice the length of the body and composed of 17–21 vertebrae, pointed bills, vestigial tails, and fluffy rear ends. Their feet are flattened, lessening resistance, and are the birds' only means of propulsion during swimming and diving. The feet are placed far back on the body, beat parallel to the water surface, and function as a rudder. The feet have become so adapted for swimming and diving that grebes are barely able to walk, doing so only for short distances and with a labored waddle and the body held upright. During relaxed swimming grebes only use one leg at a time, but during dives they move them simultaneously. Each toe has a large, unilateral swimming lobe and the three large front toes are only slightly webbed. The hind toe is small, but also lobed. The rear edge of the tarsus is serrated in adults. Serration is most pronounced in the genera *Tachybaptus* and *Rollandia*, and might serve to cut through entangling submergent vegetation. Some species have very long necks and long bills used for darting during rapid pursuit of fish, a characteristic that has evolved independently three times in the family. Most species have moderately long necks and shorter bills that also serve for feeding on small arthropods and other invertebrates. To handle fish grebes have evolved a strong bite,

A pair of western grebes (*Aechmophorus occidentalis*) "dancing" on the water at Bear River National Wildlife Refuge, Utah. (Photo by Phil Dotson. Photo Researchers, Inc. Reproduced by permission.)

but unlike several other diving birds, they have no serrated bill. Body feathers are more numerous than in any other group of bird and may amount to 20,000 or more. The feathers are downy at the base and are frequently oiled with secretion from the tufted uropygial gland. The dark skin absorbs heat during frequent sunbathing on the water, often with the bird attaining an awkward sidewise posture to expose the belly. The rectrices are strongly degenerate and barely differ from the fluffy feathers of rump and vent. The flank feathers are modified for absorbing water to decrease buoyancy during dives. The wings are fairly short and narrow, with 11 functional primaries that are somewhat curved. The number of secondaries varies interspecifically and is presumably correlated with size. There is only a single downy plumage, in which the head and neck are striped in a distinctive pattern and, in some species, also show patches of colored bare skin. The striped pattern on the downy neck is retained for months after the rest of the body has become feathered, and probably serves to appease the adults. Adults of most species shift between a breeding plumage and a duller winter plumage. The body mass changes drastically through the year in many species and is shifted between different parts of the body. Breast muscle is built up when flight is needed, leg muscle when frequent diving is needed, and during wing molt, when all the flight feathers (remiges) are shed simultaneously, enormous quantities of fat may be deposited, increasing the body mass twofold or more. Body feathers, especially the flank feathers, are molted all year round and eaten to fill half the stomach with a felt-like lining, possibly to avoid attacks by spiny-headed worms (*Acanthocephala*) and perhaps other parasites, which are common in grebes owing

to the great diversity of food ingested. It has also been suggested that the feathers help protect the stomach from puncture by fish bones. Most grebes are heavily infested, and several species of parasites are dependent on grebes. As many as 33,000 parasites, mostly flukes, tapeworms, and nematodes have been counted in a single individual. Flight is of relatively little importance, and grebes need a running start to get airborne. Young grebes cannot fly until they are six to nine weeks old. Several resident species have evolved near or total flightlessness, and the heavy build-up of fat deposits in winter quarters, together with the time of wing molt, render many other grebes flightless for a large part of the year. The oldest known grebe was 15 years old (western grebe, *Aechmophorus occidentalis*) and the little grebe is known to reach 12 years of age.

Distribution

Grebes have a nearly worldwide distribution, missing only in the Antarctic and high Arctic regions. They occur from sea level to over 13,000 ft (4,000 m). The greatest diversity is found in the Americas with 15 species representing all but one of the seven recognized genera.

Habitat

Grebes breed on shallow freshwater lakes or brackish water, often rich in submergent plants, but many winter along sea coasts. During molt and migration some species gather on larger, often saline or even super saline, lakes.

Behavior

Grebes spend a lot of time preening and sunbathing. Most species are aggressive while breeding, especially during pair formation, notably the horned grebe (*Podiceps auritus*), which will attack not only grebes, but also several other species of birds within its territory and was once observed to drive away a whole flock of the much larger greylag goose (*Anser anser*) by spiking their bellies from below. Intraspecifically grebes also keep apart, and most of those that nest colonially are territorial on the feeding grounds. Among the exceptions is the hoary-headed grebe (*Poliocephalus poliocephalus*), which both feeds and nests socially, and the Junín grebe (*Podiceps taczanowskii*) feeds in organized groups, moving forward in a lateral line and diving synchronously. The courtship of many species is spectacular and among the most complex in birds. It comprises ritualized aggressive behavior with hunched postures, head turning, running on the water side by side, synchronous diving followed by surfacing with weeds in the bill and rising breast to breast while shaking heads, and several other acts. Vocalizations accompany most displays, but are especially developed in the smaller grebes, in which the members of a pair give remarkably well-synchronized duets. Other displays are poorly developed.

When moving between breeding, staging, and winter quarters grebes fly at night. They sometimes congregate in large groups, largest in the eared grebe in North America with over two million molting on just a few supersaline lakes.

Feeding ecology and diet

The larger grebes regularly dive to depths of 80 ft (25 m), exceptionally 130 ft (40 m), but catch most of their food within 26 ft (8 m) of the surface, and some, such as the little grebe, do not dive below 7 ft (2 m). Grebes sometimes pick prey from the surface, but for the most part they feed underwater and only occasionally bring a fish to the surface before swallowing it. The importance of fish in their diet has thus been overestimated. Invertebrates, mainly insects and crustaceans but also snails and annelid worms, are important for some species, at least seasonally, and some do not eat fish at all. Other known prey include squid, tadpoles, frogs, and leeches. Even when fish form the bulk of the diet, the large numbers of invertebrate prey that may be found in their stomachs are evidence that they spend much more time catching invertebrates than fish. Prey are usually pinched, but Clark's (*Aechmorphorus clarkii*), western, and apparently sometimes great crested (*Podiceps cristatus*), and possibly great grebes (*Podiceps major*) spike fish with a quick dart of the neck during rapid pursuit. In turbid water grebes spot their prey from below. Grebes show a remarkable division of resource use between species. Generally two similar-sized species are not found in close proximity on the breeding grounds. Species that do occur in the same lakes are adapted to exploit different food items, usually in different parts of the lakes. Larger, mostly fish-eating species forage in the deeper parts, whereas smaller species that rely more on invertebrates and small fish feed closer to shore. Some of the most convincing examples of character displacement in birds is found in grebes and mainly involve change in bill size where two or three species occur in the same place.

Undigested food is regurgitated in pellets together with feathers after drinking. The hoary-headed and New Zealand grebes (*Poliocephalus rufopectus*) do not drink before regurgitating and also do not eat feathers, perhaps owing to their special diet.

Reproductive biology

The nest is almost invariably floating, but often attached to vegetation. It is built of rotting plant material and varies from small platforms to rather bulky structures, the latter most common in grebes of wind-swept habitats. Some grebes place their nest near that of a coot or other aggressive bird, presumably for protection against predators.

Besides the nest, several more platforms are often constructed. These are used for mating (which never occurs on the water), resting, and sunbathing. The fairly small, biconical eggs are light blue at first, but soon become white and then stained. They are usually 2–4 in number, at high latitudes 3–8, and have an outer layer of calcium phosphate, allowing them to breathe when wet. The incubation period is 22 to 23 days, but because of asynchronous hatching a nest may hold eggs for up to 35 days. The young can swim but

A red-necked grebe (*Podiceps grisegena*) stands at its nest in the water. (Photo by Hans Dieter Brandl/Okapia. Photo Researchers, Inc. Reproduced by permission.)

are carried on the back of the parents for weeks except during dives. After about ten days the parents may separate, more or less permanently, with one or two chicks each. The young of horned grebe are independent at two weeks, but young of most species are cared for much longer. The young can fly at six to nine weeks, but in double-brooded species they often stay even longer and help their parents feed younger siblings. Breeding success is usually two to four, but the hooded grebe (*Podiceps gallardoi*) apparently never raises more than one chick.

Conservation status

No entire genus of grebe is immediately threatened. Two species, Atitlán and Colombian (*Podiceps andinus*) grebes have gone extinct within the last two decades, and one, the Alaotra grebe, is on the brink of extinction and probably cannot be saved. Luckily all three of them have close living relatives of which they were once considered subspecies. The most critically threatened of the others is the Junín grebe, endemic to a single Peruvian lake that is subject to pollution from mining and changing water levels controlled by a hydroelectric power plant. During dry periods polluted water is fed into the lake, and with current practices a few years of drought could wipe out the entire ecosystem. The lake is declared a national reserve, but unless the intake of polluted water is cut off (which is technically possible) and the electricity for the local mines bought somewhere else, so the hydroelectric plant can cease to function, there seems to be little hope of saving the grebe from extinction.

Another species, the Madagascar grebe (*Tachybaptus pelzelnii*) is cause for concern, but its status is not critical. Its numbers have declined steadily for the last half of the twentieth century, owing to reduction of habitat and introduction of exotic herbivorous fish, which change the habitat sufficiently for a competing grebe to establish.

Significance to humans

Until the early twentieth century, grebes were hunted extensively for their silky-white belly feathers. These were used to make shoulder capes and muffs for women's clothes. Great crested grebes were nearly extirpated in Western Europe, but "grebe-fur" was then imported from other parts of the World.

Archaeological studies in the Great Salt Lake basin suggest that eared grebes, which gather there in millions to molt, were an important food in the past, but today grebes are barely hunted for food anywhere on the globe, and in many places are considered ill-tasting. Locally they are believed to harm freshwater fisheries, and are shot.

In China, the little grebe has found various uses: its fat is used as an anticorrosive; its feathered skin for fur hats; and its meat as medicine.

1. Great crested grebe (*Podiceps cristatus*); 2. Little grebe (*Tachybaptus ruficollis*); 3. Pied-billed grebe (*Podilymbus podiceps*); 4. Western grebe (*Aechmophorus occidentalis*); 5. Great grebe (*Podiceps major*); 6. Black-necked grebe (*Podiceps nigricollis*); 7. Hooded grebe (*Podiceps gallardoi*); 8. Least grebe (*Tachybaptus dominicus*); 9. Hoary-headed grebe (*Poliocephalus poliocephalus*); 10. Titicaca flightless grebe (*Rollandia microptera*). (Illustration by Barbara Duperron)

Species accounts

Titicaca flightless grebe
Rollandia microptera

TAXONOMY
Podiceps micropterus, Gould, 1868, Lake Titicaca.

OTHER COMMON NAMES
English: Short-winged grebe, flightless grebe; French: Grèbe microptère; German: Titikataucher; Spanish: Zampullín de Titicaca.

PHYSICAL CHARACTERISTICS
15.3–17.7 in (39–45 cm); 1.4 lb (635 g). Adult breeding: Above blackish brown, shaggy crown chestnut with greenish-black streaks, lower cheeks white with black streaks, throat and fore-neck white, breast and thin line on sides of neck rufous, belly mottled drab, rufous and gray, flanks with most rufous. Secondaries mostly white with black shaft streak. Bill dark suffused with yellow, eye-ring yellow, eyes dark. Nonbreeding birds with more or less whitish central underparts, immatures considerably paler both above and below.

DISTRIBUTION
Endemic to Lake Titicaca and adjacent Lake Uru-Uru at 12,100–12,600 ft (3,700–3,800 m) in southeast Peru and west Bolivia.

Rollandia microptera
  Resident

HABITAT
Breeds among patches of bulrush (*Scirpus totora*) or floating waterweeds, always with easy access to open water.

BEHAVIOR
Usually alone. Flees to open water rather than seeking cover in the vegetation.

FEEDING ECOLOGY AND DIET
Feeds nearly entirely on large fish caught in relatively open and deep water. Sometimes peers from surface.

REPRODUCTIVE BIOLOGY
Courtship elaborate, usually beginning with assembly of several birds. Breeds throughout year. Lays 2 eggs several times a year, sometimes using the same nest. Large young sometimes help feed chicks from later broods. Incubation period unknown.

CONSERVATION STATUS
Population size unknown. Estimated to between 2,000 and 10,000 birds in the 1980s, but has declined dramatically in recent years, probably mostly owing to increased use of fishing nets with fine mesh, but local eutrophication and the introduction of trout and silverside (*Odonthestes bonariensis*), which has caused disappearance of many native fish, may also have played a role. Classified as Near Threatened but situation possibly critical.

SIGNIFICANCE TO HUMANS
None known. ◆

Little grebe
Tachybaptus ruficollis

TAXONOMY
Colymbus ruficollis, Pallas, 1764, Holland. Nine subspecies.

OTHER COMMON NAMES
English: Common grebe, red-throated little grebe, dabchick; French: Grèbe castagneux; German: Zwergtaucher; Spanish: Zampullín Común.

PHYSICAL CHARACTERISTICS
9.8–11.4 in (25–29 cm); 0.26–0.53 lb (117–241 g). Adult breeding: breast, chin, lores, cap and rest of upperparts blackish, cheeks, throat and side of neck rufous. Sides and flanks dusky more or less washed with rufous, belly variable according to subspecies, ranging from silvery white to black. Most forms have no white in wing, some a small patch on inner secondaries. Bill black-tipped white and with pale yellow wattle at base, eyes red in most of range, yellow in east Asia. Non-breeding dull brownish, throat and belly whitish, immature similar but with striped neck.

DISTRIBUTION
T. r. ruficollis: Europe and northwest Africa; *T. r. iraquensis*: Iraq and southwest Iran; *T. r. capensis*: Africa south of the Sahara, Madagascar, Caucasus and eastwards through India to Myanmar; *T. r. poggei*: southeast and northeast Asia; *T. r.*

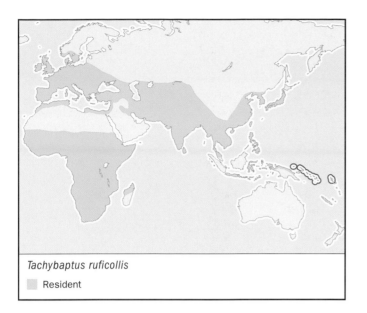

Tachybaptus ruficollis

▨ Resident

OTHER COMMON NAMES
English: American dabchick, least dabchick; French: Grèbe dominicain; German: Schwartzkopftaucher; Spanish: Zampullín Macacito.

PHYSICAL CHARACTERISTICS
8–11 in (20–27 cm); 0.25–0.40 lb (112–182 g), female decidedly smaller than male, *T. d. eisenmanni* smallest. Adult breeding: above blackish, sides of head, and neck gray. Breast and sides dusky, breast washed with buff, belly whitish mottled with gray. Eyes pale yellow to orange-yellow; bill black with pale tip. Nonbreeding duller with white throat, immature with striped head, brown eyes and pale bill.

DISTRIBUTION
T. d. dominicus: northern Caribbean; *T. d. bangsi*: Baja California; *T. d. brachypterus*: west central Mexico to Panama; *T. d. speciosus*: most of South America, including northern Argentina and southern Brazil; *T. d. eisenmanni*: western Ecuador.

HABITAT
Usually in water almost overgrown with floating vegetation. Mostly temporary ponds, but also swamps, shallow lakes and ditches. Occasionally in mangroves.

philippensis: northern Philippines; *T. r. cotabato*: southeast Philippines; *T. r. tricolor*: Sulawesi to north New Guinea; *T. r. vulcanorum*: Java to Timor; *T. r. collaris*: northeast New Guinea to Solomon Islands.

HABITAT
Mostly small and shallow lakes and ponds, but also along vegetated shores of larger lakes. When not breeding, sometimes on more open water, rarely on coast.

BEHAVIOR
Pairs may reside on the same pond all year, but non-breeding birds may assemble in loose groups of 5–30, occasionally hundreds.

FEEDING ECOLOGY AND DIET
Usually feeds within 3.3 ft (1 m) of surface, often just peering and picking with head and neck under water or picking from the surface. Diet variable, but mainly insects. Also takes small fish and, unlike most grebes, substantial numbers of snails.

REPRODUCTIVE BIOLOGY
Courtship display poorly developed and partly replaced by vocal duetting given with remarkable synchrony. Eggs 2–7, usually 4; often two, sometimes three broods a year. Incubation 20–25 days, young stay in nest for a week and can fly when 44–48 days old. They sometimes help feed older siblings.

CONSERVATION STATUS
Not threatened. Widespread and generally common.

SIGNIFICANCE TO HUMANS
None known. ◆

Tachybaptus dominicus

▨ Resident

Least grebe
Tachybaptus dominicus

TAXONOMY
Colymbus dominicus, Linnaeus, 1766, Santo Domingo. Five subspecies.

BEHAVIOR
In pairs or loose groups, territories sometimes grouped close together.

FEEDING ECOLOGY AND DIET
Mainly feeds on insects.

REPRODUCTIVE BIOLOGY
Multi-brooded, nesting at any season if conditions are suitable. Eggs usually 4–6. Incubation period 21 days.

CONSERVATION STATUS
Not threatened and locally common. Total population at least 20,000.

SIGNIFICANCE TO HUMANS
None known. ◆

Pied-billed grebe
Podilymbus podiceps

TAXONOMY
Colymbus podiceps, Linnaeus, 1758, South Carolina. Three subspecies.

OTHER COMMON NAMES
French: Grèbe à bec bigarré; German: Bindentaucher; Spanish: Zampullín Picogrueso.

PHYSICAL CHARACTERISTICS
12–15 in (30–38 cm); 0.6–1.3 lb (253–568 g). Adult breeding: above blackish, headside gray in contrast to black throat, sides of neck, breast, and sides of body grayish buff grading to mottled whitish and sooty gray on belly. Rump white. Bill short and thick, bluish white with distinct black vertical bar, eyes dark. Nonbreeding similar but throat pale, cheeks, neck, and flanks more buffy brown, bill usually fleshy pink without black bar. Immature: Head and neck boldly striped rufous, black and white, body rather uniform gray.

DISTRIBUTION
P. p. antillarum: Greater Antilles; *P. p. podiceps*: central Canada to Panama, in winter in southern part of range and Caribbean; *P. p. antarcticus*: eastern Panama and large parts of southern America.

HABITAT
Lakes, marshes, and ponds, usually with abundant reeds, floating and submergent vegetation and often with little open water.

BEHAVIOR
Alone, in pairs or family groups.

FEEDING ECOLOGY AND DIET
Eats a great variety of prey, but more than other grebes may take armored or spiny fish, crayfish, and crabs, the latter forming a substantial part of the diet in the Neotropics.

REPRODUCTIVE BIOLOGY
Courtship display poorly developed. Highly territorial. Eggs 2–10. Often double-brooded. Incubation period 21–27 days, shortest for last egg, fledging 35–37 days.

Podilymbus podiceps
■ Resident

CONSERVATION STATUS
Not threatened and common over much of its range. Often killed on rainy nights during migration when they mistake wet asphalt roads and parking lots for ponds and dive in from some height.

SIGNIFICANCE TO HUMANS
None known. ◆

Hoary-headed grebe
Poliocephalus poliocephalus

TAXONOMY
Podiceps poliocephalus, Jardine and Selby, 1827, New South Wales.

OTHER COMMON NAMES
English: Hoary-headed dabchick; French: Grèbe argenté; German: Haarschopftaucher; Spanish: Zampullín Canoso.

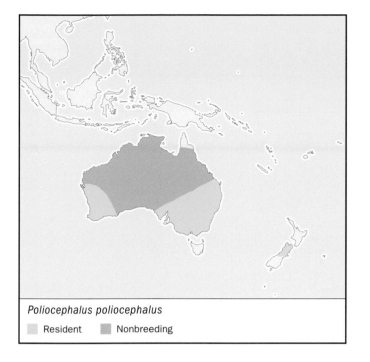

Poliocephalus poliocephalus

□ Resident ■ Nonbreeding

PHYSICAL CHARACTERISTICS
11–12 in (27–30 cm); 0.4–0.7 lb (190–311 g). Adult breeding: entire head and upperparts dark, head covered with long, fine streaks of white plumes except on black mid-crown and upper throat. Neck and breast light rusty to whitish, belly white, sides mottled with gray. Eyes buffy, bill black prominently tipped white. Nonbreeding: duller, with fewer and shorter head plumes, throat white, neck and breast whitish, bill horn; immature similar, but after shedding striped head and neck, head without any white plumes; bill pinkish with dark ridge.

DISTRIBUTION
Australia and Tasmania, recently also locally on South Island, New Zealand.

HABITAT
Mainly semi-permanent open swamps with relatively little floating vegetation, but also on open temporary ponds. In drought years non-breeders congregate in permanent wetlands and coastal lagoons.

BEHAVIOR
Gregarious, even when feeding. Semi-nomadic, sometimes appearing suddenly in groups of up to ten thousand.

FEEDING ECOLOGY AND DIET
Feeds within 6.6 ft (2 m) of surface, almost entirely on small arthropods, fish consituting less than 3% of diet.

REPRODUCTIVE BIOLOGY
Courtship display poorly developed. Nest fairly exposed, but inaccessible, in colonies with up to 400 nests. Single-brooded. Eggs 3–5.

CONSERVATION STATUS
Not threatened and locally common. Population may exceed half a million.

SIGNIFICANCE TO HUMANS
None known. ◆

Great grebe
Podiceps major

TAXONOMY
Colymbus major, Boddaert, 1783, Cayenne. Two subspecies.

OTHER COMMON NAMES
French: Grand Grèbe; German: Magellantaucher; Spanish: Somormujo Macachón.

PHYSICAL CHARACTERISTICS
22–31 in (57–78 cm), *P. m. navasi* largest, Peruvian birds smallest; 3.5 lb (1600 g). Slender with very long neck and thin, slightly upturned bill. Adult breeding: face gray to blackish, small median crest on hindcrown black, hind-neck dark gray to black, back blackish with pale feather edges, neck, breast, flanks, and vent rufous, flanks with dusky wash, belly, secondaries, inner primaries and base of outer primaries white. Eyes brown, bill black. Nonbreeding either similar, but with pale gray lores and throat, or duller, or with cap black, upper lore pale, cheeks graybrown, neck gray with some rufous, and sides gray. Immature: sides of head with bold black lines and spots, neck dull rufous, body rather uniform sooty gray, belly white.

DISTRIBUTION
P. m. major: western Peru, central Chile and all of Argentina north to southern Paraguay and southeastern Brazil, many winter on coast; *P. m. navasi*: southern Chile, in winter on coast.

Podiceps major

□ Resident

HABITAT
Large open lakes and marshes, in winter in kelp zone along coast.

BEHAVIOR
Alone or in loose groups, often nesting in colonies.

FEEDING ECOLOGY AND DIET
Mainly eats fish, but also some arthropods and mollusks, locally many crabs. Feeds in fairly deep water.

REPRODUCTIVE BIOLOGY
Courtship display rather poorly developed. Large nest, often close to each other in colonies. Sometimes double-brooded. Eggs 1–6, usually 2–3.

CONSERVATION STATUS
Common in southern part of range and not at risk. Total population estimated at 50,000.

SIGNIFICANCE TO HUMANS
None known. ◆

Great crested grebe
Podiceps cristatus

TAXONOMY
Colymbus cristatus, Linnaeus, 1758, Sweden. Three subspecies.

OTHER COMMON NAMES
French: Grèbe huppé; German: Haubentaucher; Spanish: Somormujo Lavanco.

PHYSICAL CHARACTERISTICS
18–24 in (46–61 cm), *P. c. infuscatus* smallest; 1.3–3.3 lb (568–1,490 g), heaviest while staging. Adult breeding: crown black elongated to two posterior "horns" that can be raised and spread; rest of upperparts blackish; sides of head white (upper lores and supercilium black in *infuscatus*) grading to chestnut on large posterior fan with black rear edge; underparts white, upper sides washed with dusky; secondaries, tips of inner pri-

maries, lesser wing-coverts and scapulars white. Eyes red, bill pink with dusky ridge. Nonbreeding: crest short, sides of head white with no fan, immature similar but with several black stripes on headside.

DISTRIBUTION
P. c. cristatus: Palaearctic, in winter in southern part of range, mainly on coasts; *P. c. infuscatus*: Africa locally south of Sahara; *P. c. australis*: Australia, Tasmania, and South Island, New Zealand.

HABITAT
Mainly large lakes with expanses of open water and reedy bays, but also brackish water, and tolerates heavily eutrophicated and disturbed environments such as city parks.

BEHAVIOR
Alone or on pairs, in staging areas in groups of hundreds, occasionally up to 10,000 together.

FEEDING ECOLOGY AND DIET
Mainly feeds on relatively large fish, usually in fairly deep water, but also takes frogs, crustaceans, squid and other invertebrates.

REPRODUCTIVE BIOLOGY
Courtship display well developed. Nest often placed near that of a coot. One or two broods per year. Up to 9 eggs, but usually 3–5. Incubation period 25–29 days. Young carried 3–4 weeks, associated with parents until 8–10 weeks old, able to fly at 10 weeks.

CONSERVATION STATUS
Nearly extirpated from parts of Europe in the 1800s owing to hunting for the plume trade, but now common in Palaearctic region, where increasing owing to eutrophication of lakes and where the population is estimated at around 700,000 birds. Less common in other parts of range and decreasing in parts of Africa, probably owing to drowning in monofilament gill nets. In New Zealand a drastic decline occurred since the arrival of Europeans, but population now stable.

SIGNIFICANCE TO HUMANS
Formerly extensively hunted for "grebe fur." ◆

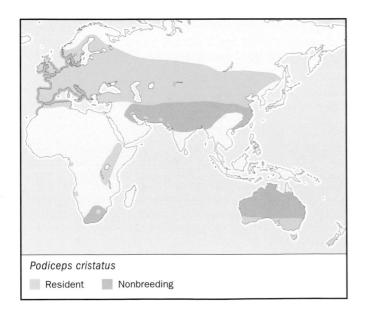

Podiceps cristatus
Resident   Nonbreeding

Eared grebe
Podiceps nigricollis

TAXONOMY
Podiceps nigricollis, C. L. Brehm, 1831, Germany. Three subspecies.

OTHER COMMON NAMES
English: Black-necked grebe; French: Grèbe à cou noir; German: Schwartzhalstaucher; Spanish: Zampullín Cuellinegro.

PHYSICAL CHARACTERISTICS
11–13 in (28–34 cm); averages 0.7 lb (325 g) while breeding, but may weigh over 1.3 lb (600 g) while staging. Females with smaller bills than males. Adult breeding: back blackish, crested head, neck and upper breast black with tuft of golden plumes behind eye, sides chestnut, rest of underparts, and secondaries white. Eyes bright red, bill black. Nonbreeding with less developed crest, blackish crown reaching to below eye where grading with white cheeks and throat, neck and sides gray;

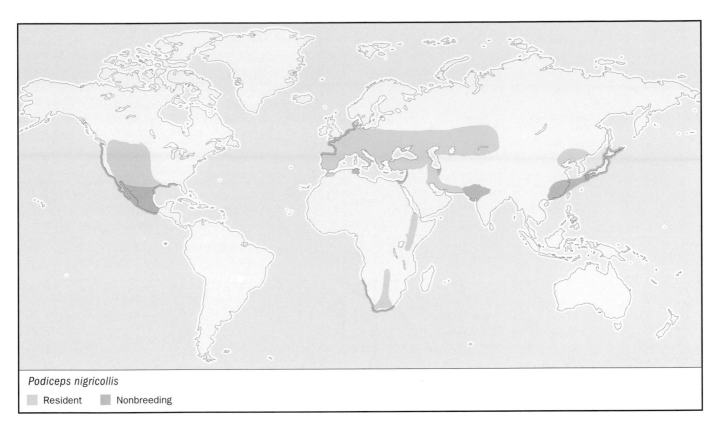

Podiceps nigricollis

☐ Resident ■ Nonbreeding

immature similar, but more brownish, especially on neck; striped pattern of head soon wearing off.

DISTRIBUTION
P. n. nigricollis: Europe and western Asia, in winter in southwestern part of range; *P. n. gurneyi*: South Africa; *P. n. californicus*: southwestern North America, in winter south to Guatemala.

HABITAT
Small, shallow, eutrophic lakes with open water and scattered patches of reed. Molts and winters on saline lakes and on coasts.

BEHAVIOR
Gregarious. In small groups where breeding, in large flocks when molting. In North America over two million stage on just a few lakes.

FEEDING ECOLOGY AND DIET
Sometimes feeds in organized groups. Eats tiny arthropods, only rarely fish.

REPRODUCTIVE BIOLOGY
Courtship display well developed. Monogamous, but nests colonially, often far from shore, usually a few together but occasionally up to 2,000 pairs in one colony, often in association with marsh terns and smaller gulls, but away from coots and other grebes. Usually single-brooded, eggs 2–4, incubation period 20–22 days.

CONSERVATION STATUS
Not threatened. The most numerous of all grebes, with a world population exceeding 5 million birds, most occurring in

North America. Numbers fluctuate greatly and species at risk while molting, when large parts of the population are concentrated on just a few lakes in flightless condition.

SIGNIFICANCE TO HUMANS
None known. ◆

Hooded grebe
Podiceps gallardoi

TAXONOMY
Podiceps gallardoi, Rumboll, 1974, Argentina.

OTHER COMMON NAMES
English: Mitred grebe; French: Grèbe mitré; German: Goldscheiteltaucher; Spanish: Zampullín Tobiano.

PHYSICAL CHARACTERISTICS
12.6 in (32 cm); 0.9–1.6 lb (420–740 g). Adult breeding: head and hindneck black with white forehead grading into orange-rufous semi-crest; back blackish, rest of body and most of wings white. Eyes bright re d, bill bluish gray. Nonbreeding similar, occasionally with some white feathers on cheeks. Immature soon loses striped head to become like adult, except for black instead of rufous on crown, and white lower cheeks and throat.

DISTRIBUTION
Patagonia, main breeding grounds on meseta between Lake Stroebel and Lake Cardiel.

Podiceps gallardoi
■ Resident

Aechmophorus occidentalis
■ Resident   ■ Nonbreeding

HABITAT
Nests in fish-free, steep-sided potholes on volcanic tablelands, water with abundant submergent and floating carpets of water milfoil (*Myriophyllum elatinoides*). After breeding gathers on large upland lakes, in winter perhaps only on coast.

BEHAVIOR
Gregarious and peaceful. In small dispersed flocks while breeding, in large flocks at other times.

FEEDING ECOLOGY AND DIET
Feeds entirely on invertebrates, perhaps mainly insects, but also snails, crustaceans and leeches, mainly caught diving.

REPRODUCTIVE BIOLOGY
Courtship display well developed, more complex and stereotyped than in any other grebe. Single-brooded. Nest large. Eggs 1–2, but only one young reared.

CONSERVATION STATUS
Population estimated at 3,000–5,000 birds. Not threatened owing to inaccessibility of habitat.

SIGNIFICANCE TO HUMANS
None known. ◆

Western grebe
Aechmophorus occidentalis

TAXONOMY
Podiceps occidentalis, Lawrence, 1858, Fort Steilacoom, Washington. Two subspecies.

OTHER COMMON NAMES
French: Grèbe élégant; German: Renntaucher; Spanish: Achichilique Común.

PHYSICAL CHARACTERISTICS
21.6–29.5 in (55–75 cm); 1.8–4.0 lb (823–1826 g), *A. o. ephemeralis* smallest, females decidedly smaller than males. Body narrow, neck very long, bill long and sharply pointed. Head with slight crest. Adult breeding: cap to below eye black, rest of upperparts blackish with faint gray scales on back; underparts white, sides spotted gray. Wings with variably sized white bar across remiges. Eyes red, bill buffy green with black ridge. Nonbreeding: similar, but crown duller, less crested and less clearly demarcated from white. Immature: like non-breeding, but crest even shorter, back without scales, and facial pattern more diffuse, sometimes with white on lores.

DISTRIBUTION
A. o. occidentalis: west to North America, in winter on coast of Texas and Pacific coast south to Baja California; *A. o. ephemeralis*: resident in central Mexico.

HABITAT
Breeds on large lakes and marshes with large expanses of open fresh or brackish water and with reedy shores. In winter mostly on salt lakes or in deep offshore waters on coast.

BEHAVIOR
Colonial, sometimes several thousand together.

FEEDING ECOLOGY AND DIET
Feeds almost entirely on large variety of fish, often spiking them, usually in fairly deep water, but on average closer to shore than Clark's grebe.

REPRODUCTIVE BIOLOGY
Courtship display well developed. Nests 3–12 ft (2–4 m) apart in colonies. Eggs 3–4, incubation period 22–24 days. Young independent at 8 weeks.

CONSERVATION STATUS
Not threatened. Population estimated at 70,000–100,000 birds, only few in Mexico.

SIGNIFICANCE TO HUMANS
None known. ◆

Resources

Books

Cramp, S., and K.E.L. Simmons, eds. *The Birds of the Western Palearctic.* Vol. 1 of *Handbook of the Birds of Europe, the Middle East and North Africa.* Oxford, London, and New York: Oxford University Press, 1977.

del Hoyo, J., A. Elliot, and J. Sargatal, eds. *Ostrich to Ducks.* Vol. 1 of *Handbook of the Birds of the World.* Barcelona: Lynx Edicions, 1992.

Ilychev, V.D., and V.E. Flint, eds. *Handbuch der Vögel der Sovjetunion.* Vol. 1. Wiesbaden: Aula-Verlag, 1985.

Marchant, S., and P.J. Higgins, Coordinators. *Handbook of Australian, New Zealand and Antarctic Birds.* Vol. 1, *Ratites to Ducks.* Oxford: Oxford University Press, 1990.

O'Donnell, C., and J. Fjeldså. *Grebes—Status Survey and Conservation Action Plan.* Gland, Switzerland, and Cambridge, United Kingdom: IUCN/SSC Grebes Specialist Group, IUCN, 1997.

Sibley, C.G., and J.E. Ahlquist. *Phylogeny and Classification of Birds. A Study in Molecular Evolution.* New Haven: Yale University Press, 1990.

Periodicals

Appert, O. "Die Taucher (Podicipedidae) der Mangokygegend in Südwest-Madagaskar." *Journal of Ornithology* 112 (1971): 61–69.

Boertmann, D. "Phylogeny of the Divers, Family Gaviidae (Aves)." *Steenstrupia* 16 (1990): 21–36.

Fjeldså, J. "Comparative Ecology of Peruvian Grebes—A Study of the Mechanisms of Evolution of Ecological Isolation." *Vidensk. Medd. Dansk Naturh. Foren.* 144 (1981): 125–249.

Fjeldså, J. "Displays of the Two Primitive Grebes *Rollandia rolland* and *R. microptera* and the Origin of the Complex Courtship Behaviour of the *Podiceps* Species." *Steenstrupia* 11 (1985): 133–155.

Geiger, W. "Die Nahrung der Haubentaucher (*Podiceps cristatus*) des Bielersees." *Ornithologische Beobachter* 54 (1957): 97–133.

Moum, T., D. Johansen, K.E. Erikstad, and J.F. Piatt. "Phylogeny and Evolution of the Auks (Subfamily Alcinae) Based on Mitochondrial DNA Sequences." *Proceedings of the National Academy of Sciences of the United States of America.* 91 (1994): 7912–7916.

Voous, K.H., and H.A.W. Payne. "The Grebes of Madagascar." *Ardea* 54 (1965): 9–31.

Organizations

IUCN/SSC Grebes Specialist Group. Copenhagen, DK 2100 Denmark. Phone: +45 3 532 1323. Fax: +45-35321010. E-mail: jfjeldsaa@zmuc.ku.dk Web site: <http://www.iucn.org>

Wetlands International. Droevendaalsesteeg 3A, Wageningen, 6700 CA The Netherlands. Phone: +31 317 478884. Fax: +31 317 478885. E-mail: post@wetlands.agro.nl Web site: <http://www.wetlands.agro.nl>

Niels K. Krabbe, PhD

Pelecaniformes

(Pelicans and cormorants)

Class Aves

Order Pelecaniformes

Number of families 5 families

Number of genera, species 9 genera; 62 species

Photo: Brown pelicans (*Pelecanus occidentalis*) with colorful gular pouches during breeding season in Baja California. (Photo by Gregory G. Dimijian. Photo Researchers, Inc. Reproduced by permission.)

Evolution and systematics

Twentieth-century ornithologists typically recognized six families in the order Pelecaniformes. Included were: the tropic birds (Phaethontidae, genus *Phaethon*) with three species, the pelicans (Pelecanidae, genus *Pelecanus*) with seven species, the cormorants and shags (Phalacrocoracidae, genera *Phalacrocorax* and *Leucocarbo*) with 34 species between them, the anhinga and darter (Anhingidae, genus *Anhinga*) with two species, the gannets and boobies (Sulidae, genera *Sula*, *Papasula*, and *Morus*) with nine species among them, and the frigatebirds (Fregatidae, genus *Fregata*) with five species. In this treatment, however, Anhingidae is classified as *Anhinga*, a genus of the family Phalacrocoracidae. At least six other families, showing characteristics similar to those living, were believed to have disappeared since the Cretaceous. Among the most spectacular of these were the pseudothorns. Bony-toothed bills and wingspans in excess of 18 ft (5.4 m) characterized these primordial predators of the Eocene (60–40 million years ago).

The tropicbirds and frigatebirds may well be the most primitive families in this diverse assemblage. Indeed, *Limnofregata azygosternon*, dating from the lower Eocene, is among the oldest known aquatic birds in the fossil record. Early additions to the cormorant (Phalacrocoracidae) and pelican (Pelecanidae) lines first appear in the Eocene-Oligocene boundary (40 mya) and early Miocene (22.5–5 mya), respectively.

Owing to the widespread geographic range of some of these birds, often occurring in discrete subpopulations, the reader should not be surprised that there is a great diversity of opinion regarding the number of extant species and subspecies. Of greater significance, however, is the accumulating body of evidence, published in the 1990s and early part of the 21st century, that casts doubt on the legitimacy of the order itself. Early avian taxonomists had qualified the group's membership largely on the basis of superficial internal or external morphology, most notable the characteristic totipalmate (webbed between all four toes) foot structure. Unlike all other water birds, families assigned to this taxon had webbing that connected all four toes. DNA-DNA hybridization data and nuclear and mitochondrial DNA sequences paint a very different picture. Results of experiments published by Charles Sibley and John Ahlquist in 1990 suggested that, while the cormorants/shags, anhingas/darters, and boobies/gannets enjoy a close genetic relationship, the other taxa are phylogenetically remote. Pelicans showed the greatest proximity to Africa's shoebill (*Balaeniceps rex*) and hammerhead (*Scopus umbretta*). Similarly, frigatebirds appeared most closely related to the petrels (Procellariiformes), penguins (Sphenisciformes), and loons (Gaviiformes). Meanwhile, tropicbirds, perennial outliers even in traditional classification systems, showed little relationship to any family. Independent phylogenetic investigations conducted later by Van Tuinen et al. supported much of the findings and caused some to wonder whether the order was actually polyphyletic and a classic example of convergent evolution.

Physical characteristics

Some of the most readily recognizable birds in the world belong in this group. Most show obvious adaptations to an aquatic lifestyle and the diagnostic characteristic shared by every one is the presence of webbing connecting all four toes, although this is much reduced in the frigatebirds.

It is their bills that have accrued the assemblage its greatest claim to fame. The enormous bill and, when distended, gular pouch of a pelican makes it difficult to confuse with any other animal. At lengths of up to 20 in (50 cm), the bill of the Australian pelican (*Pelecanus conspicillatus*) is the largest of any living bird. Much more representative of the group, however, are the more modest, often serrated, hooked-tip bills found in almost every other Pelecaniform family. The exception is the genus *Anhinga*, which is characterized by a straight, rapier-like bill of less than 4 in (10 cm).

An unfeathered gular pouch, although present in every species except the tropicbirds, is not always visible from a

A colony of nesting Australian gannets (*Morus serrator*). (Photo by T.J. Ulrich/VIREO. Reproduced by permission.)

distance. Invisible from any distance, of course, are the subcutaneous air sacs that serve to absorb the impact shock of those species that plunge dive. These air sacs account for the surprisingly light weight of birds in a group that includes some of the largest birds capable of flight.

The pelicans are the largest members of the group and may weigh up to 33 lb (15 kg). Two species have been recorded with wingspans exceeding 9 ft (3 m) and lengths of over 80 in (180 cm). By contrast, the smallest tropicbirds may weigh as little as 10.5 oz (300 g). If not for its characteristic tail feather, one species would measure as little as 15 in (38 cm) as an adult.

At first glance, these birds do not strike one as being particularly colorful. Most are primarily black or dark brown, and several are predominantly white. Upon closer inspection, however, a few colorful surprises emerge. Some of the most striking blues and greens in the animal kingdom may be seen in the eyes of some pelicans and cormorants. Colors of the bare skin of a few, most notably the boobies, is particularly striking. With the onset of the breeding season, still others develop unique pastel hues of pink or red. The plumage of the Phalacrocoracidae is unusual in that it is permeable to water.

Distribution

The greatest numbers of these birds are found in tropical and temperate regions. Overall, the group is not as tolerant

to cold as some water birds. None is found at the North or South Poles. Nevertheless, the imperial shag (*Leucocarbo atriceps*) frequents the coasts and waters surrounding the Antarctic Peninsula and four species, representing the Phalacrocoracidae and Sulidae, are present north of the Arctic Circle.

Owing to the group's heavy dependence on aquatic ecosystems, it is not surprising that it is largely absent from most arid regions. However, during migrations, even small or ephemeral water bodies may prove satisfactory for short visits.

Habitat

Open ocean, seacoasts, rivers, lakes, and ponds comprise the habitats favored by Pelecaniformes. Gannets, boobies, tropicbirds, and frigatebirds are dependent entirely on marine ecosystems, while anhingas and darters are most often found in freshwater environments. Some pelicans and cormorants are equally at home in saltwater, brackish estuaries, or freshwater.

In areas where they occur in abundance, such as the west coast of South America, these birds play a major role in the ecological processes upon which they ultimately depend. The nitrogen-rich excrement, or guano, of those species is legendary, and the microorganisms that feed upon it form the foundation upon which a spectacular web of life is spun.

One might speculate that the role these animals play in stabilizing the populations on their prey has diminished somewhat with the rise of large-scale commercial and sport fishing. Nonetheless, like so many predators, they are more apt to take organisms whose fitness has been compromised by disease, injury, age, or other factors and, in this respect, serve to enhance the overall health of those populations. By the same token, it is those birds least able to defend themselves, in particular hatchlings and eggs, that are most likely to succumb to still greater predators. Even healthy adult birds are no match for sharks, crocodilians, large birds of prey, and a host of carnivorous mammals. No species preys exclusively on these birds.

Behavior

As a general rule, these birds are colonial. They are frequently found in association with other relatively large colonial birds including, but not limited to, other families in this group. The proclivity to assemble in colonies in some families, in particular the tropicbirds, appears to be influenced more by a lack of suitable nesting sites than by any economies of scale, such as the detection of predators. Indeed, among the entire group, there are remarkably few examples of developed predator alarm calls. Croaks, grunts, and other rather uninspiring vocalizations are typical of these taxa. The exception is the shrill scream of tropicbirds. The ear-piercing utterances reminded sailors of a bosun's whistle, and they aptly named them "bosun birds."

Territoriality is most marked during the breeding season, when nest sites are at a premium. Displays of aggression may be used as an effective weapon. Ritualized threats preclude these and may involve open-bill gaping, hissing, or size-enhancing postures. There is more than ample time for such activities because, unlike many animals, these birds spend relatively little time feeding. They secure daily sustenance, often, in as little as 30 minutes although significantly more time and energy may be expended traveling to and from foraging sites. The group is a diurnal one, and several species exhibit less activity at midday than they do in the morning and late afternoon. Many species spend their lives in restricted ranges, but those dependent on temperate freshwater habitats migrate before ice makes fishing impossible.

Feeding ecology and diet

Birds representing these families feed exclusively on aquatic animals, and most eat only fish. Squid, crustaceans, and amphibian larvae serve to supplement the diet of a few species. The range of methods used to obtain prey is remarkable and includes dipping, aerial piracy, surface plunging, deep plunging, pursuit plunging, and pursuit diving. The Phalacrocoracidae, whose membership exceeds that of all the other species combined, has perfected this last technique. Unlike penguins, which propel themselves using their wings, birds in these taxa explore beneath the water's surface by foot propulsion and snatch prey with their bills. Meanwhile, anhingas and darters often impale smaller fish with their bills.

Courting blue-eyed (imperial) shags (*Phalacrocorax atriceps*) on Saunders Island, Falkland Islands. (Photo by Greg Dimijian. Photo Researchers, Inc. Reproduced by permission.)

A smaller number of species plunge dive, often from considerable heights, frequently stunning prey on impact. Tropicbirds, sulids, and the brown pelican (*Pelecanus occidentalis*) alike employ this spectacular adaptation. All other pelicans search out their quarry while cruising the water's surface. In one of the most developed forms of cooperative feeding in the avian world, the larger species assemble in a U-shaped flotilla, sometimes in excess of six individuals, that effectively drive schools of fish into shallower water and to their ultimate doom.

The low-flying Fregatidae has perfected a fourth strategy. Unable to swim with effectiveness, birds from this family take organisms occurring on or, as is often the case with flying fish, above the water's surface in the course of their aerial surveillance tours. It was another strategy, however, that earned these birds their popular name. Like the pirates of old, these airborne buccaneers will pester other seabirds to the point of disgorging their gullets. The frigatebirds then swoop down to claim their ill-gotten spoils.

Reproductive biology

Courtship and mating among Pelecaniformes is often dramatic. Courtship displays may be performed entirely in the air, as is the case with tropicbirds, or entirely on the ground. Frigatebirds use both venues; the grounded males display their brilliant inflated gular pouches for the benefit of airborne females. There is a bias toward monogamy although this trait seems, once again, largely absent among tropicbirds.

Eggs are usually chalky and may number as many as six in a clutch although several species lay only one. Those that lay more than one egg do so asynchronously, and often only the oldest chick survives. Some will breed biannually in the face of reduced food resources. Many pelagic species, among them the Guanay shag (*Leucocarbo bouganvillii*), have a reputation as being populations greatly affected by the availability of food resources. Nevertheless, there is evidence to indicate that this has frequently been overstated in the literature. Those birds occurring in temperate zones breed in the spring while tropical forms may breed at any time.

As a whole, this group is more arboreal than other aquatic birds, and most construct their nests in trees. However, a few species construct ground nests, while others use cliffs. Incubation varies from 23 to 57 days, and both parents participate in this activity. Brood patches are lacking in all but the tropicbirds; hatchlings are born naked and helpless. Nestlings take regurgitated food matter from the open bills of their parents. Fledging periods vary significantly and may take up to four months in some pelican species.

Conservation status

In 2001, the International Union for Conservation of Nature and Natural Resources (World Conservation Union) listed 22 Pelecaniform birds as being under some threat. Of these, four species were Endangered or Critically Endangered. Nevertheless, the group has fared better than many vertebrate taxa, and only one species, the spectacled cormorant (*Phalacrocorax perspicillatus*), has disappeared in modern times. Some species, such as the Guanay shag (*L. bouganvillii*) occurring in the coastal environs of western South America, may number in the several millions.

Several species have recovered from years of persecution after the establishment of adequate protection, including the elimination of toxins such as the insecticide DDT. In North America, the most notable conservation success story is that of the brown pelican (*Pelecanus occidentalis*). Once an endangered species, this animal is no longer listed by the IUCN. The range and numbers of the American white pelican (*Pelecanus erythrorynchos*) and double-crested cormorant (*Phalacrocorax auritus*) respectively, have also increased. Elsewhere in the world, meaningful action has been taken to restore such charismatic species as the Dalmatian pelican (*Pelecanus crispus*).

Loss of habitat, pollution, overfishing, and purposeful eradication, largely by those who believe the birds compete with humans for food, continue to erode some species' populations. Particularly vulnerable, of course, are those species with historically small ranges. Many seagoing species are endemic to small island clusters where breeding colonies seem especially vulnerable to predation by introduced species, such as rats. The elimination of vegetation by feral rabbits, and the accompanying reduction in the shade it produces, has adversely affected reproduction at the breeding grounds of some tropicbirds.

Significance to humans

Perhaps it is not surprising that such a diverse assemblage of bird families, as those described in the following chapters, should have had such a diverse impact on human cultures through the ages. Reverence for some species was common, and this respect has not been entirely lost in many cases. Even today, for example, the delightful tail feathers of tropicbirds play an important role in the ornaments of some South Pacific cultures. The guano produced en masse by some species is of enormous value to many agricultural economies. It has been claimed that the Guanay shag is the most economically important bird in the world. Unfortunately, with the rise of industrial and sport fishing, these birds are regarded as pests. At the turn of the 20th century, more than 1,000 double-crested cormorants were slaughtered in the northeastern United States. Those responsible, it is assumed, mistakenly believed the birds competed for game fish. In the Far East, the exceptional fishing skills of Pelecaniformes led to one of the most unusual relationships in the animal kingdom when it was discovered that cormorants and darters could be trained to fetch aquatic prey for their human masters. Many species, among them the boobies and tropicbirds, seem to enjoy the company of humans and their conveniences. Often, these birds will follow ships and even alight on them.

Resources

Books

Sibley, C.G., and J.E. Ahlquist. *Phylogeny and Classification of Birds.* New Haven: Yale University Press, 1990.

Periodicals

Van Tuinen, M., D.B. Butvill, J.A.W. Kirsch, and S.B. Hedges. "Convergence and Divergence in the Evolution of Aquatic Birds." *Proceedings of the Royal Society* 268, no. 1474 (2001): 1345–1350.

Organizations

IUCN–The World Conservation Union. Rue Mauverney 28, Gland, 1196 Switzerland. Phone: +41 22 999-0011. Fax: +41-22-999-0025. E-mail: mail@hq.iucn.org Web site: <http://www.iucn.org>

Jay Robert Christie, MBA

Tropicbirds
(Phaethontidae)

Class Aves
Order Pelecaniformes
Suborder Phaethontes
Family Phaethontidae

Thumbnail description
Medium-sized oceanic birds with pointed wings, highly elongate central tail-feathers, an overall white coloration with black markings on the wings, and excellent flight skills

Size
29–40 in (74–100 cm); includes elongated tail streamers; wingspread 37–44 in (94–112 cm)

Number of genera, species
1 genus; 3 species

Habitat
Coastal and offshore waters of warm-temperate and tropical oceans

Conservation status
Not threatened, although some populations are declining

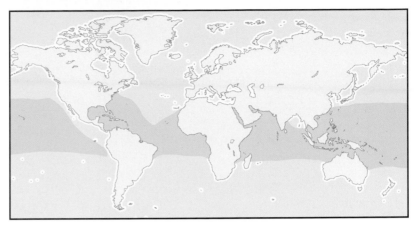

Distribution
Breed on isolated tropical islands; range widely in tropical and warm-temperate waters around the world, often far offshore

Evolution and systematics

There are only the three species of tropicbirds (genus *Phaethon*) in the family Phaethontidae. They are related to other waterbirds in the order Pelecaniformes, including pelicans, frigatebirds, cormorants, gannets, and anhingas. The Pelecaniformes lineage is ancient, with a fossil record extending to the Lower Eocene (54 million years ago).

Physical characteristics

Tropicbirds are medium-sized seabirds with a slightly decurved and pointed bill, long pointed wings, and highly elongate tail streamers (rectrices, the two central tail feathers). Body length is about 29–40 in (74–100 cm); this includes the tail streamers, which can be up to 21 in (53 cm) long and comprise about half the total body length. Wingspan is 37–44 in (94–112 cm), and the weight is 0.7–1.7 lb (0.30–0.75 kg). The plumage is overall white, sometimes with a pink flush, and black wing markings, a black eye-line, and sometimes a darker back (depending on the species). The bill is colored yellow to orange-red. Adults have elongate, narrow tail streamers. Young birds lack the tail streamers and have a gray-white banded back and wings. The short legs are placed far back on the body, making walking awkward. The feet are webbed as an aid in swimming. There is no obvious external physical difference between male and female tropicbirds.

Distribution

Tropicbirds range widely in coastal waters of tropical and warm-temperate regions, sometimes occurring far offshore.

They breed on isolated tropical islands such as Ascension Island in the Indian Ocean, Cousin Island in the Seychelles group, and Kauai in the Hawaiian Islands.

Habitat

Tropicbirds breed on isolated tropical islands. When not breeding, they range widely in coastal and offshore tropical and warm-temperate waters.

Behavior

Tropicbirds have a steady, pigeon-like flight, often settling on the surface of the water to rest. They use their webbed feet as rudders during flight, and sometimes hover in breeding displays. They do not walk well, and often shuffle on their belly when on land. They catch much of their food of small fish and marine invertebrates by making shallow plunge-dives, often from an impressive height. They have a tern-like voice and can be noisy around the breeding colonies, but are generally silent when at sea outside of the breeding season.

Feeding ecology and diet

Tropicbirds generally search for food singly or in pairs, but may also associate with large flocks of other seabirds. They often catch small flying-fish above the surface. They also feed on other species of small fish, squid, and larger

Red-billed tropicbird (*Phaethon aethereus*) in flight over Little Tobago Island in the Caribbean Sea. (Photo by Gregory G. Dimijian. Photo Researchers, Inc. Reproduced by permission.)

from the sun and rain. On some Pacific islands, they nest in trees. Ground nests are a shallow scrape and are usually located together in colonies. The social courtship display includes groups of birds flying excitedly and noisily around the nesting site in undulating flight. During this display flight, the long tail feathers wave conspicuously up and down. Tropicbirds lay only one egg, which is initially colored mottled reddish or brown but becomes paler with time as the water-soluble pigment is lost due to moisture and rubbing. The chick hatches after an incubation period of 41–45 days. The chick has a dense, silky, gray or yellow-brownish down plumage that gives protection against intense sunlight. It is fed by both parents, beginning at an age of three days, and takes 11–15 weeks to fledge. Chicks are vulnerable to being killed by adults of the same or related species of tropicbirds that are seeking a scarce nesting site. Among the red-billed and white-tailed tropicbirds of Ascension Island, such interactions within and between the two species can result in a low survival rate of young and the evolution of complex differences in breeding timetables. Red-billed tropicbirds breed every year on Ascension Island, but white-tailed trop-

crustaceans caught at the water surface, or by making a shallow plunge-dive. The typical size of prey selected is determined by the size of the bill; in areas where two species of tropicbirds occur, they tend to partition food on the basis of bill size.

Reproductive biology

Tropicbirds usually nest on ledges of coastal cliffs, in cavities under rocks, or under vegetation that gives protection

Red-tailed tropicbird (*Phaethon rubricauda*) chick on nest. (Photo by Frans Lanting. Photo Researchers, Inc. Reproduced by permission.)

Red-tailed tropicbird (*Phaethon rubricauda*) with recently hatched chick on its nest on Midway Atoll. (Photo by Frans Lanting. Photo Researchers, Inc. Reproduced by permission.)

icbirds breed every nine months and may do so at any time of the year.

Conservation status

Tropicbirds are not listed as being at risk globally by the IUCN or in the United States by the Fish and Wildlife Service. Although general population trends are not well known, some local breeding populations have declined because of disturbance and habitat loss, and perhaps because of mortality associated with commercial drift-net fishing.

Significance to humans

Tropicbirds are not of much economic importance to humans, although they are appreciated by naturalists and this can contribute to local economic benefits through ecotourism.

Formerly, tropicbird feathers were sold to the once-prominent business of millinery, or the production of women's hats and garments. The long tail feathers are still used as traditional adornments in various island cultures. The human consumption of tropicbirds, including eggs and chicks, has occurred since ancient times.

Species accounts

White-tailed tropicbird
Phaethon lepturus

TAXONOMY
Phaeton lepturus Daudin, 1802, Mauritius.

OTHER COMMON NAMES
English: Golden bosunbird, yellow-billed tropicbird; French: Phaéton à bec jaune; German: Weißchwanz-Tropikvogel; Spanish: Rabijunco Menor.

PHYSICAL CHARACTERISTICS
Adult body length (including streamers) is 29 in (74 cm), wingspan 37 in (94 cm), and weight 11 oz (0.30 kg). The overall body color is white, with black markings on the upper wings, a black eye-stripe, and a reddish (rarely yellow) bill. Juveniles have a pale-cream bill.

Phaethon lepturus

DISTRIBUTION
Tropical and warm-temperate oceans of the world.

HABITAT
Tropical and warm-temperate oceans of the world, especially in coastal waters.

BEHAVIOR
An excellent flier that commonly feeds by shallow plunge-dives and by catching flying-fish on the wing. They have a rattling call in flight, and seldom glide.

FEEDING ECOLOGY AND DIET
Small fish, squid, and larger marine invertebrates, which are caught above, at, or just under the water surface. Tends to feed closer to shore than other species of tropicbirds.

REPRODUCTIVE BIOLOGY
Breeds on remote tropical islands. A single egg is laid and incubated by both adults. Chicks have white, buff-gray, or blue-gray down. Fledging in 70–85 days.

CONSERVATION STATUS
Not threatened.

SIGNIFICANCE TO HUMANS
None known. ◆

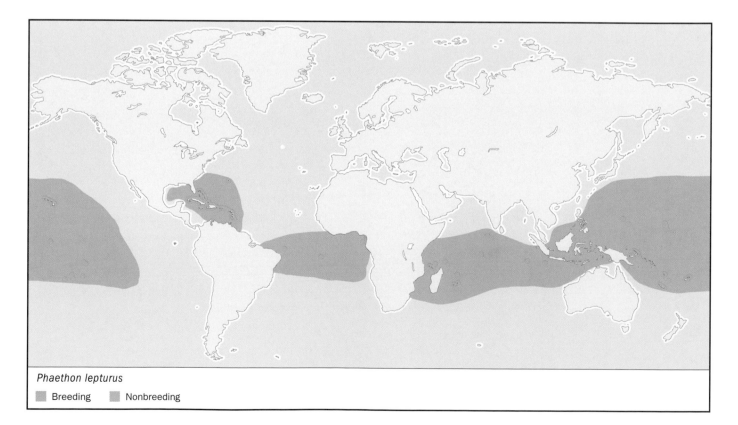

Phaethon lepturus

☐ Breeding   ☐ Nonbreeding

Red-billed tropicbird
Phaethon aethereus

TAXONOMY
Phaethon aethereus, Linnaeus, 1758, Ascension Island.

OTHER COMMON NAMES
English: Silver bosunbird; French: Phaéton à brins rouges; German: Rotschwanz-Tropikvogel; Spanish: Rabijunco Colirrojo.

PHYSICAL CHARACTERISTICS
Adult body length (including streamers) is 18 in (46 cm), wingspan 44 in (112 cm), and weight 1.6 lb (0.75 kg). The overall body color is white, with black markings on the upper wings, a darker back, a black eyestripe, red tail-streamers, and a bright red bill. Juveniles have a black bill.

Phaethon aethereus

DISTRIBUTION
Tropical and warm-temperate oceans of the western Pacific Ocean, especially off western Mexico and the Galápagos Islands, the tropical Atlantic Ocean, and the Red Sea region of the northwestern Indian Ocean.

HABITAT
Tropical and warm-temperate waters, especially in coastal areas.

BEHAVIOR
An excellent flier that feeds by shallow plunge-dives and by catching flying-fish on the wing. The voice is a harsh, clanging rattle.

FEEDING ECOLOGY AND DIET
Small fish, squid, and larger marine invertebrates, which are caught above, at, or just under the water surface.

REPRODUCTIVE BIOLOGY
Breeds on remote tropical islands. A single egg is laid and incubated by both adults. Chicks have gray down, fledge at 80–90 days.

CONSERVATION STATUS
Not threatened.

SIGNIFICANCE TO HUMANS
Not of much importance to humans, except for the economic benefits of ecotourism related to birdwatching. ◆

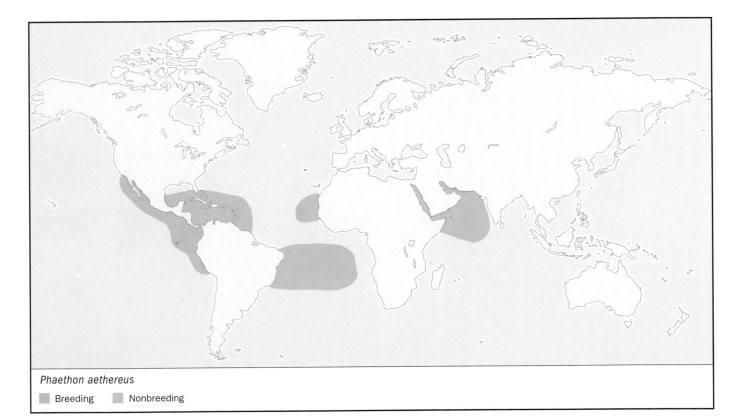

Phaethon aethereus
- Breeding
- Nonbreeding

Resources

Books

Harrison, P. *Seabirds. An Identification Guide.* Beckenham, United Kingdom: Croom Helm Ltd., 1983.

Orta, J. "Family Phaethontidae (Tropicbirds)." *Handbook of the Birds of the World.* Vol 1, edited by J. del Hoyo, A. Elliott, and J. Sargatal. Barcelona: Lynx Edicions, 1992.

Periodicals

Howell, T. R., and G. A. Bartholomew. "Experiments on Nesting Behavior of the Red-tailed Tropicbird, *Phaethon rubricauda.*" *Condor* 71 (1969): 113-119.

Stonehouse, B. "The Tropicbirds (Genus *Phaethon*) of Ascension Island." *Ibis* 103b (1962): 126-161.

Organizations

BirdLife International. Wellbrook Court, Girton Road, Cambridge, Cambridgeshire CB3 0NA United Kingdom. Phone: +44 1 223 277 318. Fax: +44-1-223-277-200. E-mail: birdlife@birdlife.org.uk Web site: <http://www.birdlife.net>

Bill Freedman, PhD

Frigatebirds
(Fregatidae)

Class Aves
Order Pelecaniformes
Suborder Pelecani
Family Fregatidae

Thumbnail description
Distinctive, dark-colored seabirds with extremely long and pointed wings, and a forked tail. Gular (throat) region is unfeathered, expanded, and colorful in breeding males. Plumage is largely black or dark brown. Bill is long and strongly hooked; nostrils absent. Middle claw is pectinate (serrated or bears projections like the teeth of a comb), and feet are webbed. Possess the lightest wing-loading of any species, giving them great flying and soaring skills

Size
Body length is 30–44 in (75–112 cm) and wingspread 69–91 in (176–230 cm); females larger

Number of genera, species
1 genus; 5 species

Habitat
Coastal waters of tropical and subtropical oceans of most of the world

Conservation status
The Christmas frigatebird (*Fregata andrewsi*) and the Ascension Island frigatebird (*F. aquila*) are listed as Critically Endangered

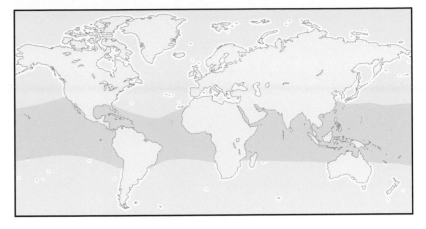

Distribution
Breed on isolated tropical islands and tend to remain fairly local to those places when feeding and during the nonbreeding season; the genus ranges worldwide in tropical and subtropical coastal waters

Evolution and systematics

The five species of frigatebirds (genus *Fregata*) are the only ones in the family Fregatidae. They are related to the pelicans, tropicbirds, cormorants, and gannets, which are also water birds in the order Pelecaniformes (characterized by four toes connected by webs, plus other traits). The Pelecaniformes lineage is ancient, with a fossil record extending to the Lower Eocene (more than 54 million years ago). The five species of Fregatidae are: the magnificent frigatebird (*F. magnificens*), the Ascension Island frigatebird (*F. aquila*), the Christmas frigatebird (*F. andrewsi*), the lesser frigatebird (*F. ariel*), and the great frigatebird (*F. minor*). A fossil frigatebird has been recovered from a deposit in England from the Lower Eocene.

Physical characteristics

Frigatebirds have a body length of 30–44 in (75–112 cm), a wingspan of 69–91 in (176–230 cm), and a weight up to 3.3 lb (1.5 kg). They have long, sweeping, narrow wings, in which the lower arm and hand bones are strongly elongated. Almost half of their body weight consists of breast muscles and feathers, and the loading per unit of wing surface area is extremely small

(proportionately to their body weight, they have the largest wings of any bird). This anatomical design allows frigatebirds to be excellent fliers and extremely efficient at soaring on rising columns of warm air. They can fly faster than 30 mph (48 kph). The tail is deeply forked and is often spread and then closed again in flight, acting as a rudder to steer the bird. The legs are short, and the small feet have only rudimentary webbing between the base of the rather short toes. The bill is long and bent into a strong hook at the tip. The gular region is nonfeathered. Females are somewhat larger than males and are marked differently, with a white throat rather than the inflatable red sac of the adult male. The body plumage is colored mostly glossy-black, with some white markings, especially on females. Young have a white head and breast.

Distribution

Frigatebirds range widely in coastal waters of tropical and subtropical regions of the world. They breed on isolated tropical islands, such as Christmas Island in the Indian Ocean, Ascension Island in the south Atlantic, and the Hawaiian Islands in the Pacific. The magnificent frigatebird is relatively widespread, but the other four species are all endemics—rare,

A lesser frigatebird (*Fregata ariel*) chick. (Photo by S. Bahrt/VIREO. Reproduced by permission.)

locally evolved species that only breed on a few highly remote, oceanic islands.

Habitat

Frigatebirds breed on isolated tropical islands and forage in coastal waters of tropical and subtropical oceans, usually fairly close to their breeding sites. They occur mainly where flying fish (*Hirundichthys*) are abundant and in regions with water of at least about 77°F (25°C).

Behavior

Frigatebirds sometimes steal food from other seabirds by harassing them relentlessly until they disgorge any fish in their gullet. The frigatebirds then scoop up such bounty as it falls through the air. This unusual, thieving behavior is known as kleptoparasitism. Frigatebirds almost never swim, as their plumage is only lightly oiled and quickly becomes wet and heavy. They are excellent fliers and can stay aloft during strong winds with little effort. Upon landing, their very long

wings can be a hindrance; this along with their short legs and tiny feet makes it awkward for them to land, perch, or walk. They often have problems when landing on their nest, particularly if strong winds are blowing. Frigatebirds tend to breed year-round and to stay near their home islands. However, they are known to sometimes fly far out over the ocean, and wander extensively when not breeding. Frigatebirds can be noisy around their nesting sites but are silent when at sea.

Feeding ecology and diet

Flying fish are a principal food of frigatebirds. They are caught in the air up to 6 or more feet (several meters) above the surface of the ocean. Frigatebirds also eat other small species of fish, as well as jellyfish, marine crustaceans, and young turtles. These foods are snatched adeptly from the surface of the water. They also eat carrion (including offal and by-catch discarded by fishing boats), eggs, and the chicks of other species of seabirds. Frigatebirds sometimes chase boobies and other marine birds, pestering them so relentlessly that they regurgitate their recently caught food in order to escape the harassment. The disgorged meal is skillfully caught by the frigatebird in the air and eaten. They take young seabirds from the ground or water surface by diving down and grabbing them with their hooked beak while in flight.

A great frigatebird (*Fregata minor*) female feeds her year-old juvenile on Wenman Island in the Galápagos. (Photo by Mark Jones. Bruce Coleman Inc. Reproduced by permission.)

A great frigatebird (*Fregata minor*) male inflates his throat pouch while standing in his nest to attract the female standing near him. (Photo by W. Wisniewski/Okapia. Photo Researchers, Inc. Reproduced by permission.)

Reproductive biology

Frigatebirds may breed at any time of the year. They usually build a nest in a low shrub or tree or sometimes on the ground. The nest is a flimsy-looking, flat platform made of interworked twigs, sticks, grasses, and reeds. They nest in colonies, which are usually close to those of other seabirds, particularly cormorants, gannets, pelicans, or terns. Frigatebirds often rob these other birds of their prey and sometimes feed on their young. During courtship, male frigatebirds use their inflatable, bright-red throat sac in a balloon-like display to impress eligible females. While trying to attract a mate, a male frigatebird occupies a suitable nesting site in the colony and shows off his outspread, glossy-black wings and inflated throat sac to any females flying above. If an interested partner approaches, the male shakes himself and conspicuously rattles his bill and wings. Frigatebirds are monogamous during each breeding attempt, but do not remain together after breeding or reunite for future attempts. Like most seabirds, frigatebirds lay only a single white egg. The egg weighs about 6% of the mother's body weight. Both parents incubate the egg over a period of 40–50 days. The young bird is naked when hatched and is fed by regurgitation by both parents. The young frigatebird is fully feathered in about 140 days and stays in the nest for a total of four to five months. The first attempt at flight occurs around 149–207 days after hatching. The fledgling depends on its parents for food for another two to six months. The young also fly about the colony in small groups and feed on scraps of food they find. Playing high in the air with feathers and bits of seaweed, young frigatebirds exercise their flight muscles and practice the techniques necessary for the highly skilled flight of adults. Frigatebirds become sexually mature at 5–7 years.

Conservation status

The Christmas frigatebird and the Ascension Island frigatebird are listed as Critically Endangered by the IUCN. Both of these extremely rare species have suffered greatly from destruction of their breeding habitat and from predation and habitat damage caused by introduced animals.

Significance to humans

Frigatebirds are not of much direct importance to humans, except for the economic benefits of tourism related to birdwatching and ecotourism. However, their ability to navigate is strong enough that these birds have been used in the past by local people to send messages between remote South Pacific islands, in the manner that carrier pigeons are used elsewhere in the world.

Several greater frigatebirds (*Fregata minor*) show their inflated throat pouches while sharing a tree. (Photo by A. Forbes-Watson/VIREO. Reproduced by permission.)

1. Magnificent frigatebird (*Fregata magnificens*); 2. Christmas frigatebird (*Fregata andrewsi*); 3. Ascension frigatebird (*Fregata aquila*). (Illustration by Patricia Ferrer)

Species accounts

Magnificent frigatebird
Fregata magnificens

TAXONOMY
Fregata magnificens Mathews, 1914, Barrington Island, Galápagos.

OTHER COMMON NAMES
English: Magnificent frigatebird; French: Frégate superbe; German: Prachtfregattvogel; Spanish: Rabihorcato Magnífico.

PHYSICAL CHARACTERISTICS
This is the largest frigatebird, with a body length of 41–44 in (103–112 cm), a wing span of 91 in (230 cm), and weight of 3.1–3.3 lb (1.4–1.5 kg). The female has a white breast and head and brownish upper-wing coverts, while the male has a mostly black body, with some white on the chest and a prominent red throat sac that is greatly inflated during sexual display.

DISTRIBUTION
Occurs in tropical and subtropical waters of the Atlantic and Pacific Oceans of the Americas. Breeds as far west as the Galápagos Islands.

HABITAT
Inhabits tropical and subtropical coastal waters, often near mangrove forest.

BEHAVIOR
Outstanding fliers, they often soar to great heights. They are silent at sea but may be noisy at the breeding colony, where they make harsh, guttural notes during courtship.

FEEDING ECOLOGY AND DIET
Feed on flying fish skillfully caught in the air, on other small fish, and on squid and other marine animals snatched at the sea's surface. They also feed on fishery offal and discarded by-catch and may predate the eggs and young of other seabirds. In addition, they feed on meals that other seabirds are harassed into disgorging in flight.

REPRODUCTIVE BIOLOGY
Females lay a single egg in a low nest, usually built in a mangrove tree or shrub. The egg is incubated by both parents for about 50 days. The chick is naked when born but fully feathered at around 140 days. It is fed regurgitated food by both parents. First flight occurs around 149–207 days after hatching. Sexual maturity is at 5–7 years.

CONSERVATION STATUS
Not threatened. Some local populations are declining because of disturbance or destruction of nesting sites and declines of food abundance caused by overfishing, but the species overall is not considered at risk.

SIGNIFICANCE TO HUMANS
Not of much importance to people, except for the economic benefits of ecotourism related to birdwatching. ◆

Christmas frigatebird
Fregata andrewsi

TAXONOMY
Fregata andrewsi Mathews, 1914, Christmas Island.

OTHER COMMON NAMES
English: Christmas Island frigatebird, Andrews' frigatebird; French: Frégate d'Andrews; German: Weissbrauch Fregattvogel; Spanish: Rabihorcado Grande.

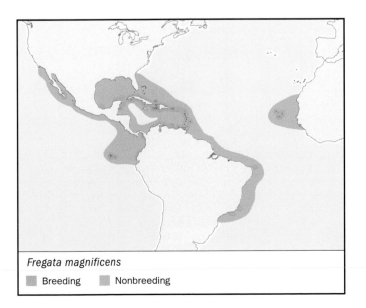

Fregata magnificens

■ Breeding ■ Nonbreeding

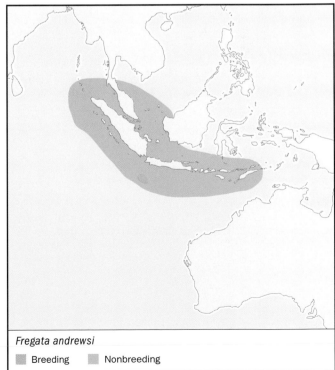

Fregata andrewsi

■ Breeding ■ Nonbreeding

PHYSICAL CHARACTERISTICS
Body length of 35–39 in (89–100 cm), a wingspan of 81–90 in (206–230 cm), and weight of about 2.6 lb (1.2 kg). Males and females have a white belly and a brown wing band. Females have a black throat, while males have a bright red, inflatable throat sac used in courtship displays.

DISTRIBUTION
Breeds on Christmas Island in the Indian Ocean but may feed more widely in tropical and subtropical coastal waters of the eastern Indian and southwestern Pacific Oceans.

HABITAT
Inhabits tropical and subtropical coastal waters, often near mangrove forest.

BEHAVIOR
Like other frigatebirds, they are outstanding fliers, often soaring to impressive heights. They are silent at sea but noisy at their breeding colony.

FEEDING ECOLOGY AND DIET
Feed on flying fish caught in the air, on other small fish, squid, and other marine food snatched at the sea's surface, and on meals they force other seabirds to disgorge in flight. They also feed on fishery offal and by-catch and predate the eggs and young of other seabirds.

REPRODUCTIVE BIOLOGY
Lays a single egg in a low nest, usually built in a mangrove tree or shrub. The egg is incubated by both parents. The chick is naked when born and is fed by both parents. Sexual maturity is at 5–7 years.

CONSERVATION STATUS
This species is listed as Critically Endangered because it breeds on only one tiny island, where its small and rapidly decreasing population is threatened by poaching, habitat destruction by mining and settlement, and an introduced ant species.

SIGNIFICANCE TO HUMANS
Not of much importance to people, except for the economic benefits of ecotourism related to birdwatching. ◆

Ascension frigatebird
Fregata aquila

TAXONOMY
Fregata aquilia Linnaeus, 1758, Ascension Island.

OTHER COMMON NAMES
English: Ascension Island frigatebird; French: Frégate aigle-de-mer; German: Adlerfregattvogel; Spanish: Rabihorcado de Ascensión.

PHYSICAL CHARACTERISTICS
Body length of 35–38 in (89–96 cm), a wingspan of 77–79 in (196–201 cm), and a weight of about 2.6 lb (1.2 kg). Males have a greenish gloss on their black plumage, while females are brownish on the upper breast, nape, and wing band. The male has a bright red, inflatable throat sac used during courtship.

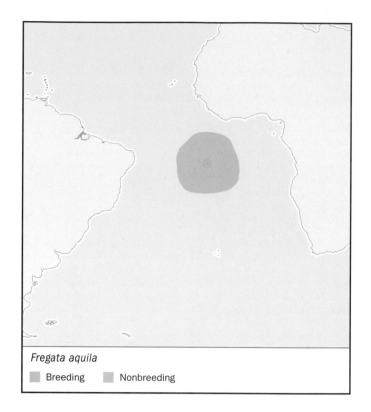

Fregata aquila
■ Breeding   ■ Nonbreeding

DISTRIBUTION
Breeds on Ascension Island in the south Atlantic Ocean. It mostly occurs near the breeding island but may also feed more widely in waters of the south Atlantic.

HABITAT
Inhabits tropical and subtropical coastal waters, often near mangrove forest.

BEHAVIOR
Outstanding fliers, they often soar to great heights. They are silent at sea but noisy at the breeding colony.

FEEDING ECOLOGY AND DIET
Feed on flying fish caught in the air, on other small fish, squid, and other marine food snatched at the sea's surface and on meals they force other seabirds to disgorge in flight. They also feed on fishery offal and by-catch and predate the eggs and young of other seabirds.

REPRODUCTIVE BIOLOGY
Lay a single egg in a low nest, usually built in a mangrove tree or shrub. The egg is incubated by both parents. The chick is naked when born and is fed by both parents. Sexual maturity is at 5–7 years.

CONSERVATION STATUS
This species is considered Critically Endangered because it breeds on only one tiny island, where its small and decreasing population is severely threatened by predation by introduced feral cats. It may also be at risk because of food depletion caused by fishing activities in its feeding habitat.

SIGNIFICANCE TO HUMANS
Not of much importance to people, except for the economic benefits of ecotourism related to birdwatching. ◆

Resources

Books

BirdLife International. *Threatened Birds of the World*. Barcelona: Lynx Edicions; and Cambridge, United Kingdom: BirdLife International, 2000.

Harrison, P. *Seabirds. An Identification Guide*. Beckenham, United Kingdom: Croom Helm Ltd., 1983.

Orta, J. "Family Fregatidae (Frigatebirds)." In Vol. 1 of *Handbook of the Birds of the World*, edited by J. del Hoyo, A. Elliott, and J. Sargatal. Barcelona: Lynx Edicions, 1992.

Organizations

BirdLife International. Wellbrook Court, Girton Road, Cambridge, Cambridgeshire CB3 0NA United Kingdom. Phone: +44 1 223 277 318. Fax: +44-1-223-277-200. E-mail: birdlife@birdlife.org.uk Web site: <http://www.birdlife.net>

IUCN–The World Conservation Union. Rue Mauverney 28, Gland, 1196 Switzerland. Phone: +41-22-999-0001. Fax: +41-22-999-0025. E-mail: mail@hq.iucn.org Web site: <http://www.iucn.org>

Other

Frigatebirds. Jan. 2002. Department of Ecology and Evolutionary Biology, Cornell University. 29 Jan. 2002. <http://www.eeb.cornell.edu/winkler/botw/fregatidae.html>

Fregatidae: Magnificent frigatebird. Jan. 2002. Florida Ecosystems. 29 Jan. 2002. <http://www.floridaecosystems.org/fregatidae.htm>

Bill Freedman, PhD

Cormorants and anhingas

(Phalacrocoracidae)

Class Aves
Order Pelecaniformes
Suborder Pelecani
Family Phalacrocoracidae

Thumbnail description
Sleek, dark-colored, large to medium-sized, long-necked waterbirds with all toes joined by webs (totipalmate); they pursue their prey (usually fish) underwater and often stand with wings spread to dry their poorly oiled, wettable plumage

Size
Variation among species ranges from 19 to 40 in (48–102 cm) and weight 1.5–7.7 lb (0.7–3.5 kg)

Number of genera, species
3 genera; 40 species

Habitat
Occur in freshwater ponds, lakes, rivers, and estuaries and in coastal marine waters

Conservation status
Endangered: 2 species; Vulnerable: 8 species; Near Threatened: 5 species

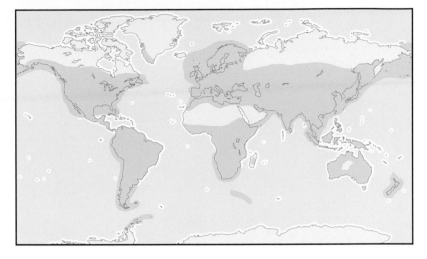

Distribution
Worldwide distribution in suitable habitat in the boreal, temperate, and tropical zones

Evolution and systematics

The 36 species of cormorants (also known as shags) and four of anhingas (also known as darters or snakebirds) make up the family Phalacrocoracidae. They are related to pelicans, frigatebirds, gannets, and tropicbirds, which are also waterbirds in the order Pelecaniformes. The Pelecaniformes lineage is ancient, with a fossil record extending to the Lower Eocene (>54 million years ago). Although cormorants and anhingas are considered here within one family, some taxonomists place anhingas in a separate family, the Anhingidae. Most taxonomists assign all of the cormorants to one genus, *Phalacrocorax*. Others, however, assign the flightless cormorant (*Nannopterum* species) and the pygmy cormorants (*Halietor* species) to separate genera.

Physical characteristics

Cormorants are sleek, large to medium-sized, long-necked waterbirds. The typical body length is 19–40 in (48–102 cm) and they weigh 1.5–7.7 lb (0.7–3.5 kg). The wings are relatively short and angular, and the spread tail is long and wedge-shaped. Cormorants are well adapted to flying and swimming, but because their legs are placed well-back on the body, they are rather clumsy when walking. When in the water, cormorants sit rather low because their bones are quite dense, with few air spaces, and their feathers are not well-oiled and so get wet when immersed. The

bill of cormorants is rather thin and tubular, hooked at the tip, and is lacking in external nares (or nostrils); the edges of the bill have tooth-like serrations. The head and upper neck have powerful muscles for closing the bill; these originate in part from special long, sesamoid bones behind the back of the head and are used to maintain a tight grip on slippery fish that have been caught (the beak serrations are also useful in this regard). For many species, the head has a plumage crest during the breeding season. Cormorant species of the Northern Hemisphere are colored glossy blackish, while those of the Southern Hemisphere tend to have a grayish body with white underparts and some black markings. Males are usually somewhat larger than females; otherwise, the sexes look alike, although they may differ in behavior, at least during the breeding season.

Anhingas are even sleeker, longer-necked waterbirds than cormorants. The typical body length is 34–36 in (86–92 cm). The bill is long, sharply pointed, and bright yellow. The wings are relatively short and rounded, and the long tail is wedge-shaped when spread. The legs are placed well-back on the body. The sexes differ in both plumage and aspects of behavior. Male anhingas have an overall black body coloration, with white markings on the wings and neck. Females also have a black body, but a light-brown neck and head. Anhingas are skilled at flying and swimming, but are clumsy on land. Like cormorants, anhingas sit low in the water because of their dense bones and feathers that get wet when immersed.

Juvenile and adult flightless cormorants (*Phalacrocorax harrisi*) at their nest in the Galápagos Islands. (Photo by Christian Grzimek/Okapia. Photo Researchers, Inc. Reproduced by permission.)

Distribution

Cormorants are widely distributed over most of the world, with species ranging from the boreal zones to the tropics (except for some Pacific islands). Anhingas occur widely in tropical and subtropical regions.

Habitat

Cormorants inhabit freshwater wetlands, swamps, lakes, rivers, estuaries, and coastal waters. Anhingas occur in freshwater wetlands, swamps, lakes, rivers, and estuaries.

Behavior

Northern species of cormorants are migratory, breeding in northern parts of their range and wintering to the south. Northern populations of anhingas are also migratory. Both cormorants and anhingas are rather gregarious, often occurring in flocks and breeding in colonies. Cormorants fly somewhat directly, often close to the water surface, using strong, steady wingbeats. They also commonly fly in groups that arrange themselves in lines or V-shaped flocks for better aerodynamic efficiency. Anhingas are also strong fliers, and they soar well, sometimes at great altitude. After swimming, both cormorants and anhingas sit on exposed perches with wings spread to the sun to dry their plumage, which lacks oily repellants and thus gets soaking wet when immersed. Cormorants and anhingas are strong swimmers and pursue prey underwater using their feet for propulsion.

Feeding ecology and diet

Cormorants and anhingas feed mostly on fish, but may also eat frogs, large crustaceans, and squid. They catch prey by an agile, underwater pursuit. Cormorants catch their prey in the bill, while anhingas often spear their quarry.

Reproductive biology

Cormorants and anhingas often breed in colonies. They build awkward stick-nests in trees or sometimes on cliff-ledges. Nests of cormorants can be rather messy, being littered with seaweed, fish remains, and other debris. Cormorants and anhingas lay two to four, elongated, chalky-surfaced eggs which are pale green or blue. Both sexes incubate the eggs (23–25 days) and rear the young. Sexual maturity is generally reached in the third or fourth year.

Conservation status

The IUCN lists 15 species of cormorants as being at risk. Of these, two are Endangered: the flightless or Galapagos cormorant (*Nannopterum harrisi*) and the Chatham Island shag (*P. onslowi*). Another eight species are Vulnerable: the Campbell Island shag (*P. campbelli*), the New Zealand king shag (*P. carunculatus*), the Stewart Island shag (*P. chalconotus*), the Aukland Island shag (*P. colensoi*), the Pitt Island shag (*P. featherstoni*), the bank cormorant (*P. neglectus*), the Socotra cormorant (*P. nigrogularis*), and the Bounty Island shag (*P. ranfurlyi*). Four species are considered at Lower Risk, Near

A double-crested cormorant (*Phalacrocorax auritus*) with its fish prey. (Photo by A. Morris/VIREO. Reproduced by permission.)

A group of double-crested cormorants (*Phalacrocorax auritus*) preens and dries their feathers after fishing at Sanibel Island, Florida. (Photo by J&L Waldman. Bruce Coleman Inc. Reproduced by permission.)

Threatened: the pygmy cormorant (*P. pygmeus*), the red-legged cormorant (*P. gaimardi*), the crowned cormorant (*P. coronatus*), and the Cape cormorant (*P. capensis*). One species, the Pallas's cormorant (*Phalacrocorax perspicillatus*), is recently extinct. Almost all of these rare cormorants are endemic species, meaning they only occur in relatively small popula-

tions on one or a few isolated, oceanic islands. Endemic species are often at an inherently high risk of extinction. The oriental anhinga (or oriental darter; *Anhinga melanogaster*) is also listed by the IUCN as being at Lower Risk.

Significance to humans

In some regions where cormorants are abundant, they may be viewed as "pests" by human fishers because they are perceived to be catching "too many fish." In almost all of these cases, however, the cormorants are feeding on smaller species or size-classes of fish than the human fishers are seeking, and so are not in much direct competition. Sometimes, cormorants nesting in colonies kill their nesting trees with their caustic excrement, which may also be perceived to be a local management problem. Several species of cormorants are extremely abundant off parts of Peru and Chile, such that their excrement and that of other abundant seabirds is collected from desert islands as a phosphorus- and nitrogen-rich fertilizer known as guano. Several local Japanese cultures have learned to use tame cormorants to catch fish for the market. In these cases, the cormorants are tethered by a leg and are prevented from swallowing fish they catch by a soft noose or collar tied loosely around their throat. Cormorants and anhingas are also sought for observation by birders and other naturalists, and so contribute to local economic benefits through ecotourism. This is especially true of the rarer species.

An adult Brandt's cormorant (*Phalacrocorax penicillatus*) feeds its young on the nest on Alcatraz Island, California. (Photo by Jerry L. Ferrara. Photo Researchers, Inc. Reproduced by permission.)

1. Double-crested cormorant (*Phalacrocorax auritus*); 2. Great cormorant (*Phalacrocorax carbo*); 3. Olivaceous cormorant (*Phalacrocorax olivaceus*); 4. American anhinga (*Anhinga anhinga*); 5. Brandt's cormorant (*Phalacrocorax penicillatus*); 6. Galapagos cormorant (*Nannopterum harrisi*); 7. New Zealand king shag (*Phalacrocorax carunculatus*); 8. Pelagic cormorant (*Phalacrocorax pelagicus*). (Illustration by Emily Damstra)

Species accounts

Great cormorant
Phalacrocorax carbo

TAXONOMY
Pelecanus carbo Linnaeus, 1758, Europe. Six subspecies.

OTHER COMMON NAMES
English: Black cormorant, white-breasted cormorant; French: Grand Cormoran; German: Kormoran; Spanish: Cormorán Grande.

PHYSICAL CHARACTERISTICS
This largest species of cormorant has a body length of about 37 in (93 cm), with a pale yellow bill, pale yellow cheek pouch bordered by a white throat, glossy blackish plumage, black legs and feet, and males somewhat larger than females (males: 5.1 lb (2.3 kg); females: 4.2 lb (1.9 kg).

DISTRIBUTION
A very widespread species in temperate regions of the world, occurring locally in the Northwest Atlantic of North America, more widely through Eurasia, and in parts of Southeast Asia, Africa, and Australia. They generally winter near their breeding grounds.

HABITAT
Nests on seacliffs, feeds in coastal waters.

BEHAVIOR
A highly social species that breeds in colonies and aggregates in flocks. Like all cormorants, it catches fish by underwater pursuit.

FEEDING ECOLOGY AND DIET
Feeds on small fish, crustaceans, and squid.

REPRODUCTIVE BIOLOGY
Lays three to four eggs in a crude stick-nest on a cliff ledge, with both sexes sharing the incubation (27–31 days) and rearing of the chicks.

CONSERVATION STATUS
Not threatened. Rather abundant over much of its range.

SIGNIFICANCE TO HUMANS
Not of great importance to humans over most of the range; however, in Japan this is one of two species (the other is the Japanese cormorant, *Phalacrocorax capillatus*) trained by human fishers to help them catch fish. ◆

Double-crested cormorant
Phalacrocorax auritus

TAXONOMY
Carbo auritus Lesson, 1831, North America. Four subspecies.

OTHER COMMON NAMES
French: Cormoran à aigrettes; German: Ohrenscharbe; Spanish: Cormorán Orejudo.

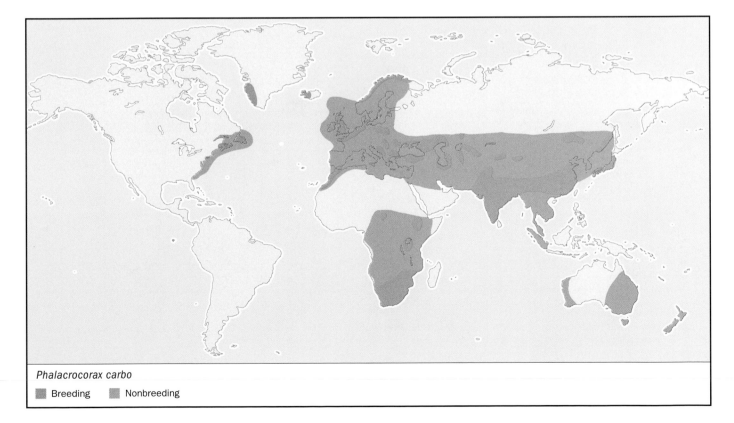

Phalacrocorax carbo

■ Breeding   ■ Nonbreeding

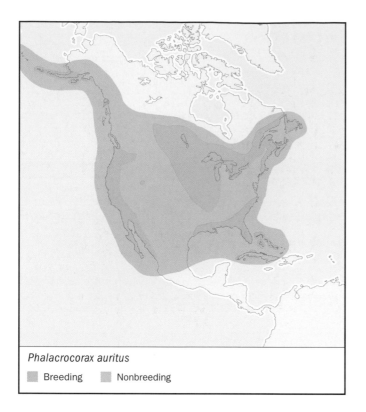

Phalacrocorax auritus
■ Breeding ■ Nonbreeding

PHYSICAL CHARACTERISTICS
Body length of 33 in (83 cm), with a bright yellow bill, yellow cheek pouch, blue eyes, glossy blackish plumage, black legs and feet, and males somewhat larger than females.

DISTRIBUTION
The most widely distributed cormorant in North America, occurring on both the Pacific and Atlantic coasts, in the Caribbean Sea, and on many larger inland lakes and rivers.

HABITAT
Usually nests on islands and feeds in coastal waters and in large lakes and rivers.

BEHAVIOR
A highly social species that breeds in colonies and aggregates in flocks; it catches its prey by underwater pursuit.

FEEDING ECOLOGY AND DIET
Feeds on small fish, crayfish, squid, and other crustaceans.

REPRODUCTIVE BIOLOGY
Lays three to four eggs in a crude stick-nest located in a tree, with both sexes sharing the incubation (c. 25–29 days) and rearing of the chick.

CONSERVATION STATUS
Not threatened. Abundant over much of its range. However, this species was considered at risk in some states in the 1970s due to organochlorine-pesticide-induced egg-shell thinning and population declines. These populations are now increasing in numbers following bans on the use of these chemicals.

SIGNIFICANCE TO HUMANS
In some parts of its range it is considered a pest for "eating too many fish" and because it kills its nesting trees with its caustic excrement. ◆

Brandt's cormorant
Phalacrocorax penicillatus

TAXONOMY
Carbo penicillatus Brandt, 1837, no locality. Monotypic.

OTHER COMMON NAMES
English: Brown's cormorant, Townsend's cormorant; French: Cormoran de Brandt; German: Pinselscharbe; Spanish: Cormorán Sargento.

PHYSICAL CHARACTERISTICS
Body length of 29 in (74 cm), with a grayish bill, blue cheek pouch, yellow throat patch, glossy blackish plumage, and black legs and feet.

DISTRIBUTION
Occurs along the Pacific coast of North America, from southern Alaska to Baja California.

HABITAT
Nests in trees and feeds in coastal waters.

BEHAVIOR
A social species that breeds in colonies and aggregates in flocks.

FEEDING ECOLOGY AND DIET
Feeds on small fish, squid, and crustaceans.

REPRODUCTIVE BIOLOGY
Lays three to four eggs in a crude stick-nest, with both sexes sharing the incubation and rearing of the chick. One or two chicks per nest normally fledge.

CONSERVATION STATUS
Not threatened. Abundant over much of its range.

SIGNIFICANCE TO HUMANS
None known. ◆

Phalacrocorax penicillatus
■ Breeding ■ Nonbreeding

Pelagic cormorant
Phalacrocorax pelagicus

TAXONOMY
Phalacrocorax pelagicus Pallas, 1811, eastern Kamachatka and the Aleutian Islands. Two subspecies.

OTHER COMMON NAMES
English: Baird's cormorant, pelagic shag; French: Cormoran pélagique; German: Meerscharbe; Spanish: Cormorán Pelágico.

PHYSICAL CHARACTERISTICS
Body length of 22 in (56 cm), with a dark bill, red cheek pouch and throat patch, glossy blackish plumage, and black legs and feet.

DISTRIBUTION
Occurs along the Pacific coast of North America, from the top of Baja California through to northwestern Alaska, across the Aleutians to eastern Siberia, and south to northern Honshu Island, Japan, plus most Beringian waters in between.

HABITAT
Nests on cliff-ledges and in trees and feeds in coastal waters.

BEHAVIOR
A social species that breeds in colonies and aggregates in flocks.

FEEDING ECOLOGY AND DIET
Feeds on small fish, squid, and crustaceans.

REPRODUCTIVE BIOLOGY
Lays three to four eggs in a crude nest, with both sexes sharing the incubation (c. 31 days) and rearing of the chick.

CONSERVATION STATUS
Not threatened. Abundant over much of its range.

SIGNIFICANCE TO HUMANS
None known. ◆

Olivaceous cormorant
Phalacrocorax olivaceus

TAXONOMY
Pelecanus olivaceus Humboldt, 1805, banks of the Magdalena River, Colombia. Two subspecies.

OTHER COMMON NAMES
English: Neotropic cormorant; French: Cormoran vigua; German: Biguascharbe; Spanish: Cormorán Biguá.

PHYSICAL CHARACTERISTICS
Body length of 25 in (63 cm), with a bright yellow bill, yellow cheek pouch, white band behind the lower mandible, glossy blackish plumage, black legs and feet, and males somewhat larger than females.

DISTRIBUTION
Occurs from the U.S. Gulf of Mexico through the Caribbean, Mexico, Central America, and almost all of South America.

HABITAT
Nests in trees near freshwater; feeds in coastal waters and in large lakes and rivers.

BEHAVIOR
A social species that breeds in colonies and aggregates in flocks.

FEEDING ECOLOGY AND DIET
Feeds on small fish, crayfish, and other aquatic animals.

Phalacrocorax olivaceus
■ Resident

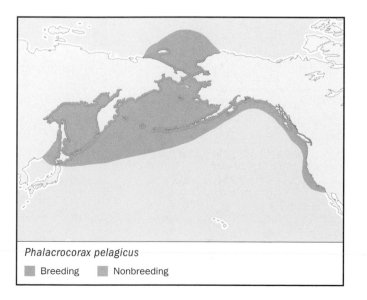

Phalacrocorax pelagicus
■ Breeding   ■ Nonbreeding

REPRODUCTIVE BIOLOGY
Lays three to four eggs in a crude stick-nest located in a tree, with both sexes sharing the incubation (c. 30 days) and rearing of the chick.

CONSERVATION STATUS
Not threatened. Abundant over much of its range.

SIGNIFICANCE TO HUMANS
None known. ◆

New Zealand king shag
Phalacrocorax carunculatus

TAXONOMY
Pelecanus carunculatus Gmelin, 1789, Queen Charlotte Sound, New Zealand, and Staten Island. Monotypic.

OTHER COMMON NAMES
English: Bronzed shag; rough-faced cormorant; rough-faced shag; French: Cormoran caronculé; German: Warzenscharbe; Spanish: Cormorán Carunculado.

PHYSICAL CHARACTERISTICS
Body length of 30 in (76 cm), with a reddish yellow bill, white throat and belly, glossy blackish plumage on the back and wings, and pink legs and feet.

DISTRIBUTION
An endemic (or local) species that only breeds on a few islands in Cook Strait between North and South Islands of New Zealand.

HABITAT
Nests on cliff-ledges and feeds in nearby coastal waters.

BEHAVIOR
A social species that breeds in colonies and aggregates in small flocks.

FEEDING ECOLOGY AND DIET
Feeds on small fish, squid, and crustaceans.

REPRODUCTIVE BIOLOGY
Lays one to three eggs in a crude nest, with both sexes sharing the incubation and rearing of the chick.

CONSERVATION STATUS
A rare species, listed as Vulnerable due to limited habitat.

SIGNIFICANCE TO HUMANS
Not of much direct importance to people, but sightings are sought after by naturalists, which brings some economic benefits through ecotourism. ◆

Galapagos cormorant
Nannopterum harrisi

TAXONOMY
Phalacrocorax harrisi Rothschild, 1898, Narborough Island, Galapagos Archipelago. Monotypic.

OTHER COMMON NAMES
English: Flightless cormorant; French: Cormoran aptère; German: Galapagosscharbe; Spanish: Cormorán Mancón.

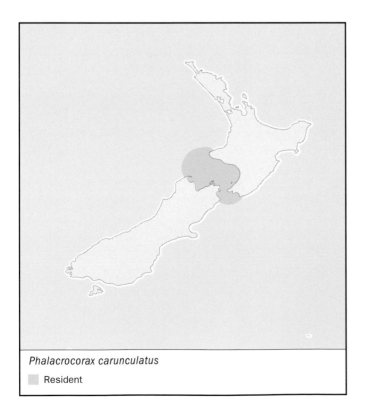

Phalacrocorax carunculatus
  Resident

Nannopterum harrisi
  Resident

PHYSICAL CHARACTERISTICS
Body length 36–39 in (91–99 cm), with short, stubby, ragged-appearing wings, an almost all-black plumage, black legs and feet, and a pinkish throat pouch.

DISTRIBUTION
An endemic (or local) species of the Galapagos Islands in the Pacific Ocean off equatorial South America.

HABITAT
Occurs in nearshore coastal waters.

BEHAVIOR
A flightless species that roosts on rocks during the night and rarely wanders far from the place where born.

FEEDING ECOLOGY AND DIET
Catches its prey of fish, squid, and crustaceans by an agile underwater pursuit.

REPRODUCTIVE BIOLOGY
Lays two to three eggs on a rocky ledge; both sexes incubate the eggs (c. 35 days) and care for the young.

CONSERVATION STATUS
An Endangered species subject to severe fluctuations in numbers in response to El Nino-related marine perturbations, with fewer than 1,000 breeding pairs surviving at only two breeding sites.

SIGNIFICANCE TO HUMANS
Not of much direct importance to people, but contributes to local economic benefits through ecotourism associated with seeing rare birds and other wildlife of the Galapagos Islands. ◆

Anhinga anhinga
■ Breeding ■ Nonbreeding

American anhinga
Anhinga anhinga

TAXONOMY
Plotus anhinga Linnaeus, 1766, Rio Tapajós, Pará, Brazil. Two subspecies.

OTHER COMMON NAMES
English: Anhinga, American darter, snakebird; French: Anhinga d'Amérique; German: Amerikanischer Schlangenhalsvogel; Spanish: Anhinga Americana.

PHYSICAL CHARACTERISTICS
Body length 34 in (85 cm), with a small head, relatively short wings, web-shaped tail, yellowish pointed beak, and yellowish legs and feet; male is colored overall black with silvery-white markings on the upper wings, while female has a brown head, neck, and upper chest.

DISTRIBUTION
Southeastern United States, Central America, South America south to northern Argentina.

HABITAT
Warm-temperate, subtropical, and tropical wetlands, rivers, lakes, swamps, and estuaries.

BEHAVIOR
Flies in a flap-and-glide manner, and often soars; often swims largely submerged, with only the head and neck exposed (this is how it got its common name, "snakebird"); roosts in trees, and dries its wet plumage by spreading its wings to the sun.

FEEDING ECOLOGY AND DIET
Catches and spears its prey of fish, crayfish, and amphibians by agile underwater pursuit. Tosses speared prey into the air with a flick of the head, then catches the prey in mid-air to swallow head-first.

REPRODUCTIVE BIOLOGY
Tends to breed in colonies; lays one to five eggs in a bulky stick-nest in a tree near water. Both sexes incubate the eggs (25–28 days) and care for the young.

CONSERVATION STATUS
Not threatened. Abundant throughout its range.

SIGNIFICANCE TO HUMANS
Not of much direct importance to humans, but often a favorite species of naturalists. ◆

Resources

Books

Cramp, S., and K.E.L. Simmons, eds. *The Birds of the Western Palearctic.* Vol. 1, *Ostrich to Ducks.* Oxford: Oxford University Press, 1977.

Orta, J. "Family Phalacrocoracidae (Cormorants)." In *Handbook of the Birds of the World.* Vol. 1, *Ostrich to Ducks,* edited by Josep del Hoyo, Andrew Elliott, and Jordi Sargatal. Barcelona: Lynx Edicions, 1992.

Orta, J. " Family Anhingidae (Darters)." In *Handbook of the Birds of the World.* Vol. 1, *Ostrich to Ducks,* edited by Josep del Hoyo, Andrew Elliott, and Jordi Sargatal. Barcelona: Lynx Edicions, 1992.

Periodicals

Hennemann, W.W. "Spread-winged Behavior of Double-crested and Flightless Cormorants, *Phalacrocorax auritus* and *P. harrisi*: Wing Drying or Thermoregulation?" *Ibis* 126 (1984): 230–239.

Jackson, J.A., and B.J.S. Jackson. "The Double-crested Cormorant in the South-central United States: Habitat and Population Changes of a Feathered Pariah. " *Colonial Waterbirds* 18 (special publ. no. 1) (1995): 118–130.

Mahoney, S.A. "Plumage Wettability of Aquatic Birds." *Auk* 101 (1984): 181–185.

Organizations

BirdLife International. Wellbrook Court, Girton Road, Cambridge, Cambridgeshire CB3 0NA United Kingdom. Phone: +44 1 223 277 318. Fax: +44-1-223-277-200. E-mail: birdlife@birdlife.org.uk Web site: <http://www.birdlife.net>

IUCN–The World Conservation Union. Rue Mauverney 28, Gland, 1196 Switzerland. Phone: +41-22-999-0001. Fax: +41-22-999-0025. E-mail: mail@hq.iucn.org Web site: <http://www.iucn.org>

Bill Freedman, PhD

Boobies and gannets
(Sulidae)

Class Aves
Order Pelecaniformes
Suborder Pelecani
Family Sulidae

Thumbnail description
Medium to large-sized seabirds with long, narrow, pointed wings and conical bill, highly adapted to catching fish by plunge-diving, often from great heights

Size
25–39 in (64–100 cm); 1.5–7.9 lb (0.7–3.6 kg)

Number of genera, species
3 genera; 9 species

Habitat
Mainly pelagic waters (open seas), breeding on offshore islands

Conservation status
Critically Endangered: 1 species; Vulnerable: 1 species

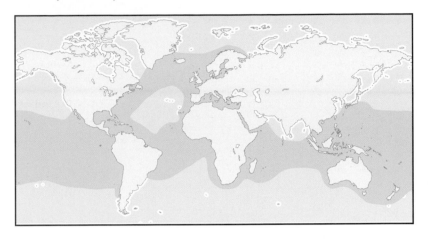

Distribution
Widespread in tropical, subtropical, and temperate oceans

Evolution and systematics

Boobies and gannets constitute a distinct family of specialist plunge-divers that are most closely related to the cormorants and anhingas (Phalacrocoracidae). They probably originated in the late Cretaceous, more than 60 million years ago. Boobies seem to have appeared first, while the gannets probably split off at a later stage, about 16 million years ago, and developed in the Northern Hemisphere. Available evidence suggests that they also occupied the North Pacific, where no gannets are currently found.

As of 2001, three genera are recognized (although some authors still keep to the classical grouping of all species under the single genus *Sula*): the genus *Papasula* comprises one species, the distinct Abbott's booby, possibly the most ancient of today's sulids. All remaining boobies are grouped under the genus *Sula*, whereas the gannets have been separated and classified in the genus *Morus*. This system reflects differences in morphology, biology, and ecology among the species, although their high degree of adaptation to the marine environment results in them sharing many characteristics.

Physical characteristics

Boobies and gannets have long, pointed wings and a characteristic cigar-shaped body. This body is medium to large in size, robust, and ends in a fairly long, wedge-shaped tail. The neck is long and thick, with strong, well-developed muscles. The head is dominated by the stout, conical bill. The bare skin around the neck and bill is often brightly colored, and

plays an active role in ritual displays. The eyes are placed at each side of the bill and are orientated towards the front, giving the birds excellent binocular vision, which is essential for active fishing from the air.

Like most fish-eating birds, boobies and gannets are predominantly light colored in the underparts, particularly the belly, but also the neck, head, and underwings. The upperparts, especially the wings, are most often dark. The white-colored underparts blend in against the brighter sky, thus rendering the predator less visible for the prey fish. The dark pigment or melanin in flight feathers protects them from ultraviolet light and salt. In some species (e.g. gannets) the white color, clearly visible from a great distance, attracts large numbers of birds to a feeding source, often a big school of fish. This may give the gannets an advantage: by attacking simultaneously in large numbers and therefore confusing the prey, the birds are able to feed on a shoal that normally is too large for a single individual to exploit efficiently.

As a further adaptation to their specialized fishing technique, boobies and gannets have subcutaneous fat and well-developed air sacs, which act as cushions and protect the birds from the violent impact of crashing into the water. For the same reason, their external nostrils are closed. Sulids also lack brood patches, which would be disadvantageous in a cold aquatic environment. Instead, they incubate by sitting on their "heels" and wrapping highly vascularized webbing of their feet around the eggs. The short and stout legs are situated far back on their bodies, allowing the birds to swim well and to maintain buoyancy even in rough seas.

A gannet dives for food. (Illustration by Patricia Ferrer)

Distribution

The family Sulidae occurs across the world's oceans, although the distribution patterns are quite distinct. The gannets are typically found in cold or temperate waters of the northern Atlantic ocean, the South African region, and Australia and New Zealand. In the off-season, they disperse over lower latitudes in the same broad areas. A further group of species, the so-called pantropical boobies (masked, red-footed, and brown), are circumpolar in distribution, occurring over most of the world's oceans between the tropics. The Peruvian and blue-footed boobies are more specialized and occur only in the eastern Pacific, along the Humboldt current area, in the Galápagos Islands, and north to southern North America. The Critically Endangered Abbott's booby is confined to tiny Christmas Island, although it had been much more widespread over the Indian Ocean.

Habitat

Boobies and gannets live primarily at sea, an environment to which they are particularly well adapted so they do not

need to set foot on land except during the breeding season. Boobies are found in tropical or subtropical waters, whereas gannets favor more temperate environments, occurring even north of the Arctic Circle. Brown and blue-footed boobies often feed in inshore waters, while red-footed and Abbott's take probably the longest foraging trips and are known to occur several hundred kilometers from the nearest land.

A variety of sites are used for breeding, although this occurs almost invariably on offshore islands and on rocky outcrops. Several species place nests directly on exposed, flat ground, others choose to do so on cliffs. The third most common nesting habitat is on top of tall tropical trees or, alternatively, on the lower scrub of oceanic islands.

Behavior

The social behavior of boobies and gannets is complex and has been the subject of various studies, most notably by J. B. Nelson. All species nest colonially in quite high densities, which has favored the development of ritualized displays, particularly to indicate site-ownership or to obtain a mate. These are most developed in aggressive species, such as the northern gannet, that live in packed-in colonies, often on cliffs or other unstable surfaces. By resorting to ritualized behavior, sulids avoid the risks implicit in physical squabbles.

Feeding ecology and diet

Boobies and gannets are highly specialized fish-eaters. They prey mostly on mobile, schooling fish that frequent open waters, such as mackerel, whiting, pilchard, and anchovy. In tropical waters, flying-fish and squid are also frequent prey. A certain degree of opportunism is common in the family; northern gannets often follow trawlers searching for fish discards.

In order to catch their prey, sulids typically plunge-dive from great heights above the water (generally from 33 to 100 ft [10–30 m], but up to 330 ft [100 m] has been recorded). Once they have located their prey, they close their wings and plunge vertically, head first, into the water. Just before entering the water, they extend their wings backwards alongside their body and thus achieve a torpedo-like shape which probably assists them in reaching a greater depth. Once under water, they may use their wings to penetrate even deeper, perhaps up to 82 ft (25 m). Prey is generally caught on the way up and is usually swallowed underwater, thus avoiding the harassment of more opportunistic feeders such as frigatebirds or gulls.

Reproductive biology

Many species, particularly of gannets, pair for life and reunite annually at the nest-site, having spent the off-season individually at sea. In that context, pair-bonding displays play an important role and can take up a significant part of the

Incubating posture of the blue-footed booby (*Sula nebouxii*), with webbed feet, as opposed to bird's breast, directly applied to the egg. (Illustration by Patricia Ferrer)

breeding season. They generally consist of exaggerated movements of the head and neck, and also of the whole body, the wings, feet, and tail. Boobies make extensive use of their wings and feet in displays, most notably the blue-footed with its unique aerial greeting, in which the incoming bird salutes with outstretched feet just as it is about to land.

Not all species are strictly seasonal. The pantropical boobies, in particular, time their breeding attempts to local conditions and food availability. Often in the same colony there may be pairs which are at different breeding stages. However, the three gannets and Abbott's and Peruvian boobies time their nesting seasons to make them coincide with the best weather and the most productive conditions at sea, both during the chick-rearing period and immediately after chicks have fledged. This increases the chances of juvenile survival when the environment is highly seasonal.

Breeding density is moderate to very high; all sulids tend to group in colonies which they sometimes share with other

A blue-footed booby (*Sula nebouxii*) protects its eggs from the sun in the Galápagos Islands. (Photo by Andrew Martinez. Photo Researchers, Inc. Reproduced by permission.)

Northern gannets (*Morus bassanus*) in a territorial dispute at the nesting site. (Photo by Hugh Clark. Photo Researchers, Inc. Reproduced by permission.)

species. Nesting colonies are particularly dense in the case of gannets, especially when located on flat ground. In those cases, nests are spaced regularly, the distance between nests being determined by the maximum length that two neighboring birds can reach with their bills while seated.

Booby and gannet nests are quite rudimentary, especially those of ground-nesting species. Often, they consist only of a slight depression or an accumulation of debris, glued together with the birds' excreta. The tree-nesting red-footed and Abbott's boobies build slightly more elaborate nests, usually a platform of sticks on one of the upper branches.

Most species lay single-egg clutches, as only one chick can be raised successfully in most environments. Only the Peruvian and blue-footed boobies, living in the exceptionally rich waters of the Humboldt current, can expect to raise more than one chick successfully and thus lay three and two eggs respectively. Other species, such as the pantropical masked and brown boobies, often lay two eggs, but subsequently reduce their brood through sibling aggression, the older chick killing its younger sibling and so ensuring that the strongest one receives all of the limited food resources.

Incubation lasts 41–45 days in most species, although Abbott's booby extends its incubation period to an average 57 days. Both sexes contribute in long stints (12–60 hours) and no feeding occurs between the breeding adults. The eggs, which are incubated by wrapping the webbed feet around them, have unusually thick shells.

The chick is born naked and is continuously guarded for the first month, until it can regulate its own body tempera-

ture. It is fed on fish remains directly from the parent's mouth and often has to reach into the parent's throat. Chicks grow rapidly, particularly so in the more seasonal gannets which soon attain the adult birds' weight or even go beyond it. In those species, however, no post-fledging care occurs and when young reach fledging age they are left unattended until they jump out to sea on their own. Booby chicks take much longer to fledge (up to five months in Abbott's) and are fed for quite a lengthy period after they fledge.

Conservation status

In its 2000 assessment of the threatened status of the world's birds, BirdLife International classed Abbott's booby as Critically Endangered and the cape gannet as Vulnerable, according to IUCN standards. The booby has its breeding range confined to Christmas Island in the Indian Ocean, and has suffered a considerable reduction of its distribution within historic times. The population on Christmas Island has declined in the past due to habitat destruction through forest clearance. The introduction in the late 1990s of an alien ant species *Anoplolepis gracilipes* is predicted to cause a rapid decline through predation on nestlings, habitat alteration, and farming of scale insects that damage the trees. The cape gannet is classified as Vulnerable due to having only six breeding colonies. This renders the species at risk from both natural disasters and human-caused hazards. Among the latter are the risk of an oil spill (one incident in 1993 killed 5,000 gannets) and the more worrying collapse of the sardine fishery in Namibia, formerly the stronghold of the species, which has caused the species to decline severely.

The remaining species are not considered to be globally threatened or in danger of extinction but their continuation depends on conservation of their nesting sites and the overall marine environment. Tourism and the overexploitation of fish stocks may adversely affect several, if not all, species of boobies and gannets. Effective protection of offshore islands where sulids place their colonies is an essential part of any conservation program for these species.

Significance to humans

Over the centuries, humans have exploited boobies and gannets, their eggs and chicks, for food. The birds were at one time an important source of protein for certain local communities in the northern Atlantic. The fact that the birds breed in quite large numbers on islands that are generally accessible has probably contributed to their exploitation by humans. This still occurs throughout the tropics, although perhaps to a lesser degree as environmental education has begun to influence human behavior. Many species have seen their numbers artificially being kept to quite low levels for centuries, and have started to recover only in the last few decades. The Peruvian booby, one of the three "guano birds," has suffered severely from direct disturbance and habitat alterations in the past, during times of the intensive exploitation of guano for agricultural purposes, which continued until well into the first part of the twentieth century.

Red-footed booby (*Sula sula*) sky pointing at Kanoehe Bay, Oahu, Hawaii. (Photo by C.K. Lorenz. Photo Researchers, Inc. Reproduced by permission.)

1. Red-footed booby (*Sula sula*); 2. Blue-footed booby (*Sula nebouxii*); 3. Masked booby (*Sula dactylatra*); 4. Northern gannet (*Morus bassanus*); 5. Australasian gannet (*Morus serrator*); 6. Cape gannet (*Morus capensis*); 7. Brown booby (*Sula leucogaster*); 8. Abbott's booby (*Papasula abbotti*); 9. Peruvian booby (*Sula variegata*). (Illustration by Patricia Ferrer)

Species accounts

Abbott's booby
Papasula abbotti

TAXONOMY
Sula abbotti, Ridgway, 1893, Assumption Island. Monotypic.

OTHER COMMON NAMES
French: Fou d'Abbott; German: Abbott-Tölpel; Spanish: Piquero de Abbott.

PHYSICAL CHARACTERISTICS
31 in (79 cm); 3.2 lb (1.46 kg). Distinctive shape with long, narrow wings. White underparts, neck and head; upperparts dark-brown. Bill slightly hooked and highly serrated, pinkish in female, blue-gray tinged pink in male.

DISTRIBUTION
Breeding currently confined to Christmas Island (Indian Ocean) from where it disperses widely for foraging. Formerly more widespread across Indian Ocean, east to western Pacific.

HABITAT
Strictly marine and pelagic. Nesting is restricted to tall forest trees in central plateau of Christmas Island. Foraging area not precisely known, but frequently seen in rich upwelling area off Java, often well away from nearest land.

BEHAVIOR
Nesting site on trees high above ground affects territorial and pair behavior, so Abbott's boobies' displays are the least fervent of all sulids. Territorial disputes are unknown in this species and even chick begging behavior is moderate in comparison with that of its congeners.

FEEDING ECOLOGY AND DIET
Not precisely known. Thought to prey mostly on flying-fish and squid. Forages well away from nesting island and presumably feeds by plunge-diving like other members of the family.

REPRODUCTIVE BIOLOGY
Fairly seasonal (laying in May through July.) but only a biennial breeder when successful. Very low reproductive success, mainly due to coincidence of breeding season with monsoons. Nest is platform of twigs and sticks high above ground; only loosely colonial. Lays only one egg, which is incubated for 57 days (longest of all sulids). Chicks fledge at 140–175 days. Post-fledging care period is also very long: 162–280 days. Does not breed until four to six years old.

CONSERVATION STATUS
Critically Endangered (BirdLife International, 2000), due to extremely reduced breeding range and small, declining population of only 2,500 active pairs as of 2000. Most highly threatened by ecological alterations on breeding island caused by introduced yellow crazy ant (*Anoplolepis gracilipes*), which is known to prey on chicks, as well as to kill the red crab (*Gecaroidea natalis*), and to farm scale insects which damage the trees. A control program for this ant has been initiated. In the past, mining of Christmas Island for phosphate extraction has reduced nesting habitat significantly. Destruction of rainforest on former breeding islands is thought to have caused their extirpation.

SIGNIFICANCE TO HUMANS
None known. Abbott's booby's secretive habits have resulted in few interactions between this species and humans. ◆

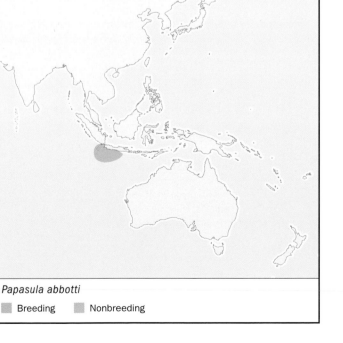

Papasula abbotti
■ Breeding   ■ Nonbreeding

Northern gannet
Morus bassanus

TAXONOMY
Pelecanus Bassanus, Linnaeus, 1758, Bass Rock, Scotland. Monotypic.

OTHER COMMON NAMES
English: (North) Atlantic gannet; French: Fou de Bassan; German: Basstölpel; Spanish: Alcatraz Atlántico.

PHYSICAL CHARACTERISTICS
34.3–39.4 in (87–100 cm); 5.1–7.9 lb (2.3–3.6 kg); wingspan 65–70.9 in (165–180 cm). Largest of sulids, a strong bird with mainly a strikingly white plumage. Compared with other gannets, bill is slightly stouter and head is paler cream. Juveniles mainly dark brown, gradually gaining white feathers of adult plumage.

DISTRIBUTION
Exclusively in the north Atlantic, where breeds on both sides 46–72° north. More widespread on eastern side, where in winter also enters the Mediterranean Sea and disperses south to subtropical waters. On western side, breeds on islands off Newfoundland and in the Gulf of St. Lawrence (Canada) and disperses south in winter to the Gulf of Mexico.

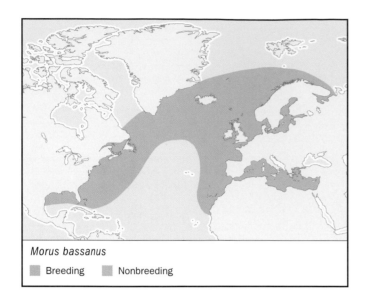

Morus bassanus
■ Breeding ■ Nonbreeding

HABITAT
Strictly marine, mainly in waters over the continental shelf. Breeds on cliffs on offshore islands or, more rarely, on mainland.

BEHAVIOR
Breeds in dense colonies where aggressiveness and intense social behavior have given way to complex repertoire of stereotyped displays. Breeding birds acquire a nest-site, which they then defend against intruders and maintain from year to year. Pair behavior is equally complex and linked to the nest-site. At sea, often occurs in groups particularly congregating around rich feeding sources but with little interaction.

FEEDING ECOLOGY AND DIET
Feeds on shoaling pelagic fish like herring (*Clupea*), mackerel (*Scomber*) and sprat (*Sprattus*), also sandeels (*Ammodytes*). Makes spectacular plunge-dives from great heights. Also regularly attends trawlers.

REPRODUCTIVE BIOLOGY
Highly seasonal, starting March through April. Forms large colonies on cliffs or on flat ground, where builds large nest of seaweed, grass, etc. and a significant amount of excreta. Lays one egg only, incubated by both parents for 44 days. Chick fledges at 90 days; on its own, after it has been deserted by parents. Does not breed until four to five years old.

CONSERVATION STATUS
Not threatened. Abundant and widespread throughout its range. Protection of breeding sites and cessation of former direct exploitation of chicks (for food) led to significant recovery over most of twentieth century. Overexploitation of fisheries remains an important threat; also suffers some degree of incidental mortality at sea.

SIGNIFICANCE TO HUMANS
Chicks used to be taken for food in some local communities, a practice that still continues in a few places (e.g., Sula Sgeir, off Scotland). Also present in literature and art. Nowadays colonies may constitute important sources of income locally, as tourist activities are developed around them. ◆

Cape gannet
Morus capensis

TAXONOMY
Dysporus capensis, Lichtenstein, 1823, Cape of Good Hope. Monotypic.

OTHER COMMON NAMES
English: African gannet; French: Fou du Cap; German: Kaptölpel; Spanish: Alcatraz del Cabo.

PHYSICAL CHARACTERISTICS
33.5–35.4 in (85–90 cm); 5.7 lb (2.6 kg). Slightly smaller than northern gannet, wings show black wingtips and secondary feathers; tail feathers also black. Black gular stripe much longer than in the other gannets; head darker cream than in northern gannet. Juveniles dark, gradually acquiring adult plumage.

DISTRIBUTION
Breeds coasts of South Africa and Namibia. Disperses north along African coasts, to the Gulf of Guinea in the Atlantic and to Mozambique, exceptionally to Kenya, in the Indian Ocean.

HABITAT
Strictly marine, mainly in waters of the continental shelf. Nests on flat offshore islands.

BEHAVIOR
Much as in northern gannet although much less aggressive and site competition less intense, despite nesting in very dense colonies on flat ground. Similarly, sexual behavior and pair-bonding displays are more moderate.

FEEDING ECOLOGY AND DIET
Feeds mostly on shoaling pelagic fish, particularly pilchard (*Sardinops*), anchovy (*Engraulis*), saury (*Scomberesox*) and mackerel (*Scomber*). Feeds by plunge-diving from 66 ft (20 m)

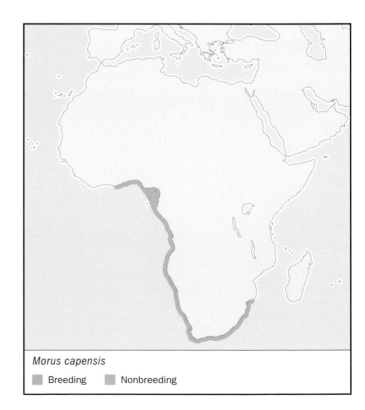

Morus capensis
■ Breeding ■ Nonbreeding

above water. Also forms large concentrations attending trawlers.

REPRODUCTIVE BIOLOGY
Highly seasonal, September through April. Nests in very dense colonies on flat ground, where nests consist of accumulation of debris with central depression. Lays one egg, exceptionally two. Incubation lasts 44 days. Young fledges at 97 days. Does not start breeding until three to four years old.

CONSERVATION STATUS
Vulnerable. Only six breeding colonies known. Population has undergone important reductions in the past and, in latter part of twentieth century, has been further reduced through overexploitation of fish stocks, particularly in Namibia. Oil pollution and mortality caused by fishing gear are also known to take a heavy toll.

SIGNIFICANCE TO HUMANS
In the past, heavily exploited for food and for fish-bait. The cape gannet is one of the guano birds, its colonies being used to extract the fertilizer until well into the twentieth century. ◆

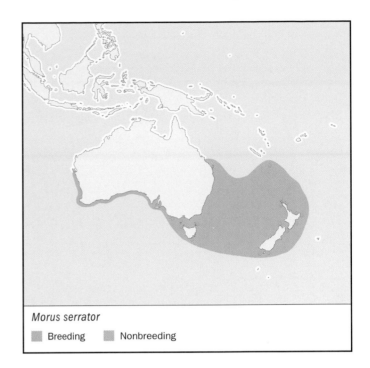

Morus serrator
■ Breeding ■ Nonbreeding

Australasian gannet
Morus serrator

TAXONOMY
Pelecanus serrator, G. R. Gray, 1843, from *Sula australis*, Gould 1841 (preoccupied), Tasmania. Monotypic.

OTHER COMMON NAMES
French: Fou austral; German: Australtölpel; Spanish: Alcatraz Australiano.

PHYSICAL CHARACTERISTICS
33.1–35.8 in (84–91 cm); 5.2 lb (2.35 kg); wingspan 63–66.9 in (160–170 cm). Resembles cape gannet but is slightly smaller, has white outer tail feathers and blue orbital ring is more intensely colored. Juveniles dark, gradually acquiring adult plumage.

DISTRIBUTION
Breeds coasts of New Zealand, Tasmania, and Australia. Disperses over those waters and along both coasts of Australia, reaching as far as Tropic of Capricorn.

HABITAT
Strictly marine, occurs mostly over continental shelf. Breeds on offshore islets.

BEHAVIOR
Much as in cape gannet, which it most closely resembles. Compared to northern gannet, less aggressive and not so competitive over nest-sites. Also complex but not so intense sexual behavior and pair-bonding displays.

FEEDING ECOLOGY AND DIET
Feeds mostly on shoaling pelagic fish, especially pilchard (*Sardinops*), anchovy (*Engraulis*), and jack mackerel (*Trachurus*). Feeds by plunge-diving. Also attends trawlers, where large numbers may concentrate.

REPRODUCTIVE BIOLOGY
Highly seasonal, October through May. Nests in rather small but dense colonies. Builds rough nest of accumulated seaweed and grass, cemented together with excreta. Lays one egg, exceptionally two. Incubation lasts 44 days. Young fledges at 102 days. Does not start breeding until five to six years old.

CONSERVATION STATUS
Not threatened. During twentieth century, population gradually recovered from earlier heavy persecution although some colonies (e.g. Tasmania) continued to decline markedly during second part of the century. Total world population is smallest of all gannets and species still suffers some degree of direct exploitation (eggs and chicks). Sometimes caught accidentally during fishing activities.

SIGNIFICANCE TO HUMANS
Breeding colonies have been traditionally raided for eggs and chicks. Species present in indigenous folklore in New Zealand. Currently some tourist activities are being developed around nesting colonies. ◆

Blue-footed booby
Sula nebouxii

TAXONOMY
Sula nebouxii, Milne-Edwards, 1882, Chile. Two subspecies recognized, *S. n. nebouxii* Milne-Edwards, 1882 and *S. n. excisa* Todd, 1948.

OTHER COMMON NAMES
French: Fou à pieds bleus; German: Blaufusstölpel; Spanish: Piquero Camanay.

PHYSICAL CHARACTERISTICS
29.9–33.1 in (76–84 cm); wingspan 59.8 in (152 cm). Upperparts generally dark brown; underparts white. Distinctive blue feet. Iris pale. Female averages larger. Race *excisa* appears larger and paler.

Sula nebouxii
◼ Breeding ◼ Nonbreeding

Peruvian booby
Sula variegata

TAXONOMY
Dysporus variegatus, Tschudi, 1843, islands off Peru. Monotypic.

OTHER COMMON NAMES
English: Variegated booby; French: Fou varié; German: Guanotölpel; Spanish: Piquero Peruano.

PHYSICAL CHARACTERISTICS
28–29.9 in (71–76 cm). Smaller version of blue-footed booby, which it most closely resembles. Lacks blue feet and averages paler, with white on head and neck. Upperwing and back are mottled white.

DISTRIBUTION
Exclusive to Humboldt current area of Pacific, breeding from northern Peru to central Chile. Occurs in Ecuador.

HABITAT
Strictly marine. Found quite close inshore, where it feeds in cool, rich waters of upwelling zones. Breeds on rocky islets and on cliff ledges.

BEHAVIOR
Behaviorally, the Peruvian booby most closely resembles the blue-footed although its displays are somewhat more moderate. It breeds in densely packed colonies, yet site-tenancy is less intense than in other species.

FEEDING ECOLOGY AND DIET
Almost an exclusive feeder on anchoveta (*Engraulis ringens*) when this was abundant. After stocks were depleted in 1970s

DISTRIBUTION
Continental coasts of east Pacific Ocean, from northwest Mexico in north to Peru in south (*S. n. nebouxii*) and Galápagos Islands (*S. n. excisa*).

HABITAT
Strictly marine. Frequents cool, rich waters in areas of upwelling, often close inshore. Breeds along rocky coasts, on cliffs and islets with little or no vegetation.

BEHAVIOR
Spectacular and elaborate pair-bonding displays, both aerial and on ground, where blue feet play important role.

FEEDING ECOLOGY AND DIET
Active fisher, preying mostly on sardine (*Sardinops*), anchovy (*Engraulis*), and mackerel (*Scomber*); also flying-fish (*Exocoetus*). Highly gregarious, often feeds in quite large groups, plunge-diving in unison. May fish with other species. Often feeds in shallow water.

REPRODUCTIVE BIOLOGY
Not markedly seasonal. Usually nests on ground, sometimes also on vegetation. Forms large colonies where nest is mere circle of accumulated excreta around a slight depression. Lays two eggs on average (one to three), which are incubated for 41 days. Chicks fledge at 102 days and afterwards are cared for 56 days on average. First breeds at two to three years of age.

CONSERVATION STATUS
Not threatened. World population quite small but quite abundant locally. Suffers predation from alien predators, at least in Galápagos. Most breeding sites currently protected.

SIGNIFICANCE TO HUMANS
Subject to exploitation for food in the past. ◆

Sula variegata
◼ Breeding ◼ Nonbreeding

and 1980s, resorted to sardine (*Sardinops*), mackerel (*Scomber*), and other fish. Feeds by plunge-diving in groups of 30–40, often more, individuals.

REPRODUCTIVE BIOLOGY
Only moderately seasonal, September through February in Peru, much later in Chile. Breeds in immense colonies where nests consist of loose pile of seaweed and debris, stuck together with excreta. Lays three eggs on average (one to four), which are incubated for 42 days. Chicks fledge at 78–105 days and are cared for a further 62 days on average. First breeding occurs at two to three years of age.

CONSERVATION STATUS
Not threatened. Has undergone significant reductions in the past (through direct exploitation, disturbance at breeding colonies, and depletion of fish stocks) but is somewhat stable at present. However, always subject to risk by regular El Niño phenomena, which can cause severe mortality of both adults and young.

SIGNIFICANCE TO HUMANS
One of the main guano birds, together with Guanay cormorant (*Phalacrocorax bouganvillii*) and Peruvian pelican (*Pelecanus thagus*). Subject to intense disturbance at breeding colonies and to direct exploitation of eggs and chicks up to middle twentieth century; practice is perhaps still maintained at a smaller scale although nesting sites are legally protected. ◆

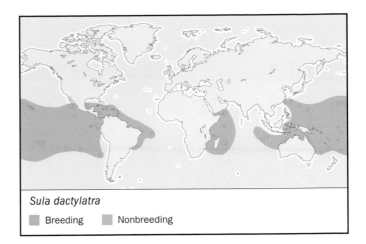

Sula dactylatra
■ Breeding   ■ Nonbreeding

Masked booby
Sula dactylatra

TAXONOMY
Sula dactylatra, Lesson, 1831, Ascension Island. Five subspecies generally recognized: *S. d. dactylatra* , Lesson, 1831; *S. d. melanops*, Heuglin, 1859; *S. d. personata*, Gould, 1846; *S. d. fullagari*, O'Brien and Davies, 1990; *S. d. granti*, Rotschild, 1902.

OTHER COMMON NAMES
English: Blue-faced booby, white booby; French: Fou masqué; German: Maskentölpel; Spanish: Piquero Enmascarado.

PHYSICAL CHARACTERISTICS
31.9–36.2 in (81–92 cm); wingspan 59.8 in (152 cm). Largest of all boobies, body feathers mostly white; flight and tail feathers black. Bare parts mostly dark, bill usually yellow in males, duller in females. Females average slightly larger in size.

DISTRIBUTION
Pantropical, race *dactylatra* occurs in Caribbean and Atlantic; *melanops* in west Indian Ocean; *personata* in east Indian Ocean and central Pacific; *fullagari* in north Tasman Sea; *granti* in east Pacific.

HABITAT
Strictly marine and fairly pelagic, prefers more offshore waters than other booby species. Nests on bare ground and cliffs on rocky offshore islands.

BEHAVIOR
Breeds in less dense colonies than other boobies. Accordingly, defends nest site less tenaciously and whole behavior is less aggressive. Much territorial behavior is based on ritualized displays. Pair-bonding behavior is also less intense than in other species.

FEEDING ECOLOGY AND DIET
Feeds mostly on shoaling fish, especially flying-fish, which it catches by plunge-diving from great heights. Feeds farther offshore than other species, also taking larger prey.

REPRODUCTIVE BIOLOGY
Only loosely colonial, very simple nest of accumulated excreta on cliff, slope or flat ground. Usually lays two eggs; brood size subsequently reduced to one chick through sibling aggression. Incubates eggs for 44 days. Chick fledges at 120 days and is further cared for another 156 days. Does not breed until two to three years old.

CONSERVATION STATUS
Not threatened. Much widespread and locally abundant, total population may number several hundred thousand individuals. Known to have undergone some declines locally, particularly as a consequence of predation by introduced animals. Eggs and chicks also taken for food locally. Booming tourist industry may pose further threat.

SIGNIFICANCE TO HUMANS
Subject to a moderate degree of exploitation for food, perhaps also for fish-bait. Some breeding colonies may be of interest for local tourist industry. ◆

Red-footed booby
Sula sula

TAXONOMY
Pelecanus Sula, Linnaeus, 1766, Barbados, West Indies. Three subspecies generally recognized: *S. s. sula*, Linnaeus, 1766; *S. s. rubripes*, Gould, 1838; *S. s. websteri*, Rotschild, 1898.

OTHER COMMON NAMES
French: Fou à pieds rouges; German: Rotfusstölpel; Spanish: Piquero Patirrojo.

PHYSICAL CHARACTERISTICS
26–30.3 in (66–77 cm); 1.9–2.2 lb (0.9–1.0 kg); wingspan 35.8–39.8 in (91–101 cm). Smallish, polymorphic sulid. Some individuals mostly white, with only flight feathers black (tail remains white in most plumages); others are wholly brown, with flight feathers always looking darker. Feet and cere around bill reddish in most plumages. Females average slightly larger.

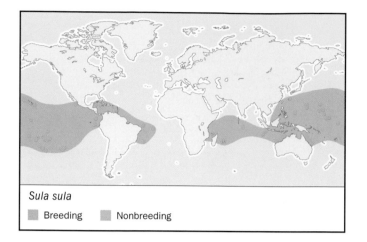

Sula sula
■ Breeding ■ Nonbreeding

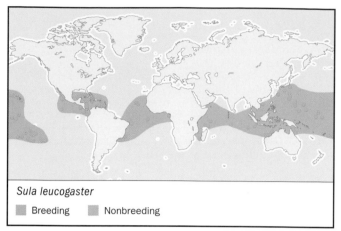

Sula leucogaster
■ Breeding ■ Nonbreeding

DISTRIBUTION
Pantropical, race *sula* occurs in Caribbean and southwest Atlantic Ocean, *rubripes* in tropical west and central Pacific and also Indian Ocean, *websteri* in east Pacific.

HABITAT
Strictly marine and largely pelagic, feeding largely offshore. Nests on offshore islands with abundant vegetation.

BEHAVIOR
Repertoire of ritualized displays more moderate than in other species, adapted to breeding habitat on trees. Role of bright-red feet largely unknown.

FEEDING ECOLOGY AND DIET
Feeds mostly offshore, preying on flying-fish and squid. Catches prey by plunge-diving from considerable height, but also takes flying-fish in flight. Partially nocturnal habits.

REPRODUCTIVE BIOLOGY
Not seasonal, may start breeding in any month. Highly colonial, builds nest of sticks on top of tree or bush. Lays one egg, incubated for 45 days. Chick fledges at 100–139 days, later cared for 190 days. First breeds at two to three years old.

CONSERVATION STATUS
Not threatened. Widely scattered, reasonably large population. Subject to direct exploitation and disturbance, most important threat comes from destruction of nesting habitat.

SIGNIFICANCE TO HUMANS
Traditionally exploited for food over much of its range. ◆

Brown booby
Sula leucogaster

TAXONOMY
Pelecanus Leucogaster, Boddaert, 1783, Cayenne. Four subspecies recognized: *S. l. leucogaster*, Boddaert, 1783; *S. l. plotus*; J. R. Forster, 1844; *S. l. brewsteri*, Goss, 1888; *S. l. etesiaca* , Thayer and Bangs, 1905.

OTHER COMMON NAMES
English: White-bellied booby; French: Fou brun; German: Weissbauchtölpel; Spanish: Piquero Pardo.

PHYSICAL CHARACTERISTICS
25.2–29.1 in (64–74 cm); 1.6–3.4 lb (0.7–1.6 kg); wingspan 52–59.1 in (132–150 cm). Wholly dark, except for white belly. Color of head and bare parts varies with race. Females average slightly larger.

DISTRIBUTION
Pantropical, race *leucogaster* occurs in Caribbean and tropical Atlantic, *plotus* in Red Sea and west Indian Ocean east to central Pacific, *brewsteri* in northeast tropical Pacific, *etesiaca* in central east Pacific.

HABITAT
Strictly marine, feeding mostly in inshore waters. Nests on cliffs, slopes, or bare ground on offshore islands or coral atolls.

BEHAVIOR
Ample repertoire of ritualized displays, including some aerial elements. Rather aggressive on breeding grounds.

FEEDING ECOLOGY AND DIET
Feeds close to the shore mostly on flying-fish and squid caught by plunge-diving from lower heights, often at an oblique angle. Uses feet and wings for underwater propulsion. Also commonly feeds on the wing, catching flying-fish or harassing other birds.

REPRODUCTIVE BIOLOGY
Only locally seasonal. Forms colonies on flat ground or among vegetation. Nest is only a small depression, sometimes lined with grass. Lays two eggs but brood size subsequently reduced through sibling aggression. Incubation lasts 43 days. Chick fledges at 85–105 days, then cared for a further 118–259 days. Does not breed until two to three years old.

CONSERVATION STATUS
Not threatened. Numerous and widespread, though numbers significantly reduced in historic times through direct exploitation. Locally threatened with alien predators, tourist development, and lack of protection at nest-sites.

SIGNIFICANCE TO HUMANS
In the past, widely taken for food and fish-bait. Such practices still persist in some areas. Due to presence of widely scattered breeding colonies, may give rise to incipient tourist activities in places. ◆

Resources

Books

Barnes, K. N., ed. *The Eskom Red Data Book of Birds of South Africa, Lesotho and Swaziland.* Johannesburg: BirdLife South Africa, 2000.

BirdLife International. *Threatened Birds of the World.* Barcelona: Lynx Edicions; and Cambridge, United Kingdom: BirdLife International, 2001.

del Hoyo, J., A. Elliott, and J. Sargatal, eds. *Handbook of the Birds of the World.* Vol. 1, *Ostrich to Ducks.* Barcelona: Lynx Edicions, 1992.

Garnett, S. T., and G. M. Crowley. *Revised Action Plan for Australian Birds.* Canberra: Environment Australia and Birds Australia, 2001.

Harrison, J. A., D. G. Allan, L. G. Underhill, M. Herremans, A. J. Tree, V. Parker, and C. J. Brown, eds. *The Atlas of Southern African Birds.* Vol. 1, *Non-passerines.* Johannesburg: BirdLife South Africa, 1997.

Nelson, J. B. *The Sulidae: Gannets and Boobies.* Oxford: Oxford University Press, 1978.

Nelson, J. B. *The Gannet.* London: T. & A. D. Poyser, 1978.

Organizations

BirdLife International. Wellbrook Court, Girton Road, Cambridge, Cambridgeshire CB3 0NA United Kingdom. Phone: +44 1 223 277 318. Fax: +44-1-223-277-200. E-mail: birdlife@birdlife.org.uk Web site: <http://www.birdlife.net>

Carles Carboneras

Pelicans

(Pelecanidae)

Class Aves
Order Pelecaniformes
Suborder Pelecani
Family Pelecanidae

Thumbnail description
Large to very large waterbirds with webbing connecting all four toes and a very long bill with a distensible pouch

Size
41–74 in (105–188 cm); 6–33 lb (2.7–15 kg)

Number of genera, species
1 genus; 7 species

Habitat
Lakes, rivers, and coastlines

Conservation status
Vulnerable: 1 species; Conservation Dependent: 1 species

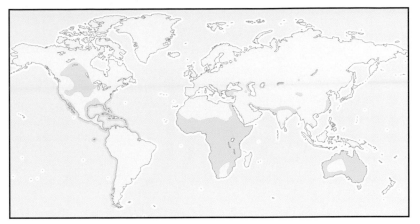

Distribution
Most temperate and tropical regions

Evolution and systematics

One of the earliest known pelicans was *Pelecanus grandis* from the Lower Miocene (22.5–5 million years ago). At least three other species have been identified in later deposits from that epoch.

The question of phylogeny among the pelecaniform birds is tricky: the Phalacrocoracidae, Sulidae, Fregatidae, and Phaetheontidae share a number of morphological characteristics with pelicans. Perhaps the most significant shared characteristic is the presence of webbing connecting all four toes. In addition, these families exhibit a greater similarity in their social displays than would be expected by evolutionary chance alone. This grouping conflicts with DNA data that suggest that pelicans are only distantly related to these taxa and are more closely related to the shoebill (*Balaeniceps rex*) and the hammerhead (*Scopus umbretta*) than to other extant birds.

Ornithologists also disagree on the phylogenetic relationships among the seven living species. Most researchers agree, however, that the brown pelican (*Pelecanus occidentalis*) assumed its own evolutionary trajectory separate from the others quite early.

Physical characteristics

Pelicans are among the most recognizable birds and the largest capable of flight (6–33 lb (2.7–15 kg)). They have long, broad wings, fairly long necks, and very long bills. Between the branches of the lower mandible there is a distensible skin pouch; the upper mandible serves merely as a flat lid to cover it. A sharp "nail" curves downward from the tip of the upper mandible. The tongue is generally small, although the tongue bone of brown pelicans is about 4 in (10 cm) long and rela-

tively large. Legs are short, feet are large, and the four toes are connected by webbing. Pelicans have 17 cervical vertebrae; the uropygial (or oil) glands have 6–9 slit-like openings. The plumage is not as water repellant as that of most other aquatic birds. With the obvious exception of the brown pelican, the feathers of adult pelicans are usually white or light silver/gray, sometimes with a pinkish tint. All of these so-called white pelicans have black wingtips.

Distribution

Pelicans can be found on every continent except Antarctica. The brown pelican is the sole neotropical species, and, being the only exclusively marine member of this family, it is not found in the South American interior. Pelicans also are absent in Asia north of Mongolia as well as in western Europe. Fossil evidence from northern Europe suggests that Dalmatian pelicans (*Pelecanus crispus*) were present in prehistoric times and vagrants of this species have been recorded there in modern times.

Habitat

Brown pelicans are the only true seabirds in the Pelecanidae. American white pelicans (*P. erythrorhynchos*) are most often found in freshwater environments, but they occur on brackish water regularly in winter. Those species living in the Old World may also be found on waters of varying salinity. In general, however, they are more apt to be found on freshwater.

Critical among the criteria required for survival is access to an ample supply of fish. Ephemeral water bodies or those subject to winter ice may be seasonally exploited by tropical

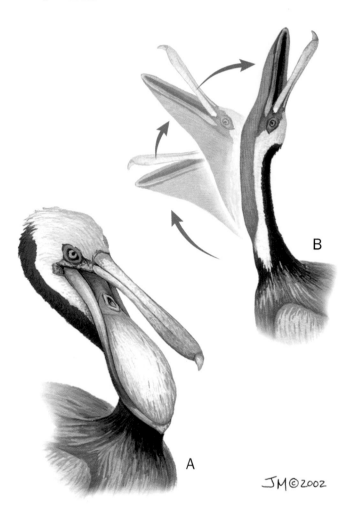

(A) glottis exposure and (B) bill toss of the brown pelican (*Pelecanus occidentalis*). (Illustration by Jacqueline Mahannah)

and Holarctic species, respectively. These birds aggregate in breeding colonies numbering, in some cases, in the thousands. For this reason, it is critical that nest sites be selected in areas where pelicans can raise their young undisturbed.

Behavior

Pelicans float high on the water and carry their wings slightly raised. The bill rests on the slightly curved neck. In flight, the head is drawn back onto the shoulders. Flight is light and elegant; gliding often alternates with wing beats. Pelicans are sociable birds that fly in small groups or larger flocks, mostly in a diagonal line with respect to the direction that they are traveling. In several species group foraging is common, and pelicans often nest in very large colonies, sometimes together with other water birds.

Adult pelicans rarely use the few calls that they have available. They make hissing, blowing, groaning, or grunting calls. Occasionally they make clattering sounds with the bill. In breeding colonies the young are much noisier; they bleat like sheep, bark or squeak, and utter grunting contact calls. These are, however, only heard if one can remain in the

colony unobserved or can approach it unseen. If one is seen by the birds, the young remain as silent as adults. Pelicans are diurnal. They spend a significant amount of time preening or just resting.

Feeding ecology and diet

In addition to fish, which constitutes the bulk of the diet of every species, researchers have recorded a number of other prey. Among these are prawns, various amphibians and their larvae, small snakes and lizards, birds, and small mammals. Pelicans use their enormous bill to secure these items. The spectacular plunge-diving strategy of brown pelicans is well known, but other pelicans display equally interesting foraging methods. Numerous observers have described the peculiar manner of fishing used by Australian pelicans (*P. conspicillatus*), American white pelicans, great or eastern white pelicans (*P. onocrotalus*), and Dalmatian pelicans. H.A. Bernatzik, who observed the behavior on Lake Malik in Albania, described it as follows: "a number of them fish in the shallow waters by arranging themselves in a semi-circle and chasing the fish towards the shore, as cormorants also do, or they gradually encircle them. They form chains of 'beaters,' scaring the fish by vigorous wing beats, blocking off large areas of the water surface. In narrow rivers they are said to occasionally divide into two parties in order to drive the prey towards one another. While doing this they may even swim in two or three rows, one behind the other." Thus they move slowly from deeper to shallower water until at last they merely have to scoop up their prey.

Reproductive biology

Mate selection seems to be an annual affair carried out by the female. Some will choose the same male every year. The pair will defend the nest site against competitors. The distance between nests is typically equal to that span whereby a neighbor's outstretched bill is just out of reach when both parties are sitting on their respective clutch.

Australian pelicans (*Pelecanus conspicillatus*) in locked combat, fighting over food in Tasmania. (Photo by Gregory G. Dimijian. Photo Researchers, Inc. Reproduced by permission.)

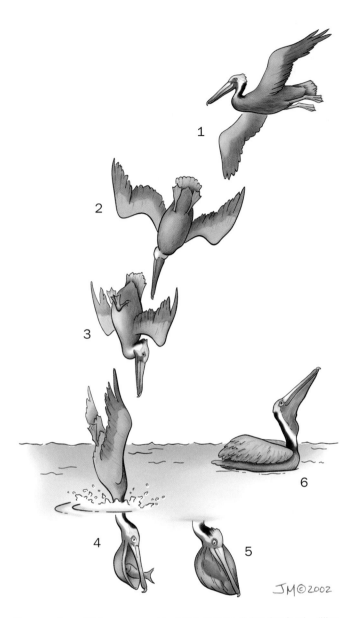

Brown pelican (*Pelecanus occidentalis*) plunge diving for food. (Illustration by Jacqueline Mahannah)

A brown pelican (*Pelecanus occidentalis*) dives for fish. (Photo by Adam Jones. Photo Researchers, Inc. Reproduced by permission.)

Female pelicans typically lay two to three eggs at two to three day intervals. Incubation may last from 29 to 35 days and is performed by both sexes. The parents relieve one another every day or two and spend off duty hours feeding themselves rather than each other. The young are born naked and helpless and regurgitated foodstuffs are extracted from the cavernous maws of the parents. Floods, cold rainy weather, and siblicide (frequently indirectly by the older offspring monopolizing mealtimes) cause great losses of eggs and young, so more than one chick being raised in a nest is rare. At three to four weeks of age, the young can escape into the reeds or water; at ten to twelve weeks they leave the colony temporarily and begin to fly and to fish on their own. Most are sexually mature at three and four years of age.

Conservation status

No pelican species has disappeared in historical times. Nevertheless, two species are of special concern to conservationists. The World Conservation Union (IUCN) lists the spot-billed pelican as Vulnerable and the Dalmatian pelican as Lower Risk/Conservation Dependent. The population of the former was estimated to be 11,500 birds and declining in 2000, whereas the latter may have stabilized somewhere between 15,000 and 20,000 birds. Ground-nesting pelicans, in particular, are sensitive to human trespassers and many colonies of several species were lost to expanding human populations in the twentieth century.

Significance to humans

Many tales and legends refer to pelicans and their strange appearance. Pelicans were known as domestic birds in ancient Egypt, as fishing helpers in India, and as reputed helpers in the building of the Kaaba in Mecca by the Muslims. The pelican was a symbol of maternal love in early Christianity; legend describes it as a bird that tears open its own breast to

Pelicans exhibit a large repertoire of displays during the brief period that precedes nest construction, an activity that is engaged in by both the male and female. Those species occurring in tropical places may breed at any time of the year while those of the Holarctic do so in the spring. All pelicans are colonial nesters and in captivity will virtually never reproduce when fewer than four pairs are present. Brown, spot-billed (*P. philippensis*), and pink-backed (*P. rufescens*) pelicans usually build their nests in trees; all others do so on the ground. To avoid terrestrial predators, those that construct ground nests often do so on islands. At the beginning of the breeding season, pelicans are very shy and sensitive to every disturbance, hence they often abandon their nests.

keep its young alive. This legend is probably based on the reddish spot that appears over the crop and gular pouch of Dalmatian pelicans during breeding season. The figure of the pelican as a martyr and a model of human mercy appears in the art of the Middle Ages, as well as on coats of arms. This belief has continued up to the present, when pelicans symbolize every form of mutual aid and Christian love of one's fellow humans.

1. Spot-billed pelican (*Pelecanus philippensis*); 2. Dalmatian pelican (*Pelecanus crispus*); 3. American white pelican (*Pelecanus erythrorhynchos*); 4. Brown pelican (*Pelecanus occidentalis*). (Illustration by Jacqueline Mahannah)

Species accounts

Dalmatian pelican
Pelecanus crispus

TAXONOMY
Pelecanus crispus Bruch, 1832, Dalmatia. Monotypic.

OTHER COMMON NAMES
English: Curly-headed pelican; French: Pélican fris; German: Krauskopfpelikan; Spanish: Pelícano Ceñudo.

PHYSICAL CHARACTERISTICS
Large birds, 63–71 in (160–180 cm); 20–29 lb (9–13 kg); male larger than female. Silvery-white shaggy or curly crest and brownish black wingtips.

DISTRIBUTION
Breeds locally from southeastern Europe to China. Winters from the Balkans through southeast China.

HABITAT
Lakes, rivers, deltas, and estuaries where human disturbance is minimal. Breeds on islands or among tall emergent vegetation.

BEHAVIOR
May display antagonistic behavior in the form of bill clattering and gaping, especially when defending nest sites. Male emits hisses and spitting sounds in concert with bowing display during courtship.

FEEDING ECOLOGY AND DIET
Less likely than other big pelicans to fish in large flotillas; usually feeds alone, in pairs, or in trios. Takes a wide variety of both freshwater and marine fish, including eels (*Anguilla*), carp (*Cyprinus* and *Carassius*), and rudd (*Scardinius*).

REPRODUCTIVE BIOLOGY
Breeds in smaller colonies than many other large pelicans. Onset of breeding varies widely depending on climate; may be as early as February or as late as August. Nests are constructed from plant material and bonded with excreta and frequently exceed 3 ft (1 m) in height and diameter. Two eggs are typically laid and incubated for 31–34 days. Chicks are hatched naked but develop white feathers within a month. Nestlings aggregate in crèches by seven weeks of age; fledge at 12 weeks; independent at 15 weeks.

CONSERVATION STATUS
Downlisted from Vulnerable to Conservation Dependent by BirdLife International at the close of the twentieth century. General population decline accelerated dangerously in the twentieth century due to reduction of wetland habitat, hunting, and overall human molestation including purposeful eradication by fishermen. Comprehensive conservation measures in Europe, including reintroduction of zoo-bred birds, are beginning to show results.

SIGNIFICANCE TO HUMANS
Prone to disturbance by tourists. Blamed for reduction in fish stocks. Bills are prized by traditional Mongolian herders who continue to hunt them. ◆

American white pelican
Pelecanus erythrorhynchos

TAXONOMY
Pelecanus erythrorhynchos Gmelin, 1789, Hudson Bay and New York. Monotypic.

OTHER COMMON NAMES
English: Rough-billed pelican, white pelican; French: Pélican à bec rouge; German: Nashornpelikan; Spanish: Pelícano Nortamericano.

PHYSICAL CHARACTERISTICS
47–70 in (120–178 cm); 8–17 lb (3.6–7.7 kg). White with yellowish gray crest and black wingtips. During breeding season, they develop a knob on the top of the orange bill.

DISTRIBUTION
Summers in western North America and southeast Texas, USA. Winters in California, Arizona, southeastern USA, and Mexico.

HABITAT
Rivers, lakes, estuaries, and seacoasts.

BEHAVIOR
Territorial during breeding season. Pair bonds strengthened by head bowing in the direction of one another and strutting walk in which male closely follows female, both with crests raised and pouches resting on chests.

FEEDING ECOLOGY AND DIET
Feed while swimming and do not dive into the water. Communal forager of various fish, typically those of little commercial value such as carp (*Cyprinus*). Also eat salamanders (*Ambystoma*) and their larvae.

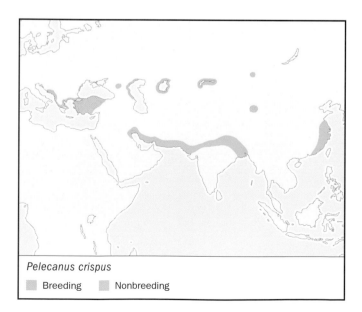

Pelecanus crispus

☐ Breeding ☐ Nonbreeding

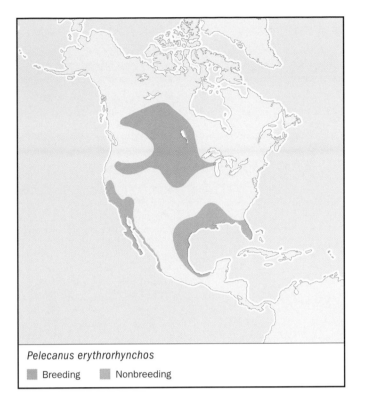

Pelecanus erythrorhynchos
■ Breeding   ■ Nonbreeding

Pelecanus occidentalis
■ Resident

REPRODUCTIVE BIOLOGY
Ground nests are constructed from plant material. Usually lays two eggs in the spring; incubates for four weeks. Nestlings aggregate in crèches by four weeks of age; fledge at nine weeks; independent at 12 weeks.

CONSERVATION STATUS
Not threatened. Population may be increasing after significant erosion throughout much of twentieth century. Several new breeding colonies recorded in the 1980s and 1990s.

SIGNIFICANCE TO HUMANS
May interact with fishermen but to a lesser degree than brown pelicans. Appears with young in logo of a North American insurance company as embodiment of mutual aid. ◆

Brown pelican
Pelecanus occidentalis

TAXONOMY
Pelecanus occidentalis Linnaeus, 1766, Jamaica. Six subspecies recognized.

OTHER COMMON NAMES
French: Pélican brun; German: Brauner Pelikan, Braunpelikan; Spanish: Peílcano Alcatraz.

PHYSICAL CHARACTERISTICS
40–60 in (102–152 cm); 6–22 lb (2.7–10 kg); male larger than female. These are the only dark-colored pelicans. Nonbreeding adults have white or yellowish head and neck and grayish brown bodies. Breeding birds have dark hindneck and a yellow patch on foreneck.

DISTRIBUTION
P. o. occidentalis: West Indies; *P. o. carolinensis*: Maryland, USA to northern Brazil; *P. o. californicus*: Oregon, USA to Panama; *P. o. murphyi*: Colombia to northern Peru; *P. o. urinator*: Galápagos; *P. o. thagus*: Peru to central Chile.

HABITAT
Seacoasts and estuaries.

BEHAVIOR
Head swaying, head turning, and bowing are among the pair bonding displays that precede breeding. Frequently roosts in trees and on manmade structures.

FEEDING ECOLOGY AND DIET
Plunge-dives from as high as 65 ft (20 m) in the air to apprehend various marine fish.

REPRODUCTIVE BIOLOGY
Breeds throughout the year; only in spring in northernmost part of range. Usually constructs nest of sticks in trees. Typically lays three eggs that are incubated for four weeks. Young fledge at 9–11 weeks.

CONSERVATION STATUS
Once Endangered, this species is no longer listed thanks largely to the elimination of harmful pesticides in North America.

SIGNIFICANCE TO HUMANS
One of three main guano birds of western South America. Among pelicans, one of the most tolerant of human activities. Frequently injured by abandoned fishing tackle. ◆

Spot-billed pelican
Pelecanus philippensis

TAXONOMY
Pelecanus philippensis Gmelin, 1789, Philippine Islands. Monotypic.

OTHER COMMON NAMES
English: Gray pelican, Philippine pelican, spotted-billed pelican; French: Pelican à bec tacheté; German: Graupelikan; Spanish: Pelícano oriental.

PHYSICAL CHARACTERISTICS
50–60 in (127–152 cm); 9–12 lb (4.1–5.4 kg); male slightly larger than female. Grayish white with dark wingtips.

DISTRIBUTION
Largest remaining populations are in India, Sri Lanka, southern Cambodia, and Sumatra. Vagrants may appear elsewhere in Southeast Asia.

HABITAT
Freshwater, brackish, and marine wetlands.

BEHAVIOR
Head bowing, head turning, and bill clapping are among the courtship displays.

FEEDING ECOLOGY AND DIET
May take small reptiles and amphibians in addition to fish. Occasionally forages communally in the manner typical of larger pelicans.

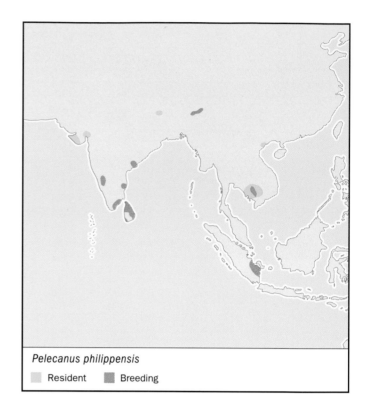

Pelecanus philippensis
▨ Resident ▨ Breeding

REPRODUCTIVE BIOLOGY
Usually lays three eggs in an arboreal nest of sticks. Incubation takes 30 days; fledging may occur between 60 and 90 days.

CONSERVATION STATUS
Listed as Vulnerable. Suffers from habitat loss, pollution, and human disturbance. Numbers decreased alarmingly in the twentieth century making it now the rarest pelican.

SIGNIFICANCE TO HUMANS
Protected by villagers in India but occasionally consumed in Cambodia. ◆

Resources

Books
BirdLife International. *Threatened Birds of the World.* Cambridge: BirdLife International, 2001.

Johnsgard, P.A. *Cormorants, Darters, and Pelicans of the World.* Washington, DC: Smithsonian Institution Press, 1993.

Sibley, C.G., and J.E. Ahlquist. *Phylogeny and Classification of Birds.* New Haven: Yale University Press, 1990.

Periodicals
Kennedy, M., H. Spencer, and R. Gray. "Hop, Step and Gape: Do the Social Displays of the Pelecaniformes Reflect Phylogeny?" *Animal Behavior* 51 (1996): 272–291.

Van Tuinen, M., D.B. Butvill, J.A.W. Kirsch, and B. Hedges. "Convergence and Divergence in the Evolution of Aquatic Birds." *Proceedings of the Royal Society* 268, no. 1474 (2001): 1345–1350.

Jay Robert Christie, MBA

Ciconiiformes

(Herons, storks, spoonbills, ibis, and New World vultures)

Class Aves
Order Ciconiiformes
Number of families 6
Number of genera, species 43 genera; 120 species

Photo: A male great egret (*Casmerodius albus*) in a breeding plumage display at Ding Darling National Wildlife Refuge, Florida. (Photo by M.H. Sharp. Photo Researchers, Inc. Reproduced by permission.)

Evolution and systematics

Among the six families that make up the order Ciconiiformes, the New World vultures (Cathartidae) occupy the most incongruous of positions. Historically, they were classified as belonging to the order Falconiiformes. Physical and behavioral similarities to carrion-feeding, hook-beaked, bare-headed Old World vultures made their inclusion in this order seem obvious. But similarities between New and Old World vultures are actually a case of convergent evolution. DNA hybridization studies show that the New World vultures' closest relatives are storks and thus they are generally accepted by taxonomists as Ciconiiformes, placed in the suborder Cathartae.

Yet ornithologists re-classify this family in its new home with great reluctance. Even in *Raptors of the World*, published in 2001, Ferguson-Lees and Christie admit that "however they are treated taxonomically, they also continue to be thought of as raptors" and unashamedly include the New World vultures in this authoritative species guide.

Their closest relatives, the true storks, share their place in the suborder Ciconiae with the ibis and spoonbill family. Fossil remains of storks are extensive, with the first identifiable stork dating to the Upper Eocene.

The shoebill (*Balaeniceps rex*) is, as of 2001, generally placed in the same suborder as storks, but like the hammerhead (*Scopus umbretta*), which sits on its own in the suborder Scopi, it defies convenient classification. The shoebill has physical characteristics that could place it among storks, herons, flamingos, or plovers.

The heron family, in the suborder Ardeae, is a very old group whose origins are in the Lower Eocene, 55 million years ago. It consists of four subfamilies, broadly defined as dayherons, nightherons, tigerherons, and bitterns.

Physical characteristics

All Ciconiiformes share the basic characteristics: a long bill and neck, a bulky body with a short tail, large, broad wings, and long legs and toes. Members of the family Ardeidae share a pectinate, or comblike, middle claw. Most herons, egrets, and bitterns have dagger-shaped bills. Those of storks are thicker and generally considerably longer. In Threskiornithidae, the bill is the defining morphological characteristic: the ibis bill is decurved, while the spoonbill does indeed have a flattened, spoon-shaped bill. New World vultures have developed meat-eating beaks, with hooked tips and sharp edges.

Without exception, Ciconiiformes are medium to very large birds. Herons show the widest variation among genera, ranging from the *Ardea* day herons, with the largest, the goliath heron (*Ardea goliath*), reaching 55 in (140 cm), to the *Ixobrychus* bitterns. But even the smallest of these, the dwarf bittern (*Ixobrychus sturmii*), is a respectable 11 in (28 cm) in length. The closely related stork and New World vulture families include some of the largest birds in the world. The male marabou stork (*Leptoptilos crumeniferus*) stands 4.9 ft (1.5 m) high; *Leptoptilos* storks and cathartids have very large wings adapted for soaring flight; the Andean condor (*Vultur gryphus*) has the largest wingspan at 10.5 ft (3.2 m).

This order tends to lack brightly colored plumage, with most species possessing a combination from gray, brown, black, or white. Most show no sexual dimorphism other than size differences, with males up to 10% larger. Those species which feed diurnally and gregariously, including many egrets, storks, and ibises, tend to be light in color. This may be so that they are less visible to prey looking up from water into the light, or for thermoregulation, since white plumage does not absorb heat as quickly. Also, waterbirds with striking light upper parts, easily seen at great distances, attract other gre-

Heron in flight. (Illustration by Wendy Baker)

garious feeders to exploit feeding opportunities. Conversely, nocturnal, and shade-feeding species, such as bitterns and night herons, have dark underparts or cryptic plumage, and blend with murky surroundings. These solitary feeders have no wish to attract attention from other Ciconiiformes.

Coloration of bare parts is often of great importance in the Ciconiiformes. In many, color changes in the legs, bill, and lore (small area between the eye and bill) take place immediately before courtship, with these bare parts becoming brighter or even—most frequently in the egrets—changing altogether, generally from yellow to red or black. The colors fade or alter again after eggs are laid, suggesting that these changes are an important part of breeding display.

The effects of color changes are most pronounced in those families where bare parts can extend to include the face, throat, neck and even the breast. Some storks, most ibises and all vultures have extensive featherless areas that usually intensify in color during courtship. A few storks and vultures can exaggerate the display by inflating large air pouches in the neck.

A feature of many heron species before courtship is the development in both males and females of ornamental plumes. Night herons and some day herons such as the great blue (*Ardea herodias*) and capped heron (*Pilherodius pileatus*), gain backward-facing head plumes which are rounded at the base and tapered at the ends (lanceolate). Some grow filoplumes, feathers resembling fine, thin hair, down the neck and breast. Most egrets have long, delicate aigrettes, ornamental plumes which trail loosely from their backs. Herons, bitterns, and egrets have powder downs, a type of feather that is not shed but becomes powder, conditioning and protecting the other feathers.

Distribution

While Ciconiiformes are found in all but the northern and southernmost parts of the earth, most species are found in tropical or sub-tropical regions. A number of those in northern temperate zones are partial or true migrants.

Habitat

With the possible exception of cathartid vultures, this order predominantly, but not exclusively, favors wetlands (103 out of 113 species), from tidal creeks, rivers, and forest streams to swamps, marshes, paddy fields, and damp meadows. Some species also forage on land-neighboring waterbodies. A few are adapted to feeding wholly in dry habitats such as savannah, light woodland, or moorland.

The range of New World vultures is governed by their ability to soar and find carrion, rather than by temperature, so they can occupy cold, mountainous areas as well as tropical forests. Scavenging abilities have enabled some species to spread into areas of dense human habitation.

Behavior

Most Ciconiiformes exhibit gregarious behavior, although the extent varies considerably among species and in function. Roosting is one of the most sociable activities within this order. Conspecifics may roost in the same colony as other Ciconiiformes and Pelicaniiformes. Roosts of gregarious herons, ibises, and storks are usually in trees, occupied year after year, and may number hundreds or even thousands of birds. Most New World vultures roost communally; American black vultures (*Coragyps atratus*) roost together in large numbers and reports from the 1940s of 20 or more California condors (*Gymnogyps californianus*) together suggest that this species' roosts were also sizeable in previous centuries. Even the otherwise solitary hammerheads roost colonially. Studies on American black vultures suggest that colonial roosting has at least one important function; inexperienced or less successful birds leave the roost to follow good hunters to sources of food. For most species, it may also be an antipredator measure.

Potential threats from predators may influence the choice of many species to nest colonially too. A total of 61% of Ciconiiformes are known to nest colonially. Various other hypotheses have been put forward; for example, social nesting may be a way of ensuring that birds have a good choice of mates for breeding.

Ciconiiformes have a limited vocal repertoire. At one extreme, cathartid vultures have no syrinx, the "voice box" that enables most birds to call and sing. The soft wheezes and whistles given at the approach of an intruder are probably only air passing through their mouth and nasal passages. Likewise, storks, shoebills, ibises, and spoonbills have little to say. Storks are silent for most of the year. Only during the breeding season do they emit a range of whistles, croaks, squeals, and grunts—usually when one birds greets another at the nest. Storks and shoebills indulge in noisy bill snapping and clattering as part of their courtship display. Ibises and spoonbills engage in bill clattering too, generally during confrontations.

Herons are the most vocal of Ciconiiformes, with most producing a range of grunts, honks, and croaks. These are emitted in greeting ceremonies during courtship, when disturbed, during antagonistic encounters, or in flight. The bitterns are exceptional in having a modified esophagus, which enables them to produce a booming call to attract a mate and proclaim a territory.

Since many Ciconiiformes must cope with extremes of temperature and humidity, they have evolved a number of thermoregulatory patterns of behavior. The extensive bare parts in many species have a large number of blood vessels close to the skin, enabling birds to radiate heat from the body. Conversely, these species are at risk from heat loss in colder conditions. Some storks of the genus *Ciconia* spend part of the year in temperate areas. In cold conditions, they stand on one leg, or tuck the bill beneath body plumage in an effort to reduce heat loss. Storks and vultures engage in sunning to warm up, standing with the wings outstretched. Among the various hypotheses for wing-stretching—including nest-shading, exposure of plumage parasites to ultraviolet light, social displays, skin conditioning during molting, maintenance of balance, and straightening of feathers after long periods of soaring—the two foremost are wing-drying and thermoregulation.

Storks, shoebills and vultures share the unusual characteristic of cooling down by urohydrosis—excreting urine on the legs to increase evaporation. Most herons, ibises, and spoonbills use the more common method of heat reduction in tropical areas, by gaping or fluttering their gular pouches.

Few Ciconiiformes are truly sedentary. A number, which spend part of the year in north temperate areas are migratory. Most species show some degree of dispersive behavior or irregular movements, usually to exploit food sources.

Most migratory species follow a north-south post-breeding route from temperate latitudes to tropical or subtropical areas, with cold weather the factor that drives migration. Some, such as American wood storks (*Mycteria americana*), move from southern breeding areas to more northern areas, to feed following breeding. Populations in warmer southern parts of the breeding range of species such as great blue herons (*Ardea herodias*), turkey vultures (*Cathartes aura*) and white-faced ibises (*Plegadis chihi*) do not migrate.

Ciconiiformes' large, broad wings are adapted for soaring on thermals and this is the most common method involved in distance travel over land among vultures, storks, ibises, and spoonbills: some of the world's biggest concentrations of these migrants are recorded passing over narrow strips of land between continents, such as Panama, or short sea crossings, including the Straits of Gibraltar and the Bosporus. Herons, however, have a slow, strong, flapping flight, enabling them to fly large distances over oceans. Purple herons (*Ardea purpurea*), for example, migrate over the widest parts of the Mediterranean Sea between Africa and Europe.

Feeding ecology and diet

The Ciconiiformes are mostly carnivorous. Those feeding in aquatic habitats eat predominantly fish, amphibians, crustaceans, insects, and mollusks. Other terrestrial and more catholic feeders take small mammals and birds, reptiles and, in a very few species, fruit and berries. The cathartid vultures,

Newly hatched roseate spoonbill (*Ajaia ajaja*) chicks. (Photo by Dan Guravich. Photo Researchers, Inc. Reproduced by permission.)

marabou stork (*Leptoptilos crumeniferus*) and greater adjutant (*Leptoptilos dubius*) are unique among Ciconiiformes in obtaining most of their food by scavenging.

All species forage for food by sight or by touch. Vultures also use their strong sense of smell. Most herons and storks stand still or wade slowly through shallow water to stalk prey and rely on striking quickly. Ibises, spoonbills, and storks of the genus *Mycteria* hunt by touch, either probing in water with slightly open bills, or moving them from side to side.

Solitary feeders defend feeding territories, which can be large in bigger species. Goliath heron (*Ardea goliath*) territories in South Africa average one bird per 3.7 mi² (6 km²). Gregarious feeders, often in mixed groups together with other Ciconiiformes and Pelicaniiformes, may exploit a temporary glut of food; the presence of a large number of birds attracts more to share the bounty. Birds on the edges of a feeding group spend longer looking out for predators. Flocks move seasonally to take advantage of optimal conditions—as one feeding area dries out or floods, they move to find more suitable habitat.

Reproductive biology

The pattern of nesting behavior is similar for most colonial and solitary nesting species. Most are monogamous. Nesting is timed to coincide with the period of peak prey availability. In temperate areas this is spring and summer. Subtropical species tend to nest during the dry season to avoid the threat of flooding. Tropical species usually nest in the wet season, when food is more plentiful.

The nest is nearly always a rough stick construction made by the female using material brought by the male. In most species, it is built in a tree, bush, low vegetation, or on the ground. The male arrives at the nest site first and defends his territory, often with stretching displays or wing flapping. The arrival of the female prompts greeting ceremonies that can be complex and may include bill snapping and tapping, mutual preening, and presenting of nest material.

In colonial-nesting species, egg-laying is almost simultaneous. Incubation is usually carried out by both species. Young are hatched blind and almost naked. Some ibis and spoonbill chicks hatch with a fine down. Often, both adults brood the young continuously for the first few weeks. The parent bird returning to the nest with food either regurgitates it onto the nest floor, or waits for the young to reach into its mouth and fish out a regurgitated meal.

After fledging, young of some species begin feeding with their parents, but return to the nest to rest. When the young finally abandon the nest, they often disperse over significant distances. Young black-crowned night herons (*Nycticorax nycticorax*) travel distances of up to 621 mi (1,000 km) from the colony where they were hatched. This reduces risk of overcrowding and enables species to exploit new habitats.

Vultures are exceptional, since they build no nest. The egg or eggs are laid on the ground in a cave, under a bush, in a large tree hole, or in an abandoned building. Hatchlings are covered with white down (or buff in black vultures). Both parents feed the young regurgitated food. Vulture young remain dependent on their parents for long periods—in the case of condors, it is six months before they even learn to fly.

Conservation status

Just over a fifth of Ciconiiformes (21 species) are under a serious level of threat, according to the IUCN, ranging in increasing degree from Endangered and Vulnerable to Critically Endangered. All but three of these species show downward population trends. An additional seven species are classified as Near Threatened.

The vulnerability of Ciconiiformes is due to a number of factors. Their large size, relatively sedentary habits, and slow flight make them easy targets for human persecution. Noisy, colonial nesting in regularly used sites by a number of species, and regular gathering in high concentrations at feeding sites increases their vulnerability, as chicks and eggs are taken and adults snared or shot by humans for food. In Asia, the white-shouldered (*Pseudibis davisoni*) and giant (*Pseudibis gigantea*) ibis, greater adjutant and milky stork (*Mycteria cinerea*) are among those species severely threatened by hunting. As human populations rise, pressures from hunting and disturbance increase. So too, do development threats. In the late 1990s, Japanese night herons (*Gorsachius goisagi*) lost habitat in Japan to golf courses, housing, and factories.

Waterbird species are particularly vulnerable to pollution. Throughout the world, there is evidence of its impact on Ciconiiformes. One of the main causes of the decline of the Oriental white stork (*Ciconia boyciana*) on the upper Amur in

Russia has been pollution from industrial waste, oil leakages, pesticides, and fertilizers. The black-faced spoonbill (*Platalea minor*) is threatened in China, where an estimated 63.1% of the rivers in the seven main river systems are polluted, according to Shen Maocheng in 2000.

The trend towards agricultural exploitation of ciconiiform habitats, particularly in Asia, has accelerated. Most of the Mekong floodplain in Laos, once a stronghold of the white-shouldered ibis, has been converted to rice paddy. Forest clearance and conversion of wetlands to agriculture in densely populated China, has fragmented the habitats of species such as the white-eared night heron (*Gorsachius magnificus*). In Indonesia and Malaysia, coastal wetlands, such as mangrove swamps, are increasingly logged and cleared to make way for fish farms and tidal rice cultivation.

Conservation efforts are addressing this generally bleak picture. For some species, the immediate aim is to carry out surveys to locate and quantify remaining populations and conduct research to understand breeding requirements, demography, and seasonal movements. Conservation groups and governments are working to establish protected areas, encompassing large tracts of habitat found to support populations of endangered species. The Japanese ibis (*Nipponia nippon*) is one such species, now severely restricted in range, but thanks to protection, one of only three Critically Endangered species showing an upward population trend.

In Laos and Cambodia, stricter enforcement of regulations and public campaigns to reduce the hunting of large waterbirds are underway. At Prek Toal reserve, in Cambodia, egg and chick theft of greater adjutant storks fell by 80% in 1997. Attempts at restoration of habitats include a replanting scheme in Assam to replace felled nesting trees of lesser adjutant storks (*Leptoptilos javanicus*).

Captive-breeding may offer hope for species whose declines are not habitat-related. The California condor, driven to extinction in the wild, largely by hunting pressures and poisoning by lead shot, will benefit from a successful reintroduction program in North America if such human pressures can be controlled. But this option is not open to the increasing number of species whose habitat is being obliterated.

Significance to humans

Myths and superstitions have historically safeguarded the populations of many Ciconiiformes. Native peoples of the Americas venerated condors and vultures as gods. Even today the Andean condor appears on the coats of arms of Bolivia, Ecuador, Chile, and Colombia. Positive associations also benefited the European white stork (*Ciconia ciconia*), Abdim's stork (*Ciconia abdimii*), and the sacred ibis (*Threskiornis aethiopicus*). White storks were considered lucky in Europe. To have one nesting on one's roof would increase fertility and wealth. Abdim's storks arrived on their breeding grounds in central Africa at the same time as life-giving rains. The "rain-bringer" was given the run of every village. Similarly, the sacred ibis returned annually to the Nile when the river flooded. The ancient Egyptians therefore worshipped it as a god. Fear motivated the protection of bitterns and hammerheads. The strange booming call of the Australasian bittern (*Botaurus poiciloptilus*) marked it as a bird of ill omen among Australian aboriginals. African villagers considered it unlucky to harm the crepuscular hammerhead or its inexplicably huge nest.

Ciconiiformes have suffered because of their perceived economic threat or value to humans. Heavily commercialized hunting of egrets, ibises, and spoonbills for their feathers during the nineteenth and early twentieth century saw millions of birds slaughtered annually. In North America, vultures were killed because they were seen, unjustifiably, as a threat to cattle farming. Herons were shot, principally in North America and Europe, since they were regarded as a threat to fishing interests.

Today, there is generally a greater acceptance of Ciconiiformes and a tolerance of those birds that live near humans. Night herons can be found in the center of built-up Hong Kong.

Resources

Books

Collar, N. J., et al. *Threatened Birds of Asia: BirdLife International Red Data Book.* Cambridge: BirdLife International, 2001.

del Hoyo, J., A. Elliot, and J. Sargatal, eds. *Handbook of the Birds of the World.* Vol. 1, *Ostrich to Ducks.* Barcelona: Lynx Editions, 1992.

del Hoyo, J., A. Elliot, and J. Sargatal, eds. *Handbook of the Birds of the World.* Vol. 2, *New World Vultures to Guinea Fowl.* Barcelona: Lynx Editions, 1994.

Hancock, J. A., H. Elliott, and R. T. Peterson. *The Heron's Handbook.* New York: HarperTrade, 1984.

Hancock, J. A., J. A. Kushlan, and M. P. Kahl. *Storks, Ibises and Spoonbills of the World.* San Diego: Academic Press, 1992.

Sibley, C. G., and J. E. Ahlquist. *Phylogeny and Classification of Birds: A Study in Molecular Evolution.* New Haven and London: Yale University Press, 1990.

Other

BirdLife International Saving Species. Birdlife International. 1 July 2001 (1 Feb. 2002). <http://www.birdlife.org.uk/species/index.cfm>.

Georgia Wildlife Web. The Georgia Museum of Natural History. 2 May 2000 (30 Jan 2002). <http://museum.hhm.uga.edu/gawildlife/birds/ciconiiformes/ciconiiformes.html>.

The Peregrine Fund. 23 Jan. 2002 (1 Feb. 2002). <http://www.peregrinefund.org/conserv.html>

Derek William Niemann, BA

Herons and bitterns
(Ardeidae)

Class Aves
Order Ciconiiformes
Suborder Ardeae
Family Ardeidae

Thumbnail description
Medium to very large wading birds, typically with long legs and toes, long bills, and long necks, which are folded over the back when flying

Size
9.7–58.5 in (25–150 cm); 0.16–9.9 lb (73 g–4.5 kg)

Number of genera, species
16 genera; 62 species

Habitat
Inland and coastal wetlands, lakes and streams, grasslands, wet forests, coasts and estuaries, islands and agricultural areas such as rice fields and aquaculture ponds

Conservation status
Endangered: 3 species; Vulnerable: 5 species; Near Threatened: 1 species

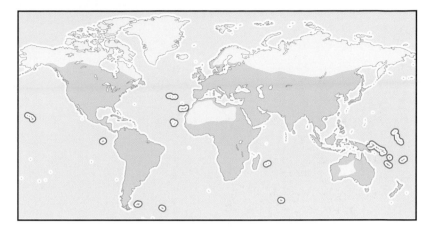

Distribution
Worldwide in tropical and temperate zones

Evolution and systematics

Herons, as far as is known, originated in the Eocene era about 60–38 million years ago. Specimens attributed to herons occur infrequently in later deposits. Herons are not well represented in the fossil record prior to the Pleistocene, probably owing to the slight structure of their bones. Missing from the record are birds clearly ancestral to what are now considered the more basal lineages of the heron family tree, the boat-billed heron (*Cochlearius cochlearius*), agami heron (*Agamia agami*), and the tiger herons. Therefore, the details of the evolutionary history of herons must be inferred primarily from modern species.

Present-day herons are a rather homogenous group of birds that have been lumped together taxonomically since Linnaeus's day. As part of the order Ciconiiformes, they are related most closely to storks, ibises, and spoonbills. They are related distantly to waterfowl, shorebirds, and pelicans, and even more distantly to such birds as loons, petrels, and albatrosses. The details of higher-level systematics among these birds may be amenable to modern molecular research techniques.

Over the decades, there has been much discussion about placement of a few relatively odd species, the shoebill (*Balaeniceps rex*), the hammerhead (*Scopus umbretta*), and the boat-billed heron. The first is most closely allied to storks, the second to pelicans, but the last is definitely a heron, but of a different sort. Together with the tiger herons, the boat-billed heron and agami heron appear to represent remnants of basal limbs of the heron evolutionary tree. In a recently proposed

taxonomic scheme, these species are considered to be representatives of distinct subfamilies—Cochleariinae, Agamiinae, and Tigrisomatinae.

The remaining herons encompass the typical herons and the bitterns. The bitterns, subfamily Botaurinae, appear to represent the herons most divergent from the primitive heron stock, a complete turnaround of systematic thinking of a few decades ago. Herons of the subfamily Ardeinae include the herons (tribe Ardeini), the egrets (tribe Egrettini) and the night herons (tribe Nycticoracini).

Within the five subfamilies and three tribes, about 62 currently recognized species are partitioned among 16 genera. Over the past several decades, anatomical and molecular studies have led to reassignment of species among infrafamilial categories, particularly among genera. Progress in heron systematics can be measured by noting that only a few decades ago, the large egrets were placed in the monotypic genus *Casmerodius* or in the genus *Egretta* and the terrestrial cattle egret was placed in a monotypic genus *Bubulcus* and later considered an *Ardeola* or an *Egretta*. It has recently been proposed that the yellow-crowned night heron (*Nyctanassa violacea*) is also more closely related to the egrets than to other birds called night herons, but additional study is called for.

Physical characteristics

Typical herons are relatively tall and thin, with relatively long necks and legs, large sharply pointed bills, large moveable

Black heron (*Egretta ardesiaca*) canopy feeding. (Illustration by Wendy Baker)

eyes, and broad wings. Their plumage is generally complex, featuring black, white, grays, or browns, and they have distinctive plumes. These fundamental heron features are primarily adaptations for wading in water in order feed on fish and other aquatic prey and for communicating with other birds.

The long neck has 20–21 cervical vertebrae, the fifth through seventh having the articulation that gives the neck its characteristic kink. The neck is long enough to be folded over the back in an "S" shape when the bird in prolonged flight, making a flying heron easily recognizable.

The long legs have feathers on the thighs but are otherwise featherless. The toes are long (including the back toe, which is on level with the rest) and slightly webbed. The claw of the middle toe has a serrated edge, which facilitates care of the plumage.

The heron bill is one of its defining characteristics. Most are elongated to effect the capture of quickly moving prey in a tweezer-like fashion following a rapid strike. Thin rapier-like bills are adaptations for fish eating, piercing through the water to capture fleeing prey. Bill color depends on species, and can vary with age and season. In some species bill color brightens and becomes more colorful during courtship.

Herons have well-developed eyes with substantial capacity for movement. Typical herons have a tall but narrow field of binocular vision that is aimed forward and includes the zone under the bill down to the feet, to aid in sighting prey. The color of the iris of some species changes seasonally, or when agitated, and so is used for social signaling. The head is fully feathered and often distinctively marked by species, except for the area between the bill and eye, which often is featherless. These bare loral patches are colored characteristically among the species, and their colors typically change during courtship and other interpersonal encounters.

Some herons are entirely white, gray, or black, while others have exceptionally complicated plumage. Plumage pattern is generally correlated with lifestyle. White-bodied herons are often highly social, feeding in flocks and nesting close together in colonies. Dark-plumaged herons tend to be more solitary, or have the capacity to be social or not as the situation requires. In other species the plumage is predominantly cryptic, featuring brown, white or buff stripes, speckles, and spots. This color scheme predominates among species that hide in reeds and bushes.

Heron plumage changes with age. The first downy plumage, which is generally gray or light brown, is immediately replaced by the first juvenal plumage. Generally an adult plumage is gained by the time of first breeding. Pond herons change plumage seasonally.

As a general rule, the sexes are not distinguishable by the coloration of their plumage. During the breeding season, however, herons use feathers on the nape, back, crest, and crown as display plumes. These are particularly functional in aggressive encounters. Plumes also occur on the front of the neck, upper neck, and back. The major types of heron plumes include filoplumes, which are elongated and hairy appearing; aigrettes, which have long shanks and few barbs so that they appear frayed; and lanceolate plumes, which are more like typical long body feathers but with frayed edges.

Herons also have patches of friable downy feathers that provide a powder used for grooming. Most herons have three pairs of down patches. The powder produced apparently keeps the plumage water-repellent and probably cleans it as well.

The wings are broad, with nine to 11 primaries. The tail is short with 12 tail feathers in most species. Large herons appear to take flight with some difficulty, holding their head out with legs dangling until they gain altitude. Smaller herons can take flight more rapidly. Once in flight, herons fly well and with endurance using slow, quiet wing beats and can travel long distances both on migration and to and from feeding grounds.

Distribution

Herons occur around the world, on all continents but Antarctica, and also on islands in all oceans. Herons occur in the greatest numbers and diversity in tropical zones. Many species range into the temperate zones, but their limits depend on the species-specific ability to nest in progressively shortened summer periods.

The herons (Ardeini) are found around the world. *Ardea* appears to be primarily an Old World genus, with 12 species there and only four in the New World, two of which are shared. The three great herons, including the goliath heron, occur in Africa and Asia. *Butorides* herons also are widely distributed continentally, and also on many islands. Pond herons (*Ardeola*) also are Old World species, mostly Asian with two representatives in Europe and/or Africa.

The egrets (Egrettini) are worldwide, with no particular concentration area. Several *Egretta* are New World species, probably originating in South America and subsequently invading North America. The Chinese egret is the only Asian species. The night herons (Nycticoracini) likely are an Old World group. The bitterns (Botaurinae) also appear to be an Old World group, with only four of 11 species occurring in the New World. The tiger herons, agami heron, and boat-billed heron (Tigrisomatinae, Agamiinae, Cochleariinae) are tropical species, four from tropical Americas and one each from Africa and New Guinea.

Temperate-zone herons are generally migratory. Some species, such as the large herons, remain rather far north such as in Canada and Great Britain, although periodic severe win-

Tricolor herons (*Egretta tricolor*) mate in their nest. (Photo by M.H. Sharp. Photo Researchers, Inc. Reproduced by permission.)

ters can cause substantial mortality in these populations. Many species that breed in the tropics also migrate regularly according to the wet and dry seasons. Many species of herons tend to wander after nesting leading to a post breeding dispersal away from their nesting areas. Due to both postbreeding dispersal and overshoots on return migrations, herons often wander far from their normal range. As a result, herons stray to high latitudes, deserts, and mountains, as well as to far off islands and ships at sea.

The ranges of some species are changing. Some in the Northern Hemisphere are expanding their ranges northward. In contrast, the large bitterns are experiencing range contraction. Some species are changing their ranges transcontinentally.

Habitat

Herons are generally aquatic birds, and the typical heron is seen feeding by standing or walking in the shallow water of a marsh or pool. However, herons use a wide array of wet and dry habitats. They may be habitat specialists or habitat generalists.

Inland wetlands are typical habitats for herons. Tree swamps are particularly favored because they provide not only foraging habitat, but also trees and bushes for roosting and for nesting. Herbaceous marshes also are used by herons worldwide. Some species, such as bitterns and the purple heron, are specialists in living among dense emergent vegetation.

Herons also feed in more open areas, such as the shallow edges of lakes, ponds, pools, and lagoons, where they tend to feed along the edges in shallow water. In these situations it is not unusual to see species arranged out from shore according to leg length, with taller birds foraging in deeper water and shorter birds at shallower sites.

Great blue heron (*Ardea herodias*) breeding pair in courtship display in Venice, Florida. (Photo by C.K. Lorenz. Photo Researchers, Inc. Reproduced by permission.)

Running water is exploited by species that feed along the banks, either by perching on overhanging trees or by feeding from the bank itself. Herons also perch on rocks, and the larger herons can withstand current sufficiently to wade into running water, although within limits. Placid streams and ditches are more commonly used by many species.

Tidal environments are of critical importance for many species. Tidal creeks, tidal mudflats and bars, salt marshes, mangrove swamps, coastal lagoons, and beaches are all used by herons. The tidal cycle determines the daily schedule of species feeding in tidal environments. They feed when conditions are appropriate, usually around the outgoing tide, and then move to nontidal habitats to continue feeding or to roost during the high tide periods. Species depending on tidal flux, especially the night herons and large herons, may also feed at night.

Artificial environments have become essential habitats for many populations. Reservoirs, farm ponds, and ditches provide aquatic or semi-aquatic habitat. Lands designated for agriculture and aquaculture are of even more importance. Rice fields have become critical habitat for herons around the world. Aquaculture provides concentrated prey of the sort that herons customarily eat. Young birds particularly may be attracted by these sources.

Despite the aquatic origins of the group, herons also use terrestrial habitats either occasionally or predominantly. Some species, such the cattle egret and black-headed heron (*Ardea melanocephala*), are predominantly terrestrial and inhabit natural grasslands and pasturelands. Many other species of herons and egrets feed on dry land at least occasionally.

Herons nest in many sorts of habitats that afford proximity to feeding areas and protection from predators. Colonial

herons use islands off shore and in lakes or rivers. Herons nest on islands of vegetation composed of trees or bushes surrounded by marsh or swamp. They also nest in tall trees, either within expansive forests or in coppices. On safely isolated islands, herons may nest on the ground, on rocks, on in cave entrances. Many species of herons nest in reed beds.

Behavior

For most herons standing is a principal feeding behavior and they spend much of their day or night resting or roosting. Many herons are sit-and-wait predators that lie in wait for prey to make itself apparent. Waiting herons can keep quite still for many minutes. Depending on species and circumstance, they stand crouched, upright, or fully erect. Crouched postures lower visibility to prey and allow the strike to occur closer to the water. In the upright posture, the body and neck are angled above the water allowing more scanning for prey than the crouched posture.

Herons may walk in the water, on the ground, over grass, or in bushes and trees. Walking herons generally move about in search of prey or stalking individual prey. Walking may be very quick or so slow as to be nearly imperceptible. Some herons run and hop from place to place in search of better feeding opportunities.

In feeding, the heron makes maximal use of its head and neck. Fish and other prey are caught after a quick movement of the head and neck. The usual method is a bill stab in which the heron issues a downward or lateral strike with the head and neck. Shorter-necked herons capture prey by bill, thrusting forward the bill, head, neck, and body, in a sort of a head-first leap. Herons also feed by more subtle methods, such as probing into mud or vegetation, pecking the ground, or gleaning insects off plants.

In most species, herons need adequate vision to see their prey before stabbing it. Other than the challenge of identifying something in the water as edible, the heron's most substantive problem is refraction of light in the water, and they compensate for the fact that their underwater prey is not actually located where it appears to be. Herons also move their heads around to compensate for reflection off the water's surface. They move their heads side to side to better locate prey binocularly and also sway both the head and neck side to side or backward and forward, a behavior also used among land-dwellers.

Herons attract or startle prey in several ways. Several medium-sized egrets have yellow feet contrasting with darker legs. They stir, rake, and wiggle their feet to attract prey or stir it into movement. *Butorides* herons attract prey by placing food items (such as dog food or corn) or imitation food (such as sticks or feathers) in the water and catching prey that the bait attracts. Some egrets also attract fish by putting their bill in the water and opening and closing it quickly, creating ripples to attract fish.

Many herons are highly social. They tend to gather in feeding sites where prey is particularly available, forming multi-species aggregations that can number in the hundreds or even thousands of birds. Most herons, the bitterns being the primary exception, nest in colonies, often of mixed species. Other species are solitary or occur in more well-dispersed pairs. In a few instances family groups may occupy a single area. Many species choose between being solitary and aggregative depending on food availability.

All herons defend themselves and their immediate surroundings with ritualistic bill lunges called forward displays. Herons at a nest site will shake twigs vigorously. A bird may also attack another by running or flying at it to supplant it from its site. Herons fight with their bills, wings, and feet. When hundreds of herons of several species gather at feeding sites, they jockey for position, open their sharp beaks, spread their wings, and rush at their enemy. Individual fights can last half an hour.

Another behavioral interaction among herons is prey theft. Large herons steal prey from smaller ones, and other birds steal prey from herons. A heron with a large or uncooperative fish must subdue it by biting, bashing, or stabbing. This delays swallowing and allows other birds to steal the fish. The attacking heron runs or flies at a victim in its attempts to steal the prey. The threat of theft may even influence the choice of prey, and a heron may sometimes pass up a larger prey item in favor of a smaller one.

Feeding ecology and diet

Herons are primarily fish-eating birds. Most species wade about looking down into the water and capture fish they see by a rapid thrust of their long sharp bill. Given their feeding method, herons can catch fish at or near the water's surface. Fish must be visible and also shallow enough that they cannot swim away before the heron can get its beak around them. Large herons can capture large fish. Lungfish are an important prey for the largest herons. The mosquitofish (*Gambusia affinis*), introduced around the world, is taken by small and medium-sized herons in great numbers wherever it occurs. Overall, many fish species can be captured by herons.

The second most common type of prey for herons is crustaceans. Crabs occur primarily in marine and estuarine environments, although some occur well inland in the large rivers and on land. In freshwater, burrowing crayfish and small shrimp are important prey. Crayfish, shrimp, and prawns are farmed in many parts of the world; herons are attracted to these sites to partake of the easy feast.

Amphibians are another important prey for herons. Frogs and toads are frequently caught, as are their tadpoles. Salamanders are less common. Insects, especially aquatic insects, are an important food source for herons. Both adults and larval forms are picked up from the water or submerged plants and rocks. Terrestrial herons primarily eat insects. Flies, dragonflies, and similar insects are often taken. Other food may include snails, bivalves—both freshwater and marine—small mammals, birds, and reptiles. Reptiles are most often an important part of heron diet on islands.

Black-crowned night herons also feed on nesting gulls, terns, or other herons in colonies. Other herons are reported

Female (left) and male little bitterns (*Ixobrychus minutus*) with chicks at their nest. (Photo by J.C. Carton. Bruce Coleman Inc. Reproduced by permission.)

to take birds as they become available. As far as is known, herons are entirely carnivorous. However, there have been reports of herons purposefully eating fruits, and vegetation may be taken along with fish.

Many species of herons aggregate to forage for several reasons. Herons are adept at finding and then exploiting ephemeral patches of highly concentrated food. Therefore, aggregations develop at places where food supply is high. This sharing of food-finding information is called local enhancement. Feeding locations for social herons change hourly, daily, and seasonally.

In some cases of aggregate foraging, the participating herons gain an advantage in that the mass of birds stir up the prey, which makes the animals more vulnerable to capture. Herons also achieve commensal benefits from following the paths of other birds or mammals in search of prey those animals may have disturbed.

In contrast, some herons are always solitary or occur in pairs. The great herons and large herons (*Ardea*) tend to feed alone.

A black-crowned night heron (*Nycticorax nycticorax*) with a fish in Pima County, Arizona. (Photo by John H. Hoffman. Bruce Coleman Inc. Reproduced by permission.)

However, when the occasion presents itself, they will join mixed species aggregations. Birds that feed alone have freedom from the disturbance of their potential prey by other animals. These birds also defend their feeding areas from invasion.

Reproductive biology

Most species of herons are serially monogamous. Although some birds return to the same site to nest year after year, many birds tend to move nests or even change nesting areas from one season to the next. As a result, pair bonds tend to be formed anew each season. For social nesters, despite monogamous pairing, promiscuous mating behavior can be common. Extra-pair mating usually occurs among individuals nesting near each other.

Bitterns and tiger herons are generally solitary nesters. More species of herons are colonial, nesting in single-species or mixed-species aggregations that can number from a few birds to hundreds or even thousands of birds. Along with other herons, colonies may include pelicans, cormorants, ibises, spoonbills, storks, and also crows and raptors. Herons tend to partition the nesting habitat, often with the larger birds on the top of trees or bushes and the smaller species underneath. Nest defense by parents is essential in a colony. Some species, such as cattle egrets, are particularly aggressive and may take over nests of other species.

Males tend to arrive at colony sites first at the future colony and claim display sites, which they defend against other birds. They give spontaneous displays and calls that attract potential mates and defend the sites. During the advertising period, a stretch display is the most universal among typical herons and egrets. The display consists of lifting the head to vertical, perhaps calling, and then bringing it down again, perhaps with a snap of the bill. Later in the nesting season, this display is also used between the mating pair. With the snap display, the bird erects head and neck feathers and extends its neck with a snap of its bill. Herons may combine snaps and stretches, bow, shake twigs, fly about the colony in circles, flip their tails, shake their feathers loosely, preen, and mock preen. After pairing, the birds give landing calls and a greeting ceremony that permits the returning mate access to the nest site.

Following courtship, the pair builds the nest. Nests are made of sticks or reeds, depending on species and nesting site. The eggs are typically blue, but may be white, greenish, or olive-brown; a few species have spotted eggs. Clutch size can range up to 10, but for most species is three to five. Incubation lasts from two to four weeks depending on the species and size of the bird. Larger birds' eggs incubate longer. Except for large bitterns, both parents incubate, taking turns.

Newly hatched young are covered with sparse down, have closed eyes, and are unable to walk, but the birds grow quickly. Both parents tend the young. They bring food in their stomachs or throats to the nest and then regurgitate it first into the beaks of the young, and later onto the edge of the nest.

Because incubation begins before the clutch is completed, the young hatch at different times. The oldest get a head start and dominate the youngest siblings in competition for food, which the parents provide preferentially to the most persistent chick. Usually the younger chicks die, either by starvation or through harassment by older chicks. How many chicks die depends on the ability of adults to supply food and the health of older chicks.

One parent broods the young for one to two weeks and continues to guard them for a bit longer. While one parent is guarding, the other forages. Chicks learn to fly gradually within the colony. Colonial species do not feed young after they fledge.

Conservation status

Many heron species remain abundant, and some are expanding their ranges and populations. Large herons often occur in rural and even urban areas. Many species use rice paddies, farm ponds, and aquaculture facilities; in fact, some populations have become dependent on them. Herons nest in artificial lakes, urban parks, and zoos, and feed along roadside ditches. Reservoir construction has increased available habitat, especially in otherwise arid areas that are inhospitable to herons.

Some species or populations in certain areas, however, teeter on the verge of extinction. *The 2000 IUCN Red List of*

Threatened Species lists three species as Endangered (*Ardea insignis, Gorsachius goisagi,* and *G. magnificus*), five species as Vulnerable (*Ardea humbloti, A. idea, Egretta eulophotes, E. vinaceigula,* and *Botaurus poiciloptilus*), and one species as Near Threatened (*Zonerodius heliosylus*). The primary threat for all species is habitat destruction and alteration, in some cases exacerbated by hunting. The most important habitat alteration involves the widespread destruction of wetlands, lowland forests, and coastal swamps and lagoons. Hunters take eggs, chicks, or adults.

Protecting and managing habitat is by far the most critical issue in heron conservation. Colonial species require nesting substrate, usually bushes and trees. Because the nesting activity can stress the plants through breakage, defoliation, and excess nutrient deposition, colony sites degrade over time, requiring either active management or the provision of alternative sites. Most solitary nesting herons need relatively large patches of marsh or forest in which to nest, although some tiger herons nest in a single isolated tree. Herons need feeding habitat throughout the year. For migrating species, this means during nesting, migration, and wintering periods. Fortunately, the habitat needs of herons coincide with those of waterfowl and other aquatic birds, making it possible for heron habitat protection to be part of larger wetland and forest conservation strategies.

Hunting is an important issue in worldwide heron conservation. Hunting for food and for body parts continues in China and Madagascar. Herons are killed at fish farms and other aquacultural facilities. Herons are killed by accidents, sport shooting, and cases of acute chemical contamination.

Significance to humans

Humans have certainly always been aware of heronries, and were probably one of the few ground predators that could access them for easy food. Many colonies occurred in places that were relatively inaccessible to early humans. Official protection afforded herons in the recent years was probably the exception rather than the rule worldwide. Herons were seen along watercourses and other places where men fished, and in these circumstances humans fostered an attachment to the birds. This relationship probably has peaked with the people of Manchar, Pakistan, who for a thousand years or more have kept herons as honored pets.

Herons have not figured in folklore to the extent that storks or some other birds have. Mention of herons nonetheless goes back to the Old Testament, ancient Egypt, and Hindu culture. The booming of the large bitterns has long been held as a bad omen in several cultures. But the graceful heron is often used

Snowy egret (*Egretta thula*) nestlings beg food from an adult. (Photo by M.H. Sharp. Photo Researchers, Inc. Reproduced by permission.)

as a symbol of natural beauty and grace in contemporary societies, and so the bird is a subject of poems and books.

Humans have long appreciated and used heron feathers. The decorative plumes of the large white and little egrets served as expensive head decorations for the Hungarian nobility and the Turks in the Middle Ages. By the late 1800s and early 1900s, birds and their feathers were quite valuable and widely used in Europe and elsewhere for ornament, particularly on hats. London became the center of the European plume market, and colonies of birds on the Danube and Theiss Rivers were among those devastated by plume-collecting expeditions. Colonies were hunted in both southern North America and through tropical America. All birds with useful plumes were killed and their feathers plucked. Non-target birds nesting in the colonies were disrupted and orphaned young died. In 1902, 3,012 lb (1,366 kg) of egret plumes were sold in London. This meant that 192,960 egrets were killed to supply the demand.

Never in their history have herons been so dependent on another species as they are on humans today. Worldwide, colony sites are protected or threatened by people. Feeding sites are part of parks, refuges, and other protected environments. Elsewhere, human destruction of mature wet forests, wetlands, and coastal environments affect populations profoundly. It is likely that herons are more and more coming to depend on their relationships with humans for their continued survival.

1. Boat-billed heron (*Cochlearius cochlearius*); 2. Least bittern (*Ixobrychus exilis*); 3. Black-crowned night heron (*Nycticorax nycticorax*); 4. Eurasian bittern (*Botaurus stellaris*); 5. White-eared night heron (*Nycticorax magnificus*). (Illustration by Gillian Harris)

1. Great blue heron (*Ardea herodias*); 2. Cattle egret (*Egretta ibis*); 3. Black heron (*Egretta ardesiaca*); 4. Squacco heron (*Ardeola ralloides*); 5. Agami heron (*Agamia agami*); 6. Little egret (*Egretta garzetta*); 7. Great egret (*Ardea alba*); 8. Gray heron (*Ardea cinerea*); 9. Goliath heron (*Ardea goliath*). (Illustration by Brian Cressman)

Species accounts

Gray heron
Ardea cinerea

SUBFAMILY
Ardeinae

TAXONOMY
Ardea cinerea Linnaeus, 1758, Sweden. Four subspecies.

OTHER COMMON NAMES
French: Héron cendré; German: Graureiher; Spanish: Garza Real.

PHYSICAL CHARACTERISTICS
A large gray heron (35–39 in [90–98 cm]) with white and black accents, a white crown with black plumes, black belly, and white thighs. Weight is 2.2–4.6 lb (1–2.1 kg)

DISTRIBUTION
Most of the Old World, including Europe, Africa, Asia, East Indies islands.

HABITAT
Typically found in and around shallow water, generally along watercourses and shorelines, and usually in locations having roost trees nearby. They may occur in inland fresh waters, along estuaries, or in marine habitats.

BEHAVIOR
Stands or walks slowly in or around shallow water. Flies to and from roosts and nesting colonies.

FEEDING ECOLOGY AND DIET
Usually hunts solitarily, but may feed in loose aggregations or mixed species flocks. Eats mostly fish, but also small mammals and amphibians. Young birds often use fish farms.

REPRODUCTIVE BIOLOGY
Nests solitarily or in colonies. Time of nesting differs according to range. In temperate areas, breeding season is restricted to spring and summer; in the tropics nesting is more flexible, usually in the wet season. Nest is a stick platform located high in a tall tree. Clutch size is normally four to five in Europe, three in the tropics. Incubation is 25–26 days (21 in tropics); chicks fledge in 50 days.

CONSERVATION STATUS
Not threatened. A rare light-colored population nesting in coastal Mauritania deserves special conservation attention.

SIGNIFICANCE TO HUMANS
Hunted in the Middle Ages, and well appreciated as falconry targets. In the present day, humans most frequently encounter the birds along rivers and at fish farms, where young birds occur frequently. They are killed in these situations in great numbers. ◆

Great blue heron
Ardea herodias

SUBFAMILY
Ardeinae

TAXONOMY
Ardea herodias Linnaeus, 1758, Hudson Bay. Five subspecies.

OTHER COMMON NAMES
English: Great white heron (white birds), Würdemann's heron (dark-white intermediate); French: Grand héron; German: Kanadareiher; Spanish: Garza Azulada.

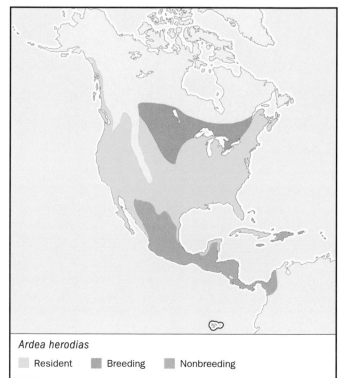

Ardea herodias

Resident ▪ Breeding ▪ Nonbreeding

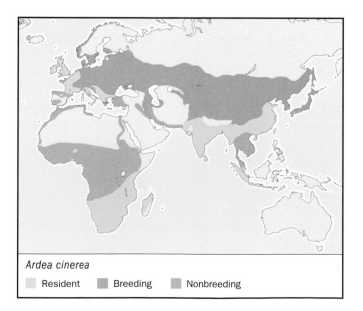

Ardea cinerea

Resident ▪ Breeding ▪ Nonbreeding

PHYSICAL CHARACTERISTICS
A large, dimorphic heron. Length is 36–54 in (91–137 cm); weight is 5–8 lb (2.3–3.6 kg). Dark gray heron has chestnut thighs and a white cap over a black eye stripe. Light birds are all white.

DISTRIBUTION
Breeds throughout much of North America except for high mountains and deserts; also in Central America and on certain islands in the Caribbean and Pacific. Nonbreeding range includes much of coastal and southern North America, West Indies, coastal Mexico, Central America, rarely to Panama and northern South America as far as Brazil.

HABITAT
Deep water to dry land. Uses freshwater and salt marshes, mangrove swamps, estuaries, meadows, flooded agricultural fields and pastures, lake and seashores, river banks, dry land pastures, coastal lagoons, mangroves, tidal flats, and sea-grass flats.

BEHAVIOR
Stands in shallow water and roosts in nearby woody vegetation. Feeds in the water or at its edge. Flies with strong slow wing beats, with its head held back. When disturbed, it gives a harsh call.

FEEDING ECOLOGY AND DIET
Eats large fish, but takes small and large animals of all sorts. Feeds mostly by stalking prey; it also feeds by diving or swimming. Commonly seen near fishing boats and at aquacultural ponds. They feed by day or night. Along the coast, the feeding schedule depends on tides. Feeding sites are often defended.

REPRODUCTIVE BIOLOGY
Begin nesting in the late winter and spring. In tropical areas, they can nest nearly year round. They nest alone, or more commonly in small colonies. Nests are in tall trees with nearby aquatic feeding areas, consisting of are stick platforms 20–39 in (0.5–1 m) across. Clutch size is two to seven, increasing from south to north. Incubation takes about 28 days. Mortality of chicks is often high; one to two are usually fledged.

CONSERVATION STATUS
Not threatened. A population found in southern Florida and the Caribbean consists of many all-white birds and is of conservation concern due to its limited range.

SIGNIFICANCE TO HUMANS
Probably the best known and appreciated heron in North America. However, it suffers conflict with aquaculture operations. Human disruption of habitat in Florida Bay has lowered the natural reproductive capacity of the highly localized white plumaged population. ◆

Great white egret
Ardea alba

SUBFAMILY
Ardeinae

TAXONOMY
Ardea alba Linnaeus, 1758, Europe. Four subspecies.

OTHER COMMON NAMES
English: Great egret; French: Grande aigrette; German: Silberreiher; Spanish: Garceta Grande.

PHYSICAL CHARACTERISTICS
A large, slender, white heron, with long neck, dark legs, and long back plumes when breeding. Length is 31–41 in (80–104 cm); weight is 1.5–3.3 lb (0.7–1.5 kg).

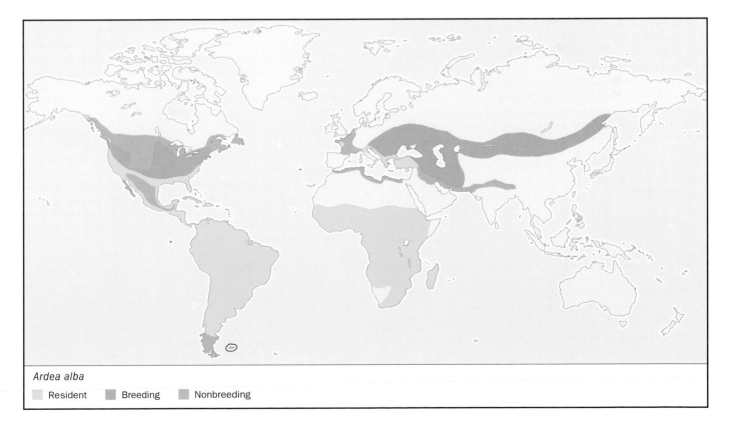

Ardea alba

☐ Resident   ■ Breeding   ■ Nonbreeding

DISTRIBUTION
Occurs through most of North, Central, and South America; east Europe, Africa, and north Asia.

HABITAT
Uses a variety of wet habitats, including marshes, swamps, river margins, lake shorelines, flooded grasslands, sea-grass flats, mangrove swamps, coastal lagoons, and offshore coral reefs. Also uses artificial sites such as drainage ditches, rice fields, crawfish ponds, and aquaculture ponds.

BEHAVIOR
Stands or walks about alone or in groups. Roosts in trees when not feeding and repairs to communal roosts at night.

FEEDING ECOLOGY AND DIET
Feeds in shallow to moderately deep water, on shore next to the water, or on dry ground. Feeds usually during the day, most actively near dawn and dusk; in tidal environments, feeds principally on outgoing tides day or night. Walks about slowly to feed, using its long neck and head to tilt, peer, and sway to better see fish. Also hops and flies, using its wings and feet to disturb prey. When feeding solitarily, it will vigorously defend its site. Highly aggressive in flocks, defending its feeding area using displays and attacking nearby birds, often stealing their prey. Principal food is fish, but in some situations insects or shrimp predominate. May also eat frogs, lizards, snakes, small mammals, and small birds.

REPRODUCTIVE BIOLOGY
Temperate birds breed in the local early spring and summer but in more tropical situations rains are more important than solar season and breeding varies from place to place and even from year to year. They usually nest in the part of the rain cycle in which food becomes maximally available. They nest in a variety of situations in trees, bushes, bamboo, reeds and other plants near water and on islands, sites that are protected from ground predators. The nest is 31–47 in (80–120 cm) wide. The eggs are pale blue and clutch size is usually three to five (range is 1–6), being smaller in the tropics. Incubation lasts about 25–26 days. Young leave the colony in 42–60 days. Brood reduction is the rule.

CONSERVATION STATUS
Not threatened. Breeding colonies are declining due to human plundering in Madagascar, however. Throughout its range the most critical conservation issue is identification, protection, and management of important nesting sites and associated feeding grounds.

SIGNIFICANCE TO HUMANS
Its long breeding plumes were some of the most sought after during the plume-hunting era. Currently it is coming into conflict with humans in aquacultural situations in North America and elsewhere. However, the species is well appreciated and has long been used as a symbol of bird conservation in North America. ◆

Goliath heron
Ardea goliath

SUBFAMILY
Ardeinae

TAXONOMY
Ardea goliath Cretzchmar, 1826, Bahr el Abiad.

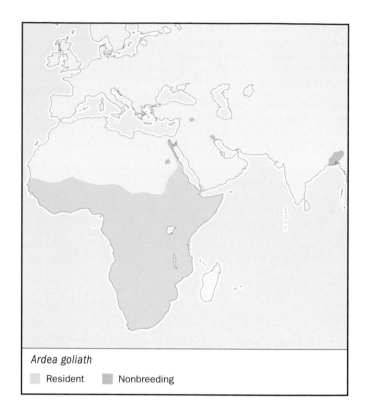

Ardea goliath

⬜ Resident ◼ Nonbreeding

OTHER COMMON NAMES
French: Héron goliath; German: Goliathreiher; Spanish: Garza Goliat.

PHYSICAL CHARACTERISTICS
The largest modern heron, it is gray with chestnut head, neck, and belly. Length is 53–55+ in (135–140+ cm).

DISTRIBUTION
Africa, the Middle East, and the Indian subcontinent.

HABITAT
Aquatic heron of both coastal and inland habitats, rarely wandering far from water. Occurs along the shallow water margins of large lakes, lagoons, and large river systems; also in tidal estuaries, reefs, and occasionally mangrove creeks and water holes in woodland savanna.

BEHAVIOR
A solitary hunter that defends large feeding territories. Stands in or near the water, or walks slowly, waiting for prey to appear. Moves to new areas by walking quickly or hopping.

FEEDING ECOLOGY AND DIET
Because of its size, this heron can wade well away from shore. Fish are caught by a lunging bill thrust that captures the fish deep in the water. It often spears them, running both mandibles through the prey. The fish is placed on the tops of floating plants and killed by restabbing, beating, and poking it with the bill. One-quarter of prey may be lost by escape or through piracy by other fish predators. Diet consists almost entirely of fish; they also will eat prawns, frogs, lizards, snakes and small mammals.

REPRODUCTIVE BIOLOGY
Breeding season coincides with the start of rains. Some populations breed year-round, and others may not breed every year.

Nesting is solitary, near colonies, and within single-species or mixed-species colonies. Solitary birds nest on riverbanks, lakeshores, and small islands. Nest sites include sedge, reeds, small trees, low bushes, mangroves, and cliffs. On islands, any tree, shrub, stone, or bare ground available can be used. The nest is a large platform made of sticks or reed stems at least 3.4–4.9 ft (1–1.5 m) in diameter. Eggs are pale blue, and the usual clutch is three or four, ranging from two to five. Young fledge at about five weeks. Older young can trample younger siblings, leading to brood reduction. Production is one or two young per successful nest.

CONSERVATION STATUS
Not threatened. However, the status of this species is currently unknown in south Iraq/Iran and the Indian subcontinent, where birds are infrequently reported.

SIGNIFICANCE TO HUMANS
Well known in its range, but little is understood of important aspects of its biology. ◆

Cattle egret
Egretta ibis

SUBFAMILY
Ardeinae

TAXONOMY
Ardea ibis Linnaeus, 1758, Egypt. Three subspecies.

OTHER COMMON NAMES
English: Buff-backed heron; French: Héron garde-boeufs; German: Kuhreiher; Spanish: Garcilla Bueyera.

PHYSICAL CHARACTERISTICS
A short-legged white egret. Has a relatively short yellow bill, and in breeding season attains a buff wash over much of its body. Length is 18–22 in (46–56 cm); weight is 12–14 oz (340–390 g).

DISTRIBUTION
Mid latitudes to warm temperate zone in North America and South America, Europe, Africa, Asia, and Australia.

HABITAT
Forages in native grasslands and in pastures alongside hoofed livestock. Also uses irrigated alfalfa fields, dumps, parks, athletic fields, golf courses, meadows, rice fields, lawns, and road margins. Nests in colonies with other wading birds, usually on islands over or surrounded by water.

BEHAVIOR
Walks slowly adjacent to moving cattle or other hoofed stock and may perch on these animals as they rest or move from place to place. Walks with the head alternately withdrawn and then pulled forward with each step, a gait characteristic of the species. Cattle egrets are among the most social of herons, forming small and large flocks on their feeding grounds. Feeds during the day, most actively in the morning and afternoon. During midday and at other times when grazing stock rest to ruminate, foraging flocks often loaf with other birds in trees or on the ground near the resting herd. At night, it roosts with other species, sometimes in the thousands.

FEEDING ECOLOGY AND DIET
Captures food made obvious by the movement of cattle, native large mammals, birds, or tractors. In Africa, its primary natural beater was probably the African buffalo (*Syncerus*) but also follows many other species. Locusts, grasshoppers, and crickets

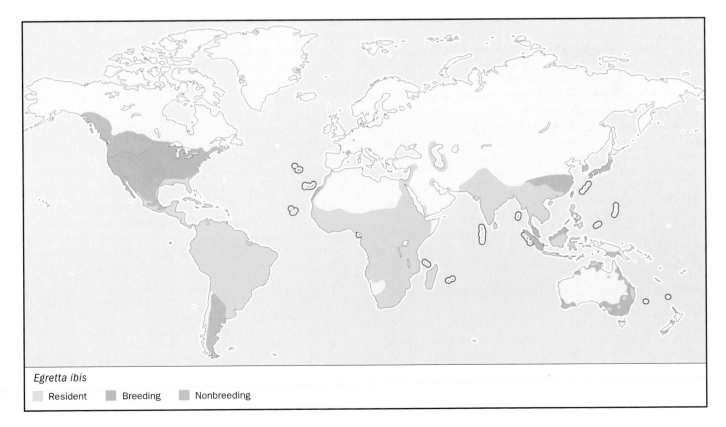

Egretta ibis

☐ Resident ☐ Breeding ☐ Nonbreeding

are the common element of its diet worldwide. Other insects also eaten include flies, beetles, caterpillars, dragonflies, mayflies, and cicadas.

REPRODUCTIVE BIOLOGY
Nesting season varies according to food availability. In the temperate north, it nests in spring and summer. In the tropics, nesting occurs at the end of the rainy season, as grasslands are drying out. The birds are highly colonial, breeding in mixed-species colonies of a few hundred pairs to several thousand pairs. The nest is made of reeds, twigs, or branches, 16 in (40 cm) wide. Eggs are white with a pale green or blue tinge, broad oval and somewhat pointed, lighter than most other medium-sized egrets. The clutch is usually four to five eggs. Incubation lasts about 24 days. Parents share care of the chicks. Young are guarded until day 10, leave the nest and climb in nearby branches at two weeks, fledge at 30 days, becoming fully independent in 15 more days. Nesting success is usually fairly high.

CONSERVATION STATUS
Probably the most abundant heron in the world.

SIGNIFICANCE TO HUMANS
Usually is easily recognized and not persecuted on its feeding ground in that it is perceived to be beneficial or neutral to cattle activities. However its tendency to develop large new colonies in and near towns and villages creates what may be perceived to be public nuisances. Efforts to control the populations can adversely impact other herons, whose conservation status may be more of a concern. ◆

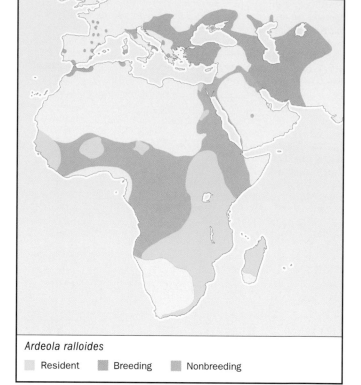

Ardeola ralloides
■ Resident ■ Breeding ■ Nonbreeding

Squacco heron
Ardeola ralloides

SUBFAMILY
Ardeinae

TAXONOMY
Ardea ralloides Scopoli, 1769, Carniola. Monotypic.

OTHER COMMON NAMES
French: Crabier chevelu; German: Rallenreiher; Spanish: Garcilla Cangrejera.

PHYSICAL CHARACTERISTICS
Tawny buff brown with a streaked head, crest, and back, and light belly. Length is 16.5–18.5 in (42–47 cm); weight is 8–13 oz (230–370 g). In breeding it develops a distinctive black and white mane. Immature birds are similar to adults in nonbreeding plumage, but drabber and lack crest and back plumes.

DISTRIBUTION
Occurs in Europe, Africa, Madagascar, and the Middle East to Iran.

HABITAT
Occurs in dense marshes—shallow fresh water with a cover of reeds and dense bushes. Its principal habitat throughout its range is now rice fields. It also occurs in ponds, canals, ditches, irrigated land, similar shallowly flooded areas. Seacoasts, reefs and islands are used on migration. For nesting, it tends to prefer dense trees and shrubs near its feeding areas.

BEHAVIOR
Often overlooked because it blends into dense vegetation. Roosts in groups, using sheltered woods or reed beds. The alarm and flight call given when disturbed or when flying to and from roosts is highly recognizable, giving the bird its name.

FEEDING ECOLOGY AND DIET
Typically feeds by searching for prey in a standing, crouched posture, either in the open or among the reeds. Usually feeds alone, defending its territory, although it also feeds in small groups or large flocks in winter and on migration. Feeding success is higher for solitary birds than those feeding in flocks. Feeds during the day, especially at twilight. Diet is relatively small prey, particularly fish, frogs, and tadpoles, as well as insects and insect larvae.

REPRODUCTIVE BIOLOGY
Herons in Europe and North Africa nest from late spring to summer. In tropical Africa, it breeds primarily in the rainy season. Nests in dense bushes or small trees, near or overhanging water, and less frequently in reed beds and papyrus swamps, using either the reed or small trees. Typically nests colonially with other species, although sometimes solitarily. Nests are small, bulky, and compact, 7–11 in (17–27 cm) in diameter made of reeds, grass, and twigs. Eggs are greenish blue. The clutch is four to six eggs in Europe, three to four in Madagascar and southern Africa. Clutch sizes have decreased in southern Europe over several decades. Incubation is 22–24 days in Europe, 18 days in Madagascar. Young begin to clamber from the nest into branches at 14 days. They are fledged at 45 days (35 days in Madagascar). Young form groups at the colony site.

CONSERVATION STATUS
Not threatened, but its populations are variable. Historic declines appear to be due to a combination of hunting, habitat

change, and perhaps climate. In some areas, the bird has increased its range in recent decades, likely due to its concentrated use of rice fields.

SIGNIFICANCE TO HUMANS
Often occurs close to humans, living in rice fields and marshes adjacent to towns and villages. It is not often noticed, but its call is distinctive. ◆

Black heron
Egretta ardesiaca

SUBFAMILY
Ardeinae

TAXONOMY
Ardea ardesiaca Wagler, 1827, Senegambia. Monotypic.

OTHER COMMON NAMES
French: Aigrette ardoisée; German: Glockenreiher; Spanish: Garceta Azabache, Garceta Gorgirroja.

PHYSICAL CHARACTERISTICS
Medium-sized (17–26 in [42.5–66 cm]), all-black plumaged African heron with yellow feet, usually seen feeding in open shallow water.

DISTRIBUTION
Occurs in Madagascar and Africa south of the Sahara.

HABITAT
Prefers shallow open waters, especially margins of fresh water lakes and ponds. Also uses marshes, river edges, rice fields, and seasonally flooded grasslands. Along the coast it feeds along

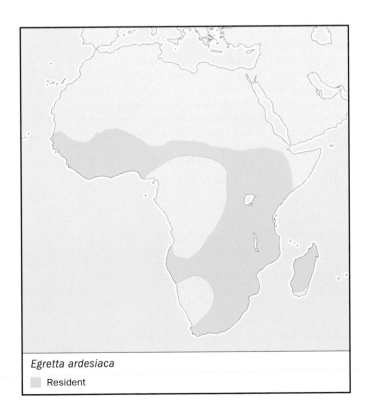

Egretta ardesiaca
▨ Resident

tidal rivers and creeks, mangroves, alkaline lakes, and tidal flats.

BEHAVIOR
Exhibits distinctive feeding behavior called canopy feeding. It spreads its wings over its head in a full umbrella, with the tips of its primaries touching the water and erect nape plumes completing the canopy. The heron forms the canopy above the potential prey over the course of a few steps. It peers under the canopy for a few seconds, perhaps also stirring with its feet. The heron then moves on a few steps to form another canopy, usually within a few more seconds. It frequently pauses to shake itself.

FEEDING ECOLOGY AND DIET
The functioning of the canopy feeding behavior remains unclear, although the canopy reduces reflection and provides better visibility in addition to obscuring the silhouette of the heron. Fish are likely attracted to the shadow or are attracted to or flee the foot stirring. Some resident black herons feed solitarily in well-defended feeding territories. They also feed in groups of up to 50 individuals, with over 200 being reported. Feeds by day, especially around dusk. Roosts communally at night and, on the coast, at high tides. Eats small fish, but also takes aquatic insects and crustaceans.

REPRODUCTIVE BIOLOGY
The nest is a solid structure of twigs placed over water in trees, bushes, and reed beds. Nests at the start of the rainy season, in single or mixed-species colonies that may number in the hundreds. Eggs are dark blue and the clutch is two to four eggs.

CONSERVATION STATUS
Threatened on Madagascar, where human interference and habitat change have led to massive population reductions. Elsewhere, the heron is patchily distributed but not uncommon. Its greatest threats are human disturbance, predation at nest sites, and threats to aquatic habitats.

SIGNIFICANCE TO HUMANS
The distinctive feeding behavior and its feeding in open areas makes it easily noticed where it occurs. ◆

Little egret
Egretta garzetta

SUBFAMILY
Ardeinae

TAXONOMY
Ardea garzetta Linnaeus, 1766, Malalbergo, Italy. Six subspecies.

OTHER COMMON NAMES
English: Lesser egret; French: Aigrette garzette; German: Seidenreiher; Spanish: Garceta Común.

PHYSICAL CHARACTERISTICS
A thin, medium-sized heron, with a long thin neck and bill, dark legs and yellow feet (in most forms). Length is 22–25.5 in (55–65 cm); weight is 10–22.5 oz (280–638 g). In breeding it has distinctive head, chest and back plumes, and red lores. Some populations are dimorphic, having both dark and white birds.

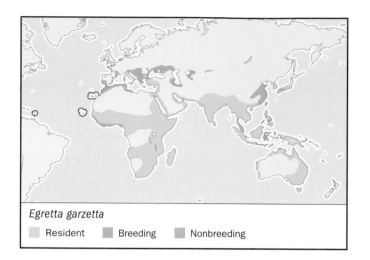

Egretta garzetta

■ Resident ■ Breeding ■ Nonbreeding

DISTRIBUTION
Occurs in Europe, Africa, Madagascar, Asia, East Indies, Australia, Pacific Ocean islands. It has recently colonized the West Indies.

HABITAT
Typically uses open or sparsely vegetated shallow to very shallow water for feeding. Frequently uses artificial feeding habitats, including rice fields, fish ponds, and irrigation pools. Occasionally feeds in pasture and other dry land situations, and is known to feed communally with cattle or other ungulates. Nests in trees, bushes, or islands that offer protection and isolation.

BEHAVIOR
Highly social, usually seen in groups, either feeding in or at the edges of shallow water bodies, roosting at midday or on high tides, or nesting. Highly aggressive and territorial when feeding. Runs or hops between feeding sites, opening its wings to startle and chase down fish. Also uses such feeding behaviors as using floating bread or their bills to attract fish, following cattle, or riding bathing water buffalo. Birds roost when not feeding, and in the evening fly in small flocks to communal roosts.

FEEDING ECOLOGY AND DIET
Feeds in shallow, open, and unvegetated sites where water levels and dissolved oxygen are fluctuating (tidally, seasonally, or daily), where fish are concentrated in pools or at the water's surface. Typically feeds by walking slowly with the neck stretched out through the water in search of fish or other prey, stirring the substrate with its feet. Feeds in deeper water by flying above the surface, dipping its bill into the water to catch fish, or dragging its feet at the surface to frighten them into movement. Switches habitats through the year, and feeds alone or in groups. Follows other birds closely, frequently robs them of prey, and is robbed in turn. Diet is mainly small fish, generally only 0.4–2.4 in (1–6 cm) long. Also takes small birds, lizards, snakes, frogs, toads and tadpoles, insects, prawns, amphipods, crayfish, crabs, and many other invertebrates.

REPRODUCTIVE BIOLOGY
Breeding season varies across its range, spring in temperate areas and most often at the peak or after the peak of the rainy season in the tropics. Nests colonially, sometimes in mixed-species colonies that can number in the thousands. Coastal birds tend to nest in smaller colonies or alone. Nests are small platforms, 12–14 in (30–35 cm) wide. The eggs are variable greenish blue, fading to off-white. Clutch size varies geographically, with a range of two to eight. Incubation period is 21–25 days. Parents attend young for 10–15 days. Nestlings compete for food, and the youngest birds typically die. Young leave the nest at 35–50 days.

CONSERVATION STATUS
Not threatened. Loss of inland and coastal wetlands has occurred throughout its range.

SIGNIFICANCE TO HUMANS
A well-known species because it occurs near and with human populations. ◆

Black-crowned night heron
Nycticorax nycticorax

SUBFAMILY
Ardeinae

TAXONOMY
Ardea nycticorax Linnaeus, 1758, Europe. Four subspecies

OTHER COMMON NAMES
English: Night heron; French: Bilhoreau gris; German: Nachtreiher; Spanish: Martinete.

PHYSICAL CHARACTERISTICS
A stocky dark gray and white heron with a distinctive glossy black bill, crown, and back. Length is 22–25.5 in (56–65 cm); weight is 18.5–28 oz (525–800 g). During breeding it develops white head plumes that may reach 10 in (25 cm) long. It has relatively short legs that do not extend much beyond the tail when in flight. Juveniles are cryptic gray-brown with buff and white spots above and stripes below.

DISTRIBUTION
Occurs across the temperate and tropical world from North and South America, Europe, Africa, and Asia to the East Indies.

HABITAT
Typically found along the vegetated margins of shallow freshwater or brackish rivers, streams, ponds, lakes, marshes, swamps, mangroves, and mud flats. Also uses grasslands and coastal habitats, especially on migration, and unlike most herons occurs on high mountains. Uses pastures, ponds, reservoirs, canals, ditches, fishponds, rice fields, wet-crop fields, and dry grasslands. Usually nests in bushes and trees but also in reeds, sedge, grass tussocks, on the ground in protected areas like islands, and in protected locations in urban areas. Large nesting colonies especially appear to be associated with protected sites in large wetlands.

BEHAVIOR
A noisy bird having a raucous "quawk" call. Also has a breeding call that is like the sound of a rubber band being plunked and followed by a buzz or hiss. It flies with wing beats that are faster than most herons. Roosts by day in trees and bushes and is most often seen flying to roost in the morning and out in the evening, giving the quawk call. Roosts commonly are in rural areas and within towns.

FEEDING ECOLOGY AND DIET
Typically feeds at night, locating prey by sight and sound, also feeds during the day when nesting. Usual method is standing in a crouched posture and making a lunging strike at prey. During daylight, it may run, dive into the water from the air,

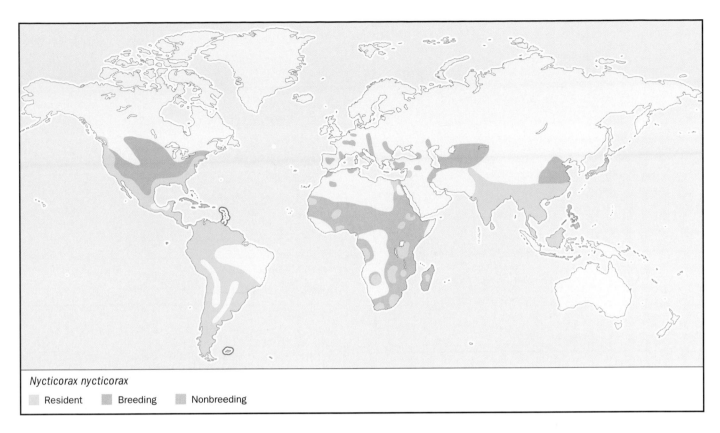

Nycticorax nycticorax

☐ Resident   ☐ Breeding   ☐ Nonbreeding

hover, swim, or use its wings to startle prey. Also attracts fish by vibrating its bill or using baits. Mostly a solitary forager, maintaining territories that it defends vigorously. Also feeds in aggregations when prey is highly concentrated. Fish, frogs, and aquatic insects predominate in the diet. Often eats the young of other colonial nesting waterbirds.

REPRODUCTIVE BIOLOGY

In temperate areas, nesting occurs in spring, often early, but in tropical and subtropical areas nesting is more variable. The species often nests in rural, suburban, and urban settings, particularly in zoos. Nesting is usually colonial, in single-species or mixed-species colonies that number sometimes in the thousands. Nests are a platform of sticks and reeds, 12–18 in (30–45 cm) wide. Eggs are green to pale blue-green. Clutch size is two to five eggs with an overall range of one to seven. Incubation averages 23 days. Parents brood the young for 10 days. The young clamber out of the nest by three weeks and fledge in six or seven weeks. Nesting success is often high.

CONSERVATION STATUS

Not threatened. However, nesting is limited to few areas in some regions, such as in Europe, so conservation of these sites is crucial. In North America, populations declined due to pesticides, particularly up to the 1960s.

SIGNIFICANCE TO HUMANS

Fairly tolerant of human activities, and often nest and roost near humans. Night herons are often killed at fish hatcheries and are still hunted for food in some places. Most human interaction has been positive for the species. ◆

White-eared night heron
Nycticorax magnificus

SUBFAMILY
Ardeinae

TAXONOMY
Nycticorax magnifica Ogilvie-Grant, 1899, Hainan. Monotypic.

OTHER COMMON NAMES
English: Magnificent night heron; French: Bihoreau superbe; German: Hainanreiher; Spanish: Martinete Magnifico.

PHYSICAL CHARACTERISTICS
A medium-brown heron with a brown streaked breast and a white patch on the side of the head. Length is about 21 in (54 cm). Juvenal plumage has brown-black feathering spotted with buff or white.

DISTRIBUTION
Occurs in Southeast Asia.

HABITAT
Occurs in dense, primary forests with streams and adjacent marshes. It currently is found only in mid-altitude mountains, but was likely originally also a lowland species.

BEHAVIOR
A poorly known species, with few having been observed in the wild. Feeds at night and roosts high in trees during the day. Has been reported as feeding singly or in isolated pairs.

FEEDING ECOLOGY AND DIET
Diet includes fish, shrimp, and insects.

Nycticorax magnificus
■ Resident   ■ Nonbreeding

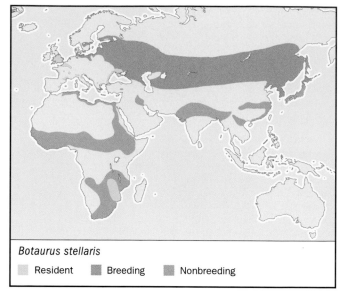

Botaurus stellaris
■ Resident   ■ Breeding   ■ Nonbreeding

REPRODUCTIVE BIOLOGY
Nearly nothing is known about the breeding biology of this species. Nests in tall trees and perhaps bamboo. May nest in mixed colonies.

CONSERVATION STATUS
Endangered. Only two breeding sites are known, a third was submerged by a reservoir. The principal threat to the species is habitat loss caused by deforestation, reforestation with pine monoculture, reservoir construction, and gold mining. Hunting is also a threat, even in nature reserves. The species is nationally protected in China, but the ability of local people to save the birds may be limited unless education and local conservation initiatives are undertaken.

SIGNIFICANCE TO HUMANS
This species is severely threatened due to human habitat alteration and hunting. Local people eat herons, and young herons of several species are captured for the market. Balancing the needs of the local population for food with conservation of this species is a huge challenge. ◆

Eurasian bittern
Botaurus stellaris

SUBFAMILY
Botaurinae

TAXONOMY
Ardea stellaris Linnaeus, 1758, Sweden. Two subspecies.

OTHER COMMON NAMES
English: Great bittern, common bittern; French: Butor étoilé German: Rohrdommel; Spanish: Avetoro Común.

PHYSICAL CHARACTERISTICS
A thick-necked, medium-sized, golden brown heron with a black head and moustache. Length is 25–31 in (64–80 cm);

weight is 1.9–4.3 lb (0.9–1.9 kg). Its back is cryptic mottled and mottled. It is the largest of the four species of large bitterns. Males are significantly larger than females.

DISTRIBUTION
Occurs in the Old World temperate and tropical zones of Europe, Asia, and Africa.

HABITAT
Occurs in densely vegetated wetlands. During the breeding season, it is found only in reed beds characterized by dense plants, stable shallow flooding, and with intermittent clearings or channels. During the nonbreeding season, it uses more varied and open aquatic habitats such as small ponds, gravel pits, wet grassy meadows, ditches, tall rice fields, fish ponds, floating leafed plant beds, and sewage lagoons.

BEHAVIOR
Hunts by walking with stealth in a crouched posture, the bill pointed forward, and the feet lifted high with each step. Moves about by climbing over the emergent stems, using long toes to grasp the stems. Can hold a concealing behavior, called the bittern posture, for hours. In this posture, it raises its bill to the sky and peers directly forward, swaying as if in the breeze and turning slowly to keep eyes on a moving intruder.

A solitary feeder that fiercely defends its feeding and nesting area during breeding using its booming call. The call consists of two to four deep, resonant booms preceded by a few short grunts or pumps, sometimes accompanied by clappering the bill. It is aggressive in physically defending its site, and also flies to supplant intruders and fights in the air, even to the death.

FEEDING ECOLOGY AND DIET
Feeds at the edge of emergent reeds and open water, such as along a pool, channel, or ditch, avoiding unflooded ground. Primarily feeds in the morning and evening but is known to hunt during the day and at night. Fish, amphibians, and insects usually dominate the diet. Small mammals, birds, and snakes are also taken.

REPRODUCTIVE BIOLOGY
Nests solitarily in spring and summer or in the rainy season in the tropics. Non-migrating birds begin to call as early as late

winter. Nest is a pad of matted reeds and other marsh vegetation that is built by the female. A polygamous species, males may have up to five mates, each of which has a nest within the male's territory. Eggs are olive brown with spotting. The normal clutch is four to five eggs; range is three to seven. Incubation begins immediately, and the range of hatching dates for a large clutch may stretch over two weeks. Only the female incubates, lasting 25–26 days. Young can leave the nest after about two weeks and fledge at 50–55 days.

CONSERVATION STATUS

The Eurasian bittern was formerly widespread and abundant, but suffered significant population changes. In Europe, it declined steadily since as early as the nineteenth century, being extirpated from England in 1868. It began a comeback through Europe in early 1900s, increased into the 1960s and then began a second decline, in some cases very rapidly. It is now regionally Vulnerable in Europe. The southern African population is even more at risk, given its rapid decline over the past several decades.

SIGNIFICANCE TO HUMANS

The Eurasian bittern occurs in reed beds and marshes throughout its range. It is a skulking species that stays hidden, at least during the day. It nonetheless is well known owing to its booming call. This call has entered into folklore wherever large bitterns occur, generally as a harbinger of evil. ◆

Least bittern
Ixobrychus exilis

SUBFAMILY
Botaurinae

TAXONOMY
Ardea exilis Gmelin, 1789, Jamaica. Five subspecies.

OTHER COMMON NAMES
English: Nitlin, gaulin; French: Petit blongios; German: Amerikanische Zwergdommel; Spanish: Avetorillo Panamericano.

PHYSICAL CHARACTERISTICS
The least bittern is the smallest heron (11–14 in [28–36 cm]), a pale buff bird with a dark crown and back and buff-colored wing patches. The female averages larger than the male. It has chestnut, rather than black, upperparts, a less prominent crown, darker neck stripes, dark brown chest streaks, and paler wing patch. Juveniles are similar to females.

DISTRIBUTION
The least bittern occurs in North America, Central America, West Indies, and north, west and east South America.

HABITAT
The habitats typically are very dense marsh vegetation in water with both woody growth and open water patches. These include fresh water marshes, lake edges, salt marshes in temperate areas, and mangroves in the tropics.

BEHAVIOR
The least bittern feeds by stalking through the reeds or along the edge of dense reed stands or on branches over the water. It walks in very crouched posture, with its neck extended and its bill nearly touching the water. It may also feed by standing in

Ixobrychus exilis

■ Resident ■ Breeding ■ Nonbreeding

one place and may build feeding platforms. The bittern posture is often assumed as a defensive display. The least bittern is very vocal, giving a low pitched, dove like advertising call and a rattling disturbance call.

FEEDING ECOLOGY AND DIET
It feeds within dense emergent vegetation. The principal prey is small fish, but its overall diet is much broader including crabs, crayfish, insects, frogs, tadpoles, salamanders, small mammals, and even small birds.

REPRODUCTIVE BIOLOGY
In the north, it nests in the spring and summer and at more varied times in the tropics. Nests are placed in thick herbaceous marshes, most commonly in cattail. It nests solitarily or in small groups. The male constructs the nests and advertises with a distinctive cooing call and defends its territory. The eggs are white. Clutch size is four or five eggs, with fewer in the tropics. Unlike in the large bitterns, both sexes incubate and care for young. Chicks develop quickly, being able to leave the nest temporarily by day five, wandering by two weeks. They fledge in about three or four weeks.

CONSERVATION STATUS

Not threatened. Conservation of this species depends on marsh preservation. Water impoundments and wetland construction for various purposes increase the potential habitat for the species as it often nests in cattail marshes created by human activities.

SIGNIFICANCE TO HUMANS

None known. Least bitterns are seldom noticed, due to their small size and secretive ways. They are charming small birds that well deserve additional attention. ◆

Agami heron
Agamia agami

SUBFAMILY
Agamiinae

TAXONOMY
Ardea agami Gmelin, 1789, Cayenne. Monotypic.

OTHER COMMON NAMES
English: Chestnut-bellied heron; French: Héron agami; German: Speerreiher; Spanish: Garza Agamí.

PHYSICAL CHARACTERISTICS
The agami heron is a strikingly colored medium-sized heron. The rapier-like bill averages 5.5 in (140 mm) but sometimes

reaches 6.4 in (163 mm), about one-fifth the bird's total length (24–30 in [60–76 cm]). The neck is very long and snake-like. Its back is bottle green, upper neck is chestnut with a central white stripe bordered by black contrasting with a gray lower neck, which sports a distinctive mat of shaggy, light gray feathers. The belly is chestnut. In the breeding season it has ribbon-like light blue crest feathers, up to 5 in (125 mm) long, and also broad slaty blue plumes on the lower back.

DISTRIBUTION
The agami heron occurs in Central and northern South America, especially in the Orinoco and Amazon basins.

HABITAT
This heron occurs in dense tropical lowland forest along margins of streams, small rivers, and swamps. They are also found less commonly along the margins of pools, oxbow lakes, and other small bodies of water.

BEHAVIOR
The agami heron typically is seen standing in crouched posture on banks, dykes, bushes, or branches overhanging the water. It also walks slowly in shallow water at the edge of streams or ponds. It has a distinctive, low-pitched, rattling alarm call.

FEEDING ECOLOGY AND DIET
The agami heron is a specialized bank fisher. Its short legs and long neck permit a long lunging strike. It feeds alone, with individuals scattered along water courses. With its long neck and bill, it is primarily a fish-eating heron.

REPRODUCTIVE BIOLOGY
Nesting is during the wet season. It nests in small single species or mixed-species colonies. Nests are in isolated clumps of mangroves, dead branches of drowned trees in an artificial lake, trees standing in water, and bushes within marshes, well hidden within the vegetation. The nest is a loose, thick platform of sticks or twigs, rather deeply cupped. The eggs are pale blue-green or dull blue. Clutch size is two to four eggs. Nothing is reported on incubation. Young gain weight quickly, more than doubling in the first week.

CONSERVATION STATUS
It is likely not at risk over the entirety of its large range and is readily seen along rivers and streams in certain parts of its range. Given its known habits, it would be threatened by deforestation and damming of rivers.

SIGNIFICANCE TO HUMANS
None known. This is a little known, highly specialized species of the deep tropical forest. It is infrequently seen. ◆

Agamia agami

Resident

Boat-billed heron
Cochlearius cochlearius

SUBFAMILY
Cochleariinae

TAXONOMY
Cancroma cochlearius Linnaeus, 1766, Cayenne. Five subspecies.

OTHER COMMON NAMES
French: Savacou huppé German: Kahnschnabel; Spanish: Martinete Cucharón, Pato Pico de Barco.

Cochlearius cochlearius

▨ Resident

HABITAT
It primarily uses wooded or mangrove fringes of freshwater creeks, lakes and marshes. It roosts in the day time in bushes or trees overhanging water.

BEHAVIOR
Despite its great bill, this species for the most part feeds like a typical heron by standing, sometimes for many minutes, usually in a crouched posture. It also walks in its crouched, hunched-backed posture with very slow deliberate steps on the ground or along branches and roots. It sometimes uses a rhythmic movement of the body but not the head. It sometimes walks very quickly or runs about. It also feeds non-visually by wading along with its bill partially submerged thrusting it forward in a scooping motion with each step. During the day it perches in dense trees and bushes, and also retreats there when disturbed. When roosting, its large bill rests on its breast or under a wing. It uses its crest for communication, raising it in response to disturbance and as a greeting display. This is a noisy heron, having a raucous laughing call that is a commonly heard sound along the tropical mangroves and inland forests. Also makes a popping noise with its bill.

FEEDING ECOLOGY AND DIET
It feeds nocturnally and crepuscularly although it occasionally feeds during the day. It feeds alone, flying from the communal roost to independent, probably defended, feeding areas. Occasionally it will feed in aggregations. The diet is broad and includes insects, shrimps, fish, amphibians, and small mammals.

REPRODUCTIVE BIOLOGY
Breeding timing is variable, generally in the rainy season. The heron nests solitarily or in small groups of a few to a dozen pairs but also joins mixed heronries. The eggs are pale blue to green, often with spotting. Clutch size is usually three eggs, range one to four eggs. Incubation is 26 days. The young are at first fed entirely at night. The adult is aggressive in defending the young from all intruders, a behavior not typical of herons.

CONSERVATION STATUS
Not threatened. The species is widespread and found in suitable habitat throughout its range. There is little information available on population sizes and status, but it is not rare.

SIGNIFICANCE TO HUMANS
This is the most unusual of the herons, with its huge bill, unusual behaviors, and evolutionary distinctiveness from other herons. It is well known locally where it occurs, especially due to its calls from nesting colonies. ◆

PHYSICAL CHARACTERISTICS
The boat-billed heron is a stocky, medium-sized (18–20 in [45–51 cm]) mostly black and white and sometimes buff heron, with a huge black bill. The head is black, with a crest of long, black, lanceolate plumes that are most extravagant during the nesting season. The huge eyes bulge out from the face. The upper back is black, the rest of the back and upper wings are gray. The underparts are a rich rufous. During breeding the mouth lining, lores, and gular area turn black.

DISTRIBUTION
The boat-billed heron occurs in South and Central America.

Resources

Books

Brown, Leslie. *The Birds of Africa.* Vol. 1. San Diego: Academic Press, 1983.

del Hoyo, J., A. Elliot, and J. Sargatal, eds. *Handbook of the Birds of the World.* Vol. 1, *Ostrich to Ducks.* Barcelona: Lynx Edicions, 2001.

Hancock, J.A. *Birds of the Wetlands.* London: Academic Press, 1999.

Hancock, J.A., J.A. Kushlan, and M.P. Kahl. *Storks, Ibises and Spoonbills of the World.* London: Academic Press, 1992.

Kushlan, J.A., and H. Hafner. *Heron Conservation.* London: Academic Press, 2001.

Kushlan, J.A., and J.A. Hancock. *The Herons.* Oxford: Oxford University Press, 2002.

Periodicals

Draulans, D., and J. van Vessem. "Some Aspects of Population Dynamics and Habitat Choice of Gray Herons (*Ardea cinerea*) in Fish-pond Areas." *Gerfault* 77 (1987): 43–61.

Resources

Kushlan, J.A. "Feeding Behavior of North American Herons." *Auk* 93 (1976): 86–94.

Kushlan, J.A. " Feeding Ecology of Wading Birds." *Wading Birds, National Audubon Society Research Report* 7 (1978): 249–297.

Maddock, M., and G.S. Baxter. "Breeding Success of Egrets Related to Rainfall, A Six Year Australian Study." *Colonial Waterbirds* 14 (1991): 133–139.

Marion, L. "Territorial Feeding and Colonial Breeding are Not Mutually Exclusive: The Case of the Gray Heron (*Ardea cinerea*)." *Journal of Animal Ecology* 58 (1989): 693–710.

Organizations

Herons Specialist Group. Station Biologique de la Tour du Valat, Le Sambuc, Arles 13200 France. Phone: +33-4-90-97-20-13. Fax: 33-4-90-97-29-19. E-mail: hafnerh@aol.com Web site: <http://www.tour-du-valat.com>

Waterbird Society. National Museum of Natural History, Smithsonian Institution, Washington, DC 20560 USA. Web site: <http://www.nmnh.si.edu/BIRDNET/cws>

Other

Kushlan, J.A., and L. Garrett. *A Bibliography of Herons.* <http://www.pwrc.usgs.gov/library/bibs.htm>

James A. Kushlan, PhD

Hammerheads

(Scopidae)

Class Aves
Order Ciconiiformes
Suborder Scopi
Family Scopidae

Thumbnail description
Large uniform-brown wading bird with a
distinctive large, backward-pointed crest

Size
Length: 20–24 in (50–60 cm); wing: 11.6–12.4
in (297–316 mm); weight: 0.91–0.95 lb
(415–430 g)

Number of genera, species
1 genus; 1 species

Habitat
Most freshwater habitats, even small temporary
ponds

Conservation status
Not threatened; common in appropriate habitat

Distribution
Sub-Saharan Africa: Senegal to southern Somalia south to southern South Africa;
Madagascar, and southwest Arabian Peninsula

Evolution and systematics

The origins of the hamerkop or hammerhead (*Scopus umbretta*) are obscure. Discovered by Gmelin in Senegal, Africa, in 1789, it has been traditionally placed in the Ciconiiformes with other large, long-legged, wading birds.

The hammerhead shares affinities with different groups. It shares a pectinated claw on the middle toe with herons; it shares a free hind toe with flamingos; and its ectoparasites are related to those of plovers. Its egg-white proteins place it close to the storks; DNA suggests that it should be between the herons and flamingos, close to the storks.

It is distinctive enough that it has been placed in its own suborder Scopi.

Physical characteristics

There are two subspecies, which vary in size and appearance. In the nominate race (*S. u. umbretta*), the hammerhead stands about 22 in (56 cm) tall; the more restrictive West

African race (*S. u. minor*) is smaller and darker. The entire body is dull brown, paler on the chin and throat. The head is strongly crested, and the crest points backward. The crest and the long, heavy bill suggest the name "hammerhead." The female is similar to the male but slightly larger.

Distribution

The species is found in sub-Saharan Africa from Senegal in the west, east to southern Somalia and south to southern South Africa. It is also found in Madagascar and the southwest Arabian Peninsula.

The hammerhead is nonmigratory, although in drier areas there may be some seasonal dispersal as rains lead to additional temporary feeding sites.

The nominate race (*S. u. umbretta*) is found through most of the range in tropical Africa, Madagascar, and the southwest Arabian peninsula. The West-African race (*S. u. minor*) is found in the coastal belt from Sierra Leone to eastern Nigeria.

Hammerhead (*Scopus umbretta*). (Illustration by Joseph E. Trumpey)

Habitat

The hammerhead is found in almost all types of wetlands. For feeding it requires shallow wetlands from lakeshores to the banks of large rivers to small temporary ponds.

For breeding, the species requires a foundation to support its large, complicated nest. Both male and female hammerheads participate in building nests, which can weigh 100 times more than the bird. Almost always these are located in large trees but rarely cliffs or rocky hillsides are used. These areas are also used for roosting.

Behavior

Hammerheads are generally active during the day or are crepuscular, but are not active at night as some have suggested. As with other tropical birds, they are less active during the heat at mid-day. They are usually found alone or in small groups, though occasionally large groups (up to 50 birds) may roost together.

Feeding ecology and diet

Hammerheads feed by wading in shallow water and picking prey from among vegetation. They may stir the water with their feet ("foot stirring") or open their wings ("wing flicking") to encourage prey to move. They also capture prey while flying, often catching tadpoles or small fish when in small pools. They can fly slowly over the water because they have low wing loadings (i.e., large wing area compared to low mass).

Their major prey varies geographically. They are particularly known for taking clawed frogs (*Xenopus* sp.) or their tadpoles in south and east Africa. In other areas (e.g., Mali), they concentrate on small fish. They also take shrimp, crustaceans, and even small mammals.

Reproductive biology

Hammerheads are monogamous and territorial, although territories often overlap. The breeding time varies geographically, but is primarily in the dry season or late in the wet season. At this time, decrease in water leads to a concentration of prey, making it easier for the adults to collect adequate food for their young.

They are famous for their very large and elaborate nests. The domed nests may include over 8,000 items, be as much as 5 ft (1.5 m) deep, weigh up to 55 lb (25 kg), and fully support an adult man or woman. The materials include mostly sticks, leaves, and mud. The structures include a chamber up to 16 in (40 cm) wide and 24 in (60 cm) high, and an entrance 4–6 in (10–15 cm) in diameter and 16–24 in (40-60 cm) long. Nests are most often built in a fork in a tree, usually about 30 ft (9 m) above the ground, but occasionally on cliffs or rarely on the ground.

Both members of a pair share in nest construction, often working together. Working primarily in the morning and evening, they construct the nest platform, followed by the walls, and then the roof.

While a nest may be used for one or many seasons, a pair may build 3–5 nests in a single season. Some are abandoned before completion; some are used for roosting; one may be used for breeding, or other animals may take advantage of these nests. Other animals observed using nests, include: Verreaux's eagle owl (*Bubo lacteus*), barn owl (*Tyto alba*), Egyptian goose (*Alopochen aegyptiacus*), genets (*Genetta* spp.) monitor lizard (*Veranus* spp.), and spitting cobra (*Naja nigricollis*).

After the nest is complete, three to seven white eggs, measuring 1.6–2.1 in by 1.3–1.5 in (41–53 mm by 32–37 mm) and

This hammerhead (*Scopus umbretta*) works on its huge nest that may take six months to build in Kruger National Park, South Africa. (Photo by Nigel J. Dennis. Photo Researchers, Inc. Reproduced by permission.)

weighing about 1 oz (25 g) are laid in one to two day intervals. Both parents incubate and care for the young. Hatching occurs after about 30 days following the completion of the clutch. At hatching, the downy young are pale brown. They begin fledge at about 50 days old.

Conservation status

The hammerhead is common to abundant throughout its range.

Significance to humans

Throughout much of its range, it is considered a bird with supernatural powers and even evil powers. It is thought that if improperly treated, a hammerhead can cause a house to melt, an epidemic among cattle, or even death. It has been given great respect and distance.

This respect may be due to the bird's strange appearance, its gigantic nest, and the many other birds and animals that may occupy abandoned nests, including deadly cobras.

Resources

Books

Brown, L.H., E.K. Urban, and K. Newman. *The Birds of Africa*. Vol. 1, *Ostrich to Falcons*. London: Academic Press, 1982.

del Hoyo, J., A. Elliott, and J. Sargatal, eds. *Handbook of the Birds of the World*. Vol. 1, *Ostrich to Ducks*. Barcelona: Lynx Edicions, 1992.

Hancock, J.A., J.A. Kushlan, and M.P. Kahl. *Storks, Ibises and Spoonbills of the World*. London: Academic Press, 1992.

Maclean, G.L. *Roberts' Birds of Southern Africa*. 5th ed. London: New Holland Publishers, 1985.

Wilson, R.T., and M.P. Wilson. *Breeding Biology of the Hamerkop in Central Mali*. Proceedings of the Fifth Pan-African Ornithological Congress, edited by J. Ledger.

Periodicals

Liversidge, R. "The Nesting of the Hamerkop, *Scopus umbretta*." *Ostrich* 34 (1963): 55-62.

Wilson, R.T. "Nest Sites, Nesting Seasons, Clutch Sizes and Egg Sizes of the Hamerkop *Scopus umbretta*." *Malimbus* 9 (1987): 17-22.

Wilson, R.T., M.P. Wilson, and J.W. Durkin. "Aspects of the Reproductive Ecology of the Hamerkop *Scopus umbretta* in Central Mali." *Ibis* 129 (1987): 382-388.

Malcolm C. Coulter, PhD

Storks

(Ciconiidae)

Class Aves
Order Ciconiiformes
Suborder Ciconiae
Family Ciconiidae

Thumbnail description
Distinctive medium to large wading birds with long legs, long necks, and large powerful bills

Size
30–60 in (75–152 cm); 2.9–19.7 lb (1.3–8.9 kg)

Number of genera, species
6 genera; 19 species

Habitat
Wide variety of mainly lowland habitats, generally in warm climates. Many species prefer to be in or near wetlands, though some occur in drier areas

Conservation status
Endangered: 3 species; Vulnerable: 2 species; Near Threatened: 2 species

Distribution
Widely distributed; found on all continents except Antarctica.

Evolution and systematics

There are 19 species of stork in six genera. Taxonomists place the birds in three "tribes": the Mycteriini (including the wood stork (*Mycteria americana*) and the openbills (*Anastomus* spp), the Ciconiini (including the European white stork (*Ciconia ciconia*) and black stork (*Ciconia nigra*) and the Leptoptilini (including large storks such as the marabou (*Leptoptilos crumeniferus*) and jabiru (*Jabiru mycteria*).

Stork remains have been identified from the Upper Eocene (about 40 million years ago) in France, and the group was distinct in the early part of the Tertiary (about 65 million years ago). Traditionally storks are placed taxonomically with other long-legged wading birds such as herons, but their nearest relatives may be New World vultures such as the ubiquitous turkey vulture. Although the similarities are not immediately apparent, DNA analysis supports this conclusion. Interestingly, both New World vultures and storks share the rather unpleasant habit of defecating on their own legs to facilitate heat loss, and this has been cited as a behavioral similarity to support the biochemical findings.

Physical characteristics

Storks are distinctive medium to large wading birds. They have long legs, long necks, and large powerful bills. The only birds with which they might be confused are herons, but in general herons are of a much slighter build and characteristically fly with neck retracted, as opposed to storks who fly mostly with their necks outstretched. Plumages are combinations of white, black, and gray. Strikingly colored bills in various combinations

of red, black, and yellow often complement these plumages. Some species, such as the North American wood stork and the African marabou, lack feathers on their head and neck, a response to their habit of feeding in muddy pools and on carcasses, situations in which feathers would soon become soiled.

Distribution

Storks have a wide distribution and are found on all continents except Antarctica. They reach their greatest diversity in tropical regions and show a strong preference for warmer climates; indeed the few species that breed in colder temperate areas migrate to warmer countries after nesting. North America has the least diversity, with the wood stork as the region's only, and very marginal, representative.

Habitat

Storks are found in a wide variety of mainly lowland habitats. Many species prefer to be in or near wetlands, although some, such as the marabou, occur in drier areas. The stork with possibly the most atypical habitat is the black stork. In the northern summer, this bird inhabits the extensive forests of Eastern Europe and Asia, albeit within easy reach of small pools and rivers for feeding.

Behavior

The social behavior of storks is varied. Many species, such as the painted stork, nest in colonies and are highly gregari-

Wood storks (*Mycteria americana*), like some other wading birds, feed by touch rather than sight. Known as "tacto-location," this method of feeding enables the birds to take fish and other food items from cloudy water. (Illustration by Emily Damstra)

ous during the breeding season. Others nest in smaller, much looser, groups, and a few species, such as the black stork and saddlebill (*Ephippiorhynchus senegalensis*), nest alone. Outside breeding season, storks are either solitary or congregate in small groups.

Storks are adept at soaring in flight and regularly exploit warm currents of rising air (thermals) to gain height before gliding down to their destination. Most fly with necks outstretched, although those with particularly heavy bills, such as the marabou, may retract them to keep their aerial balance. Storks rarely fly in formation.

Although storks are not very vocal, they can produce a variety of croaks, honks, hisses, and wheezes. They are also well known for their noisy bill-clattering displays during the breeding season. In the towns and villages where the stork often breeds, the clattering can go on well into the night, to both the chagrin and delight of residents.

Feeding ecology and diet

Storks are carnivores and consume a wide variety of animals, from small aquatic invertebrates, amphibians, and fish to more unlikely items such as young crocodiles and young birds. Two closely related species, the marabou and the greater adjutant, are at home scavenging at carcasses and even on human waste.

Such a varied diet elicits a similarly varied range of feeding techniques. Some species, such as the wood stork, hunt almost entirely by touch, capturing small fish the moment they chance to touch the bird's sensitive bill, which is purposely held open in readiness. In experiments, wood storks

have been recorded reacting in 25 milliseconds, the fastest-known response rate of any vertebrate.

The gap between the mandibles of the bill of the openbill has prompted much speculation as to its purpose in relation to the bird's feeding technique. Some observers have speculated that it might be used to break the shells of the openbill's preferred prey, apple snails (*Pomacea*); others have thought that the opening might help the birds carry the snails. Neither of these appears to be the case. It is perhaps more likely that the curvature of the lower mandible was originally a simple deformity that had the advantage of enabling some birds to extract snails from their shells more efficiently. Natural selection then favored these birds and the trait was perpetuated.

Other species, such as members of the Ciconiini tribe, are more opportunistic, and simply take what is available. Their typical feeding method involves slowly pacing their feeding grounds looking for prey which, when located, is seized with a sudden forward lunge.

Reproductive biology

Storks are either highly colonial, loosely colonial, or solitary breeders. Solitary breeders form monogamous pairs. *Mycteria*, *Anastomus*, and *Leptoptilos* are decidedly colonial, their chosen breeding sites sometimes consist of thousands of nests, often in the company of other storks, as well as wading birds such as herons and egrets. European white and maguari storks (*Ciconia maguari*) are much less colonial, breeding in smaller groups or, occasionally, alone. A number of storks, such as the black stork, woolly-necked stork (*Ciconia episcopus*), and jabiru always nest alone.

Almost invariably storks choose to nest in trees, and often at quite a height. Some species, such as the wood stork, pre-

A saddlebill (*Ephippiorhynchus senegalensis*) swallows a bream (a fish) after catching it in the Khwai River, Botswana. (Photo by Gregory G. Dimijian. Photo Researchers, Inc. Reproduced by permission.)

White stork (*Ciconia ciconia*) adult and young in nest (Turkey). (Photo by U. Walz/Okapia. Photo Researchers, Inc. Reproduced by permission.)

fer the security of islands. Abdim's stork (*Ciconia abdimii*) will nest on cliffs or on the top of village huts, and the European white stork is renowned for nesting on structures such as telegraph poles, chimney stacks, and pylons.

Stork nests are made from sticks and twigs, with other plant materials occasionally woven into the final construction. As with some other wading birds, nest building is shared between male and female. Often the tasks are split, with the male collecting sticks and the female arranging them. The final nests, especially if built on older nests, can be huge. In the case of the European white stork, they have been known to be as much as 9 ft (2.7 m) in depth.

The eggs are oval and white, the average clutch size is five, and incubation lasts between 25 to 38 days, depending on species. After hatching, the young are completely dependent on their parents, who attentively bring and regurgitate food on the nest floor for the young to pick at. Chick development is rapid. Once the young have fledged they leave the nest, but may still remain dependent on their parents for support for some weeks. Most storks only reach breeding condition at between three and five years.

Conservation status

Birdlife International lists three species as Endangered (Oriental white stork, Storm's stork, and greater adjutant) and two as Vulnerable (lesser adjutant and milky stork). The painted stork and the black-necked stork are listed as Near Threatened. Many other species are suffering regional declines in the face of ever-increasing pressure for land for agriculture and building development. The wood stork suffered catastrophic declines in the southeastern United States following the wholesale drainage of wetlands such as the Everglades in Florida. However, the numbers of marabou are increasing, perhaps in part due to their fondness for feeding around human garbage.

Asian openbill (*Anastomus oscitans*) in greeting display in Wat Phai Lom, Thailand. (Photo by M.P. Kahl. Photo Researchers, Inc. Reproduced by permission.)

Significance to humans

Storks are frequently held in great affection by local people across the world. In western countries the stork is often cited as the bird that brings babies. The roots of this myth are unclear, but it may be linked to the notion that storks nesting on houses will ensure fertility in the household. The welcome white storks receive is mirrored in other species, and many colonies are afforded special protection. In Thailand, Asian openbills (*Anastomus oscitans*) nesting in the grounds of a Buddhist temple at Wat Phai Lom have been protected by the monks for many years.

1. Asian openbill (*Anastomus oscitans*); 2. Painted stork (*Mycteria leucocephala*); 3. Marabou (*Leptoptilos crumeniferus*); 4. Black stork (*Ciconia nigra*); 5. Saddlebill (*Ephippiorhynchus senegalensis*); 6. White stork (*Ciconia ciconia*); 7. Jabiru (*Jabiru mycteria*); 8. Wood stork (*Mycteria americana*). (Illustration by Emily Damstra)

Species accounts

Wood stork
Mycteria americana

SUBFAMILY
Tribe Mycteriini

TAXONOMY
Mycteria americana Linnaeus, 1758, Brazil ex Marcgraf. Monotypic.

OTHER COMMON NAMES
English: Wood ibis; French: Tantale d'Amérique; German: Waldstorch; Spanish: Tántalo Americano.

PHYSICAL CHARACTERISTICS
Length 33–40 in (83–102 cm), wingspan 59 in (150 cm). Weight 4.4–6.6 lb (2–3 kg). White with gray featherless neck and head; long, slightly downcurved bill.

DISTRIBUTION
Southeastern United States, through tropical Central and South America to northern Argentina.

HABITAT
Wetlands with shallow water.

BEHAVIOR
Highly social, nests in colonies, and feeds and roosts in flocks.

FEEDING ECOLOGY AND DIET
Mostly fish, found almost entirely by sense of touch.

REPRODUCTIVE BIOLOGY
Nests colonially in trees. Clutch size three eggs; incubation 28–32 days; fledging 60–65 days.

CONSERVATION STATUS
Not threatened, but suffered historical declines in the United States.

SIGNIFICANCE TO HUMANS
Regarded as a "barometer" of wetland quality in the United States. ◆

Painted stork
Mycteria leucocephala

SUBFAMILY
Tribe Mycteriini

TAXONOMY
Tantalus leucocephalus Pennant, 1769, Ceylon. Monotypic.

OTHER COMMON NAMES
English: Painted wood stork, Indian wood ibis; French: Tantale Indien; German: Buntstorch; Spanish: Tántalo Indio.

PHYSICAL CHARACTERISTICS
Length 3–3.3 ft (93–102 cm), wingspan 4.9–5.2 ft (150–160 cm); 4.4–7.8 lb (2–3.5 kg). Black and white with orange/red face and yellow bill slightly downcurved at the tip.

DISTRIBUTION
India and Indochina.

Mycteria americana
☐ Resident   ☐ Nonbreeding

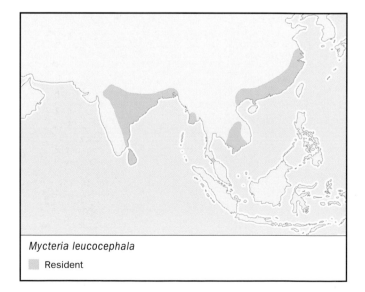

Mycteria leucocephala
☐ Resident

HABITAT
Shallow freshwater lakes, marshes, and flooded fields.

BEHAVIOR
Gregarious. Flies with neck extended and slightly lowered. Generally quiet, but performs "wing-woofing" and bill-clattering during courtship displays.

FEEDING ECOLOGY AND DIET
Mostly fish, but also frogs, small reptiles, and invertebrates. Locates prey by touch, stalking shallow water with an open bill, using feet and wing flaps to disturb prey.

REPRODUCTIVE BIOLOGY
Colonial, up to 100 nests together. Clutch size three to four, incubation 28–32 days, fledging 60 days.

CONSERVATION STATUS
Not threatened. Local declines have occurred though through hunting and capture for zoos.

SIGNIFICANCE TO HUMANS
Popular species whose colonies are actively supported and protected by locals. ◆

Asian openbill
Anastomus oscitans

SUBFAMILY
Tribe Mycteriini

TAXONOMY
Ardea oscitans Boddaert, 1783, Pondicherry. Monotypic.

OTHER COMMON NAMES
English: White openbill; French: Bec-ouvert Indien; German: Silberklaffschnabel; Spanish: Picotenaza Asiático.

PHYSICAL CHARACTERISTICS
Length 31 in (81 cm), wingspan 58–59 in (147–149 cm). Small pale gray or white stork with black wings and black forked tail. Distinctive "open" bill formed by lower mandible curving down, then back, to meet upper mandible at tip.

DISTRIBUTION
India, Indochina.

HABITAT
Shallow marshes and flooded fields.

BEHAVIOR
Social. In flight soars on thermals, then glides to destination. Call a mournful "hoo-hoo."

FEEDING ECOLOGY AND DIET
Mainly apple snails and occasionally other small aquatic animals. Prey located by touch and sight. Snails extracted from shells using sharply pointed lower mandible.

REPRODUCTIVE BIOLOGY
Highly social, nests in large tree colonies with other waterbirds such as herons. Clutch size two to five eggs, incubation 27–30 days, fledging 35–36 days.

CONSERVATION STATUS
Not threatened. The most common Asian stork.

SIGNIFICANCE TO HUMANS
Generally well regarded. Specially protected in Thailand, where colonies are located in the grounds of Buddhist monasteries. ◆

Black stork
Ciconia nigra

SUBFAMILY
Tribe Ciconiini

TAXONOMY
Ardea nigra Linnaeus, 1758, Sweden. Monotypic.

OTHER COMMON NAMES
French: Cigogne noire; German: Schwarzstorch; Spanish: Cigüeña Negra.

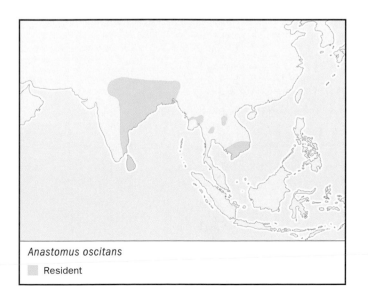

Anastomus oscitans
■ Resident

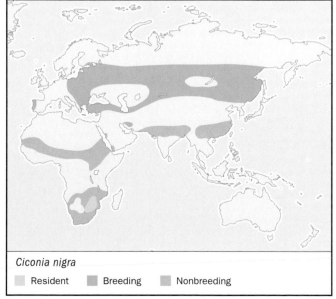

Ciconia nigra
■ Resident ■ Breeding ■ Nonbreeding

PHYSICAL CHARACTERISTICS
Length 37–39 in (95–100 cm), wingspan 57–61 in (144–155 cm); 6.6 lb (3 kg). Glossy black except for white feathering on belly. Red bill can appear slightly recurved.

DISTRIBUTION
Largest breeding range of any stork, nesting from eastern Europe through central Asia. Winters in Africa and Asian tropics. Separate resident population occurs in southern Africa.

HABITAT
Wooded areas with access to water.

BEHAVIOR
More solitary than some other storks. Agile flier, can fly through the forest canopy. More vocal than other storks, communicates with variety of hisses and whistles.

FEEDING ECOLOGY AND DIET
Fish and occasionally aquatic invertebrates. Locates prey visually, grabbing food items with forward lunge of the head. Has been observed shading water with outstretched wings while hunting.

REPRODUCTIVE BIOLOGY
Monogamous. Solitary nester in trees, the same nest often used over many seasons. Sometimes "adopts" other bird nests, such as those of black eagles and hammerheads. Clutch size three to four eggs, incubation 32–38 days, fledging 63–71 days.

CONSERVATION STATUS
Declining locally from persecution and deforestation, especially in Europe.

SIGNIFICANCE TO HUMANS
Heavily hunted, especially during migration through southern Europe and Asia. ◆

European white stork
Ciconia ciconia

SUBFAMILY
Tribe Ciconiini

TAXONOMY
Ardea ciconia Linnaeus, 1758, Sweden. Two subspecies.

OTHER COMMON NAMES
English: White stork; French: Cigogne blanche; German: Weissstorch; Spanish: Cigüeña Blanca.

PHYSICAL CHARACTERISTICS
Length 39–40 in (100–102 cm), wingspan 61–65 in (155–165cm); 5.1–9.7 lb (2.3–4.4 kg). Mostly white with black on wings and red/orange bill and legs.

DISTRIBUTION
Summer breeding population in Europe and western Asia, wintering in tropical Africa and India. A resident population also in South Africa.

HABITAT
Open spaces without tall and thick vegetation, frequently in or near wetlands. Will nest in towns and villages.

BEHAVIOR
Less gregarious than other storks, but migrates in groups. Adept at soaring on thermals during long migrations along well-defined routes. Uses bill-clattering in displays.

Ciconia ciconia
▪ Resident ▪ Breeding ▪ Nonbreeding

FEEDING ECOLOGY AND DIET
Varied diet of animal matter, from insects and earthworms, to lizards, snakes, and amphibians. Locates prey by sight.

REPRODUCTIVE BIOLOGY
In temperate north, nesting starts between February and April. Nests are large constructions of sticks lined with a variety of soft natural or human-made objects located in trees or on suitably tall human-made structures. Loosely colonial, but may nest alone. Clutch size averages four eggs, incubation 33–34 days, fledging 58–64 days.

CONSERVATION STATUS
Significant local declines in Western Europe, where it is Threatened. Declines linked to the reduction of swarms of locusts (an important source of food) in west African wintering grounds, and reduction of food-rich habitats in breeding areas as a result of the intensification of agriculture. Also threatened by hunting and collisions with power lines.

SIGNIFICANCE TO HUMANS
Traditionally a popular bird, nesting on houses is welcomed as conferring good fortune and fertility to householders. Also some economic value as pest controllers. ◆

Saddlebill
Ephippiorhynchus senegalensis

SUBFAMILY
Tribe Leptoptilini

TAXONOMY
Mycteria senegalensis Shaw, 1800, Senegal. Monotypic.

OTHER COMMON NAMES
English: African jabiru, saddlebilled stork; French: Jabiru de Sénégal; German: Sattelstorch; Spanish: Jabirú Africano.

PHYSICAL CHARACTERISTICS
Length 55–59 in (140–150 cm), wingspan 94–106 in (240–270 cm); 11–16.1 lb (5–7.3 kg). One of the largest storks. Mostly

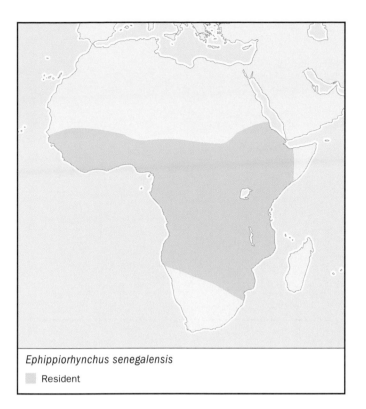

Ephippiorhynchus senegalensis

Resident

Jabiru mycteria

Resident

orange/red bill divided by a black band, surmounted by patch of yellow. Black neck and flight feathers, white body feathers; dark legs with red "knees." Males and females differ both in size (the male is larger) and iris color (brown in male, yellow in female).

DISTRIBUTION
Tropical Africa south of the Sahara.

HABITAT
Open wetlands.

BEHAVIOR
Mostly solitary. Flies with heavy wing-beats and neck out-stretched; makes use of thermals.

FEEDING ECOLOGY AND DIET
Mainly fish. Hunts by sight and, occasionally, touch.

REPRODUCTIVE BIOLOGY
Monogamous. Nests alone toward end of rainy season. Nest a platform of sticks. Clutch size two to three eggs, incubation 30–35 days, fledging 70–100 days.

CONSERVATION STATUS
Not threatened.

SIGNIFICANCE TO HUMANS
Popular with ecotourists on wildlife holidays in East Africa. ◆

Jabiru
Jabiru mycteria

SUBFAMILY
Tribe Leptoptilini

TAXONOMY
Ciconia mycteria Lichtenstein, 1819. Monotypic.

OTHER COMMON NAMES
English: American jabiru, jabiru stork; French: Jabiru d'Amérique; German: Jabiru; Spanish: Jabirú Americano.

PHYSICAL CHARACTERISTICS
Length 48–55 in (122–140 cm), wingspan 90–102 in (230–260 cm); weight 17.6 lb (8 kg). Mostly white with dark bill and neck, colored red at base.

DISTRIBUTION
Tropical Central and South America to northern Argentina.

HABITAT
Freshwater wetlands.

BEHAVIOR
May retract neck in flight due to having heavy bill that, if out-stretched, would cause problems with balance.

FEEDING ECOLOGY AND DIET
Fish and other aquatic animals. Uses both sight and touch to locate prey. Will splash bill in shallow water to disturb prey prior to capture.

REPRODUCTIVE BIOLOGY
Nests alone or in small groups in trees. Nests are large plat-forms of sticks and mud, may be built upon season after sea-son. Clutch size three to four eggs, fledging 80–95 days.

CONSERVATION STATUS
Not threatened.

SIGNIFICANCE TO HUMANS
Hunted for food in some areas. ◆

Marabou
Leptoptilos crumeniferus

SUBFAMILY
Tribe Leptoptilini

TAXONOMY
Ciconia crumenifera Lesson, 1831. Monotypic.

OTHER COMMON NAMES
English: Marabou stork; French: Marabout d'Afrique; German: Marabu; Spanish: Marabú Africano.

PHYSICAL CHARACTERISTICS
Length 3.3–5 ft (115–152 cm), wingspan 7.4–9.4 ft (225–287 cm); 8.8–19.6 lb (4–8.9 kg). Black and white, with featherless pink neck spotted black, and heavy greenish yellow bill.

DISTRIBUTION
Tropical Africa south of the Sahara.

HABITAT
Arid or semiarid open country within flying distance of rivers and lakes.

BEHAVIOR
Fairly social, colonial during breeding season. Feeds in flocks, often with other species such as vultures. Voice a variety of whistles, clatters bill as display prior to mating.

FEEDING ECOLOGY AND DIET
Wide variety of animals including fish, reptiles, amphibians, and invertebrates; also carrion.

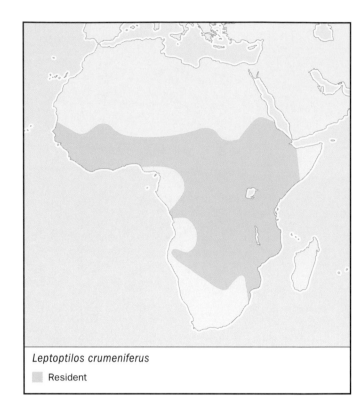

Leptoptilos crumeniferus
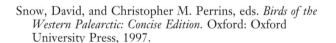
☐ Resident

REPRODUCTIVE BIOLOGY
Long breeding period, starting in the dry season through subsequent rains. Nest in trees, woven from sticks and lined with softer plant material. Clutch size two to three eggs, incubation 29–31 days, fledging after 95 days.

CONSERVATION STATUS
Not globally threatened.

SIGNIFICANCE TO HUMANS
Scavenge on carrion from waste dumps and other areas around human settlements.

Resources

Books

Collar, N. J., M. J. Crosby, and A. J. Stattersfield. *Birds to Watch 2: The World List of Threatened Birds.* Cambridge: BirdLife International, 1994.

del Hoyo, J., A. Elliot, and J. Sargatal, eds. *Handbook of the Birds of the World.* Vol. 1, *Ostrich to Ducks.* Barcelona: Lynx Edicions, 1992.

Hancock, J. A., J. A. Kushlan, and M. P. Kahl. *Storks, Ibises and Spoonbills of the World.* London: Academic Press, 1992.

Snow, David, and Christopher M. Perrins, eds. *Birds of the Western Palearctic: Concise Edition.* Oxford: Oxford University Press, 1997.

Organizations

BirdLife International. Wellbrook Court, Girton Road, Cambridge, Cambridgeshire CB3 0NA United Kingdom. Phone: +44 1 223 277 318. Fax: +44-1-223-277-200. E-mail: birdlife@birdlife.org.uk Web site: <http://www.birdlife.net>

Tony Whitehead, BSc

New World vultures
(Cathartidae)

Class Aves
Order Ciconiiformes
Suborder Cathartae
Family Cathartidae

Thumbnail description
Medium large to very large, highly social birds that feed primarily on carrion with very few or no feathers on their heads; plumage color is generally black, gray, or brown with some portion of white in three of the largest species

Size
2.2–33.1 lb (1–15 kg)

Number of genera, species
5 genera; 7 species

Habitat
Forests, savannas, woodland, pastures, mountains, deserts, river ways, and seashores

Conservation status
Critically Endangered: 1 species; Near Threatened: 1 species

Distribution
Southern Canada to Tierra del Fuego

Evolution and systematics

The Cathartidae include seven species of vultures that range exclusively in North and South America. Also referred to as "Neotropical" or New World vultures, they were once thought to be closely related to Old World vultures, found in Europe, Asia, and Africa, but any similarities of the two groups can best be attributed to convergence or parallel evolution.

As recently as the last quarter of the twentieth century it was argued successfully on the basis of physiology, behavior, and genetics, that the cathartid vultures are not descended from the Accipitidiae as is accepted for the Old World vultures, but descended instead from a common ancestor with the Ciconidae, or storks. For instance, all of the cathartids and ciconids use "urohydrosis" or the method of cooling themselves by emitting liquid waste on the bare portion of their legs where densely packed blood vessels close to the skin are cooled by evaporation and, in turn, body core temperature is reduced. They also never rest on one foot as do birds of prey but instead lie down. While both New and Old World vultures express a distinct social hierarchy, that of Old World vultures is based more on the hunger level of a bird

arriving at a carcass while the social status of New World vultures is based primarily on an individual's personal status that is determined by the individual's species, age, sex, and experience.

The evolutionary history of the Old and New World vultures is comparatively good. The earliest New World vulture was reported in England dating from late Paleocene deposits. Several "cathartid type" fossils are known from middle and late Eocene deposits in France and Germany but no remains after early Miocene have been located in the Old World. Fossil records show the New World vultures first appearing in America in the early Oligocene, flourishing along side Old World type vultures that became extinct toward the end of the Pleistocene, only 10,000–20,000 years ago. When the mass mammalian extinctions occurred, both Old and New World type vultures followed the fate of their prey.

The California condor appears to be the only large vulture to survive. Its Pleistocene range that spanned southern North America and included both coasts, was reduced to the west coast from British Columbia to northern Baja by modern times. Like the Andean condor today, it relied heavily on carrion found along the coast.

Turkey vultures (*Cathartes aura*) feeding on a mule deer carcass in California. (Photo by Richard R. Hansen. Photo Researchers, Inc. Reproduced by permission.)

Physical characteristics

The family Cathartidae consists of five genera with seven species. The smallest species, by weight, is the 2.1 lb (0.94 kg) lesser yellow-headed vulture (*Cathartes burrovianus*). The other two species in the genus *Cathartes*, the turkey vulture (*C. aura*) weighing 3.3 lb (1.5 kg) and the greater yellow-headed vulture (*C. melambrotus*) at 2.6 lb (1.2 kg), are not much heavier yet all give the appearance of being much larger than their actual weight. The physical effect of their large flying surface area to weight ratio, called a "light wing loading," makes these three species comparatively more buoyant in air and able to take advantage of the slightest thermal to stay aloft close to the ground. The black vulture (*Coragyps atratus*) on the other hand has a relatively short wingspan, or flying surface, for its heavier weight of 4.4 lb (2.0 kg). These flight characteristics join with other behavioral and anatomical factors to help define niche separation in each species of this scavenger guild. The colorful king vulture (*Sarcoramphus papa*) is built more like a black vulture but larger, weighing about 7.5 lb (3.4 kg). The California condor (*Gymnogyps californianus*), like the smaller vultures, is sexually monomorphic in size and color, but is much heavier weighing 17–24 lb (7.7–10.9 kg) with a wingspan of 114 in (290 cm). The Andean condor (*Vultur gryphus*) is one of the largest flying birds in the world. It is sexually dimorphic in shape, color, and size. Females range in weight from 18 to 23 lb (8.3–10.5 kg), have dark gray skin on the head which has no caruncle, similar to the male. Their iris color in the adult is a deep red while that of the male is tan, plumages are the same. The larger male ranges from 24 to 33 lb (10.9–15 kg).

While the Cathartidae have only a rudimentary syrinx and cannot call or sing as other birds can they are able to communicate with a suprising array of grunts, growls, and hisses.

Distribution

The most widely distributed cathartid species is the turkey vulture, ranging from the Canadian border to the southern end of South America. There are also four subspecies of turkey vultures usually recognized based on slight differences in head color and distribution. The most migratory appears to be *C. aura aura* which has been extending its summer breeding range over the last few decades north though New England. It spends winters from the southern United States to northern South America competing with the more sedentary subspecies of that region. The movement patterns of the other cathartid species appear to be less latitudinally migratory but are more regional, associated with weather patterns, food supply, and breeding season.

Roosting areas are particularly important in influencing the distribution pattern in these vultures particularly for the rarer, larger condors where roosting conditions are more specific. Where hundreds of black and turkey vultures may roost in particular groves of trees or on the supports of man-made towers, it is more difficult for condors to find appropriate cliff roosts with the right climatic conditions. Dozens of condors may use a network of these traditional roosts as secure bases from which to forage in a particular area and they are as im-

A juvenile Andean condor (*Vultur gryphus*) sunning with its wings extended in the Andes of South America. (Photo by Kenneth W. Fink. Photo Researchers, Inc. Reproduced by permission.)

portant as adequate nesting sites in delineating the distribution of these vultures.

Habitat

From the northern to southern portions of where cathartid vultures range they are found in every habitat where their carrion food supply can be effectively exploited, including but not limited to deserts, coastlines, water ways, open grasslands and savannas, forests, cities, and mountain and canyon regions.

Unlike Old World vultures that do not forage in closed forests, three cathartid species in the New World are well adapted to successfully exploit forest as well as open habitat. They accomplish this through a sense of smell that is so acute they can find even small bird, reptile, and mammal carcasses in a forest shortly after they begin to decompose. Both condor species use winds generated off mountain slopes and thermal activity to move distances of hundreds of kilometers at thousands of meters of altitude within a few days covering several habitat types.

Behavior

New World vultures are highly gregarious roosting nightly and foraging communally by day. Unless there is already wind, in which case they can and will fly before dawn, vultures and condors typically wait until later in the morning when rising thermals can assist soaring flight to where the most recent carcass is located. When the first rays of sun hit the roost or throughout the day when it peaks from behind a cloud, vultures will spread their wings and orient themselves at right angles to the sun. Seen also after bathing, this pose, called sunning, dries and straightens flight feathers and functions to assist with preening in reducing ectoparasites.

Like the accipitrid vultures in Africa, the Cathartidae partition the food resources of the available carrion in an area through timing, anatomical differences, and a relatively ordered hierarchy. Black vultures, with a heavier wing loading, are most efficient at foraging when flying at higher altitudes where they can best observe the behavior and activities of their own and other species. Lacking the ability to use olfaction, like the three smaller *Carthartes* species, they rely on vision to hunt. When flying conditions allow, a foraging flock of black vultures will disperse over miles, where they can effectively scan for resource opportunities to exploit over a large area. When activity of interest is noted by one or more individuals, the adjusted flying pattern appears to signal the attention of other flock mates. As more and more birds gather over a carcass, the inadvertent signal produced by the mass of large black birds at varying altitudes can persist over many days attracting the larger, less common species of scavengers such as king vultures and Andean condors. The same scenario is carried out on the west coast of North America involving turkey vultures, ravens, and California condors.

Where black vultures occur, their population levels can get quite large, even into the thousands sustained by some type of consistent and predictable food source like a city dump. So behaviorally astute are black vultures that they will adjust their flight distance to humans depending on the circumstances. The same marked bird feeding without fear within a few feet of people at the Panama City market place one day will flush away the next day when approached within 164 ft (50 m) in the countryside. When black vultures first arrive at a natural carcass, not part of the usual system, they have usually followed one of the *Cathartes* species. At low, equal numbers, black vultures generally cannot dominate turkey vultures, but as their group size increases, the *Cathartes* species often move to the side of the main feeding activity or leave altogether.

Feeding ecology and diet

King vultures, and especially condors, can go longer periods between meals than the smaller vultures as long as water is available. Reintroduced California condors re-trapped after 26 days of unsuccessful foraging showed no behavioral or clinical signs of stress. They generally fly higher, faster, and over larger areas, foraging away from roost site and nesting territories. When they arrive at a carcass, they dominate any of the smaller species. Observations of marked wild Andean condors indicate that birds with high status are less reluctant to approach the carcass. The smaller bills of the *Cathartes* and *Coragyps* vultures are not sufficient to tear through the hide of large carcasses and usually are confined to natural openings of the mouth, eyes, ears, and anus. Three to five of the most dominant individuals of the 100 member black vulture flock can defend the few holes in a fresh, large carcass. Sometimes hundreds of birds wait off to one side for the social dynamics to change. When condors or even king vultures arrive, the waiting, lower status vultures become alert and begin to crowd the carcass as the first condor approaches. With a bill every bit as powerful as the largest Old World vulture, one or more condors soon open several access points in the tough hide, making it difficult or impossible for a few dominant black vultures to successfully defend the carcass. With a breakdown in the hierarchy, a feeding frenzy ensues and even young, normally submissive birds can race into the confusion and successfully dash out with food.

Reproductive biology

The black vulture and the three *Cathartes* species all lay three eggs on a yearly basis. Although sexually mature by age two they may take several more years to acquire a mate and successfully breed. The king vulture and the two condors lay only one egg per season. Condors have a very slow reproductive rate and may take two or more years to produce one young. Parental dependence is months long in the larger species but short to non-existent in the genus *Cathartes* possibly due to their almost immediate success at finding food through olfaction. Monogamy is typical, with pair bonds that are life long, but shifts in mates may occur if the pair is un-

productive over several years. Territory defense in condors is males against intruding males and females against females.

Conservation status

Where turkey vultures and black vultures are gradually expanding their range in some areas, the king vulture and condors have had significant declines. The California condor that narrowly escaped extinction at the end of the Pleistocene had been declining since the early 1800s when the growing human population reduced its coastal food supply and directly shot and poisoned the species. By the early 1980s, only 21 had survived and by 1987 all of the wild flock had to be brought into captivity to insure the species survival. Captive breeding programs at the Los Angeles Zoo and the San Diego Wild Animal Park and later, the World Center for Birds of Prey in Idaho have been highly successful in producing numbers of birds while preserving the remaining founding genetic lines. The reintroduction of the species into its former habitat began in 1992 and by the close of 2001 there were 183 birds with nearly 60 of those in the wild at three release sites; two in California and one in Arizona. Attempts to breed in the wild began in 2002. Andean condors have also been reintroduced into parts of their former range where they had been extirpated. North American Zoos, through their Species Survival Plan for Andean Condors, raised and released over 80 Andean condors in Venezuela and Colombia where they now breed in the wild.

Significance to humans

Condors were important in the mythology and featured in the rituals of the pre-Columbian cultures in the Andes. Their image is found incorporated in the textile designs, pottery, and carvings of these peoples. Even today the Andean condor appears on the coats of arms of Colombia, Bolivia, Ecuador, and Chile. Native North American groups greatly respected the California condor. It was buried with the dead, and its image was incorporated into their artistic motifs.

Many early human cultures associated vultures with death and these birds became important symbols in burial rituals. Today vultures are not respected as they were by earlier cultures, but they are tolerated for the valuable environmental service they provide. Vultures have suffered in recent years from human-generated pollution. Positioned as they are at the end of the food chain, vultures are likely to accumulate toxins and contaminants, which can kill them outright or damage their reproductive success. Other human-made hazards, such as power lines, can also pose a danger to flying birds.

1. California condor (*Gymnogyps californianus*); 2. Andean condor (*Vultur gryphus*); 3. Black vulture (*Coragyps atratus*); 4. Turkey vulture (*Cathartes aura*); 5. Lesser yellow-headed vulture (*Cathartes burrovianus*); 6. King vulture (*Sarcoramphus papa*); 7. Greater yellow-headed vulture (*Cathartes melambrotus*). (Illustration by Jonathan Higgins)

Species accounts

Turkey vulture
Cathartes aura

SUBFAMILY
Catharnae

TAXONOMY
Vultur aura Linnaeus, 1758, Veracruz, Mexico. Four sup-species.

OTHER COMMON NAMES
French: Urubu à tête rouge; German: Truthahngeier; Spanish: Aura Gallipavo.

PHYSICAL CHARACTERISTICS
Depending on race 1.9–4.4 lb (0.85–2 kg); 25.2–31.9 in (64–81 cm); sexually monomorphic. Brownish black plumage with bare head and neck; skin color varies from pink to bright red.

DISTRIBUTION
Southern border of Canada in North America to Tierra del Fuego, Chile, winter migration from northern portions of United States to Central America and north and central South America. Has been expanding its range north over the last three decades.

HABITAT
Woodland, savanna, desert, and seashore.

BEHAVIOR
Roosts socially and hunts for carrion in small groups or singly using olfaction. Lowest status at large carcasses generally subordinate to larger vultures. When caught will regurgitate and feign death.

FEEDING ECOLOGY AND DIET
Will feed on dead animals of any size but specializes on smaller carcasses found by relying on olfaction. Has been seen to feed on rotten fruit and vegetables.

REPRODUCTIVE BIOLOGY
Two eggs are brown blotched over a cream base; nests in shallow cave, on the ground in dense vegetation, or in hollow log; 38–41 days of incubation. Both parents tend eggs and young. Chick "down" color white gradually changing overall to dark brown as contour feathers emerge. Fledging from nest area at about 67 days with a relatively short parental dependency period of a few weeks.

CONSERVATION STATUS
Not threatened. Populations seem stable in most areas unless roost trees are destroyed. Expanded use of road kills and laws against shooting large birds are possible reasons for recent northward expansion.

SIGNIFICANCE TO HUMANS
Never used as food. ◆

Cathartes aura

Resident ▢   Breeding ▢

Lesser yellow-headed vulture
Cathartes burrovianus

SUBFAMILY
Catharnae

TAXONOMY
Cathartes burrovianus Cassin, 1845, near Veracruz City, Mexico. Two subspecies sometimes recognized.

OTHER COMMON NAMES
English: Savanna vulture; French: Urubu à tête jaune; German: Kleiner Gelbkopfgeier; Spanish: Aura Sabanera.

PHYSICAL CHARACTERISTICS
Smallest of the *Cathartes* on average, 23–26 in (58–66 cm), 2.1–3.3 lb (0.94–1.5 kg); sexually monomorphic. Black plumage; bare skin on head and neck varies from bright yellow to orange and blue tones.

DISTRIBUTION
South Central America to Uruguay, east of the Andes Mountains.

HABITAT
Open savanna and water courses, open flat grasslands to the edge of forest but generally not over forest.

Cathartes burrovianus
▨ Resident

Cathartes melambrotus
▨ Resident

BEHAVIOR
Uses olfaction as other members of its genus.

FEEDING ECOLOGY AND DIET
Will feed on dead animals of any size but specializes on smaller carcasses found by relying on olfaction.

REPRODUCTIVE BIOLOGY
Very little known but that in general it is similar to *C. aura*.

CONSERVATION STATUS
Status and distribution poorly known but not threatened.

SIGNIFICANCE TO HUMANS
Never considered food. Like other small vultures generally tolerated because of feeding habits that reduce the presence of animal carcasses decreasing the likelihood of disease transmission to humans. ◆

Greater yellow-headed vulture
Cathartes melambrotus

SUBFAMILY
Catharnae

TAXONOMY
Cathartes melambrotus Wetmore, 1964, Kartabo, Guyana. Monotypic.

OTHER COMMON NAMES
English: Forest vulture; French: Grand urubu; German: Großer Gelbkopfgeier; Spanish: Aura Selvática.

PHYSICAL CHARACTERISTICS
29–32 in (74–81 cm), 2.6–3.6 lb (1.2–1.65 kg) but with larger wings than the heavier *C. aura*. Similar head color to *C. burrovianus*.

DISTRIBUTION
Not well known. Southern Venezuela and the Guianas and other parts of Amazonia.

HABITAT
Typically low tropical forest and less inclined to use open or disturbed forest.

BEHAVIOR
Uses olfaction as other members of its genus.

FEEDING ECOLOGY AND DIET
Specializes on smaller carcasses found in forested areas by relying on olfaction.

REPRODUCTIVE BIOLOGY
Not known, but probably similar to *C. avra* in most respects.

CONSERVATION STATUS
Status and distribution poorly known but in undisturbed forests can be common.

SIGNIFICANCE TO HUMANS
None known. ◆

American black vulture
Coragyps atratus

SUBFAMILY
Catharnae

TAXONOMY
Vultur atratus Bechstein, 1793, St. John's River, Florida. Three subspecies.

Coragyps atratus
■ Resident

Finds carcasses by observing the behavior of other species and checking areas where food was historically more predictable. Cannot use olfaction as can members of the genus *Cathartes*, relies primarily on sight.

REPRODUCTIVE BIOLOGY
Two brown mottled eggs laid in darkened sheltered cave, thick vegetation, or abandon buildings. No nest material used. Incubation takes 38–45 days. Chicks covered with reddish brown down; fledging at about 90 days. Extended parental dependency of several months.

CONSERVATION STATUS
Not threatened. Common in most areas of its range.

SIGNIFICANCE TO HUMANS
Never considered food. Like other small vultures generally tolerated because of feeding habits that reduce the presence of animal carcasses decreasing the likelihood of disease transmission to humans. ◆

King vulture
Sarcoramphus papa

SUBFAMILY
Catharnae

TAXONOMY
Vultur papa Linnaeus, 1758, Suriname. Monotypic.

OTHER COMMON NAMES
French: Sarcoramphe roi; German: Königsgeier; Spanish: Zopilote Rey.

OTHER COMMON NAMES
English: Black vulture; French: Urubu noir; German: Rabengeier; Spanish: Zopilote Negro.

PHYSICAL CHARACTERISTICS
22–27 in (56–69 cm), 2.4–4.2 lb (1.1–1.9 kg). Entirely black with light gray ventral wing patches at the base of the primary feathers. Sexes alike. Smooth dark gray head skin of young birds becomes fleshy and warty with age.

DISTRIBUTION
Southern North America to southern South America.

HABITAT
Open areas and water ways.

BEHAVIOR
Roosts, forages and feeds more socially than other species of Cathartidae except for the condors. Flocks at roosts can number several hundred. Heavy wingloading causes tendency to fly at higher altitudes and flap more compared to *Cathartes* species. Most gregarious of the species.

FEEDING ECOLOGY AND DIET
Will feed on dead animals of any size. Have been known to kill injured or highly compromised animals on rare occasions.

Sarcoramphus papa
■ Resident

PHYSICAL CHARACTERISTICS

28–32 in (71–81 cm), 6.6–8.3 lb (3–3.8 kg). Most brilliantly colored of the New World vultures with varying hues ranging from purple and blue to red and orange on its head. Its contrasting black and white plumage is opposite that of condors, with a white body and black primary and secondary wing feathers.

DISTRIBUTION

Southern Mexico to northern Argentina.

HABITAT

Usually associated with lowland tropical forests but can also be found in savannas, grasslands, and desert margins.

BEHAVIOR

Seldom seen in large groups, usually visits carcasses as a pair with their single offspring where they easily dominate the smaller vultures. Forages for food at high altitudes. They are less gregarious and roost in pairs or family threesomes.

FEEDING ECOLOGY AND DIET

Has no apparent sense of smell and finds carcasses through the activities of other vultures. Its bill, which is more powerful than the bills of the smaller cathartids, enables it to feed more easily on large carcasses.

REPRODUCTIVE BIOLOGY

Territorial pairs lay a single white egg in hollow trees, sometimes high off the ground. As with other carthatids, no nesting material was used in the few nests that have been found. The 53–58 day incubation is shared equally by both sexes, as regularly seen in captivity. Down of young chick is white. Fledging is at three months with an extended parental dependency period of several more months.

CONSERVATION STATUS

CITES III status in Honduras, but not globally threatened. Relatively rare compared to smaller vultures, but appear to be naturally uncommon even in undisturbed forests.

SIGNIFICANCE TO HUMANS

Indigenous cultures depicted this striking species in artwork. ◆

California condor

Gymnogyps californianus

SUBFAMILY

Catharnae

TAXONOMY

Vultur californianus Shaw, 1798, Monterey, California. Monotypic.

OTHER COMMON NAMES

French: Condor de Californie; German: Kalifornischer Kondor; Spanish: Cóndor Californiano.

PHYSICAL CHARACTERISTICS

46–53 in (117–134 cm) 17–24 lb (7.7–10.9 kg). Wingspan of nearly 10 ft (3 m). Entirely black plumage except for conspicuous triangle of white feathers in the ventral portion of both wings in the adult. The grayish triangle of juvenile birds gradually becomes whiter by five years of age. The grayish head color of the juveniles also gradually changes to reddish orange as they mature.

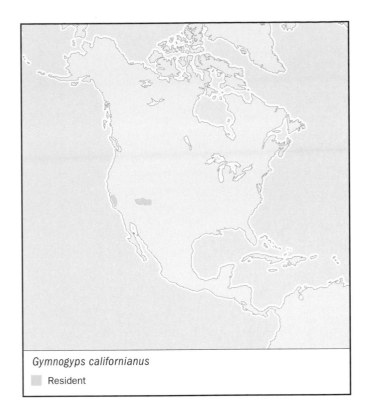

Gymnogyps californianus

■ Resident

DISTRIBUTION

While the species once ranged from British Columbia to northern Baja in the early 1800s, it became extinct in the wild in 1987. As a result of reintroduction efforts, its 2002 range includes the coastal mountains in California from Monterey to just north of Los Angeles, and a disjunct population north of the Grand Canyon in Arizona.

HABITAT

Roosting and nesting occur in mountainous areas where winds allow the birds to range widely. Foraging occurs in open areas of savanna, grasslands, and coastal beaches where food is located through the activities of other scavenger species.

BEHAVIOR

Highly curious and intelligent, it finds food through observing the behaviors and activities of other species. Condors can travel hundreds of miles in a single day, foraging alone or in well dispersed groups at high altitudes. Strict hierarchy at a carcass reduces aggression to a minimum.

FEEDING ECOLOGY AND DIET

Scavenger of large carcasses that include marine mammals as well as wild and domestic ungulates. Their large and powerful bill enables them to tear open thick hide.

REPRODUCTIVE BIOLOGY

Courtship displays generally begin in October with egg laying beginning in mid-January to late April. The single white egg is incubated equally by both parents through its 57 day incubation period. The chick takes about six months to fledge, with a lengthy parental dependency period that appears to vary with food availability. Sexual maturity is at five to six years of age. Pair bonds are for life as long as the pair remains productive.

CONSERVATION STATUS
Critically Endangered. Wild population was as low as 19 birds. In 1987 the remaining birds were brought into captivity to re-build the population. Through strict out-breeding of pairs within the 14 family lines and multiple clutching techniques, the species has rebounded to 183 birds by the end of 2001.

SIGNIFICANCE TO HUMANS
Native American cultures along the California and Oregon coasts have had a long and intimate history with this species. They have used feathers and bones for ceremonies, depicted the birds in artwork, and incorporated them in their legends. ◆

Andean condor
Vultur gryphus

SUBFAMILY
Catharnae

TAXONOMY
Vultur gryphus Linneaus, 1758, Chile. Monotypic.

OTHER COMMON NAMES
French: Condor des Andes; German: Andenkondor; Spanish: Cóndor Andino.

PHYSICAL CHARACTERISTICS
39–51 in (100–130 cm) 18.1–23.1 lb (8.2–10.5 kg). with a wingspan of as much as 10 ft (3 m). Black plumage with large white patches on the dorsal portion of the wings. Neck ruff of short white feathers. Sexually dimorphic with males exhibiting a large fleshy crest, tan-colored irises, and head skin ranging from dark gray to yellow. Females have red irises, grayish skin on their head, and can be 2.2–11 lb (1–5 kg) lighter.

DISTRIBUTION
Andes Mountains from Venezuela to Tierra del Fuego, rang-ing to the coast from northern Peru to southern Chile and Argentina.

HABITAT
Roosting and nesting occur in mountainous areas where winds allow the birds to range widely. Foraging occurs in open areas of savanna, grasslands, deserts, and beaches along the coast.

BEHAVIOR
Highly curious and intelligent, it finds food through observing the behaviors and activities of other species. Condors can travel hundreds of miles in a single day, foraging alone or in well dispersed groups at high altitudes. Strict hierarchy at a carcass reduces aggression to a minimum. At Andean condor roosts in South America, groups consisting of several imma-ture birds seem to gain the same information advantage as in the black vulture system. In areas where large roosts are not convenient, however, juvenile bands move between territory pairs led by the older more experienced individuals of the group. Their brown, immature stage plumages, which gradu-ally change to black and white by seven years of age, afford them safe passage between nesting cliffs defended against adults and gain them the advantages of associating with knowledgeable adult birds.

Vultur gryphus
◼ Resident

FEEDING ECOLOGY AND DIET
Scavenger of large carcasses that include marine mammals as well as wild and domestic ungulates. Have been known to feed on sea bird nestlings on offshore guano islands. Their large and powerful bills enable them to tear open thick hide.

REPRODUCTIVE BIOLOGY
Courtship displays have been seen throughout the year near the equator. To the far south nesting occurs from May through August. Incubation from 58 to 62 days (captive) and fledging takes about six months. While pairs may hold territo-ries, nesting may occur every other year and may be postponed for several years if food availability is low. Parental dependency period is lengthy and may vary with food supply.

CONSERVATION STATUS
Listed as Near Threatened but still abundant in the Chile/ Argentina Andes where thousands of birds exist. Fewer birds as one progresses north along the Andes until Colombia and Venezuela where the population consists of mostly reintro-duced birds.

SIGNIFICANCE TO HUMANS
Pictographs and legends found throughout native cultures. "The sun rose and set by the wings of the condor," according to an Inca belief. Festivals and rituals involving this species still exist. A modern Andean festival involves tying a condor to a bull's back and sending the bull running through the town. If the condor survives, it symbolizes the successful resistance of the native South Americans against the Spanish conquistadors, and it is released back to the wild. ◆

Resources

Books

Blake, E.R. *Manual of Neotropical Birds.* Vol. 1: *Spheniscidae (Penguins) to Laridae (Gulls and Allies).* Chicago & London: University of Chicago Press, 1977.

Snyder, N.F.R., and H.A. Snyder. *Birds of Prey: Natural History and Conservation of North American Raptors.* Stillwater, Minnesota: Voyageur Press, 1991.

Wallace, M.P., and W. Toone. "Captive Management for the Long Term Survival of the California Condor." In *Wildlife 2001: Populations,* edited by D. R. McCullough and R.H. Barrett. Elsevier Applied, 1992.

Wilbur, S.R., and J.A. Jackson, eds. *Vulture Biology and Management.* University of California Press, 1983.

Periodicals

Audubon, J.J. "Account of the Habits of the Turkey Buzzard *Vultur aura* Particularly with the View of Exploding the Opinion Generally Entertained of Its Extraordinary Powers of Smelling." *Edinb. New Phil. Journal* 2(1826):172–184.

Bang, B.G. "The Nasal Organs of the Black and Turkey Vultures: A Comparative Study of the Cathartid Species *Coragyps atratus atratus* and *Cathartes aura septentrionalis* (With Notes on *Cathartes aura falklandica, Pseudogyps bengalensis* and *Neophron percnopterus*)." *Journal of Morphology* 115(1972): 153–184.

Bernal, L.G., D.C. Houston, and P. Cotton. "The Role of Greater Yellow-headed Vultures as Scavengers in Neotropical Forests." *Ibis* 136(1994).

Clinton-Eitniear, J. "King Vulture Research Report." *Vulture News* 6(1981):7–8.

Cox, C.R., V.I. Goldsmith, and H.R. Engelhardt. "Pair Formation in California Condors." *Amer. Zool.* 33(1993): 126–138.

Davis, D. "Morning and Evening Roosts of Turkey Vultures at Malheur Refuge, Oregon." *Western Birds* 10(1979): 125–130.

Gailey, J., and N. Bolwig. "Observations on the Behaviour of the Andean Condor *Vultur gryphus.*" *Condor* 75(1973):60–68.

Graves, G.R. "Greater Yellow-headed Vulture (*Cathartes melambrotus*) Locates Food by Olfaction." *Journal of Raptor Research* 26(1992): 38–39.

Houston, D.C. "Competition for Food Between Neotropical Vultures in Forest." *Ibis* 130(1988): 402–417.

Kiff, L.F. "An Historical Perspective on the Condor." *Outdoor California* 44(1983): 5–6, 34–37.

Kiff, L.F. "To the Brink and Back: The Battle to Save the California Condor." *Terra* 28(1990): 6–18.

Meretsky, V., and N.F.R. Snyder. "Range Use and Movements of California Condors." *Condor* 94(1992): 313–335.

Pattee, O.H "The Role of Lead in Condor Mortality." *Endangered Species Bulletin* Vol. 12, No. 9 (1987): 6–7.

Rabenold, P.P. "Family Associations in Communally Roosting Black Vultures." *Auk* 103(1986): 32–41.

Snyder, N.F.R., R.R. Ramey, and F.C. Sibley. "Nest Site Biology of the California Condor. " *Condor* 88(1986): 228–241.

Snyder, N.F.R., and H.A. Snyder. "Biology and Conservation of the California Condor." *Current Ornithology* 6(1989): 175–267.

Stewart, P.A. "The Biology and Communal Behaviour of American Black Vultures." *Vulture News* 9/10(1983): 14–36.

Toone, W., and A.C. Risser. "Captive Management of the California Condor *Gymnogyps californianus.*" *International Zoology Yearbook* 27(1988): 50–58.

Michael Phillip Wallace, PhD

Shoebills
(Balaenicipitidae)

Class Aves
Order Ciconiiformes
Suborder Ciconiae
Family Balaenicipitidae

Thumbnail description
This is a large gray wading bird with a very large bulbous bill tipped with a hooked nail; legs are long with long toes allowing the bird to walk on submerged aquatic vegetation; in flight the neck is retracted like herons; not outstretched as in storks

Size
4 ft (120 cm); wing: 31 in (780 mm); bill: 7.5 in (191 mm); male slightly larger than female

Number of genera, species
1 genus; 1 species

Habitat
Swamps, primarily papyrus swamps or cattail marshes

Conservation status
Near Threatened

Distribution
Central Africa: Sudan, Uganda, Tanzania, Democratic Republic of Congo, Central African Republic, and Rwanda

Evolution and systematics

The shoebill (*Balaeniceps rex*), with one genus and species, has been placed within its own family and has traditionally been allied with the storks and/or herons. It was first classified in 1850 by Gould, who spotted the creature along the banks of the upper White Nile and called it "the most extraordinary bird I have seen for many years". Suggestions that it is related to pelicans based on skull structure have been discounted, although recent DNA analyses support this relationship.

Physical characteristics

The shoebill stands about 43–55 in (110–140 cm) tall. The bird is gray to blue-gray with an ashy gray crown. Most prominent is the enormous, almost bulbous, prominently hooked bill, which resembles a shoe and lends the bird its common name. The bill is yellowish with irregular dark patches. The toes are long: 6.6–7.3 in (16.8–18.5 cm) long and completely divided; the claw of the hind toe is larger than those on the fore-toes. In flight, the neck is retracted as in herons; not outstretched as in storks. The female is similar in all respects but slightly smaller than the male.

Distribution

Central Africa: Most populous in southern Sudan and northern Uganda, but also found in Tanzania, Democratic Republic of Congo, Central African Republic, and Rwanda.

Habitat

Swamps and marshy lakeside, usually where papyrus (*Cyperus papyrus*) or cattails (*Typha* spp.) are dominant.

Behavior

Generally solitary. At favored locations, several birds may fish near each other. Shoebills are slow-moving, sedentary birds that are silent, except during nest building, when they make clapping noises with their bills.

Feeding ecology and diet

Shoebills use a stand-and-wait approach, although they occasionally feed by walking slowly. A shoebill stands with bill

Shoebill (*Balaeniceps rex*). (Illustration by Joseph E. Trumpey)

pointed downwards, almost motionless, sometimes for over thirty minutes. When prey is sighted, the shoebill thrusts its whole body forward, wings outstretched, as the bird attempts to seize its prey. The vegetation is separated from the prey. Prey is often decapitated. The shoebill usually swallows water after feeding. If the feeding attempt is unsuccessful, it usually moves to new location.

Shoebills feed primarily on lungfish (*Protopterus aethiopicus*), bichirs, catfish and other fish. Because shoebills feed in stagnant swamps, fish prey are caught when they come to the surface for air.

Reproductive biology

Nests singly in monogamous pairs. The nest is made of aquatic vegetation and measures up to 8 ft (2.5 m) across. When nests are built in the swamp, supplementation may be necessary to counteract sinking; sometimes the nest is built on a solid mound.

One to three (usually two) dull white eggs are laid. The eggs measure 3–3.5 in (80–90 mm) by 2–2.5 in (55–63 mm) and weigh about 6 oz (165 g). Eggs are laid at intervals of up to five days. Both birds incubate for about 30 days. During hot weather the parents may bring water that they regurgitate to cool the eggs.

At hatching, the young are downy, white, or silvery gray. The young stay in the nest for up to 105 days. During the first 35 days, they cannot stand and are brooded by the adults. Following this, parents spend less time at the nest while collecting food for the chicks. Feathers develop. By 95 days, the young begin to wander off the nest and they can fly about ten days later.

Both parents feed and care for the young. In addition, parents cool their young by bringing water and pouring this over the chicks. The parents may do this as many as five times on hot days.

Usually only a single chick fledges.

Conservation status

Near Threatened. Total population is estimated at 12,000–15,000 birds.

Significance to humans

None known.

Shoebill (*Balaeniceps rex*) preening. (Photo by Tom McHugh. Photo Researchers, Inc. Reproduced by permission.)

Resources

Books

BirdLife International. *Threatened Birds of the World.* Barcelona: Lynx Edicions; and Cambridge, United Kingdom: BirdLife International, 2000.

Brown, L.H., E.K. Urban, and K. Newman. *The Birds of Africa.* Vol. 1, *Ostrich to Falcons.* London: Academic Press, 1982.

Collar, N.J., and S.N. Stuart. *Threatened Birds of Africa and Related Islands.* Cambridge, United Kingdom: ICBP, 1985.

del Hoyo, J., A. Elliott, and J. Sargatal, eds. *Handbook of the Birds of the World.* Vol. 1, *Ostrich to Ducks.* Barcelona: Lynx Edicions, 1992.

Hancock, J.A., J.A. Kushlan, and M.P. Kahl. *Storks, Ibises and Spoonbills of the World.* London: Academic Press, 1992.

Sibley, S.G., and J.E. Ahlquist. *Phylogeny and Classification of Birds.* New Haven: Yale University Press, 1990.

Periodicals

Collar, Nigel J. "Shoebill." *Bulletin of the African Bird Club* 1 (March 1994).

Sibley, C. G., and J. E. Ahlquist. " Phylogeny and Classification of Birds Based on the Data of DNA-DNA Hybridization." *Current Ornithology* 1(1983): 245–292.

Malcolm C. Coulter, PhD

Ibises and spoonbills
(*Threskiornithidae*)

Class Aves
Order Ciconiiformes
Suborder Ciconiae
Family Threskiornithidae

Thumbnail description
Medium-sized wading and terrestrial birds of temperate and tropic regions, with prominent bills (decurved in ibises, broad and flat in spoonbills), long neck and legs, anterior toes, and highly social habits

Size
19–43 in (48–110 cm): 1–5.5 lb (0.5–2.5 kg)

Number of genera, species
13 genera; 32 species

Habitat
Wetlands, forests, grassland, arid or semi-arid areas

Conservation status
Extinct: 1 species; Critical: 4 species: Endangered: 2 species; Vulnerable: 1 species; Near Threatened: 2 species

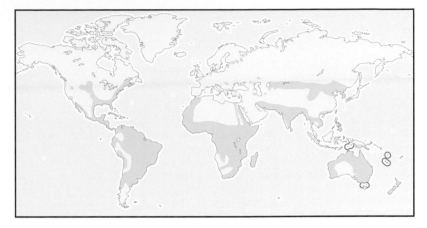

Distribution
Worldwide distribution in temperate and tropical zones. All continents except Antarctic have representatives of this family

Evolution and systematics

Two groups of Ciconiiformes, both with peculiarly-shaped beaks, make up the family of the ibises (Threskiornithidae). They are related to the storks, with the wood ibises (*Mycteria* also known as wood storks) forming a slight link with true ibises. With their slender curved beaks, the ibises differ strikingly from the flat-billed spoonbills but are nevertheless closely related. Spoonbill-ibis hybrids have been successfully raised in zoos. Hybridization raises some questions about the usual division of these birds into two subfamilies, but this division is retained here for practical purposes.

The two subfamilies are readily distinguishable by external characteristics: the ibises (Threskiornithinae), with their long, narrow, and markedly down-curved beak, probe for insects, mollusks, crustaceans, and worms in mud and soil; occasionally they also catch larger prey. Wing beats alternate with periods of gliding; when in flocks all birds alternate from one form of flight to the other at more or less the same time. There are 12 genera with 26 species.

The spoonbills (subfamily Plataleinae), with a beak that is flattened and widened at the tip, seize prey in side-to-side movements of the bill. They do not interrupt wing beats by gliding. This subfamily is comprised of one genus and six species.

Physical characteristics

All members of the family Threskiornithidae are medium to large in size. The face and throat are bare of feathers in most species; the medium-length legs are sturdy. The vocal apparatus is only feebly developed; they only utter low sounds or are almost mute, although a few species utter far-reaching calls. Spoonbills can also clatter with the beak. Both sexes are similar in color, the females generally being somewhat smaller than the males. Most plumage is white, brown, or black. Uniform coloration is the rule, sometimes with adornments such as display plumes. The standout exceptions in the family are the roseate spoonbill (*Ajaia ajaja*), whose shaded pink plumage is offset by a strange-looking head with bare greenish skin, and the scarlet ibis (*Eudocimus ruber*), with its striking uniform red plumage broken only by black wingtips. Most species have some areas of bare skin on the face. The sacred ibis (*Threskiornis aethiopicus*), has no feathers anywhere on the head or neck.

The fossil record of this family goes back 60 million years. It appears that, several times over the course of this long history, flightless species developed on islands. Of these, only the reunion flightless ibis (*Threskiornis solitarius*), survived into historical times.

Distribution

Ibises and spoonbills can be found almost everywhere in the world that moderate or warm temperatures prevail. They marginally inhabit the edges of deserts like the Sahara. With the exception of some regions of northern Africa and the Arabian Peninsula, most of the non-Antarctic world south of 45° North latitude is home to at least one species.

Scarlet ibises (*Eudocimus ruber*) feeding in Los Llanos area, Venezuela. (Photo by François Gohier. Photo Researchers, Inc. Reproduced by permission.)

Habitat

Ibises and spoonbills can adapt to a surprising variety of habitats. Some species live on arid plateaus and mountains, while most inhabit savannas, forests, and wetlands of all types. Agricultural areas often attract these birds: in Asia, ibises often live near rice paddies, which provide excellent hunting grounds.

Behavior

Most species are very sociable, often breeding in large colonies and wandering about or migrating in flocks, often mingling with other Ciconiiformes such as storks and herons. Migration is common, especially in species living in areas such as sub-Saharan Africa, where food is highly dependent on seasonal rainfall patterns. Their social behavior extends to relationships between species: mixed flocks are common. As many as seven species have been counted in a roosting area.

In flight, the neck is extended forward, similar to that of storks. During the day, ibises and spoonbills will often leave foraging sites to drink and bathe in freshwater ponds. Preening is common and can take a considerable amount of time.

Feeding ecology and diet

Ibises and spoonbills generally obtain their food in shallow water and on the banks, catching small fish, crustaceans, insects, and miscellaneous other invertebrates. Occasionally, they will feed on the eggs of reptiles or other birds. Feeding in the water is done primarily by the sense of touch provided by the sensitive bill.

Reproductive biology

Trees and bushes are popular nest sites for the species in this family, although a few species build nests on the ground or on cliffs. Males often find a suitable nest site and advertise their presence to females, making a show of pointing their bills in the air, bowing, and other movements. They often snap their bills shut to make a popping sound, and will sometimes pick up a twig and shake it. When a female lands nearby, the male may initially reject her: if he accepts her, they join in a display of preening and bowing. Copulation normally takes place at the nesting site, and the male gathers the nesting materials. Both parents incubate the eggs, and share in the task of gathering and regurgitating food for the hatchlings. Clutch size is two to five eggs. White and blue

Ibis feeding its young. (Illustration by Brian Cressman)

A roseate spoonbill (*Ajaia ajaja*) makes a call by clacking the top and bottom of its bill together. (Photo by Lawrence E. Naylor. Photo Researchers, Inc. Reproduced by permission.)

are the predominant egg colors, and in some species, the eggs have dark spots. The incubation period averages 20–31 days, with the chicks remaining in the nest for a fledgling period of 28–56 days.

Conservation status

The Reunion flightless ibis met a premature extinction, apparently at the hands of humans, around 1705. Several existing species are perilously close to following it. The four species classed as Critically Endangered by the IUCN are the dwarf olive ibis (*Bostrychia bocagei*), the hermit ibis or waldrapp (*Geronticus eremita*), the white-shouldered ibis (*Pseudibis davisoni*), and the giant ibis (*Pseudibis gigantea*). Considered Endangered are the black-faced spoonbill (*Platalea minor*) and the Japanese or crested ibis (*Nipponia nippon*), whose population in 2002 (wild and captive) was counted at 48 birds. The black-faced, or Australian, spoonbill (*Platalea minor*) nests only on islands off the east cost of the Korean Peninsula. Destruction of the tidal zones that are the species' preferred feeding grounds is the suspected cause of the birds' decline, and as few as 700 individuals remain. The bald ibis (*Geronticus calvus*) is considered Vulnerable, while species subject to lesser threat are the Madagascar crested ibis (*Lophotibis cristata*) and the black-headed ibis (*Threskiornis melanocephalus*). Ibises and spoonbills are under pressure mainly due to wetland reduction by human activity and direct hunting. Pesticides, especially DDT (which is still used in many areas of the world and is blamed for thin, easily broken eggshells) are another source of concern.

Significance to humans

Large-scale trade of bird feathers has dwindled, and with it the hunting that drove many species into peril. However, in many parts of the world, local species are still hunted as a source of food. Ibises in particular have taken on religious significance in some areas. The sacred ibis (*Threskiornis aethiopicus*) has been a part of cultural history for 5,000 years; in ancient Egypt, it was revered as the embodiment of Thoth, the god of wisdom, as well as the scribe of the gods.

Australian white ibis (*Threskiornis molucca*) rookery in Healesville Sanctuary, Australia. (Photo by Tom McHugh. Photo Researchers, Inc. Reproduced by permission.)

1. Scarlet ibis (*Eudocimus ruber*); 2. Japanese ibis (*Nipponia nippon*); 3. Hadada ibis (*Bostrychia hagedash*); 4. Sacred ibis (*Threskiornis ibis*); 5. Roseate spoonbill (*Ajaia ajaja*); 6. Hermit ibis (*Geronticus eremita*); 7. Spoonbill (*Platalea leucorodia*); 8. White-faced glossy ibis (*Plegadis chihi*). (Illustration by Brian Cressman)

Species accounts

Sacred ibis
Threskiornis aethiopicus

SUBFAMILY
Threskiornithinae

TAXONOMY
Tantalus aethiopicus Latham, 1790, Egypt. Three subspecies.

OTHER COMMON NAMES
French: Ibis sacré; German: Heiliger ibis; Spanish: Ibis Sagrado.

PHYSICAL CHARACTERISTICS
25.5–35 in (65–90 cm); 3.3 lbs (1,500 g). Plumage is mostly white; primary and secondary wing feathers tipped in black. Head and neck are bare, skin is black. Legs are black. Thickest bill of its genus.

DISTRIBUTION
Lives in most of the African continent south of 15° North latitude. There is an isolated colony at the southern tip of Iraq. Apparently the bird once was common in Egypt but it has not bred there since the first half of the nineteenth century.

HABITAT
Mainly coastal lagoons, marshes, damp lowlands, and agricultural areas (when flooded), but sometimes will travel far from water. Also garbage dumps and recently burned areas.

BEHAVIOR
These birds commonly fly in staggered lines, with each bird slightly ahead and to one side of the bird behind. A relatively quiet bird, the sacred ibis will make grunting and croaking noises during the breeding season, but these are the only vocalizations. Juveniles lack the long bill of the adults, and eat by reaching their bills into the parents' throats and removing food.

FEEDING ECOLOGY AND DIET
The sacred ibis is specialized by nature for aquatic prey like small fish and invertebrates, but is an opportunistic eater that will take anything available, such as carrion, bird eggs and nestlings, or small mammals.

REPRODUCTIVE BIOLOGY
The nest is a platform made of sticks and twigs, lined with vegetation. Clutch size is usually three to four eggs, from which the young hatch after 21 days and are fledged in five to six weeks. Their downy plumage is dark.

CONSERVATION STATUS
While its range has decreased, the species is not in imminent danger. Not threatened.

SIGNIFICANCE TO HUMANS
Thoth, the ibis-headed god of wisdom, is shown in many murals and sculptures, such as in the temple of Sethos, where it hands the hieroglyph of life to Osiris. It was used as a hieroglyphic symbol, and entire "cemeteries" of ibis mummies have been found at Sakkara near Cairo and at Hermopolis in middle Egypt. ◆

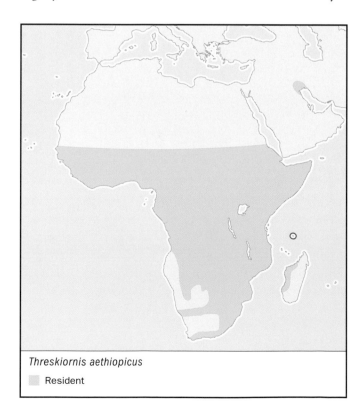

Threskiornis aethiopicus
■ Resident

White-faced glossy ibis
Plegadis chihi

SUBFAMILY
Threskiornithinae

TAXONOMY
Numenius chihi Viellot, 1817, Paraguay. Monotypic.

OTHER COMMON NAMES
English: White-faced ibis; French: Ibis à face blanche; German: Brillensichler; Spanish: Morito Cariblanco.

PHYSICAL CHARACTERISTICS
17–25.5 in (43–65 cm); 1.3 lb (610 g). Deep chestnut plumage with metallic green and purple gloss on back, wings, head, and neck. A border of white feather surrounds the pinkish to red facial skin. Legs are reddish.

DISTRIBUTION
The range forms a broad band across South America, reaching as far north as southern Peru and Brazil and south to include the northern thirds of Chile and Argentina. The range is markedly discontinuous, with the bird being absent north of this band until it reappears in central and western Mexico, northern California, and a large area of the midwestern and western United States, plus the western half of the United States Gulf coast.

Plegadis chihi

☐ Resident   ■ Breeding   ■ Nonbreeding

SIGNIFICANCE TO HUMANS
None known. ◆

Hermit ibis
Geronticus eremita

SUBFAMILY
Threskiornithinae

TAXONOMY
Upupa eremita Linnaeus, 1758, Switzerland. Monotypic.

OTHER COMMON NAMES
English: Waldrapp, northern bald ibis; French: Ibis chauve;
German: Waldrapp; Spanish: Ibis Eremita.

PHYSICAL CHARACTERISTICS
27.5–31.5 in (70–80 cm); 2.5 lb (1,280 g). Plumage is dark with
metallic green and purple gloss. Front portion of head is bare
(skin is reddish orange with black marks over the eyes), back
portion features a crest of dark feathers.

DISTRIBUTION
The species breeds only in three colonies in Morocco, all in
the Souss-Massa National Park. Formerly, the hermit ibis had
widespread distribution in Africa and Europe, but populations
dwindled due to loss of habitat and mass hunting in the seventeenth century.

HABITAT
Rocky plateaus, high-altitude meadows and streams, and arid
or semi-arid plains within foraging range of riverbeds or ocean
beaches.

HABITAT
Inhabits wetlands and all types of agricultural land. Congregates
around streams, creek beds, lakes, and other water sources.

BEHAVIOR
Some populations migrate, moving between breeding and wintering grounds, but others, especially those in the southern
part of the range, stay in one place throughout the year.

FEEDING ECOLOGY AND DIET
Feeds in the shallows of lakes, ponds, streams, rivers, and wetlands. Also forages in rice and alfalfa fields when flooded.
Takes fish and various other small aquatic vertebrates and invertebrates. In some areas, earthworms collected in irrigated
fields are a dietary staple.

REPRODUCTIVE BIOLOGY
Nests can be found in swamps, marshes, bushes, or trees, especially on vegetated islands. Nests built on the ground are usually woven from dry reeds, while those in trees are built of
sticks and twigs. Clutch size is three or four eggs, with an incubation period of about three weeks.

CONSERVATION STATUS
Some local populations are threatened, mainly by habitat destruction.

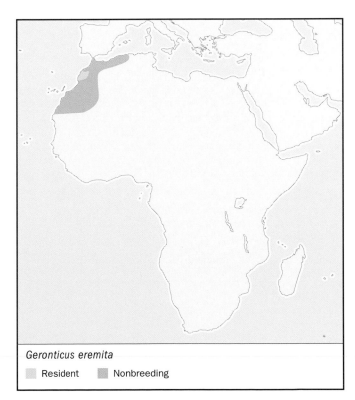

Geronticus eremita

☐ Resident   ■ Nonbreeding

BEHAVIOR
Except for the habit of nesting on cliffs, the hermit ibis is a typical member of its family. It is colonial, not given to loud calls, and spends its non-breeding days wading in the shallows.

FEEDING ECOLOGY AND DIET
Primarily insects, larvae, spiders, worms, and small reptiles and amphibians.

REPRODUCTIVE BIOLOGY
The species breeds colonially. Clutch size is usually two to three eggs. Adults feed not only their own young but also the young of other pairs from beak to beak. Incubation is 27 to 28 days, and the young are fledged after 46 to 51 days.

CONSERVATION STATUS
Critically Endangered. A captive breeding experiment failed to save the Turkish population, leaving only the Moroccan colonies. The hermit ibis has undergone a long-term decline and now has an extremely small range and population. The major reasons for the shrinkage of the species' range include agriculture, development, and hunting. In 2001, the World Conservation Monitoring Centre estimated the entire population at 220 birds in the wild, plus 700 in captivity.

SIGNIFICANCE TO HUMANS
No economic significance. In ancient times, the bird's return to the Euphrates River was a harbinger of spring, and was celebrated with a festival. ◆

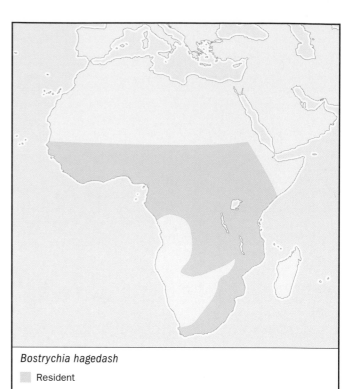

Bostrychia hagedash
▨ Resident

Hadada ibis
Bostrychia hagedash

SUBFAMILY
Threskiornithinae

TAXONOMY
Tantalus hagedash Latham, 1790, Cape of Good Hope. Three subspecies.

OTHER COMMON NAMES
English: Hadeda, Hadedah; French: Ibis hagedash; German: Hagedasch; Spanish: Ibis Hadada.

PHYSICAL CHARACTERISTICS
25.5–30 in (65–76 cm); 2.4 lb (1,250 g). General tone of plumage is gray to olive-brown (depending on subspecies) with metallic green gloss. Culmen has a distinctive red base. No crest of feathers on the head.

DISTRIBUTION
Senegal and Gambia across the continent to Ethiopia and southern Somalia, and south to include most of South Africa.

HABITAT
Primarily in savanna, grassland, and along wooded rivers and streams. Also in gardens and cultivated land.

BEHAVIOR
Hadadas are not as social as most ibises. They gather in flocks for breeding, but nest alone, not in colonies. Most populations are sedentary except for the normal radiation of young pushing out from the breeding area and local moves to adapt to environmental conditions. Hadadas do not hesitate to colonize areas of human habitation within their range, and are also known to display aggression toward domestic dogs and cats.

FEEDING ECOLOGY AND DIET
Insects and other small invertebrates, along with small fish and reptiles.

REPRODUCTIVE BIOLOGY
Hadadas nest most often in trees, and occasionally in telephone poles. Pairs generally breed on their own in wooded ravines up to elevations of 6,600 feet (2,000 m), but the birds descend to agricultural areas for feeding. Both partners incubate and feed the young. Eggs hatch after 26 days, and the young stay in the nest for about 33 days.

CONSERVATION STATUS
Not threatened. While other species have suffered from human activity, the hadada appears to have profited. The population is gradually rising.

SIGNIFICANCE TO HUMANS
None known. ◆

Japanese ibis
Nipponia nippon

SUBFAMILY
Threskiornithinae

TAXONOMY
Ibis nippon Temminck, 1835, Japan. Monotypic.

OTHER COMMON NAMES
English: Japanese crested ibis, crested ibis; French: Ibis nippon; German: Nipponibis; Spanish: Ibis Nipón.

Nipponia nippon
■ Resident

breeding, but the attempt was unsuccessful. The last Japanese
ibis in Japan died in 1995 at the estimated age of 26. ◆

Scarlet ibis
Eudocimus ruber

SUBFAMILY
Threskiornithinae

TAXONOMY
Scolopax ruber Linnaeus, 1758, Bahamas. Monotypic.

OTHER COMMON NAMES
French: Ibis rouge; German: Scharlachsichler; Spanish: Coro-
coro Rojo.

PHYSICAL CHARACTERISTICS
24 in (60 cm); 2 lb (900 g). Scarlet plumage and black recurved
bill; non-breeding adults have pink or reddish bills.

DISTRIBUTION
Coastal Brazil to north Venezuela, Colombia, and eastern
Ecuador.

HABITAT
Mangrove swamps, lagoons, estuaries, wetlands, and mudflats.

BEHAVIOR
Often gather in large flocks, feeding during the day and roosting in trees at night in large numbers. Has a plaintive, high-
pitched call.

PHYSICAL CHARACTERISTICS
22–31 in (56–79 cm); 2.2 lbs (1,000 g). Mostly white, this ibis
has orange-brown flight and tail feathers, a bare, orange-red
face, and a crest of long, white feathers extending backward
from the head. Legs are orange-red.

DISTRIBUTION
Before the twentieth century, this species bred in large areas of
eastern China and Japan, and existed in Korea until the 1940s.
Today, the remaining birds live in a reserve in southern
Shaanxi, a province in east-central China.

HABITAT
Forested hills and adjoining wetlands, rice paddies, lakes, and
ponds.

BEHAVIOR
The Japanese ibis does not migrate. The known birds only
travel from their breeding ground to foraging areas and back.

FEEDING ECOLOGY AND DIET
Frogs, newts, fish, crustaceans, and insects.

REPRODUCTIVE BIOLOGY
Breeding takes place in a colonial setting. The nest is a simple
platform of sticks built in a tree. Three eggs are normally laid.

CONSERVATION STATUS
Endangered and on the edge of extinction, with a total of 48
individuals recorded in 2001. In recent years, the sole wild
colony has never totaled higher than 22 birds, although an average of five fledglings per year was recorded over the last
decade. One bird per year is taken from the wild to add to a
captive breeding project in the Beijing Zoo, where several birds
have been hatched and raised successfully. Hunting (once
widespread, although now illegal), habitat destruction, and pesticides are blamed for the species' decline.

SIGNIFICANCE TO HUMANS
Revered as a Japanese national symbol. When only two were
left in Japan, in 1994, a pair was brought from China for

Eudocimus ruber
■ Resident

FEEDING ECOLOGY AND DIET
Hunts fish, frogs, newts, insects, crabs, and other invertebrates. Forages in both saltwater and freshwater ecosystems.

REPRODUCTIVE BIOLOGY
Pair formation takes place in the small nest territory which is defended by both partners. Nest material is brought mainly by the male, and is used for building by the female. Clutch size is usually two eggs. The young hatch after 21–23 days, are tended by both parents, and are fledged at 35–42 days.

CONSERVATION STATUS
Not threatened.

SIGNIFICANCE TO HUMANS
None known. ◆

Spoonbill
Platalea leucorodia

SUBFAMILY
Plataleinae

TAXONOMY
Platalea leucorodia Linnaeus, 1758, Sweden. Three subspecies.

OTHER COMMON NAMES
English: Eurasian spoonbill, common spoonbill; French: Spatule blanche; German: Löffler; Spanish: Espátula Común.

PHYSICAL CHARACTERISTICS
27.5–37.5 in (70–95 cm); 3.3 lb (1,500) g. Overall white plumage with varying amounts of yellow (from small patch to ring) at the base of the neck. Crest of white feathers on the back of the head. Black bill tipped in yellow and black legs. Males somewhat larger than females.

DISTRIBUTION
Has the largest modern range of any species in its family. Found across the Eurasian mainland, from the Atlantic coast of the Netherlands east across the Caspian and Black Seas, over

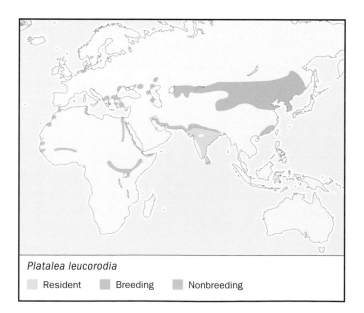

Platalea leucorodia
☐ Resident　☐ Breeding　☐ Nonbreeding

most of China, to Mongolia, southern Siberia, and the Korean Peninsula.

HABITAT
Marshes, lakes, ponds, rivers, lagoons, flooded areas, and mudflats.

BEHAVIOR
Flies with the head and legs extended, using majestic, slow beats of its wings. Groups may fly in single file or in a loose V formation. Spoonbills rarely utter any cries. On the ground, it often rests standing on one leg. It will swim for short distances to reach suitable areas of shallow water.

FEEDING ECOLOGY AND DIET
Mainly insects, crustaceans, and fish. The spoonbill holds its straight, flattened bill slightly open while foraging, sweeping it through shallow water and picking up prey items disturbed by the motion. Experiments have determined that the bill's shape lets it act as a hydrofoil, setting up water currents which affect objects up to four inches (10 cm) away from the bill itself.

REPRODUCTIVE BIOLOGY
The spoonbill breeds, like most birds of the ibis family, in colonies of varying size. Clutch size is about three to five eggs. The young hatch after 21 days and are cared for by both parents.

CONSERVATION STATUS
Not threatened. Some local pressures due to hunting and habitat destruction.

SIGNIFICANCE TO HUMANS
None known. ◆

Roseate spoonbill
Ajaia ajaja

SUBFAMILY
Plataleinae

TAXONOMY
Ajaia ajaja Linnaeus, 1758, Brazil. Monotypic.

OTHER COMMON NAMES
English: Pink curlew, rosy spoonbill; French: Spatule rosée; German: Rosalöffler; Spanish: Espátula Rosada.

PHYSICAL CHARACTERISTICS
31 in (80 cm); 3.3 lbs (1,500 g). The only pink spoonbill.

DISTRIBUTION
Range covers most of South America, excluding some western areas such as Chile, most of Argentina, and almost all of Peru. Also Central American nations up to northern Mexico and east along the Gulf Coast to Louisiana and Florida. Also occurs in Cuba, Haiti, and the Dominican Republic.

HABITAT
Mangrove stands, lagoons, swamps, rivers, lakes, and ponds.

BEHAVIOR
Roseate spoonbills are colonial birds but are nonetheless territorial, with the male staking out and defending nesting areas. Unlike some members of its family, the roseate spoonbill sometimes feeds at night. The birds fly with neck and legs ex-

Ajaia ajaja

▨ Resident

tended, flapping the wings and then gliding. The flight is described as more leisurely than that of ibises.

FEEDING ECOLOGY AND DIET
Primarily small fish, although other types of small aquatic prey, such as crayfish and crustaceans, are also taken. Like the nominate species of spoonbill, the roseate spoonbill swings its flattened beak from side to side, disturbing prey species. When the sensitive nerve endings in the inner linings of the bill report contact, the bill claps shut. The birds toss their heads backward to swallow prey.

REPRODUCTIVE BIOLOGY
Roseate spoonbills nest in colonies. Copulation takes place on the nest, which is loosely woven of sticks and twigs. Eggs are laid at the rate of one every two days. Clutch size averages three eggs, and incubation lasts an average of 22–23 days. Hatchlings have pink skin covered with short, sparse white down.

CONSERVATION STATUS
Currently, the roseate spoonbill is not threatened. Before World War II, the species suffered a considerable decline in the areas of its range populated by humans, due to hunting for meat and feathers as well as habitat destruction. At one point, the population in the United States may have numbered as few as 20 to 25 nesting pairs. Before modern conservation efforts began on the species' behalf, safety was afforded only by the remote areas of South and Central America.

SIGNIFICANCE TO HUMANS
Once widely hunted for plumes and meat, the birds today have no economic significance. ◆

Resources

Books

del Hoyo, J., A. Elliot, and J. Sargatal, eds. *Handbook of the Birds of the World*. Vol. 1, *Ostrich to Ducks*. Barcelona: Lynx Editions, 1992.

Elphick, Chris, John B. Dunning, Jr., and David Allen Sibley. *The Sibley Guide to Bird Life and Behavior*. New York: Alfred A. Knopf, 2001.

Hancock, J. A., J. A. Kushlan, and M. P. Kahl. *Storks, Ibises and Spoonbills of the World*. San Diego: Academic Press, 1992.

Sibley, C. G., and J. E. Ahlquist. *Phylogeny and Classification of Birds: A Study in Molecular Evolution*. New Haven and London: Yale University Press, 1990.

Periodicals

Martinez, Carlos, and Antonio Rodrigues."Breeding Biology of the Scarlet Ibis on Cajual Island, Northern Brazil." *Journal of Field Ornithology* 70 (4)(1999): 558–566.

Other

United Nations Environment Programme World Conservation Monitoring Centre. *Crested Ibis*. <http://www.unep-wcmc.org/index.html> <http://www.unep-wcmc.org/species/data/species_sheets/crestedi.htmmain> 30 October 2001.

United Nations Environment Programme World Conservation Monitoring Centre. *Waldrapp (Northern Bald Ibis)*. <http://www.unep-wcmc.org/index.html> <http://www.unep-wcmc.org/species/data/species_sheets/waldrapp.htmmain> 30 October 2001.

Matthew A. Bille MSc
Cherie McCollough, MS

Phoenicopteriformes
Flamingos
(Phoenicopteridae)

Class Aves
Order Phoenicopteriformes
Family Phoenicopteridae
Number of families 1

Thumbnail description
Large, very long-legged and long-necked water-birds with specialized down-curved, filter-feeding bills, and pink, black, and white plumage

Size
31.5–63 in (80–160 cm); 5.5–7.7 lb (2.5–3.5 kg)

Number of genera, species
3 genera; 5 species

Habitat
Shallow saline, brackish, and alkaline waters

Conservation status
Near Threatened: 2 species

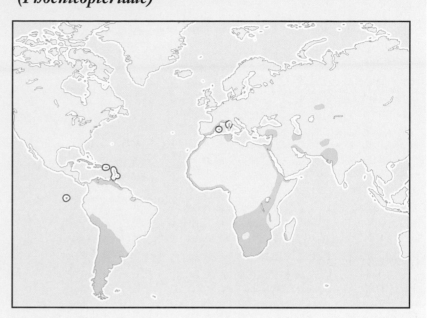

Distribution
South America including Galápagos, the Caribbean, Africa, southern Europe, southwest Asia, the Middle East, Indian subcontinent

Evolution and systematics

The taxonomic status of flamingos continues to be the subject of much debate. Current placement in a separate order between storks and herons (Ciconiiformes) and wildfowl (duck, geese, and swans; Anseriformes) has been followed most often and, despite recent challenges, still seems to be the most suitable. Certain skeletal features and courtship displays are similar to those of storks, while egg-white proteins show similarities with herons. There are several characteristics shared with wildfowl, including bill structure, webbed feet, voice, other courtship displays, chick behavior patterns, and external parasites. Flamingos are set apart from these families by, for example, the existence of communal displays, an absence of territorial behavior, and creching of young.

A study of muscle and bone structure, eggs, and internal parasites has suggested an evolutionary origin from long-legged wading birds (Charadriiformes), e.g., stilts and avocets. However, DNA-DNA hybridization techniques suggest a closer kinship with not just storks, but pelicans and New World vultures. Work using bile-gland acids puts flamingos firmly into the Anseriformes.

Primitive flamingos exist as fossils from c. 50 million years ago (Middle Eocene); fossils from the Oligocene period (about 30 million years old) appear identical with present-day genera. Flamingos were then more widespread than they are today, occurring in both North America and Australia, whence they are now absent, as well as over a much larger area of Europe. Their range then as now would have been linked to availability of shallow wetlands in a warm-temperate to tropical climate and their fossils therefore suggest the extent of such habitats in the past.

The separation of flamingos into three genera and five species, plus one sub-species, is based on relatively minor differences. The greater (*Phoenicopterus ruber*) and Chilean (*Phoenicopterus chilensis*) flamingos are placed in one genus, the former with two subspecies. The other three species, Lesser (*Phoeniconaias minor*), Andean (*Phoenicoparrus andinus*), and James' (*Phoenicoparrus jamesi*) flamingos are placed in two further genera, the difference between them being the presence of a hind toe in the lesser flamingo.

Physical characteristics

Flamingos are unmistakable in size, shape, and coloring, made more so by their habit of flocking, sometimes in exceptional numbers, e.g. gatherings of lesser flamingos in excess of one million birds. All five species are similar in shape and have common plumage features. They are separable in the field by size and the coloring of plumage and soft parts. The body is oval, with exceptionally long legs and a long neck. The relatively small size of the head is emphasized by the

Lesser flamingos (*Phoeniconaias minor*) run on water as they take flight in Lake Nakuru National Park, Kenya. (Photo by Adam Jones. Photo Researchers, Inc. Reproduced by permission.)

large bill, which is sharply decurved in the middle. Very unusual among birds, the upper bill is smaller than the lower. Both are lined with lamellae for filter-feeding, as is the large, fleshy tongue. The extremely long, spindly legs are an adaptation for wading, and the three front toes are webbed, supporting the birds on mud and allowing them to swim.

The largest of the flamingos, the greater, stands up to 60 in (150 cm) tall; the smallest, the lesser, stands about 35 in (90 cm). The dominant colors are pink and crimson-red. These colors are vital to their display, contrasting with black flight-feathers. The bill and legs are brightly colored, too, in red, pink, and yellow. Sexes are similar in plumage, but the male is slightly larger than the female. Small young are covered in whitish-grayish down, and their first set of feathers is gray-brown. At about one year old, they molt into a very pale version of adult plumage, acquiring full breeding coloring at three to four years. They breed for the first time from about four years old.

Distribution

Flamingos are found in South America, including the Galápagos Islands; in the Caribbean; and throughout much of Africa, including the north and west coasts, the full length of the Rift Valley in East Africa, and in South Africa. In Europe, they are confined to the extreme south, around the Mediterranean; and are found in Turkey and east into southwest Asia, and from the Middle East to Pakistan, India, and Sri Lanka.

Three species, the Chilean, Andean, and James', occur only in South America, the Chilean having the largest range, from central Peru through Chile, Bolivia, and Argentina south to Tierra del Fuego. The Andean and the James' Flamingos are confined to much the same area of the high Andes, encompassing southern Peru, western Bolivia, northern Chile, and northwestern Argentina. The lesser flamingo has a large range in East and South Africa, with much smaller numbers occurring in West Africa, and in Pakistan and India.

The most widespread species is the greater flamingo, of which one subspecies (*Phoenicopterus ruber ruber*) breeds on the Galápagos; on several Caribbean islands, including the Bahamas, Cuba, and the Netherlands Antilles; and on the coasts of northern South America and eastern Mexico. The nominate subspecies (*Phoenicopterus ruber roseus*) has a very extensive, if scattered, range in West, East, and South Africa; on the northern and southern coasts of the Mediterranean; in the Middle East; southwest Asia; and Pakistan, India, and Sri Lanka.

There is little information on historical ranges, but the greater flamingo formerly bred in Kuwait, Egypt, Algeria, and the Cape Verde Islands.

Habitat

Flamingos are specialized feeders requiring a very specialized habitat, consisting of shallow lakes and lagoons, which can be inland or coastal, including tidal, and ranging from strongly saline (up to twice or even more the salinity of seawater) to strongly alkaline (with a pH in excess of 10). While quite small waterbodies, including artificial saltpans, may be used for feeding, breeding normally takes place on much larger waters, including lakes in East Africa up to hundreds of square miles in extent. In southern France, the greater flamingo also feeds in fresh water, on rice-paddies, but this is a relatively new habit and probably related to the greatly increased numbers in the area.

The waterbodies used by flamingos extend from sea-level to nearly 14,000 ft (3,500 m) in the Andes. At these altitudes, the birds are able to live throughout the year in the presence of hot springs that keep the water from freezing. The birds' tolerance to conditions shared by only a handful of other organisms, e.g., aquatic invertebrates, diatoms, and algae, all of which they feed on, is astonishing. They can cope not only with water temperatures of up to 155°F (68°C), but with the extremely caustic nature of the water, containing chlorides and sulfates often in very high concentrations. Their adaptation to living in such conditions has allowed them to exploit an abundant food supply in the absence of any competitors. That the absence of food competitors is important has been demonstrated for the Chilean flamingo, which lives almost exclusively on lakes without indigenous fish and avoids those with them.

Behavior

The main collective display of flamingos starts well before the breeding season and consists of ritualized stretching and preening movements, including self-explanatory "head-flagging"; the "wing-salute," when the wings are briefly opened to expose their bright colors; and "marching," when the entire tightly packed group of birds walks rapidly in one direction before abruptly turning about and walking back again.

Vocalizations form an important part of the ritualized displays. Loud honking calls are given during head-flagging and lower-pitched grunts during the wing-salute. The voice is important for keeping flocks together, particularly during movements, and for communication between the breeding pair and their chick.

While all five species of flamingo indulge in some movements, many of these are adaptations to changes in their habitat, rather than true seasonal migrations. The more northerly European and Asian populations of the greater flamingo make regular southerly movements in autumn, returning in spring. However, the movements of both greater and lesser flamingos over their extensive ranges in Africa are dictated by irregular patterns of drought and rainfall and associated water-level changes, which in turn affect both the food supply and availability of suitable nesting areas. Chilean flamingos breeding in the high Andes descend to the coast for the winter, but such vertical movements are rare among both the Andean and James' flamingos. The population of greater flamingos in the Galápagos is sedentary, as are at least some of those in the Caribbean.

Feeding ecology and diet

Flamingos have three main foods: algae; diatoms; and small aquatic invertebrates, including, in different areas, brine-shrimp, brine-flies, and snails. The greater and Chilean flamingos are generalist feeders, taking a wide variety of available invertebrates, some seeds, algae, and diatoms. The other three species are specialists, lesser flamingos feeding exclusively on blue-green algae, while the Andean and James' flamingos take mainly diatoms.

Bills have broad areas of filtering lamellae. The smaller upper portion of the bill fits onto the larger lower one like a lid. A sharp angle about the middle of the bill ensures that, when the flamingo lowers its head into the water to sieve for food, the upper portion of the bill faces downward with the bill upside down. It also means that the lamellae-lined cleft between the upper and lower portions remains small along its entire length when the bill is open.

This mechanism permits particles only up to a certain maximum size to be sucked into the beak with the water. The bird creates suction by retracting its thick fleshy tongue with the beak slightly opened, reducing the pressure in the bill and causing water to enter. Closing the bill and moving the tongue forward, expels the water leaving the food particles caught on the lamellae (thin flat membranes). At the next retraction of the tongue, these food particles are carried into the oral cavity by the bristle-like projections of the tongue, and simultaneously water once more enters the bill.

The differences among species in their major foods causes species with the finer lamellae to sift just beneath the water surface, while species with the coarser weave largely work in the mud beneath the water. This makes it possible for multiple flamingo species to live in the same area and even to feed in the same lake without competing. Thus ranges of the greater and lesser flamingos and of the Chilean, Andean, and James' flamingos, overlap. Species with similar filtration apparatus, and hence similar food, always have separate areas of distribution.

Reproductive biology

Flamingos are among the most gregarious of birds, feeding and breeding in flocks and colonies that may contain more than a million individuals. However, even though they live in these huge assemblies, the birds are monogamous and probably pair for life. The social stimulation of group displays, which can involve hundreds or thousands of birds, is a vital factor in initiating breeding attempts and bringing them into close synchronization.

Pair formation displays are similar to the group displays already described, but take place slightly apart from the large flocks, as does copulation. Actual nest sites within the colony site are selected by the female shortly before egg-laying, and she commences nest construction, though both birds will complete it. The nest is a truncated cone of heaped mud, with a shallow depression in the top for the single elongated, chalky white egg. Nest-building continues for several days after egg-laying, the birds using available materials, including mud, stones, shells, etc., within their reach while standing or sitting on the nest site, piling them up around themselves. The height of the mud cone varies according to the nature of the

The feeding action of a greater flamingo (*Phoenicopterus ruber*), as it retrieves its food from the water. (Photo by Kenneth W. Fink. Photo Researchers, Inc. Reproduced by permission.)

ground; it may be up to 16 in (40 cm) high or be altogether absent on rocky ground.

Incubation is by both parents and lasts 27–31 days. The newly hatched chick has a white, downy plumage; a straight, red bill; and thick, red legs, which become black after 7–10 days. The chick leaves the nest when it is 4–7 days old and is, to start with, accompanied by the parents and defended against other birds that come too close. Soon after this, the parents leave the chick alone for longer and longer periods, and it joins others to form loose groups, or creches. At about 2–3 weeks, the chick grows a second gray, downy plumage, and the bill begins to bend. At about 4 weeks, the first contour feathers appear on the shoulders.

The bill lamellae are not yet fully functional in 70-day-old young, which can already fly. Up to this age, young depend largely on a highly nutritious liquid secreted by the parents in the region of the esophagus and proventriculus. This secretion has a nutritional value comparable to that of milk; its content of carotenoids and blood give it a bright red color and are the same pigments synthesized by the parents to color their feathers. Parents know their own young by their voices and will feed no others, even when the young are gathered in groups.

The breeding season of flamingos in tropical and subtropical areas is dictated mainly by rainfall providing suitable shallows and food abundance, and therefore may take place at any time of the year, or not at all, for several years in succession. In temperate areas, such as southern Europe, breeding occurs mainly in spring, but an attempt may be abandoned if conditions become too dry, or the birds may miss a year altogether. This irregularity extends to the colony sites chosen. These sites may be used several years in succession, or the birds may shift to a new site almost every year.

Conservation status

The Andean and James' flamingos are both classified as Near Threatened, having been changed from Threatened in recent years, as a result of better information. Both are thought to have populations of up to 50,000 birds, but both are concentrated on relatively few waterbodies where some habitat destruction through pollution, mining, and diversions of streams has taken place. Increasing access through road construction has also led to increased egg-harvesting by humans and colony disruption by foxes. However, the creation of reserves should provide some relief from these problems.

The Chilean flamingo is thought to number around 200,000 birds, and despite some habitat loss and egg-harvesting, it probably has a favorable conservation status, as have both the greater and lesser flamingos. The former has populations of about 400–500 in the Galápagos, up to 90,000 in the Caribbean. and perhaps 800,000 in its European-African-Asian range. Numbers in the Caribbean have declined sharply in recent years, through habitat destruction by drainage and reclamation, but have increased strongly in southwest Europe, as a result of conservation measures on its principal breeding localities in France and Spain. The lesser flamingo is the most numerous and certainly numbers 3–4 million birds, perhaps as many as 6 million.

Significance to humans

The flamingo was known to Neolithic man, who illustrated it around 5000 B.C. in cave paintings in southern Spain. Egyptians used the flamingo as one of their hieroglyphic symbols to indicate the color red. They also regarded it as a living embodiment of the sun god Ra, and there may also have been a link with the mythical Phoenix.

Flamingos have long been eaten, with the Romans regarding the tongue as a special delicacy. Hunting of flamingos has always been constrained by the remote and difficult terrain in which so many of them live and by the extreme water conditions where they breed, though this hasn't stopped some native peoples from regular egg-harvesting, which has probably existed for many centuries.

The only known instance of flamingos becoming pests has been in the Camargue region of southern France where birds from this increasing population started to feed in newly sown rice paddies in the vicinity, taking the rice grains as food. The problem has been solved largely by systematic scaring during the critical period before the rice sprouts.

1. Greater flamingo (*Phoenicopterus ruber*); 2. Lesser flamingo (*Phoeniconaias minor*); 3. Chilean flamingo (*Phoenicopterus chilensis*); 4. Andean flamingo (*Phoenicoparrus andinus*); 5. James' flamingo (*Phoenicoparrus jamesi*). (Illustration by Patricia Ferrer)

Species accounts

Greater flamingo
Phoenicopterus ruber

TAXONOMY
Phoenicopterus ruber Linnaeus, 1758, Bahamas. Two subspecies: *P.r. ruber* and *P.r. roseus*.

OTHER COMMON NAMES
English: Caribbean, West Indian flamingo, rosy flamingo; French: Flamant rose; German Rosaflamingo; Spanish: Flamenco Común.

PHYSICAL CHARACTERISTICS
47–57 in (120–145 cm) 4.6–9.0 lb (2.1–3.4 kg); female approximately 10–20% smaller than male. Largest of the flamingos, adults are rosy red (Caribbean population) or whitish tinged with pink (European-African-Asian population) with brighter pink on the wings. The flight feathers are black. The bill is pink with a black tip, and legs are pink with darker pink joints. Hatchling is dark or light gray down with bright red legs and straight red bill. Juvenile is gray-brown, acquiring pale pink upper wing panel and pink tinge to gray legs and bill at 11 months; at four years, body plumage and lower portion of bill still grayer than adults.

DISTRIBUTION
P. r. ruber: Galápagos and Caribbean; *P. r. roseus*: North, West, East, and South Africa, southern Europe, Middle East, southwest Asia and Pakistan, India, and Sri Lanka.

HABITAT
Shallow saline and alkaline lakes and lagoons.

BEHAVIOR
Gregarious, with group displays involving ritualized movements of head and wings, accompanied by loud calls. In flocks of a few hundred to over one million.

FEEDING ECOLOGY AND DIET
Sieves aquatic invertebrates, seeds, algae, and diatoms from shallow water and mud.

REPRODUCTIVE BIOLOGY
Lays single egg (large, elongated, white, and chalky with reddish yolk) on mud nest close to or in shallow water, the time of breeding being dictated by rainfall rather than seasons. Nests in dense colonies, up to tens or hundreds of thousands of pairs. Incubation period 27–31 days; fledging 65–90 days. Both parents incubate and care for young, which gather into groups. Productivity very variable, with complete failures in some years. Age of first breeding normally five or six years.

CONSERVATION STATUS
Not threatened. Has declined in the Caribbean but increased in southwestern Europe. Elsewhere, very numerous, though subject to wide fluctuations in numbers based on rains and breeding success.

SIGNIFICANCE TO HUMANS
Sometimes hunted for food or sport, e.g., in Egypt. ◆

Chilean flamingo
Phoenicopterus chilensis

TAXONOMY
Phoenicopterus chilensis Molina, 1782, Chile.

Phoenicopterus chilensis
☐ Resident

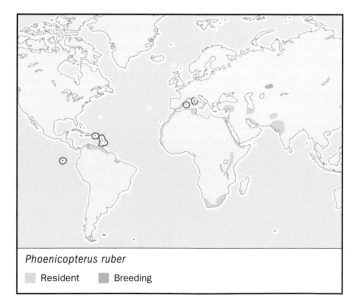

Phoenicopterus ruber
☐ Resident ☐ Breeding

OTHER COMMON NAMES
French: Flamant du Chili; German: Chileflamingo; Spanish: Flamenco Chileno.

PHYSICAL CHARACTERISTICS
38–42 in (96–107 cm); c. 5.0 lb (2.3 kg); female approximately 10% smaller than male. Smaller than the greater flamingo, with overall coloring similar to that of the European-African-Asian population, though pinker on the neck and breast. The inner third of the bill is pink, the remainder black, while legs are pink with darker pink joints. Chicks covered in gray down when born; may retain gray markings, at least in part, or develop white plumage that remains until two to three years of age. The juvenile is gray-brown.

DISTRIBUTION
Central Peru south through Chile, Bolivia, and Argentina to Tierra del Fuego.

HABITAT
Shallow saline and alkaline lakes and lagoons.

BEHAVIOR
Gregarious, with group displays involving ritualized movements of head and wings, accompanied by loud calls. In flocks of a few hundred to tens of thousands.

FEEDING ECOLOGY AND DIET
Sieves aquatic invertebrates, seeds, algae, and diatoms from shallow water and mud.

REPRODUCTIVE BIOLOGY
Lays single egg (chalky-white, goose-sized) on mud nest close to or in shallow water, the time of breeding being dictated by rainfall rather than seasons. Nests in dense colonies, up to several thousands of pairs. Incubation period 27–31 days; fledging 70–80 days. Both parents incubate and care for young, which gather into groups. Productivity very variable, with complete failures in some years.

CONSERVATION STATUS
Not threatened. Egg-harvesting and habitat destruction have caused declines at some colonies, but overall status probably stable.

SIGNIFICANCE TO HUMANS
Egg-harvesting and perhaps some hunting. ◆

Lesser flamingo
Phoeniconaias minor

TAXONOMY
Phoeniconaias minor Geoffroy, 1798, no locality = Senegal.

OTHER COMMON NAMES
French: Flamant nain; German: Zwergflamingo; Spanish: Flamenco Enano.

PHYSICAL CHARACTERISTICS
31–35 in (80–90 cm); 3.3–4.4 lb (1.5–2.0 kg); female approximately 10% smaller than male. Smallest of the flamingos, the adults are similar to the greater flamingo of the European-African-Asian population but with a proportionately longer

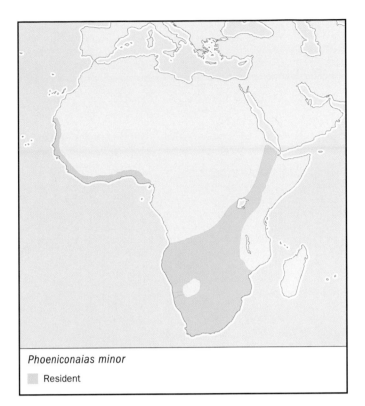

Phoeniconaias minor
▢ Resident

bill, which is much darker red, paler toward the tip, which is black. Legs are dark pink. Hatchling has bright coral-red legs and straight reddish pink bill. Juvenile is gray-brown.

DISTRIBUTION
East and South Africa, with small numbers in West Africa and in Pakistan and India.

HABITAT
Shallow saline and alkaline lakes and lagoons.

BEHAVIOR
Gregarious, with group displays involving ritualized movements of head and wings, accompanied by loud calls. In flocks of a few hundred to over one million.

FEEDING ECOLOGY AND DIET
Sieves blue-green algae from shallow water and mud.

REPRODUCTIVE BIOLOGY
Lays single egg (chalky-white, elongated; large but slightly smaller than *P. ruber*) on mud nest close to or in shallow water, the time of breeding being dictated by rainfall rather than seasons. Nests in dense colonies, up to hundreds of thousands of pairs. Incubation period 28 days; fledging 70–75 days. Both parents incubate and care for young, which gather into groups. Productivity very variable, with complete failures in some years. Age of first breeding normally three or four years.

CONSERVATION STATUS
Not threatened. Although numbers show huge fluctuations at individual sites, overall numbers believed more or less stable.

SIGNIFICANCE TO HUMANS
None known. ◆

Andean flamingo
Phoenicoparrus andinus

TAXONOMY
Phoenicoparrus andinus Philippi, 1854, salt lake near Altos de Pingopingo, Antifagasta, Chile.

OTHER COMMON NAMES
English: Greater Andean flamingo; French: Flamant des Andes; German: Andenflamingo; Spanish: Parina Grande.

PHYSICAL CHARACTERISTICS
40–43 in (102–110 cm); 4.4–5.3 lb (2.0–2.4 kg); female approximately 10% smaller than male. The head and neck are suffused with wine-red, the rest of the body whitish, tinged with pink, brightest on the wings. The flight feathers are black. The inner third of the bill is yellow, with a reddish patch between the nostrils, the remainder black. Legs and feet are yellow. The juvenile is grayish, streaked darker.

DISTRIBUTION
Found only in the high Andes of southern Peru, Bolivia, northern Chile, and northwestern Argentina.

HABITAT
Shallow high-altitude saline and alkaline lakes and lagoons.

BEHAVIOR
Gregarious, with group displays involving ritualized movements of head and wings, accompanied by loud calls. In flocks of a few hundred to several thousand.

FEEDING ECOLOGY AND DIET
Sieves diatoms from shallow water and mud.

REPRODUCTIVE BIOLOGY
Lays single egg on mud nest close to or in shallow water, the time of breeding being dictated by rainfall rather than seasons. Nests in dense colonies, up to several thousands of pairs. Incubation period c. 28 days; fledging probably 70–80 days. Both parents incubate and care for young, which gather into groups. Productivity very variable, with complete failures in some years. Age of first breeding probably four or five years.

CONSERVATION STATUS
Not threatened, but has declined in some areas in recent years through habitat destruction and egg-harvesting. Recent establishment of reserves should benefit the species.

SIGNIFICANCE TO HUMANS
Egg-harvesting. ◆

James' flamingo
Phoenicoparrus jamesi

TAXONOMY
Phoenicoparrus jamesi P.L. Sclater, 1886, Sitani, at foot of Isluga volcano, Tarapacá, Chile.

OTHER COMMON NAMES
English: Lesser Andean flamingo, Puna flamingo; French: Flamant de James; German: Jamesflamingo; Spanish: Parina Chica.

PHYSICAL CHARACTERISTICS
35–26 in (90–92 cm); c. 4.4 lb (2.0 kg); female approximately 10% smaller than male. Adults are overall whitish, tinged with

Phoenicoparrus andinus
Resident

Phoenicoparrus jamesi
Resident

pink, with a band of carmine streaks across the breast; bright red on the wings. Flight feathers are black. Bill is yellow, with red at the base and a broad black tip. Legs and feet are pink. The juvenile is brownish, streaked darker.

DISTRIBUTION
Found only in the high Andes of the extreme south of western Bolivia, northern Chile, and northwestern Argentina.

HABITAT
Shallow high-altitude saline and alkaline lakes and lagoons.

BEHAVIOR
Gregarious, with group displays involving ritualized movements of head and wings, accompanied by loud calls. In flocks of a few hundred to several thousand.

FEEDING ECOLOGY AND DIET
Sieves diatoms from shallow water and mud.

REPRODUCTIVE BIOLOGY
Lays single egg on mud nest close to or in shallow water, the time of breeding being dictated by rainfall rather than seasons. Nests in dense colonies, up to a few thousands of pairs. Incubation period probably c. 28 days; fledging probably 70–80 days. Both parents incubate and care for young, which gather into groups. Productivity very variable, with complete failures in some years. Age of first breeding probably four or five years.

CONSERVATION STATUS
Has declined in some areas in recent years, through habitat destruction and egg-harvesting. Recent establishment of reserves should benefit the species.

SIGNIFICANCE TO HUMANS
Egg-harvesting. ◆

Resources

Books

Cramp, S., and K.E.L. Simmons, eds. Vol. 1 of *The Birds of the Western Palearctic*. New York: Oxford University Press, 1977.

del Hoyo, J., A. Elliot, and J. Sargatal, eds. *Handbook of the Birds of the World*. Vol. 1, *Ostrich to Ducks* Barcelona: Lynx Edicions, 1992.

Kear, J., and N. Duplaix-Hall, eds. *Flamingos*. Berkhamsted, United Kingdom: Poyser, 1975.

Ogilvie, M., and C. Ogilvie. *Flamingos*. Gloucester, England: Alan Sutton Publishing Ltd., 1986.

Periodicals

Johnson, A.A. "Greater Flamingo." *BWP Update* 1 (1997): 15–24.

Olson, S.L., and A. Feduccia. "Relationships and evolution of Flamingos (Aves: Phoenicopteridae)." *Smithsonian Contributions to Zoology* No. 316 (1980).

Sibley, C.G., K.W. Corbin, and J.H. Haavie. "The relationships of the Flamingos as indicated by the egg-white proteins and hemoglobins." *The Condor* 71 (1969): 155–179.

Organizations

Wetlands International/Survival Service Commission Flamingo Specialist Group. c/o Station Biologique de la Tour du Valat, Le Sambuc, Arles 13200 France.

Other

Chilean Flamingos Page. Decatur (Illinois) Public Schools. 6 Dec. 2001. <http://www.dps61.org/jh/ZooWeb/Flame/webpage.htm>.

Chilean Flamino Page. Rolling Hills Refuge Wildlife Conservation Center, Salina, Kansas. 6 Dec. 2001. <http://www.rhrwildlife.com/theanimals/f/flamingochilean>.

Flamingo, Chilean Page. Phoenix Zoo. 6 Dec. 2001 <http://www.phoenixzoo.org/zoo/animals/facts/flamchil.asp>.

The Roberts VII Project. Draft species texts. *Greater Flamingo* 6 Dec. 2001 <http://www.uct.ac.za/depts/fitzpatrick/docs/r096.html>. *Lesser Flamingo* 6 Dec. 2001 <http://www.uct.ac.za/depts/fitzpatrick/docs/r097.html>.

Malcolm Ogilvie, PhD

Falconiformes

(Diurnal birds of prey)

Class Aves

Order Falconiformes

Number of families 3

Number of genera, species 76 genera; 306 species

Photo: Crested caracara (*Polyborus plancus*) in Llanos, Venezuela. (Photo by Art Wolfe. Photo Researchers, Inc. Reproduced by permission.)

Evolution and systematics

Raptors are potentially as old as more "primitive" families, such as the loons (Gaviidae), with the oldest claimed fossil record, a tiny falcon-like bird from the British Isles, dating from the lower Eocene some 55 million years ago. Well-documented finds from the late Eocene and early Oligocene, some 30–50 million years ago, are all from Europe, with raptors showing up from the New World only in the Miocene, about 23 million years ago. However, there is no clue as to the order's geographic origin, with modern-day representatives found on every continent except Antarctica and the greatest diversity found in the Neotropics.

The oldest fossils are of forms unrelated to any modern-day species, though the osprey (*Pandion haliaetus*) has been around for 10 million years. There is also nothing to show that the different families of the Falconiformes share a common ancestor and, although this is a comparatively well-studied order, understanding of its systematics is limited. The raptors have been traditionally grouped in respect of their similar behavior, external morphology, moult patterns, and internal anatomy. Comparisons of feather proteins and DNA seem to confirm the relationships between the families, including the closer relationship between the monotypic Sagittariidae and the rest of the family than with the cranes (Gruidae) or bustards (Otididae), with which it shares some behavioral features.

The Falconiformes are divided into three families: the Accipitridae (hawks, eagles, and allies), the Falconidae (falcons, caracaras, and allies), and the unique Sagittariidae (secretary bird). Of the three, only the Falconidae is further divided, into the Polyborinae (caracaras and forest-falcons) and the Falconinae (the "true" falcons and falconets). The Accipitri-

dae is numerically dominant and one of the largest avian families with more than 200 species, though the changing nature of taxonomy means that there may be up to 250, and the dividing lines between the genera and species are not especially precise.

Physical characteristics

Raptors have a strong, compact body and a large, generally rounded, head, joined by a strong neck that is very short in most species. The smallest species is the black-thighed falconet (*Microhierax fringillarius*), with a 12 in (30 cm) wingspan and weighing as little as 1.1 oz (28 g). At the other end of the scale, the Himalayan griffon vulture (*Gyps himalayensis*) has a wingspan of over 9 ft (3 m) and can weigh up to 26 lb (12 kg). In most species, there is significant sexual dimorphism, with males significantly smaller than females, enabling a pair to exploit a greater size range of prey. Although body length is a less useful measure in falconiformes than in other orders, it is notable that the secretary bird (*Sagittarius serpentarius*) stands at 4 ft (1.2 m) tall.

With a few exceptions, raptors excel in the air, and each family is well adapted to particular hunting techniques: the flight and tail feathers are large, with 10 primaries and 12–16 secondaries on each wing, and the bill and feet are designed for catching or ripping open the skin of prey. Even in those species that are not primarily carnivorous, such as the honey-buzzards, this sharp, hooked bill is essential to its lifestyle. Cutting edges of the upper mandible project over those of the lower mandible to form a scissors-like instrument. The legs are generally short, with long toes and bent, sharp claws, used to grasp prey. Often overlooked are the bristles at the base of

Gray morph gyrfalcon (*Falco rusticolus*) in flight over Seward Peninsula, Alaska. (Photo by Jim Zipp. Photo Researchers, Inc. Reproduced by permission.)

the bill of most species, which may protect the eyes when feeding, but may also provide sensory information on wriggling prey. In honey-buzzards, these are replaced by flattened, scale-like feathers, which may provide protection from bee and wasp stings, or may simply be to prevent honey from soiling the head plumage, in the same way that some kites and vultures have bare facial skin around the bill.

Most raptor species look relatively dull, with shades of brown, gray, and buff dominating the plumage. None is brightly colored and relatively few are predominantly black or chestnut and white as adults. In most orders, plumage color, especially of females, is camouflage to evade predators. This is less important for many adult raptors, which have few natural enemies, but the plumage has evolved to reduce detection as they hunt. For example, species that catch live prey, such as harriers (*Circus*) and falcons (*Falco*), tend to sport paler underparts, which make them less visible from below.

Distribution

The Falconiformes are a global order, with the Accipitridae and the Falconidae found in all continents except Antarctica. Only the monotypic Sagittariidae is limited to a single zoogeographical region, the Afrotropics (though secretary bird-like fossils have been found in Europe and North America). At least one or two breeding species can be found in every major habitat type around the globe, from urban—where scavenging vultures and kites can thrive—to the high arctic tundra, where the gyr falcon (*Falco rusticolus*) raises its chicks on the abundant seabirds, waders, and lemmings.

The greatest number of species is found at lower latitudes and altitudes, especially among the Accipitridae, many of which require thermals to hunt. The Falconidae are more adaptable and able to exploit some of the harsher environments. The peregrine falcon (*Falco peregrinus*) may have the widest distribution of any breeding bird, and is now found in

the center of some of the busiest cities, including London and New York.

Habitat

At the top of the food chain in many habitats, birds of prey are good indicators of habitat quality—without sufficient food or nest sites, these long-lived birds cannot survive. Many species occur at low densities, requiring large home ranges for feeding, especially in higher latitudes. They are terrestrial birds, although Steller's sea-eagles (*Haliaeetus pelagicus*) will fish from drifting icebergs several miles from shore.

Tropical rainforests contain the greatest abundance of Falconiformes (especially the Accipitridae, caracaras, and forest-falcons), many of which nest or roost in trees, even though some forage in open, agricultural landscapes. The true falcons tend to reside in more open habitats, while some species—such as harriers—are more adapted to grasslands or even reedbeds.

Behavior

Pairs of most species live a solitary lifestyle, especially those at higher latitudes, where resources are often scarcer and home range can be several dozen square miles. Some species are more social, particularly those that are less predatory, feeding on invertebrates. The adults of a few species, such as Eleonora's falcons (*Falco eleonorae*) and some vultures, nest colonially, while many roost and feed together outside the breeding season. Many migratory species also make their transcontinental journeys *en masse*. Social groups of immature birds are probably important in developing the skills for later breeding success.

Raptors have simple calls, usually repeated notes, often high pitched and harsh. Calls are used for many social situations, including maintaining contact between a pair or family. Kites and buzzards are the most vocal, with a variety of mewing and screeching calls, which peak during courtship.

Raptor migration is among the greatest spectacles of the avian world. Raptors with a low wing-loading are unable to fly for a long distance over water, so require thermals to make the distance. Thus, narrow peninsulas and isthmuses—such as Panama, Gibraltar, and Sinai—are the meeting points for the raptors from a whole region and, in the right conditions, thousands can pass every hour, for days or weeks on end. The principal long distance migrants are the Accipitridae species that breed in the northern hemisphere, with relatively small numbers of the Falconidae making such journeys, though a few make seasonal altitudinal movements.

Feeding ecology and diet

Most species are exclusively carnivorous, feeding on every major group of vertebrates and most of the invertebrates. Some, especially the larger Accipitridae, are generalist scavengers of carrion, but many have specialized diets, such as the eponymous bat falcon (*Falco rufigularis*) and the snail kite (*Ros-*

trhamus sociabilis). Visual acuity is critical to success, especially among the high-speed falcons, and is up to eight times better than human eyesight.

In general, the Falconidae fly rapidly, taking prey on the wing at speed, catching it in their talons, and killing it with a bite to the back of the neck. By contrast, small Accipitridae tend to hunt from a perch, making a short flight to catch small prey on the ground, squeezing it to death with their strong feet, and often taking it a short distance for plucking or ripping. Larger hawks and eagles search for prey from thermals, so do not hunt until the air has warmed up, several hours after sunrise. Some genera have developed specialized methods, such as the hovering utilized by kestrels and the kicking adopted by the secretary bird. The soft organs, with the highest protein levels, are extracted first. Aerial feeders, such as Eurasian hobby (*Falco subbuteo*) swallow invertebrates alive and whole, while flying. Indigestible material is regurgitated in a pellet through the bill, some 16–18 hours later.

Reproductive biology

Most raptors are monogamous, though polygyny is known in three harrier species when there is an abundant food supply, and polyandry (where one female mates with several males, which help to rear the brood) is recorded occasionally from a few species. Small falcons breed at one year old, whereas large vultures and forest eagles do not mature until six to nine years, though it can be significantly sooner where the population is well below carrying capacity for the habitat.

Courtship is usually a simple, soaring display flight by the male to advertise his ownership of a territory, though some species undertake a dramatic "rollercoaster" flight. Most of the Falconidae use a shallow scrape on a cliff face or a hole in a tree, whereas the Accipitridae and the caracaras build a nest platform that can attain a height of several feet over several years. Territories are evenly spaced and often traditional—those of some larger species are used by different pairs over many decades.

As long-lived birds, raptors have more breeding opportunities than many bird families, but rearing such large young is costly on resources, so most species successfully rear just one young each year, and some of the larger species do not breed annually. Typically, males hunt on behalf of females during incubation and while the chicks are young. The sexual dimorphism comes into its own as the nestlings grow, with the male and female able to hunt for a wide range of prey. The time taken to fledge is related to the ultimate size of the adult, with young sea-eagles and vultures remaining in the nest for several months after hatching.

Conservation status

Raptor populations are, by nature, stable, with population density remaining remarkably constant over many decades. Their history during the last 400 years is indicative of the pressures facing birds of prey and their habitats, though—perhaps surprisingly—none has become extinct (though one falcon subspecies has been lost from central America). Thirty-

Bald eagle (*Haliaeetus leucocephalus*) nest in western North America. (Photo by Tom & Pat Leeson. Photo Researchers, Inc. Reproduced by permission.)

four species are listed as Threatened or Near Threatened by BirdLife International and IUCN, though the global populations are not known for most species.

Many species are believed to be less abundant than in the recent past as a result of habitat changes that have altered the prey base. In particular, deforestation and agricultural monocultures have had a dramatic effect on the densities of many species, with most finding it harder to survive, though a few—such as kestrels—have probably benefited from the expansion of low intensity farming into former forests. An indication in reverse comes from the post-Soviet abandonment of collective farms which resulted in reduced grazing and thus dramatic falls in the populations of susliks, followed rapidly by that of saker falcons (*Falco cherrug*).

As well as loss of wooded and wetland habitat, modern agriculture also brought organochlorine pesticides that significantly reduce breeding success, by preventing eggshells from thickening. During the 1950s and 1960s, populations of several falcon and hawk species fell dramatically in Europe and North America, ultimately resulting in the prohibition of the compounds and the subsequent recovery of most species. However, compounds such as DDT remain in widespread use in many parts of the world, with little knowledge of the deleterious effects on raptors.

Direct persecution remains a serious problem for some species, especially those that come into conflict with the land-uses adopted in their favored habitats. Depredation by raptors of livestock, particularly sheep, and small gamebirds can elicit a lethal response from farmers and gamekeepers. Carrion feeders are especially vulnerable to poison baits, targeted either at them or mammalian predators. In addition, some falcon species are targeted by collectors for sale or falconry, while in Britain raptors have long been the target of egg-collectors. The longevity and low breeding success of raptors means that recovery from any decline is slow.

The most dramatic decline occurred among the *Gyps* vultures of the Indian subcontinent in the late 1990s, in which an epidemic probably killed hundreds of thousands of birds in the region. Disease appears to be the cause, but by 2002, it remained unclear whether an environmental factor had made the birds more prone than previously.

The pressures on raptors have been recognized in many parts of the world by protective legislation. Indeed, in many parts of the world, members of the order have the highest levels of protection, enabling the populations of many species to recover to former levels. Their readiness to breed in captivity has enabled the reintroduction of several species into former parts of their range.

Significance to humans

Although most species do not live close to people, raptors have had a strong role in popular culture since prehistoric times, some being worshipped by earlier religions. Their im-age is used to symbolize power, freedom, and agility on the flags or arms of many nations and on many logos in the corporate world. Birds of prey have a special place in the hearts of conservation biologists for their role as an environmental indicator brought to the fore as the flagship for the campaign to ban certain pesticides in the 1960s. Popular with bird-watchers, raptors are a draw worth millions of dollars to many tropical areas.

Some species have always had a close relationship with people. At least 30 species can be found in large cities, and many carrion-feeders make regular use of conurbations. In Elizabethan London, red kites (*Milvus milvus*) scavenged on the streets, while vultures and kites are a regular sight on the refuse tips surrounding Asian and African cities. Their role as nature's cleaners has long been respected by society, especially by the Parsi sect in India.

Others have a more ambivalent relationship, seeing raptors as vermin that threatens their livelihood or sport, be it livestock farming, hunting, or pigeon-racing, though the perception of the damage caused is often much greater than the reality. The hooked bill is enough to warrant the blame. It is, therefore, ironic that in some parts of the world, hunters use raptors in falconry and hawking to hunt small game.

Resources

Books

BirdLife International. *Threatened Birds of the World*. Cambridge: BirdLife International, 2000.

Cade, T. J. *Falcons of the World*. London: Collins, 1982.

Chancellor, R.D., and B.-U. Meyburg. *Raptors at Risk*. Berlin and London: World Working Group on Birds of Prey/Hancock House, 2000.

del Hoyo, J., A. Elliott, and J. Sargatal. *Handbook of the Birds of the World*. Vol. 2, *New World Vultures to Guineafowl*. Barcelona: BirdLife International and Lynx Edicions, 1994.

Ferguson-Lees, J., and D.A. Christie. *Raptors of the World*. London: Christopher Helm, 2001.

Forsman, D. *The Raptors of Europe and the Middle East*. London: T & D Poyser, 1999.

Sibley, C.G., and B.L. Monroe, Jr. *Distribution and Taxonomy of Birds of the World*. New Haven: Yale University Press, 1990.

Snow, D.W., and C.M. Perrins. *The Birds of the Western Palearctic, Concise Edition*. Vol. 1, *Non-Passerines*. Oxford and New York: Oxford University Press, 1998.

Organizations

BirdLife International. Wellbrook Court, Girton Road, Cambridge, Cambridgeshire CB3 0NA United Kingdom. Phone: +44 1 223 277 318. Fax: +44-1-223-277-200. E-mail: birdlife@birdlife.org.uk Web site: <http://www.birdlife.net>

The Hawk and Owl Trust. 11 St Marys Close, Abbotskerswell, Newton Abbot, Devon TQ12 5QF United Kingdom. Phone: +44 (0)1626 334864. Fax: +44 (0)1626 334864. E-mail: hawkandowl@aol.com Web site: <http://www.hawkandowltrust.org>

Raptor Research Foundation. 1752 Robin Hood Road, Mt. Bethel, PA 18343 USA. Phone: (570) 897-6863. E-mail: ednjudy@juno.com Web site: <http://biology.boisestate.edu/raptor/>

World Center for Birds of Prey, The Peregrine Fund. 566 West Flying Hawk Lane, Boise, Idaho 83709 USA. Phone: (208) 362-3716. Fax: (208) 362-2376. E-mail: tpf@peregrinefund.org Web site: <http://www.peregrinefund.org>

Other

USGS Raptor Information System. <http://ris.wr.usgs.gov/>

Julian Hughes

▲

Hawks and eagles

(Accipitridae)

Class Aves
Order Falconiformes
Suborder Accipitres
Family Accipitridae

Thumbnail description
Powerful predators with broad wings, hooked beaks, strong legs and feet, sharp talons, and keen sight; carnivorous hunters and scavengers

Size
7.9–59 in (20–150 cm); 2.6–441 oz
(75–12,500 g)

Number of genera, species
64 genera, 236 species, and 535 taxa (species or subspecies)

Habitat
Most habitats from seacoasts to mountains, deserts to wetlands, woodlands and lush forests, remote wilderness and isolated islands to farmlands, suburbs, and cities

Conservation status
Critical: 8, Endangered: 4, Vulnerable: 22, Not Threatened: 24, Data Deficient: 1

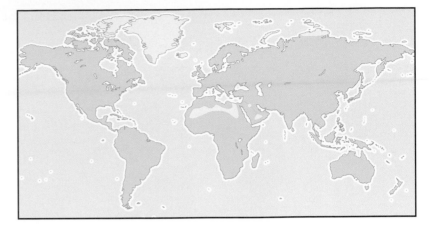

Distribution
Worldwide, except the Antarctic and extreme northern Arctic

Evolution and systematics

There are two diurnal raptor families, Accipitridae (hawks, eagles, and allies) and Falconidae (falcons, caracaras, and allies), which have no obvious relatives among the other birds; it is not even agreed that they are closely related to each other. Their similarities, which include strong, sharp beaks, feet, and talons, and forward-directed eyes for stereoscopic vision, may well be evolutionary convergent adaptations to similar lifestyles rather than indicative of taxonomic affinity. This is almost certainly so for the New World vultures, which are now thought to be allied with the storks.

The accipitrid family can be split into two subfamilies. The subfamily Pandioninae has only one representative, the osprey, whose relationship to the other hawks and eagles remains controversial. Hence it is sometimes placed in its own monospecific family. Its fossil record extends back at least to the Miocene—10–15 million years ago (mya). By the late Miocene, the osprey was already widespread with virtually the same form and distribution as today.

Fossil representatives of the other subfamily, the Accipitrinae (the hawks, eagles, and allies), have been found in Tertiary deposits (30–50 mya). These were buzzard-like raptors that bear no obvious relationship to any living raptor. They first appeared in South America and were widespread by the Miocene. Also widely distributed at this time were the Old World vultures, which no longer occur in the Americas, a clear indication that present day distributions do not neces-

sarily reflect evolutionary origins. By the end of the Miocene (5 mya), when the fossil record improves, many of the modern raptor forms had already appeared.

The number of species and subspecies in taxonomic lists of the family varies depending on the views of the author, and molecular studies have been of limited use in clarifying relationships. Some genera such as *Morphnus* have only one representative (monotypic). At the other extreme, the genus *Accipiter* contains about 50 species (polytypic). Some species have been split into a multitude of subspecies—the aptly named variable goshawk (*Accipiter novaehollandiae*) has about 23—others are monotypic. A good many of these arrangements do not stand close scrutiny, particularly for little known taxa.

For convenience, the accipitrids are often split into groups of like species. These "natural" groups attempt to reflect general evolutionary trends within the family, from the so-called "primitive," less predatory species to the more "advanced," highly predatory forms:

1. The kites lack the bony eye shield ("brow") which gives many of the other accipitrids a fierce expression. The (a) white-tailed kites (*Elanus, Chelictinia, Gampsonyx*) have ungrooved talons, unlike the other accipitrids, and eat mammals and insects. Arguably, the most primitive group of kites are (b) unusual specialist feeders (*Aviceda, Macheiramphus,*

A female northern harrier (*Circus cyaneus*) at the nest feeding its young. (Photo by Edgar T. Jones. Bruce Coleman Inc. Reproduced by permission.)

Pernis, Leptodon, Chondrohierax) that eat caterpillars and mantids (*Aviceda*), bats (*Macheiramphus*), wasp nests (*Pernis, Leptodon*), and arboreal snails (*Chondrohierax*). On the one hand, they link to the (c) Australasian endemic kites, which include *Lophoictinia*, *Hamirostra*, and *Henicopernis*, and, on the other, to (d) the South American forms *Rostrhamus, Ictinia*, and *Harpagus*. In turn these (c and d) are linked to the next group of kites, (e) the typical kites *Milvus* and *Haliastur*, by the fact that that they all have the basal joint of the middle toe fused with the next joint. The (f) fish-eagles (*Haliaeetus, Ichthyophaga*) are basically large typical kites represented by 10 species which replace each other geographically. Several other, large, powerful Australasian/South American species (g) including *Erythrotriorchis, Megatriorchis, Harpia, Harpyopsis*, and *Pithecophaga* may also be offshoots or relict forms of the kite radiation.

2. Old World vultures, 15 species, all scavengers. For example, *Gypaetus, Gyps*, and *Torgus*, whose closest relatives are probably the fish-eagles.

3. Serpent-eagles, made up of 15 species of snake eaters, such as *Circaetus, Terathopius*, and *Spilornis*, possibly also have close links with the kites.

4. Harriers (*Circus*) and harrier-hawks (*Polyboroides*), the latter most closely related to the serpent-eagles.

5. Goshawks, containing about 58 species of "true" hawks, including chanting goshawks (*Melierax*), goshawks, and sparrowhawks (*Accipiter*).

6. Buzzard-like hawks, a grab-bag of species that may not all belong together, including *Parabuteo, Buteogallus, Butastur*, and *Geranoaetus*.

7. Typical buzzards (*Buteo*) with 28 species.

8. Typical eagles containing 33 species, including *Aquila, Hieraaetus*, and *Spizaetus*, which, like *Buteo*, have feathered legs.

Physical characteristics

The familiar characteristics of the birds of prey include the strongly hooked beak and, at is base, the bare, often brightly colored cere in which the nostrils are situated. Features that distinguish the hawks and eagles (accipitrids) from the other raptorial family, the falcons and caracaras, include several skeletal differences, yellow, red, or hazel eyes (vs. brown), well-developed nest-building behavior (vs. absent or poor), and the forceful squirting of excreta (vs. dropping of excreta).

Members of the family range in size from tiny active hunters, the South American pearl kite (*Gampsonyx swainsonii*) and African little sparrowhawk (*Accipiter minullus*), both weighing less than 3.5 oz (100 g) and with wingspans of 21 and 39 in (54 and 39 cm), respectively, to the Himalayan vulture (*Gyps himalayensis*), a hulking scavenger with a wingspan exceeding 9 ft (3 m) and weight reaching to 26 lb (12.5 kg), and the fearsome harpy eagle (*Harpia harpyja*) and Steller's sea-eagle (*Haliaeetus pelagicus*), both reaching 20 lb (9 kg) and the largest of all flying predators.

The hawks and eagles occur in a great variety of forms. The basic types are kites and vultures, hawks and eagles, and within each there are weaker and stronger forms, and mildly predatory and highly predatory species. For example, all the vultures are scavengers but some have immensely robust beaks to tear tough skin and tendons, whereas others have long, lightweight beaks to reach deep into the carcass to nibble tender parts and still others rely on scraps left by large vultures and other predators. The vultures have rather weak feet and stubby, flattish talons that are not used to clutch prey, compared with the eagles with their powerful grasp and dagger-like talons that hold and squeeze the prey. The more powerful eagles that hunt large, difficult prey have deep powerful bills and stout legs, whereas those that eat smaller, more easily captured prey have quite gracile bills and slender legs. The long, double-jointed legs of the harrier-hawks (*Polyboroides*) allow them to reach deep in to tree holes to extract nestling birds.

Form also reflects function in wings and tail shapes. Short-winged, long-tailed hawks are adept at flight through the confines of forest; long, broad-winged, broad-tailed forms are soarers that ride wind currents to great height and cover vast distances effortlessly. Both the African bateleur (*Terathopius ecaudatus*) and Australian black-breasted buzzard (*Hamirostra melanosternon*) have long, broad wings and very little tail, characteristics of hawks that glide low for long distances and have little need of agility. Migrants are longer-winged than non-migrants, particularly those that use active flight to travel across the world. Even within a species, such as the osprey, migratory populations are longer-winged than those that are sedentary.

Vultures and other scavenging raptors that delve into carcasses have parts of their face, head, and even neck bare of feathers, presumably for cleanliness. Most species are rather cryptic shades of gray, brown, or whitish, often streaked or barred ventrally, depending upon their typical habitat. A few hawks have plumage morphs such that two or more color forms occur and interbreed.

In most species the sexes share similar plumage, although the male may be slightly brighter; exceptions include several harriers in which the female is brown and the male gray (dichromatism). Immatures tend to be more brown or heavily marked than adults or, in species with sexual dichromatism, most like the adult female. An interesting feature of the family is that females are larger than males (dimorphism). This is most obviously so in the species that kill relatively large prey that is difficult to catch. Hence the vultures are only slightly dimorphic, whereas in many of the sparrowhawks the female is twice as heavy as the male.

Eurasian sparrowhawk (*Accipiter nisus*) in flight over England. (Photo by Stephen Dalton. Photo Researchers, Inc. Reproduced by permission.)

All raptors have keen eyesight, with particular sensitivity to movement. To help them distinguish their green insect prey from green vegetation, the eyes of bazas (*Aviceda*) have red oil droplets that act like filters. Bazas and other similar kites, which are relatively non-predatory, have quite laterally placed eyes, whereas active pursuers such as accipiters have more forward placed eyes for greater stereoscopic vision. Crepuscular hunters and the few species that are truly nocturnal, such as the letter-winged kite (*Elanus scriptus*), also depend on sight and have relatively large eyes and hunt by moonlight. A few species, including the bat hawk and harriers, are quite reliant on hearing to help them locate concealed prey and have a facial disc of stiff feathers that funnels sounds to their large ear openings. The sense of smell does not seem to be particularly important. Unlike some of the New World vultures, the accipitrid vultures do not have well-developed sense of smell to lead them to carrion.

Distribution

The family has an almost world-wide distribution although only one species occurs in the high Arctic and none in the Antarctic. Some genera, such as *Accipiter*, are extremely widespread, occurring on many islands and all continents except Antarctica. Others have a much more restricted distribution; for example, the great Philippine eagle (*Pithecophaga jefferyi*) is found only on large islands of the Philippines.

Many species with Arctic breeding grounds vacate them after breeding for more benign climates, sometimes flying across the globe, from North to South America or Europe to Africa. Where the climate is stable or moderate all year species tend to be resident. Elsewhere, they may escape the harshest

A lappet-faced vulture (*Torgos tracheliotus*) challenging another near a zebra carcass on the shortgrass plains of the Serengeti, Tanzania. (Photo by Gregory G. Dimijian. Photo Researchers, Inc. Reproduced by permission.)

season: in some species the population simply shifts its range slightly southwards, other species stay put in some regions and all or part of the population migrates from other regions.

Only one species has been successfully introduced to a part of the world where it was not endemic. The Pacific marsh harrier (*Circus approximans*) was taken to Tahiti in about 1885 to control rodents, and self-spread to other islands in the Society Group. Races of several other species have been translocated as part of reintroduction programs for conservation purposes.

Habitat

The hawks, eagles, and their allies are found in most habitats throughout the world. Forests and woodlands support the most species, especially in tropical areas. Poorer, less varied habitats such as tundra, desert steppe, and intensive agriculture support few species. Small or isolated oceanic islands may have one or no species. Even the large islands of New Zealand have only one species, the Pacific marsh harrier.

Many raptors prefer particular habitats, for example, the goshawks and sparrowhawks, genus *Accipiter*, favor forest and woodland, and the harriers (*Circus*) require flat, treeless areas. Other species, such as Swainson's hawk (*Buteo swainsoni*), are more generalized, ranging over many habitats.

Sea coasts from Asia to the arctic support sea-eagles, and they are joined by ospreys (*Pandion haliaetus*) and Brahminy kites (*Haliastur indus*) on warmer Asian-Australasian coasts, and also frequent large inland waterbodies. Several species frequent ecotones, where two or more habitats meet. The

Australian black-breasted buzzard may nest and roost along broad, dry inland watercourse but hunts far out into the surrounding desert and savanna. Sea-eagles require trees or cliffs for nesting but hunt along shoreline and in-shore waters. Some species, such as the kites (*Elanus*) and harriers, can hunt where groundcover is long, others need lower groundcover to hunt successfully. Indeed structure seems to be more important than vegetation composition.

Migrants tend to occupy similar habitats at either end of their migration path. The pallid harrier (*C. macrourus*) moves from breeding grounds in the grassy plains and dry steppes of middle Europe to similar "wintering" habitats in Africa and India. Its congener, the western marsh harrier (*C. aeruginosus*), makes the same trip, but favors reedy wetlands.

Towns and cities with parks, open spaces, and abundant prey can support a number of species including sparrowhawks. Where sanitation is poor and rubbish dumps common a number of species live communally with humans. For example, black kites (*Milvus migrans*), hooded vulture (*Necrosyrtes monachus*), and Indian white-backed vulture (*Gyps bengalensis*) thrive around settlements and cities in parts of Africa and India.

Behavior

Most hawks and eagles are active by day, usually during the period when their prey is most mobile. Some of the largest species are dependent on the heat of the day to create thermals to help them get them airborne and carry them high and far. When resting they perch quietly, often in a sheltered position on a cliff or among foliage. At the perch, they spend considerable time in feather maintenance, keeping their plumage clean, parasite free, and well aligned. Most species have an oil gland at the base of the tail from which they spread oil through the feathers, although a few, such as the *Elanus* kites, have powder down (special feathers that flake into a fine powder that is spread through the plumage to clean it). Many species bathe, sometimes by flying through wet foliage but mostly by wading into water.

The majority of hawks and eagles are solitary but several are colonial and hunt, roost, and breed in numbers. Even some solitary species, such as the steppe eagle (*Aquila nipalensis*), become more gregarious outside the breeding season and forage and roost with other individuals or species where food is plentiful.

Even though they are capable predators, aggressive encounters between individuals seldom progress beyond displays and bluffing. The naked skin on the head of the lappet faced vulture (*Torgos tracheliotus*) "blushes" with emotion. Several species, for example, the long-crested eagle (*Lophaetus occipitalis*) and booted eagle (*Hieraaetus pennatus*), have crests which they raise when threatened.

Hawks and eagles make greatest use of their voices during the breeding season, to defend and advertise territories and in courtship and breeding. For any particular species the range of calls used is usually very limited: often simply repeated whistles, mews, barks, cackles, yelps, or chitters. A few species have far-carrying melodic calls. The whistling kite (*Haliastur sphenurus*) throws its head back to make a single leisurely, descending whistle followed by a staccato series rising in pitch. In contrast, its larger cousin the white-bellied sea-eagle (*Haliaeetus leucogaster*), which also throws its head back, emits a series of goose-like honks that echo across the landscape. The vultures hiss and spit when squabbling but are otherwise silent. In fact, outside the breeding season, most hawks and eagles are seldom vocal.

Especially in the northern hemisphere, species or populations in cooler areas often vacate their breeding grounds to travel to milder climates, where food is more plentiful, in the non-breeding season. Broad-winged species, such as the broad-winged hawk (*Buteo platypterus*), use thermals and updraughts to fly long distances with few stops. Other species, including Eurasian sparrowhawks (*Accipiter nisus*) and pallid harrier (*C. macrourus*), are more active fliers and frequently stop to feed in suitable habitat between breeding and wintering grounds. Where migration routes are channeled at narrow sea crossings, land bridges, or between mountains, migrating raptors may fill the sky. The autumn movement of raptors of several species between North and South America concentrates spectacularly at the isthmus of Panama; 2.6 million have been counted passing through.

Feeding ecology and diet

All the hawks and eagles are carnivorous and most eat only freshly caught prey. Some eat carrion at times or almost exclusively, although rarely truly putrid flesh, and a few eat vegetable and other organic matter.

A bald eagle (*Haliaeetus leucocephalus*) diving for fish in Montana. (Photo by Alan & Sandy Carey. Photo Researchers, Inc. Reproduced by permission.)

Crabs gathered from coastal mangroves are almost the exclusive diet of the crab hawk (*Buteogallus aequinoctialis*), whereas the white-necked hawk (*Leucopternis lacernulata*) appears to specialize on insects, especially those flushed by ants, monkeys, birds, and humans, and only takes a few vertebrates. The bat hawk (*Macheiramphus alcinus*) has a wide gape to swallow bats whole. Wasps and hornets, larvae, pupae, and adults, plucked from the comb, are the favored food of the honey-buzzards (*Pernis apivorus*). Palm nuts are the main food of the palm nut vulture (*Gypohierax angolensis*), although it does eat some invertebrates, fish, crabs, and carrion, and the bearded vulture (*Gypaetus barbatus*) lives on bones left by other scavengers. The osprey (*Pandion haliaetus*) rarely eats anything but fish. The dynamics of *Elanus* kite populations are closely tied to cyclic populations of the rodents on which they are dependent. At the other extreme, the generalist feeder, the red kite (*Milvus milvus*) hunts small animals and eats almost anything organic, alive or dead.

Some of the larger eagles are among the most potent of predators, regularly overpowering prey as large or larger than themselves. The most powerful hunt big, dangerous prey: the South American harpy eagle (*Harpia harpyja*) takes adult monkeys, sloths, porcupines, and the largest of the massive-billed parrots; in Africa, the crowned eagle (*Stephanoaetus coronatus*) weighs 6–8 lb (3–4 kg) but hunts monkeys, small antelope, and other animals up to 40 lb (20 kg) and the 2–3 lb (1–1.5 kg) ornate hawk-eagle (*Spizaetus ornatus*) hunts toucans, macaws, squirrels, and agoutis. The 6–8 lb (3–4 kg) Australian wedge-tailed eagle (*Aquila audax*) eats a range of prey, including medium-sized birds, mammals, reptiles, and carrion, and occasionally hunts cooperatively to exhaust and kill adult kangaroos and dingoes many times its own weight. In contrast, the very similar Verreaux's eagle (*Aquila verreauxii*) of Africa specializes on hyrax, which it takes by surprise from

A square-tailed kite (*Lophoictinia isura*) with chick at their nest. (Photo by Michael Morcombe. Bruce Coleman Inc. Reproduced by permission.)

rock outcrops. Several of the accipiters (goshawks and sparrowhawks) are among the swiftest and most agile of aerial hunters, overtaking birds after a brief pursuit through forest or woodland.

Within species, prey captured varies seasonally, young individuals take easier prey (especially abundant invertebrates) and, during the breeding season, adults tend to take more vertebrates than in other seasons.

Searching and hunting methods vary according to the main prey types and habitats. Vultures and some larger open country eagles soar to great height and search over great distances. The bateleur spends long hours on the wing effortlessly gliding low (about 16 ft [50 m] above ground) to search for, surprise and flush prey from the ground. The harriers quarter open country, flapping slow and low, up and down a field or wetland. Still other species sit and wait at a perch and make short sallies out to pursue passing prey. The *Elanus* kites hover into the slightest breeze and drop onto small mammals below. Especially in the non-breeding season, many species gather opportunistically at termite agates and rodent and locust plagues. Species with the most generalized of diets usually use the most diverse range of hunting techniques. The sea-eagles wade into water after fish, swoop from the air to scoop it from the water, pursue rabbits across the land, scavenge along the shoreline, and frequently harry other predators for their kill.

Some species use assisted hunting, either with their own kind or with other animals, machinery, or fire. Pairs of the great Philippine eagle (*P. jefferyi*) hunt cooperatively: one bird distracts the monkey troop while the other strikes. The plumbeous kite (*Ictinia plumbea*) feeds in association with marmosets as they move through the forest, catching the cicadas they flush. Attracted for miles by the smoke, roadside hawks (*Buteo magnirostris*) gather at fires to catch animals fleeing the flames.

Small hawks and eagles must feed more often than large ones, and species that eat easily captured prey may have to

spend more time gathering or locating prey than those that eat large, difficult prey. Sparrowhawks hunt at least daily, eagles typically hunt every few days, and a large vulture may need to gorge only once a fortnight or so.

Hunters of live prey mostly capture and kill with their feet, and the bill is used in opposition to the feet to dissect prey. The snail eaters have their own characteristic techniques: the hook-billed kite (*Chondrohierax uncinctus*) uses its robust upper mandible to forcefully break open the whorls of the shell, whereas, with their long modified bills, the snail kites (*Rostrhamus*) sever the columellar muscle that attaches the snail to the shell.

Reproductive biology

Most raptors defend a breeding territory from conspecifics and other intruders. This may be the area immediately around the nest or a wider area. Spacing between nests tends to be quite regular, where nest sites allow. Pairs of colonial species, such as Rüppell's vulture (*Gyps rueppellii*) and letter-winged kite, nest on the same cliff or share a tree with other pairs. The larger predatory species space more widely, tens of kilometers apart, closer where food is most available. Territorial activity is usually most vigorous as the breeding season approaches, when boundaries are advertised in some species by spectacular display flights.

Most species are monogamous and only a handful vary from this. Polygamy is known to be common only in three harrier species, including Montagu's harrier (*Circus pygargus*), for which breeding resources (food and nest sites) are concentrated. Experienced males are able to defend and support two or more females, although the primary female and her brood get the larger share of food captured by the male. In parts of their range where food is less available, Harris' hawks (*Parabuteo unicinctus*) breed in cooperative groups, where the core pair are assisted by unrelated helpers and young from the previous breeding attempt.

Typically, in the more predatory species the male feeds the female as part of courtship and continues to supply food during incubation and when the chicks are small; once the nestlings can maintain their own body heat, the female also hunts. In all but the vultures, which share nest duties, the female tears up the food and distributes it among the young.

All accipitrids build a nest of sticks lined with softer material. Nest sites are usually in a commanding position on a cliff or in a tree, but a few species such as the harriers nest on the ground. In some species successive generations return to reuse a traditional site for decades.

Most species breed annually, in the season when food is predictable and abundant, usually spring. The largest eagles attempt to breed every second year. Species that depend on prey that has extremes of population size, such as the plaguing rodents, tend to breed when the opportunity arises, regardless of season, and continue to breed until prey numbers subside.

The eggs are oval, mainly white marked with shades of brown, red, and purplish gray. Larger species tend to lay one or two eggs and smaller species three or more; and clutch sizes tend to be larger in harsher climates. Rodent specialists, such as the rough-legged buzzard (*Buteo lagopus*), show the greatest extremes; they lay very large clutches in years of plenty and have small clutches or do not breed in poor rodent years.

The length of the incubation period ranges from about three-and-a-half weeks in small sparrowhawks to 21 weeks in the harpy eagle. Chicks stay in the nest for a similar period and, once fledged, are dependent for several more weeks as they gradually learn to hunt.

A feature of some species is siblicide, in which the first hatched chick kills its sibling. In a few species, including Verreaux's eagle, this is obligatory and no second chicks survive. In other species it depends on the availability of food and brood size is adjusted to suit the conditions: when food is scarce weak nestlings are killed and eaten so that there is no wastage and the chances of survival of the remaining chick(s) improve.

Conservation status

In the 2000 IUCN world listing 34 species are assessed as Threatened—8 Critically Endangered, 4 Endangered, and 22 Vulnerable. None have become extinct since 1600, but several species were lost from large islands in historical times, following early colonization by humans, among them Haast's eagle (*Harpagornis moorei*), a huge moa-eating eagle of New Zealand.

Not surprisingly, species with small distributional ranges tend to be most vulnerable and these are often on islands. Among the most endangered species, two are in Cuba, where less than 250 individuals of the Cuban kite (*Chondrohierax wilsonii*) and about 300 of Gundlach's hawk (*A. gundlachii*) survive and continue to decline from deforestation and persecution. The Reunion harrier (*C. maillardi*) is under pressure from increasing urbanization, persecution, and poaching but since protection was tightened in the 1970s, public awareness has increased and numbers have increased to 200–340. On continents, species with very small distributions include the white-collared kite (*Leptodon forbesi*) of the humid forests of north-east Brazil, where there has been massive logging; this population is thought to number less than 250 individuals. The Spanish imperial eagle (*Aquila adalberti*), with an estimated total population of 252, is declining from deforestation and persecution, in this case by poisoning at game hunting reserves.

The large island of Madagascar has the greatest number of Threatened species. The Madagascar harrier (*Circus macrosceles*), serpent-eagle (*Eutriorchis astur*), and fish-eagle (*Haliaeetus vociferoides*)—the latter two of which are thought to have populations of less than 250 individuals—are all declining and threatened by habitat loss and degradation. The fish-eagle also suffers from human hunting and persecution. Another two species are Near Threatened: Henst's goshawk (*A. henstii*) and the Madagascar sparrowhawk (*A. madagascariensis*) are confronted by problems from widespread deforestation.

More specialized species suffer more particular threats. The huge vultures—white-rumped (*Gyps bengalensis*), long-billed (*G. indicus*), Cape griffon (*G. coprotheres*), and lappet-faced—fall victim to poison left in carcasses for control of other predators and, at the same time, find fewer carcasses and refuse on which to feed, in part because of competition from humans. Human pressure—from such threats as overfishing, coastal development, and hydroelectric schemes—is a common problem to another three sea-eagles Sanford's, Steller's (*H. pelagicus*), and Pallas's (*H. leucoryphus*). Yet, human caused habitat loss to development and natural resource harvesting threatens by far the greatest number of species.

Remedial action has re-established or stabilized some species. The organochlorine pesticides (including DDT, which causes raptors to lay thin-shelled eggs, and dieldrin, which causes direct mortality) caused massive population decreases in species such as the Eurasian sparrowhawk (*Accipiter nisus*) in the 1960s and 1970s; populations began to recover when until the chemicals were banned in developed countries. Many countries give raptors full legal protection, which has lessened persecution and disturbance. Some populations have been assisted by hands-on conservation efforts. The white-tailed eagle (*H. albicilla*), which last bred in the British Isles in 1908, has been successfully reintroduced to Scotland. Northern goshawk (*A. gentilis*) and red kite, absent from Britain for more than a century, have also been re-established, the former aided by escaped falconers birds. By protection, provision of food, and careful attention to their sociable nature, the Eurasian griffon (*Gyps fulvus*) has been returned to the mountains of south-central France. Nest protection, erection of artificial nest sites, and breeding manipulation by double-clutching and egg and nestling translocation has benefited the recovery of species such as the bald eagle (*Haliaeetus leucocephalus*) in the United States and the osprey in Britain. A handful of species has adapted to life in cities, croplands, and plantations, but for the majority of accipitridae, particularly those that are large or highly specialized, the long-term prognosis is poor.

Significance to humans

Birds of prey have long been admired for their hunting prowess and powers of flight and sight. Paradoxically, they are also despised for their depredations on livestock and perceived cruelty; their fortunes fluctuating with the times.

In ancient Egypt, the Eurasian griffon was worshipped, appearing in the crown of the goddess of childbirth Nekhebet, and a vulture headdress was the privilege of a queen. On victory steles, vultures carried away bodies of vanquished. In some parts of the world supernatural hawks took on a part-human form or were fearful messengers of more human-like gods. Greek legend has Zeus's eagle stealing the beautiful Ganymede to make him cupbearer to the gods of Olympus. Garuda, the steed of the Hindu god Vishnu, has the red wings and white face of the Brahminy kite; its image guards the sacred temples at Angkor and elsewhere. The thunderbird of many tribes from the Americas to the Cook Islands often took the form of a terrifying, eagle-like bird that brought rain or created the world. From Tibet to the Nile Valley mythical eagles were worshipped as harbingers of the wind, stars, sun, and rain and protectors of the gods and their earthy representatives, including the Inca emperors. In early Christianity the eagle signified escape and fulfillment. Because of such associations, they were often an essential part of the trappings of nobility. The German imperial two-headed eagle of medieval heraldry signified the power of emperors. Even today images of eagles appear on coats of arms and company logos, symbols of strength and reliability.

One of the closest relationships between humans and raptors is falconry or hawking. As early as 2000 B.C. in Asia, humans were hunting with trained hawks. The practice flourished in Europe and the middle East from A.D. 500 to 1600 and was practical, providing fresh meat for the table, as well as recreational. In these feudal societies there was often strict hierarchy: the upper classes were allowed the more prestigious species including the larger falcons and eagles and the middle classes less desirable birds such as Eurasian sparrowhawk and northern goshawk. There are still devotees of the sport, particularly among Arabian royalty, and the custom continues in some central Asian tribes, where golden (*Aquila chrysaetos*) and imperial eagles (*A. heliaca*) are used to hunt wolves, foxes, and gazelles from horseback. However, in many developed countries falconry is not allowed, mainly for conservation and ethical (animal rights) reasons. Where it is legal, there are restrictions on the keeping of birds and taking from the wild and many are now bred in captivity.

The modern era brought a lessening of superstitions and with the widespread introduction of firearms hawks were no longer a useful means of hunting. Fear and loathing replaced the general reverence for raptors and they were persecuted in their millions, sometimes fueled by government sponsored bounties. Although a few larger species occasionally prey on young livestock such as cattle, sheep, reindeer, or poultry, their impact on healthy herds is invariably exaggerated. Even harmless species, tarred with the same brush, have suffered. Today, livestock is better managed, and as many raptor populations dwindle there is increasing concern for their conservation and appreciation of their beauty and role in nature. Large-scale egg and specimen collecting, popular for much of the 1900s and which had a local impact on thinly scattered, already beleaguered raptor populations, is no longer fashionable. Nevertheless, many hawks are still are trapped, shot, or poisoned to protect livestock and thousands are destroyed (sometimes for food or medicine) where they gather in numbers on migration through Europe, China, and elsewhere. Body parts of the critically endangered Madagascar fish-eagle (*Haliaeetus vociferoides*) continue to be used in traditional medicines. Conversely, in the Solomon Islands, breakdown of traditional taboos now allows hunting that threatens Sanford's sea-eagle (*H. sanfordi*).

1. African little sparrowhawk (*Accipiter minullus*); 2. Lappet-faced vulture (*Torgos tracheliotus*); 3. Gurney's eagle (*Aquila gurneyi*); 4. Harris' hawk (*Parabuteo unicinctus*); 5. Andaman serpent-eagle (*Spilornis elgini*). (Illustration by Barbara Duperron)

1. Rough-legged buzzard (*Buteo lagopus*); 2. Northern goshawk (*Accipiter gentilis*); 3. Harpy eagle (*Harpia harpyja*); 4. Hen harrier (*Circus cyaneus*). (Illustration by Barbara Duperron)

1. Long-tailed honey-buzzard (*Henicopernis longicauda*); 2. Hook-billed kite (*Chondrohierax uncinatus*); 3. White-rumped vulture (*Gyps bengalensis*); 4. Madagascar cuckoo-hawk (*Aviceda madagascariensis*); 5. Letter-winged kite (*Elanus scriptus*). (Illustration by Barbara Duperron)

1. Black-breasted buzzard (*Hamirostra melanosternon*); 2. Black kite (*Milvus migrans*); 3. Osprey (*Pandion haliaetus*); 4. Steller's sea-eagle (*Haliaeetus pelagicus*); 5. Egyptian vulture (*Neophron percnopterus*). (Illustration by Barbara Duperron)

Species accounts

Osprey

Pandion haliaetus

SUBFAMILY
Pandioninae

TAXONOMY
Falco haliaetus Linnaeus, 1758, Sweden. Four subspecies.

OTHER COMMON NAMES
English: Fish hawk; French: Balbuzard pêcheur; German: Fischadler; Spanish: Aguila Pescadora.

PHYSICAL CHARACTERISTICS
21.7–22.8 in (55–58 cm); male 2.6–3.5 lb (1.2–1.6 kg), female 3.5–4.4 lb (1.6–2 kg). Brown upperparts with white legs and chest, accented with speckled necklace.

DISTRIBUTION
P.h. haliaetus: Scandinavia to Japan, the Mediterranean, Red Sea, and Cape Verde Islands; wintering in South Africa, India, Indonesia, and the Philippines. *P.h. carolensis*: Labrador to Alaska to Florida and Arizona, wintering in Peru and South Brazil. *P.h. ridgwayi*: Caribbean. *P.h. cristatus*: Australia to New Caledonia to New Guinea, Java, and Sulawesi.

HABITAT
Low altitude inland and shallow marine waters, including marshes, lakes, reservoirs, bays, sea coasts and islands, estuaries, and, less often, rivers. Almost exclusively coastal and subcoastal in Australasia and much of Asia.

BEHAVIOR
Solitary, pairs, or family groups, occasionally larger groups; northern populations, where winter sends fish to deeper water, are migratory, southern populations are sedentary.

FEEDING ECOLOGY AND DIET
Feeds almost exclusively on live fish, rarely turtles and seabirds, and fish found dead or dying.

REPRODUCTIVE BIOLOGY
Breeds in solitary pairs (e.g., Australia and Britain) or loose colonies (e.g., Mediterranean and United States), sometimes of hundreds of birds; usually monogamous but polygynous trios found; large stick nest lined with flotsam, seaweed, dead grass or leaves, near water, on islet, sea-cliff, mangrove or other tree, man-made structure, or on ground on predator-free island. Annual breeding season, usually starting winter-spring (into summer in the north). Usual clutch is three eggs; incubation about five weeks; fledge at about seven weeks; fledglings remain with adults two to eight weeks until migration in northern populations, longer in resident populations.

CONSERVATION STATUS
Not threatened. Generally common and locally abundant throughout much of range.

SIGNIFICANCE TO HUMANS
Occasionally regarded as a competitor for fish and can be a nuisance at inland fisheries/hatcheries and when nesting on powerpoles (shorting-out electricity). ◆

Madagascar cuckoo-hawk

Aviceda madagascariensis

SUBFAMILY
Accipitrinae

TAXONOMY
Pernis madagascariensis A. Smith, 1834, Madagascar. Monotypic.

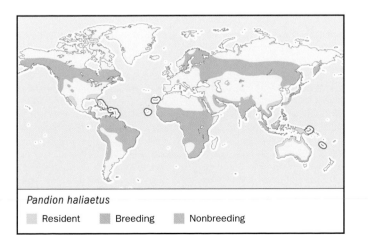

Pandion haliaetus

　▢ Resident 　▣ Breeding 　▦ Nonbreeding

Aviceda madagascariensis

　▢ Resident

OTHER COMMON NAMES
English: Madagascar baza, Madagascar cuckoo-falcon; French: Baza malgache; German: Lemurenweih; Spanish: Baza Malgache.

PHYSICAL CHARACTERISTICS
15.7–17.7 in (40–45 cm). Dull brown wings, barred tail, mottled brown and white underparts.

DISTRIBUTION
Much of Madagascar.

HABITAT
Evergreen and dry deciduous forest interior and edge; clearings in forest, villages within forest and palm plantations.

BEHAVIOR
Poorly known. Apparently non-migratory. By night, roosts in the canopy. Hunts by day and, perhaps, crepuscularly.

FEEDING ECOLOGY AND DIET
Main prey is large insects and small reptiles and frogs snatched from foliage. Perches in canopy to glide down onto prey; sometimes flies low over canopy in search of prey or hawks aerial insects.

REPRODUCTIVE BIOLOGY
Little known. Distinctive rocking with wings held high and tumbling courtship flight. Breeds in solitary pairs, laying in October to December. Builds small, flimsy nest lined with green leaves, high in the canopy. Clutch size unknown, probably two to three eggs. Incubation probably about 32 days and fledging about five weeks as in other Bazas.

CONSERVATION STATUS
Not threatened. Fairly common in forested areas but deforestation an increasing threat.

SIGNIFICANCE TO HUMANS
None known. ◆

Chondrohierax uncinatus
▨ Resident

Hook-billed kite

Chondrohierax uncinatus

SUBFAMILY
Accipitrinae

TAXONOMY
Falco uncinatus Temminck, 1822, Brazil. Three subspecies.

OTHER COMMON NAMES
French: Milan bec-en-croc; German: Langschnabelweih; Spanish: Milano Picogarfio.

PHYSICAL CHARACTERISTICS
15–16.5 in (38–42 cm); male about 8.8 oz (250 g); female 9–12.7 oz (255–360 g). Large hooked bill with green and yellow cere. Extreme variation in plumage, with males typically bluish gray.

DISTRIBUTION
C.c. uncinatus: western Mexico and extreme southern United States, southwards to northern Argentina. *C.c. wilsonii*: eastern Cuba.

HABITAT
Lower canopy and dense understorey of rainforest, seasonally flooded forest and montane tall forest. Also low forest on Grenada and acacia thorn-scrub in Mexico, forest edge and clearings.

BEHAVIOR
Apparently sedentary. Unobtrusive, most often seen as it soars over forest.

FEEDING ECOLOGY AND DIET
Feeds mainly on tree snails in the understorey. Occasionally takes lizards, frogs, salamanders, freshwater crabs, slugs, and insects. Hops about in the canopy or glides down from a perch to snatch prey.

REPRODUCTIVE BIOLOGY
Monogamous. Builds a rather small, flimsy nest of sticks high, and often precariously, in the canopy. Lays one or two eggs in late dry season. Chicks fledge in the rainy season to take advantage of the plentiful tree snails.

CONSERVATION STATUS
Not yet considered globally threatened. Continental subspecies *C. u. uncinatus* is widespread and generally uncommon. Cuban subspecies (which has yellow bill), now confined to eastern Cuba, is Critically Endangered and on the verge of extinction due mainly to habitat destruction by logging; some persecution because of mistaken belief that it preys on poultry; harvesting of snails has also depleted its prey. Grenadan subspecies *C. u. mirus* is also Endangered because of habitat loss and introduced snails, thought to be too large for the kite to prey on, which feed on the native snail. Recommendations for conservation action include protection by law, protection of remaining habitat, public awareness campaigns to reduce persecution and protection of snails on which the species preys.

SIGNIFICANCE TO HUMANS
None known. ◆

Long-tailed buzzard
Henicopernis longicauda

SUBFAMILY
Accipitrinae

TAXONOMY
Falco longicauda Garnot, 1828, New Guinea. Monotypic.

OTHER COMMON NAMES
English: Long-tailed honey-buzzard; French: Bondrée à longue queue; German: Langschwanzweih; Spanish: Abejero Colilargo.

PHYSICAL CHARACTERISTICS
19.7–23.6 in (50–60 cm); male 15.9 oz (450–630 g), female 20.1–25.7 oz (570–730 g). Mottled brown and honey colored upperparts with barred tail, and neck and chest streaked with white.

DISTRIBUTION
New Guinea and western Papuan and Aru islands.

HABITAT
Tropical rainforest and forest edge from lowlands to mid-mountain (c. 9,200 ft [2800 m]).

BEHAVIOR
Usually seen singly, in pairs or trios. Thought to be sedentary.

FEEDING ECOLOGY AND DIET
Hunts by day and at dusk. Preys on wasps and their larvae, ants, grasshoppers, mantids, and other invertebrates, small birds and their eggs and nestlings, and small lizards.

REPRODUCTIVE BIOLOGY
In display, the pair wheel over the forest; as they pass, one bird rolls on back to present talons to other. Monogamous. Builds a stick nest high in a tree, less often on a cliff ledge. Laying recorded May and August. Little else known.

CONSERVATION STATUS
Not threatened. Quite common and widespread although deforestation and hunting for traditional uses have caused it to become scarce in some areas. New Britain honey-buzzard is poorly known and may be declining due largely to clearing to establish oil palm plantations.

SIGNIFICANCE TO HUMANS
The buzzard's flight (wing and tail) feathers feature in ceremonial headdresses of some New Guinea tribes. ◆

Black-breasted buzzard
Hamirostra melanosternon

SUBFAMILY
Accipitrinae

TAXONOMY
Buteo melanosternon Gould, 1841, inland New South Wales. Monotypic.

OTHER COMMON NAMES
English: Black-breasted kite, black-breasted buzzard-kite; French: Milan à plastron; German: Schwarzbrustmilan; Spanish: Milano Pechinegro.

PHYSICAL CHARACTERISTICS
19.7–23.6 in (50–60 cm); male c. 1.3 lb (1.3 kg); female c. 3.3 (1.5 kg). Heavy build. Short legs with large white feet. Black and brown body with white accents.

DISTRIBUTION
Mainly arid central and tropical northern Australia.

Henicopernis longicauda

▢ Resident   ▨ Nonbreeding

Hamirostra melanosternon

▢ Resident   ▨ Nonbreeding

HABITAT

Arid deserts, grasslands and plains, especially along wooded creek lines, and tropical woodlands, grasslands, and savannas.

BEHAVIOR

Usually solitary or in pairs or family groups but gathers in small numbers (up to nine recorded) at large carcasses. Movement by part of the population northward in winter for the dry season, but many birds resident year round. In summer, escapes tropical coasts (wet season) and hottest deserts.

FEEDING ECOLOGY AND DIET

Spends much time on wing in search of prey, soaring, gliding, or low quartering; also walks across the ground in search of prey. Main prey is medium-sized mammals (such as young rabbits), birds, large lizards, and nestlings of other birds including raptors. One of the few raptors to use a tool (see also Egyptian vulture): mainly, picks up a rock in the beak and hurls it at large eggs (e.g., Emu eggs) to gain access to their contents. Also breaks egg directly with bill and by throwing egg itself. Scavenges carrion and occasionally catches snakes and large insects. Not a powerful predator like the superficially similar Wedge-tailed eagle (*Aquila audax*).

REPRODUCTIVE BIOLOGY

Usually nests as solitary pair but one polyandrous trio recorded. Builds a large platform of sticks lined with leaves in a living or dead tree in the open. Typically, lays a clutch of two eggs in August–October. Incubation about 36 days; chicks fledge after about seven or eight weeks.

CONSERVATION STATUS

Not threatened. Widely but thinly distributed across range and generally uncommon. Declined in south-east of range due to habitat degradation and loss of prey species.

SIGNIFICANCE TO HUMANS

Susceptible to poisoning when scavenging on baited carcasses, but harmless to stock. Traditionally hunted by some aboriginal tribes and feathers used in hairbelts and other decorative products, but custom largely lapsed. ◆

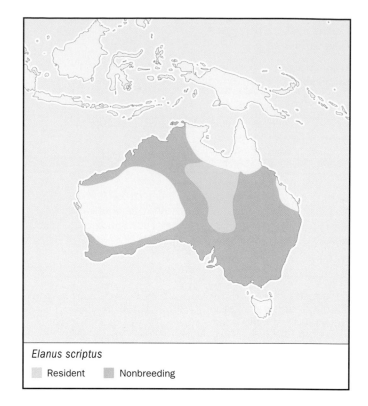

Elanus scriptus
■ Resident   ■ Nonbreeding

Letter-winged kite

Elanus scriptus

SUBFAMILY

Accipitrinae

TAXONOMY

Elanus scriptus Gould, 1842, Cooper Creek, South Australia.

OTHER COMMON NAMES

French: Élanion lettré; German: Schwarzachselaar; Spanish: Elanio Escrito.

PHYSICAL CHARACTERISTICS

13.4–14.6 in (34–37 cm); male about 9.2 oz (260 g); female 11.3 oz (320 g). Distinctive black band around eyes.

DISTRIBUTION

Mainly Central Australia.

HABITAT

Arid and semi-arid grasslands and tree-lined watercourses. Following irruptions may reach more coastal grasslands and open woodlands.

BEHAVIOR

One of the few truly nocturnal accipitrids. Roosts by day in leafy trees, sometimes in colonies of hundreds when not breeding. Follows cycles of main rodent prey, especially long-haired rats (*Rattus villosissimus*), which plague irregularly every five to 10 years following good rains that fill inland waterways. Breeds when rats abundant, then disperses widely, often reaching coastal areas as rat numbers wane, then usually perish. Presumably a core of adults remains inland to repopulate when conditions allow.

FEEDING ECOLOGY AND DIET

A rodent specialist, mostly long-tailed rat, but also takes other small mammals and lizards, and large insects. Usually hunts by night when main prey active; quarters the ground, hovers, and drops vertically onto prey.

REPRODUCTIVE BIOLOGY

Typically, breeds in loose colonies in coolibahs along inland (arid zone) watercourses whenever food is abundant. Monogamous. Builds a nest of small sticks lined with leaves or dung. Egg-laying mostly in late-winter to spring and autumn. Clutch size is usually four or five incubation about 31 days; nestlings fledge at about five weeks.

CONSERVATION STATUS

Not threatened. Generally rare and rather mysterious due to poor knowledge of movements (here today, gone tomorrow habits) and boom and bust breeding strategy. Some threat from overgrazing of already fragile landscape and breeding colonies sometimes invaded by feral cats.

SIGNIFICANCE TO HUMANS

None known. ◆

Black kite
Milvus migrans

SUBFAMILY
Accipitrinae

TAXONOMY
Falco migrans Boddaert, 1783, France. Seven subspecies.

OTHER COMMON NAMES
French: Milan noir; German: Schwarzmilan; Spanish: Milano negro.

PHYSICAL CHARACTERISTICS
21.7–23.6 in (55–60 cm); 19.8–33.5 oz (560–950 g) (measurements varies with race) female larger and heavier than male. Mostly reddish brown. Plumage and bill color vary with race.

DISTRIBUTION
M.m. migrans: northwest Africa, Europe to central Asia and south to Pakistan; winters in Africa, south of the Sahara. *M.m. lineatus*: Siberia to Amurland, Japan, India, Burma, and China; winters in Iraq, India, and southeastern Asia. *M.m. formosanus*: Taiwan and Hainan, China. *M.m. govinda*: Pakistan to India, Sri Lanka, Indo-China, and the Malay Peninsula. *M.m. affinus*: Sulawesi to New Guinea, New Britain, and Australia. *M.m. aegyptius*: Egypt, Arabia, coastal eastern Africa to Kenya. *M.m. parasitus*: Africa south of Sahara to Madagascar.

HABITAT
Desert to grassland, savanna and woodland, but avoids dense forests. Often near wetlands and found in suburbs and towns, around rubbish tips, abattoirs.

BEHAVIOR
Migratory or partly so, particularly in Europe and Asia, from which it migrates after breeding to sub-Saharan Africa, the Middle East, southeastern Asia and Indian subcontinent. Migrates in flocks and gathers to cross sea straits in tens of thousands. Elsewhere, such as Australia, New Guinea, and Egypt, some populations resident, movements less regular or nomadic. Gregarious, often in forages in large flocks, sometimes roosts communally (in trees), and may breed in very loose colonies.

FEEDING ECOLOGY AND DIET
Feeds on a wide variety of prey, live or dead, and scraps. Offal, garbage, excrement, fish, invertebrates, some vegetable matter such as oil palm nuts. Steals from other raptors and waterbirds. Also catches small mammals, birds, reptiles, and amphibians snatched from the ground, foliage, or water.

REPRODUCTIVE BIOLOGY
Often returns to traditional nest sites on return from migration. Monogamous. Nests as solitary pair or in loose colonies of tens of pairs. Usually builds a stick nest in a tree, less often on a cliff, lined with rubbish such as rags, dug, and fur. Timing depends on region, usually the dry season. Clutch size two or three eggs. Incubation about 31 days; chicks fledge after six or seven weeks.

CONSERVATION STATUS
Not threatened. Nevertheless, sometimes poisoned or shot because it steals young poultry, feeds on stock carcasses.

SIGNIFICANCE TO HUMANS
Traditionally trapped by several indigenous people for food and decoration and feature in their legends. For example, thought to spread fire by some Australian aboriginal tribes, presumably because of their habit of travelling from far and wide to congregate at fires and swooping at prey among the flames. ◆

Steller's sea-eagle
Haliaeetus pelagicus

SUBFAMILY
Accipitrinae

TAXONOMY
Aquila pelagicus Pallas, 1811, islands in the sea of Okhotsk.

Haliaeetus pelagicus
■ Resident   ■ Breeding   ■ Nonbreeding

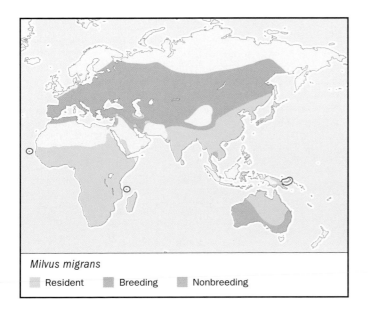

Milvus migrans
■ Resident   ■ Breeding   ■ Nonbreeding

OTHER COMMON NAMES
English: White-shouldered sea-eagle; French: Pygargue empereur; German: Riesenseeadler; Spanish: Pigargo Gigante.

PHYSICAL CHARACTERISTICS
33.5–94 in (85–94 cm); 10.8–19.8 lb (4.9–9 kg); female larger and heavier than male. Blackish brown all over, except white tail and shoulders. Morph "niger," found in Korea, is all black.

DISTRIBUTION
Coastal west Bering Sea and Sea of Okhotsok, wintering further south as far as Korea. Breeds mainly Kamchatka Peninsula, Sea of Okhotsk the lower reaches of the Amur River, and on northern Sakhalin and Shatar, Russia.

HABITAT
Coast and lower reaches of rivers, less often inland along rivers and lakes where fish are abundant. Most often in forested river valleys which provide trees for nesting.

BEHAVIOR
Shift in population southward for the winter. Some stay at Kamchatka and on the Okhotsk coast; most winter in Japan, reaching north-east China, North and South Korea.

FEEDING ECOLOGY AND DIET
Mostly large fish, alive or dead, especially Pacific salmon, but will catch a variety of other prey and scavenge.

REPRODUCTIVE BIOLOGY
Monogamous. Mostly nests in large trees, but also sea cliffs and cliffs far inland near lakes and larger rivers. Lays in April–May in a large stick nest. Clutch size usually two; incubation about seven weeks, fledging about 10 weeks.

CONSERVATION STATUS
Vulnerable. Total world population is estimated at 5,000 birds and declining. Main threats are felling of old forest and building of hydroelectric plants, over-fishing and lead-poisoning from shot in deer carcasses left by hunters. Recommendations for alleviation of threats include minimizing the impact of industrial development in Russia, establishing artificial feeding sites, encouraging sustainable management of fishing stocks and protection of salmon spawning grounds.

SIGNIFICANCE TO HUMANS
None known. ◆

Egyptian vulture
Neophron percnopterus

SUBFAMILY
Accipitrinae

TAXONOMY
Vultur perenopterus [sic] Linnaeus, 1758, Egypt. Two subspecies.

OTHER COMMON NAMES
English: Scavenger vulture; French: Vautour percnoptère; German: Schmutzgeier; Spanish: Alimoche Común.

PHYSICAL CHARACTERISTICS
22.8–27.6 in (58–70 cm); 3.5–4.9 lb (1.6–2.2 kg). Distinctive contrasting coloration between white head and body and black flight feathers.

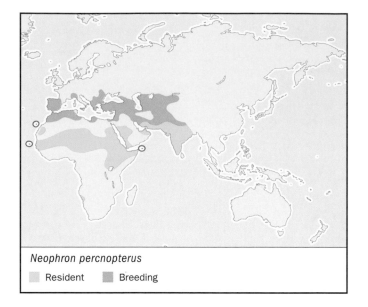

Neophron percnopterus
☐ Resident ■ Breeding

DISTRIBUTION
N.p. percnopterus: Europe to central Asia and northwest India, south to Tanzania, Angola, and Namibia; also Canary and Cape Verde Islands and Socotra. *N.p. ginginianus*: India and Nepal.

HABITAT
Frequents extensive open country of dry, arid regions: steppe, scrub, desert, pastures, and cereal crops. Also in flat mountainous areas usually at low to moderate altitudes, cities and towns (especially Africa and India). Nests in rocky areas.

BEHAVIOR
Usually solitary or in pairs but a hundred or more may congregate where food is abundant and at roosts on cliffs, trees or on buildings. In north of range migrate to Africa just south of Sahara and north of the equator. In India, Arabia, sub-Saharan Africa, Balearic and Canary Islands apparently sedentary or make local movements.

FEEDING ECOLOGY AND DIET
Opportunistic feeder, dependent on rubbish dumps and carcass disposal sites; carrion and refuse is main food. Less often, catches live prey, usually sick or otherwise vulnerable. Also insects, crustaceans lifted from the water and birds' eggs; large eggs broken by throwing a stone.

REPRODUCTIVE BIOLOGY
Usually, breeds as solitary pair but occasionally two nests in close proximity. Monogamous. Builds a substantial, untidy nest of sticks lined with wool, rags and hair in a cleft, cave or narrow ledge at height on a cliff, often overhung; also on ruins, date palms and other trees where no cliffs. Typically, lays two eggs in March–May (earlier in some areas); incubation 42 days; fledges at about 11 weeks. Unlike most raptors, regurgitates food for chicks.

CONSERVATION STATUS
Not threatened. Population has undergone a general decline but may now be stable. Main European population is now Spain; main population is Ethiopia. Fewer carcasses, reductions in small prey species, poisoning and persecution all thought to be factors in decline.

SIGNIFICANCE TO HUMANS
Its image was carved into Egyptian monuments but apparently the species was never worshipped, as was the more powerful Eurasian griffon (*Gyps fulvus*). ◆

White-rumped vulture
Gyps bengalensis

SUBFAMILY
Accipitrinae

TAXONOMY
Vultur bengalensis Gmelin, 1788, Bengal. Monotypic.

OTHER COMMON NAMES
English: Indian white-backed vulture, white-backed vulture;
French: Vautour chaugoun; German: Bengalengeier; Spanish:
Buitre Dorsiblanco Bengalí.

PHYSICAL CHARACTERISTICS
29.5–33.5 in (75–85 cm); 7.7–13 lb (3.5–6 kg). Blackish bird,
distinguished by white lower back and underwing coverts.

DISTRIBUTION
From south-east Iran to Pakistan, through India to south-
central China, Indochina, and the northern Malay Peninsula.

HABITAT
Mainly open plains near villages, towns, and parks. Also into
hilly woodlands of Himalayan foothills to 4,900 ft (1,500 m).

BEHAVIOR
Apparently sedentary. A social species, usually found in non-
specific flocks. Also roosts in large flocks in trees.

FEEDING ECOLOGY AND DIET
Feeds on carrion, largely dead livestock, and human remains.
Gorges then rests for an extended period on ground or in tree
while heavy load of food is digested.

REPRODUCTIVE BIOLOGY
Breeds in small colonies, often in tall trees near human habitation, along canals or streams. Monogamous. Builds a large nest
of sticks. Lays a single egg clutch in about October-November.
Incubation 45 days and fledging after about three months.

CONSERVATION STATUS
Critically Endangered. Previously widespread and abundant
across its distributional range. East of India the species has
been all but extinct since the early 1900s probably due to the
rarity of wild large mammals and consumption of dead livestock by humans. Now rare in China and remaining strong-
holds are Pakistan and India. However, recently (2000)
upgraded to Critically Endangered because of rapid population
decline: in mid-2000, across Nepal, Pakistan and India, large
numbers of *Gyps* vultures were found dead and dying. The
cause is unknown but may have been viral. Other threats include poisoning, pesticides, and changes in processing of dead
livestock and other waste.

SIGNIFICANCE TO HUMANS
Traditionally, the Parsee of India dispose of their dead by leaving bodies on special towers so that the vultures can carry the
remains heavenward. The vultures' habit of roosting habitually
in large flocks at the same site can kill trees through accumulation of excrement and can be a problem in coconut plantations
and mango groves. ◆

Lappet-faced vulture
Torgos tracheliotus

SUBFAMILY
Accipitrinae

TAXONOMY
Vultur tracheliotus J. R. Forster, 1791, South Africa. Three subspecies.

OTHER COMMON NAMES
English: African black vulture, African king vulture, Nubian
vulture; French: Vautour oricou; German: Ohrengeier; Spanish: Buitre Orejudo.

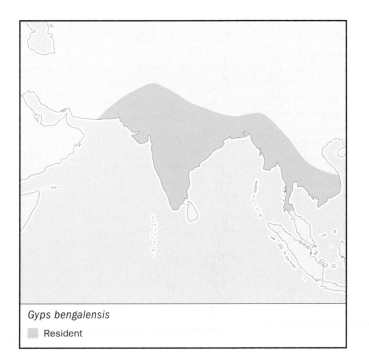

Gyps bengalensis
▓ Resident

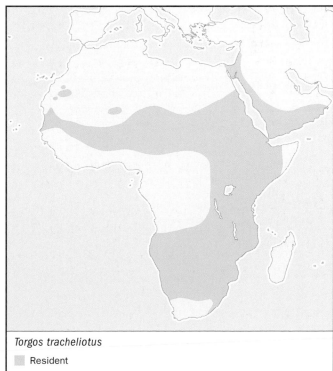

Torgos tracheliotus
▓ Resident

PHYSICAL CHARACTERISTICS
45.3 in (115 cm); 11.9–20.7 lb (5.4–9.4 kg). Very large bird, with bald pinkish head and lappet, wings dark brown and chest white with brown accents.

DISTRIBUTION
T.t. tracheliotus: southwest to Morocco, southern Mauritania to Ethiopia, Kenya, and South Africa. *T.t. nubicus*: Egypt and northern Sudan. *T.t. negevensis*: Israel and Arabian peninsula.

HABITAT
Semi-arid areas and desert with scattered trees and short grass. Occasionally into mesic open savanna and grassland.

BEHAVIOR
No regular migration known but some local movement to avoid the rainy season. Sociable, congregates at carcasses (up to 50 recorded in company of other vultures) but often in pairs.

FEEDING ECOLOGY AND DIET
Mainly a scavenger, feeds on carrion, skin, and bone fragments from large carcasses. Dominant to other vultures when hungry, aggressively bounding at them, but often socializes around carcass before feeding.

REPRODUCTIVE BIOLOGY
Monogamous. Nests as solitary pair in flat-topped thorny trees. Builds a large platform of sticks lined with grass. Lays a single egg in the dry season, beginning about October–December, depending on region. Incubation about 55 days; fledging at about four months.

CONSERVATION STATUS
Vulnerable. Formerly thinly scattered throughout wide range. In 2000 only a small, declining population remained, estimated at about 8,500 individuals. Accidental poisoning from baits left by farmers for predators and persecution in the mistaken belief that the vulture preys on livestock are problems. Increasing numbers of recreational off-road vehicles may also be a threat because of the species' sensitivity to nest disturbance.

SIGNIFICANCE TO HUMANS
None known. ◆

Andaman serpent-eagle
Spilornis elgini

SUBFAMILY
Accipitrinae

TAXONOMY
Haematornis elgini Blyth, 1863, South Andaman Island. Monotypic.

OTHER COMMON NAMES
English: Andaman dark serpent eagle; French: Serpentaire des Andaman; German: Andamanenschlangenweihe; Spanish: Culebrera de Andamán.

PHYSICAL CHARACTERISTICS
19.3–21.3 in (49–54 cm); 27.9–35.3 oz (790–1,000 g). Plumage mainly dark brown with small white spots.

DISTRIBUTION
Andaman Islands.

HABITAT
Mainly forests and forest clearings of inland, occasionally on hillsides with scattered trees.

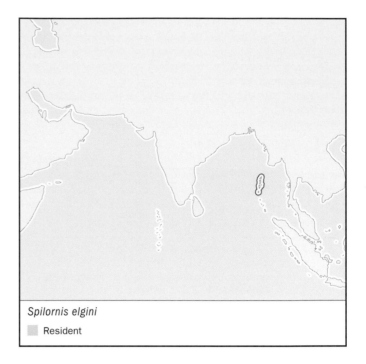

Spilornis elgini
▮ Resident

BEHAVIOR
Sedentary.

FEEDING ECOLOGY AND DIET
Not well known. Takes a variety of prey, including birds, frogs, lizards, snakes, and rats; perhaps catches mainly reptiles, as do other serpent-eagles.

REPRODUCTIVE BIOLOGY
Mutual soaring and calling over territory. No other information. Perhaps a small clutch, of one egg, as *S. cheela*.

CONSERVATION STATUS
Near Threatened. Most numerous raptor in the Andaman Islands but listed as rare or Near Threatened because of very small distributional range and anticipated increasing threats. Hunting is common and may also be a problem for the eagle.

SIGNIFICANCE TO HUMANS
None known. ◆

Hen harrier
Circus cyaneus

SUBFAMILY
Accipitrinae

TAXONOMY
Falco cyaneus Linnaeus, 1766, Europe. Two subspecies.

OTHER COMMON NAMES
English: Northern harrier, marsh harrier; French: Busard Saint-Martin; German: Kornweihe; Spanish: Aguilucho Pálido.

PHYSICAL CHARACTERISTICS
16.9–20.5 in (43–52 cm); male approx. 12.3 oz (350 g); female 18.7 oz (530 g). Pale gray upperparts, with blackish gray band on secondary feathers.

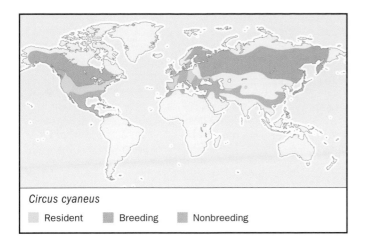

Circus cyaneus

☐ Resident ☐ Breeding ☐ Nonbreeding

DISTRIBUTION
C.c. cyaneus: Europe and northern Asia to Kamchatka, wintering from Europe to northern Africa, southern Asia, southeastern China, and Japan. *C.c. hudsonius*: North America, wintering as far south as northern South America.

HABITAT
Open country with grasses, shrubs, or young trees, grassland, steppe, swamps and other wetlands, young plantations, croplands, and meadows.

BEHAVIOR
Sits tall and slender, often on the ground, but also posts, rocks, or trees. Flaps low, on upswept wings, over open country. Roosts communally in winter on the ground, often at traditional roosts with tens of other individuals, occasionally hundreds. At northerly latitudes, entire population migrates, on a broad front, southwards for the winter.

FEEDING ECOLOGY AND DIET
Hunts by day but also quite crepuscular, active into dusk. Feeds mainly on mammals such as mice, rats, voles, and young rabbits and hares, which it often locates in vegetation by sound, also on birds (usually passerines), frogs, birds' eggs, and insects.

REPRODUCTIVE BIOLOGY
Nests as solitary pair in a loose colony around a marsh or similar, also polygamous, two or three females to a male, rarely up to seven. Lays in the northern spring-summer, mainly May; earlier at more southern latitudes. Nests on the ground in dense grass, rushes, shrubs, crops or young pine plantations in a nest of grasses and small sticks. Clutch of three to six eggs; incubation about 30 days. Fledges at four to five weeks.

CONSERVATION STATUS
Not threatened. Main threats include habitat loss to intensified agriculture, drainage of wetlands, reforestation, and, locally, severe persecution by gamekeepers.

SIGNIFICANCE TO HUMANS
None known. ◆

African little sparrowhawk
Accipiter minullus

SUBFAMILY
Accipitrinae

TAXONOMY
Falco minullus Daudin, 1800, Gamtoos River, South Africa. Monotypic.

OTHER COMMON NAMES
English: Little sparrowhawk; French: Épervier minule; German: Zwergsperber; Spanish: Gavalancito Chico.

PHYSICAL CHARACTERISTICS
9.1–10.6 in (23–27 cm); male 2.6–3 oz (74–85 g); female 2.4–3.7 oz (68–105 g). Small gray hawk with lightly barred underparts.

DISTRIBUTION
Africa: southern Sudan and Ethiopia, south to South Africa, and west to Angola and Namibia.

HABITAT
Woodland and forest patches, often along rivers or in valleys. Occasionally, small plantations of exotics in savanna.

BEHAVIOR
Apparently sedentary.

FEEDING ECOLOGY AND DIET
A tiny but bold hunter. Typically, flies at speed from perch, winding agilely through foliage, to catch prey on wing. Specializes on small birds from 0.4–1.4 oz (10–40 g). Occasionally takes small bats, lizards, and insects.

REPRODUCTIVE BIOLOGY
Breeds as solitary pair in March–April in northeast Africa, mostly October–November in southern Africa. Monogamous. Builds a small stick nest of twigs lined with green leaves, high in a tree fork. Usually two eggs; incubation 31 days; fledging about 26 days.

CONSERVATION STATUS
Not threatened. Widespread and common in appropriate habitat and quickly colonizes new habitat such as plantation.

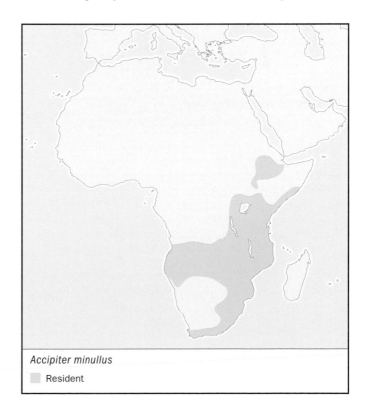

Accipiter minullus

☐ Resident

SIGNIFICANCE TO HUMANS
None known. ◆

Northern goshawk
Accipiter gentilis

SUBFAMILY
Accipitrinae

TAXONOMY
Falco gentilis Linnaeus, 1758, Alps. Eight subspecies.

OTHER COMMON NAMES
English: European goshawk; French: Autour des palombes;
German: Habicht; Spanish: Azor Común.

PHYSICAL CHARACTERISTICS
18.9–27.2 in (48–69 cm); male 18.2–41.3 oz (515–1170 g); female 28.9–53.3 oz (820–1510 g). Brownish gray upperparts and
barred underparts with geographical variation among subspecies in size, plumage, and color.

DISTRIBUTION
A.g. gentilis: Europe and northwest Africa. *A.g. arrigonii*: Cor-
sica and Sardinia. *A.g. buteoides*: Northern Eurasia from Sweden to River Lena, wintering south to central Europe and
central Asia. *A.g. albidus*: Siberia and Kamchatka. *A.g. schvedowi*:
Asia from the Urals to Amurland and south to central China,
wintering south to the Himalayas and Indochina. *A.g.
fujiyamae*: Japan. *A.g. atricapillus*: North America. *A.g. laingi*:
Queen Charlotte and Vancouver Islands, British Columbia.

HABITAT
Mature woodlands—mainly coniferous, also deciduous and
mixed—especially edges and clearings; from lowlands to the
treeline. Occasionally in small isolated woods and town parks.

BEHAVIOR
Mainly sedentary. Migratory in northernmost parts of range,
departs mainly October–November and returns March–April.
Irruptions of goshawks from Arctic, some reaching southern
limits of distribution, following seasons of superabundant prey,
about every 10 years.

FEEDING ECOLOGY AND DIET
Hunts by day; takes small to medium-sized birds and mammals
as large a grouse or hare, mainly on the ground. Prey varies
geographically.

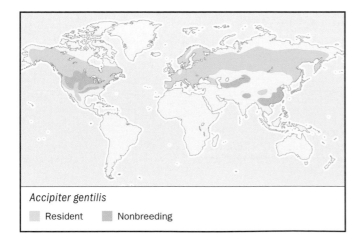

Accipiter gentilis

Resident   Nonbreeding

REPRODUCTIVE BIOLOGY
Nests as solitary pair in large territory. Monogamous. Builds a
stick nest, lined with fresh leaves, in the fork, or on a branch
near the trunk, of a large tree. Lays in April–May; most
common clutch three or four eggs; incubation about 36 days;
fledging at about five or six weeks.

CONSERVATION STATUS
Not threatened. Decline in Europe since nineteenth century
but populations now mostly stable and some recovering. Extinct in Britain since 1800s because of pesticides, persecution,
nest robbing for falconry, and deforestation; re-established in
the late 1960s apparently from escaped falconers' birds. Population stable in North America, increasing in Russia. Still killed
in places (e.g., Finland) by hunters and vulnerable to poisoning
from baits left for other predators. Reforestation is beneficial.

SIGNIFICANCE TO HUMANS
Used by falconers for centuries. Remains the most popular
hawk among falconers. ◆

Harris' hawk
Parabuteo unicinctus

SUBFAMILY
Accipitrinae

TAXONOMY
Falco unicinctus Temminck, 1824, western Minas Gerais, Brazil.
Two subspecies.

OTHER COMMON NAMES
English: Bay-winged hawk; French: Buse de Harris; German:
Wüstenbussard; Spanish: Busardo Mixto.

PHYSICAL CHARACTERISTICS
19–22 in (48–56 cm); male: 25 oz (725 g), female: 34 oz (950 g).
Sooty brown body, with rufous accents on shoulders, thighs,
and underwings, and black tail.

DISTRIBUTION
P.u. harrisi: southwest United States to Mexico, Central Amer-
ica, western Colombia, Ecuador, and Peru. *P.v. unicinctus*:
Northeastern Colombia and western Venezuela to Bolicia,
Brazil, Chile, and southern Argentina.

HABITAT
Seasonally dry desert, Chaco and savanna, occasionally swamp-
land. In more arid regions, near large waterbodies.

BEHAVIOR
Largely sedentary.

FEEDING ECOLOGY AND DIET
Hunts large prey for its size, mostly mammals, up to the size
of rabbits and jackrabbits, also birds including flickers and rails.
Also reptiles (snakes and lizards) and insects. Hunts larger prey
co-operatively, social groups of two to six gather at dawn to
work through territory to flush, ambush, and sequentially attack rabbits.

REPRODUCTIVE BIOLOGY
Typically monogamous, usually nesting as solitary pair. Builds a
stick nest, lined with moss, grass and leaves, in a tree. Lays one
to four eggs in June–July. Incubation about 34–35 days; fledging about 40 days. Some pairs renest in late summer or early

Parabuteo unicinctus
  Resident

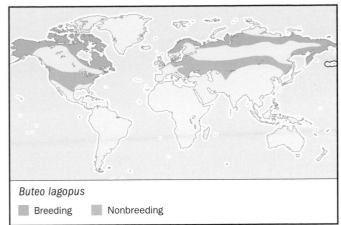

Buteo lagopus
  Breeding   Nonbreeding

autumn, even following a successful first (winter) nesting attempt. Cooperative breeding reported in United States but not elsewhere: one to five juvenile or adult helpers bring food and defend the nest of the dominant (alpha) pair. The beta birds appear to be unrelated to the breeding pair and the gamma birds are often young from the previous breeding attempt.

CONSERVATION STATUS
Not threatened. Occasionally poisoned by strychnine-baited carcasses left by sheep farmers for other predators. Reintroduced to California, where small population established.

SIGNIFICANCE TO HUMANS
None known. ◆

Rough-legged buzzard
Buteo lagopus

SUBFAMILY
Accipitrinae

TAXONOMY
Falco lagopus Pontoppidan, 1763, Denmark. Four subspecies.

OTHER COMMON NAMES
English: Rough-legged hawk; French: Buse pattue; German: Rauhfulßbussard; Spanish: Busardo Calzado.

PHYSICAL CHARACTERISTICS
19.7–23.6 in (50–60 cm); male 21.2–48.7 oz (600–1380 g); female 27.5–58.6 oz (780–1660 g). Brown and white mottled

plumage varies in intensity among subspecies. White tail with dark subterminal band.

DISTRIBUTION
B.l. lagopus: northern Eurasia from Scandinavia to River Yenisey, wintering south to central Europe and central Asia. *B.l. menzbieri*: northeastern Asia, wintering south to central Asia, northern China, and Japan. *B.l. kamtschatkensis*: Kamchatka, wintering south to central Asia. *B.l. sanctijohannis*: Alaska and northern Canada, wintering south to central and southern United States.

HABITAT
Mainly treeless tundra, but also wooded tundra and extreme northern taiga when lemmings and voles are abundant. Usually flat low country. Wintering grounds are also mainly flat, open country, including prairie, cropland, and marsh.

BEHAVIOR
Clear migrant with separate breeding and wintering grounds. Depart breeding grounds about September–October and return about April–May. Timing and extent of migration depends on seasonal prey abundance at either end.

FEEDING ECOLOGY AND DIET
Mainly preys on mammals, especially voles and lemmings. Also takes birds, other vertebrate including fish, insects and carrion, particularly when main prey scarce. Hunts by day, but occasionally crepuscular.

REPRODUCTIVE BIOLOGY
Breeds as solitary pair, laying in May–June. Monogamous. Usually nests on a protected ledge, high on a riverbank, cliff or rocky outcrop, rarely in tree. Builds a bulky nest of sticks lined with grass and prey remains; three to five eggs; greater number (up to seven) in good seasons when food abundant. Incubation about 30 days; fledging about five or six weeks.

CONSERVATION STATUS
Not threatened. No obvious threats in breeding grounds but winter quarters are subject to habitat disturbance and other human pressures.

SIGNIFICANCE TO HUMANS
None known. ◆

Gurney's eagle
Aquila gurneyi

SUBFAMILY
Accipitrinae

TAXONOMY
Aquila (? Heteropus) gurneyi G.R. Gray, 1860, Bacan, Moluccas. Monotypic.

OTHER COMMON NAMES
French: Aigle de Gurney; German: Molukkenadler; Spanish: Aguila Moluqueña.

PHYSICAL CHARACTERISTICS
29.1–33.9 in (74–86 cm); female 107.9 oz (3,060 g); males are smaller than females. Chocolate brown plumage.

DISTRIBUTION
New Guinea and larger surrounding islands including Misool, Waigeo, Salawati, Aru, Yapen, Normandy and Goodenough, West Papuan, and Aru Islands, and the Moluccas, including Morotai, Halmahera, Ternate, Bacan, Ambon, and Seram.

HABITAT
Hillside and lowland primary rainforest and swamp forest. Hunts into nearby littoral zone, cultivated farmland and grassland. Inland but usually within 9.3 mi (15 km) of coast.

BEHAVIOR
Uses uplifts to soar along hillsides and cliffs; soars to great height on thermals. Usually solitary in pairs or trios, the latter possibly family groups. Adults apparently sedentary.

FEEDING ECOLOGY AND DIET
Reported to take cuscus and other arboreal mammals. Slowly quarters forest canopy or ground, patrols seashore.

REPRODUCTIVE BIOLOGY
Not known.

CONSERVATION STATUS
Not threatened. Uncommon and seldom encountered. Deforestation of lowlands may be a threat.

SIGNIFICANCE TO HUMANS
None known. ◆

Harpy eagle
Harpia harpyja

SUBFAMILY
Accipitrinae

TAXONOMY
Vultur harpyja Linnaeus, 1758, Mexico. Monotypic.

OTHER COMMON NAMES
French: Harpie féroce; German: Marpyie; Spanish: Arpía Mayor.

PHYSICAL CHARACTERISTICS
35–41.3 in (89–105 cm); male 8.8–10.6 lb (4–4.8 kg); female 16.8–19.8 lb (7.6–9 kg). Large, regal raptor with gray head, white breast, and long barred tail.

DISTRIBUTION
Southern Mexico through Central America to Columbia, east through Venezuela and south through Bolivia, Brazil, and north-east Argentina.

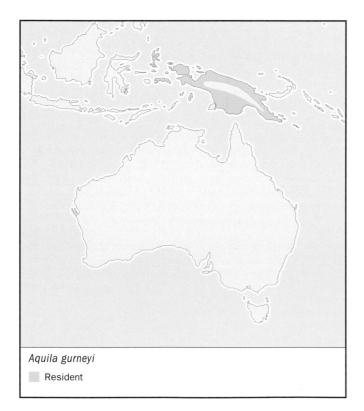

Aquila gurneyi
▨ Resident

Harpia harpyja
▨ Resident

HABITAT
Lowland tropical forest, mostly up to about 2,950 ft (900 m). Occurs in uninterrupted forest, but will nest where high-grade trees have been logged and hunt through forest remnants intermixed with pasture.

BEHAVIOR
Occasionally, in the early morning sunbathes on prominent perches emerging from the forest. Rarely, if ever, soars, unlike typical eagles. Thought to be largely sedentary but suggestion that the population in southern Atlantic forests may be migratory.

FEEDING ECOLOGY AND DIET
One of the most powerful of avian predators. Preys on large, difficult vertebrates including howler, capuchin and saki monkeys, sloths, opossums, porcupines, and anteaters. Also reptiles, such as snakes and iguanas, and ground mammals, such as agoutis, domestic pigs and young deer. Bird prey include curassows, macaws, and seriemas. Hunts from a perch at the forest edge or clearing, at rivers and beside salt licks.

REPRODUCTIVE BIOLOGY
Monogamous. Lays in June in Guyana, September–November in Brazil. Builds a bulky nest of large sticks, usually in enormous, emergent tree. Clutches of incubation is 56 days; fledge at about give months. Unusually, male brings prey to nest only twice a week during first half of nestling period.

CONSERVATION STATUS
Not globally threatened but considered Near Threatened. Uncommon and sparsely distributed throughout range. Has all but disappeared from large parts of former range, notably north and central South America. Extensive deforestation is a significant and continuing threat.

SIGNIFICANCE TO HUMANS
None known. ◆

Resources

Books

BirdLife International. *Threatened Birds of the World.* Barcelona and Cambridge: Lynx Edicions and BirdLife International, 2000.

Brown, L. H., E. K. Urban, and K. Newman. *The Birds of Africa.* Vol. 1. London: Academic Press, 1982.

Coates, B. J. *The Birds of New Guinea.* Vol. I, *Non-Passerines.* Dove Publications: Alderley, 1985.

Cramp, S., ed. *The Birds of the Western Palearctic.* Vol. II, *Hawks to Bustards.* Oxford: Oxford University Press, 1980.

del Hoyo, J. A., A. Elliot, and J. Sargatal, eds. *Handbook of the Birds of the World.* Vol. 2, *New World Vultures to Guineafowl.* Barcelona: Lynx Edicions, 2000.

Ferguson-Lees, J. *Raptors: An Identification Guide to the Birds of Prey of the World.* Academic Press: New York, 2001.

Fox, N. *Understanding the Bird of Prey.* Surrey: Hancock House, 1995.

Long, J. L. *Introduced Birds of the World.* Sydney: Reed, 1981.

Marchant, S. and P. J. Higgins, eds. *Handbook of Australian, New Zealand and Antarctic Birds.* Vol. 2, *Raptors to Lapwings.* Oxford: Melbourne, 1993.

Newton, I., and P. Olsen, eds. *Birds of Prey.* London: Merehurst, 1990.

Olsen, P. *Australian Birds of Prey.* University of Sydney and Baltimore: New South Wales Press and Johns Hopkins, 1995.

Poole, A. F. *Ospreys: A Natural and Unnatural History.* Cambridge University Press: Cambridge, 1989.

Organizations

The Hawk and Owl Trust. 11 St Marys Close, Newton Abbot, Abbotskerswell, Devon TQ12 5QF United Kingdom. Phone: +44 (0)1626 334864. Fax: +44 (0)1626 334864. E-mail: hawkandowl@aol.com Web site: <http://www.hawkandowltrust.org>

Raptor Research Foundation. P.O. Box 1897, 810 E. 10th Street, Lawrence, Kansas 66044-8897 USA. Web site: <http://biology.biosestate.edu/raptor>

World Center for Birds of Prey, The Peregrine Fund. 566 West Flying Hawk Lane, Boise, Idaho 83709 USA. Phone: (208) 362-3716. Fax: (208) 362-2376. E-mail: tpf@peregrinefund.org Web site: <http://www.peregrinefund.org>

World Working Group on Birds of Prey and Owls. P.O. Box 52, Towcester, NN12 7ZW United Kingdom. Phone: +44 1 604 862 331. Fax: +44 1 604 862 331. E-mail: WWGBP@aol.com Web site: <http://www.raptors-international.de>

Penny Olsen, PhD

▲
Secretary birds
(Sagittariidae)

Class Aves
Order Falconiformes
Suborder Sagittarii
Family Sagittariidae

Thumbnail description
Very large predatory bird with hooked bill; long stork-like legs; bare facial skin orange; and long, black feathers forming erectile crest on nape. Pale gray and white but with black leggings and flight feathers; central tail feathers elongated

Size
49–59 in (125–150 cm); 7.5–9.5 lb (3.4–4.3 kg)

Number of genera, species
1 genus; 1 species

Habitat
Open woodland, savanna, and semi-desert steppe

Conservation status
Not threatened

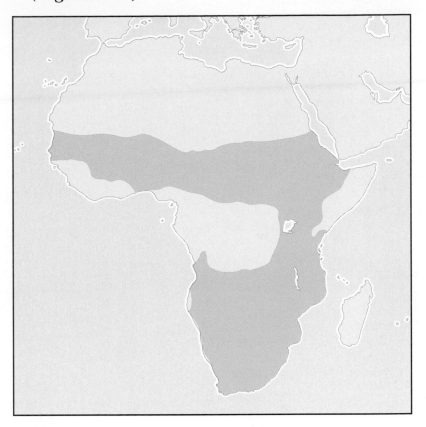

Distribution
Sub-Saharan Africa

Evolution and systematics

The secretary bird (Sagittariidae) is sole member of a unique African family, a status shared only with the hammerhead (*Scopus umbretta*). Prehistorically, however, fossil remains of at least two secretary bird-like species with shorter legs are known from 20 million-year-old Miocene and even older Oligocene deposits in France. The special attributes of the leg morphology and karyotype of the secretary bird have always been recognized by placing it in its own family and suborder and sometimes even its own order. It is generally considered an aberrant bird of prey, in the order Falconiformes, but with some stork-like features that probably represent an earlier common ancestry with ciconiiform waterbirds, as is also shown by the New World vultures (Cathartidae). DNA-DNA hybridization studies also confirmed the diurnal birds of prey as nearest relatives. Proposed relationships with gruiform birds, such as seriemas (Cariamidae), cranes (Gruidae), or bustards (Otididae) seem to represent convergence of morphology for a terrestrial lifestyle rather than genetic relationship. Overall, the secretary bird behaves and has the skull anatomy of a large "marching" eagle, although aspects of its breeding biology are most similar to storks.

Physical characteristics

The secretary bird stands about 4 ft (1.2 m) tall on long pink legs. Stubby toes are armed with thick claws and used to kick prey into submission. Plumage of the upper parts is pale gray and of the underparts white, with the exception of the thighs and abdomen, which are black. Flight feathers on the broad wings are black, contrasting with the gray upperwing and white underwing and undertail coverts. Tail feathers are dark gray, with a broad black subterminal band and white tip; they are graduated in length from shortest on the outside to the exaggerated length of the central pair. The head is striking with a hooked, pale gray beak and broad cere; large bare areas of orange facial skin; and elongated, black nape feathers that form either a droopy crest or are erected to form a spiky halo. The iris is dark brown or yellow, although the geographical details of where this difference occurs are unclear. Sexes seem identical in size and plumage. Juveniles have a paler orange face, a brown wash to the plumage, fine gray bars across the white underwing and undertail coverts, and an iris that changes from brown to pale gray before attaining the adult color.

Secretary bird (*Sagittarius serpentarius*). (Illustration by Joseph E. Trumpey)

Distribution

The secretary bird occurs throughout sub-Saharan Africa except for areas of tropical forest along the West African coast and across the Congo River Basin. It is most common in areas of open savanna and steppe but wanders widely to open areas within woodlands of central Africa and less arid areas of southwest and northeast Africa.

Habitat

Open woodland, savanna, and steppe comprise the optimum habitat of the secretary bird. It is absent from stands of dense forest and woodland, although it may enter larger clearings, and it wanders into true desert only after exceptional rainfall. Its main requirements are a few low trees on which to roost and nest and an adequate food supply of small animals.

Behavior

In more productive areas, the secretary bird is resident and sedentary, but in areas with fluctuating conditions, it is a nomadic visitor during times of plenty. It normally occurs in pairs, each within a defended territory, at densities of 7.7 –193 mi² (20–500 km²) per pair depending on local conditions and abundance of food. At night, it roosts on top of a low tree, either standing or lying down on an old nest platform if available. At dawn, it flies to the ground to begin its daily walk across the veld. Depending on terrain and cover, it either strides briskly along and looks from side to side or shuffles slowly along and searches just in front of its feet. The normal rate is about 120 paces/min, which, with a stride of about 16 in (40 cm), covers about 2 mi/hr (3 km/hr). Prey is captured and eaten where- and whenever encountered. Members of a pair may hunt alone or together. Once satiated, the birds

may rest or, using thermals in the heat of the day, rise up on broad wings, which span 7 ft (2.12 m), and soar to their nest, water, or alternative hunting areas. By dusk, they usually return to their roost site. Encounters with neighbors may lead to chases, bouts of kicking, or aerial pursuit, accompanied by deep croaking calls. For most of the day, however, the species is silent and pedestrian.

Feeding ecology and diet

Any small animals that can be kicked into submission are killed and eaten. These range in size from moths and grasshoppers to mongooses, hares, and gamebirds. Very small insects, such as termites or wasp larvae, may be picked up in the bill, but any large or active prey, including poisonous snakes, are killed after fast, active pursuit, kicking, and disablement. The bird will only stoop to pick up its prey after it is completely immobile. The exact diet depends on locality and availability: locusts and rodents predominate in one area with beetles and lizards in another. The hunting technique allows a wide range of prey types and sizes to be captured, most of which are swallowed whole or, more rarely, torn to pieces or cached under a bush for later consumption. Undigested remains are regurgitated as large pellets, 1.6–1.8 in (40–45 mm) in diameter and up to 4 in (100 mm) long, mainly below roost and nest sites, where they provide a quick indication of prey consumed.

Reproductive biology

Members of a pair share breeding duties, from building a nest of weeds, sticks, and grass on top of a tree; through incubation; to brooding and feeding chicks. Courtship includes high flights above the nest area, with pendulum-like displays of repeated diving and swooping up, accompanied by deep croaking calls. The nest is built into a stable platform of 3.3–6.6 ft (1–2 m) diameter, with a bed of dry grass in the

Secretary birds (*Sagittarius serpentarius*) building a nest in Kenya. (Photo by Renee Lynn. Photo Researchers, Inc. Reproduced by permission.)

center to accommodate the clutch of 1–3 white eggs. Parents take turns incubating or brooding while the off-duty mate goes off to feed. Nest relief includes a greeting display. Once chicks have hatched, a cropful of food is regurgitated onto the nest floor for their consumption. Incubation takes 42–46 days and the nestling period varies from 65–106 days. Animal food is at first torn up and fed to small chicks, but within a few weeks of hatching, they are able to gulp down whole prey almost immediately. Where available, parents also swallow water before coming to the nest and dribble this into a chick's bill, along with partly digested food.

Breeding normally takes place during summer rains when food is most abundant but can occur at any time of year if prey numbers persist. Food availability seems to determine the number of eggs laid, rate of chick growth, age at fledging, and interval between broods. In some years, three broods may be attempted, in others none. A full brood of three chicks is raised only rarely since food shortages during breeding often lead to the starvation of the youngest and smallest sibling. Flexibility in breeding biology, along with nomadic movements to areas of abundant food, allow the secretary bird to be unusually productive for such a large bird in such a variable environment.

Conservation status

The secretary bird is widespread and common across its range in Africa, including larger national parks and other conservation areas. It has disappeared from areas of high human density or intensive agriculture, but clearing of bush and planting of pastures has extended its range. Its flexible breeding biology enables it to capitalize on good habitats and conditions, whereever and whenever available, although the low nest and terrestrial habits make it vulnerable to casual persecution.

Significance to humans

No particular significance seems to be attached to the secretary bird by Africans, although it is a much sought-after species by visiting birders. Its reputation as a predator of poisonous snakes often earns it protection.

Resources

Books

Brown, L.H., E. Urban, and K. Newman. *The Birds of Africa.* Vol. 1. London: Academic Press.

del Hoyo, J., A. Elliott, and J. Sargatal, eds. *Handbook of Birds of the World.* Vol.2, *New World Vultures to Guineafowl.* Barcelona: Lynx Edicions, 1994.

Sibley, C.G., and J.E. Ahlquist. *Phylogeny and Classification of Birds: A Study in Molecular Evolution.* New Haven and London: Yale University Press, 1990.

Steyn, Peter. *Birds of Prey of Southern Africa, Their Identification and Life Histories.* Cape Town, South Africa: David Philip, 1982.

Periodicals

Kemp, M.I., and A.C. Kemp. " Bucorvus and Sagittarius: Two Modes of Terrestrial Predation." *Proceedings Symposium on*

African Predatory Birds, edited by Alan Kemp. Northern Transvaal Ornithological Society, Pretoria, 1977.

Kemp, A.C. "Aspects of the Breeding Biology and Behaviour of the Secretary Bird *Sagittarius serpentarius* near Pretoria, South Africa." *Ostrich* 66 (1995): 61–68.

Organizations

Raptor Conservation Group, Endangered Wildlife Trust. Private Bag X11, Parkview, Gauteng 2122 South Africa. Phone: +27-11-486-1102. Fax: +27-11-486-1506. E-mail: ewt@ewt.org.za Web site: <http://www.ewt.org.za>

World Working Group on Birds of Prey and Owls. P.O. Box 52, Towcester, NN12 72W United Kingdom. Phone: +44 1 604 862 331. Fax: +44 1 604 862 331. E-mail: WWGBP@aol.com Web site: <http://www.raptors-international.de>

Alan Charles Kemp, PhD

Falcons and caracaras
(Falconidae)

Class Aves
Order Falconiformes
Suborder Falcones
Family Falconidae

Thumbnail description
Small to medium-sized largely diurnal raptors with strong feet, usually pointed wings, and sharply curved beaks; from scavengers to among the swiftest and, for their size, most powerful of avian predators

Size
5.5–25.6 in (14–65 cm); 0.06–4.6 lb (28–2,100 g)

Number of genera, species
10 genera; about 62 species

Habitat
Most habitats, from treeless desert and tundra to dry forest and rainforest, featureless plains to rugged gorges and escarpments

Conservation status
Vulnerable: 4 species; Near Threatened: 6 species; one subspecies Extinct

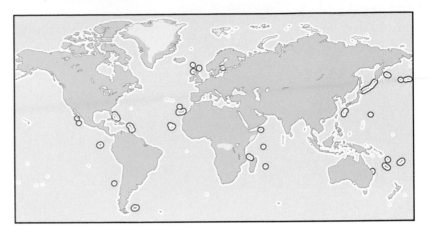

Distribution
Worldwide, except the Antarctic

Evolution and systematics

The relationship of falconids to other birds has long been unclear. Their similarity to the accipitrids (family Accipitridae, hawks and eagles) is obvious, and it has been suggested that they have morphological and anatomical links with the owls. Yet molecular evidence suggests that in both cases the resemblance is convergent; that they have evolved to look and behave in a similar fashion because of their similar lifestyles. One classification, based on DNA hybridization, groups them within the Ciconiiformes (herons, storks, and ibises). Nevertheless, in most modern classifications they are still placed with the accipitrids in the order Falconiformes.

There is no good fossil evidence to suggest where the family might have originated geographically. Africa now has the greatest diversity of falcons, but that does not necessarily reflect their past distribution or give clues about where they might have arisen. Contemporary caracaras and forest falcons (Polyborinae) are South and Central American, and there is no indication the subfamily was ever otherwise, so it seems reasonable to suggest they arose in the neotropics.

The earliest fossil attributed to the family was found in England and dated to 55 million years ago. Better-substantiated fossil members of the family have been reported from 36-million-year-old deposits in France. In the Americas, falconids have been identified from 23 million years ago, including a caracara-like form. The falcons, genus *Falco*, the largest group of falconids, are clearly closely related and are thought to have

undergone rapid radiation and expansion early in the Pleistocene (about 1.85 million years ago), when the grasslands opened up.

Within the family, division into the two subfamilies is well supported but there is much argument over relationships within each. Nevertheless, each major region has its group of similar, presumably closely related forms (gray falcon [*Falco hypoleucos*], black falcon [*F. subniger*], brown falcon [*F. berigora*] and New Zealand falcon [*F. novaeseelandiae*] in Australasia) and there are a few groups of more widespread sister species (the desert falcons, kestrels, peregrines, and hobbies).

Physical characteristics

Falconids are readily recognizable as birds of prey and are likely to be confused only with the other family of diurnal raptors, the Accipitridae. Both have a fleshy cere atop a strong hooked beak, and a strong hallux (hind toe) that opposes three forward toes. Both families capture animals with their feet and, unlike most bird species, females are larger than males. The difference between the sexes is greatest in bird-eaters (female around 150% of male weight), intermediate in mammal-eaters, and least in insect-eaters and scavengers (females marginally heavier).

Differences between the families include several anatomical features, such as the structure of the syrinx, the fact that falconids kill with the beak or by the blow when they strike

A peregrine falcon (*Falco peregrinus*) with its prey. (Photo by Tim Davis. Photo Researchers, Inc. Reproduced by permission.)

(accipitrids squeeze with their feet), and eggs that have reddish translucence when held up to the light (accipitrid eggs are bluish or greenish).

Most falconids have strong needle-sharp talons. The exception is the caracaras, which have heavy but flatter talons for their more terrestrial, vulture-like lifestyle. Other structures also vary with lifestyle, form reflecting function. Hence, toes of bird-catching species are long and slender, mammal eaters have thicker toes, and snake eaters the shortest and thickest. Beaks also vary with prey: short and robust in most species, particularly deep in species that take large prey; longer and less powerful in carrion eaters. Fast aerial falconids of open country or those that migrate tend to have the longest, most pointed wings. Slower species tend to have slightly broader wings. Forest-dwelling species have short rounded wings and a long tail that gives them maneuverability in tight spaces.

Plumages tend to be fairly cryptic: shades of brown, black, and gray or white, often mottled or streaked. The caracaras, which tend to be mostly scroungers with no need for camouflage, are more colorful and some have raven-like, glossy black feathers with a green sheen. Most species have yellow soft parts (cere, narrow eye-ring, legs, and feet), occasionally gray or red. Caracaras and forest falcons are distinguished by obvious areas of featherless bright red or yellow skin around the face, more extensive in the more scavenging species.

The family is noted for its powers of sight, with an ability to detect the smallest movement at great distance. Most have rather large brown eyes; those living in dimly lit habitats or that regularly hunt into the evening tend to have the largest eyes.

Distribution

Except for the Antarctic and parts of the far high Arctic, the falcons alone are distributed worldwide. The remaining Falconinae, the falconets, are mainly tropical *Spiziapteryx* (1

South American species), *Polihierax* (1 African species; 1 Indian), and *Microhierax* (5 Asian species).

All 16 species of Polyborinae are neotropical, mainly South American, and do not venture beyond that region. These include caracaras *Daptrius* (2 species), *Phalcoboenus* (4), *Polyborus* (1), *Milvago* (2), laughing falcon *Herpetotheres* (1), and forest falcons *Micrastur* (6).

Of the Falconinae, several falcons migrate from one region to another (most from Eurasia to Africa; none to Australia).

Among breeding grounds, the species are distributed as follows, and some occur in more than one region: Australasia (8 falcons), African region (1 pygmy falcon *Polihierax*, 14 falcons), Central and South America (3 falcons), North America (9 falcons), and Eurasia (6 falconets, 11 falcons).

Individual species vary widely in the extent of their distribution. The peregrine falcon (*Falco peregrinus*, meaning *the wanderer*) is the most widespread of all, occurring almost everywhere habitat is suitable, with a few notable gaps such as New Zealand. The Seychelles (*F. araea*) and Mauritius (*F. punctatus*) kestrels are probably the most restricted in distribution, confined to some tiny islands of the Seychelles and Mauritius, Indian Ocean. The crested caracara (*Polyborus plancus*), found throughout South and Central America, has the most extensive range of the Polyborinae. The striated caracara (*Phalcoboenus australis*) and plumbeous forest falcon (*Micrastur plumbeous*) are very localized. Perhaps not surprisingly, all species with very limited distributions are listed as being threatened with extinction.

Several species extended their distribution in historical times because extra habitat was created for them or they adapted to disturbed habitats. One example is the Australian kestrel (*F. cenchroides*), which has established populations on Christmas Island, Indian Ocean, and Norfolk Island, southwest Pacific Ocean. The kestrel self-colonized both islands; the former by hitching a ride on warships operating in the area during World War II, the latter after the rainforested island was turned to farmland.

One species was purposely introduced and became established: the chimango caracara (*Milvago chimango*), endemic to southern South America, was translocated to Easter Island, South Pacific, in 1928 and is now common.

Habitat

As a group, the falconids occur in most major habitats around the world. They reach their greatest diversity near the tropics and only two species are found in the high Arctic (gyrfalcon *F. rusticolis*, peregrine falcon). Many species are quite adaptable and for them the structure of the habitat is more important than its individual components. For these species, habitat disturbance may not be important provided prey and nest sites are still available. For example, conversion of forest and woodland to farmland appears to have favored the Australian kestrel, but recent agricultural intensification has rendered some areas unsuitable again. Another example is the use of city canyons by peregrine falcons and several species of kestrel; cities and suburbs can offer nest sites, abundant prey (pigeons, starlings, rats, insects), and refuge from persecution.

Other species are more specialized. For example, several forest falcons need large tracts of more-or-less intact forest. Eleanora's falcon (*F. eleanorae*) nests on a few small quiet islands and islets, pockmarked with holes and ledges and sited on the migratory routes of small birds in the Mediterranean. In the nonbreeding season the population migrates to a very different habitat in the open woodlands and forests of Madagascar.

Typically, forest falcons and falconets require forest and forest clearings and edges. In contrast, falcons and caracaras favor open country, from grasslands to open woodlands and scrub. As always there are exceptions; the bat falcon (*F. rufigularis*) depends on forest edge, and the Seychelles kestrel lives in dense secondary forest.

In the nonbreeding season many species, such as Eleanora's falcon and several from high latitudes in the Northern Hemisphere, make drastic habitat changes. Annually, the entire population leaves breeding grounds made barren by the winter desertion of small birds or by snowfalls to fly across the globe to more productive zones. Other species make less drastic, more unpredictable shifts in distribution, such as local movements down from the mountains to winter in milder coastal areas. The African pygmy falcon (*Polihierax semitorquatus*) leaves more arid parts of its range when the season is particularly dry. In some species there may be no shift in mild winters, or only part of the population may leave to winter elsewhere. In many species, such as several kestrels, females are most likely to depart the breeding grounds for the winter; males will stay if they can support themselves.

Behavior

Typically, falconids are active by day, but several species hunt crepuscularly. For example, the Eurasian hobby (*F. subbuteo*) chases moths after dusk, and peregrine falcons hunt shearwaters as they return to their holes well after dark. At night they usually roost at a regular roost, sheltered from the prevailing elements and safe from predators. Even on migration, the same roosts are often used year after year.

Some species are resident, others are migratory or partially migratory (in areas where winter is harsh, part of the population migrates). Distances traveled range from 6,000–12,000 mi (10,000–20,000 km) (Arctic peregrine falcons) down to very local shifts (e.g., Australian peregrine falcons), both to areas where prey is more available. One of the most dramatic examples is the movement of several falcons from Eurasia and the Mediterranean to Africa and Madagascar, to feed on vast swarms of termites and ants that arise during the wet season. Some species gather in flocks and move en masse (red-footed falcon *F. vespertinus*), others find their way individually. Juveniles of most species disperse after their post-fledgling dependency, except for juveniles of the red-throated caracara, that may join the communal group. In all species studied, juveniles are more dispersive than adults, and females tend to depart earlier and move farther than males.

Crested caracara (*Caracara plancus*) pair preening and bonding in south Texas. (Photo by Larry Ditto. Bruce Coleman Inc. Reproduced by permission.)

Although falcons are typically viewed as solitary creatures, many species are gregarious and may flock together to feed, roost, breed, or migrate. Some flock opportunistically (many kestrels), others live in cooperative groups as a way of life (red-throated caracara *Daptrius americanus*). Still others live in loose groups but don't directly assist each other (Eleanora's falcon). At the other end of the spectrum are purely solitary species that are highly territorial year-round (collared falconet *Microhierax caerulescens*).

Falconids have rather harsh voices, at least for longer-distance communication. For more intimate contact they make softer sounds. Forest falcons and laughing falcons call at dawn (and again at dusk), presumably to advertise that they are still present on their territory in the dense forest. Caracaras are also rather vocal, particularly when annoyed over food or other territorial disputes, throwing their heads back to release the far-carrying calls for which they are named. By contrast, falcons can be rather silent, mainly heard when humans disturb a nest and the falcons call in defense (usually a strident *kak kak kak kak* that is higher pitched in smaller species).

Feeding ecology and diet

Feeding ecology is a better-known aspect of falconoid ecology, mainly because they eat large prey and produce pellets (wads of indigestible parts of the prey—fur, feathers, scales, bone—that are regurgitated once a day or so) that can be analyzed for the prey they contain. Pellets build up under nests and roosts and make interesting study.

Falconids are predatory meat eaters, yet some also scavenge and the caracaras eat some vegetable matter. Several, especially the more aggressive species (black falcons *F. subniger*, and caracaras), are not above stealing prey from other rap-

Striated caracara (*Phalcoboenus australis*) female near the nest where her young sits. (Photo by F. Pölking/Okapia. Photo Researchers, Inc. Reproduced by permission.)

tors, herons, and others. Some species are specialized; for example, the peregrine feeds mainly on birds and the laughing falcon on snakes, but both take a variety of species in those prey groups. The gyrfalcon takes ground-dwelling creatures such as medium-sized mammals and terrestrial birds. Other species spread their diet over a range of prey groups. Caracaras can kill prey as large as lambs and capture birds on the wing, but mostly they scrounge for carrion (especially its maggots) and horse dung, and take easier prey such as earthworms, maggots, lizards, insects, nestling birds, frogs, and fish. Forest caracaras (*Daptrius*) eat fruit, and wasp and bee nests, for which they seem to have a repellant that protects them from stings from the angry insects. The black caracara (*D. ater*) picks ticks off tapirs, which appear to solicit the caracaras by calling and then lying down to have the ticks removed. Many species hawk moths in the evening and insects on the wing, especially when the insects swarm, such as after rain and around water. In the breeding season most species concentrate on live prey. Juveniles tend to capture easier prey (such as insects) than do adults.

Hunting methods are as varied as the diet. Bird-catching falcons (even diminutive falconets) are the most spectacular hunters, sometimes catching prey as large as themselves at breathtaking speed, either in a direct tail chase or by diving from height to strike with great force. Other species capture

prey on the ground or as it flushes. Most kestrels hover (giving them a fixed aerial vantage point from which to spot and drop onto prey) or swoop down from a perch. Some longer-legged species run over the ground after insects.

Falcons tend to hunt in the open, by sight. Forest falcons make greater use of sound to locate prey in the dim dense forest, and have a circle of stiff feathers around the ear openings to funnel sound to the ears. All species use the element of surprise to improve their success rate.

Most species hunt alone, occasionally as cooperative pairs (several falcons), and a few forage cooperatively in groups (red-throated caracara). Several species use herds, tractors, fires, cars, and other birds to flush prey. Most if not all falcons cache a little prey in excess of their immediate needs, placing it in a rock crevice or under a grass tussock for later retrieval.

Reproductive biology

Most falconids breed once a year in a traditional breeding territory. Territories range from a few square yards (meters) for colonial species like the lesser kestrel (*Falco naumanni*) and Eleanora's falcon, to 400 mi² (1,000 m²) for the solitary nesting gyrfalcon. Spacing between pairs depends on food and nest availability. Typically falconids are monogamous, nesting as solitary faithful pairs. Yet many variations in breeding arrangements found in other birds are present in the family. About 10% of falconids are colonial or semicolonial and nest in loose groups. The red-throated caracara has a more cooperative approach to group living; only the alpha pair breeds and it is assisted in feeding and protection by other group members. Some peregrine falcon pairs in France had an extra female who appeared to help with feeding, leading to greater breeding success at assisted nests. Pygmy falcons are sometimes polyandrous, with more than one male attending the nest and presumably having a chance to father young.

Most species have courtship rituals involving fancy flying and increased calling. Except for the caracaras, falconids do not build nests. Rather they use a tree hole or cliff cavity, an open stick or enclosed woven nest of another species, an epiphyte, or some other suitable situation. Caracaras build an untidy nest of sticks or dried grass, sometimes lined with wool or grass.

Caracaras, forest falcons, and many of the falcons lay two to three red-brown blotched buff eggs. The laughing falcon lays a single dark egg; the falconets and some species that nest in harsher environments lay three to four or more eggs. Some falcons (gyrfalcon, many kestrels) lay a much larger clutch when conditions are good than when they are poor; other species are more conservative in the size of their clutch (brown falcon, peregrine falcon, island kestrels). Widespread species have smaller clutches in milder parts of their range.

Typically, there is a division of roles: the female does most of the incubation and the male catches prey for her and the chicks, at least for the first part of the nestling period when she is brooding. Incubation is quite fixed for individual species and ranges from about 28 days for smaller species to 35 days for the gyrfalcon. The length of the nestling period is also related to size, with smaller species (falconets and small falcons) taking four weeks, larger species seven or eight weeks (gyrfalcons and caracaras). The nestlings grow rapidly and at fledgling are as heavy as the adults but with flight feathers (wing and tail) not quite fully grown. Fledglings are dependent on adults for variable periods, seemingly longer in species that hunt difficult or scarce prey.

Conservation status

Falcons and caracaras have a long history of persecution for their perceived impact on livestock. Particularly in the 1960s and 1970s, organochlorine pesticides (DDT, dieldrin, and others) had a very real impact on eggshell thickness and the mortality of some species, leading to massive population declines. Yet the group's remarkable resilience is illustrated by the fact that no species are currently listed as Endangered or Critically Endangered, the two highest-risk categories used by the World Conservation Union (IUCN).

Only one falconid has become extinct in historical times: the Guadeloupe caracara (*Polyborus plancus lutosus*) of Isla Guadeloupe, Mexico. It is generally regarded as a subspecies of the crested caracara (*P. plancus*), which is widespread on the Central and South American mainlands, but some taxonomists consider it a full species. Regardless, settlers persecuted the caracara and it has not been seen since 1900. Its demise must have been accelerated by the denuding of the island's dense vegetation by goats.

The main threat for falconids in general is widespread habitat destruction by logging, clearing and burning of forests and woodlands, and intensifying land-use by humans for agriculture, grazing, subsistence hunting, or housing. Local impacts include introduced predators, pesticides, and trade in birds and eggs. Threats to survival of the IUCN-listed species are typical of those facing most falconids.

Four species, the Plumbeus forest falcon and three kestrels, are listed as Vulnerable; that is, they face extinction in the medium-term future if factors causing their vulnerability are not reversed or held at bay. All three kestrels—Seychelles, Mauritius, and lesser kestrels—are found in the African region. The lesser kestrel is a winter migrant to the region, South Africa in particular. In 2000 its world population was estimated to have declined by about 50% in the past 50 years, recovering to 80% of former numbers in the past decade. Deforestation and urbanization of its extensive breeding grounds in western Europe and elsewhere, and intensification of agriculture in its breeding and wintering grounds are believed to be major causes of decline. The other two kestrels are confined to islands where their forest habitat is decimated and introduced predators raid their nests. Conservation actions such as restoring and predator-proofing nest sites are helping all three species. As a result of a captive breeding-for-release program and management in the wild, the Mauritius kestrel was recently down-listed from Endangered because its population has undergone a spectacular recovery, from about four wild birds in the 1970s to more than 500 in 2000.

Another six species are considered Near Threatened. The striated caracara occurs in low numbers on islands and islets off southern Argentina and Chile, including the Falklands where it was persecuted for its occasional attacks on weak or stranded sheep. At present it seems to be free of major threats but remains vulnerable because of its small population. The five remaining species are not well known but are presumed to have a small total population. The two falconets are Asian in distribution, where habitat clearing is a major problem. The white-rumped falcon (*Polihierax insignis*) was once widespread and common in the grasslands and deciduous forests of Thailand, Laos, Cambodia, and Vietnam, and may still survive in extensive areas that remain uncleared. The white-fronted falconet (*Microhierax latifrons*) has a very restricted distribution centered on Sabah, Malaysia; an ability to survive in disturbed habitats may allow it to persist. Past habitat clearance, persecution by farmers and pigeon and poultry keepers, and egg destruction by possums introduced from Australia have caused concern for the New Zealand falcon, but population trends are unknown. The gray falcon is thinly scattered across vast spaces of arid and semiarid Australia and, while it may have suffered some small past contraction in range from habitat loss to grazing by livestock, its population is thought to be stable. Similarly, the taita falcon (*F. fasciinucha*) occurs in naturally low densities in scattered gorges and escarpments from Ethiopia to South Africa, and no major threats are apparent.

Regional conservation initiatives have had some success; for example, reserves such as the Snake River Bird of Prey Conservation Area in Idaho. There, 800 pairs of 15 species nest and another nine use the park to winter or stop over on migration. Guarding nests at risk from egg collectors and falconers, and fostering nestlings into safer nests bolsters populations. In recent decades, education and promotion has generally increased tolerance of predatory birds and appreciation of their place in nature. Many falconids, such as the peregrine falcon, bat falcon (*F. rufigularis*), merlin *F. columbarius*), and various kestrels have moved into suburbs and cities where they hunt in the urban spaces and city skies, and nest in parklands and on buildings.

Significance to humans

Falconids have had a long association with people, and tend to elicit a strong response as friend or foe. They are much admired for their hunting prowess, flying skills, and keen sight, and appear as icons and in the legends and folklore of many cultures. In ancient Egypt, Horus the falcon god was ubiquitous, and as a hieroglyph, Horus represented the king. Falcons (Lanner *F. biarmicus*, Barbary falcon *F. peregrinus pele-*

grinoides, a desert race of the peregrine, common *F. tinnunculus* and perhaps lesser kestrels) were buried in tombs, often mummified. At about the same time (2000 B.C.) or earlier, falcons were used to capture meat for humans. The first known falconry was practiced in Asia, from which it spread. It was most popular in Europe in medieval times (twelfth century), when kings, merchants, and even nuns kept trained falcons. Some of the jargon, such as haggard (a wild-caught adult falcon), has become part of everyday language. Today falconry is still popular where it is not outlawed by wildlife-protection legislation, particularly in the Middle East. Falconids are still significant to some indigenous peoples as totems and in legend. For similar reasons, they appear as symbols of strength, courage, or speed on crests and in logos and product names.

Paradoxically, often falconids are hated because of their depredations on livestock, poultry, pigeons, and even native birds. This is most often a case of misunderstanding of their role in nature and an exaggeration of their impact.

Because pesticide use caused it to vanish from vast areas of its natural range, the peregrine falcon became a flagship of the conservation movement. A concerted effort restored the peregrine and other affected species to much of their range. Although many falconids have adapted to altered environments and human activities, they remain charismatic symbols of a natural world.

1. Gyrfalcon (*Falco rusticolis*); 2. Peregrine falcon (*Falco peregrinus*); 3. Crested caracara (*Polyborus plancas*); 4. Brown falcon (*Falco berigora*); 5. Fox kestrel (*Falco alopex*); 6. Amur falcon (*Falco amurensis*); 7. Plumbeous forest-falcon (*Micrastur plumbeus*); 8. Spot-winged falconet (*Spiziapteryx circumcinctus*); 9. Laughing falcon (*Herpetotheres cachinnans*); 10. White-fronted falconet (*Microhierax latifrons*). (Illustration by Wendy Baker)

Species accounts

Crested caracara
Polyborus plancus

SUBFAMILY
Polyborinae

TAXONOMY
Polyborus plancas J. F. Miller, 1777, Tierra del Fuego. About three or four extant subspecies.

OTHER COMMON NAMES
English: Common caracara, Guadeloupe caracara; French: Caracara huppé; German: Schopfkarakara; Spanish: Caracara Carancho.

PHYSICAL CHARACTERISTICS
19–23 in (49–59 cm); in Panama male 1.8 lb (835 g), female 2.1 lb (955 g); larger in Chile and Peru 2.5–3.5 lb (1,150–1,600 g). A distinctive heavy-beaked falconid with a laterally flattened head, black cap with slight crest, and a bare red face; long-legged and somewhat terrestrial. Mantle, rump, throat, and vent are white or buff and finely marked with black; rest of back is brown-black and finely marked with white or buff; breast is brown-black; tail is barred black and buff with black wide-barred tip. Juvenile is browner and more streaked. Races differ somewhat in size and color.

DISTRIBUTION
Central and South America from the southern United States (Texas to Arizona, and Florida) to Tierra del Fuego and the Falkland Islands.

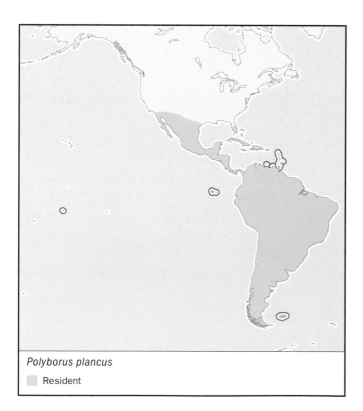

Polyborus plancus
■ Resident

HABITAT
Typically lowland open and semiopen country: grazing land, pasture, river edge. Also Patagonian river valleys, shrub steppe, and grassy foothills. Increasingly moving into uplands of Colombia and Ecuador.

BEHAVIOR
Gregarious, gather at carcasses. Outside the breeding season they roost communally in the tops of isolated trees. Quite vocal, with a loud raucous territorial call made in characteristic pose with head thrown far back. Some local movements, but largely sedentary.

FEEDING ECOLOGY AND DIET
Among the most opportunistic of raptors. Walks over ground and through shallow water in search of prey; arrives at carcasses before vultures. Typically feeds on carrion, road kills, and dead and dying fish and livestock. Occasional live prey includes turtles, iguanas, snakes, crabs, earthworms, caterpillars, and beetles. Raids nests of passerines and other caracaras, and egret and spoonbill colonies. Aggressive; steals prey from other caracaras and raptors; pursues pelicans until the pelicans regurgitate. Eats coconut flesh at harvest in Guyanan plantations.

REPRODUCTIVE BIOLOGY
Breeds as solitary pair over a prolonged season; occasionally two clutches in same year. Builds a large rough lined or unlined nest of sticks on top of cactus or palms, on dense tangles of tree branches, or on the ground. The usual clutch is two eggs; incubation is about a month. Fledglings remain with adults for perhaps three months.

CONSERVATION STATUS
Not threatened. Generally common and locally abundant. Some local persecution and population decreases where they are poisoned and shot for predation on lambs. Habitat loss of farmland to citrus plantation threatens Florida population, but elsewhere conversion of forest to ranch land has helped species expand its distribution.

SIGNIFICANCE TO HUMANS
The caracara is the national bird of Mexico, a nod to Aztec legend. ◆

Laughing falcon
Herpetotheres cachinnans

SUBFAMILY
Polyborinae

TAXONOMY
Herpetotheres cachinnans Linnaeus, 1758, Surinam. Three subspecies usually recognized.

OTHER COMMON NAMES
French: Macagua rieur; German: Lachfalke; Spanish: Halcón Reidor.

PHYSICAL CHARACTERISTICS
17.7–20.9 in (45–53 cm); male 1.2–1.5 lb (565–690 g), female 1.4–1.8 lb (625–800 g). A distinctively patterned, large-headed,

Herpetotheres cachinnans
☐ Resident

bats and, in disturbed areas, reptiles, rodents, and fish. Hunts from a perch where it sits in wait for long periods with head slightly bowed.

REPRODUCTIVE BIOLOGY
Nests as solitary pair that duets ("laughs") near nest at dawn and dusk. Lays single dark brown egg in trees or cliff cavities, stick nests of another species, or on epiphytes. Nestlings fledge at about eight weeks and stay with parent for some months.

CONSERVATION STATUS
Not threatened. Uncommon or fairly common throughout much of their extensive range.

SIGNIFICANCE TO HUMANS
Of traditional significance to local Indians. ◆

Plumbeous forest falcon
Micrastur plumbeus

SUBFAMILY
Polyborinae

TAXONOMY
Micrastur plumbeus W. L. Sclater, 1918, Río Bogotá, Ecuador. Monotypic.

OTHER COMMON NAMES
English: Sclater's forest falcon; French: Carnifex plombé; German: Einbinden-Waldfalke; Spanish: Halcón-montés Plomizo.

black-masked falconid unlike any other, with short wings and long tail. Head and underparts cinnamon-buff to white, with a wide black mask from eyes to hindneck. Upperparts black-brown. Stout legs and short toes. Juvenile has dark feathers edged rufous or buff. Races vary in size and color.

DISTRIBUTION
Central and South America, from Mexico to Paraguay and northern Argentina.

HABITAT
Tropical and subtropical forest, near openings, tracks, or edge, and open forest. Mainly in lowlands.

BEHAVIOR
Has very large home range for its size, estimated at 6,200 acres (2,500 ha) in continuous forest, less in more disturbed habitats. Thought to be sedentary.

FEEDING ECOLOGY AND DIET
Feeds almost exclusively on snakes that are terrestrial and arboreal, venomous and harmless. Occasionally takes birds and

Micrastur plumbeus
☐ Resident

PHYSICAL CHARACTERISTICS
11.8–14.6 in (30–37 cm). Goshawk-like falcons with short
wings, long tail, and long legs. A bare area of bright yellow
skin surrounds the eyes and links the species with the
caracaras. Upperparts slate gray. Underparts: gray throat
grades to a white breast and belly barred with black. Tail distinctive: black tipped with white and a single white band mid-
tail. Juvenile has fainter barring to underparts.

DISTRIBUTION
Restricted in distribution: from Cauca and Nariño of southwest
Colombia to Esmereldas, northwest Ecuador.

HABITAT
Wet forest interiors of the lowlands and foothills of the Pacific
slope.

BEHAVIOR
Little known. Presumed to be sedentary.

FEEDING ECOLOGY AND DIET
Almost unknown. Land crab and lizard in one stomach. Hunts
accipiter-like within forest, with short bursts of great speed and
agility, and may also run over ground. May follow army-ant
swarms, catching small animals that they flush, as do other
Micrastur.

REPRODUCTIVE BIOLOGY
Unknown. Nests as solitary pair probably in tree holes, as do
other *Micrastur*.

CONSERVATION STATUS
Vulnerable. Logging increasingly removes undisturbed closed-
canopy forest on which the species depends. Infrastructure,
particularly roads, has opened the region to agriculture, mining, and exploitation by international interests. Few pairs have
ever been found and the species' secretive nature makes it impossible to know the population size.

SIGNIFICANCE TO HUMANS
None known. ◆

Spiziapteryx circumcinctus
▨ Resident

Spot-winged falconet
Spiziapteryx circumcinctus

SUBFAMILY
Falconinae

TAXONOMY
Harpagus circumcinctus Kaup, 1852, Chile; error, Mendoza, Argentina. Monotypic.

OTHER COMMON NAMES
French: Carnifex à ailes tachetées; German: Tropfenfalke;
Spanish: Halconcito Argentino.

PHYSICAL CHARACTERISTICS
11.0–13.0 in (28–33 cm); about 3.5 oz (100 g). Female larger
than male. A slight short-winged, long-tailed forest dweller.
Head, back, and wings gray-brown; a pale streak that extends
back from above the eye, black ear coverts and moustachial
streak form a characteristic pattern. Shoulders and wing spotted conspicuously with white. Underparts pale gray, narrowly
streaked with brown. Central tail feathers black, remainder

barred with white to form wide black bars. No distinctive juvenal plumage (unlike other falconids).

DISTRIBUTION
South America: eastern Bolivia, through Paraguay to north and
central Argentina.

HABITAT
Savanna, scrubby woodland, and semidesert.

BEHAVIOR
Sedentary. Roosts in communal nests of monk parakeets
(*Myiopsitta monachus*) in winter, even when the nests are occu-
pied by parakeets.

FEEDING ECOLOGY AND DIET
Hunts mainly birds, some as large as they are, and lizards and
insects such as locusts and cicadas. Bird prey includes nestling
and adult monk parakeets.

REPRODUCTIVE BIOLOGY
Nests as solitary pair in about November–December in woven
nests of other species such as cachalotes (Furnariidae) and
colonial monk parakeets; falconets enlarge the nest and entrance. In Argentina, 15 of 70 parakeet nests (21%) had nesting
pairs of falconets. Lays clutches of two to four eggs; incubation
period is unknown; young fledge after about 33 days.

CONSERVATION STATUS
Not threatened. Status is unknown but habitat is not seriously
degraded.

SIGNIFICANCE TO HUMANS
None known. ◆

White-fronted falconet

Microhierax latifrons

SUBFAMILY
Falconinae

TAXONOMY
Microhierax latifrons Sharpe, 1879, Lawas River and Lumbidan, Borneo. Monotypic.

OTHER COMMON NAMES
French: Carnifex à ailes tachetées; German: Tropfenfalke; Spanish: Halconcito Argentino.

PHYSICAL CHARACTERISTICS
5.9–6.7 in (15–17 cm); 1.2–2.3 oz (35–65 g). Female larger than male. The second smallest falconid. Small, swift, and powerful for its size. Upperparts mainly black; underparts white with light tan wash to belly. Black eye stripe. Forehead, crown, and cheek white in male, chestnut in female. Juvenile similar to adult female.

DISTRIBUTION
Northern Borneo: Sabah and far northeast Sarawak.

HABITAT
Open forest and forest clearings and edges, up to about 3,900 ft (1,200 m).

BEHAVIOR
Sedentary.

FEEDING ECOLOGY AND DIET
Hunts from perch to snatch prey from the air and tree canopies. Mainly large insects such as bees, moths, cicadas, butterflies; sometimes birds and perhaps lizards and bats. May hunt in family parties and feed communally, sharing prey.

REPRODUCTIVE BIOLOGY
Little known. Nests as solitary pairs in sites such as barbet and woodpecker holes in March–April.

CONSERVATION STATUS
Listed as Near Threatened because of its restricted range and uncertain status. Forest clearing is a threat. Nevertheless, the species appears to be fairly common and copes with some disturbance of its habitat.

SIGNIFICANCE TO HUMANS
None known. ◆

Fox kestrel

Falco alopex

SUBFAMILY
Falconinae

TAXONOMY
Tinnunculus alopex Heuglin, 1861, Gallabat, Sudan. Monotypic.

OTHER COMMON NAMES
French: Crécerelle renard; German: Fuchsfalke; Spanish: Cernícalo Zorruno.

PHYSICAL CHARACTERISTICS
13.8–15.4 in (35–39 cm); 8.8–10.6 oz (250–300 g). A large deep-chestnut kestrel with long broad wings and long tail. Chestnut all over, streaked with black. Juveniles are more heavily marked. Unlike most kestrels, sexes indistinguishable by color but female larger than male.

DISTRIBUTION
Central Africa: from Senegambia east to Red Sea, Ethiopia, south to northeast Zaire, northwest Kenya, and northeast Uganda.

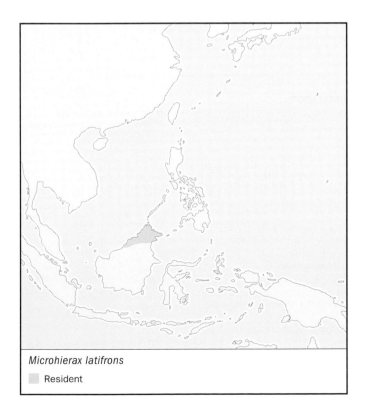

Microhierax latifrons

■ Resident

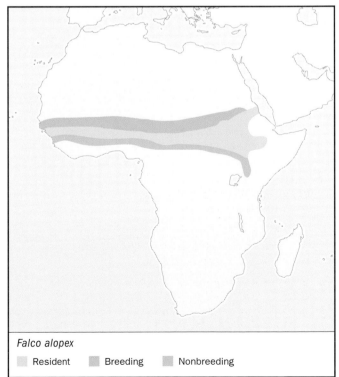

Falco alopex

■ Resident   ■ Breeding   ■ Nonbreeding

HABITAT
Cliffs, rocky outcrops, and hills adjoining open arid savanna.

BEHAVIOR
Moves south from more arid parts of range in dry season (October to March) to attend bush fires; with rains, returns north to nest. Elsewhere appears sedentary.

FEEDING ECOLOGY AND DIET
Hunts large insects, small mammals and birds, and lizards. Drops onto prey from perch or from hover, or snatches from the air in direct flight (termites). Said to remain on the wing for long periods. Follows fires to catch animals flushed by the flames.

REPRODUCTIVE BIOLOGY
Nests semicolonially, 20–25 pairs in proximity on a rock face. Lays about March to May in a shallow scrape on a ledge. Clutch size two or three.

CONSERVATION STATUS
Not threatened. Species little known but thought to be secure. Human exploitation and savanna degradation may be a threat.

SIGNIFICANCE TO HUMANS
None known. ◆

Amur falcon
Falco amurensis

SUBFAMILY
Falconinae

TAXONOMY
Falco vespertinus var. *amurensis* Radde, 1863, Zeya River, Amurland. Sister species to very similar red-footed falcon *F. vespertinus*. Monotypic.

OTHER COMMON NAMES
English: Manchurian, Amur, or eastern red-footed falcon; French: Faucon de l'Amour; German: Amurfalke; Spanish: Cernícalo del Amur.

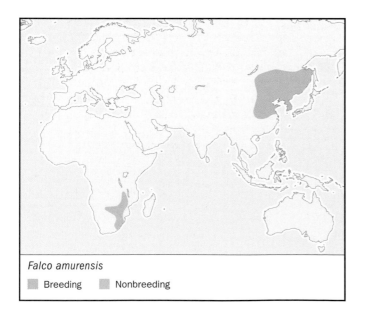

Falco amurensis
■ Breeding   ■ Nonbreeding

PHYSICAL CHARACTERISTICS
11.0–11.8 in (28–30 cm); male 3.4–5.5 oz (95–155 g), female 3.9–6.7 oz (110–190 g). A colorful little falcon with unusual red legs, cere, and eye ring. Male mostly slate gray with contrasting chestnut lower belly, undertail coverts, and thighs. Female very different: upperparts gray; underparts whitish streaked and chevroned with black. Juvenile similar to adult female but with rufous edging to upperparts.

DISTRIBUTION
Breeds in southeast Siberia and from northeast Mongolia to Amurland, south into northern and eastern China and North Korea, occasionally northeast India. Winters in southern Africa, mainly from Malawi to Transvaal.

HABITAT
Breeding grounds: open woodland and woodland (coniferous and deciduous) margins and marshes, likes mature trees and avoids treeless areas. Wintering grounds: savanna and grassland with clumps of trees.

BEHAVIOR
Whole population migrates from eastern Asia to southern Africa, often in large groups, sometimes with other small falcons. Leaves Asia in September, begins arriving in Africa in late November. In Africa, roosts at traditional sites in groups of trees, used by hundreds to thousands of birds. Returns to Asia, departing mainly March, arriving mainly April.

FEEDING ECOLOGY AND DIET
Feeds on insects such as locusts, grasshoppers, and beetles, and on small birds and occasionally amphibians. Hunts, often around dusk, from perch to capture prey in air, less often on ground, sometimes hovers. Outside breeding, gathers in flocks to feed on termite and ant swarms often associated with rain storms.

REPRODUCTIVE BIOLOGY
Nests annually as solitary pair or in small colonies, in old corvid nests or tree holes. Lays in May–June; usually three or four eggs; incubation about 28–30 days; fledges at about one month.

CONSERVATION STATUS
Not threatened. Thought to be common, at least locally.

SIGNIFICANCE TO HUMANS
None known. ◆

Brown falcon
Falco berigora

SUBFAMILY
Falconinae

TAXONOMY
Falco berigora Vigors and Horsfield, 1827, New South Wales.

OTHER COMMON NAMES
English: Brown hawk; French: Faucon bérigora; German: Habitchfalke; Spanish: Halcón Berigora.

PHYSICAL CHARACTERISTICS
16.1–20.1 in (41–51 cm); male 0.7–1.3 lb (316–590 g), female 0.9–1.9 lb (430–860 g). A medium-sized long-legged, buteo-like falcon. Adults extremely variable: from tan and buff to

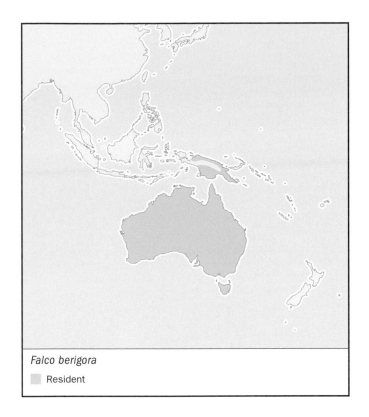

Falco berigora
▨ Resident

chocolate brown with variable white underparts (males tend to have more white), to near black all over (with some barring visible in wing and tail). Juveniles brown with buff touches to forehead, nape, and vent. Some regional variation in predominant color. Tend to be darker in humid areas, paler in arid areas, smaller in tropics, larger in temperate zone.

DISTRIBUTION
Australia and New Guinea, except highlands.

HABITAT
Open woodland, savanna, grassland, farmland, and desert up to about 6,600 ft (2,000 m).

BEHAVIOR
Adults sedentary. Gathers in sometimes large flocks post-breeding, especially at fires and locust and mouse plagues, sometimes with other raptors.

FEEDING ECOLOGY AND DIET
Versatile and opportunist hunter: takes prey from perch, hover, in direct flight, or running over the ground. Occasionally robs other raptors. Attracted to fires, cattle herds, farm machinery, and livestock for the animals they flush. Pairs occasionally hunt cooperatively. Feeds on fresh carrion but takes mostly live prey: mammals, birds, reptiles (especially snakes), amphibians, and large insects; rarely crabs and fish.

REPRODUCTIVE BIOLOGY
Breeds annually as solitary pair in stick nest of mainly corvids, and other raptors and magpies *Gymnorhina tibicen*; rarely on a man-made structure, tree fern, tree hole, cliff, or termitarium. Lays mainly August–September in south, earlier in north. Clutch 2–3, mostly three; incubation 33 days; young fledge at five to six weeks.

CONSERVATION STATUS
Not threatened. Generally common and extremely widespread. Expanded into forested areas turned to farmland and open woodland.

SIGNIFICANCE TO HUMANS
Traditional significance and food to some aboriginal tribes, but practice largely lapsed. ◆

Gyrfalcon
Falco rusticolis

SUBFAMILY
Falconinae

TAXONOMY
Falco rusticolus Linnaeus, 1758, Sweden. Monotypic (no subspecies).

OTHER COMMON NAMES
English: Gyrfalcon; French: Faucon gerfaut; German: Gerfalke; Spanish: Halcón Gerifalte.

PHYSICAL CHARACTERISTICS
Male 18.9–24.0 in (48–61 cm), female 20.1–25.2 in (51–64 cm); male 1.8–2.9 lb (800–1,325 g), female 2.2–4.6 lb (1,000–2,100 g). Largest of the falconids and the only white falcon. Highly variable plumage: from nearly pure white through various barred, chevroned, and streaked gray plumages to nearly uniform dark gray-brown. Adults have bright yellow legs and feet. Juveniles tend to be slightly browner and more heavily streaked; pale gray legs and feet. White form usual in high Arctic; dark form in Labrador; gray forms predominate in Iceland; mostly gray individuals grading to equal numbers of white individuals from west to east across Russia and Siberia.

DISTRIBUTION
The most northern of all diurnal raptors. Breeds around the Arctic circle: Iceland, Greenland, North America, and Eurasia; winters farther south.

HABITAT
Fairly uniform habitat: tundra and taiga, from sea level to about 4,600 ft (1,400 m), ice bound and snow covered much of the year. Favors rivers and seacoasts, also mountains. Winter migrant to ice edge, farmland, agricultural land, and steppe.

BEHAVIOR
In populations below 70° north many birds are resident, especially adult males. Migratory above 70° north, moves mainly but not only south to over-winter in warmer areas where prey is plentiful, mostly north of 40° north. Juvenile tracked with satellite transmitter from Alaska to Russia and back.

FEEDING ECOLOGY AND DIET
Hunts mostly ground-dwelling birds and mammals, such as ptarmigan, grouse, ground squirrels, and lemmings. Mostly flies low and fast to surprise and flush prey; occasionally takes birds after pursuit on the wing, and lifts waterfowl and shorebirds from water.

REPRODUCTIVE BIOLOGY
Breeds annually as solitary pairs from March to July. Lays eggs in depression on cliff ledge, large stick nest of another species, or man-made structure. High variation in clutch size and nesting success, depending on prey availability; clutch usually three

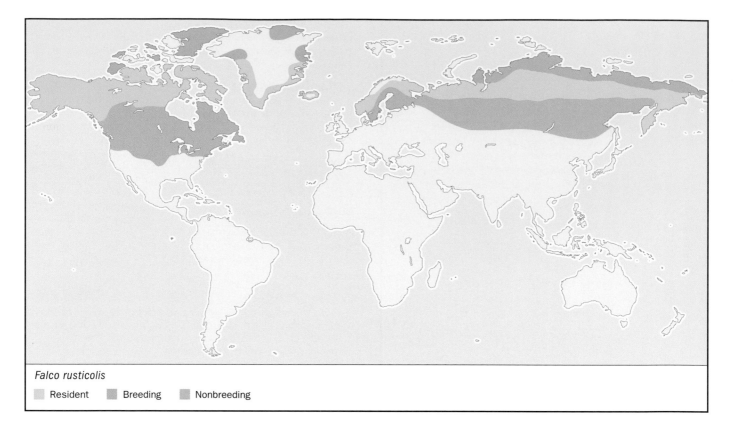

Falco rusticolis

▢ Resident   ▨ Breeding   ▨ Nonbreeding

or four but up to seven; incubation 33–36 days; fledges at about seven weeks.

CONSERVATION STATUS
Not threatened. Naturally uncommon but can be locally common. Its preference for remote habitat gives it some protection from threats to many other raptors. Fur trappers in Arctic Russia may kill 1,000–2,000 annually, some are taken by egg collectors and falconers.

SIGNIFICANCE TO HUMANS
Prized for falconry, but probably small numbers taken from wild and are now bred in captivity for that purpose. ◆

Peregrine falcon
Falco peregrinus

SUBFAMILY
Falconinae

TAXONOMY
Falco peregrinus Tunstall, 1771, Great Britain. About 18 subspecies.

OTHER COMMON NAMES
English: Duck hawk; Barbary falcon, Kleinschmidt's falcon, Peale's falcon; French: Faucon pèlerin; German: Wanderfalke; Spanish: Halcón Peregrino.

PHYSICAL CHARACTERISTICS
13.4–19.7 in (34–50 cm); 1.1–3.3 lb (500–1,500 g); female considerably larger than male (15–20% larger; 50% heavier). Size

and color varies according to subspecies: largest is Peale's falcon (*F. p. pealei*) of coastal North America and the Aleutian Islands; smallest are desert forms of Eurasia and Africa. A handsome, powerful falcon with pointed wings, a rather short tail and a black-helmeted head. Upperparts blue-black to charcoal; underparts white to rufous with black broken bars. Juveniles (first year) are browner, with streaking to underparts. Subspecies tend to be darkest in humid areas, palest in deserts.

DISTRIBUTION
Perhaps the widest breeding distribution of any bird. Almost worldwide, on all continents except the Antarctic and many oceanic islands (such as Fiji); notably absent from the high Arctic (Iceland, Newfoundland) and New Zealand.

HABITAT
Among the most variable of all birds, from the hot tropics to cold coasts and islands, dry deserts and rugged cliff lines to forest and flat treeless tundra, from sea level to about 13,000 ft (4,000 m).

BEHAVIOR
Many Northern hemisphere populations are long-distance migrants, moving south for the winter, stopping to rest and eat en route, with the most northern birds often travelling farthest south (for example, from Greenland to extreme southern South America). In Southern hemisphere and at mid-latitudes, adults are mainly resident year round.

FEEDING ECOLOGY AND DIET
A specialist hunter of birds caught in flight. Attacks from a perch or the air, sometimes in a spectacular stoop at great speed to strike prey in mid-air, less often on the ground or in water. Favors flock species, particularly pigeons and doves,

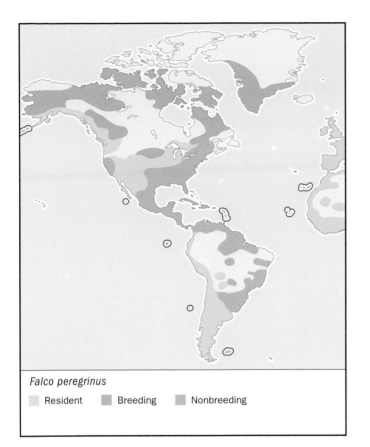

Falco peregrinus

☐ Resident   ■ Breeding   ■ Nonbreeding

parrots, auks and petrels, European starling (*Sturnus vulgaris*). Occasionally also mammals such as rabbits, voles, bats, large insects, and (rarely) reptiles and fish. Takes prey up to 4.4 lb (2 kg). Pairs occasionally hunt cooperatively, one splitting a flock and the other following to catch the scattering birds unawares.

REPRODUCTIVE BIOLOGY

Breeds annually as solitary pair, details depending on region. Lays eggs usually in late winter to spring. Does not build a nest, rather makes a shallow scrape in the substrate of the nest, which is most often on a cliff ledge or in a pothole or cave; also on buildings and bridges, on the ground (in predator-free parts of its range such as the Baltic bogs), in large abandoned stick nests of another species (Australia, Germany), or a large tree hole (mainly Australia). Clutch is larger in Northern hemisphere: usually three or four eggs in the Arctic and two or three in Africa and Australia. Incubation lasts about 30–33 days and chicks fledge when they are about five to six weeks old.

CONSERVATION STATUS

Not globally threatened. Populations plummeted in the 1960s and 1970s from ill effects of chlorinated hydrocarbon pesticides used in griculture. After widespread banning of the chemicals beginning in the 1970s, most populations have recovered, some assisted by releases of captive-bred birds. Population numbers typically very stable. A few island subspecies are rare but may always have been so. Still persecuted by keepers of racing pigeons.

SIGNIFICANCE TO HUMANS

Where falconry is legal, the peregrine is a bird of choice. During the pesticide era, the species became a flagship species for the conservation movement and remains a charismatic reminder of that time. ◆

Resources

Books

BirdLife International. *Threatened Birds of the World.* Barcelona and Cambridge: Lynx Edicions and BirdLife International, 2000.

Brown, L. H., E. K. Urban, and K. Newman. *The Birds of Africa.* Vol. 1. London: Academic Press, 1982.

Cade, T. J. *Falcons of the World.* London: Collins, 1982.

Cade, T. J. "Progress in Translocation of Diurnal Raptors." In *Raptors at Risk,* ed. R. D. Chancellor and B.-U. Meyburg. Berlin and London: World Working Group on Birds of Prey and Hancock House, 2000.

Clum, N. J., and T. J. Cade. "Gyrfalcon *Falco rusticolus.*" *Birds of North America.* No. 114. 1995.

Cramp, S., ed. *The Birds of the Western Palearctic.* Vol. II, Hawks to Bustards. Oxford: Oxford University Press, 1980.

del Hoyo, J. A., A. Elliot, and J. Sargatal, eds. *Handbook of the Birds of the World.* Vol. 2. *New World Vultures to Guineafowl.* Barcelona: Lynx Edicions, 2000.

Eastham, C. P., J. L. Quinn, and N. C. Fox. "Saker *Falco cherrug* and Peregrine *Falco peregrinus* Falcons in Asia: Determining Migration Routes and Trapping Pressure." In *Raptors at Risk,* ed. R. D. Chancellor and B.-U. Meyburg.

Berlin and London: World Working Group on Birds of Prey and Hancock House, 2000.

Fox, N. *Understanding the Bird of Prey.* Surrey: Hancock House, 1995.

Love, O. P., and D. M. Bird. "Raptors in Urban Landscapes: A Review and Future Concerns." *Raptors at Risk,* ed. R. D. Chancellor and B.-U. Meyburg. Berlin and London: World Working Group on Birds of Prey and Hancock House, 2000.

Mindell, D. P. *Avian Molecular Evolution and Systematics.* San Diego: Academic Press, 1997.

Newton, I., and P. Olsen, eds. *Birds of Prey.* London: Merehurst, 1990.

Olsen, P. *Australian Birds of Prey.* Sydney and Baltimore: New South Wales Press and Johns Hopkins, 1995.

Sibley, C. G., and J. E. Ahlquist. *Phylogeny and Classification of Birds.* New Haven: Yale University Press, 1990.

Wink, M., and P. Heidrich. "Molecular Evolution and Ssystematics of the Owls." In *Owls: A Guide to Owls of the World,* ed. C. Konig, F. Weick, and J.-H. Becking. Sussex: Pica, 1999.

Resources

Organizations

The Hawk and Owl Trust. 11 St Marys Close, Abbotskerswell, Newton Abbot, Devon TQ12 5QF United Kingdom. Phone: +44 (0)1626 334864. Fax: +44 (0)1626 334864. E-mail: hawkandowl@aol.com Web site: <http://www.hawkandowltrust.org>

World Center for Birds of Prey, The Peregrine Fund.. 566 West Flying Hawk Lane, Boise, Idaho 83709 USA.

Phone: (208) 362-3716. Fax: (208) 362-2376. E-mail: tpf@peregrinefund.org Web site: <www.peregrinefund.org>

World Working Group on Birds of Prey and Owls. P.O. Box 52, Towcester, NN12 7ZW United Kingdom. Phone: +44 1 604 862 331. Fax: +44 1 604 862 331. E-mail: WWGBP@aol.com Web site: <www.raptors-international.de>

Penny Olsen, PhD

Anseriformes

(Ducks, geese, swans, and screamers)

Class Aves

Order Anseriformes

Number of families 2

Number of genera, species 41 genera; 147 species

Photo: Mute swans (*Cygnus olor*) in Munich, Germany. (Photo by Frank Krahmer. Bruce Coleman Inc. Reproduced by permission.)

Evolution and systematics

Although the flamingos (Phoenicopteriformes) have webbed feet, goose-like calls, and complex bill laminae indicating a close relationship to the Anseriformes, there is compelling evidence that waterfowl more directly evolved from a galliform ancestor allied to the guans and curassows (Cracidae) of the New World. This theory is supported by the existence of the Neotropical screamers (Anhimidae), which share a number of anatomical features with waterfowl, and even more strongly with the unique magpie goose (*Anseranas semipalmata*) of Australia and New Guinea. The latter exhibits so many peculiar features that set it apart from all other ducks, geese, and swans and link it with the screamers that Sibley and Monroe have allocated its own monotypic family, the Anseranatidae.

The ancestors of present day waterfowl probably began their evolution in tropical swamps prior to the Eocene age more than 50 million years ago. However, the earliest undoubted fossil anserine fragments come from the latter part of the Eocene epoch discovered in deposits in Colorado, Utah, and France. Although over 100 extinct species of waterfowl have been described, many remains are indistinguishable from those of present-day species.

As of 2001, it is generally recognized that the 147 species and 41 genera of the Anatidae fall into seven sub-families: magpie goose (Anseranatinae), whistling-ducks (Dendrocyginae), stiff-tails (Oxyurinae), swans and geese (Anserinae), shelducks (Tadorninae), dabbling ducks (Anatinae), and sea ducks (Merginae). An alternative systematic interpretation of the number of taxa recognized as species was proposed by Bradley Livezey in 1997; based on phylogenetic (cladistic) analysis of 157 morphological features, Livezey suggested that the number of species be raised to 175. However, limits between several of these forms remain ill-defined; particularly difficult to interpret are relationships between the various forms of torrent duck (*Merganetta armata*), common eider (*Somateria mollissima*), Canada goose (*Branta canadensis*), and brant (*Branta bernicla*).

Physical characteristics

Members of the Anatidae show considerable variation in size and structure from the smallest, the tropical pygmy-geese (*Nettapus* spp.) measuring only 12 in (31 cm) and weighing a meager 10 oz (269 g), to the immense trumpeter swan (*Cygnus buccinator*) which is 72 in (183 cm) in length and can weigh well over 38 lb (17 kg). Waterfowl have relatively large, compact bodies, long or prominent necks, relatively short, stout tarsi, and, unlike the magpie goose, full webbing between the three forward-pointing toes (the hind toe or hallux is insignificant). To keep their plumage in good waterproofed condition, they are endowed with a particularly large preen gland. They have a distinctive bill structure, with the lower mandible flat and the upper roughly cone-shaped tapering to a drop-like hard process, the "nail," at the tip. Waterfowl have horny lamellae along the interior of the bill near the cutting edges, which provide a sifting apparatus. Waterfowl cannot soar or glide to any extent but fly directly and rapidly with their necks outstretched. Only five surviving species are habitually flightless: three of the four species of steamer-ducks (*Tachyeres* spp.), and the closely-related Auckland Island teal (*Anas aucklandica*) and Campbell Island teal (*Anas nesiotis*), these include the southernmost of all duck species, and all have very restricted ranges. The Anatidae have a relatively broad, usually pointed, wing composed of 10–11 primaries. Their tails are flat with 12–24 feathers and usually quite concealed by the folded wing. The young are nidifugous, have a dense downy plumage, and, except in some of the Oxyurinae, are tended for a long time by one or both parents.

Plumage coloration varies from the unpatterned white of most of the swans, through the drab brown of many of the geese to the brightly patterned and colored nuptial plumage of drakes of many northern ducks. The inconspicuous plumage of most female ducks, as with many other primarily ground-nesting birds, helps camouflage them while incubating. This does not apply to the males, most of which take no part in incubation. On the contrary, their conspicuous colors effectively enhance their species-specific display movements involved in pair formation and the maintenance of the pair bond. A metallic luster is especially developed on the secondaries of many ducks in the form of a colorful speculum, the pattern of which varies between species and no doubt helps as a species-recognition signal when mixed flocks of ducks take to the air. What might appear to be bright patterning in such colorful species as the harlequin duck (*Histrionicus histrionicus*) and wood duck (*Aix sponsa*) can be surprisingly cryptic, breaking up the shape of the bird and providing remarkable concealment when amongst the light and shade of dappled water or wave-splashed rocks.

In the true geese and swans, the sexes are similar in plumage and there is a single annual molt. However, the ducks—with few exceptions—are strongly sexually dimorphic and molt their contour feathers twice a year. In both groups there are exceptions, with several duck species undergoing only a partial second molt, while others like the long-tailed duck (*Clangula hyemalis*) have four partial plumage changes in a year. As a rule, female ducks have an inconspicuous plumage all year round, in which dull browns and grays predominate. Males of most migratory duck species have a pre-basic molt into a drab female-like "eclipse" plumage soon after breeding and undergo a pre-alternate molt back into their bright nuptial colors in the late fall and early winter when pair formation commences for many species.

In all ducks, swans, and geese the flight feathers are molted only once a year and dropped simultaneously as a completion of pre-basic molt, rendering the bird incapable of flight for a short period. The exception to this is the magpie goose (now allocated its own monotypic family) which has a gradual molt and retains the ability to fly throughout. The time of shedding the flight feathers is usually related to the reproductive cycle. In ducks, where only the female incubates and tends the young, males shed the flight feathers earlier than females. The latter shed these feathers only when the young are half grown. In the swans, the sequence is reversed; females become flightless soon after breeding and the males later, at a time when females can already make use of their wings to some extent. During the slow development of the young, the parents molt one after the other so that one of them is always ready to guide and defend the young. The ability to fly is reached by the young and regained by the adult male in late summer, the family is then ready to begin the fall migration or move to other waters.

Distribution

Waterfowl are extremely widespread in suitable habitats, being absent only from Antarctica and many of the oceanic islands. Some 50% of the 150 species breed in the northern hemisphere, with half of these executing long distance migrations. This enables them to take advantage of the rich food sources and nesting space available during the short boreal summers, but forces them to migrate to more temperate regions to avoid the inhospitable winters. Such birds have been responsible for establishing isolated populations on some oceanic islands. In some cases these have already evolved to acceptable species levels, for example Laysan duck (*Anas laysanensis*) and Hawaiian duck (*A. wyvilliana*) of the central Pacific evolved from a relatively recent mallard ancestor, and Eaton's pintail (*A. eatoni*) of the southern Indian Ocean from northern pintail (*A. acuta*). Many southern dabbling ducks are endemic to various islands, having presumably arisen from a common ancestor; most remarkable among these compose the eight species within the austral teal complex: Madagascar teal (*A. bernieri*), Andaman teal (*A. albogularis*), Sunda teal (*A. gibberifrons*), gray teal (*A. gracilis*), chestnut teal (*A. castanea*), brown teal (*A. chlorotis*), Auckland Island teal, and Campbell Island teal, the last four forming the more southerly "brown teal" group of species, and the first four the "gray teal" group. Of the many genera the most widespread is *Anas* (the dabbling ducks) which contains almost one third of all waterfowl species and is found worldwide. Surprisingly only one (the monotypic *Lophodytes*, hooded merganser) is endemic to north America and only two to Palearctic Eurasia (monotypic *Marmaronetta* [marbled duck} and *Mergellus* [smew]) but the two combined as the Holarctic region have 11 (of which five are monotypic but two are polytypic—*Anser*, with some 11 species and *Branta*, with at least five). In comparison, the southern hemisphere has no less than 10 genera endemic to the Neotropics (of which eight are monotypic), this preponderance of endemic genera to the southern hemisphere is supported by the eight monotypic Australasian genera and five monotypic genera in Africa. Some distribution patterns indicate that well-separated land-masses were once connected: the fulvous whistling-duck (*Dendrocygna bicolor*) and comb duck (*Sarkidiornis melanotos*) are found across South America, Africa, and Asia and the southern pochard (*Netta erythrophthalma*) and white-faced whistling-duck (*Dendrocygna viduata*) in both South America and Africa. Australasia has several very strange and presumably primitive species, apart from the peculiar magpie goose mentioned earlier, including the bizarre musk duck (*Biziura lobata*) and the weird freckled duck (*Stictonetta naevosa*), which is believed to be allied to the stifftails and is probably not far removed from the black-headed duck (*Heteronetta atricapilla*) of South America.

Habitat

Waterfowl can be found by almost any sort of wetland, from Arctic shores to mountain streams, steppe lakes, steamy rainforests, rivers, and tidal estuaries. Even some of the most familiar species, the Canada goose and mallard for example, can be found in their various forms in some of the most remote places on earth, yet appear equally at home on park lakes in cities. However, the majority of species are extremely specialized and fussy about their environment. Some, such as the pink-headed duck (*Rhodonessa caryophyllacea*), have become extinct through being over-specialized, the latter almost certainly having been a bird of the now fragmented swamp

The spectacled eider (*Somateria fischeri*) is an endangered species. This male and female are in the Yukon Delta National Wildlife Refuge in Alaska. (Photo by Stephen J. Krasemann. Photo Researchers, Inc. Reproduced by permission.)

grasslands of north-eastern India. Arctic and subarctic coastal waters are the abode of the various species of eiders, which have a marine existence matched only by the ecologically similar but quite unrelated steamer-ducks in southern South America. Although spending most of their lives on the sea, the scoters (*Melanitta* spp.) move inland to nest by freshwater lakes and pools, as does the diminutive long-tailed duck. Fast-flowing mountain streams are favored by the torrent ducks of the Andes, and flowing water features as the habitat preference of the Brazilian merganser (*Mergus octosetaceus*), blue duck (*Hymenolaimus malacorhynchos*) of New Zealand, and Salvadori's duck (*Salvadorina waigiuensis*) of New Guinea. The beautiful harlequin duck seems to be equally at home by mountain rivers and rocky coasts. By complete contrast a number of species are closely associated with forests, at least in the breeding season, the goldeneyes (*Bucephala* spp.), most mergansers, maned duck (*Chenonetta jubata*), wood duck, and mandarin all nest in tree holes in temperate regions. Much of the interior of Australia is very arid; sudden rainfall can create huge temporary shallow lakes, which can trigger-off a sudden colonization following the arrival of large numbers of nomadic pink-eared ducks (*Malacorhynchus membranaceus*).

Behavior

The majority of species are diurnal; in winter, flocks of most species of northern geese and swans feed in fields during the daytime, and fly to the safety of lakes or tidal flats at dusk, returning to their feeding areas at daybreak. However, if persecuted by man the reverse often happens. Most waterfowl are sociable when not nesting, with large winter gather-

ings of geese and swans being composed of many family units, which stay together through the first year of the young bird's life. Winter gatherings of waterfowl can be enormous, particularly notable being the thousands of snow geese (*Chen caerulescens*) that winter on farmland in the southern United States and the very dense concentrations, totaling some 300,000, of Baikal teal (*Anas formosa*) estimated wintering in South Korea. Most waterfowl are relatively solitary when nesting, although bar-headed geese (*Anser indicus*) form colonies composed of hundreds of nests by some Tibetan lakes and some colonies of black swans (*Cygnus atratus*) in Australia and New Zealand are numbered in their thousands.

Vocal communication between waterfowl is particularly important with the larger species that habitually flock prior to migration and keep together in family units during the ensuing winter. Males of the Anatidae have a sac-like dilatation, the bulla, at the bifurcation of the trachea. The whistling-ducks have a symmetrically placed bulla in both sexes. Looped trachea found in the magpie goose and the northern swans allow these birds to call with a far-carrying bugling quality. The mute swan (*Cygnus olor*) lacks such a resonating chamber; its trachea is straight and its voice is restricted to grunts, snorts, and hisses, which have given it its English name. Instead of vocalizations, mute swans keep in contact during flight by producing a "singing" wing sound, audible for more than a 109 yd (100 m). In geese, swans, and whistling-ducks, the voices of the two sexes are very similar; they do, however, substantially differ between the sexes of most species of ducks and sheldgeese. Males of most duck species are relatively quiet, some such as the marbled duck (*Marmaronetta angustirostris*) are virtually silent. The majority however do produce some

Mute swans (*Cygnus olor*) at their nest. (Photo by Manfred Danegger/ OKAPIA. Photo Researchers, Inc. Reproduced by permission.)

form of rasping or whistled sounds during courtship, which is more strongly developed in some species than in others: the far-carrying yodeling of spring long-tailed ducks is a very evocative sound and the whistled cries from drake wigeons are far from insignificant. Most female ducks utter a variety of abrupt grunts or quacks given for different occasions, and several Anas species have a well-known descending mocking cackle uttered by females when left alone by their partners. Shrill plaintive peeping cries are given by the small young of most species which help keep the family in contact while foraging.

Most geese and swans migrate in flocks along relatively narrow traditional fly-ways and have regular stop-over points en route where they can rest and feed in large numbers. Geese and swans form life-long pair bonds and migrate together with their youngsters of that year as family units; thus the young birds seemingly learn the route from older, experienced birds. Movements of many southern hemisphere waterfowl are closely linked with seasonal rainfall, those of Australia being particularly vulnerable to dramatic dispersive movements into the interior following rains and rapid range contractions towards the south-east following droughts. Several species also perform post-breeding molt-migrations, this may allow the birds to gather in large numbers for safety at a time when they are flightless and vulnerable and presumably produces a rich food source to sustain them at a difficult time.

Feeding ecology and diet

The Anatidae can be divided into several groups according to their methods of obtaining food. Many geese effectively graze on land, cropping grass and similar vegetation, and gleaning fallen seeds and grain. Swans, shelducks, and surface-feeding ducks either dabble at the surface of the water or mud, swinging their bills and sifting for small invertebrates and plant materials, or up-end by immersing their heads and necks with the rear of the body projecting upright above the water. In this way they can reach the bottom of shallow waters with their bills. Diving ducks also generally get their food from the bottom, but they can reach greater depths by diving completely below the surface. Most diving ducks rely on their feet for propulsion when underwater, but some, the long-tailed duck for example, use their wings as well. Perhaps the greatest maneuverability is achieved by the mergansers, who use only their feet to execute remarkably quick twists and turns in the pursuit of fish. In most waterfowl (the mergansers are the exception) there are rows of horny lamellae inside the mandibles which, together with the tongue, form a fine sifting apparatus. The tongue serves to push food items between the ridges of the bill in grazing birds, in dabbling ducks and swans it functions as a suction piston sucking the muddy water in through the tip of the beak and squeezing it out again at the base. This creates a characteristic chattering sound as food particles are filtered through the mandibles.

Reproductive biology

The young of most duck species are sexually mature when nine to 11 months old; thus they breed in the second summer of their life. Whistling-ducks generally breed when one year old, the true geese at three years, but the swans do not do so until they reach four to five years of age. The duration of the pair bond varies; in most species of ducks it ends with egg-laying, while in others, the males remain near the nest until the young hatch, and only then gather into continually growing flocks of males. In the swans and true geese, the males always guard the nest and take a share in leading the family after hatching. Thus a permanent pair bond results that is often so strong that a widowed bird will not pair up again for the rest of its life. Interestingly, such species show little or no plumage difference between the sexes, whereas the majority of ducks, which have short-term pair bonds lasting only through a single season, are strongly sexually-dimorphic.

Colorful male plumage, aided by vocalizations and display postures, is an important factor in selecting a mate of the correct species and in building pair bonds. This is particularly true of migratory northern species which have a relatively short period of time to complete the breeding process. In complete contrast, many southern, more resident species, even within the same genus, show little or no sexual dimorphism. The peculiar stiff-tails have a different strategy in which males select a territorial patch from which they vigorously display to attract females, the grotesque musk duck of Australia may mate with several females in a season, with the oldest and biggest males faring better than younger novices. Unlike many other families of birds courtship feeding is virtually unknown amongst waterfowl, and mutual preening is only normally undertaken by the whistling-ducks. Copulation almost always occurs on water, only the Hawaiian goose (*Branta sandvicensis*) and Cape Barren goose (*Cereopsis novae-hollandiae*) like the unique magpie goose, copulate on land.

Despite its aversion to water, the Hawaiian goose's pre-copulatory display includes head-dipping and washing movements, indicating that its ancestors formerly copulated on water. Males of all ducks, geese, and swans have a copulatory organ that is relatively long and curves to the left when evaginated from the cloaca during copulation. Such an erectile pseudo-penis is particularly long in the freckled duck and the stiff-tails, but is found in only a few other orders of birds.

Many species of waterfowl build their nests as close as possible to water; but a number of surface-feeding ducks, able to walk with relative ease, will nest over a mile away from water. Although most waterfowl nest on the ground, some of the shelducks prefer disused mammal burrows, while a number of other ducks nest in tree cavities. Mallards nest almost anywhere, sometimes even utilizing nests of gamebirds, or disused tree nests of other large birds such as crows. The nest is a hollow, lined with stems and leaves pulled in from around the nest site. Shortly before the last egg is laid, the females pluck nest-down from their underparts to complete the lining. The color of the down differs between species, hole-nesting ducks tend to have whitish down (presumably to help the incubating bird locate its clutch in darkness) while ground-nesting species have dark down as an aid to camouflage. The eggs of nearly all waterfowl are pale and unmarked, although they do become stained by the body of the incubating bird. An exception is the relatively large rich brown eggs laid by the peculiar white-backed duck (*Thalassornis leuconotus*) of Africa. The largest eggs, at 12.5 oz (355 g), are produced by the trumpeter swan, while those of the smallest ducks weigh little over 0.7 oz (20 g).

In general, only the females incubate; however, males of the magpie goose, black swan, and the whistling-ducks help with these duties. Sometimes a female will dump her eggs in strange nests before she has constructed her own. In this way, almost unbelievably large clutches may arise: as many as 87 eggs have been counted in a single redhead (*Aythya americana*) nest. The black-headed duck of South America always lays its eggs in strange nests and does not concern itself about its offspring. It prefers to lay into the nests of the rosybill (*Netta peposaca*) but often lays in the nests of quite unrelated birds, including those of gulls or birds of prey.

In most waterfowl species the female lays one egg per day until the clutch is complete; in the swans there is an interval of two to three days between each egg. Clutch size varies considerably, from as few as two in the musk duck, to as many as 22 in the mallard.

The young of the Anatidae are among the best developed precocial birds. They are covered with a dense downy feathering that becomes water repellent by becoming greased during contact with their mother's abdominal plumage. Only a few hours after the hatching the mother leads the brood to water. Leaving the nest site is difficult for the young of tree nesters because of the drop involved, but the ducklings are light, their bones are soft and pliable, allowing them to fall without harm. Ducklings feed independently from their first day; the adults, normally the female, merely warm and protect their offspring. In some swans, the care of the young includes carrying tired youngsters on the parent's back. Goose

Mallard duck (*Anas platyrhynchos*) taking off in flight. (Photo by Manfred Danegger/OKAPIA. Photo Researchers, Inc. Reproduced by permission.)

families often return to their nests in the evening during the first few days. Ducks tend their youngsters until they are capable of flight (a period of some 40–70 days according to species). Geese and swans, as already mentioned, lead their young until the following spring.

Conservation status

As of 2001, 26 waterfowl species were listed as Endangered or Threatened. Three of the five species classed as Critically Endangered are probably already Extinct (crested shelduck, *Tadorna cristata*), pink-headed duck, and Madagascar pochard (*Aythya innotata*); however still surviving with tiny populations is the Brazilian merganser and Campbell Island teal. Four of the six Endangered waterfowl are also confined to islands, Madagascar teal and Meller's duck (*Anas melleri*) on Madagascar, brown teal (New Zealand) and Hawaiian duck, while the other two, swan goose (*Anser cygnoides*) and white-winged duck (*Cairina scutulata*), are found as fragmented populations in eastern Asia. A further 15 species are treated as Vulnerable, with another seven classed as Near Threatened, these are not yet considered rare enough to cause major concern.

Although hunting is a major factor, there is little doubt that habitat loss is the prime threat to populations of many species. Introduced predators have had a disastrous impact on the eco-systems of many island faunas in various parts of the world. Wetland drainage and methods of riverine management have been particularly harmful, with the building of dams making conditions unsuitable for species favoring faster flowing water. Many ports are situated near the mouths of large rivers, the accompanying spread of industrial development results in the pollution and loss of extensive areas of marshland. Although there is much international concern and legislation over oil pollution, accidents involving tankers do occur and their huge oil slicks spread death and destruction among waterbirds. As a counter-measure to these changes of the habitat, the management of seasonal flood plains and creation of freshwater marshes and shallow lakes as waterfowl refuges and reserves have been of much benefit to many species. The importance of protecting key wetlands cannot be overestimated, indeed an international conference on wetland conservation held in Iran in 1971, the Ramsar Convention, encouraged all countries to protect their key wetlands and submit them as sites of international importance; as of 2001, there were 126 countries that had signed up a total of 1,140 wetland sites. The work of many international conservation organizations, especially those of the Wildfowl and Wetlands Trust and Wetlands International, have done much to save rare species both by creating captive stock and in promoting wetland conservation.

Significance to humans

Humans have has historically had a close association with waterfowl as a food source. The greylag goose (*Anser anser*), swan goose, and mallard have long been known in domestication (for at least 4,000 years in the case of the greylag); all three were probably first domesticated in eastern Asia. Through selective breeding a number of local breeds, often bearing little resemblance to their wild ancestor, have been developed. By contrast the Muscovy duck came into domestication in South America. The mute swan was farmed in Europe during the middle ages. In Iceland colonial nesting common eiders have long been farmed for the excellence of the down that the females line their nests with. Swans have a number of myths and legends associated with them, these formed the basis for Tchaikovsky's ballet "Swan Lake." In Greek mythology, Zeus transformed into a swan to seduce the King of Sparta's beautiful wife, Leda (famed also as being the mother of Helen of Troy).

Resources

Books

del Hoyo, J., A. Elliott, and J. Sargatal, eds. *Handbook of Birds of the World*. Vol. 1, *Ostrich to Ducks*. Barcelona: Lynx Edicions, 1992.

Madge, S., and H. Burn. *Waterfowl*. New York: Houghton Mifflin, 1988.

Marchant, S. and P. J. Higgins, eds. *Handbook of Australian, New Zealand and Antarctic Birds*. Vol. 1, *Pelicans to Ducks*. Melbourne: Oxford University Press, 1990.

Todd, F. S. *Natural History of the Waterfowl*. Vista: Ibis Publishing Company, 1996.

Periodicals

Daugherty, C. H., M. Williams, and J. M. Hay. "Genetic Differentiation, Taxonomy and Conservation of Australasian Teals (*Anas* spp.)" *Bird Conservation International* 9 (1999): 29–42.

Kang, H.-Y., and S.-R. Cho. "Wintering Ecology of the Baikal Teal (*Anas formosa*) and Carrying Capacity of Their Habitats." *Korean Journal of Ornithology* 3 (1996): 33–41.

Livezey, B. C. "A Phylogenetic Classification of Waterfowl (Aves: Anseriformes), Including Selected Fossil Species." *Annals of Carnegie Museum* 66 (1997): 457–496.

Organizations

Wetlands International (the Americas). 7 Hinton Avenue North, Suite 200, Ottawa, Ontario K1Y 4P1 Canada. Phone: (613) 722-2090. E-mail: davidson@wetlands.org Web site: <http://www.wetlands.org>

Wildfowl and Wetlands Trust. Slimbridge, Glos GL2 7BT United Kingdom. Phone: +44 01453 891900.

Steve Madge

Ducks, geese, and swans
(Anatidae)

Class Aves

Order Anseriformes

Suborder Anseres

Family Anatidae

Thumbnail description
Medium to very large-sized birds of predominantly brown, white, black,and metallic green colors, with a broad stocky body, webbed feet, and a flattened bill

Size
13–71 in (34–180 cm); 0.56–29.8 lb (255 g–13.5 kg)

Number of genera, species
41 genera, 147 species

Habitat
Marshlands, coastal waters, lakes, rivers, and streams

Conservation status
Extinct: 5 species; Endangered: 5 species; Critically Endangered: 5 species; Vulnerable: 11 species; Near Threatened: 7 species

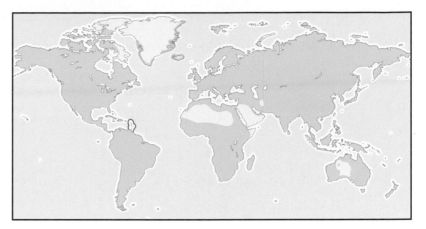

Distribution
All continents except Antarctica

Evolution and systematics

The taxonomic history of the Anatidae began early with the work of F. Willughby and J. Ray who, in 1676, wrote the first comprehensive classification of aquatic birds. The systematics of the Anatidae have evolved considerably since then; however, at the beginning of the twenty-first century much controversy remains. The different classificatory schemes by J. Delacour, R. Verheyen, B. C. Livezey, and others, recognize anywhere from two up to 14 subfamilies. Here we follow P. A. Johnsgard who divided the Anatidae into seven subfamilies: the Anseranatinae (magpie goose), Dendrocygninae (whistling-ducks), Anserinae (geese and swans), Tadorninae (shelducks), Anatinae (wood ducks, dabbling ducks, pochards), Merginae (sea ducks), and Oxyurinae (stiff-tailed ducks).

The earliest fossils that can be identified as anseriform were those of *Anatalavis rex*. Two bones recovered from the Hornerstown Formation of New Jersey may date back to the Late Cretaceous or early Paleocene (80–50 million years ago). Similar bones, about 40 million years old, were found in England and helped identify these fossils as a member of the Anseranatinae. The most common anseriform in the fossil record is *Presbyornis* from the Paleocene and early Eocene (65–50 million years ago). According to S. L. Olson, *Presbyornis* may have looked like "a duck-like skull on the body of a long-legged wading bird." On the evolutionary tree of the anseriformes, *Presbyornis* branched off between the Anseranatinae and the other six subfamilies.

Physical characteristics

The Anatidae range in size from the minute African pigmy geese (*Nettapus auritus*), measuring only up to 13 inches (33 cm) and weighing no more than 0.51 lb (230 g), to the large trumpeter swan (*Cygnus buccinator*), which reaches a body length of 70.87 in (1.80 m) and a weight of 30 lb (13.5 kg). Some mute swans (*Cygnus olor*) may even weigh as much as 49.6 lb (22.5 kg). Plumages range from dull and inconspicuous, as in the greylag goose (*Anser anser*), to spectacularly colorful as in the mandarin duck (*Aix galericulata*).

Despite their variety, the Anatidae have a characteristic "Bauplan," or external morphology, which makes them readily distinguishable from all other groups of birds. The most obvious characteristics are a somewhat flattened bill with horny lamellae, a broad body, and partially webbed feet. The members of this family also share a hard process, the "nail," at the tip of the bill, long necks, a large preen gland crowned by a tuft of feathers, and a large external penis in males.

The structure of the bill and the lamellae are perfectly suited to the bird's diet and feeding methods. True geese are mostly herbivorous and feed by grazing. The bills are therefore strong, the "nail," used to grasp vegetation, is wide and the lamella stout and flat. Ducks that strain food particles from the water or mud have blade-like lamellae. These are tightly packed in filtering specialists such as shovelers. Shovelers also have a very broad, spatula-like bill to enhance straining efficiency. The "nail" is small in dabblers and filter-feeders. A

White-faced whistling ducks (*Dendrocygna viduata*) allopreening. (Illustration by Emily Damstra)

muscovy duck (*Cairina moschata*) and crests as in the mergansers. A mane on the neck adorns the maned duck (*Chenonetta jubata*). Even more striking are the sickle feathers of the falcated teal (*Anas falcata*) and the elongated lancet-like flank feathers of the plumed whistling-duck (*Dendrocygna eytoni*). The wide, upward-bent inner vanes of the shoulder feathers of the mandarin duck (*Aix galericulata*) look like orange sails. In the male oldsquaw (*Clangula hyemalis*) the tail feathers are greatly elongated. Some members of the Anatidae are able to produce acoustic signals with their flight feathers. Lesser whistling-ducks (*Dendrocygna javanica*) produce a whistling sound, and golden-eyes (*Bucephala*) and the black scoter (*Melanitta nigra*) produce ringing sounds.

In some Anatidae, parts other than the plumage contribute to the decorative effect of the nuptial plumage. A highly colored fleshy hump at the base of the beak is present during the breeding season in many species, such as in some swans, scoters, and others. The bill may also be intensely colored in black, yellow, red, and blue.

In all ducks and geese, the flight feathers are molted only once a year. They are lost simultaneously so that the bird is incapable of flight for a short period of time. Only the magpie goose sheds its primaries sequentially and never loses its ability to fly. The contour feathers are usually molted twice a year, but three times in the oldsquaw (*Clangula hyemalis*). The "nest-down" on the abdomen grows before the onset of the breeding season. Before the clutch is complete, the incubating bird plucks down with its bill and pads the nest with it.

Distribution

From the Arctic Circle to Tierra del Fuego, South Georgia Island, New Zealand, and the Cape of Good Hope, ducks can be found on all continents except Antarctica.

striking adaptation to feeding can be found in the mergansers (*Lophodytes, Mergus*). The backward pointing lamellae are serrated, almost tooth-like, and the horny bill is unusually narrow. Thus mergansers are able to get a firm grasp on slippery fish, their main diet.

The broad cross-section of the body is a result of powerful, bulging pectoral muscles needed for continuous beating of the narrow, pointed wings during flight. Except for a few island endemics and three species of steamerducks (*Tachyeres brachypterus*), all waterfowl are strong flyers.

Short thighs and tarsi form a powerful lever arm that makes waterfowl good swimmers. With the exception of the magpie goose (*Anseranas semipalmata*) and the Hawaiian goose (*Branta sandvicensis*), all Anatidae have webs between their front toes. Trapped air in the plumage and in the respiratory system provides buoyancy. Because of the wide body and relatively short legs, ducks and geese have a distinctive waddling gait.

Males of all ducks, geese, and swans have a copulatory organ which is evaginated from the cloaca for copulation. The sperm do not flow through a central canal of this erectile penis, as in mammals, but along grooves on the outside.

The palette of colors found in the Anatidae is very varied and spans all colors from red to blue and white to black. Nuptial plumages are also decorated by modified feathers or whole regions of the plumage. There are curled hoods, as in the

An emperor goose (*Anser canagica*) takes a defensive posture over its hatching nest in Yukon Delta National Wildlife Refuge, Alaska. (Photo by Stephen J. Krasemann. Photo Researchers, Inc. Reproduced by permission.)

Snow geese (*Anser caerulescens*) fly in a V-formation over the St. Lawrence River in Quebec, Canada. (Photo by Jeff Lepore. Photo Researchers, Inc. Reproduced by permission.)

The magpie goose lives in Australia and New Guinea. The Dendrocygninae are mostly restricted to tropical or subtropical regions of North, Central, and South America, the West Indies, Africa, Asia, and Australia. They are absent in Europe. The Anserinae are limited in their distribution to mostly temperate and Subarctic regions and, with the exception of a few vagrants, are absent from Africa. The Hawaiian goose lives only on Hawaii Island. The Tadorninae have South American (10 species), African (five species), and Australasian representatives (five species). The Anatinae (86 species) are cosmopolitan. With 33 resident and nonresident species, Central and South America have the highest species richness, followed by Asia (28 species) and Africa (27 species). Europe and North America harbor 16 species each and Australia 11. This group includes the well-known mallard (*A. platyrhynchos*) and the American black duck (*A. rubripes*). Five species and 2 subspecies are limited entirely to islands, such as New Zealand, Madagascar, and Hawaii. The Merginae are a group of the north temperate and Arctic regions. Most species can be found in North America. The Oxyurinae are distributed on all continents, but are more prevalent in South America, Africa, and Australia.

Habitat

The Anatidae can be found anywhere as long as some wetland or body of water is present. For instance, the marbled teal (*Marmaronetta angustirostris*) lives in arid regions of the Mediterranean, the oldsquaw in the high Arctic tundra, and the Brazilian teal (*Amazonetta brasiliensis*) in the lush rainforests of South America. Five species require fast flowing streams and two hardly depend on aquatic habitats at all. The Cape Barren goose (*Cereopsis novaehollandiae*) enters the water only while molting its wing feathers or rearing young and the Hawaiian goose inhabits the lava fields of volcanoes. In contrast, the scoters (*Melanitta*) rarely go ashore except during reproduction. Several species can be found at high elevation. A good example is the bar-headed goose (*Anser indicus*) which breeds on the highland plateaus of the central Asia, between 13,100 and 16,400 ft (4,000–5,000 m).

Behavior

Almost half of the species (47.6%) in this family are either completely or partially migratory. Of the remaining species the vast majority wander over wide areas in response to changing water levels. During migration, waterfowl may fly at high altitudes, as high as 32,800 ft (10,000 m) in some geese.

One of the most notable aspects of waterfowl behavior are the highly ritualized displays. The specific functions of displays are to aid in family group cohesion, convey information about the reproductive status, establish pair bonds, defend a mate or territory, and prevent hybridization. The displays may have originated from common behavioral patterns as they bear a remarkable resemblance to movements used in plumage maintenance (e.g. preening, bathing, and shaking), feeding, and locomotion.

Typical for the Anserinae is the so-called "triumph ceremony," which is used to establish dominance status in wintering flocks as well as to advertise territories in the breeding season.

The Dendrocygninae are gregarious and to large degree nocturnal or crepuscular. Auditory signals are therefore very important. Within the Tadorninae, the highly aquatic steamerducks (*Tachyeres*) stretch their necks and cock their tails as the most conspicuous means to signal from the water surface. Paired females in *Tadorna* (also in *Aythya*) perform a display called "inciting." It shows an intruding male that she already has a partner and stimulates her mate to attack the intruder.

The voices of the Anatidae vary from whistling sounds in whistling-ducks and sea ducks, to a variety of typical "quack" sounds in dabbling ducks and deep honks in swans. Most species posses a bulla, a more or less ossified enlargement at the union between that trachea and the bronchi, that acts as a resonating body.

The Anatinae are a very diverse group with a variety of habitat-specific displays. The torrent duck (*Merganetta*

A female mallard duck (*Anas platyrhynchos*) feeds in the water. (Photo by Wayne Lankinen. Bruce Coleman Inc. Reproduced by permission.)

A mute swan (*Cygnus olor*) cob (adult male) and cygnets (young swans) feed in Devon, United Kingdom. (Photo by Michael Black. Bruce Coleman Inc. Reproduced by permission.)

armatta) is especially interesting because it does not share its display repertoire with any other species of ducks. For instance, it uses its legs in hostile displays in which it also reveals painful wing spurs.

The Merginae have complex signaling systems as courtship occurs mainly on the water. Vocal repertoires are diverse and many species use courting flights and/or underwater pursuits. The precopulatory behaviors in common eiders (*Somateria mollisima*) are extremely, and puzzlingly, complex.

The Oxyurinae have many distinct behavioral features not shared with any other group. Birds may be seen contorting their necks and tails into bizarre postures.

Feeding ecology and diet

Although most swans, geese, and ducks require wetlands or other water-bodies for their survival, not all species forage in the water. The subfamilies Anseranatinae, Dendrocyginae, Anserinae, and some Tadorninae (especially *Chloephaga*) feed mostly on land on a largely vegetarian diet. *Tachyeres* steamerducks eat almost exclusively on aquatic invertebrates. The Anatinae are largely omnivorous and the Merganinae are largely piscivorous and insectivorous. The Oxyurinae, complement a diet based on aquatic invertebrates with seeds and green parts of aquatic plants.

The methods used in feeding are as varied as the diets. Vegetarians often graze or browse on land. Most species that feed in the water dabble, dip their heads into the water, or upend by immersing half of their body head down. The more aquatic species mostly forage by diving. The duration of the dives is usually 60–70 seconds, but may be as long as 2 min in some scoters (*Melanitta*).

Reproductive biology

Most species of Anatidae are monogamous. The Anserinae, Dendrocygninae, and Tadorninae may stay paired for several seasons, mute swans even for life. With the exception of some promiscuous stifftails, most ducks are seasonally monogamous. Pair bonds often last only halfway through the breeding season until midincubation or hatching of the young. The Anseranatinae are unusual among the waterfowl in that they form polygamous "trios."

Breeding in the temperate regions usually begins in the spring. In those species that are not perennially monogamous, pair formation is most often achieved before the breeding season. In other species, such as the ruddy duck (*Oxyura jamaicensis*), pair formation takes place on the breeding grounds. In tropical latitudes the beginning of the breeding season is variable, depending to large degree on water levels.

The nests are made on the ground with surrounding vegetation or in cavities. Most species also line their nests with down and feathers. The magpie goose builds its nest on a mound of floating vegetation. Most Dendrocygninae, Anserinae, and Anatinae make ground nests. The Tadorninae has both ground and cavity nesters. The Merginae nest on the ground, oftentimes in nothing more than a small depression. The Oxyurinae build their nests of vegetable matter on the ground although some stiff-tails reuse nests of other species.

On average, clutch size is of 4 to 16 eggs. Should a clutch fail, relaying occurs after 4–20 days. Incubation lasts between 22 and 40 days. Sexual maturity is usually attained after the first to third year. The eggs are large, and the young hatch completely covered with down and are able to swim and dive within a few hours.

An interesting aspect of anatid reproductive behavior is that many species parasitize the nests of other birds, most often of the same species. Up to 30 eggs may accumulate in so-called "dump nests." The black-headed duck (*Heteronetta atricapilla*) is entirely parasitic.

Male parental care (incubation and protection of young) is uncommon in the Anatidae. It can be found mostly in species that are perennially monogamous and in several tropical Anatinae. In the latter, an unpredictable environment and breeding season force the drake to secure a mate before environmental conditions are favorable. The male magpie goose and the comb duck even feed the young.

Conservation status

Five species have already become extinct due to overhunting in historic times and at least 14 more in prehistory. The latter are almost all island species that disappeared as people colonized the Pacific islands and New Zealand. Thirty-three species (22%) and 5 subspecies are under some category of threat as defined by the IUCN. Most of these species come from Asia or islands, such as Hawaii, New Zealand, and Madagascar. Of the four species listed as Critically Endangered, three, the crested shelduck (*Tadorna cristata*), the pink-headed duck (*Rhodonessa caryophyllacea*),

A female Cape Barren goose (*Cereopsis novaehollandiae*) sits on her nest while the male stands guard nearby, on Kangaroo Island, off the south coast of Australia. (Photo by Gregory G. Dimijian. Photo Researchers, Inc. Reproduced by permission.)

and the Madagascar pochard (*Aythya innotata*) may already have become Extinct. Five species and two subspecies are Endangered. The Vulnerable list includes 11 species and two subspecies. An additional seven species are considered Near Threatened.

The greatest threat comes from overhunting and wetland drainage. Large quantities of waterfowl are still hunted during migration, especially in Siberia. In Asia, a largely poor rural population also puts pressure on waterfowl by subsistence hunting. Wetland drainage results in a loss of suitable breeding habitat. The destruction of habitat surrounding the main breeding areas can also be detrimental. The river specialists, for instance, suffer from increased siltation caused by deforestation.

Significance to humans

The Anatidae have always had a close relationship with humans. At least four species have been domesticated: the greylag goose, swan goose (*Anser cygnoides*), muscovy duck, and mallard (*Anas platyrhynchos*). More than 23 variants of mallard are known. The mute swan and Canada goose (*Branta canadensis*) are kept in semi-liberty.

Waterfowl have been hunted ever since *Homo sapiens* walked the earth. Considered a delicacy by almost all cultures, at the end of the twentieth century their hunting and observation has also become a lucrative business with significant economic impact. In America, outdoor stores sell millions of dollars worth of hunting gear and many states also derive benefit through hunting licenses. More importantly, they are an important element in the maintenance of wetland ecosystems.

Ducks, geese, and swans figure prominently in myths and stories. For example, in Greek mythology Zeus took on the form of a swan in order to conquer Leda. In the Germanic Lohengrin saga, the knight is pulled over a lake by swans, which symbolize purity and love.

In science, observations on waterfowl displays have advanced our knowledge on animal behavior in general. Waterfowl still promise to be great model species for studies on sexual selection, the origin of mating systems, and the mechanistic basis of behavior.

1. Mandarin duck (*Aix galericulata*); 2. Comb duck (*Sarkidiornis melanotos*); 3. Cape Barren goose (*Cereopsis novaehollandiae*); 4. African pygmy goose (*Nettapus auritus*); 5. Magpie goose (*Anseranas semipalmata*); 6. Brown teal (*Anas aucklandica*); 7. Canada goose (*Branta canadensis*); 8. Ruddy shelduck (*Tadorna ferruginea*); 9. Magellanic steamerduck (*Tachyeres pteneres*); 10. White-faced whistling duck (*Dendrocygna viduata*); 11. Mute swan (*Cygnus olor*); 12. Salvadori's teal (*Anas waigiuensis*). (Illustration by Emily Damstra)

1. Black-headed duck (*Heteronetta atricapilla*); 2. Torrent duck (*Merganetta armata*); 3. American widgeon (*Anas americana*); 4. Madagascar pochard (*Aythya innotata*); 5. Harlequin duck (*Histrionicus histrionicus*); 6. Male oldsquaw (*Clangula hyemalis*) in summer plumage; 7. Brazilian merganser (*Mergus octosetaceus*); 8. Mallard (*Anas platyrhynchos*); 9. Marbled teal (*Marmaronetta angustirostris*); 10. Northern shoveler (*Anas clypeata*); 11. Musk duck (*Biziura lobata*); 12. King eider (*Somateria spectabilis*). (Illustration by Emily Damstra)

Species accounts

Magpie goose
Anseranas semipalmata

SUBFAMILY
Anseranatinae

TAXONOMY
Anas semipalmata Latham, 1798, Hawkesbury River, New South Wales, Australia. Monotypic.

OTHER COMMON NAMES
English: Pied goose; Semipalmated goose; French: Canaroie semipalmé; German: Spaltfußgans; Spanish: Ganso Urraco.

PHYSICAL CHARACTERISTICS
27.6–35.4 in (70–90 cm); female 4.4 lb (2.0 kg); male 6.2 lb (2.8 kg); webs on the toes reduced; hind toe very long.

DISTRIBUTION
Southern New Guinea and Queensland, Australia; reintroduced in Victoria, southeastern Australia.

HABITAT
Swamps and grasslands in riverine floodplains.

BEHAVIOR
Often bigamous. Parental care extensive. Gregarious.

FEEDING ECOLOGY AND DIET
Feeds on grasses and seeds using its feet to bend down taller plants. Also digs out roots and bulbs with its hooked bill.

REPRODUCTIVE BIOLOGY
Often bigamous, with one male paired with two females. One or both females lay 1-16 eggs in same nest, Feb.–Apr. in north, Aug.–Sept. in south. The nest is a mound of floating vegetation. Incubation 23–25 days; fledging c. 11 weeks; sexual maturity after two years in females and 3–4 years in males.

CONSERVATION STATUS
Size of populations increasing. Generally common in its habitat. Protected from hunting except during open season in Northern Territory, Australia.

SIGNIFICANCE TO HUMANS
Hunted traditionally by aboriginies. Considered a pest by rice growers. ◆

White-faced whistling duck
Dendrocygna viduata

SUBFAMILY
Dendrocygninae

TAXONOMY
Anas viduata Linnaeus, 1766, Cartagena, Colombia. Monotypic.

OTHER COMMON NAMES
English: White-faced tree duck; French: Dendrocygne veuf; German: Witwenpfeifgans; Spanish: Suirirí Cariblanco.

PHYSICAL CHARACTERISTICS
15–19 in (38–48 cm); 1.1–1.8 lb (502–820 g). Brown and gray feathers with dark neck and eponymous white face.

DISTRIBUTION
Throughout tropical America from Costa Rica south through northern and eastern Colombia and Guyana in east; in south

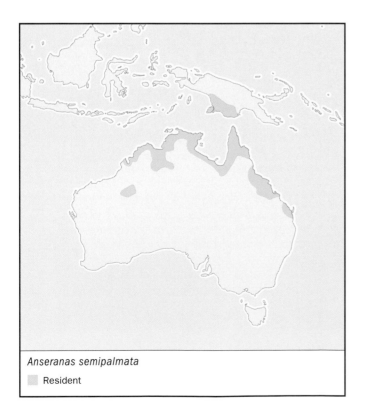

Anseranas semipalmata
▨ Resident

Dendrocygna viduata
▨ Resident

from eastern Bolivia east to Uruguay; Trinidad. In Africa south of the Sahara to Namibia and Natal; Madagascar; Comoro Islands.

HABITAT
Freshwater marshes, grassy lagoons, and flooded fields.

BEHAVIOR
Not territorial. Move regionally as a response to varying water levels.

FEEDING ECOLOGY AND DIET
Forages mainly at night by diving; also wades and dabbles near surface. Feeds on grasses, seeds, rice, and invertebrates.

REPRODUCTIVE BIOLOGY
Perennially monogamous. Breeds during rainy season. Well-concealed nests made on the ground. Lay 4–13 eggs; incubation 26–28 days; fledging c. 8 weeks.

CONSERVATION STATUS
Not threatened. Common throughout range.

SIGNIFICANCE TO HUMANS
None known. ◆

Mute swan
Cygnus olor

SUBFAMILY
Anserinae

TAXONOMY
Anas olor Gmelin, 1789, "Russia, Sibiria, Persico etiam littore maris Caspii." Monotypic.

OTHER COMMON NAMES
French: Cygne tuberculé; German: Höckerschwan; Spanish: Cisne Vulgar.

PHYSICAL CHARACTERISTICS
49–63 in (125–160 cm); 14.6–33 lb (6.6–15.0 kg). Characteristic knob on bill.

DISTRIBUTION
Central and northern Europe, locally in Russia and Siberia, patchily from Turkey to eastern China. Winters in northern Africa, Black Sea, northwestern India, and Korea. Populations anywhere else are introduced.

HABITAT
Freshwater marshes, lagoons, and rivers; artificial lakes and canals.

BEHAVIOR
Territorial. The 11.12 acres (4.5 ha) large territories defended very aggressively, even by killing other birds. Clap feet on water and perform rotation displays to advertise their territories. Migratory.

FEEDING ECOLOGY AND DIET
Feeds mostly on aquatic vegetation, seeds, grasses, small amphibians, and invertebrates. Dabbles and rarely dives.

REPRODUCTIVE BIOLOGY
Perennially monogamous. Breeds in spring. Nests are large platforms of vegetation built on floating mats or reeds. Lay

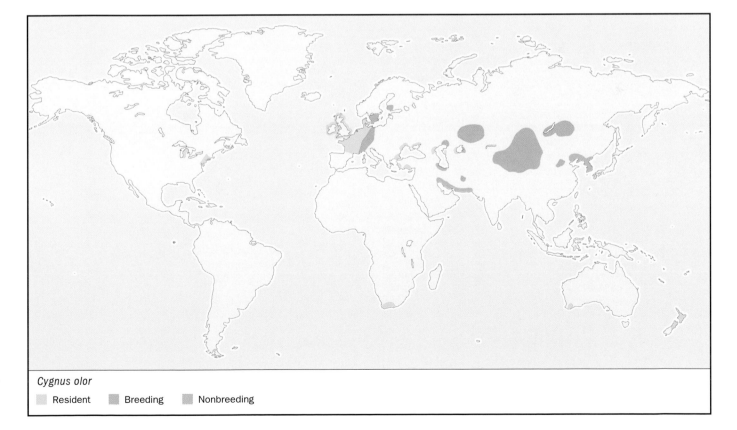

Cygnus olor

☐ Resident   ■ Breeding   ■ Nonbreeding

5–7 eggs. Incubation 35–36 days; fledging 120–150 days; become sexually mature at the earliest during the third year.

CONSERVATION STATUS
Range expanding and abundances increasing.

SIGNIFICANCE TO HUMANS
Symbolizes purity, love, and elegance in many cultures making it the stuff of myths. Common in parks. ◆

Canada goose
Branta canadensis

SUBFAMILY
Anserinae

TAXONOMY
Anas Canadensis Linnaeus, 1758, City of Quebec, Canada. Subspecies considered by some as full species.

OTHER COMMON NAMES
French: Bernache du Canada; German: Kanadagans; Spanish: Barnacla Canadiense.

PHYSICAL CHARACTERISTICS
21.7–43.3 in (55–110 cm); 4.5–14.4 lb (2.06–6.52 kg). Large long-necked goose with dark solid-colored neck and mottled plumage.

DISTRIBUTION
B. c. leucopareia: Aleutian Islands; winters in California. *B. c. minima*: west coast of Alaska; winters in California. *B. c. taverneri*: Alaska to western Northwest Territory; winters in

Washington south to northern Mexico. *B. c. occidentalis*: southwestern Alaska; winters from Prince William Sound south to northern California. *B. c. fulva*: coastal southern Alaska and western British Columbia. *B. c. parvipes*: north central Canada; winters from California south to Louisiana and northern Mexico. *B. c. moffitti*: British Columbia east to Manitoba and in south from northern California to eastern Colorado; winters in southern part of its range and northern Mexico. *B. c. maxima*: Alberta to Manitoba. *B. c. hutchensii*: arctic tundra of Canada; winters in New Mexico, Texas, and northeastern Mexico. *B. c. interior*: central and eastern Canada; winters from Wisconsin east to New York, and south to Gulf and Atlantic coasts. *B. c. canadensis*: northeastern Canada; winters along eastern coastal provinces of Canada south to North Carolina. Introduced into United Kingdom, northwestern Europe, and New Zealand.

HABITAT
Mostly open habitats such as tundra, semi-desert, wooded areas, and agricultural lands.

BEHAVIOR
Territorial to colonial. Migratory.

FEEDING ECOLOGY AND DIET
Diet mostly vegetarian. Grazes on land and submerges head when on water.

REPRODUCTIVE BIOLOGY
Perennially monogamous with mate guarding. Breeds in spring. Usually lay 4–7 eggs in shallow ground nest; incubation 24–30 days; fledging 40–86 days; Sexually mature at 2–3 years.

CONSERVATION STATUS
Not threatened, though some populations have declined due to over-hunting and habitat alteration.

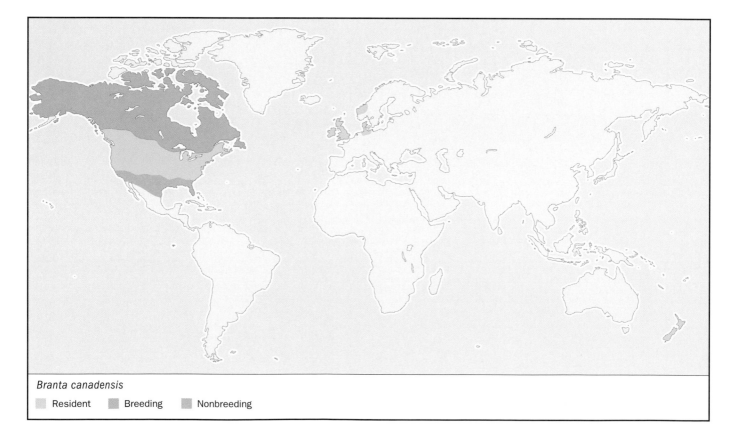

Branta canadensis

■ Resident   ■ Breeding   ■ Nonbreeding

SIGNIFICANCE TO HUMANS
Hunted for game. ◆

Cape Barren goose
Cereopsis novaehollandiae

SUBFAMILY
Anserinae

TAXONOMY
Cereopsis n. hollandiae Latham, 1801, New South Wales = islands
of Bass Strait. Monotypic.

OTHER COMMON NAMES
English: Cereopsis goose; French: Céréopse cendrée; German:
Hühnergans; Spanish: Ganso Cenizo.

PHYSICAL CHARACTERISTICS
29.5–39.4 in (75–100 cm); 7–14 lb (3.17–6.80 kg). Pale gray,
with distinctive dark spots on wings.

DISTRIBUTION
Islands off southern Australia from Recherche Archipelago to
Tasmania.

HABITAT
Scrub and grassy areas near coast; edges of lakes and lagoons.

BEHAVIOR
Territorial or colonial during the breeding season, but disperse
after breeding season. Perform triumph ceremonies.

FEEDING ECOLOGY AND DIET
Feeds on grasses, seeds of grasses, sedges, and leaves by grazing.

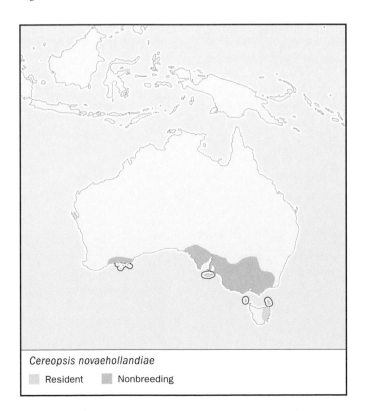

Cereopsis novaehollandiae
☐ Resident ▨ Nonbreeding

REPRODUCTIVE BIOLOGY
Perennially monogamous. Breeds in southern winter May–Jun.
Lays 3–6 eggs into shallow ground nest; incubation c. 34–37
days; fledging 70–76 days.

CONSERVATION STATUS
Not threatened. Populations stabile due to favorable conditions
on agricultural lands and a stop of intensive hunting.

SIGNIFICANCE TO HUMANS
Limited hunting allowed in Tasmania. ◆

Ruddy shelduck
Tadorna ferruginea

SUBFAMILY
Tadorninae

TAXONOMY
Anas ferruginea Pallas, 1789, no locality = Tartary. Monotypic.

OTHER COMMON NAMES
English: Brahminy duck; French: Tadorne casarca; German:
Rostgans; Spanish: Taro Canelo.

PHYSICAL CHARACTERISTICS
24.8–26 in (63–66 cm); 2.0–3.6 lb (925–1,640 g). Golden head,
ruddy brown body, and iridescent green secondary feathers.

DISTRIBUTION
Patchy from southern Spain and northwestern Africa east to
Mongolia; winters south of breeding range in Africa, Arabia,
and East Asia.

HABITAT
Brackish lakes, lagoons, and wetlands in mostly open landscapes.

BEHAVIOR
Territorial during breeding season. Asian populations are
mostly migratory.

FEEDING ECOLOGY AND DIET
Feeds on plants and invertebrates on land and water by grazing,
dabbling and upending.

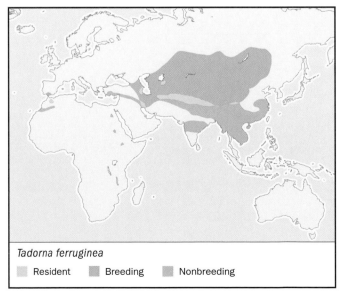

Tadorna ferruginea
☐ Resident ▨ Breeding ▨ Nonbreeding

REPRODUCTIVE BIOLOGY
Seasonally monogamous, but pair bonds may last several seasons. Breeding begins in Mar. or Apr. Nests in cavities. Lays usually 8–9 eggs; incubation c. 28–29 days; fledging c. 55 days; becomes sexually mature at 2 years.

CONSERVATION STATUS
Has declined (now recuperating?) in western part of the range due to hunting and wetland drainage. Asian populations appear healthy.

SIGNIFICANCE TO HUMANS
Hunted for food. ◆

Magellanic steamerduck
Tachyeres pteneres

SUBFAMILY
Tadiorninae

TAXONOMY
Anas pteneres Foster, 1844, Tierra del Fuego. Monotypic.

OTHER COMMON NAMES
English: Magellanic flightless steamerduck; French: Brassemer cendré; German: Magellan-Dampfschiffente; Spanish: Patovapor de Magellanes.

PHYSICAL CHARACTERISTICS
29.1–33.1 in (74–84 cm); 8.0–13.6 lb (3.63–6.18 kg); small wings render it almost flightless.

Tachyeres pteneres
Resident

DISTRIBUTION
Coastal Chile from Chiloé Island south to Cape Horn, and coastal Argentina from Chubut south to Tierra del Fuego and Staten Island.

HABITAT
Rocky shores and sheltered bays, often several miles from coast.

BEHAVIOR
Highly territorial and very aggressive. Both males and females defend territory. Perform triumph ceremonies. Moves over water by flapping wings like a steamer boat. Sedentary.

FEEDING ECOLOGY AND DIET
Feeds by diving in shallow kelp beds for marine mollusks and crustaceans.

REPRODUCTIVE BIOLOGY
Perennially monogamous. Breeding starts Sept. or Oct.. Lay 5–8 eggs into well hidden nests; incubation c. 30–40 days.

CONSERVATION STATUS
Common in limited range.

SIGNIFICANCE TO HUMANS
None known. ◆

Comb duck
Sarkidiornis melanotos

SUBFAMILY
Anatinae

TAXONOMY
Anser melanotos Pennant, 1769, Ceylon. Two subspecies.

OTHER COMMON NAMES
English: Knob-billed duck, black-backed goose; French: Canard-à-bosse bronzé; German: Glanzente; Spanish: Pato Crestudo.

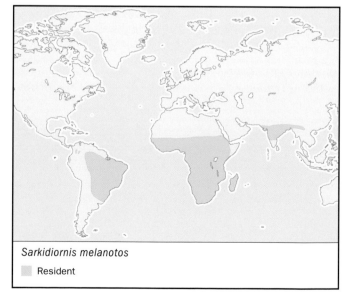

Sarkidiornis melanotos
Resident

PHYSICAL CHARACTERISTICS
22.1–29.9 in (56–76 cm); 2.7–5.8 lb (1.23–2.61 kg); males have a prominent black fleshy crest on upper bill.

DISTRIBUTION
S. m. melanotos: Africa south of Sahara and Madagascar; from Pakistan through tropical India to southern China. *S. m. sylvicola*: eastern Panama, western Ecuador, eastern Colombia east through Guyana and south to northern Argentina and Uruguay; occasionally Trinidad.

HABITAT
Open swamps, rivers, and lakes with sparse trees.

BEHAVIOR
Very sociable. Males defend 17.3 acres (7 ha) large territories, threatening intruders with *wing flap* display. Female/female aggression common while searching for nest cavities.

FEEDING ECOLOGY AND DIET
Vegetarian, but also feeds on some invertebrates. Grazes on land; swims, dabbles, and wades in shallow water.

REPRODUCTIVE BIOLOGY
Seasonally monogamous, but sometimes polygynous. Forced copulations are common. Breeding season depends very much on the local rainy season. Nests in tree cavities or occasionally on ground. Lays 6–20 eggs; incubation c. 28–30 days; fledging c. 10 weeks.

CONSERVATION STATUS
Listed on Appendix II of CITES. South American subspecies is uncommon and threatened by deforestation, hunting, and poisoning by rice farmers. African and Asian populations are not threatened.

SIGNIFICANCE TO HUMANS
Considered a pest by rice farmers. Hunted for food. ◆

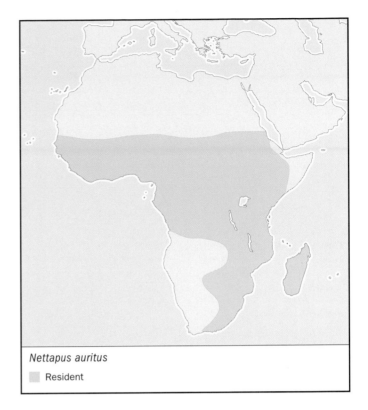

Nettapus auritus

■ Resident

African pygmy goose
Nettapus auritus

SUBFAMILY
Anatinae

TAXONOMY
Anas aurita Boddaert, 1783, Madagascar. Monotypic.

OTHER COMMON NAMES
English: Dwarf goose, pygmy goose; French: Anserelle naine; German: Afrikanische Zwergente; Spanish: Gansito Africano.

PHYSICAL CHARACTERISTICS
11.8–13.0 in (30–33 cm); 0.57–0.63 lb (260–285 g). Iridescent green back with ruddy flanks and white underparts.

DISTRIBUTION
South of Sahara except southwestern Africa; islands of Pemba, Zanzibar, Mafia, and Madagascar.

HABITAT
Shallow wetlands and slow flowing rivers with abundant floating aquatic vegetation, preferrably water lilies.

BEHAVIOR
Territorial and sedentary, but moves in response to habitat changes.

FEEDING ECOLOGY AND DIET
Feeds on insects and quatic plants, especially seeds of water lillies, while swimming. Also dives for small fish and aquatic insects.

REPRODUCTIVE BIOLOGY
Seasonally monogamous in captivity and in the wild. Nests in tree cavities. Lays 6–12 eggs; incubation c. 23–24 days; sexually mature probably at 1 year.

CONSERVATION STATUS
Not threatened. Locally common to abundant. Listed on Appendix III of CITES for Ghana.

SIGNIFICANCE TO HUMANS
Hunted for food. ◆

Mandarin duck
Aix galericulata

SUBFAMILY
Anatinae

TAXONOMY
Anas galericulata Linnaeus, 1758, China; monotypic.

OTHER COMMON NAMES
French: Canard mandarin; German: Mandarinente; Spanish: Pato Mandarín.

PHYSICAL CHARACTERISTICS
16.1–20.1 in (41–51 cm); 0.98–1.10 lb (444–500 g). Multicolored upperparts of gray, green, black, and ruddy brown. White underparts.

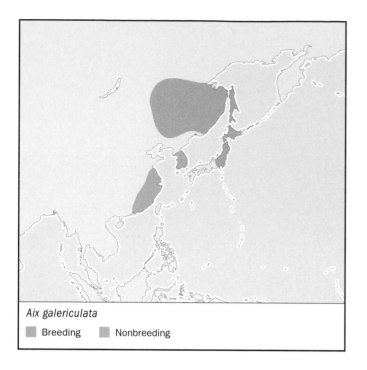

Aix galericulata
- Breeding
- Nonbreeding

Merganetta armata
- Resident

DISTRIBUTION
Eastern Siberia and eastern Chinasouth to South Korea and
Japan; winters in southeastern China below 40° north and
Taiwan.

HABITAT
Fast flowing rocky streams and wooded ponds.

BEHAVIOR
Not known if territorial. Migratory.

FEEDING ECOLOGY AND DIET
Feeds on grains, seeds, and acorns as well as land snails, little
fish, and insects. Forages by dabbling, dipping head into the
water, and upending.

REPRODUCTIVE BIOLOGY
Seasonally to perennially monogamous; some forced copulations may occur. Breeding begins in Apr. Nests in tree holes
up to 30 ft (10 m) high. Lays 9–12 eggs; incubation 28–30
days; fledging c. 40–45 days. Sexually mature at 1 year.

CONSERVATION STATUS
Not threatened, but of special concern as populations have declined over several decades during the twentieth century.

SIGNIFICANCE TO HUMANS
Bred by aviculturalists. In China and Japan symbolizes happiness
and marital fidelity. Not hunted for food because distasteful. ◆

Torrent duck
Merganetta armata

SUBFAMILY
Anatinae

TAXONOMY
Merganetta armata Gould, 1842, Andes of Chile. Six subspecies.

OTHER COMMON NAMES
French: Merganette des torrents; German: Sturzbachente;
Spanish: Pato de torrente.

PHYSICAL CHARACTERISTICS
16.9–18.1 in (43–46 cm); 0.69–0.97 lb (315–440 g); slender
bodied.

DISTRIBUTION
M. a. colombiana: Andes of western Venezuela south to southern Ecuador. *M. a. leucogenis*: Andes of southern Ecuador south
to northwest Argentina. *Merganetta a. armata*: Andes of Mendoza, Argentina, and Atacama, Chile, south to Tierra del
Fuego. Found between 984 and 15,100 ft (300–4,600 m), in
south down to sea level.

HABITAT
Fast-flowing, rocky mountain streams with clear water in
páramo grasslands and humid montane forest.

BEHAVIOR
Very territorial and aggressive year round.

FEEDING ECOLOGY AND DIET
Feeds on aquatic invertebrates and possibly fish. Forages by
diving, dipping head into water, and upending; searches
crevices among rocks and boulders.

REPRODUCTIVE BIOLOGY
Perennially monogamous. May mate for life. Grassy nest lined
with down can be found in cavities, among rocks, or in dense
vegetation. Breeding season depends on locality. Lays 3–4
eggs; incubation c. 43–44 days.

CONSERVATION STATUS

M. a. colombiana and *M. a. leucogenis* appear to be declining and may soon be considered Near Threatened. Main causes for the decline are deforestation, river siltation, hunting, and mining.

SIGNIFICANCE TO HUMANS

Hunted for food. ◆

Salvadori's teal

Anas waigiuensis

SUBFAMILY

Anatinae

TAXONOMY

Salvadorina waigiuensis Rothschild and Hartert, 1894, Waigeo. Monotypic.

OTHER COMMON NAMES

English: Salvadori's duck; French: Canard de Salvadori; German: Salvadoriente; Spanish: Anade Papúa.

PHYSICAL CHARACTERISTICS

15.0–16.9 in (38–43 cm); 0.88–1.2 lb (400–550 g). Dark head, barred wings, and stippled feathers.

DISTRIBUTION

Mountains of New Guinea.

HABITAT

Fast flowing mountain streams and brooks as well as slower flowing small rivers and small lakes; from 1,640–11,800 ft (500–3,600 m) elevation.

BEHAVIOR

Territorial year round.

FEEDING ECOLOGY AND DIET

Feeds mostly on aquatic invertebrates and possibly tadpoles and small fish.

REPRODUCTIVE BIOLOGY

Perennially monogamous. Possibly produces two broods a year. Lays 3–4 eggs; incubation longer than 28 days.

CONSERVATION STATUS

Not threatened. Locally common. May become threatened due to increased hunting, river pollution, predation by introduced mammals, and competition with introduced predatory fish.

SIGNIFICANCE TO HUMANS

Hunted for food. ◆

American wigeon

Anas americana

SUBFAMILY

Anatinae

TAXONOMY

Anas americana Gmelin, 1789, Lousinana and New York. Monotypic.

OTHER COMMON NAMES

English: Baldpate; French: Canard d'Amérique; German: Nordamerikanische Pfeifente; Spanish: Silbón Americano.

PHYSICAL CHARACTERISTICS

17.7–22.1 in (45–56 cm); 1.5–1.7 lb (680–770 g). Male has dark green band along side of head.

DISTRIBUTION

Western and Central North America; some populations breed from New Brunswick south to Massachusets. Winter from Alaska south to Central America, Gulf and Atlantic coasts.

Anas waigiuensis

Resident

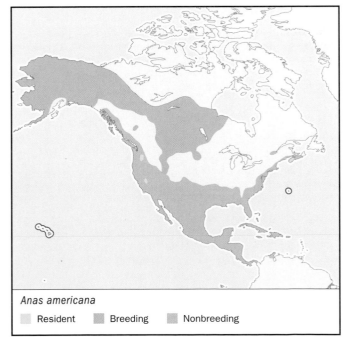

Anas americana

Resident   Breeding   Nonbreeding

HABITAT
Shallow lakes and open wetlands surrounded by meadows or agricultural lands; during winter mostly in coastal wetlands.

BEHAVIOR
Territorial during the early breeding season. Paired males swim 'in tandem' with lone intruding males. Migratory.

FEEDING ECOLOGY AND DIET
Vegetarian, feeds by walking, wading, or swimming in shallow water.

REPRODUCTIVE BIOLOGY
Seasonally monogamous until midincubation. Breeding begins Apr.–May. Lays 4–8 eggs into a concealed shallow depression; incubation 23–25 days; fledging c. 37–48 days. Becomes sexually mature at 1–2; years.

CONSERVATION STATUS
Not threatened. Locally abundant. Critical habitats protected.

SIGNIFICANCE TO HUMANS
Hunted for sport and food. ◆

Brown teal
Anas aucklandica

SUBFAMILY
Anatinae

TAXONOMY
Nesonetta aucklandica G. R. Gray, 1844, Auckland Islands; the three subspecies may deserve species status.

OTHER COMMON NAMES
English: Brown duck, New Zealand teal, Pateke; French: Sarcelle brune; German: Auklandente; Spanish: Cerceta Maorí.

PHYSICAL CHARACTERISTICS
14.2–18.9 in (36–48 cm); 0.8–1.5 lb (375–700 g); *A. a. aucklandica* and *A. a. nesiotis* are flightless.

DISTRIBUTION
New Zealand and nearby islands. *A. a. aucklandica*: Auckland Islands. *A. a. chlorotis:* patchily distributed on North Island and southwestern South Island. *A. a. nesiotis*: Dent Island to the northwest of Campbell Island; 12 individuals have been released on Codfish Island in 1999.

HABITAT
Sheltered coastlines with kelp beds, inland wetlands with some tree cover.

BEHAVIOR
Male participates in brood rearing. Very territorial during the early breeding season. Males patrol territories using aggressive displays and chest-to-chest fighting. Often nocturnal.

FEEDING ECOLOGY AND DIET
Feeds by probing, dabbling, upending, and diving. Searches for food in kelp beds, washed-up algae on beaches, marshes, ponds, and slow flowing waters. Diet consists invertebrates, roots, and tips of shoots.

REPRODUCTIVE BIOLOGY
Seasonally monogamous. Nest is well hidden in thick vegetation. *A. a. chlorotis* breeds mostly from Jun. to Oct. and *A. a. aucklandica* begins breeding season in Dec.–Jan.

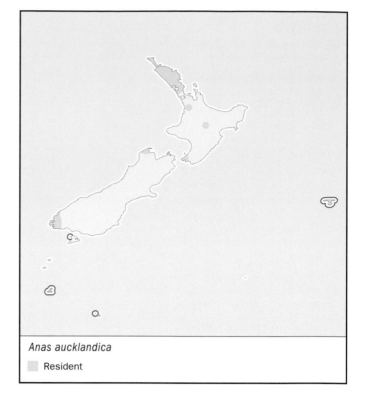

Anas aucklandica
■ Resident

Probably lays 4–8 eggs; incubation c. 29–30 days; fledging c. 50–55 days.

CONSERVATION STATUS
A. a. aucklandica considered Vulnerable and *A. a. chlorotis* Endangered. Only about 25 pairs remain of *A. a. nesotis* which is considered Critically Endangered. The species appears to be declining due to introduced predators, habitat destruction, and hunting. Listed on Appendix I of CITES.

SIGNIFICANCE TO HUMANS
Sometimes hunted. ◆

Mallard
Anas platyrhynchos

SUBFAMILY
Anatinae

TAXONOMY
Anas platyrhynchus Linnaeus, 1758, Europe. Seven subspecies.

OTHER COMMON NAMES
English: Greenhead, koloa; French: Canard colvert; German: Stockente; Spanish: Anade Azulón.

PHYSICAL CHARACTERISTICS
19.7–25.6 in (50–65 cm); 1.7–3.5 lb (750–1,580 g). Green head, brown chest, blue speculum.

DISTRIBUTION
A. p. platyrhynchus: widespread throughout the nearctic, in palearctic from Iceland to Kamchatka and south to the Mediterrean; winters on most of Pacific coast from the Aleu-

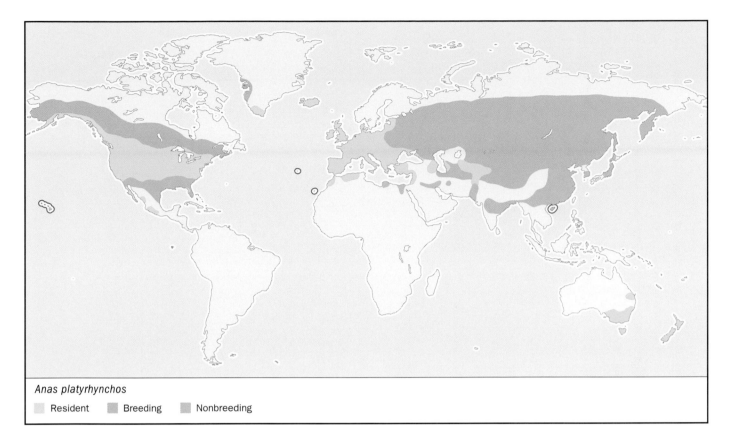

Anas platyrhynchos

■ Resident   ■ Breeding   ■ Nonbreeding

tians to California, southern half of the United States, north-eastern Mexico, the West Indies, northern Africa, and from Iraq west to southeastern China. *A. p. conboschas*: coasts of southern Greenland. *A. p. fulvigula*: coastal Alabama west to Florida peninsula. *A. p. maculosa*: Gulf coast from Mississippi south to central Tamaulipas, Mexico; winters south to Veracruz. *A. p. diazi*: southeastern Arizona to western Texas and south to Mexico. *A. p. wyvilliana*: Kauai, Oahu, and Hawaii, Hawaiian Islands. *A. p. laysanensis*: Laysan Island, Hawaiian Islands. Populations elsewhere are introduced.

HABITAT
Shallow and calm waters of all types of natural or artificial wetlands and saltwater and brackish water. Prefers some vegetative cover.

BEHAVIOR
Territorial to midincubation. Males then abandon their mates and territory. The c. 39.54–274.29 acres (16–111 ha) large, overlapping territories are defended aggressively. Forced copulations occur. Typical behaviors are the *grunt-whistle* and *head-up-tail-up* displays used in courtship. Migratory.

FEEDING ECOLOGY AND DIET
Feeds on water by dabbling, head-dipping, upending, and rarely diving, and on land forages by grazing and probing. Omnivorous diet includes terrestrial and aquatic invertebrates, fish, amphibians, and various plant parts.

REPRODUCTIVE BIOLOGY
Seasonally monogamous. Breeding begins Feb.–Jun., depending on locality. The nest is made in a cavity or the ground. Generally lays 9–13 eggs; incubation c. 27–28 days; fledging 50–60 days. Becomes sexually mature at 1 year.

CONSERVATION STATUS
Common. Only Hawaiian subspecies are rare. *A. p. laysanensis* considered Vulnerable and listed on Appendix I of CITES. *A. p. wyvilliana* considered Critically Endangered. The other subspecies are common. *A. p. diazi* may be threatened by hybridization with southward spreading *A. p. platyrhynchus*.

SIGNIFICANCE TO HUMANS
Hunted for sport. ◆

Northern shoveler
Anas clypeata

SUBFAMILY
Anatinae

TAXONOMY
Anas clypeata Linnaeus, 1758, coasts of Europe. Monotypic.

OTHER COMMON NAMES
English: Shoveler; French: Canard souchet; German: Löffelente; Spanish: Cuchara Común.

PHYSICAL CHARACTERISTICS
16.9–22.1 in (43–56 cm); 0.9–2.4 lb (410–1,000 g). Green head, white breast, ruddy underparts.

DISTRIBUTION
Throughout most of the Nearctic and Palearctic. Winters in southern United States south to Colombia, in Mediterranean east to southeastern Asia and south to tropical Africa.

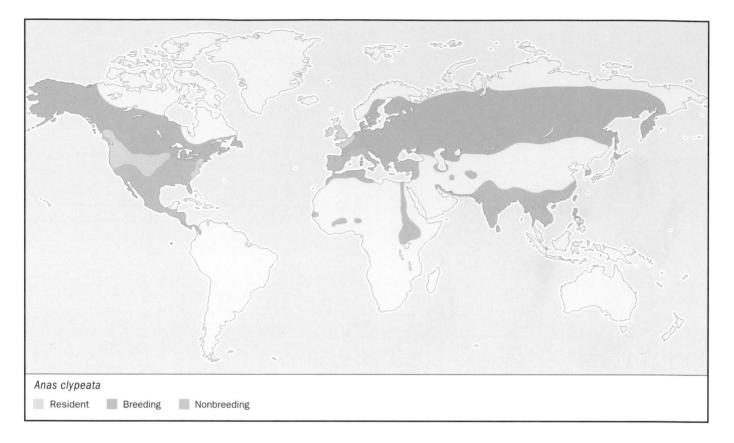

Anas clypeata

☐ Resident   ■ Breeding   ■ Nonbreeding

HABITAT
Open, shallow freshwater wetlands and, in winter, brackish waters and tidal mudflats.

BEHAVIOR
Territorial during early breeding season. Males aggressive; use wing noises to advertise their presence on a territory and chase intruders. Migratory.

FEEDING ECOLOGY AND DIET
Omnivorous diet consisting of small aquatic invertebrates, seeds, and vegetative parts of plants. Feed by dabbling.

REPRODUCTIVE BIOLOGY
Seasonally monogamous. Pair bonds last through brood rearing. Breeding season begins in Apr–May. Lays 9–11 eggs in a depression on the ground; incubation c. 22–23 days; fledging 40–45 days; becomes sexually mature at 1 year.

CONSERVATION STATUS
Not threatened. Locally common. Listed on Appendix III of CITES for Ghana.

SIGNIFICANCE TO HUMANS
Hunted for food and sport. ◆

Marbled teal
Marmaronetta angustirostris

SUBFAMILY
Anatinae

TAXONOMY
Anas angustirostris Ménétriés, 1832, Lenkoran. Monotypic.

OTHER COMMON NAMES
English: Marbled duck; French: Sarcelle marbrée; German: Marmelente; Spanish: Cerceta Pardilla.

PHYSICAL CHARACTERISTICS
15.4–18.9 in (39–48 cm); 1.0–1.3 lb (450–590 g). Plumage mainly grayish brown with white marbling.

DISTRIBUTION
From southern Spain and northwestern Africa east to Pakistan; winters south of breeding range to the northern Sahara and east to northwestern India.

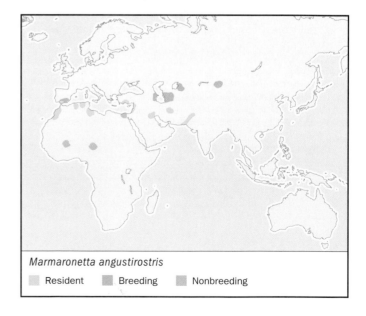

Marmaronetta angustirostris

☐ Resident   ■ Breeding   ■ Nonbreeding

HABITAT
Heavily vegetated brackish marshes within arid regions.

BEHAVIOR
Not territorial, but male guards female until she finishes laying eggs. Migratory.

FEEDING ECOLOGY AND DIET
Feeds on plant parts and aquatic invertebrates by dabbling, upending, diving, mud-filtering, and wading on shore.

REPRODUCTIVE BIOLOGY
Seasonally monogamous. Breeding begins in Apr.–Jun., and May–Jul. in northern Africa. Lays 7–14 eggs into a small concealed depression on ground; incubation 25–27 days; becomes sexually mature at 1 year.

CONSERVATION STATUS
Populations are declining due to hunting and wetland drainage. Considered Vulnerable.

SIGNIFICANCE TO HUMANS
Hunted for food. ◆

Madagascar pochard
Aythya innotata

SUBFAMILY
Anatinae

TAXONOMY
Nyroca innotata Salvadori, 1894, Madagascar. Monotypic.

OTHER COMMON NAMES
English: Madagascan white-eye; French: Fuligule de Madagascar; German: Madagascarmoorente; Spanish: Porrón Malgache.

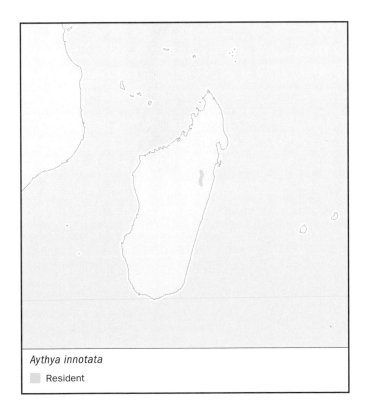

Aythya innotata

▨ Resident

PHYSICAL CHARACTERISTICS
17.7–22.1 in (45–56 cm). Brown duck with white belly.

DISTRIBUTION
Occurs only in and around Lake Aloatra in eastern Madagascar.

HABITAT
Freshwater wetlands with a mixture of open water and vegetation islands.

BEHAVIOR
Not known.

FEEDING ECOLOGY AND DIET
Mostly dives for aquatic invertebrates and seeds of aquatic plants.

REPRODUCTIVE BIOLOGY
Probably monogamous. Nesting has been observed Mar.–Apr. Hidden nests are slightly raised on a bank or a clump of vegetation.

CONSERVATION STATUS
Considered Critically Endangered, but possibly extinct. Last seen in 1991. Declined due to habitat conversion and excessive hunting.

SIGNIFICANCE TO HUMANS
Hunted for food. ◆

King eider
Somateria spectabilis

SUBFAMILY
Merginae

TAXONOMY
Anas spectabilis Linneaus, 1758, Canada, Sweden. Monotypic.

OTHER COMMON NAMES
French: Eider à tête grise; German: Prachteiderente; Spanish: Eider Real.

PHYSICAL CHARACTERISTICS
16.9–24.8 in (43–63 cm); 3.3–4.4 lb (1.50–2.01 kg). Male has colorful blue, yellow, and white head.

DISTRIBUTION
Coasts of the Arctic. Winters off the coast of Iceland, Norway, and Kuril and Aleutian Islands in the Old World, and as far south as California and Long Island in the New world.

HABITAT
Open marine waters. Breeds on nearby land in Arctic freshwater wetlands.

BEHAVIOR
Sometimes semi-colonial with the male guarding his mate. Migratory.

FEEDING ECOLOGY AND DIET
Feeds mostly on marine invertebrates. Dives in deep water and upends and head-dips in shallow water. Also eats the green parts of tundra vegetation.

REPRODUCTIVE BIOLOGY
Seasonally monogamous until midincubation. Some forced copulations occur. Breeding begins in Jun. Lays 4–5 eggs into a deep depression on the ground with little nesting material; incubation 22–24 days; becomes sexually mature at 3 years.

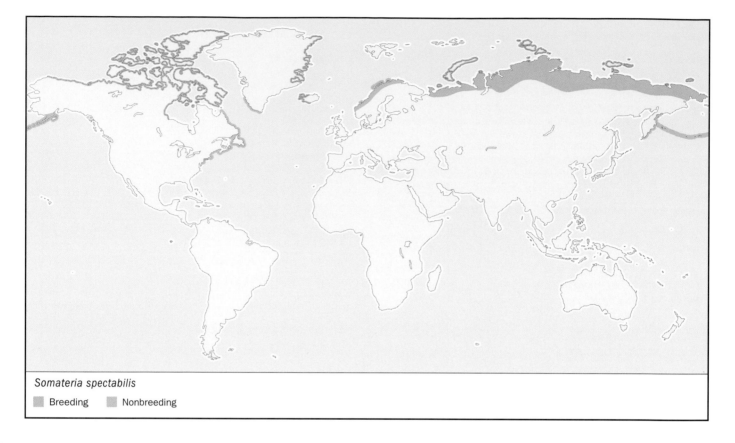

Somateria spectabilis

■ Breeding   ■ Nonbreeding

CONSERVATION STATUS
Not threatened. Common throughout its range. Has disappeared from Kuril Island. Intense hunting pressure during migration and possible oilspills during the nonbreeding season may pose future threats.

SIGNIFICANCE TO HUMANS
Hunted for food and game. ◆

Harlequin duck
Histrionicus histrionicus

SUBFAMILY
Merginae

TAXONOMY
Anas histrionicus Linnaeus, 1758, America = Newfoundland *ex* Edwards. Monotypic.

OTHER COMMON NAMES
English: Harlequin; French: Arlequin plongeur; German: Kragenente; Spanish: Pato Arlequín.

PHYSICAL CHARACTERISTICS
15.0–20.1 in (38–51 cm); 1.2–1.5 lb (540–680 g). Distinctive white markings on head, chest, and back.

DISTRIBUTION
Eastern Siberia from Lake Baikal north to about 68° and east to central western Alaska and Yukon. South in North America to California and east to southern Baffin Island and Quebec. Greenland, and Iceland. Winters along coasts of Kamchatka, Bearing Sea islands, Japan, Korea, China, California, and from southern Labrador south to Long Island.

HABITAT
Fast flowing rocky rivers during the breeding season and rocky coastlines during the nonbreeding season.

BEHAVIOR
Loosely territorial and aggressive. Males guard their mate. Return to the same breeding area each year. Migratory.

FEEDING ECOLOGY AND DIET
Has mostly an animal diet of invertebrates and some fish. Mostly dives for food, but also dabbles, up-ends, and dips its head in shallow water.

REPRODUCTIVE BIOLOGY
Seasonally monogamous until midincubation. Same birds may re-pair in the following season. Breeding begins May–Jun. The nest is well hidden on the ground. Commonly lays 5–7 eggs; incubation 27–29 days; fledging c. 60–70 days; becomes sexually mature at 2 years.

CONSERVATION STATUS
Not threatened. Locally common with stable populations.

SIGNIFICANCE TO HUMANS
None known. ◆

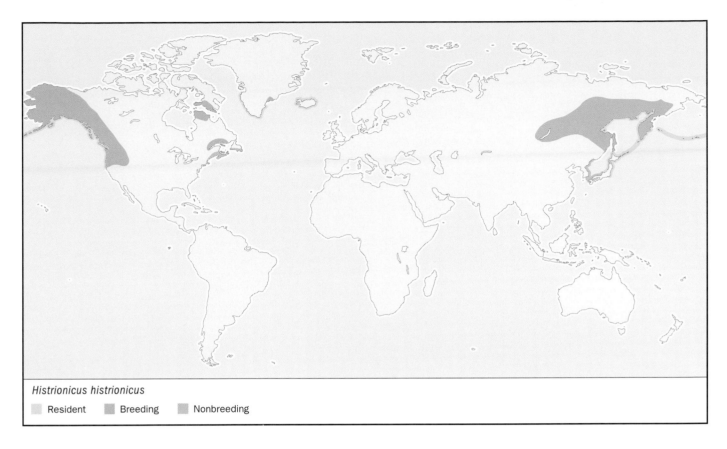

Histrionicus histrionicus

Resident   Breeding   Nonbreeding

Oldsquaw
Clangula hyemalis

SUBFAMILY
Merginae

TAXONOMY
Anas hyemalis Linnaeus, 1758, Arctic Europe and America; monotypic.

OTHER COMMON NAMES
English: Long-tailed duck; French: Harelde borréale; German: Eisente; Spanish: Pato Havelda.

PHYSICAL CHARACTERISTICS
15.0–22.8 in (38–58 cm); 1.4–1.8 lb (650–800 g). Plumage brighter in winter. Characteristic long streamers.

DISTRIBUTION
Coasts of the high Arctic. Winters south of breeding area along coastlines of Bearing Sea Islands, southern Alaska to California, Great Lakes, Atlantic coast from southern Labrador to the Carolinas, the British Isles, Northern and Baltic Seas, Caspian Sea, various lakes in Turkistan, Japan, Korea, and northeastern China.

HABITAT
Breeds in wetlands of tundra and Arctic coasts. Winters in open sea and large deep freshwater lakes.

BEHAVIOR
Territorial until hatching of young. Male defends a 1.24 acres (0.5 ha) small territory on which the female forages, but does not nest. Intruders are threatened with various displays and vocalizations before being chased in flights. Migratory.

FEEDING ECOLOGY AND DIET
Feeds by diving on marine invertebrates and fish. Rarely consumes plant matter.

REPRODUCTIVE BIOLOGY
Seasonally monogamous. Breeding begins May–Jun. Lays 6–9 eggs into a small depression; incubation 24–29 days; fledging 35–40 days; becomes sexually mature at 2 years.

CONSERVATION STATUS
Not threatened. Common to abundant throughout its range. Threats include excessive hunting and oil spills which may kill thousands of birds aggregating in large rafts.

SIGNIFICANCE TO HUMANS
Intensively hunted. ◆

Brazilian merganser
Mergus octosetaceus

SUBFAMILY
Merginae

TAXONOMY
Mergus octosetaceus Vieillot, 1817, Brazil. Monotypic.

OTHER COMMON NAMES
French: Harle huppard; German: Dunkelsäger; Spanish: Serreta Brasileña.

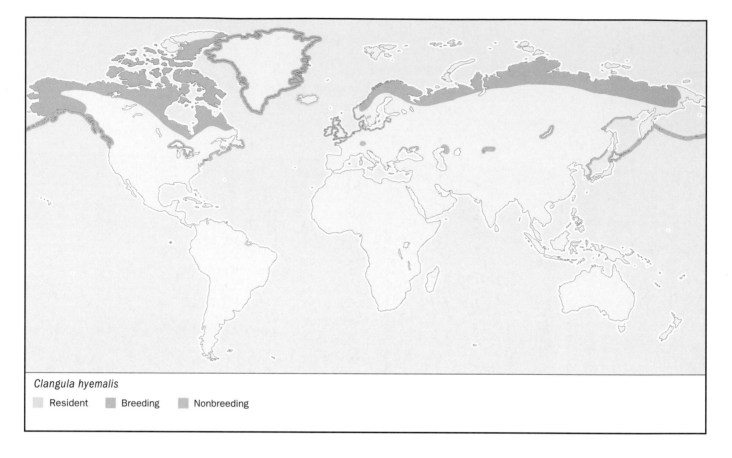

Clangula hyemalis

⬜ Resident ⬛ Breeding ◩ Nonbreeding

Mergus octosetaceus

⬜ Resident

PHYSICAL CHARACTERISTICS
19.3–22.1 in (49–56 cm). Iridescent green crest and head on stippled gray-brown body.

DISTRIBUTION
Southeastern Brazil, eastern Paraguay, and northeastern Argentina

HABITAT
Fast flowing rivers and streams surrounded by tropical forest.

BEHAVIOR
Territorial with pairs defending stretches of river.

FEEDING ECOLOGY AND DIET
Feeds mostly on fish and some invertebrates by diving.

REPRODUCTIVE BIOLOGY
Apparently seasonally monogamous. Breeds during rainy season (beginning Jun.). Nest in tree cavities.

CONSERVATION STATUS
Considered Critically Endangered. Only few populations are known. Major threats include habitat destruction and hunting.

SIGNIFICANCE TO HUMANS
Hunted for food. ◆

Black-headed duck
Heteronetta atricapilla

SUBFAMILY
Oxyurinae

TAXONOMY
Anas atricapilla Merrem, 1841, Buenos Aires. Monotypic.

OTHER COMMON NAMES
French: Hétéronette à tête noire; German: Kukkucksente;
Spanish: Pato Rinconero.

PHYSICAL CHARACTERISTICS
13.8–15.8 in (35–40 cm); 1.1–1.3 lb (513–565 g). Black head
and upperparts, mottled brown underparts.

DISTRIBUTION
Santiago, Chile, in west, to Paraguay and Buenos Aires
province, Argentina, in east.

HABITAT
Freshwater wetlands with abundant emergent vegetation in
open or sparsely vegetated regions.

BEHAVIOR
Not known.

FEEDING ECOLOGY AND DIET
Feeds on seeds, plants, and some aquatic invertebrates by diving, dabbling, head-dipping, upending, and mud-filtering.

REPRODUCTIVE BIOLOGY
Probably seasonally monogamous for only short periods of time.
Breeding mostly between Sept. and Dec. Only completely parasitic anatid. Female lays on average 2 eggs in the nests of other
waterfowl, especially coots (*Fulica*) and rosy-billed porchard.

CONSERVATION STATUS
Common throughout its range, but may be threatened by habitat loss, hunting, and pollution.

SIGNIFICANCE TO HUMANS
Hunted for food. ◆

Musk duck
Biziura lobata

SUBFAMILY
Oxyurinae

TAXONOMY
Anas lobata Shaw, 1796, New South Wales = King George
Sound, Western Australia. Monotypic.

OTHER COMMON NAMES
English: Lobed duck; French: Erismature à barbillons; German: Lappenente; Spanish: Malvasía de Papada.

PHYSICAL CHARACTERISTICS
Male 26.0 in (66 cm), female 21.7 in (55 cm); male 4.0–6.9 lb
(1.81–3.12 kg), female 2.2–4.1 lb (993 g–1.84 kg). Male has
fleshy lobe underneath mandible.

DISTRIBUTION
Southwestern and southeastern Australia, Tasmania.

HABITAT
Freshwater wetlands with abundant reedbeds. Favors deep waters
of lagoons, estuaries, and coastlines in nonbreeding season.

Heteronetta atricapilla
☐ Resident ☐ Breeding

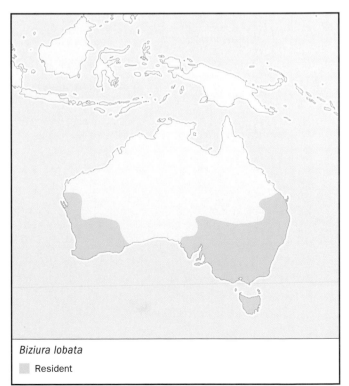

Biziura lobata
☐ Resident

BEHAVIOR
Males are very territorial early during the breeding season.

FEEDING ECOLOGY AND DIET
Feeds mostly by diving on aquatic invertebrates, amphibians, some fish, and small ducklings.

REPRODUCTIVE BIOLOGY
No strong pair bond develops, probably polygamous. Breeding mostly during Sept. to Oct. Commonly lays 2–3 eggs into a concealed cup nest made of sticks and stems; incubation 24 days in captivity.

CONSERVATION STATUS
Not threatened. Locally abundant, but generally at low abundances. May be threatened in future by habitat loss, fishing nets, and hunting.

SIGNIFICANCE TO HUMANS
Hunted for food. ◆

Resources

Books
Batt, D. J., Alan D. Afton, Michael G. Anderson, C. Davison Ankney, Douglas H. Johson, John A. Kadlec, and Gary L. Krapu, eds. *Ecology and Management of Breeding Fowl.* Minneapolis: University of Minnesota Press, 1992.

del Hoyo, J., A. Elliot, and J. Sargatal, eds. *Ostrich to Ducks.* Vol. 1 *Handbook of the Birds of the World.* Barcelona: Lynx Edicions, 1992.

Johnsgard, P. A *Ruddy Ducks and Other Stifftails: Their Behavior and Biology.* Norman: University of Oklahoma Press, 1996.

Madge, S., and H. Burn. *Waterfowl: An Identification Guide to the Ducks, Geese and Swans of the World.* Boston: Houghton Mifflin, 1988.

Schneck, M. *Ducks and Waterfowl.* New York: Todtri Book Publishers, 2000.

Todd, Frank S. *Ducks, Geese, and Swans of the World.* San Diego: Seaworld, 1979.

Periodicals
Callaghan, Des A. "Conservation status of the torrent ducks *Merganetta.*" *Wildfowl* 48 (1997–1998): 166–173.

Johnson, Kevin P., Frank McKinney, and Michael D. Sorenson. "Phylogenetic constraint on male parental care in dabbling ducks." *Proceedings of the Royal Society of London, Series B* 266 (1999): 759–763.

Livezey, Bradley C. "A phylogenetic classification of waterfowl (Aves: Anseriformes), including selected fossil species." *Annals of Carnegie Museum* 66 (1997): 457–496.

Olson, Storrs L. "The anseriform relationships of Antalavis Olson and Parris (Anseranatidae), with a new species from the Lower Eocene London clay." *Smithsonian Contributions to Paleobiology* 89 (1999): 231–243.

Organizations
Ducks Unlimited, Inc. One Waterfowl Way, Memphis, Tennessee 38120 USA. Phone: (800) 453-8257. Fax: (901) 758-3850. E-mail: mlafarge@ducks.org Web site: <http://www.ducks.org>

Japanese Association for Wild Geese Protection. Minamimachi 16, Wakayangi, 989-5502 Japan. Phone: +81 228 32 2004. Fax: +81 228 32 2004. E-mail: secretariat@jawgp.org Web site: <http://www.japwgp.org>

Other
Miyabayashi, Yoshihiko, and Taej Mundkur "Atlas of key sites for Anatidae in the East Asian flyway" <http://www.jawgp.org/anet/aaa1999/aaaendx.htm>. (1999)

Sauer, J. R., J. E. Hines, and J. Fallon. *The North American Breeding Bird Survey, Results and Analyses 1966–2000. Version 2001.2.* Laurel, Maryland: USGS Patuxent Wildlife Research Center, 2001.

The International Union for Conservation of Nature and Natural Resources Species Survival Commission "2000 IUCN Red List of Threatened Species" <http://www.redlist.org/>. (2000).

Markus Patricio Tellkamp, MS

Screamers
(Anhimidae)

Class Aves
Order Anseriformes
Suborder Anhimae
Family Anhimidae

Thumbnail description
Large, heavy-bodied, fowl-like, semi-aquatic birds with two distinctive bone spurs at the bend of each wing

Size
Body length is 28–36 in (71–92 cm); 5–7 lb (2–3 kg)

Number of genera, species
2 genera, 3 species

Habitat
Occur in the vicinity of tropical lowland lakes, ponds, marshes, and rivers

Conservation status
The northern screamer (*Chauna chavaria*) is Near Threatened

Distribution
Tropical regions of South America, from Venezuela and Colombia south to Uruguay and northern Argentina

Evolution and systematics

Although screamers (family Anhimidae) are rather fowl-like in appearance, they are placed in the same order (Anseriformes) as the geese, swans, and ducks (family Anatidae). Prior to clarification of the evolutionary proximity of the screamers with the waterfowl, which was largely based on aspects of their internal anatomy, they had been considered more closely related to the rails (family Rallidae) or the storks and their allies (order Ciconiiformes). Because some of the characters of screamers are primitive, in particular their wing-spurs, they had even been thought to be the closest living relatives of the ancient bird progenitor, *Archaeopteryx*. The common ancestry of the screamers and the anatids is reinforced by aspects of the biology of the Australian magpie goose (*Anseranas semipalmata*), which shows similarities to the screamers in its manner of molting, the absence of webbing on the feet, plates on the tarsus, the formation of the sternum, and the tendency to perch in trees. Fossil remains of screamers are known from deposits of the Quaternary Period in Argentina. There are three living species: the horned screamer (*Anhima cornuta*),

northern horned screamer (*Chauna chavaria*), and southern horned screamer (*Chauna torquata*).

Physical characteristics

Screamers are large, semi-aquatic, goose-like birds that swim well and occur near and in fresh water. They have a stout body, a crest or horn-like projection on the top of the head, a short, downcurved, fowl-like bill, rather thick, long legs, and stout feet lacking webbing between the strong toes. Their body length is 28–36 in (71–92 cm) and they weigh 5–7 lb (2–3 kg). Screamers have two sharp, spur-like outgrowths of fused carpal bones on each of their wings. The spurs are covered externally by keratin, a tough polysaccharide similar to the material in fingernails. The presence of these two spurs is unique in screamers. Like most birds, the interior of most skeletal bones is permeated by abundant air sacs, which continue into a well-developed air-containing tissue in the dermis. Male screamers do not have a copulatory organ. The general body coloration is black or gray on top,

Southern screamers (*Chauna torquata*) near the water. (Photo by K. Schafer/VIREO. Reproduced by permission.)

and somewhat lighter below. The two genera of screamers have marked differences in their internal anatomy, and in addition the horned screamers (genus *Anhima*) have 14 tail feathers, while the crested screamers (*Chauna*) have 12.

Distribution

Screamers occur in tropical regions of South America, from Venezuela and Colombia south to Uruguay and northern Argentina.

Habitat

Screamers occur in the vicinity of tropical lowland lakes, ponds, marshes, and rivers.

Behavior

Screamers are non-migratory birds, remaining all year within their breeding area. They are somewhat gregarious. Outside the breeding season, horned screamers live in groups of five to 10 birds, while crested screamers are found in larger flocks that circle above water bodies in the evenings, calling vociferously. Screamers can fly well but slowly, and they may soar for extended periods of time. They swim well and may walk on dense mats of floating vegetation. They roost in trees. They sometimes use the sharp spurs on their wings as weapons in fights connected with pair formation. The spurs are also

used as a defense against predators. Crested screamers have a goose-like call, and also a gargled throaty sound that resembles drumming. The horned screamer is less vocal, uttering a loud, shrill hoot. The cries of screamers are among the loudest of any birds.

Feeding ecology and diet

Screamers are vegetarians that feed on aquatic plants and seeds.

Reproductive biology

The nests of screamers are rather large and are located near or in marshy vegetation in shallow water. Nests are constructed of plant materials. Screamers lay two to seven smooth, yellowish-white eggs, which are incubated by both parents. The down-covered young are nidifugous, meaning they leave the nest almost immediately after hatching. Both parents care for the young.

Conservation status

The World Conservation Union (IUCN) lists the northern screamer as Near Threatened. This species has declined greatly in range and abundance, and further deterioration of its circumstances would render it threatened.

Significance to humans

Screamers are sometimes hunted as food. Young screamers are sometimes caught and tamed by local people. They readily take to captivity and can be kept with chickens in farmyards, where they defend their companions against birds of prey and other enemies. Sometimes they are allowed to walk about at liberty in South American parks and zoos.

A southern screamer (*Chauna torquata*) brooding chicks. (Photo by Kenneth W. Fink. Bruce Coleman Inc. Reproduced by permission.)

1. Horned screamer (*Anhima cornuta*); 2. Northern screamer (*Chauna chavaria*). (Illustration by Bruce Worden)

Species accounts

Horned screamer
Anhima cornuta

TAXONOMY
Palamedea cornuta Linneaus, 1766, eastern Brazil. Monotypic.

OTHER COMMON NAMES
French: Kamichi cornu; German: Hornwehrvogel; Spanish: Chajá Añuma.

PHYSICAL CHARACTERISTICS
Body length of 34–37 in (86–94 cm). Body is colored greenish black, with a white belly. A long, quill-like "horn" protrudes from its forehead.

DISTRIBUTION
Widespread but local in Amazonian regions of Venezuela, the Guianas, Colombia, Ecuador, Peru, Bolivia, and Brazil.

HABITAT
Inhabits wetlands in flooded tropical forest, such as oxbow lakes, marshes, and swamps. It occurs as high as about 3,300 ft (1,000 m).

BEHAVIOR
Has an extremely loud and distinctive set of calls. It swims or walks on aquatic vegetation while feeding, and often roosts in shrubs and trees.

FEEDING ECOLOGY AND DIET
Feeds on aquatic vegetation.

REPRODUCTIVE BIOLOGY
Builds a nest of plant materials, floating but anchored among marsh vegetation. It lays two yellowish-white eggs, which are incubated by both parents. The down-covered young leave the nest almost immediately after hatching. Both parents care for the young.

CONSERVATION STATUS
Not threatened. Relatively widespread and abundant species, although decreasing in abundance where heavily hunted.

SIGNIFICANCE TO HUMANS
Hunted as a source of meat and sometimes kept in captivity. ◆

Northern screamer
Chauna chavaria

TAXONOMY
Parra chavaria Linneaus, 1766, lakes near Río Sinú, south of Cartagena, Colombia. Monotypic.

OTHER COMMON NAMES
English: Black-necked screamer; French: Kamichi chavaria; German: Weisswangen-Tschaja; Spanish: Chajá Chicagüire.

Anhima cornuta
■ Resident

Chauna chavaria
■ Resident

PHYSICAL CHARACTERISTICS
Has a body length of 30–36 inches (76–91 cm). A large, stout body colored dark gray, with the throat and sides of head white. Has a crest of feathers on the top of its small head.

DISTRIBUTION
Occurs in tropical regions of northern Colombia and north-western Venezuela.

HABITAT
Inhabits wetlands in lowland tropical forest, such as oxbow lakes, marshes, lagoons, and swamps. It occurs as high as about 650 ft (200 m).

BEHAVIOR
Has a loud, bugled call. Outside of the breeding season it occurs in loose groups.

FEEDING ECOLOGY AND DIET
Feeds on aquatic vegetation.

REPRODUCTIVE BIOLOGY
Builds a nest of plant materials, floating but anchored among marsh vegetation. It lays two to seven yellowish-white eggs, which are incubated by both parents. The down-covered young leave the nest almost immediately after hatching and are cared for by both parents.

CONSERVATION STATUS
Near Threatened. This species has declined greatly in range and abundance, and further deterioration of its circumstances would render it Threatened.

SIGNIFICANCE TO HUMANS
Hunted as a source of wild meat and sometimes kept in captivity. ◆

Resources

Books

BirdLife International. *Threatened Birds of the World*. Barcelona and Cambridge, U.K.: Lynx Edicions and BirdLife International, 2000.

Blake, E. R. *Manual of Neotropical Birds*. Vol. 1, *Spheniscidae (Penguins) to Laridae (Gulls and Allies)*. Chicago: University of Chicago Press, 1977.

del Hoyo, Josep, Andrew Elliot, and Jordi Sargatal, eds. *Handbook of the Birds of the World*, Vol. 1. *Ostrich to Ducks*. Barcelona: Lynx Edicions, 1992.

Hilty, S. L., and W. L. Brown. *A Guide to the Birds of Colombia*. Princeton, N.J.: Princeton University Press, 1986.

Organizations

BirdLife International. Wellbrook Court, Girton Road, Cambridge, Cambridgeshire CB3 0NA United Kingdom. Phone: +44 1 223 277 318. Fax: +44-1-223-277-200. E-mail: birdlife@birdlife.org.uk Web site: <http://www.birdlife.net>

IUCN–The World Conservation Union. Rue Mauverney 28, Gland, 1196 Switzerland. Phone: +41-22-999-0001. Fax: +41-22-999-0025. E-mail: mail@hq.iucn.org Web site: <http://www.iucn.org>

Bill Freedman, PhD

Galliformes
(Chicken-like birds)

Class Aves

Order Galliformes

Number of families 5

Number of genera, species 77 genera; 281 species

Photo: Plain chachalacas (*Ortalis vetula*) in Texas. (Photo by Erwin & Peggy Bauer. Bruce Coleman Inc. Reproduced by permission.)

Evolution and systematics

Few birds have such a long relationship with people as the Galliformes, but their own history is even older. Fossils show that their predecessors date back to the Eocene period (50 to 60 million years ago), when northern latitudes were tropical. The earliest known cracid ancestor was found in the United States in Wyoming, although the megapodes are probably more primitive. All the Galliformes are of a similar, standard design, perfected for a terrestrial lifestyle that has been little modified by millions of years of evolution.

There are two tribes: the Craci (the megapodes, chachalacas, guans, and curassows) and the Phasiani (the turkeys, grouse, New World quails, pheasants, partridges, and guinea fowls). The two are distant in evolutionary terms, and no examples are known of a bird from one tribe hybridizing with one from the other. They are distinguished by the hallux, the hind toe, which in the Craci is in line with the other toes, but in the Phasiani is above the others.

The Phasianidae is numerically dominant, accounting for 155 species. Work on the mitochondrial DNA of birds in the late twentieth century has resulted in the New World quails (Odontophoridae) being split from the pheasants (Phasianidae). Discoveries continue to change our understanding. For example, the Udzungwa forest-partridge (*Xenoperdix udzungwensis*), discovered in southern Tanzania in 1991, is more closely related to the Asian hill-partridge (*Arborophila torqueola*).

Physical characteristics

Several characteristics are common to the Galliformes, all of which can be seen in the domestic chicken, derived from the red junglefowl (*Gallus gallus*) of Southeast Asia. Most gallinaceous species are medium to large in size, with a stocky

body, small head, and short wings. The Old World quails are the smallest, the most diminutive being the Asian blue (*Coturnix chinensis*) at just 5–6 in (12–15 cm) and weighing less than 1 oz (20 g). By contrast, the wild turkey (*Meleagris gallopavo*) weighs 17–22 lb (8–10 kg); only the domesticated forms destined for the table can attain 44 lb (20 kg); while a large, male green peafowl (*Pavo muticus*) is up to 98 in (250 cm) long, although its immense tail accounts for more than half of this. Pheasants in particular show a significant size difference between males and females, with the tail often responsible for one-third of the total.

In many cases, males and females are mottled brown or black, adapted to camouflage in the forest or scrub. In a few species, however, males are colorful, with iridescent colors that have long made them attractive to humans. The male Indian peafowl (*Pavo cristatus*), the "peacock" of art and the movies, whose fanned tail has hundreds of "eyes" on the tips, is perhaps the most well known.

Feeding ecology and diet

Galliformes are terrestrial, spending their day foraging for food in grasslands or the understory of the forests. Birds have a short, often downcurved bill, used to peck plant material from the ground or from short vegetation, though several species in northern latitudes, such as the capercaillie (*Tetrao urogallus*), depend on the stiff needles of coniferous trees (*Pinus*) to see them through the long winter when snow covers the ground. They also have large, strong feet, a crucial attribute that allows them to expose seeds and roots that are inaccessible to most other animals (the name "megapode" is derived from the Greek words for "big foot"). These feet are capable of moving heavy branches or stones; the orange-footed scrubfowl (*Megapodius*

A great curassow (*Crax rubra*). (Photo by Kenneth W. Fink. Bruce Coleman Inc. Reproduced by permission.)

reinwardt) can move a stone up to eight times its own weight. Their heavy build indicates a diet based on bulky, vegetable matter, although the chicks of many species depend on insects and larvae during their first few weeks.

Gallinaceous birds have a roomy, flexible crop, which can be extended to cache food before beginning to digest it. They also have a very strong gizzard, used to grind down the hard exterior of seeds and nuts, and the tough fibers in green vegetation. To aid digestion, birds regularly swallow small stones. Even the most secretive forest Galliformes visit roads and tracks early in the morning in search of grit before their day's feeding. Some make only occasional visits to water, even during dry periods, but a few species visit salt licks, where they ingest claylike soil to supplement their diet with minerals.

Reproductive biology

Galliformes display a wide variety of breeding strategies. In general, species with the least sexual dimorphism in size and color are monogamous, and those in which the male has more resplendent plumage are polygynous. In many grouse species, males display at communal leks, seeking to be the dominant male to attract a harem of females.

Male Galliformes have a range of adornments to attract females: bright colors, crests, unusually shaped tail feathers, or markings. Some have additional modifications, such as long, pendant wattles, dewlaps, combs, or "eyebrows." Most species display one of these "badges," or white patches on the wings or tail, although the curassows are the most highly evolved family, with colorful knobs, or ramphothecae, on their bill, which grow larger as the bird ages.

Unlike many nonpasseriform birds, calls play an important part in display and territorial ownership, and also simply to keep in touch with a mate. This is not surprising, since many species are solitary, living deep in scrub or forest, and crepuscular, being most active at dawn and dusk. In tropical areas, the wailing calls of tinamous and guans travel across the forest in the fading daylight, up to 4 mi (6.4 km) in some species. This is possible because a modification in the length of the trachea and a loop between the skin and the pectoral muscles enables some cracids to produce calls at a lower pitch and a higher volume than most other species, although swans (*Cygnus*) and cranes (*Grus*) have a similar modification.

The breeding strategy of the megapodes, which do not use body heat to incubate their eggs, is unique in the avian world, although it does not demonstrate a link to reptiles, as some have suggested. The male builds a huge mound of sand or plant material or constructs a burrow, invites a female to lay her eggs, then tends to the nest, regulating the temperature for many weeks until the young hatch. Indeed, some male megapodes are attached to their mound for 11 months of the year. In monogamous species, however, both birds help to rear the young, maintaining the pair bond through mutual preening or activities such as wing-drumming.

In most species, the young are precocious, able to feed semi-independently within a few hours of hatching. Generally, the first downy feathers are subdued in color to reduce the risk of a predator seeing them.

Distribution and habitat

Gallinaceous birds are found in a wide variety of habitats, in semideserts, steppes, savannas, forests, mountains, and farmland. The cracids are the most arboreal family, with most species spending at least part of the time in the forest canopy, but even some of the chachalacas feed in more open habitats. Members of other families are more specialized; for example, the British race of the red grouse (*Lagopus lagopus scoticus*) lives only on upland *Calluna* heather moorlands, devoid of any trees, whereas the other 18 races live around dwarf trees!

The Galliformes occur on every continent except Antarctica, with some families found on a single continent—megapodes in Australasia, cracids in Central and South America, turkeys in North America, New World quails in North and South America, and guinea fowl in sub-Saharan Africa. Only two families are spread across more than one continent—grouse in North America and Eurasia, and pheasants and partridges in Africa, Eurasia, and Australasia. There is relatively little geographic overlap between the families, perhaps not surprising given their sedentary nature; most birds moving only a few miles from where they hatched.

Behavior

The social behavior of the Galliformes is complex, and the commoner species have been the subject of many studies by ornithological and hunting interests. Many species are solitary or spend the year in pairs; the males being strongly territorial, charging intruders with their neck raised and wings spread open. In some grouse species, this has developed into a mating display, in which males demonstrate their defensive prowess to females gathered nearby, a behavior known as "lekking." Outside the breeding season, some species, such as chachalacas, brush-turkeys, and pheasants, feed communally where there is a good supply. Some will also roost communally, flying to the tree canopy where they are safer from predatory ground mammals.

Many Galliformes have cryptic plumages. Birds sit tight in thick vegetation, hoping not to be noticed, and only when the threat is almost upon a bird will it move. A few species, such as the Nicobar scrubfowl (*Megapodius nicobariensis*), evade predation by running swiftly away, but most species explode into the air in a rush of wings. This is possible because Galliformes have strong breast muscles and strong legs, enabling a near vertical take-off. In flight, many are bulletlike, especially the partridges, their wings beating rapidly, although only over short distances. Birds fly close to the ground, although this brings its own problems. Many capercaillies and black grouse (*Tetrao tetrix*) are killed against high deer fences around European forestry plantations.

Most gallinaceous birds bathe, often visiting the same sites repeatedly, squatting in a shallow pit and beating their wings to shower sand or dust across their plumage to maintain the feathers and remove parasites. In tropical species, bathing and preening usually takes place during the middle of the day when birds are resting, whereas birds are more active, displaying and feeding, during the three hours around sunrise and sunset.

Most species are sedentary, but a few are altitudinal migrants, moving down the mountainsides outside the breeding season; four Old World quails (*Coturnix*) are true long-distance migrants, traveling from breeding grounds in Eurasia to sub-Saharan Africa.

Conservation status

Of the 281 species, 104 are Threatened or Near Threatened, far above the average of 10% for all bird species. The pheasants and partridges are under the greatest threat, with 71 listed by the IUCN as Critically Endangered, Endangered, Vulnerable, or Near Threatened; one, possibly two, pheasant species have become extinct since 1600.

Gaps in our knowledge constrain the development of conservation measures for many of the species in the remotest habitats. There are many species, such as the Bruijn's brush-turkey (*Aepypodius bruijnii*), for which further research is crucial to their conservation.

Hunting of adults or the collection of eggs for food remains a problem for several of these "game" birds. Historically, cracids were an important sustainable source of protein

Profile of a vulturine guineafowl (*Acryllium vulturinum*). (Photo by H. & J. Eriksen/VIREO. Reproduced by permission.)

for native Amerindians, but rapid human colonization since 1492 has led to over-exploitation of birds, as well as the destruction of their tropical rainforest habitat. Some species are now on the brink of extinction, with the Alagoas curassow (*Mitu mitu*) only known in captivity since the 1980s, the horned guan (*Oreophasis derbianus*) limited to a few isolated mountain ranges in Mexico and Guatemala, and the endemic Trinidad piping-guan (*Pipile pipile*) now restricted to a few square miles of montane forest on a single island. It is doubtful whether the harvesting of eggs from some species is sustainable, and over-exploitation has probably caused the extinction of several megapode species on Pacific Islands. Such gathering continues today, with, for example, an estimated five million eggs taken every year from a single site in New Britain where Melanesian scrubfowl (*Megapodius eremita*) gather to breed.

However, habitat destruction is the principal threat. Galliformes that depend on primary, tropical forests are under the greatest threat, with logging of timber or intensive burning to clear the land for agriculture being major problems, especially in Southeast Asia. As if the removal of the old forest was not enough, herbivorous livestock, such as cattle, sheep, and goats, compete with Galliformes for seeds and vegetation. The Galliformes' dependence on certain habitats makes them good indicators of environmental change. As consumers of a large biomass of seeds and roots, they also play a critical role in dispersing seeds, especially in tropical forests.

Our fascination with these amazing birds provides a potential, sustainable solution to the need of local people to earn an income. In parts of South America, for example, ecotourism can be more important to the local economy than logging or beef production, and is certainly better for the Galliformes that birdwatchers come to see.

Significance to humans

We can only guess when *Homo sapiens* first discovered that some Galliformes were relatively easy to catch and that their meat tasted good and was high in protein. Some time later, as people moved from being hunter-gatherers to farmers, they learned to domesticate several species, including turkeys, chickens, and guinea fowl, which remain part of human diet across the world. Galliform eggs are much sought after, with a high yolk content that provides a rich source of protein. The word "fowl," which is applied generically to game and domesticated birds, has its origins in the Old English *fugol*, the Old Norse *fogl*, and the modern German *vogel*.

During the last 200 years, many gallinaceous species have been moved between countries and continents for decoration or for shooting on vast, private estates. Introductions, whether deliberate or accidental, are the third most serious threat to global avifauna, after habitat destruction and degradation. Introductions are most problematic on islands, and although Galliformes are not known to be an special threat, they have been widely transferred. At least 45 of the 281 species have been introduced to two archipelagoes, Hawaii and New Zealand, though only two-thirds colonized successfully, the rest failing either because of poor stock or an inability to deal with predators that they were not used to facing.

The sedentary nature of most Galliformes makes them popular for shooting. Even in countries where strict legislation makes it illegal to shoot birds, many Galliformes are excluded. At the height of the British Empire, aristocrats and civil servants spent their leisure time in Africa and Asia shooting small game (Galliformes), and brought some of the most numerous back to Europe, most notably the ring-necked pheasant (*Phasianus colchicus*). Each year, tens of millions are reared and released for shooting. These and European game species, such as the gray partridge (*Perdix perdix*) and rock partridge (*Alectoris graeca*), were subsequently introduced into North America for sport.

Resources

Books

BirdLife International. *Threatened Birds of the World.* Cambridge: BirdLife International, 2000.

Brooks, D. M., and S. D. Strahl, eds. *Curassows, Guans and Chachalacas: Status Survey and Conservation Action Plan for Cracids 2000–2004.* Cambridge: International Union for Conservation of Nature and Natural Resources (IUCN), 2000.

del Hoyo, J., A. Elliott, and J. Sargatal, eds. *Handbook of the Birds of the World.* Vol. 2, *New World Vultures to Guineafowl.* Barcelona: Lynx Edicions, 1994.

Hudson, P. J., and M. R. W. Rands. *Ecology and Management of Gamebirds.* Oxford: BSP Professional Books, 1988.

Johnsgard, P. A. *The Grouse of the World.* Lincoln: University of Nebraska, 1983.

Sibley, C. G., and B. L. Monroe, Jr. *Distribution and Taxonomy of Birds of the World.* New Haven: Yale University Press, 1990.

Snow, D. W., and C. M. Perrins. *Birds of the Western Palearctic: Concise Edition.* Vol. 1, *Non-Passerines.* Oxford: Oxford University Press, 1998.

Storche, I., ed. *Grouse: Status Survey and Conservation Action Plan 2000–2004.* Cambridge: IUCN and the World Pheasant Association, 2000.

Urban, E. K., H. C. Fry, and S. Keith, eds. *Birds of Africa* Vol. 2, *Gamebirds to Pigeons.* London: Academic Press, 1986.

Periodicals

Buchholz, R. "Older Males Have Bigger Knobs: Correlates of Ornamentation in Two Species of Curassow." *Auk* 108 (1991): 153–160.

Vaurie, C. "Taxonomy of the Cracidae." *Bulletin of the American Museum of Natural History* 138 (1968): 135–243.

Webre, A., and J. Webre. "Ecotourism and the Plain Chachalaca *Ortalis vetula* in Texas." *Bulletin of the IUCN/Birdlife/WPA Cracid Specialist Group.* 6 (1998): 13–14.

Organizations

BirdLife International. Wellbrook Court, Girton Road, Cambridge, Cambridgeshire CB3 0NA United Kingdom. Phone: +44 1 223 277 318. Fax: +44-1-223-277-200. E-mail: birdlife@birdlife.org.uk Web site: <http://www.birdlife.net>

IUCN Species Survival Commission. Rue Mauverney 28, Gland, 1196 Switzerland. Phone: +41 22 999 01 53. Fax: +41 22 999 00 15. E-mail: lwh@hq.iucn.org Web site: <http://www.iucn.org>

World Pheasant Association. P. O. Box 5, Lower Basildon St, Reading, RG8 9PF United Kingdom. Phone: +44 1 189 845 140. Fax: +44 118 984 3369. E-mail: office@pheasant.org.uk Web site: <http://www.pheasant.org.uk>

Julian Hughes

Moundbuilders

(Megapodiidae)

Class Aves
Order Galliformes
Suborder Craci
Family Megapodiidae

Thumbnail description
Medium-sized to large chicken-like birds with long legs and toes

Size
11–27 in (28–70 cm); 1.1–5.5 lb (0.5–2.5 kg)

Number of genera, species
7 genera; 22 species

Habitat
Rainforest, subtropical and tropical closed forest, mallee

Conservation status
Endangered: 1 species; Critical: 1 species; Vulnerable: 7 species; Near Threatened: 1 species

Distribution
Islands of Southeast Asia and southeast Pacific, Australia, and Papua New Guinea

Evolution and systematics

The moundbuilders, also known as incubator-birds, brush-turkeys, and megapodes, are chicken-like birds, closely resembling other galliforms such as pheasant and quail in general body shape, plumage, and forest-floor habitat. Their unique method of incubating eggs using environmental heat sources, rather like crocodiles and some reptiles, led some to suggest that these birds were extremely primitive. While this idea has now been thoroughly discredited, there has been considerable discussion about the relationship between the moundbuilders and other galliforms. The most authoritative accounts regard moundbuilders (Megapodiidae) as a sister group of all other galliforms, including the cracids (Cracidae), the South American family they most closely resemble. This classification is followed by the major monograph, *The Megapodes*, published in 1995.

Although there is a relatively large collection of remains of extinct species from this family, most are from very recent times. They paint a picture of progressive extermination as humans moved through the South Pacific during the last few thousand years. Fossil material from earlier periods is limited, making it difficult to determine the origins of this family. The

oldest megapode remains were initially thought to have been found in France, a claim now discredited. This material was used in part to suggest that the earliest moundbuilders had originated in the Northern Hemisphere and subsequently moved south to their present stronghold in the Australasian region. The alternative view is that moundbuilders evolved in the south, probably on the ancient super-continent of Gondwana, and later moved north into Southeast Asia and the Pacific. Although the Southern Origin theory is currently favored, both theories remain speculative due to the lack of reliable fossil material.

Among the Pleistocene remains of moundbuilders are several species much larger than extant species, including *Progura gallinacea*, from southern Australia, and *Megapodius molistructor*, from New Caledonia. These and other species, exterminated either directly or indirectly by humans, suggest that as recently as 3,000 years ago, as many as 23–33 additional species of moundbuilders were alive throughout the islands of the South Pacific.

Many of the species exterminated recently belonged to the genus *Megapodius*, the largest and most widely distributed of all the taxonomic groups within the family. Even excluding these extinct species, the number of species within this genus has puzzled experts for centuries. Harold Frith, the author of this section in the first edition, recognized only four species, which gave a total of 12 species for the entire family. The authors of *The Megapodes* proposed that there are 22 extant species, which include: 14 species of megapodes, sometimes called the scrubfowl (13 *Megapodius* and one species in the very similar *Eulipoa*); three species of brush-turkeys (one *Alectura* and two *Aepypodius*); three species of talegallas (*Talegalla*); and two species belonging to single-species genera, the malleefowl (*Leipoa*) and the maleo (*Macrocephalon*).

Physical characteristics

Moundbuilders have large, strong legs and feet and a short down-curved bill. Generally, most moundbuilders resemble other galliforms in body shape and drab plumage. In those few species in which the plumage is patterned or colorful, the effect is one of camouflage rather than conspicuousness. Most species are mainly drab brown, gray or blackish in color without detailed patterning, the main exception being the malleefowl (*Leipoa ocellata*). In this species the plumage is heavily spotted, with the patterns resembling eye shapes (ocellated). The most distinctive plumage of any moundbuilder is that of the maleo (*Macrocephalon maleo*) from Indonesia, a relatively large species whose dark blackish upperparts and head starkly contrast with its elegant salmon breast.

Although most moundbuilders have fairly drab plumage, many have naked patches of red, yellow, or blue skin on the face and neck, though this is rather inconspicuous. The three species of brush-turkey, however, each possess brightly colored skin on a virtually naked neck and head. The males in these species also possess a variety of sometimes oddly shaped wattles, combs, and neck sacs. These ornaments, vividly colored in yellows, reds, and blues, are used in sexual and dominance displays. In species such as the Australian brush-turkey (*Alectura lathami*), the large bright-yellow neck sac develops only among breeding adult males and is used as a signal of status. It may also be inflated for use as a device for enhanced vocalization.

Distribution

Moundbuilders are found mainly in Australia and New Guinea and on a large number of islands throughout the southeastern Pacific and Southeast Asia. They extend from the remote island of Niuafo'ou, Tonga, in the west, to the Palau and Mariana islands in the north, and Nicobar and Andaman islands in the Bay of Bengal to the east. Within this region, they occur west of Wallace's Line (between Borneo and Sulwesi) and are, therefore, largely an Australasian rather than an Oriental group. This is supported by the absence of moundbuilders from most of the larger land masses of the region, including most of Borneo, Java, Sumatra, and the mainland of Southeast Asia.

This largely island-based distribution has been explained as resulting either from competition for food with other galliforms present in the area (mainly pheasant and jungle fowl), or from predation pressure from carnivores, primarily civets

and civet cats. The latter hypothesis suggests that these predators have prevented the expansion of moundbuilders onto the larger landmasses for reasons directly related to their methods of incubation; the use of incubation mounds requires attendance for prolonged periods of time, a situation making them vulnerable to predators. In the small number of locations in which feline predators and moundbuilders occur together, the birds involved do not build mounds but are burrow-nesters, laying in soil heated by the sun or volcanic activity.

Habitat

Moundbuilders are largely birds of tropical and subtropical rainforests. Although some species venture into more open habitats and certain locations with seasonally dry climates, the moundbuilding habits of most require conditions that enhance the decomposition of organic matter. Even the burrow-nesting species, which may travel to exposed beaches or geothermal areas for egg-laying, typically return promptly to forests when finished.

Only two species are routinely found in habitats other than dense, moist forests. The malleefowl is the most aberrant of all species in this regard, living in the dry scrub of dwarf eucalyptus trees, or mallee, of south Australia. The only other species found in drier habitats are relict populations of the Australian brush-turkey. In general, this bird is an edge-species, preferring the margins of rainforests for its mounds, but some populations occur in the dry scrub forests far inland.

Behavior

Almost every aspect of life among moundbuilders is influenced by their incubation methods. Because eggs are laid in individual holes deep within the incubation site, each chick hatches separately, and, without assistance from parents, digs for 2–15 hours to reach the surface. Adult moundbuilders appear not to recognize the young of their own species. All hatchlings must, therefore, be able to leave the incubation site, find food and water, recognize and avoid predators and, even thermoregulate, alone. Malleefowl chicks (*Leipoa ocellata*) can run swiftly within one hour and fly within 24 hours of their hatching. Chicks of moundbuilders are certainly among the most precocial of all birds, and some experts now define their behavior as "supercocial"—meaning the young are completely independent at hatching. Studies of juvenile Australian brush-turkeys by Ann Göth have confirmed that, despite the absence of assistance by adults, these birds are able to respond differently and appropriately to aerial versus ground predators

The incubation sites used by these birds, both mounds and burrows, are the only sites in which eggs can be laid. In most species, construction and maintenance of these sites is undertaken jointly by a mated male and female. In these pairings, members are rarely seen separately and appear to be monogamous. However, detailed investigations of the Nicobar megapode (*Megapodius nicobariensis*) revealed examples of mate-exchanges and extra-pair copulations, and at least one

A malleefowl (*Leipoa ocellata*) nests at the top of the mound it has built. (Photo by R. Brown/VIREO. Reproduced by permission.)

case of polygyny has been recorded in the otherwise monogamous malleefowl.

In the three species of brush-turkeys, monogamy clearly does not occur. Males construct large incubation mounds alone and defend these from other males. Males allow access only to females willing to mate with them. In Australian brush-turkeys, males may construct several mounds or may take over another male's mound nearby. Females have unrestricted choice among all the mound-defending males, although there may be considerable competition for particular mounds among them. Although it is unclear how females choose among males, some individual males attract a large proportion of the eggs while other males attract very few.

All moundbuilders produce a variety of typical galliform-like clucks, grunts, and squawks, all of a low frequency, deep in pitch and generally fairly low in volume. Two species produce a distinctive, low-frequency boom by inflating either the external neck sac (Australian brush-turkey) or chambers in the throat (malleefowl) and then forcing air out through the nostrils. Many species—all *Megapodius* as well as the maleo—also produce a series of remarkably loud and conspicuous calls during both the day and night. Male brush-turkeys communicate with booms to advertise their mounds to females or to assert themselves while engaged in aggressive male-to-male behavior. In some places, the effect of roosting groups of these birds answering one another in these very loud and raucous choruses for prolonged periods, can be a

memorable, if sleep-disturbing, experience of being in the tropical jungle.

Feeding ecology and diet

Most moundbuilders appear to be generalist forest-floor foragers, eating a very wide variety of food types but taking advantage of seasonal abundances of items such as fallen fruits. Although the few studies on feeding ecology mention large amounts of small leaf-litter invertebrates along with fruit and seeds being eaten, unusual items such as ants, scorpions, and even small snakes are also mentioned.

While most of these items will be encountered as the birds rake their way through the leaf-litter, moundbuilders also seek out specific types of food. For example, Australian brush-turkeys excavate large holes in the forest floor in order to gain access to the succulent tubers of certain rainforest plants. This species also feeds on fruit within the forest canopy.

Reproductive biology

While many of these birds may look like large and somewhat dull galliforms, it is their unique approach to incubation that sets the moundbuilders apart from all other birds. The discovery of their unusual incubation sites was discussed by Bernard Grzimek himself in the first edition of this work. It was Navarette, a monk on Magellan's ill-fated circumnavigation of

Australian brush-turkey (*Alectura lathami*) incubation mound. The male (left) watches as the female digs down to lay her eggs with those previously laid by her and perhaps other females. Afterward the male fills the hole and monitors the mound for proper temperature and humidity. (Illustration by Dan Erickson)

the world (1519–1522), who gave us the first European account of a moundbuilder, probably the Philippine megapode (*Megapodius cumingii*). Unfortunately, Navarette's less-than-precise description did not aid the scientific reception of these ideas; in Grzimek's words, "[the monk] brought back . . . the story of fowl which laid eggs larger than themselves in heaps of leaves. The eggs, then, hatched without any further care. The matter of the size of the eggs was not quite correct; in regard to the second point, people believed more in mermaids and giant sea serpents than in such skills in fowl."

It is now known that this family of birds exploits three different sources of natural heat: solar radiation (primarily on beaches), geothermal activity (especially the soil near volcanic areas), and the decomposition of organic matter (in mounds). In the third process, decomposition, heat results from the respiration of the microorganisms that break down the material gathered together into a mound. Most species construct some form of incubation mound and this is the activity that gives the family its common name.

Incubation mounds vary greatly in size and composition but all consist of moist organic matter, mainly leaf-litter and soil, in which microbes can proliferate. The birds maintain a reliable incubator for many months by carefully and regularly adding fresh material to conserve moisture and prevent drying. The male brush-turkey and malleefowl maintain the temperature of the mounds at 91°F (33°C) by opening the mounds to cool them or piling on more sand. They probe the mounds with their bills to check the temperature. The amount of work required to construct and maintain a mound varies greatly among species and with climatic conditions. By far the most complex mounds are those made by malleefowl. Living in arid areas, this species faces the challenge of gathering together large amounts of damp leaf-litter and preventing the material from drying out. It does so by excavating a shallow depression, filling this with plant material, and waiting for rain. This debris is then piled into a mound and buried beneath a huge layer of sand and dry soil, which effectively insulates the mound. Whenever the birds need to gain access to the inner egg chamber for egg laying or temperature testing, they must remove and replace up to 1,875 lb (850 kg) of sand.

The incubation mounds of most other moundbuilders require much less effort to construct and maintain, many being little more than a large pile of leaf-litter, or simply a gathering of material around the base of a rotting log. Such mounds are made annually and often decompose to a small soil hump. Species such as the orange-footed megapode (*Megapodius reinwardt*), however, use the same mound year after year, with the resulting construction often growing to hillocks 14.8 ft (4.5 m) in height and more than 29.5 ft (9 m) in diameter.

Some species, including certain populations of those constructing mounds, are actually burrow-nesters rather than mound-builders, incubating their eggs in material that is heated by the sun or geothermal sources. In most cases, the eggs are laid in the loose soil at the bottom of long burrows excavated into the substrate.

Moundbuilders lay their eggs at intervals of several days, unlike most birds that lay eggs at smaller intervals. Each egg begins to develop immediately upon being deposited in the warm material. A female may lay a total of 12–30 very large (2.6–8.1 oz [75–230 g]) eggs per season, an egg mass that may comprise 120–180% of her body weight. Moundbuilder eggs contain an extremely high (32–49%) proportion of weight as yolk, a feature than correlates with their prolonged incubation period of 45–70 days.

Conservation status

Of the 22 extant species of moundbuilders, almost half face some level of threat. The Polynesian megapode (*Megapodius pritchardii*), which is limited to a single tiny island and may number less than 300 individuals, is Critically Endangered; and the Micronesian megapode (*Megapodius laperouse*), which is threatened by a variety of natural and human-based disturbances, is Endangered. A further seven species are classified as Vulnerable, due mainly to habitat loss, over-exploitation of eggs by humans, and predation by introduced animals such as foxes.

Significance to humans

The moundbuilders have had a close relationship with indigenous peoples for millennia throughout their range. Wherever they occur, humans have harvested their eggs, an extremely important source of protein and portable food. While humans have undoubtedly been responsible for the disappearance of numerous species, especially in the South Pacific, in other places, the exploitation of megapode eggs seems to have been practiced carefully for thousands of years. More recently, however, sustainable harvesting appears to have broken down through human population growth and internal migration. Today the conservation of several threatened species is being undertaken by indigenous programs that aim to restore habitat and prevent over-harvesting.

1. Australian brush-turkey (*Alectura lathami*); 2. Maleo (*Macrocephalon maleo*); 3. Malleefowl (*Leipoa ocellata*). (Illustration by Dan Erickson)

Species accounts

Australian brush-turkey
Alectura lathami

TAXONOMY
Alectura lathami Gray, 1831, Sydney, Australia. Two subspecies recognized.

OTHER COMMON NAMES
English: Scrub turkey, bush turkey, pouched talegallus; French: Talégalle de Latham; German: Bruschhuhn; Spanish: Talégalo Cabecirrojo.

PHYSICAL CHARACTERISTICS
23.6–27.6 in (60–70 cm); female 4.4–5.5 lb (1.98–2.51 kg), male 4.6–6.4 lb (2.12–2.9 kg). Large, mainly black, ground-dwelling bird with bright red head and neck, males with either yellow or light purple extendable neck sac during breeding season. Chicks, born fully feathered, have a uniform color buff-brown to sooty brown, closely resembling quail.

DISTRIBUTION
East Australia, from Cape York to northern New South Wales.

HABITAT
Rainforest and closed forest.

BEHAVIOR
Loosely social, males building and defending incubation mounds. Roosts communally in trees.

FEEDING ECOLOGY AND DIET
Generalist ground-forager, feeding on leaf-litter invertebrates and fruits.

REPRODUCTIVE BIOLOGY
Mounds constructed in July and are maintained until about December. Males polygynous, maintaining mounds (often two) in which females lay several eggs before moving to another mound. Up to 18–24 eggs; white and elliptical; laid by each female although a mound may have incubated up to 50 by the end of the season. Young extremely precocial.

CONSERVATION STATUS
Abundant and locally common throughout most of range, especially in southern Queensland where it is often a nuisance in urban gardens.

SIGNIFICANCE TO HUMANS
Eggs harvested. ◆

Malleefowl
Leipoa ocellata

TAXONOMY
Leipoa ocellata Gould, 1840, Swan River, Western Australia. Monotypic.

OTHER COMMON NAMES
French: Léipoa ocellé; German: Thermometerhuhn; Spanish: Talégalo Leipoa.

PHYSICAL CHARACTERISTICS
23.6 in (60 cm); female 3.3–4.5 lb (1.52–2.05 kg), male 4.0–5.5 lb (1.81–2.50 kg). Large and distinctive, upper parts boldly

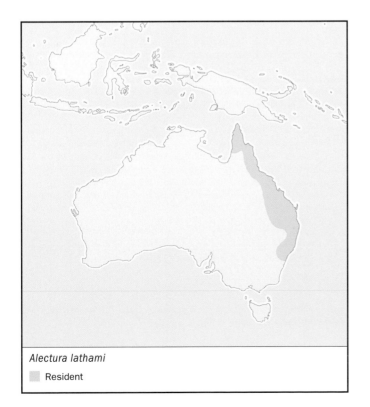

Alectura lathami
▭ Resident

Leipoa ocellata
▭ Resident

barred, streaked and fringed with gray, white, black, and rufous. Sexes similar though males slightly larger.

DISTRIBUTION
Originally found widely throughout the inland of southern Australia, now restricted to small patches of suitable habitat in southern states of Australia. Recently desert populations rediscovered in central Australia.

HABITAT
Arid and semi-arid low eucalypt and acacia woodland (mallee) and heath.

BEHAVIOR
Territorial, pairs defending area of incubation mound. Often solitary, with male spending long periods near mound while female wanders widely. Roost in trees and rarely fly. Three main calls: three-syllable booming (territorial), loft lowing call (communication), and sharp grunt (alarm).

FEEDING ECOLOGY AND DIET
Very broad diet of largely plant materials, especially seeds, fruit, and buds although up to 20% of food taken is ground-dwelling invertebrates.

REPRODUCTIVE BIOLOGY
Breed at two to four years of age. Mound-building species, with male spending up to 11 months of each year in mound construction and attendance. Mounds may be used for several generations. Usually strictly monogamous though some cases of polygyny are known. Females lay 2–34 eggs at intervals of 5–10 days. Each egg weighs 10% of female's body weight, the egg's pale pink color changes to dark beige during incubation. Incubation takes 55–77 days, depending on temperature of mound.

CONSERVATION STATUS
Classified by IUCN as Vulnerable, species having undergone 20% decline during last 45 years due to habitat destruction and impact of introduced predators.

SIGNIFICANCE TO HUMANS
Species has strong totemic significance for Central Australia indigenous people. ◆

Maleo
Macrocephalon maleo

TAXONOMY
Macrocephalon maleo S. Müller, 1846, Sulawesi. Monotypic.

OTHER COMMON NAMES
English: Gray's brush-turkey, maleofowl; French: Mégapode maléo; German: Hammerhuhn; Spanish: Talégalo Maleo.

PHYSICAL CHARACTERISTICS
21.7 in (55 cm); females 3.3–3.9 lb (1.50–1.76 kg), males 2.9–3.5 lb (1.34–1.59 kg). Large, striking bird with upperparts deep black and underparts white with strong salmon-pink tinge. Head topped with distinctive black casque. Sexes very similar, though males slightly larger.

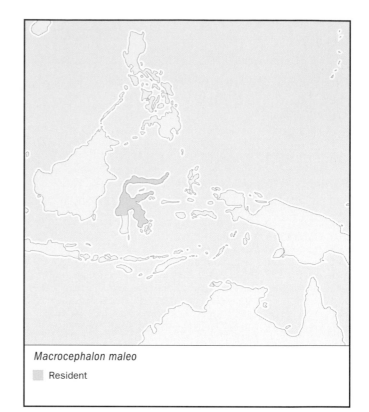

Macrocephalon maleo
▨ Resident

DISTRIBUTION
Endemic to island of Sulawesi, Indonesia, though absent from all deforested areas.

HABITAT
Tropical forests, as well as secondary vegetation and plantations.

BEHAVIOR
Shy and wary when approached. Pairs inseparable and will drive other pairs away from favored egg-laying positions.

FEEDING ECOLOGY AND DIET
Omnivorous, feeds on wide variety of fruits, seeds, and invertebrates encountered while foraging on forest floor.

REPRODUCTIVE BIOLOGY
A burrow-nesting species, using geothermally heated soils in the forest or sun-exposed beaches. Pairs travel to nesting grounds together and share effort in excavating laying hole and chasing away other maleos. Eggs laid at depths of 4–40 in (10–100 cm), at intervals of 10–12 days, the longest interval of any bird.

CONSERVATION STATUS
Classified as Vulnerable by the IUCN due to rapid population decline and over-exploitation and habitat degradation.

SIGNIFICANCE TO HUMANS
This charismatic species is an important species to the people of Sulawesi, where its eggs have been harvested for centuries. ◆

: Moundbuilders

Resources

Books

Bird, David M. *The Bird Almanac: The Ultimate Guide to Essential Facts and Figures of the World's Birds.* Buffalo: Firefly Books, 1999.

Dekker, R., R. Fuller, and G. Baker. *Megapodes: Status Survey and Conservation Action Plan 2000–2004.* Cambridge, United Kingdom: IUCN, 2000.

Harrison, Colin, and A. Greensmith. *Birds of the World.* New York: Dorling Kindersley Inc., 1993.

Jones, D., R. Dekker, and C. Roselaar. *The Megapodes.* Oxford: Oxford University Press, 1995.

Whitfield, P., ed. *The MacMillan Illustrated Encyclopedia of Birds.* New York: Collier Books, 1988.

Periodicals

Göth, A. "Innate Predator-recognition in Australian Brush-turkey (*Alectura lathami*, Megapodiidae) Hatchlings." *Behaviour* 138 (2001): 117–136.

Organizations

WPA/BirdLife/SSC Megapode Specialist Group. c/o Department of Ornithology, National Museum of Natural History, P.O. Box 9517, Leiden, 2300 RA The Netherlands.

Other

Malleefowl Conservation Group, Inc., 25 Nov. 2001. 9 Dec. 2001. <http://www.malleefowl.com.au>

Wong, Sharon. "Development and Behaviour of the Australian Brush-turkey *Alectura lathami*," Griffith University, 1999. Australian Digital Theses Project. 9 Dec. 2001. <http://www.gu.edu.au/ins/lils/adt/approved/adt>

Darryl Noel Jones, PhD

Grzimek's Animal Life Encyclopedia 411

Curassows, guans, and chachalacas

(Cracidae)

Class Aves
Order Galliformes
Suborder Craci
Family Cracidae

Thumbnail description
A group of Neotropical gamebirds, long-tailed and medium to large in size

Size
16.5–36.2 in (42–92 cm); 0.8–9.5 lb (385–4,300 g)

Number of genera, species
11 genera; 50 species

Habitat
Predominately tropical forest and woodland

Conservation status
Critically Endangered: 3 species; Endangered: 4 species; Vulnerable: 7 species; Near Threatened: 8 species; Extinct in the Wild: 1 species

Distribution
South Texas to central Argentina

Evolution and systematics

Cracids are a primitive, ancestral family of gamebirds (Galliformes), probably originating in Central America and southern North America. The earliest known cracid is recognized by a fossil approximately 50 million years old, found in Wyoming. This primitive bird appeared to be primarily arboreal, living at a time when most of North America was tropical. Younger fossils (around 30 million years old) similar to chachalacas have been found in nearby South Dakota. Recent fragments (approximately 20,000 years old) of more contemporary cracid fossils (e.g., *Crax*, *Penelope*) have been found within their existing range. The cracids are most closely related to the moundbuilders (Megapodiidae).

In Peter's checklist, the cracids form a family with 8 genera and 44 species. In other, more recent classifications, the family has 11 genera and 50 species. The chachalacas (*Ortalis*), with 12 species, are the smallest members of the family, and characterized by more drab coloration, lack of sexual dimorphism, and the ability to occupy a variety of habitats including secondary growth. The medium-sized guans are represented by six genera: true guans (*Penelope*) with 15 species, piping guans (*Pipile*) and sickle-winged guans (*Chamaepetes*)

with four and two species, respectively, and three monotypic genera: the wattled guan (*Aburria aburri*), the highland guan (*Penelopina nigra*), and the horned guans (*Oreophasis derbianus*). Coloring in guans ranges from drab to dramatic, and sexual dimorphism is present in only a few species. Curassows are the largest cracids, represented by four genera: the monotypic nocturnal curassow (*Nothocrax urumutum*), the razor-billed curassows (*Mitu*) with four species, helmeted curassows (*Pauxi*) with two species, and the true curassows (*Crax*) with seven species.

Physical characteristics

The Cracidae range in size from scarcely as large as a black grouse (*Lyrurus tetrix*) to almost turkey-sized. They replace the pheasants of Asia in tropical America and are in some respects also reminiscent of the American turkeys. Spanish-speaking Latin Americans therefore call them *pavos* or *pavones* (turkeys), or *faisanes* (pheasants). Their length is 16.5–36.2 in (42–92 cm), and weight is 0.8–9.5 lb (385–4,300 g). These birds are slim, long-legged, and have short, rounded wings and a fairly long tail. The beak is strong but fairly short and lightly curved, often with a conspicuous cere at the base. The

A plain chachalaca (*Ortalis vetula*) displays its tailfeathers. (Photo by R. Robles Gil/VIREO. Reproduced by permission.)

feet are like those of the moundbuilders and, in contrast to the Phasianidae, they have long, well-developed hind toes on the tarsus at the same level as the other toes ("pigeon-footed"). Cracids have a long caecum; in many species the trachea is prolonged. The plumage is mainly a shiny black or olive-brown to reddish brown, often with white marks, which form a crest on the head of some species. The feathers lack an aftershaft. The great curassow's head generally has an erectile crest of stiff, forward-curled feathers. In many curassow species the males have a fleshy knob or hump on the root of the beak, or bare, brightly colored areas of skin on the head. The cere of the beak is often colorful.

Distribution

Cracids are restricted to the New World, distributed from south Texas through most of tropical South America as far as central Argentina. Perhaps one of the most puzzling and intriguing patterns of cracid distribution is that some of the highland species show a strongly disjunct (separate) distribution, while many of the lowland forms are strongly parapatric (their

distributions adjoin each other rather than overlap). Riverine barriers may be a cause of the strong parapatric distribution of many lowland forms; other more disjunct species may have displayed more continuous distributions historically.

The countries harboring the highest diversity of cracids are Colombia and Brazil, with 24 and 22 species, respectively. In contrast, the United States harbors only a single species: the plain chachalaca (*Ortalis vetula*).

About 13 of the cracids are regionally restricted (six restricted to Brazil, three to Colombia, two to Mexico, and one each to Peru and Trinidad). The species with the widest distributions are perhaps the rusty-margined guan (*Penelope superciliaris*), which ranges through most of eastern tropical South America, and Spix's guan (*Penelope jacquacu*), which ranges through most of the western Amazon Basin. In contrast, the most range-restricted species is perhaps the Alagoas curassow (*Mitu mitu*) which is likely extinct in the wild and occurs only in captivity.

Habitat

The large members of the family are birds of dense tropical forest country. Some live in areas with a long dry season where the trees periodically shed their leaves, or in the "fringing forests" that extend along waterways in otherwise treeless country. Most guans are dependent upon primary forest, but a few *Penelope* species are able to tolerate secondary habitat. Curassows are almost invariably dependent upon tropical forest. The smaller chachalacas avoid the interior of dense forests and are at home in light secondary growth, such as secondary growth forest on formerly cultivated land. Chachalacas may live on plantations or near houses as long as there are trees and shrubs nearby and they are not disturbed. Most species inhabit warmer lowlands, though some occur in cooler mountain forests up to altitudes exceeding 9,800 ft (3,000 m).

All cracids are adapted to tree life. They walk about lightly and skillfully on thin branches in the tops of trees. Before crossing a woodland clearing, they jump and fly from branch to branch until they have reached the top of the highest tree at the edge of the clearing. From there they launch themselves into the air and as soon as they have gained enough speed in their fall, they spread their wings and glide downwards, often over distances far greater than 330 ft (100 m). Only when this glide will not enable them to reach their goal do they beat their wings to gain height. However, occasionally even heavy species like the great curassow undertake more distant aerial journeys. Cracids roost in trees overnight.

Behavior

Since the Cracidae are so shy and generally avoid humans, little is known about their social life. There are indications that at least some species live in polygamy; male yellow-knobbed curassows (*Crax daubentoni*), for example, may be found in the breeding season with three, four, or occasionally more hens. Guans are found at all seasons more or less socially in flocks, and their nests sometimes stand together in groups.

Chachalaca flight pattern. (Illustration by Brian Cressman)

The vocalizations of the Cracidae are loud and rarely pleasant to the human ear. In a number of species, the vocal power of the cocks, and sometimes of the hen as well, is increased by a prolonged trachea. It runs far back between skin and breast muscles, and then turns and runs forward to the point of entry into the chest cavity.

Apart from vocal utterances, several species of this family also produce drumming or clattering sounds with the wings during special display flights. The drumming sound of the black guan sounds "sharper" than the crested guan's wing drumming, because in the latter the ends of the two outermost primaries have almost no vanes, but only bare shafts.

Altitudinal migration apparently occurs in some of the montane species.

Feeding ecology and diet

Cracids are mainly plant eaters, but eat insects and other small animals to a lesser extent. They feed mainly on fruit and seeds that ripen during the course of the year in the tropical forests. They swallow berries and other small fruits whole,

however, they will bite into larger fruit like mangoes and guavas. They also bite off soft leaves and opening buds. Insects, snails, and other small animals form only very small parts of the menu. Now and then the birds come down from the trees to eat. When feeding on the ground, cracids, in contrast to many other Galliformes, do not appear to scratch the ground.

The general trend in diet appears to be more leaves and flowers and less fruit in smaller species (chachalacas), to more fruit and fewer leaves in larger species (curassows). The large curassows gather large numbers of fallen fruit on the forest floor. Similarly, animal matter seems to be more prevalent in the diets of smaller species (e.g., insects in the diet of chachalacas, snails in the diet of piping guans) than in curassows.

A crop is present as a dilatation of the gullet in curassows and in the horned guan. Other members of the family, which lack a crop, have a distensible gullet, so that food can be stored in it before digestion begins. The stomach is emptied first, and then food enters it from the crop or the gullet. Smaller seeds are passed through, whereas larger seeds are regurgitated on the spot. As the birds disperse the seeds of

A Salvin's curassow (*Mitu salvini*) adult and chick. (Photo by J. Alvarez A./VIREO. Reproduced by permission.)

incubation twice a day, once in the early morning, and again late in the afternoon, staying off the eggs from 60 to 75 minutes each time. Incubation periods vary according to species. The eggs of chachalacas may hatch in 21–23 days, but incubation lasts 34–36 days in the horned guan.

Cracid chicks are well developed at hatching, and the young are very soon able to fly or at least to flutter over short distances. They leave the nest very soon after their down is dry, sometimes even before. Skutch observed a chachalaca feeding its chicks soon after hatching. After the last chick had become dry, it was another three hours before the mother got down to the ground from the nest, which was 5 ft (1.5 m) high. The young left the nest with her. Cracids can move at birth, and fly, hop, and walk along twigs at quite a respectable height when only a few days old.

Conservation status

Cracids are the most threatened family of birds in the Americas. Of the 50 species, about half are of conservation concern. This includes nine of the 14 species (64%) of the curassows, but only two of the 12 species (16%) of the chachalacas; the other threatened species are guans, of which approximately 50% are threatened. The main threats are a combination of overhunting and habitat destruction.

Young cracids do not reproduce until they are two years old, and they breed only once a year with each hen rearing few young. Compared to other gallinaceous birds, cracids thus have a very low rate of reproduction. Only when humans appear on the scene and pursue them with firearms does their low rate of reproduction become insufficient and consequently dangerous. Wherever these birds are not protected by sensible and strictly enforced laws, they are in danger of extermination. Only the small chachalacas, which can adapt to the plants of cultivated country and are less desirable as food for humans, seem able to flourish in more densely settled and cultivated areas.

Significance to humans

Indigenous tribes may use tail or wing feathers in their ornamentation. Additionally, cracids constitute an important protein source in the diets of hunters in Latin America and often represent a substantial portion of the prey base. Unfortunately, the life history of cracids will often not permit intensive hunting pressure due to their low reproductive rate, long generation time, dependence upon specific habitat, and poor dispersal qualities. Because cracids are so heavily affected by both hunting and habitat destruction, they can be used effectively as bioindicator species for managing parks and protected areas in the Neotropics. By monitoring the population status of cracids in a particular area, wildlife managers can determine whether or not the forest resources in a given region are being over-exploited.

their preferred food plants throughout the forest in this way, they help regenerate the tropical forests in which they live.

Reproductive biology

The Cracidae build their nests in trees or in bushes in the forest or in thickets. The nest is a rough, disorderly structure shaped like a flat dish or a platform with a depression, which is often longer than it is wide. It is built from twigs, climbing plant stems, leaves, grass, palm frond pieces, and similar items. Larger species may use branches of 0.8–1.2 in (2–3 cm) in diameter for the nest base. Often they pluck leaf-bearing twigs or grasses that they bring to the nest while still fresh and green.

Hens appear to lay only two eggs in curassows, often three in chachalacas, and three or four eggs in guans. A clutch of nine plain chachalaca eggs found by R. J. Fleetwood was evidently derived from three hens. As far as is known, all Cracidae lay white eggs. The thick shell is usually rough and grainy or has markings that resemble pin pricks. Often the eggs are surprisingly large. During incubation, particularly in wet weather, the white eggs become stained by the leaves on which they lie.

For most species it appears that only hens incubate the eggs. A female chachalaca observed by Skutch interrupted her

1. Rufous-vented chachalaca (*Ortalis ruficauda*); 2. Horned guan (*Oreophasis derbianus*); 3. Plain chachalaca (*Ortalis vetula*); 4. Black guan (*Chamaepetes unicolor*); 5. Alagoas curassow (*Mitu mitu*); 6. Wattled curassow (*Crax globulosa*); 7. Northern helmeted curassow (*Pauxi pauxi*); 8. Crested guan (*Penelope purpurascens*). (Illustration by Brian Cressman)

Species accounts

Plain chachalaca
Ortalis vetula

SUBFAMILY
Penelopinae

TAXONOMY
Penelope vetula Wagler, 1830, Mexico. Four subspecies.

OTHER COMMON NAMES
English: Common chachalaca; French: Otalide chacamel; German: Blauflügelguan; Spanish: Chachalaca Norteña.

PHYSICAL CHARACTERISTICS
19–22.8 in (48–58 cm); 15.5–28 oz (440–794 g). Plain coloration, races vary in size and color.

DISTRIBUTION
This species ranges from south Texas to eastern Mexico and Costa Rica.

HABITAT
Scrub and tall brush vegetation. Also occurs in lowland and pre-montane forest in Central America.

BEHAVIOR
The full morning chorus of the plain chachalaca (*Ortalis vetula*) is unforgettable. One of these birds, sitting in a tree above dense secondary growth, calls with a rough, unmelodic, but remarkably strong voice, "cha cha lack, cha cha lack." The neighbors take part and a real din of loud calls arises. When those nearby have become quiet, one hears other more distant voices. The chorus seems to decline until from a distance of over half a mile (1 km), one can hear no more. Then the noise surges back with increasing strength and finally an earsplitting din is produced by a group of six to eight of the birds situated vertically above the observer.

FEEDING ECOLOGY AND DIET
Fleshy fruit comprises much of their diet. Also green leaves, buds, shoots, and twigs. Some insects.

REPRODUCTIVE BIOLOGY
Nests are most often built in trees 3.3–33 ft (1–10 m) off the ground. The nest itself virtually nonexistent, with birds often laying eggs on bare limbs. The clutch size is typically two to four eggs with an average incubation of 25 days.

CONSERVATION STATUS
The Utila Island subspecies (off north Honduras) population has declined, and is possibly extinct.

SIGNIFICANCE TO HUMANS
Sometimes consumed for food. ◆

Rufous-vented chachalaca
Ortalis ruficauda

SUBFAMILY
Penelopinae

TAXONOMY
Ortalida ruficauda Jardine, 1847, Tobago. Two subspecies.

Ortalis ruficauda
▪ Resident

Ortalis vetula
▪ Resident

OTHER COMMON NAMES
English: Rufous-tailed chachalaca, rufous-tipped chachalaca;
French: Otalide à ventre roux; German: Rotschwanzguan;
Spanish: Chachalaca Culirroja.

PHYSICAL CHARACTERISTICS
20.1–24 in (53–61 cm); 15.2–28.2 oz (430–800 g). Rufous undertail coverts, grayish belly. Immature resembles adult.

DISTRIBUTION
This species occurs in northeastern Colombia, northern
Venezuela, and Tobago.

HABITAT
Lowland forest and thorny brush. Often found near water in
gallery forest and along rivers.

BEHAVIOR
Its native name, "guacharaca," is a good reproduction of its
call. It calls loudly during moonlight, but its fullest choruses
are heard at daybreak. During the rainy season, it calls on and
off all morning.

FEEDING ECOLOGY AND DIET
Leaves, shoots, fruits, and flowers. Usually forages in trees in
groups of 4–20.

REPRODUCTIVE BIOLOGY
Nests are often built in trees 3–10 ft (1–3 m) off the ground.
The nest itself is typically made of twigs and leaves. The clutch
size is typically three to four eggs with incubation lasting approximately 28 days.

CONSERVATION STATUS
Not threatened.

SIGNIFICANCE TO HUMANS
Sometimes consumed for food. ◆

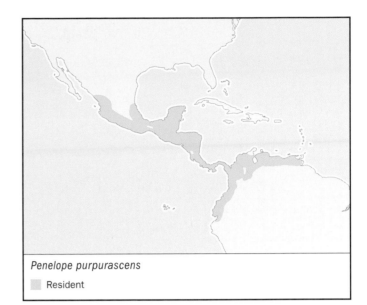

Penelope purpurascens
▨ Resident

Crested guan
Penelope purpurascens

SUBFAMILY
Penelopinae

TAXONOMY
Penelope purpurascens Wagler, 1830, Mexico. Three subspecies.

OTHER COMMON NAMES
English: Purple guan; French: Pénélope panachée; German:
Rostbauchguan; Spanish: Pava Cojolita.

PHYSICAL CHARACTERISTICS
This is the largest species of guan, with some specimens reaching over 3 ft (90 cm), and nearly 5.5 lb (2.5 kg).

DISTRIBUTION
Ranges from Mexico to Venezuela, northern Colombia, and
southwestern Ecuador.

HABITAT
Generally in humid forest and hilly lowlands. Occasionally
found in gallery forest.

BEHAVIOR
Pairs of crested guans have territories in which they may remain with their young until the next breeding season.

Crested guans are particularly noisy when disturbed. They
perch high in trees and continually protest with a very loud
prolonged shrieking which sounds peculiarly high for such relatively large birds.

The crested guan may climb to the top of a high tree at the
edge of a clearing and fly with slow measured beats over the
open space. When it has gained enough speed, it will beat its
wings much more rapidly, producing a loud drumming noise.
Then it may glide for a stretch, drum again, and continue its
flight across the clearing into the trees on the opposite side. This
peculiar drumming is heard only rarely, just at dawn or dusk and
on moonlit nights. The display is likely related to breeding.

FEEDING ECOLOGY AND DIET
Fruits, figs, and berries; also seeds, leaves, shoots, and occasionally ground insects. Usually forages in the high branches of trees.

REPRODUCTIVE BIOLOGY
Monogamous, possibly maintaining a permanent pair bond.
The birds mature after two to three years. Nests are most often built in trees, and are made of sticks and twigs and lined
with leaves. The clutch size is typically two eggs.

CONSERVATION STATUS
Not threatened.

SIGNIFICANCE TO HUMANS
This species is often consumed for food. ◆

Black guan
Chamaepetes unicolor

SUBFAMILY
Penelopinae

TAXONOMY
Chamæpetes unicolor Salvin, 1867, Veragua. Monotypic.

OTHER COMMON NAMES
English: Black sickle-winged guan; French: Pénélope unicolore;
German: Mohrenguan; Spanish: Pava Negra.

Chamaepetes unicolor

■ Resident

PHYSICAL CHARACTERISTICS
24–27 in (62–67 cm); 2.4–2.6 lb (1.1–1.2 kg). Male is completely black with a bare blue facial patch.

DISTRIBUTION
This species is restricted to Costa Rica and western Panama.

HABITAT
Found in montane forest above 3,300 ft (1,000 m) in southern Central America.

BEHAVIOR
Black guans are one of the few cracids known to live singly outside the breeding season and are found in pairs only during the breeding season.

The black guan's wing sound is quite different compared to the crested guan. When this bird, in its long glide, has reached the middle of a clearing, it beats its wings rapidly over a short stretch in such a way that the longer feathers alternately separate and beat together. Thus it produces a wooden-sounding clatter of surprising loudness, which can be imitated by drawing a thin, narrow piece of wood over an iron grating or by holding it against the spokes of a wheel.

FEEDING ECOLOGY AND DIET
Generally fruits, also seeds and several species of plants. Forages singly, in pairs, or in small groups.

REPRODUCTIVE BIOLOGY
Clutch size is typically two to three eggs. Downy chicks are brown with striped pattern on the head.

CONSERVATION STATUS
Considered Near Threatened, this species is threatened with habitat destruction, and is vulnerable because it is restricted to montane forests at 3,300–8,200 ft (1,000–2,500 m). Furthermore, it is endemic to a small region of Costa Rica and Panama. Nonetheless it might be locally common in at least some regions where it is protected. For example, in the early

1990s densities were approximately 7.4 birds per km² at the Monteverde Cloud Forest Reserve in Costa Rica.

SIGNIFICANCE TO HUMANS
None known. ◆

Horned guan
Oreophasis derbianus

SUBFAMILY
Penelopinae

TAXONOMY
Oreophasis derbianus G.R. Gray, 1844, Guatemala. Monotypic.

OTHER COMMON NAMES
English: Lord Derby's mountain pheasant; French: Oréophase cornu; German: Zapfenguan; Spanish: Pavón Cornudo.

PHYSICAL CHARACTERISTICS
29.5–33.5 in (75–85 cm). This species is quite unique, and believed by some authorities to be a link between guans and curassows because it shares some characteristics of both. Most prominent is the large red horn and the white scalloping on the breast. Sexes are identical.

DISTRIBUTION
The horned guan is found in Guatemala and Chiapas (the most southern province of Mexico).

HABITAT
Found in montane forest above 4,900 ft (1,500 m) in northern Central America.

BEHAVIOR
When disturbed, horned guans give off a throaty (guttural) shriek, which, in its suddenness and intensity, has the effect of a loud explosion. Then they threaten the intruder from a high perch by clattering their yellow beaks like castanets.

Oreophasis derbianus

■ Resident

FEEDING ECOLOGY AND DIET
Generally fruits and green leaves from a vast assortment of plant species. Forages primarily in tree branches.

REPRODUCTIVE BIOLOGY
This species is likely one of the few cracids (and perhaps the only guan) where polygyny is observed, where one male might mate with several females, one after another, during the breeding season (serial polygyny). Nests are often built in very high trees, up to 66 ft (20 m) off the ground. The nest itself is typically made of twigs and epiphyte roots. The clutch size is typically two eggs, with one of the longest incubations documented for any cracid, up to 36 days.

CONSERVATION STATUS
Considered Endangered. The remaining populations are small, fragmented, and only partly protected. Habitat destruction and hunting continue to threaten this species.

SIGNIFICANCE TO HUMANS
May be hunted for food. These birds have been successfully raised in captivity, which may be important to the future survival of this species. ◆

Alagoas curassow
Mitu mitu

SUBFAMILY
Cracinae

TAXONOMY
Crax mitu Linnaeus, 1766, Brazil and Guiana = northeastern Brazil. Monotypic.

Mitu mitu
■ Resident

OTHER COMMON NAMES
English: Bare-eared curassow; French: Hocco mitou; German: Mituhokko; Spanish: Paují de Alagoas.

PHYSICAL CHARACTERISTICS
About 32.5 in (83 cm) in length; males weigh about 104 oz (2.960 g); females weigh about 97 oz (2,745 g). Shorter head crest and tall, narrow red comb running along the forehead and upper mandible of the beak. Lacks dense feathering in the auricular region, hence its alternate name of bare-eared curassow.

DISTRIBUTION
This species was endemic to Alagoas, Brazil, but is now extinct in the wild and survives only in captivity.

HABITAT
Extinct in the wild, but formerly found in lowland forest.

BEHAVIOR
Little is known, though the birds have been raised successfully in captivity.

FEEDING ECOLOGY AND DIET
Little known. Fruit was found in the stomach of a specimen collected in 1951; and three birds were spotted eating fruit in the late 1970s.

REPRODUCTIVE BIOLOGY
One nest from the late 1970s was found in a tree amidst much foliage cover. Captive birds are apparently sexually mature after a couple of years, and the clutch size is often two eggs, as is the case with most curassows.

CONSERVATION STATUS
This species is Extinct in the Wild, and most of the region it lived in before has been converted to soybean plantations. A small population persists at two collections in Brazil, jointly exceeding 50 birds in the year 2000.

SIGNIFICANCE TO HUMANS
These birds have been hunted for food. Their survival depends upon the populations currently in captivity. ◆

Northern helmeted curassow
Pauxi pauxi

SUBFAMILY
Cracinae

TAXONOMY
Crax pauxi Linnaeus, 1766, Mexico, error = Venezuela. Two subspecies.

OTHER COMMON NAMES
English: Helmeted curassow; French: Hocco á pierre; German: Helmhokko; Spanish: Paují de Piedras, Paují de Yelmo.

PHYSICAL CHARACTERISTICS
Helmeted curassows are characterized by a bony outgrowth on the forehead. This species has a unique ornamentation, a large globose bluish grey knob above the bill. Length is 33.5–36 in (85–92 cm). Males weigh 7.7–8.3 lb (3.5–3.75 kg); females weigh 5.8 lb (2.65 kg).

DISTRIBUTION
This species is found in northern Venezuela and northeastern Colombia.

HABITAT
Found in dense, cool montane cloud forest above 1,650 ft (500 m).

Pauxi pauxi
▢ Resident

Wattled curassow
Crax globulosa

SUBFAMILY
Cracinae

TAXONOMY
Crax globulosa Spix, 1825, Rio Solimes, Brazil. Monotypic.

OTHER COMMON NAMES
English: Red wattled curassow, Yarrell's curassow; French: Hocco globuleux; German: Karunkelhocco; Spanish: Pavón Carunculado.

PHYSICAL CHARACTERISTICS
32.2–35 in (82–89 cm); roughly 5.5 lb (2.5 kg). This is the only species where the male has red globose ornamentation both above and below the bill. The female has a red cere and rufous venter. Male is black with a white belly, females have reddish brown belly feathering.

DISTRIBUTION
They are found from Colombia and western Brazil to Bolivia.

HABITAT
This species is strongly tied to *várzea* (seasonally flooded) forest in the Amazon basin. It is almost entirely arboreal.

BEHAVIOR
The wattled curassow has a soft whistle, "yeeeeeee," which lasts four to six seconds.

FEEDING ECOLOGY AND DIET
Not known.

BEHAVIOR
The call of the helmeted curassow is a prolonged, low-pitched grunting or groaning which sounds like "mm-mm-mm-mm." The cock produces it by breathing out with a closed beak.

FEEDING ECOLOGY AND DIET
Generally fallen fruits and seeds; also leaves, grasses, and buds. Forages on the ground singly, in pairs, or in family groups.

REPRODUCTIVE BIOLOGY
Nests are often built in trees 13–20 ft (4–6 m) off the ground. The nest itself is typically rudimentary, made of twigs and lined with leaves. The eggs are among the largest of any cracid, 3.3–3.7 in (8.5–9.5 cm) long and 2.4–2.5 in (6–6.5 cm) wide. The clutch size is typically two eggs with incubation lasting approximately 34 days. The hen leaves her eggs once a day, usually between eight and ten in the morning, and she stays off them from one to 2.5 hours. If it rains the whole day, she may omit this "outing."

CONSERVATION STATUS
Vulnerable. The northern helmeted curassow is endemic to a small region of northern South America, and is threatened with habitat destruction and overhunting. The montane forest this species is restricted to is being converted into cattle grazing ranchland at higher altitudes, and narcotics plantations at lower altitudes. Its geographic range is estimated at 13,900 mi^2 (36,000 km^2). Its numbers are estimated at fewer than 10,000, with populations declining. It does however occur in a number of Venezuelan reserves, as well as the Cocuy National Park in Colombia.

SIGNIFICANCE TO HUMANS
This species is often hunted for food, unsustainably in many situations. ◆

Crax globulosa
▢ Resident

REPRODUCTIVE BIOLOGY

No information available about breeding in the wild. In captivity, clutch size is two eggs; chicks brownish above, light buff below.

CONSERVATION STATUS

Vulnerable. The wattled curassow is rarely encountered, and very patchily distributed throughout its geographic range in the western Amazonian basin. However, this species may be more difficult to detect than other species because its call is less easily detected, it is restricted to the higher strata of the forest, and it often flies away silently in retreat. This species is threatened by overhunting, and the total population is suspected of being small and declining. Only future research will shed more light on the status of this curassow.

SIGNIFICANCE TO HUMANS

This species is often hunted for food, unsustainably in many situations. ◆

Resources

Books

Birdlife International. *Threatened Birds of the World.* Barcelona: Lynx Edicions, 2000.

Brooks, Daniel M., and Fernando Gonzalez-Garcia. *Biology and Conservation of Cracids in the New Millenium.* Houston: Miscellaneous Publications of the Houston Museum of Natural Science, Number 2, 2001.

Brooks, Daniel M., et al. *Biology and Conservation of the Piping Guans (Pipile).* Houston: Special Monograph Series of the Cracid Specialist Group, Number 1, 1999.

Brooks, Daniel M., and Stuart D. Strahl. *Cracids: Status Survey and Conservation Action Plan.* Switzerland: IUCN, 2000.

Delacour, Jean, and Dean Amadon. *Curassows and Related Birds.* New York: The American Museum of Natural History, 1973.

Strahl, Stuart D., et al. *The Cracidae: Their Biology and Conservation.* Washington: Hancock House Publications, 1997.

Organizations

The Cracid Specialist Group. PO Box 132038, Houston, TX 77219-2038 USA. Phone: (713) 639-4776. E-mail: dbrooks @hmns.org Web site: <http://www.angelfire.com/ca6/cracid>

Neotropical Bird Club. c/o The Lodge, Sandy, Bedfordshire SG19 2DL United Kingdom. E-mail: secretary @neotropicalbirdclub.org

Daniel M. Brooks, PhD

Guineafowl

(Numididae)

Class Aves
Order Galliformes
Suborder Phasiani
Family Numididae

Thumbnail description
Medium-sized, stocky, highly vocal birds with short, stout bills, and naked heads normally topped by casque or crest of feathers

Size
15.5–28.2 in (40–72 cm); 1.5–3.5 lb (0.7–1.6 kg)

Number of genera, species
4 genera; 6 species

Habitat
Forest, woodlands, and savanna

Conservation status
Vulnerable: 1 species

Distribution
Sub-Saharan Africa, introduced to extreme southern Arabia, Madagascar, and many tropical islands

Evolution and systematics

Guineafowl have sometimes been incorporated as a subfamily within the family Phasianidae, which includes pheasant-like birds such as turkeys, grouse, quails, and partridges. However, evidence from an examination of their internal anatomy, plumage, behavior, breeding biology, egg-white proteins, chromosomes, and DNA suggests that they be accorded family status. Their evolution, as with that of the Australasian megapodes (Megapodiidae) and the largely South American cracids (Cracidae), was almost certainly influenced by the break-up of Gondwana (the southern mega-continent comprising Australia, India, Africa and South America) over 100 million years ago. Guineafowl essentially bridge the evolutionary gap between the relatively advanced, pheasant-like birds (Superfamily Phasianoidea) and the anatomically primitive megapodes and cracids. There is no unequivocal fossil record of guineafowls before the Pleistocene.

Effects of climate change due to long-term fluctuations in rainfall and temperature and movement of landmasses have had a profound influence on guineafowl evolution. Expanding and contracting forests and open-country vegetation, and the de-

velopment of the Rift Valley, have caused repeated fragmentation of the ranges of species tied to forests or open-country habitats. This process has promoted speciation and sub-speciation within the genera *Agelastes*, *Guttera*, and *Numida*.

As of 2001, the scientific community recognizes six species grouped within four genera. However, the crested guineafowl (*Guttera pucherani*) may be further sub-divided into a second species, *G. edouardi*.

Each species has a largely unfeathered head and neck; background body plumage is largely black, adorned (in most species) with tear-drop-sized white spots. The naked head and neck is thought to be an adaptation for cooling relatively warm arterial blood flowing upwards to the brain from the heart within a car-radiator-like nexus of micro-veins carrying relatively cool blood downwards. This "radiator" allows guineafowl to forage and socialize throughout the heat of the day.

Physical characteristics

Guineafowl are medium-sized, chicken-like birds with short, stout bills. They have strong legs, either without leg

Helmeted guineafowl (*Numida meleagris*) gathered at a watering hole. (Photo by P. Craig-Cooper/VIREO. Reproduced by permission.)

spurs (genera *Guttera* and *Numida*), or with short, blunt, multiple spurs (genus *Acryllium*), or long, sharp, single spurs (genus *Agelastes*); these spurs grow directly out of the tarsometatarsus as in the junglefowl *Gallus* rather than the hypotarsus (as in most spurred galliforms). The tail is generally short and points downward, with 14 feathers (genus *Agelastes*) or 16 feathers (the rest of the genera). Guineafowl in the genera *Guttera* and *Numida* have cartilaginous wattles that hang from the base of the bill. The males, although larger and heavier on average, resemble the females in all species.

The two species of lowland forest guineafowl lack the characteristic white spotting on their plumage and have no adornments on their naked, red heads. The plumage of the black guineafowl (*Agelastes niger*) has a delicate, brownish wave design on the otherwise black plumage. The white-breasted guineafowl (*Agelastes meleagrides*), as its name implies, has white breast feathers. The rest of the body is black with fine, white, wavy bars. The body plumage of the remaining species of guineafowl is black with white spots.

The crested and plumed guineafowl (genus *Guttera*) each have a trachea, which forms a loop, embedded in a kind of pocket at the base of the wishbone. Plumed guineafowl (*Gut-*

tera plumifera) have a brush-like tuft of straight feathers on the crown. Crested guineafowl (*Guttera pucherani*) have slightly to markedly curled feathers. There are five subspecies of crested guineafowl, but only two subspecies of plumed guineafowl, that vary mainly in the color of the facial skin on the head and the form of the feathered crest.

The head of the vulturine guineafowl (*Acryllium vulturinum*) is mostly featherless except for a chestnut-brown patch of short feathers on the sides and back of the head. This species has a hackle of spear-shaped feathers which have lengthwise stripes of black, white, and blue. The breast is blue, the edges of the wings are violet, and the rest of the plumage, except for the black ventral feathers, shows the usual spotted design. The central tail feathers are elongated and pointed. This guineafowl is slender and has longer legs than the helmeted guineafowl.

The helmeted guineafowl (*Numida meleagris*) has a bony casque or helmet on top of its head covered with horny cartilage. Each of the nine subspecies of helmeted guineafowl has a characteristic helmet shape. Each is also characterized by different coloring of the bare parts of the head, wattle, and neck feathers, as well as by the absence or presence of conspicuous bristles near the nostrils.

Distribution

With the exception of one (almost certainly extinct) sub-species of helmeted guineafowl that inhabited western Morocco, the original range of all guineafowl species is sub-Saharan Africa. Humans have introduced the helmeted guineafowl to southern Arabia, Madagascar, and as far afield as Cuba and other tropical islands, where they have acclimatized.

Black guineafowl occur in forests from Cameroon to the lower Congo and eastward into the Ituri territory. White-breasted guineafowl replace black guineafowl in the forests of Upper Guinea.

Plumed guineafowl inhabit forests in Cameroon, northern Congo, and the Central African Republic, extending southwards to Gabon and northern Angola. Crested guineafowl are distributed in less well-developed forests, from Guinea Bissau, eastward through Congo, the Democratic Republic of Congo, and as far as western Kenya, and southwards as far as Angola and northeastern South Africa.

Vulturine guineafowl inhabit the dry steppes of southern Ethiopia and southern Somalia as far as northwestern Tanzania.

Helmeted guineafowl occupy open-country areas virtually continuously from Senegal, eastwards to Somalia, and south to South Africa.

Habitat

Black guineafowl and white-breasted guineafowl inhabit primary, unspoiled, rainforests. Plumed guineafowl occur in primary rainforest and very mature secondary re-growth. Crested guineafowl prefer secondary forest, forest edge, and gallery forest. Helmeted guineafowl occupy virtually all open-country habitats in sub-Saharan Africa outside of deserts; they can thrive on cultivated land within which aggregations of more than 1,000 individuals (which are comprised of many flocks) may be encountered.

Behavior

All species of guineafowl live in sedentary groups when not breeding and in monogamous pairs when breeding.

Black guineafowl and white-breasted guineafowl live in groups of 15–20 and roost in trees at night. Their advertising calls are very different from other species of guineafowls. The calls consist of short, soft, low-pitched whistling sounds reminiscent of the cooing of doves.

The advertising calls of plumed and crested guineafowl are low-pitched "kuk-kuk-kuks" given in series and increasing in volume. Little is known about the behavior of plumed guineafowl.

Crested guineafowl occur in flocks averaging fewer than 20 birds. During early morning after descending from their nightly roosts, crested guineafowl flocks will move into forest glades to preen and socialize in the warmth of the morn-

A vulturine guineafowl (*Acryllium vulturinum*) preens in East Africa. (Photo by Erwin & Peggy Bauer. Bruce Coleman Inc. Reproduced by permission.)

ing sun. In dim light before sunrise and after sunset, flocks will venture onto bush tracks and thus be vulnerable to collisions with motor vehicles, particularly because crested guineafowl will fly up from the track only to land in front of the vehicle due to the impenetrable vegetation on either side. Crested guineafowl will fly up into trees to feed on fruit rather than eat maize scattered on the ground. This species has also been observed, primarily in the late afternoon, foraging in association with vervet monkeys (*Cercopithecus aethiops*), feeding below on scraps of fruit and feces that fall to the forest floor. This relationship may also take advantage of a two-way alarm system: the monkeys warn of intruders from the trees above and the guineafowl scout out the forest floor. Although deeper than the call of the helmeted guineafowl, the crested guineafowl alarm call is also a staccato "chuk-chuk-chukchuk-err." The lower pitch may be due, in part, to its windpipe being housed in the hollowed-out blade of its wishbone. Crested guineafowl are noisy, sometimes calling well into the night and during the quiet pre-dawn hours. Flock members keep in contact by emitting a low-pitched "chuk" call. Male crested guineafowl apparently do not have a hump-backed display as

The vulturine guineafowl (*Acryllium vulturinum*) sometimes perches in bushes to feed on fruits. (Photo by Nigel Dennis. Photo Researchers, Inc. Reproduced by permission.)

in the helmeted guineafowl, but rather cock their tails like bantams when alarmed.

Vulturine guineafowl occur in flocks of 20–30 birds. Their advertising call is similar to that of the crested and plumed guineafowl, but is a much higher-pitched "keek-keek-keek," also given in series.

The helmeted guineafowl is by far the most social guineafowl species and may be seen in aggregations of hundreds of birds. At dawn, flock members fly down from their roost sites to a watering place. During early morning after drinking, they often seek patches of dry, light soil to dust bathe, presumably to rid themselves of feather parasites. During late morning and early afternoon, they stay in the shade. By late afternoon, they start looking for food, and at nightfall they return to their roosts, generally in tall trees.

Feeding ecology and diet

Black, white-breasted, plumed, and crested guineafowl feed on a range of plants and small invertebrates obtained by scratching in the leaf litter. In the case of white-breasted guineafowl, when one flock member finds a concentration of food, others immediately converge on it and try to crowd it away, pressing with their shoulders and pushing with their legs. However, birds competing for food do not use their bills

aggressively during these encounters. Crested guineafowl may also feed in trees on berries and other fruits, and on bulbs from below the ground.

Vulturine guineafowl also feed on a range of plant and animal items and will sometimes perch in bushes to feed on fruits. Unlike all other guineafowl, this species appears to survive without readily-available drinking water.

Helmeted guineafowl are the most omnivorous of the guineafowl species, taking a broad range of plant and animal material (even small toads) from above and below the ground, switching their preferences to whatever appears to be abundant at the time, but focusing on insects and other arthropods during the breeding season.

Reproductive biology

Guineafowl appear to be monogamous breeders. Black guineafowl may be found breeding in any month of the year, but generally do so during drier months (December–February). The nest has not been described. The clutch size is unknown, and the eggs are pale reddish brown, sometimes with a violet or yellowish tinge. The downy keet (guineafowl chick) has dark-rufous back plumage with black markings.

White-breasted guineafowl may also breed during any month of the year, but breeding probably peaks at the end of the rainy season (November–January). The nest has not been described. Clutch size is about 12 eggs that are reddish buff with white pores. The back plumage of the downy keet is grayish brown with reddish brown highlights. The crown of the head and the mid-line of the neck are black.

Plumed guineafowl also appear not to have a fixed breeding season, but tend to nest during the rainy season(s). The nest is a simple scrape in the ground lined with leaves. Clutch size is 10 eggs that are pale buff with darkened pores. The downy keet has dark-buff back plumage with dark brownish black longitudinal stripes.

Crested guineafowl may also be found breeding during any month of the year in the equatorial parts of its range, but this species breeds during the rainy season in areas away from the equator. The nest is a simple scrape in the ground lined with leaves. Clutch size is normally four to five eggs, ranging up to seven eggs, that are dark buff to pinkish or white. The downy keet has dark-buff back plumage with dark brownish black longitudinal stripes. Chicks can flutter-fly at 12 days and fledge in downy plumage at about 30 days.

Vulturine guineafowl breed primarily during June and December–January, after the rainy seasons. The nest is a simple scrape in the ground, generally in thick grassy areas. Clutch size is 7–10 eggs, ranging up to 15 eggs, that are creamy white to pale brown. The incubation period is 22–25 days. The downy keet has yellowish buff back plumage with dark brownish black mottling.

For helmeted guineafowl, the breeding season is strongly tied to the timing and amount of rainfall. In west Africa, they breed during May–July. In southern Africa, the most intense

A helmeted guineafowl (*Numida meleagris*) in a dust bath. (Photo by Kenneth W. Fink. Photo Researchers, Inc. Reproduced by permission.)

breeding activity is during summer (October–March) in the predominantly summer-rainfall regions of eastern and southern South Africa, and during late summer and early autumn (January–March) in the north (in Botswana and Namibia). Breeding activity is, however, determined largely by the timing of regular heavy rainfall—in the winter-rainfall regions (the Western Cape and the western half of the Eastern Cape), peak breeding is between September and December, to take advantage of food fostered by winter rains. The nest is a simple scrape in the ground lined with grass, protected by a bush or tall grass. Clutch size is 12–23 eggs that are extremely hard-shelled, brown or yellowish brown in color, and pointed at one end. The downy keet has dark buff back plumage with blackish brown longitudinal stripes. Chicks can flutter-fly at 14 days and fledge in downy plumage at about 30 days.

Conservation status

The white-breasted guineafowl is Vulnerable, according to the IUCN. It is extremely sensitive to habitat destruction and illegal hunting by humans. On a global scale, none of the other species of guineafowl appears to be threatened.

Significance to humans

Members of the subspecies of helmeted guineafowl from the vicinity of what was then known as the Gulf of Guinea (hence the family name) in western Africa were domesticated in Europe and selected artificially for rapid growth to a heavier body mass (i.e., for meat production). Additional effects of this selection have been a change in color of the body plumage from black to white (or partly white) and of the legs and toes from black to yellow-orange.

Helmeted guineafowl are also highly prized, but grossly undervalued, objects of wingshooting (gamebird hunting).

Species accounts

Helmeted guineafowl
Numida meleagris

TAXONOMY
Phasianus meleagris Linnaeus, 1758, Africa = Nubia, Upper Nile. Nine subspecies.

OTHER COMMON NAMES
French: Pintade de Numidie; German: Helmperlhuhn; Spanish: Pintade Común.

PHYSICAL CHARACTERISTICS
20–25 in (50–63 cm), 2.5–3.5 lb (1.15–1.6 kg) male and female. Plumage mainly blackish gray with white spots and lines. Widespread variation among the subspecies in the head adornments (filoplumes, cere, wattles, and casque).

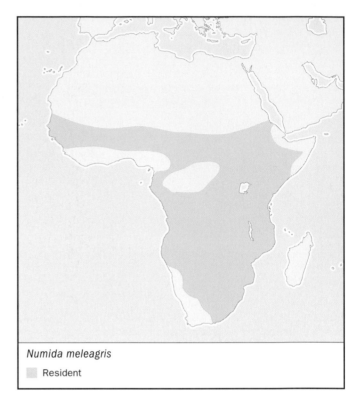

Numida meleagris

Numida meleagris
▨ Resident

DISTRIBUTION
N. m. sabyi (probably extinct): northwestern Morocco; *N. m. galeata*: sub-Saharan, to northern Senegal and into northern Angola; *N. m. meleagris*: sub-Saharan, north of equator to Sudan and western Ethiopia; *N. m. somaliensis*: eastern Ethiopia to Eritrea and Somalia; *N. m. reichenowi*: Kenya and central Tanzania; *N. m. mitrata*: western and eastern Tanzania to northern Mozambique; *N. m. marungensis*: southern Congo to Angola and western Zambia; *N. m. damarensis*: extreme southwestern Angola to Namibia and western Botswana; *N. m. coronata*: Zimbabwe, southern Mozambique and South Africa.

HABITAT
Open savannas, woodland and dry thorn-scrub.

BEHAVIOR
Highly social and vocal; sexually dimorphic. Usually in flocks of up to 35 individuals but may form aggregations of more than 1,000 birds in feeding areas.

The timing of roosting is strongly correlated to the time of sunset, with later roosting on longer days. In May 1970 on the island of Nantucket, Massachusetts, a flock of domesticated helmeted guineafowl took to their roost during a total eclipse of the sun at midday.

FEEDING ECOLOGY AND DIET
Active forager, mainly on the ground. Forages opportunistically on what is abundant. Omnivorous.

REPRODUCTIVE BIOLOGY
Apparently monogamous. 12–23 eggs with 24–27 day incubation time. Both parents feed and care for newly hatched chicks, known as downy keets. Breeding period varies depending on subspecies and rainfall patterns.

CONSERVATION STATUS
Widespread and common at the continental scale, but recent documentation indicates significant declines in parts of the range, probably due to habitat destruction.

SIGNIFICANCE TO HUMANS
Probably Africa's (and certainly southern Africa's) most popular terrestrial gamebird. Domesticated in Europe and introduced elsewhere. ◆

Resources

Books

Little, R.M., and T.M. Crowe. *Gamebirds of Southern Africa.* Cape Town, South Africa: Struik, 2000.

Martínez, I. "Numididae (Guineafowl)." In *Handbook of the Birds of the World.* Vol. 2, *New World Vultures to Guineafowl*, edited by Josep del Hoyo, Andrew Elliott, and Jordi Sargatal. Barcelona: Lynx Edicions, 1994.

Organizations

Gamebird Research Programme, Percy FitzPatrick Institute, University of Cape Town. Private Bag, Rondebosch, Western Cape 7701 South Africa. Phone: +27 21 6503290. Fax: +27-21-6503295. E-mail: tmcrowe@botzoo.uct.ac.za Web site: <http://www.uct.ac.za/depts/fitzpatrick>

Timothy Michael Crowe, PhD

Fowls and pheasants
(Phasianidae)

Class Aves
Order Galliformes
Suborder Phasiani
Family Phasianidae

Thumbnail description
Plump, ground-based birds of a great size range, with short, broad wings and stout bills and feet; males of larger species are often heavier with striking plumage and elaborate displays

Size
6–49 in (15–125 cm); 1.5 oz–24.2 lb (43 g–11.0 kg)

Number of genera, species
46 genera; 179 species

Habitat
Forest, woodland, bogs, tundra, mountains, savanna, desert fringes

Conservation status
Critically Endangered: 3 species; Endangered: 9 species; Vulnerable: 38 species; Near Threatened: 21 species

Distribution
North America, Europe, Asia, Africa, and Australasia

Evolution and systematics

From studies based on DNA comparisons as well as traditional morphological work, it seems clear that the turkeys (Meleagridinae), grouse (Tetraoninae), and pheasants and Old World partridges (Phasianinae) form an assemblage distinct from the other Galliformes. The current DNA evidence confirms that grouse and turkeys are closely related and probably evolved alongside the other main types within this complex family. When more such data becomes available and the resulting pattern of relationships among these species stabilizes, a completely new taxonomy for the Phasianinae is likely to emerge.

Unraveling the taxonomic affinities of some individual species in this family has produced some real surprises. The DNA evidence shows that the Congo peafowl (*Afropavo congensis*) is not closely related to either the guineafowls or partridges with which it shares its continent, Africa, having the most in common with the other peafowls (*Pavo* spp.) in South and Southeast Asia. The Udzungwa forest-partridge (*Xenoperdix udzungwensis*), first discovered in 1991, appears to be more closely related to the Southeast Asian hill-partridges (*Arborophila*) than to any African species. The Gunnison sage grouse (*Centrocercus minimus*) was only recognized as a full species in 2000, and lives in just eight localities at the southern limit for this genus in southwest Colorado and southeast Utah.

Some long-established species have also come under renewed scrutiny, including Edwards's pheasant (*Lophura ed-*

wardsi) and the imperial pheasant (*L. imperialis*) from central Vietnam. During 2000 and 2001, the Berlioz's silver pheasant (*L. nycthemera berliozi*) was deliberately crossed with Edwards's pheasant to produce birds that appear identical to imperial pheasants, thus suggesting that the few imperials recovered from the wild were rare interspecific hybrids. Yet another pheasant had been discovered in the same Annamese lowlands area in 1964, and was adopted as a new species, the Vietnamese pheasant (*L. hatinhensis*) in 1975. It differs from the Edwards's pheasant only in having white central tail feathers in the male.

Physical characteristics

The species in this family vary enormously in size, from tiny quails (*Coturnix* spp.) weighing less than 2 oz (43 g) to the wild turkey (*Meleagris gallopavo*) weighing up to 24.2 lb (11 kg). Their dominant common feature is a heavy rounded body, which is the result of extreme development of the flight muscles over the sternum. This in turn has evolved in parallel with their typical flight behavior: an explosively energetic take-off to gain height followed by a fast glide. Generally, the legs and neck are short, the head and tail small, although in some of the large species, longer necks and tails have evolved.

Facial adornments are many and various, again particularly in the larger species and especially in the males. Male turkeys have a naked red crop and the fleshy and flexible caruncle, which can change rapidly in color from red to blue and dangles beside the beak. The ring-necked pheasant (*Phasianus* colchicus) has a bright red skin patch around the eye, while in the crested fireback (*Lophura ignita*) it is blue. Fleshy and brightly colored wattles are characteristic of Bulwer's pheasant (*Lophura bulweri*) and the junglefowls (*Gallus* spp.). Garishly colored bib-like air sacs and paired erectile horns are unique to the tragopans (*Tragopan* spp.). Other specialties include feathery ears in the eared-pheasants (*Crossoptilon*), head crests as in the koklass (*Pucrasia macrolopha*), monals (*Lophophorus*), and crested wood-partridge (*Rollulus rouloul*), and neck-ruffs in Lady Amherst's pheasant (*Chrysolophus amherstiae*) and the ruffed grouse (*Bonasa umbellus*).

Many grouse species have bright yellow to red fleshy combs above the eyes, which in males especially become engorged and more prominent in the mating season. In those species believed to have polygynous or promiscuous mating systems, the males are more extravagantly adorned and may be up to twice the size of the more dowdy but well-camouflaged females. In the prairie grouse (*Tympanuchus* spp.), males have a pair of large and brightly colored air sacs on their necks, which they inflate during their breeding season displays. Grouse are adapted to live in cold climates by having an exceptionally thick and heavy plumage, as well as feathering right down to the toes. The ptarmigans (*Lagopus* spp.) adopt a special white winter plumage so that they remain well camouflaged.

A number of species have long tails, which can be held in fans during displays. In grouse, these are plain, as in the blue grouse (*Dendragapus obscurus*) or barred, as in sage grouse (*Centrocercus urophasianus*). In the peafowls (*Pavo* spp.) and peacock-pheasants (*Polyplectron* spp.), they bear numerous eye-like ocelli.

Distribution

The relict distribution of the ocellated turkey (*Meleagris ocellata*) in the Yucatán peninsula of Central America is the southern-most point for this family in the New World, with the wild turkey (*M. gallopavo*) and prairie grouse (*Tympanuchus* spp.) originally occupying much of the United States between them. Other grouse species occur over large parts of Canada, including most of the islands in the extreme north. The range of the ruffed grouse (*Bonasa umbellus*) spans the continent, as does its counterpart in the Old World, the hazel grouse (*B. bonasia*). The willow grouse (*Lagopus lagopus*) and rock ptarmigan (*L. mutus*) have circumpolar distributions in the northern tundra.

Between them, the Old World partridges and pheasants occupy almost all of Europe, Africa, and Australasia. The genus *Francolinus* contains numerous African species, extending to almost every corner of that continent. The bush-quails (*Perdicula* spp.) and spurfowl (*Galloperdix* spp.) occur only in southern Asia, while the hill-partridges (*Arborophila* spp.) are Southeast Asian and Chinese in distribution.

Habitat

Most grouse species inhabit northern tundra or boreal forests, and the equivalent habitats in isolated mountainous regions further south, as in the case of the capercaillie (*Tetrao urogallus*) in the Cantabrian Mountains of northern Spain.

A male wild turkey (*Meleagris gallopavo*) gobbling. (Photo by John Shaw. Bruce Coleman Inc. Reproduced by permission.)

Several North American grouse species and the two turkeys have evolved to occupy relatively open temperate and subtropical habitats over much of the continent, the niche occupied by the bustards (Otidae) in the Old World.

Most pheasant species (Phasianini) are forest specialists, with exceptions such as the cheer (*Catreus wallichi*), which occupies open grass and scrub habitats in the western Himalayan foothills, and the Chinese monal (*Lophophorus lhuysii*), which lives in alpine scrub and grassland. The great argus (*Argusianus argus*) occurs in lowland tropical rainforests in Southeast Asia, while the koklass is a temperate forest species of the Himalayas and China.

Partridge (Perdicini) occur in all habitats except for the northern boreal forests and tundra, where they are replaced by grouse. Conversely, in the high alpine areas of central Asia, where there are no grouse, the snowcocks (*Tetraogallus*), snow partridge (*Lerwa lerwa*), and Tibetan partridge (*Perdix hodgsoniae*) occupy their niche. The temperate grasslands of Europe are home to the gray partridge (*Perdix perdix*) and several rock partridges (*Alectoris* spp.). Many species occur in tropical grasslands and savanna such as the yellow-necked francolin (*Francolinus leucoscepus*) in southeastern Africa and the painted francolin (*F. pictus*) in India. Philby's rock partridge (*Alectoris philbyi*) lives on rocky slopes in southwestern Arabia. Lowland tropical rainforest species include the crested wood-partridge (*Rollulus rouloul*) of Southeast Asia and Latham's francolin (*F. lathami*) from equatorial Africa. Montane forests harbor the Hainan hill-partridge (*Arborophila ardens*) from China and the red-billed hill-partridge (*A. rubirostris*) of Sumatra. The swamp francolin (*F. gularis*) and Manipur bush-quail (*Perdicula manipurensis*) inhabit wet grasslands south of the Himalayas.

Behavior

Out of the breeding season, open country species such as the wild turkey, snowcocks, African savanna francolins, and the alpine Tibetan eared-pheasant (*Crossoptilon harmani*) form groups of 20–100, presumably as a protection against predation. Forest species are in general much less gregarious. However, in some species, female gregariousness in the pre-breeding period has apparently enabled males to defend small groups of them and thereby acquire a harem, as in the wild turkey and the red junglefowl (*Gallus gallus*). An alternative tactic is to defend territories in strategically important habitats for females: thus males of the introduced ring-necked pheasant (*Phasianus colchicus*) in Britain defend territories on woodland edge in farmland where groups of females nest.

Studies of territorial behavior in various grouse species reveal that loud and often repeated calling by the males is the routine method for establishing and maintaining territories. Fighting is also frequent, particularly if there is a large surplus of birds still seeking territories in a limited area of habitat, as for the red grouse (*L. l. scoticus*) in the British uplands.

The daily routine of birds in this family is rather universal. They emerge from their roosts at dawn for an intensive period of feeding activity, in order to refill their crops following the night fast. After an hour or two, birds retreat to cover, presumably avoiding inclement weather and predators. In winter, arctic grouse need to feed in the open for as little time as possible to avoid excessive heat loss. They have an unusually large crop in which to store rapidly gathered food for later digestion in warmth and safety. Towards the end of the day, there is usually another burst of feeding activity, sometimes followed by calling as families or larger groups gather to roost for the night.

Migration is not a prominent feature of these species. Despite their small size, quails are unusual among them in making long-distance spring and autumn movements between wintering and breeding grounds. The only other species that migrate regularly are the northern-most populations of the

A male common pheasant (*Phasianus colchicus*) dust bathing near Arundel, United Kingdom. (Photo by Roger Wilmshurst. Photo Researchers, Inc. Reproduced by permission.)

rock and willow ptarmigan (*Lagopus mutus*) that move several hundred miles/kilometers south in fall to escape the worst winter conditions. Some montane pheasant species such as the Himalayan monal (*Lophophorus impejanus*) have been found to mimic this kind of movement by undertaking altitudinal migrations of up to 4,900 ft (1,500 m) in fall, returning to breed in the sub-alpine scrub in spring after the snow has receded.

Feeding ecology and diet

High-montane snowcocks feed almost exclusively on vegetation. Boreal forest species such as the spruce grouse (*Falcipennis canadensis*) subsist on the oily buds and needles of conifers throughout the winter months. Grouse also have exceptionally long caecae, the blind-ended tubes in the gut where symbiotic bacteria digest cellulose into sugars. Forest pheasants such as tragopans thoroughly dig over areas of litter and soil with their feet, the monals also using their stout beaks, to depths of at least 12 in (30 cm) in order to feed on tubers, bulbs, and roots.

Tropical forest species take a huge range of items as food, with ants, termites, and other invertebrates of the forest floor and understory being prominent alongside fruits, seeds, and leaves. Prairie chickens also take a significant amount of insect food as adults, mainly in the form of grasshoppers in summer on the North American plains. In the dry grasslands of southern Africa, the gray-winged fracolin (*F. africanus*) takes a varied mixture of roots and bulbs, and seeds, fruits, and invertebrates. All these open-country species, as well as the wild turkey and the quails, have been able to adapt well to agricultural expansion by feeding in fields of crops and on seeds left after harvesting.

Newly hatched chicks in almost all species rely completely on a protein-rich diet of invertebrates, although they rapidly

This wild turkey (*Meleagris gallopavo*) hen is dust bathing in west Texas. (Photo by John Snyder. Bruce Coleman Inc. Reproduced by permission.)

switch to a less exclusive diet within their first month. An exception is provided by snowcock chicks, which take legumes as a major part of the diet.

Reproductive biology

Nests are usually simple scrapes on the ground, lined with only a little vegetation, and camouflaged by grasses, shrubs, or rocky overhangs. Clutch size in some open-country perdicines and the ring-necked pheasant can be a high as 15–20. In contrast, the Malaysian peacock-pheasant (*Polyplectron malacense*) lays just one egg.

Incubation is usually carried out entirely by the female. Chicks immediately leave the nest at hatching and feed themselves. The care provided by one or both parents is normally limited to protecting chicks from predators, and sheltering them under their wings in inclement weather and at night. Rudimentary flight is achieved in just 7–10 days in many grouse and partridges, although families may stay together for two to three months.

Females generally come into breeding condition each year including their first, although wild turkeys defer breeding for a year if their fat reserves are low in spring. Males in many larger species do not mate in their first season. Male silver pheasants and tragopans only molt into full adult colors in their second fall, while the male great argus does not achieve adult plumage until the third year. The train of the Indian peafowl takes four years to develop fully.

The promiscuous grouse and pheasants species gather at collective display and mating grounds known as leks. Here, 5–50 males defend territories just a few yards wide, with a very small percentage of the males obtaining the vast majority of copulations. Males with the largest, most decorated, or least damaged tails are most successful. Males with these characteristics also father high-quality chicks, are in better body condition, carry lower parasite loads, and survive better themselves. This suggests that sexual selection via both male-male competition and female choice can cause the evolution of elaborate plumage, bizarre displays, and the much greater body size of the males in such species.

Conservation status

In 2000, of the 179 extant species in the Phasianidae, 50 (28%) were included on the IUCN Red List as being threatened with extinction, a proportion nearly three times that for all birds (11%). A further 21 (12%) were classified Near

A male great argus pheasant (*Argusianus argus*) shows his courtship display, hoping to interest the female next to him. (Photo by Kenneth W. Fink. Photo Researchers, Inc. Reproduced by permission.)

Only the female ruffed grouse (*Bonasa umbellus*) incubates and raises the young. This female is incubating eggs in Pennsylvania in early spring. (Photo by Jeff Lepore. Photo Researchers, Inc. Reproduced by permission.)

Threatened. Of the 108 species native to Asia, 70% of the forest specialists are threatened, compared to only 18% of those living in open habitats, implying that forest degradation and fragmentation are important threats. However, being ground-based and often either large or gregarious, these species are universally harvested as a source of food. Over 90% of the threatened species in Asia are suspected of being over-hunted.

Only one species in this family has become Extinct relatively recently: the New Zealand quail (*Coturnix novaezelandiae*) was common on both North and South Islands in 1850, but was last seen in 1876; it was probably wiped out by a disease carried by an introduced bird species. One of the Critically Endangered species is the Djibouti francolin (*Francolinus ochropectus*), which is only known from one small area of highly disturbed juniper forest in this politically unstable corner of Africa. The other two species in this highest threat category are the Himalayan quail (*Ophrysia superciliosa*) and the gorgeted wood-quail (*Odontophorus strophium*).

The nine Endangered species are: the newly-described Gunnison sage grouse (*Centrocercus minimus*); Edwards's pheasant (*Lophura edwardsi*); Vietnamese pheasant (*L. hatinhensis*); Bornean peacock-pheasant (*Polyplectron schleiermacheri*); Nahan's francolin (*Francolinus nahani*) on the border between Uganda and the Democratic Republic of Congo; Mount Cameroon francolin (*F. camerunensis*); Sichuan hillpartridge (*Arborophila rufipectus*) from central southern China;

orange-necked hill-partridge (*A. davidi*) from south Vietnam; and chestnut-headed hill-partridge (*A. cambodiana*).

Significance to humans

It is easy to suggest that this family of birds is of greater importance to the human race than any other, as it contains the wild ancestors of both domestic chickens and turkeys, as well as many species that have been hunted in the wild for food over millennia. The precise origins of the domestic chicken are uncertain, but there is a common consensus that the progenitor is the red junglefowl. Archaeological investigations at the sites of cities dating from the third millennium B.C. in the Indus valley of south Asia indicate that their sophisticated inhabitants kept domesticated fowl as well as a variety of hoofed livestock. By 1500 B.C., the chicken was being used in China, Egypt, and northwest Europe. It subsequently achieved global distribution, even reaching many islands in the South Pacific where it became feral.

Bones found during investigations of pre-Columbian settlements in North America suggest that the wild turkey was an important source of meat in the diet of Native Americans. By around A.D. 500–700, it was being kept as a domesticated bird by people living in northern New Mexico and Arizona. The domestic turkey arrived in Europe as a result of the 1519 Cortés expedition to Mexico by the Spaniards, and in 1607 it was taken back to the New World by the first European settlers. It now takes a place of pride on tables at both Thanksgiving in the United States and Christmas throughout the Western world.

It has been judged that at the end of the 1970s around 8.5 million grouse were hunted each year in North America. One species takes the brunt of this harvest: the ruffed grouse (*Bonasa umbellus*) at around six million per year, but more than half a million sharp-tailed grouse and sage grouse are also taken. In the Old World the willow grouse is cropped at a rate of about eight million per year, mainly in Russia, and in Fenno-Scandinavia the annual total for all the species is more than half a million birds. Since the late nineteenth century in Britain, the habitat of the red grouse has been nurtured in order to provide sport hunting of the highest caliber. In most species, it is believed that hunting has little effect on populations, because a high proportion of the birds shot would die through some other cause.

In the European Alps, hunting revenues are sufficient to support the costs of habitat preservation and improvement for capercaillie. Clear felling of old open woodland and its replacement by forests in which the trees are much closer together, reduces or obliterates the understory layer on which this species depends for food and nesting cover. Efforts to reverse these effects for capercaillie have had a wider impact on the utility of the habitat for other wildlife: areas where these grouse occur have more woodpeckers and a greater number of songbird species than places not yet re-colonized.

1. Edwards's pheasant (*Lophura edwardsi*); 2. Brown eared-pheasant (*Crossoptilon mantchuricum*); 3. Male satyr tragopan (*Tragopan satyra*) in display; 4. Red junglefowl (*Gallus gallus*); 5. Male Palawan peacock-pheasant (*Polyplectron emphanum*) in lateral display; 6. Ring-necked pheasant (*Phasianus colchicus*); 7. Wild turkey (*Meleagris gallopavo*); 8. Udzungwa forest-partridge (*Xenoperdix udzungwensis*); 9. Chinese monal (*Lophophorus lhuysii*). (Illustration by Emily Damstra)

1. Capercaillie (*Tetrao urogallus*); 2. Ruffed grouse (*Bonasa umbellus*); 3. Greater prairie chicken (*Tympanuchus cupido*); 4. Tibetan snowcock (*Tetraogallus tibetanus*). (Illustration by Bruce Worden)

1. King quail (*Coturnix chinensis*); 2. Chinese grouse (*Bonasa sewerzowi*); 3. Crested wood-partridge (*Rollulus roulroul*); 4. Willow ptarmigan (*Lagopus lagopus*); 5. Gray partridge (*Perdix perdix*); 6. Sichuan hill-partridge (*Arborophila rufipectus*); 7. Red-necked francolin (*Francolinus afer*). (Illustration by Bruce Worden)

Grzimek's Animal Life Encyclopedia

Species accounts

Wild turkey
Meleagris gallopavo

SUBFAMILY
Meleagridinae

TAXONOMY
Meleagris gallopavo Linnaeus, 1758, North America = Mexico.
Six subspecies.

OTHER COMMON NAMES
French: Dindon sauvage; German: Truthuhn; Spanish: Gua-
jolote Gallipavo.

PHYSICAL CHARACTERISTICS
Male 39–49 in (100–125 cm); female 30–37 in (76–95 cm); male
11.0–24.2 lb (5.0–11.0 kg); female 6.6–11.0 lb (3.0–5.0 kg).
Males have a bare blue and pink head, red wattles, dark
plumage with iridescent green and bronze highlights, white-
barred flight feathers, a blackish breast tuft, and pinkish spurred
legs. Females are duller in color and smaller than males.

DISTRIBUTION
Native of central and northern Mexico, throughout the United
States ranging from southern Vermont to Florida, and west to
Washington, Oregon, and California.

HABITAT
Prefers mix of hardwood forest, scrub, grass, and agricultural
land, but tolerant of dry scrub and subtropical forest.

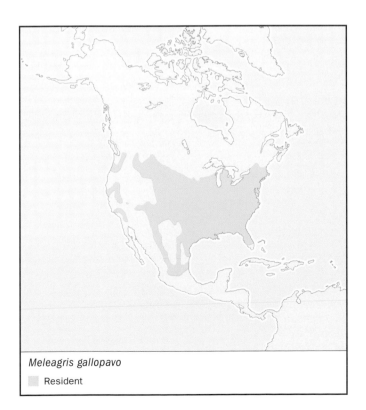

Meleagris gallopavo
☐ Resident

BEHAVIOR
Gathers into large flocks of 50, exceptionally 500, in winter
with males usually separate from females and young; no territorial defense; will run to cover, only flying routinely to reach
communal roosts in trees.

FEEDING ECOLOGY AND DIET
Feeds mainly by picking at ground; takes leaves, shoots, small
seeds, acorns, buds, fruits, as well as grasshoppers, crabs, and
small vertebrates.

REPRODUCTIVE BIOLOGY
Male display starts in February, and later takes up most of
their time; successful males attract a group of four to five females for mating; nest in dense cover; clutch size usually
10–12; incubation by female alone, 27–28 days; female and
young remain together until following spring.

CONSERVATION STATUS
Until 1940s, over-hunting in the United States was reducing
its range and population; following many successful translocations and systematic management of hunting, it is again common and widespread; it is local and much less common in
Mexico.

SIGNIFICANCE TO HUMANS
A hunted game bird; the subject of turkey-calling contests; a
flagship for its habitats. ◆

Willow ptarmigan
Lagopus lagopus

SUBFAMILY
Tetraoninae

TAXONOMY
Tetrao lagopus Linnaeus, 1758, Swedish Lapland. Nineteen subspecies.

OTHER COMMON NAMES
English: Red grouse, willow grouse; French: Lagopéde des
saules; German: Moorschneehuhn; Spanish: Lagópodo Común.

PHYSICAL CHARACTERISTICS
14–17 in (36–43 cm); male 0.9–1.8 lb (405–795 g); female
0.9–1.5 lb (405–700 g). Males have a rusty head and upperparts, bright red eye combs, white underparts, and a black tail.
Females are grayer, are more heavily barred on the breast and
flanks, and lack the bright red eye combs. Both sexes are all
white in winter except for black tail.

DISTRIBUTION
Circumpolar between 47° and 76°N.

HABITAT
Arctic tundra, sub-arctic scrub, and boreal forest edge; preferring moister areas with dwarf deciduous trees.

BEHAVIOR
In large groups of variable sex ratio in winter; males highly territorial in spring through calling from landmarks and in flight.

Lagopus lagopus
▨ Resident

FEEDING ECOLOGY AND DIET
Willow and birch buds and twigs in winter; invertebrates taken especially by young chicks in summer; berries in fall.

REPRODUCTIVE BIOLOGY
Mostly monogamous. Pairs occupy exclusive territories; nesting starts April–June depending on latitude; clutch size eight to 11; incubation 22 days; male broods chicks; families remain together until fall.

CONSERVATION STATUS
Not threatened. Locally common and widespread.

SIGNIFICANCE TO HUMANS
Locally managed to provide a substantial hunted surplus; keenly hunted in the United Kingdom, Scandinavian countries, Finland, and Russia, with 2.4 million birds taken annually in these areas together. ◆

Capercaillie

Tetrao urogallus

SUBFAMILY
Tetraoninae

TAXONOMY
Tetrao urogallus Linnaeus, 1758, Sweden. Twelve subspecies.

OTHER COMMON NAMES
French: Grand tétras; German: Auerhuhn; Spanish: Urogallo Común.

PHYSICAL CHARACTERISTICS
Male 32–35 in (83–90 cm); female 23–25 in (59–64 cm); male 7.3–14.3 lb (3.3–6.5 kg); female 3.3–5.5 lb (1.5–2.5 kg). Males

are mostly slate gray with a blackish head and neck, red eye combs, glossy greenish black breast, dark brown wings with white carpal patch, varying amounts of white on upper wings and underparts, and long, rounded tail. Females are mottled and barred in gray, buff, and black with a large rusty breast patch.

DISTRIBUTION
Northern Britain and Scandinavia to eastern Russia; more fragmented in eastern and southeastern Europe, the Alps; isolated populations in northern Spain and Pyrenees.

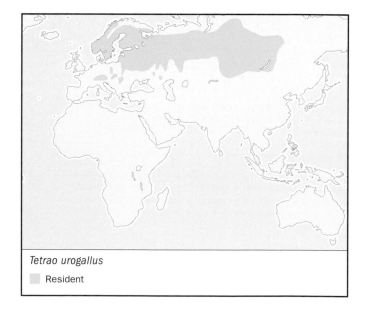

Tetrao urogallus
▨ Resident

HABITAT
Mainly in old conifer forest with moderate understory usually of bilberry, interspersed with bogs; up to 6,600 ft (2,000 m) in Pyrenees.

BEHAVIOR
Male usually alone, while females and young form wintering groups of up to 10; males gather loosely in lek areas to defend territories and attract females using calls, erect strutting, and tail-fanning displays.

FEEDING ECOLOGY AND DIET
Pine needles, holly leaves, birch buds, berries; leaves of heath plants; young chicks especially take invertebrates.

REPRODUCTIVE BIOLOGY
Promiscuous. Laying in April–June; nest in thick cover in forest; clutch size six to nine; incubation 26 days; chicks able to fly after three weeks; males defer mating until third year.

CONSERVATION STATUS
Small and fragmented populations are threatened and prone to extinction; continued hunting a particular threat in central and south Europe, where increased predator numbers, alpine tourism, and collisions with power lines and deer fences all cause problems.

SIGNIFICANCE TO HUMANS
Attracts trophy hunters; hunted for food mainly in fall. ◆

Chinese grouse
Bonasa sewerzowi

SUBFAMILY
Tetraoninae

TAXONOMY
Tetrastes sewerzowi Przevalski, 1876, Gansu, China. Monotypic.

OTHER COMMON NAMES
English: Severtzov's grouse, black-breasted hazel grouse; French: Gélinotte de Severtzov; German: China-haselhuhn; Spanish: Grévol Chino.

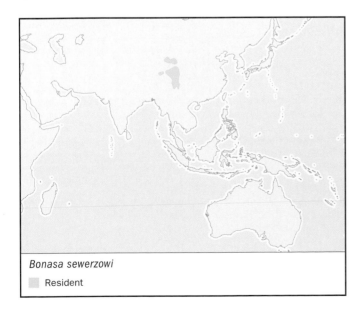

Bonasa sewerzowi
░ Resident

PHYSICAL CHARACTERISTICS
13–14 in (33–36 cm); 0.6–0.7 lb (270–310 g). Brownish gray with black bars on upperparts, black chin bordered in white, a chestnut upper breast, and underparts spotted and barred in dark gray and ochre. Small erectile crest on head. Barred pattern on the tail distinguishes this species from *B. bonasia*.

DISTRIBUTION
China: central Gansu to southern Quinghai, eastern Tibet, northwestern Yunnan, and northern and western Sichuan.

HABITAT
Montane forests at 3,300–13,100 ft (1,000–4,000 m); conifer near treeline, birch and conifer below, willow thickets on riverbanks.

BEHAVIOR
Forms flocks of up to 15 for fall-winter; spring dispersal for breeding; males repeat noisy display for most of day, and fight in treetops.

FEEDING ECOLOGY AND DIET
Forages on ground and in trees for buds and shoots of willow and birch, also taking various flowers, seeds, and berries.

REPRODUCTIVE BIOLOGY
Monogamous. Nests on ledges and stumps in May–June; clutch size five to eight; incubation 25 days.

CONSERVATION STATUS
Near Threatened as forest clearance and fragmentation causes local extinctions, with hunting and egg-collecting a problem outside protected areas.

SIGNIFICANCE TO HUMANS
A hunted resource locally. ◆

Ruffed grouse
Bonasa umbellus

SUBFAMILY
Tetraoninae

TAXONOMY
Tetrao umbellatus Linnaeus, 1766, Pennsylvania, United States. Fourteen subspecies.

OTHER COMMON NAMES
French: Gélinotte huppée; German: Kragenhuhn; Spanish: Grévol Engolado.

PHYSICAL CHARACTERISTICS
17–19 in (43–48 cm); male 1.3–1.4 lb (600–650 g); female 1.1–1.3 lb (500–590 g). Cryptic plumage; gray and brown color morphs with gray commoner in northern parts of range and brown commoner in southern parts. Small crest on head, erectile black ruff on sides of neck, and fan-shaped tail with distinctive subterminal dark band.

DISTRIBUTION
North America from Alaska to Labrador and Nova Scotia, south to California and Utah in west and through Appalachians to northern Georgia in east; Nevada and Newfoundland.

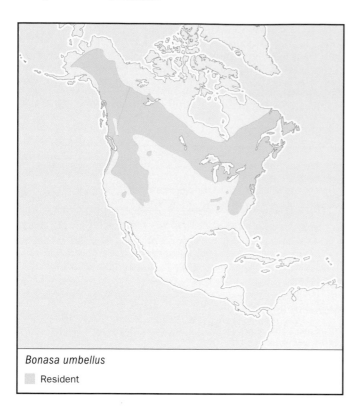

Bonasa umbellus

■ Resident

Tympanuchus cupido

■ Resident

HABITAT
Pacific Coast rainforest, boreal, and dry deciduous woodlands; prefers a mixed-age forest mosaic with aspen and brushwood clearings.

BEHAVIOR
Roosts in conifers; drums with wings while in upright posture year round, but most intensely at dawn in March–June when males defend territories and sometimes form a loose lek.

FEEDING ECOLOGY AND DIET
Buds and twigs from aspens and other deciduous trees; herb flowers and catkins; berries and some invertebrates; newly hatched chicks depend on insects; also fungi and acorns.

REPRODUCTIVE BIOLOGY
Promiscuous. Nests in May on forest floor; clutch size 10–12; incubation 23–24 days; chicks can fly at 10–12 days.

CONSERVATION STATUS
Widespread and common in north of range; elsewhere range contractions have been reversed by restocking.

SIGNIFICANCE TO HUMANS
Most hunted grouse in North America. ◆

Greater prairie chicken
Tympanuchus cupido

SUBFAMILY
Tetraoninae

TAXONOMY
Tetrao cupido Linnaeus, 1758, Virginia, United States. Two extant subspecies.

OTHER COMMON NAMES
English: Pinnated grouse, prairie grouse; French: Tétras des prairies; German: Präriehuhn; Spanish: Gallo de las Praderas Grande.

PHYSICAL CHARACTERISTICS
16–18 in (41–47 cm); male 2.2 lb (990 g); female 1.7 lb (770 g). Brown overall with extensive barring on both upperparts and underparts; short, rounded, blackish tail; elongated feathers on the sides of the neck (pinnae) are erect during courtship; golden yellow cervical sacs; yellow to orange eye combs. Females similar to males but with smaller pinnae and smaller, paler cervical sacs.

DISTRIBUTION
North America, mainly from Oklahoma to North Dakota.

HABITAT
Prairie remnants amid arable cropland.

BEHAVIOR
Classic lek-forming species; spectacular display, audible over a mile or more; gather into large mixed-sex flocks in winter.

FEEDING ECOLOGY AND DIET
Takes acorns, smaller seeds, leaves, buds; cultivated grains (corn, soya); grasshoppers and other invertebrates.

REPRODUCTIVE BIOLOGY
Promiscuous. Mating at leks, generally excluding yearling males; nests in thick grass cover in April–June; clutch size eight to 13; incubation 23–25 days.

CONSERVATION STATUS
Not threatened overall, but the heath hen (*T. c. cupido*) of New England is Extinct, and Atwater's prairie chicken (*T. c. atwateri*) persists as two populations in southeastern Texas.

SIGNIFICANCE TO HUMANS
Enjoyed by bird-watchers for the spring spectacle and used as a flagship species for prairie conservation. ◆

Tibetan snowcock
Tetraogallus tibetanus

SUBFAMILY
Phasianinae (Tribe Perdicini)

TAXONOMY
Tetraogallus tibetanus Gould, 1854, Tibet = Ladakh, India. Four subspecies.

OTHER COMMON NAMES
French: Tetraogalle du Tibet; German: Tibetkönigshuhn; Spanish: Perdigallo Tibetano;.

PHYSICAL CHARACTERISTICS
19–22 in (50–56 cm); male 3.3–4.0 lb (1.5–1.8 kg); female 2.6–3.5 lb (1.2–1.6 kg). Distinguished from other snowcocks by white underparts with heavy blackish streaks on flanks.

DISTRIBUTION
Tibetan plateau including northern India, Nepal, Bhutan, eastern Tajikistan, Tibet, and adjacent parts of Qinghai, Gansu, Sichuan, and Yunnan in China.

HABITAT
Mainly on bare and grassy slopes at 16,400–19,700 ft (5,000–6,000 m); stays above tree-line year round.

BEHAVIOR
Gathered in groups of up to 50 out of breeding season; pairs formed by April and males call from vantage points at dawn; runs if disturbed, only taking flight reluctantly; roosts in rocky scree, flying downhill early to start feeding.

FEEDING ECOLOGY AND DIET
Roots, shoots, seeds, and berries; chicks consume legumes, but few invertebrates.

REPRODUCTIVE BIOLOGY
Probably monogamous; nests late May concealed by shrub or boulder; clutch size four to seven.

CONSERVATION STATUS
Not threatened, having an extensive range; relatively common in Nepal.

SIGNIFICANCE TO HUMANS
None known. ◆

Red-necked francolin
Francolinus afer

SUBFAMILY
Phasianinae (Tribe Perdicini)

TAXONOMY
Tetrao afer Muller, 1776, Benguela, Angola. Seven subspecies.

OTHER COMMON NAMES
English: Bare-throated francolin, red-necked spurfowl; French: Francolin à gorge rouge; German: Rotkehlfrankolin; Spanish: Francolín Gorgirrojo.

PHYSICAL CHARACTERISTICS
10–15 in (25–38 cm); 1.0–1.7 lb (440–770 g). Distinguished from *F. rufopictus* and *F. swainsonii* by scarlet throat and red bill and legs.

DISTRIBUTION
Africa, from Angola, across Congo basin to coastal Kenya, and from Rift Valley to east coast, south to southeast South Africa.

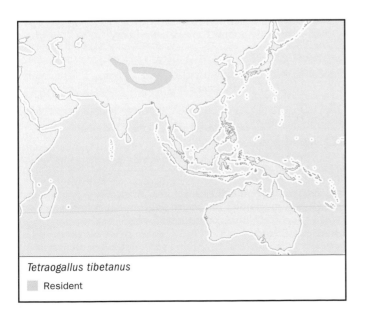

Tetraogallus tibetanus
◻ Resident

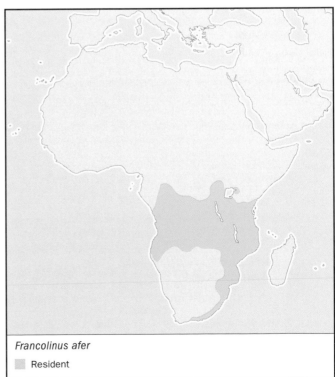

Francolinus afer
◻ Resident

HABITAT
Moist evergreen forest in west and north of range, but in drier grass and scrub elsewhere.

BEHAVIOR
Reluctant to fly, runs if disturbed; males call early in the day and are territorial near nests; roost at night and midday in bushes and trees; single or in small parties, sometimes with other francolin species.

FEEDING ECOLOGY AND DIET
Tubers, roots, bulbs, shoots, berries, crops, and invertebrates, including ticks and termites.

REPRODUCTIVE BIOLOGY
Most breeding late in rainy seasons with two clutches in some places; probably monogamous; clutch size three to nine; incubation 23 days; chicks fly at 10 days and fully grown in four months; families together for most of year.

CONSERVATION STATUS
Not threatened and covering a large range; locally numerous but hunting may reduce numbers significantly in places.

SIGNIFICANCE TO HUMANS
None known. ◆

Gray partridge
Perdix perdix

SUBFAMILY
Phasianinae (Tribe Perdicini)

TAXONOMY
Tetrao perdix Linnaeus, 1758, Sweden. Seven subspecies.

OTHER COMMON NAMES
English: Common partridge; French: Perdrix grise; German: Rebhuhn; Spanish: Perdiz Pardilla.

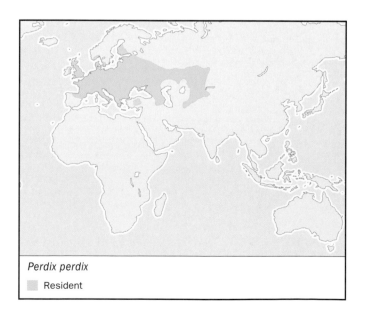

Perdix perdix
▨ Resident

PHYSICAL CHARACTERISTICS
11–12 in (29–31 cm); male 0.7–1.3 lb (325–600 g); female 0.7–1.3 lb (310–570 g). Appearance varies especially in terms of grayer or browner plumage. Races in the western parts of the range tend to be more rufous brown, while those in the east are generally grayer and paler.

DISTRIBUTION
West to southeastern Europe, including south Scandinavia to Siberia and south to Kazakhstan and western Xinjiang.

HABITAT
Native of temperate grasslands and steppe, but now mainly in less intensively managed croplands; up to 8,500 ft (2,600 m) in Spain and Caucasus.

BEHAVIOR
In parties of 5–25, consisting of one or more families, in overlapping home ranges for winter; pairs form and live in more exclusive areas in spring.

FEEDING ECOLOGY AND DIET
Weed and cereal seeds, grass, and clover leaves, with chicks completely dependent on insects for first two weeks.

REPRODUCTIVE BIOLOGY
Usually monogamous. Nests April–June depending on locality; clutch size usually 15–17; incubation 23–25 days; chicks fly at two weeks and reach adult weight in three months.

CONSERVATION STATUS
Still widespread and locally abundant, but adversely affected by intensive farming because of removal of nesting cover, herbicide treatment for weeds, and reduction in insect availability through pesticide applications to crops; 80% reduction estimated overall.

SIGNIFICANCE TO HUMANS
An important game bird for hunting in Europe and United States; flagship for conservation in lowland agricultural landscapes. ◆

King quail
Coturnix chinensis

SUBFAMILY
Phasianinae (Tribe Perdicini)

TAXONOMY
Tetrao chinesis Linnaeus, 1766, China and Philippines = Nanking, China. Ten subspecies.

OTHER COMMON NAMES
English: Asian blue quail, painted quail, Chinese quail, blue-breasted quail; French: Caille peinte; German: Zwergwachtel; Spanish: Codorniz China.

PHYSICAL CHARACTERISTICS
5–6 in (12–15 cm); male 1.2–1.7 oz (35–48 g); female 1.1–1.4 oz (31–41 g). Males are a dark brownish blue with a lighter bluish gray breast and chestnut belly; the face and throat are black and white. Females are a mottled brown overall and lack the black-and-white coloration on the face and throat.

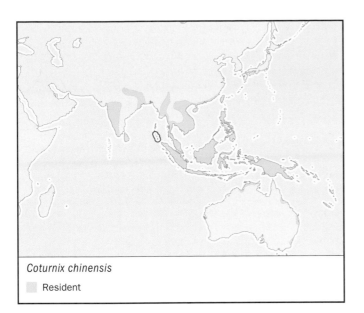

Coturnix chinensis
Resident

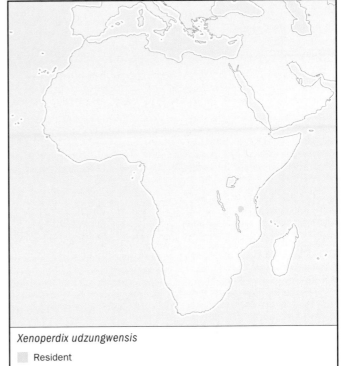

Xenoperdix udzungwensis
Resident

DISTRIBUTION
Southwestern and northeastern India, Sri Lanka, eastern Myanmar, Indochina to Hainan and Taiwan, Malaysian peninsula, Philippines and Indonesia, northern and eastern coast of Australia.

HABITAT
Wet shrubland, swampy grassland, rice paddy, mainly in lowlands and coastal areas, but up to 6,600 ft (2,000 m) in Sri Lanka and India.

BEHAVIOR
Crouches or runs rather than flying if disturbed; dust-bathes in open drier areas; usually in pairs or families.

FEEDING ECOLOGY AND DIET
Leaves, grass, seeds, and invertebrates, especially termites.

REPRODUCTIVE BIOLOGY
Strong pair bond and assumed monogamous; nest often domed with grasses and sedges; clutch size four to eight; incubation 18–19 days by female; two broods per year in good conditions; chicks mature in eight weeks.

CONSERVATION STATUS
Not threatened, widespread but cryptic; likely to be declining as favored swampy grasslands are drained for agriculture throughout range.

SIGNIFICANCE TO HUMANS
None known. ◆

Udzungwa forest-partridge
Xenoperdix udzungwensis

SUBFAMILY
Phasianinae (Tribe Perdicini)

TAXONOMY
Xenoperdix udzungwensis Dinesen et al., 1994, Ndundulu Mountains, Tanzania. Monotypic.

OTHER COMMON NAMES
English: Udzungwa partridge; French: Xénoperdrix de Tanzanie; German: Udzungwawachtel; Spanish: Perdiz de Udzungwa.

PHYSICAL CHARACTERISTICS
11 in (29 cm); 8–9 oz (220–239 g); male slightly larger. Barred upperparts and blotched underparts; red bill.

DISTRIBUTION
First found in 1991 and known only from the eastern Udzungwa highlands and the Rubeho Mountains in southern Tanzania, Africa.

HABITAT
Montane and sub-montane evergreen forest with open understory at 4,400–6,200 ft (1,350–1,900 m).

BEHAVIOR
Usually in small flocks of up to eight; roosts in trees and shrubs.

FEEDING ECOLOGY AND DIET
Forages for invertebrates and seeds by searching litter on forest floor.

REPRODUCTIVE BIOLOGY
Adults with chicks seen in November–December. No other information is available.

CONSERVATION STATUS
Vulnerable. Only known from four populations, but these populations appear to be stable.

SIGNIFICANCE TO HUMANS
None known. ◆

Sichuan hill-partridge
Arborophila rufipectus

SUBFAMILY
Phasianinae (Tribe Perdicini)

TAXONOMY
Arboriphila rufipectus Boulton, 1932, west Sichuan, China. Monotypic.

OTHER COMMON NAMES
English: Boulton's hill-partridge, Sichuan partridge; French: Torquéole de Boulton; German: Boultonbuschwachtel; Spanish: Arborófila de Sichuán.

PHYSICAL CHARACTERISTICS
12 in (29–31 cm); male 14–17 oz (410–470 g); female 12–13 oz (350–380 g). Distinguished from the common hill-partridge by a white throat and russet breast patch.

DISTRIBUTION
China: Southern Sichuan and adjacent northern Yunnan within a fragmented range.

HABITAT
Primary subtropical broadleaf forest, and adjacent disturbed or broadleaf plantation areas, with relatively open understory, at 3,600–7,400 ft (1,100–2,250 m).

BEHAVIOR
Pairs in spring, with family parties staying together only until late fall.

FEEDING ECOLOGY AND DIET
Seeds and fruits from forest floor and shrubs, some invertebrates.

REPRODUCTIVE BIOLOGY
Breeding pairs widely separated, probably monogamous; nests concealed among tree roots in April–May; clutch size five to six.

CONSERVATION STATUS
Endangered; threatened by clear-felling of primary forest, agricultural encroachment, bamboo-shoot collection and livestock browsing; only in one small protected area.

SIGNIFICANCE TO HUMANS
None known. ◆

Crested wood-partridge
Rollulus rouloul

SUBFAMILY
Phasianinae (Tribe Perdicini)

TAXONOMY
Phasianus rouloul Scopoli, 1786, Malacca, Malaysia. Monotypic.

OTHER COMMON NAMES
English: Crested partridge, Roulroul; French: Rouloul couronné, German: Straußwachtel; Spanish: Perdiz rulrul.

PHYSICAL CHARACTERISTICS
10 in (25 cm); male 8–11 oz (225–300 g); female 8–10 oz (225–275 g). Males have a spectacular reddish crest, dark plumage, and bright red bare parts. Females are mostly green with a gray head.

DISTRIBUTION
Southern Myanmar and Thailand through Malaysian peninsula to Sumatra and Borneo.

HABITAT
Lowland tropical rainforest, including disturbed areas, up to 3,900 ft (1,200 m).

BEHAVIOR
Encountered as singles, pairs, families, and larger groups of up to 15; scatters on foot if disturbed, then regroups using calls; paired birds remain close together calling in soft whistles; roosts in low shrubs and trees.

FEEDING ECOLOGY AND DIET
Seeds, fruits, and invertebrates.

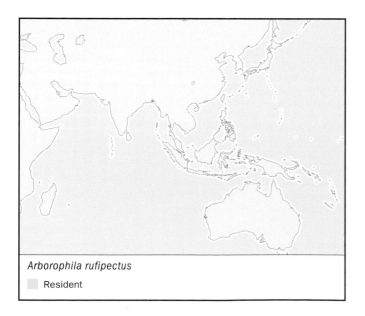

Arborophila rufipectus
▨ Resident

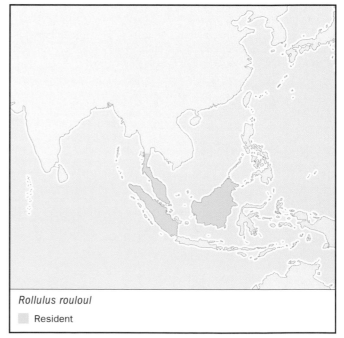

Rollulus rouloul
▨ Resident

REPRODUCTIVE BIOLOGY
Breeds mainly in first half of year; nest concealed on forest floor under pile of leaves; clutch size five to six; incubation 18 days; unusually for Galliformes, both parents actively feed chicks.

CONSERVATION STATUS
Near Threatened due to continuing degradation and clearance of lowland tropical forests in Southeast Asia.

SIGNIFICANCE TO HUMANS
None known. ◆

Satyr tragopan
Tragopan satyra

SUBFAMILY
Phasianinae (Tribe Phasianini)

TAXONOMY
Meleagris satyra Linnaeus, 1829, Bengal = Sikkim, India. Monotopic.

OTHER COMMON NAMES
English: Crimson tragopan, Indian tragopan, crimson horned pheasant; French: Tragopan satyre; German: Satyrtragopan; Spanish: Tragopán Sátiro.

PHYSICAL CHARACTERISTICS
Male 26–28 in (67–72 cm); female 22–23 in (57–59 cm); male 3.5–4.6 lb (1.6–2.1 kg); female 2.2–2.6 lb (1.0–1.2 kg). Males have deep red underparts, bare blue facial skin, and brown plumage on lower back and rump; upperwing-coverts also brown. Females are dull brown to rufous with bars and lance-shaped markings.

DISTRIBUTION
Himalayas from northern India through Nepal, Sikkim, Bhutan to western Arunachal Pradesh, including southeastern Tibet.

HABITAT
Temperate montane forest with dense understory at 5,900–14,100 ft (1,800–4,300 m).

BEHAVIOR
Usually in pairs or singles early in year, and family parties in July–September; male's repeated wailing call at dawn from roost in April–June heralds breeding season.

FEEDING ECOLOGY AND DIET
Roots, bulbs, and invertebrates, and plucking leaves from cover and trees.

REPRODUCTIVE BIOLOGY
Monogamous in captivity; usually crude stick nests are constructed up to 20 ft (6 m) in trees; clutch size two to three; incubation 28 days.

CONSERVATION STATUS
Not threatened and still widespread, but hunted (except in Bhutan).

SIGNIFICANCE TO HUMANS
Used as a flagship for central Himalayan forest conservation campaigns. ◆

Chinese monal
Lophophorus lhuysii

SUBFAMILY
Phasianinae (Tribe Phasianini)

TAXONOMY
Lophophorus lhuysii Geoffrey St. Hilaire and Verreaux, 1866, Moupin, China. Monotypic.

OTHER COMMON NAMES
English: Chinese monal pheasant; French: Lophophore de Lhuys; German: Grünschwanzmonal; Spanish: Monal Coliverde.

PHYSICAL CHARACTERISTICS
Male 30–31 in (76–80 cm); female 28–29 in (72–75 cm); 6.2–7.0 lb (2.8–3.2 kg). Larger and more heavily built than other pheasants in this genus.

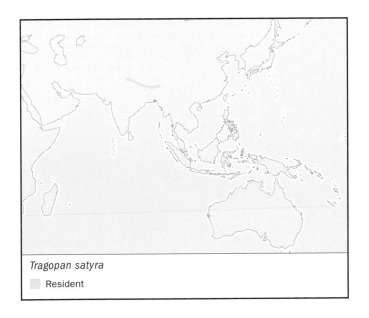

Tragopan satyra
 Resident

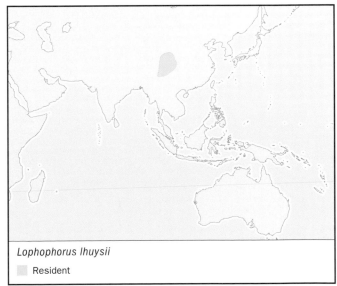

Lophophorus lhuysii
 Resident

DISTRIBUTION
China: centered on western Sichuan, but extending into south-eastern Quinghai, southern Gansu, northeastern Tibet, and northwestern Yunnan.

HABITAT
Alpine meadows and sub-alpine scrub adjacent to highest conifer forests, mostly at 9,200–16,100 ft (2,800–4,900 m).

BEHAVIOR
Groups of two to eight individuals common in winter, with single- and mixed-sex flocks being seen in spring; a vocal species at roost in spring and summer, and when alarmed.

FEEDING ECOLOGY AND DIET
Tubers and bulbs, but also takes moss, leaves, flowers, and some invertebrates.

REPRODUCTIVE BIOLOGY
Breeding starts during March in snow; nests at 12,500–13,100 ft (3,800–4,000 m); clutch size three to five; incubation 28 days.

CONSERVATION STATUS
Vulnerable and on CITES Appendix I, prohibiting trade in wild birds.

SIGNIFICANCE TO HUMANS
None known. ◆

Red junglefowl
Gallus gallus

SUBFAMILY
Phasianinae (Tribe Phasianini)

TAXONOMY
Phasianus gallus Linnaeus, 1758, Poulo Condor, Vietnam. Five subspecies.

OTHER COMMON NAMES
English: Wild junglefowl; French: Coq bankiva; German: Bankivahuhn; Spanish: Gallo Bankiva.

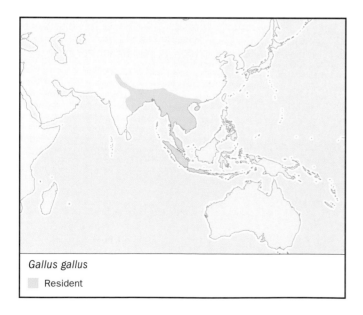

Gallus gallus
  ☐ Resident

PHYSICAL CHARACTERISTICS
Male 25–30 in (65–78 cm), female 16–18 in (42–46 cm); male 1.5–3.2 lb (0.7–1.5 kg); female 1.1–2.3 lb (0.5–1.1 kg).

DISTRIBUTION
Northern India, Nepal, and Bangladesh, southern Yunnan to Hainan Island, Southeast Asian peninsula, and Sumatra, Java, and Bali.

HABITAT
Woodland edge and secondary scrub in tropical and sub-tropical areas from sea level to 6,560 ft (2,000 m).

BEHAVIOR
Often seen in groups consisting of one male, several females, and offspring; roosts socially in trees or bushes.

FEEDING ECOLOGY AND DIET
Seeds, including rice, and invertebrates, including eggs.

REPRODUCTIVE BIOLOGY
Polygamous. Breeds almost year round in India and Malaysia; nests typically under bushes or in bamboo; clutch size five to six; incubation 18–21 days; chicks can fly at seven days.

CONSERVATION STATUS
Widespread and locally common.

SIGNIFICANCE TO HUMANS
The supposed progenitor of all domestic fowl, and therefore arguably the most important bird species of all to humans. ◆

Edwards's pheasant
Lophura edwardsi

SUBFAMILY
Phasianinae (Tribe Phasianini)

TAXONOMY
Gennaeus edwardsi Oustalet, 1896, Quangtri, Vietnam. Two subspecies.

Lophura edwardsi
  ☐ Resident

OTHER COMMON NAMES
English: Annam pheasant; French: Faisan d'Edwards; German: Edwardsfasan; Spanish: Faisán de Edwards.

PHYSICAL CHARACTERISTICS
23–25.5 in (58–65 cm); 2.4 lb (1.1 kg). Males have black plumage with a blue sheen and metallic green fringes on up-perwing-coverts; small white crest on head. Females are chestnut brown with darker flight feathers and tail and no crest.

DISTRIBUTION
Very restricted range in adjacent parts of Quangtri, Thua Thien, and Quang Binh provinces, east of the Annamese mountains in central Vietnam.

HABITAT
Primary and secondary forest on level lowlands below 1,300 ft (400 m).

BEHAVIOR
Little known, but shy and fond of dense understory.

FEEDING ECOLOGY AND DIET
Nothing known.

REPRODUCTIVE BIOLOGY
Clutch size four to seven; incubation 21–22 days in captivity.

CONSERVATION STATUS
Endangered and on CITES Appendix I, prohibiting trade in wild birds.

SIGNIFICANCE TO HUMANS
None known. ◆

Brown eared-pheasant
Crossoptilon mantchuricum

SUBFAMILY
Phasianinae

TAXONOMY
Crossoptilon mantchuricum Swinhoe, 1863, vicinity of Peking = Beijing, China. Monotypic.

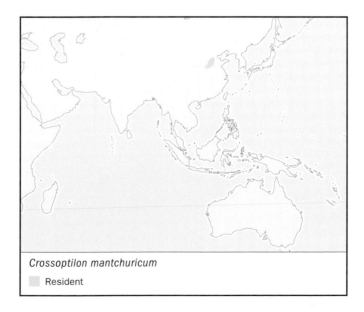

Crossoptilon mantchuricum
▨ Resident

OTHER COMMON NAMES
English: Manchurian eared-pheasant; French: Hokki brun; German: Brauner ohrfasan; Spanish: Faisán Orejudo Pardo.

PHYSICAL CHARACTERISTICS
37–39 in (96–100 cm); male 3.7–5.5 lb (1.7–2.5 kg); female 3.3–4.4 lb (1.5–2.0 kg). The only brownish member of its genus.

DISTRIBUTION
Formerly widespread in China; now mainly in Shanxi.

HABITAT
Mixed montane forest with shrub understory at 3,600–8,500 ft (1,100–2,600 m).

BEHAVIOR
Encountered in large groups (10–30) moving through forest for much of the year.

FEEDING ECOLOGY AND DIET
Roots, bulbs, and tubers, but also a wide variety of leaves, some fungi, and invertebrates.

REPRODUCTIVE BIOLOGY
Males issue long calls in April–June; nests on ground in conifer forest; clutch size usually five to eight; incubation 26–27 days.

CONSERVATION STATUS
Vulnerable and on CITES Appendix I, prohibiting trade in wild birds.

SIGNIFICANCE TO HUMANS
Tail feathers have been worn on military uniforms since about 500 B.C., reflecting battles fought by males in the mating season. ◆

Ring-necked pheasant
Phasianus colchicus

SUBFAMILY
Phasianinae (Tribe Phasianini)

TAXONOMY
Phasianus colchicus Linnaeus, 1758, Africa, Asia = Rion River. Thirty-one subspecies.

OTHER COMMON NAMES
English: Common pheasant; French: Faisan de Colchide; German: Fasan; Spanish: Faisán Vulgar.

PHYSICAL CHARACTERISTICS
Male: 29–35 in (75–89 cm); female 21–24 in (53–62 cm); male 2.6 lb (1.2 kg); female 2.0 lb (0.9 kg).

DISTRIBUTION
Japan, Taiwan, central and eastern China, with apparently isolated populations spread across central Asia to the Caucasus; possibly into southeastern Europe; introductions worldwide.

HABITAT
Mixed temperate scrub, riverine and woodland edge, adjacent to cultivation, avoiding dense forest, dry areas, and high mountains.

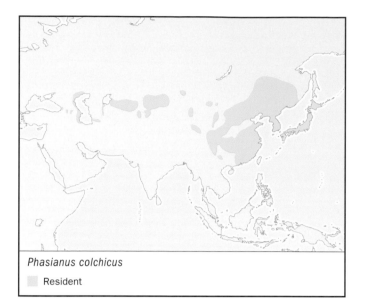

Phasianus colchicus
◼ Resident

Polyplectron emphanum
◼ Resident

BEHAVIOR
Often in large feeding groups with wide variation in sex ratio; tends to run for cover rather than fly if disturbed.

FEEDING ECOLOGY AND DIET
Leaves, cereal grains, tree seeds, buds, fruits and some invertebrates as adults; invertebrates essential to young for first month after hatching.

REPRODUCTIVE BIOLOGY
Single males crow loudly in spring and some associate with a group of females, suggesting harem polygyny; nests in ground cover; clutch size eight to 14; incubation 22–25 days.

CONSERVATION STATUS
The most widespread and common pheasant in the world, although some of its isolated western subspecies may be threatened.

SIGNIFICANCE TO HUMANS
Tens of millions of birds are artificially reared each year for release in sport hunting enterprises, especially in Europe and North America. ◆

Palawan peacock-pheasant
Polyplectron emphanum

SUBFAMILY
Phasianinae (Tribe Phasianini)

TAXONOMY
Polyplectron emphanum Temminck, 1831, Sunda Islands or Moluccas (=error: Palawan Island, the Philippines). Monotypic.

OTHER COMMON NAMES
English: Napoleon's peacock-pheasant; French: éperonnier napoléon; German: Napoleonfasan; Spanish: Espolonero de Palawan.

PHYSICAL CHARACTERISTICS
Male 20 in (50 cm); female 16 in (40 cm); male 1.0 lb (0.4 kg); female 0.7 lb (0.3 kg). Male has a long, pointed crest on head, solid black underparts, shiny blue and green on mantle, and a distinctive black and white face pattern. Female is brown with buff markings.

DISTRIBUTION
Palawan Island, southwestern Philippines.

HABITAT
Previously typical of lowland coastal forest below 2,000 ft (600 m); this now mostly logged and species found inhabiting montane forest and bamboo scrub at 4,900 ft (1,500 m) in 2000.

BEHAVIOR
No information available.

FEEDING ECOLOGY AND DIET
No information available.

REPRODUCTIVE BIOLOGY
Males call and keep small areas clear of leaves, for use as courtship display grounds; clutch size two; incubation 18–20 days.

CONSERVATION STATUS
Vulnerable and on CITES Appendix I, prohibiting trade in wild birds.

SIGNIFICANCE TO HUMANS
None known. ◆

Resources

Books

del Hoyo, J., A. Elliot, and J. Sargatal, eds. *Handbook of the Birds of the World.* Vol. 2. *New World Vultures to Guineafowl.* Barcelona: Lynx Edicions, 1994.

Delacour, J. *The Pheasants of the World.* 2nd edition. Hindhead: Spur/Saiga/ World Pheasant Association, 1977.

Fuller, R. A., J. P. Carroll, and P. J. K. McGowan, eds. *Partridges, Quails, Francolins, Snowcocks, Guineafowl, and Turkeys. Status Survey and Conservation Action Plan 2000-04.* WPA/BidLife/SSC Partridge, Quail and Francolin Specialist Group. Gland and Cambridge: IUCN/Reading: World Pheasant Association, 2000.

Fuller, R. A. and P. J. Garson, eds. *Pheasants. Status Survey and Conservation Action Plan 2000-04.* WPA/BidLife/SSC Pheasant Specialist Group. Gland and Cambridge: IUCN/Reading: World Pheasant Association, 2000.

Hill, D. A., and P. A. Robertson. *The Pheasant.* Oxford: BSP Professional Books, 1988.

Johnsgard, P. A. *The Grouse of the World.* Lincoln: University of Nebraska Press, 1983.

Johnsgard, P. A. *The Pheasants of the World.* 2nd ed. Washington, DC: Smithsonian Institution Press, 1999.

Johnsgard, P. A. *The Quails, Patridges and Francolins of the World.* London: Oxford University Press, 1986.

Madge, S., and P. McGowan. "Pheasants, Partridges, and Grouse." *Helm Identification Guides.* London: Christopher Helm, 2002.

Potts, G. R. *The Partridge.* London: Collins, 1986.

Storch, I., ed. *Grouse. Status Survey and Conservation Action Plan 2000-04.* WPA/BidLife/SSC Grouse Specialist Group. Gland and Cambridge: IUCN/Reading: World Pheasant Association, 2000.

Periodicals

Dinesen, L., T. Lehmberg, J. O. Svendsen, L. A. Hansen, and J. Fjeldså. "A New Genus and Species of Perdicine Bird (Phasianidae, Perdicini) from Tanzania: A Relict Form with Indo-Malayan Affinities." *Ibis* 136 (1994): 2–11.

Kimball, R. T., E. L. Braun, and J. D. Ligon. "Resolution of the Phylogenetic Position of the Congo Peafowl, *Afropavo congensis:* A Biogeographic and Evolutionary Enigma." *Proceedings of the Royal Society of London* Series B 264 (1997): 1517–1523.

Kimball, R. T., E. L. Braun, P. W. Zwartes, T. M. Crowe, and J. D. Ligon. "A Molecular Phylogeny of the Pheasants and Partridges Suggests that these Lineages are Not Monophyletic." *Molecular Phylogenetics and Evolution* 11 (1999): 38–54.

Young, J. R., C. E. Braun, S. J. Oyler-McCance, T. W. Quinn, and J. W. Hupp. "A New Species of Sage Grouse (Phasianidae: *Centrocercus*) from Southwestern Colorado, USA." *Wilson Bulletin* 112 (2000): 445.

Organizations

Game Conservancy Trust. Fordingbridge, Hampshire SP6 1EF United Kingdom. Phone: +44 1425 652381. Fax: +44 1425 651026. E-mail: info@gct.org.uk Web site: <http://www.gct.org.uk>

Ruffed Grouse Society. 451 McCormick Rd, Coraopolis, PA 15108. Phone: (888) 564-6747. Fax: (412) 262-9207. Web site: <http://www.ruffedgrousesociety.org>

World Pheasant Association. P.O. Box 5, Lower Basildon, Reading RG8 9PF United Kingdom. Phone: +44 1 189 845 140. Fax: +44 118 963369. E-mail: office@pheasant.org.uk Web site: <http://www.pheasant.org.uk>

Peter Jeffery Garson, DPhil

New World quails
(Odontophoridae)

Class Aves
Order Galliformes
Suborder Phasiani
Family Odontophoridae

Thumbnail description
Plump, medium-sized birds with short, powerful wings and strong running and scratching legs and feet; all have a characteristic "toothed" bill and lack tarsal spurs in all species

Size
7–15 in (17–37 cm); 4–16 oz (125–465 g)

Number of genera, species
9 genera; 32 species

Habitat
Forest, woodlands, savanna, grasslands, and agricultural

Conservation status
Critically Endangered: 1 species; Vulnerable: 4 species; Near Threatened: 3 species

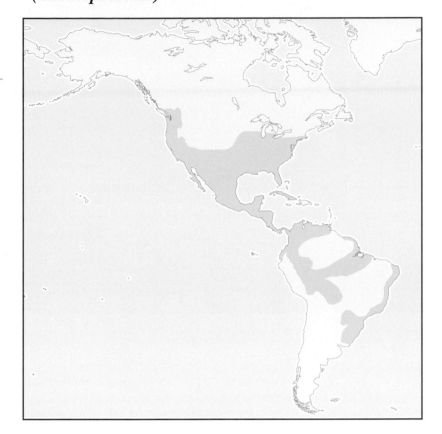

Distribution
Southern Canada, eastern United States, west coast and southwestern United States, much of Mexico and Central America, northern South America through the Amazon basin and Andes south to Bolivia, and east coast of Brazil and Uruguay; North American species introduced widely to Argentina, Chile, parts of Europe, and New Zealand

Evolution and systematics

There have been various groupings of the New World quail based on morphological characteristics, generally combining them with the pheasants as a subfamily within Phasianidae. However, DNA evidence suggests that family status is warranted and that the New World quail are not particularly closely related to the pheasants or Old World quail. Earliest fossils of quail-like birds have been found in Saskatchewan, Canada dating back 37 million years. Evidence appears to be pointing toward a divergence of the New World quail in South America 35–63 million years ago.

Within Odontophoridae there are 32 recognized species, although this number is subject to change as the lesser studied Latin American species are more thoroughly researched. There are nine genera including four (*Oreortyx, Philortyx, Dactylortyx,* and *Rhynchortyx*) containing only one species each. The remaining genera, *Dendrortyx, Callipepla, Colinus, Cyrtonyx,* and *Odontophorus* have three, four, four, two, and 15 species, re-

spectively. Even taxonomy within these genera has varied especially in the *Callipepla* and *Colinus.* In the past, *Callipepla* has been split into two genera, *Callipepla* and *Lophortyx,* but it is now generally agreed that these species represent only a single genus. Among the better studied species, there is still confusion regarding taxonomy. The northern bobwhite (*Colinus virginanus*) may represent a super-species that can be split into several species. There is some suggestion that the masked bobwhite (*C. v. ridgwayi*) should be considered a distinct species. The *Odontophorus* wood-quails, which contains the largest number of species, are made up of several complexes of species. For example, the rusty-breasted complex and the black-throated complex are both found in the northern Andes. These nine species may be grouped or split in the future as more research is completed.

Physical characteristics

The New World quail are smallish Galliformes and, like most members of the order, have plump bodies and short wings.

Northern bobwhite (*Colinus virginianus*) covey in a sleeping circle, Virginia. (Photo by Nell Bolen. Photo Researchers, Inc. Reproduced by permission.)

Odontorphoridae are much less variable in size compared to the other families of Galliformes ranging from the smallest bobwhites (*Colinus* spp.), barred quail (*Philortyx fasciatus*), and tawny-faced quail (*Rhynchortyx cinctus*)—reported to be as little as 7 in (17 cm) in length and 4 oz (125 g)—to the long-tailed wood-partridge (*Dendrotyx macroura*) which may be 16 in (37 cm) in length and weigh 16 oz (454 g). Plumage tends to be more subtle than many of the other Galliformes. Also, sexual dimorphism which tends to be dramatic in the Phasianidae is much less distinct. Often there are slight size and plumage coloration differences between males and females. Some species appear to have no distinct external differences between males and females. A number of species have distinct crests ranging from small tufts to very long plumes. Many species have distinct, often red, fleshy rings around the eyes.

The serrated edge of the bill is a distinct characteristic of this family. None of the New World quail have tarsal spurs, unlike many of their Old World counterparts. In some species the legs and feet are very thick and strong for digging. All others still have strong legs and feet for running and scratching.

Distribution

Outside of human introductions, the New World quail are restricted to North, Central, and South America. The greatest number of genera and species are found in the vicinity of southern Mexico and Guatemala with the number of species decreasing outwards north and south. The genus *Odontophorus* is found mainly in southern Central America and northern South America. *Colinus* is distributed from eastern United States and Canada through Central America to Colombia,

Venezuela, and the Guianas. *Callipepla* is restricted to western United States and Mexico. *Oreortyx* is restricted to western United States and just a small part of Mexico. *Dendrortyx* is restricted to Mexico and the vicinity of Honduras and Guatemala. *Philortyx* is restricted to Mexico. *Cyrtonyx* is found in southwestern United States, and Mexico, through to western Guatemala. *Dactylortyx* is found in southern Mexico through Honduras. *Rhynchortyx* is found along the eastern coast of Honduras to Panama where it is then found on the west coast down to Ecuador.

The northern bobwhite and California quail (*Callipepla californica*) have been widely introduced by humans. For the bobwhite this includes established populations in the northwestern United States and British Columbia, Puerto Rico, Cuba, Hawaii, New Zealand, Italy, and Germany. The California quail has been introduced to Chile, Argentina, New Zealand, Australia, and Hawaii.

Habitat

The New World quail are ground birds inhabiting a wide range of tropical, subtropical, and temperate ecosystems. The bobwhites inhabit grassland, savannas, rangeland, and agricultural lands, but are also considered woodland edge species. The *Callipepla* species are mainly scrub to desert inhabitants. The two *Cyrtonyx* species are found in open pine or oak woodlands, but also inhabit scrub. The three *Dendrortyx* species inhabit montane and cloud forests. *Oreortyx* is found in mixed forest, forest edge, and chaparral. *Philortyx* inhabits xeric scrub and farmland. The wood-quails (*Odontophorus*) are found in a variety of tropical, subtropical, and montane forests. *Rhyn-*

Gambel's quail (*Callipepla gambelii*) dust bathing in the Sonoran Desert, southwestern United States. (Photo by Stephen J. Krasemann. Photo Researchers, Inc. Reproduced by permission.)

chortyx is restricted to lowland tropical forest. Finally, *Dactylortyx* is found in a variety of montane forests, but some populations are found in lowland scrub and woodland edge.

Behavior

The most notable behavior of the New World quail is the covey, which has been reported in almost all species. It is interesting to note that these were formerly thought of as family groups, but now covey membership is thought to be much more complex. In some of the *Odontophorus* wood-quails there is some suggestion that coveys are family groups including adults pairs and helpers from previous clutches. Most species are diurnal and spend most of their time on the ground. Tree roosting has been observed in a number of the forest species. None of the New World quail are true migrants, although some appear to be altitudal migrants in mountainous regions.

Although the vocal repertoire of the New World quail is rather limited, most use a variety of calls and whistles to communicate. The bobwhites probably have the largest number of calls with at least 19 distinct calls. The *Dendrortyx* wood-partridges will give loud hooting calls; whereas the wood-quails exhibit loud guttural choruses.

Feeding ecology and diet

Most of the New World quail are gleaners and scratchers for seeds. Most eat a variety of seeds including those from grasses, forbs, shrubs, and trees. Many species will move into agricultural lands and eat waste grain seeds. Commonly reported crops eaten by various species include corn, wheat, sorghum, peanuts, and black beans. Some of the tropical forest wood-quails have been observed digging for fleshy roots. The two species of *Cyrtonyx* feed extensively on tubers and bulbs, especially the tubers of wood sorel (*Oxalis* spp). When

studied, the diet of chicks is comprised mainly of invertebrates, and as they age they become more granivorous.

Reproductive biology

Breeding biology is not well studied in most species. Generally the New World quail were thought to be monogamous. However, data is mounting to the contrary. The northern bobwhite quail is now thought to have a flexible mating system that includes monogamy, polygyny, polyandry, and promiscuity. New genetic techniques combined with field methodologies are now being applied to this species to clarify its mating system. Anecdotal evidence suggests that some of the wood-quails may have adult pairs with helpers at nests and rearing young.

The temperate and grassland adapted species have the largest clutch sizes. For example, clutch size of the California quail ranges up to 17. Clutch size tends to decrease among the more tropical and forest adapted species. Where described, clutch sizes of three to five have been reported for some of the wood-quails. Nests are usually constructed on the ground forming a small bowl. Many species cover the nest with vegetation and form an "igloo-like" structure with an opening at one end. Incubation period is quite variable, although not well described in many species. Reports among the New World quail range from 16 to 30 days. Chicks are precocial and are capable of leaving the nest within hours of hatching. They grow rapidly and are capable of flight in less than two weeks. Mortality rates of nests and chicks is reported to be quite high in the better studied temperate and grassland species. Nest failure rates of 40–80% are quite common. Chick mortality rates of 20–50% are often reported. These species are persistent nesters with up to four nesting attempts reported in a breeding season. Although not reported for the more tropical and forest adapted species, the mortality rates of nests and chicks is probably lower.

Northern bobwhite (*Colinus virginianus*) nest in a tussock of wire grass (*Aristida stricta*), New Hanover County, North Carolina. (Photo by Jack Dermid. Bruce Coleman Inc. Reproduced by permission.)

Bobwhite quail (*Colinus virginianus*) with young in south Texas. (Photo by John Snyder. Bruce Coleman Inc. Reproduced by permission.)

Conservation status

Conservation status of the New World quail varies widely among species. The more temperate and grassland adapted species of *Colinus* and *Callipepla* tend to be very common. Some of these species are increasing in population and distribution as a result of human activity. Many of the forest adapted species seem to tolerate some human impact on their habitat, therefore seem to be maintaining reasonable populations. Most of the species with a conservation status are forest wood-quails in the genus *Odontophorus*. One species, the gorgeted wood-quail (*Odontophorus strophium*), found in oak forests in Colombia is considered critically endangered because almost all of its mid-elevational habitat in the Central Andes has been destroyed. The conservation status of many of the Latin American species is tentative because of lack of significant research to assess their status. For example, the bearded wood-partridge (*Dendrortyx barbatus*) was considered critically endangered in 1995. However, subsequent surveys of its montane forest habitat in the vicinity of Veracruz, Mexico identified a number of small and disjunct populations. This species is now considered to be Vulnerable because of continued threat to the remnant forest patches it inhabits.

Significance to humans

Like many of the Galliformes the New World quail are important to humans. Some species such as the northern bobwhite quail are among the most studied birds in the world. This species is widely hunted and contributes greatly to local economies in parts of the United States and Mexico. Management for hunting in some areas has a significant impact on land use, oftentimes reducing the negative impact of grazing and farming on other wildlife. It is also widely raised in captivity to be released for hunting, as well as for the restaurant market. Most species are hunted either for sport or subsistence. This is done sustainably for a few species; however, the impact of hunting is not known for most of the Latin American species. There are a few cases of crop depredation by some species.

1. Northern bobwhite (*Colinus virginianus*); 2. Bearded wood-partridge (*Dendrortyx barbatus*); 3. Venezuelan wood-quail (*Odontophorus columbianus*). (Illustration by John Megahan)

Species accounts

Bearded wood-partridge
Dendrortyx barbatus

TAXONOMY
Dendrortyx barbatus Gould, 1846, Jalapa, Veracruz, Mexico.
Monotypic.

OTHER COMMON NAMES
English: Bearded partridge, Bearded tree-quail; French: Colin
barbu; German: Bartwachtel; Spanish: Colin barbudo, Chiviz-
coyo.

PHYSICAL CHARACTERISTICS
This is one of the larger species ranging 9–13 in (22–32 cm) in
length and estimated weights of males about 16 oz (459 g) and
for females about 14 oz (405 g). This bird has a grayish head
with a small crest. Most notable are the bright red eye-ring,
bill, and legs. The body has an overall reddish brown coloration with a darker rump and wings.

DISTRIBUTION
Restricted distribution to 14 fragmented populations ranging
from west-central Veracruz to extreme northeast Queretaro,
Mexico.

HABITAT
It is found in montane pine-oak, cloud, and older second
growth forests. It is also associated with forest edges, shade
coffee, and some agricultural land. Most of the forest, especially in Veracruz, Mexico is highly fragmented. The recently
discovered population in Queretaro inhabits the most remote
and intact forests.

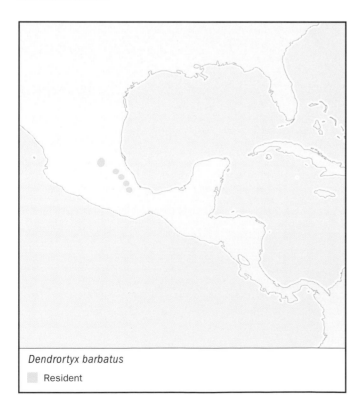

Dendrortyx barbatus
■ Resident

BEHAVIOR
Little is known. They form coveys like most of the other quails
and have quite raucous chorus calls in the morning.

FEEDING ECOLOGY AND DIET
Little information available, but is known to feed on a variety
of seeds and fruits. Will eat domestic crops such as black beans.
Captive birds readily eat beans, corn, bananas, and grapes.

REPRODUCTIVE BIOLOGY
Little is known, but the breeding season is suspected to be
April–June. Broods of five have been seen in the wild. Captive
birds constructed nests in shallow depressions in the ground
and lined them with palms.

CONSERVATION STATUS
Until 1995 this species was considered Critically Endangered.
Surveys undertaken in late 1990s resulted in discovery of addi-
tional populations in remnant forests. Now downgraded to
Vulnerable. However, rapid human growth and encroaching
agriculture further threaten this species.

SIGNIFICANCE TO HUMANS
Because of its limited distribution and population, it is not an
important species to humans. However, even in the remaining
populations birds are still hunted for sport, and trapped as cage
birds and for food. Local farmers have poisoned them when
they are found depredating black bean fields. ◆

Venezuelan wood-quail
Odontophorus columbianus

TAXONOMY
Odontophorus columbianus Gould, 1850, Caracas, Venezuela.
Monotypic.

OTHER COMMON NAMES
French: Tocro du Venezuela; German: Venezuelawachtel;
Spanish: Perdiz montañera.

PHYSICAL CHARACTERISTICS
This is a medium sized species ranging 11–12 in (28–30 cm) in
length and with estimated weight of males about 12 oz (343 g)
and 11 oz (336 g) for females. This species has a low brownish
crest and general body color of reddish brown. The throat is
white with black streaks. The breast is covered by distinct
white teardrops outlined in black. The bill is black and the legs
dark gray.

DISTRIBUTION
Restricted distribution in remnant forests in the northern
coastal mountains west of Caracas, Venezuela and the northern-most tip of the Andes in northwestern Venezuela. Found
in Henri Pittier and San Esteban National Parks.

HABITAT
Found in montane subtropical forests at altitudes of
2,950–7,900 ft (900–2,400 m). A recent study found that foraging habitat was usually associated with areas containing high

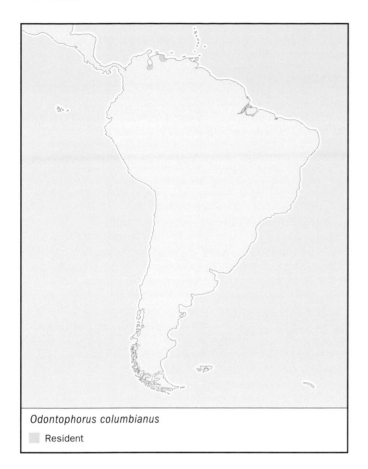

Odontophorus columbianus
■ Resident

numbers of monocots, high vertical foliage density and low frequency of palms.

BEHAVIOR
Like most of the other New World quails this species is found in coveys which are thought to be family groups. Group size in ranges from two to five individuals with one survey reporting an average group size of 4.5. Coveys are quite vocal and loud during early morning choruses. This might be for territory establishment and maintenance. When feeding they make a quiet, "güp-güp" call. Although almost always observed on the ground even when roosting, they have been observed roosting on palm fronds above the ground.

FEEDING ECOLOGY AND DIET
Little information available, but they have are suspected to feed on seeds, fruits, insects, and worms. They typically forage as a group working through the litter on the forest floor. They have been observed digging along the edges of surface roots and feeding on fleshy bits of the root.

REPRODUCTIVE BIOLOGY
Little is known and only one nest has been described. Breeding season appears to run from March to July which corresponds to the wet season. The single nest was found at the base of palms and contained a roof of vegetation. It contained a clutch of six eggs and was incubated for 30 days.

CONSERVATION STATUS
Considered Near Threatened because it occurs in several large national parks, most notably Henri Pittier National Park. However, populations and distribution have probably declined

significantly as forests outside of the parks have been converted to agriculture and urban uses. Limited research suggests that there might be a decline in the population in the vicinity of Rancho Grande in Pittier National Park due to increasing tourism and vehicle traffic.

SIGNIFICANCE TO HUMANS
Because of its limited distribution and population, it is not an important species to humans. It may be opportunistically harvested by subsistence hunters who are hunting in remnant forest patches, and in the vicinity of the national parks. ◆

Northern bobwhite quail
Colinus virginianus

TAXONOMY
Tetrao virginianus Linnaeus, 1758, America (=South Carolina). Twenty-two subspecies.

OTHER COMMON NAMES
English: Bobwhite, northern bobwhite; French: Colin de virginie; German: Baumwachtel, Virginiawachtel, Wachtel; Spanish: Codorniz; Colín de virginia.

PHYSICAL CHARACTERISTICS
Among the smaller Galliformes ranging in length from 8 to 10 in (20–25 cm) and 4 to 8 oz (129–233 g) in weight. Largest birds are found in the northern part of the range and the smallest in southern Mexico. Females are slightly smaller than males. Quite variable male plumage usually contains a combination of black, gray, and white. Males of the *virginianus* group usually have a white throat, as do the birds in the *graysoni* and *pectoralis* groups. *Graysoni* males tend to be more rufous col-

Colinus virginianus
■ Resident

ored; whereas the *pectoralis* males tend to look most like the *virginianus* group, except the black collar is wider and more streaky on breast. Males of the *coyolcos* group have a solid black head and throat.

In females, the head and throat are mostly buff. The black parts of the head tend to be more chestnut in color.

DISTRIBUTION
Widespread distribution ranging from southern New England west through southern Ontario, Canada to southeastern Minnesota. The distribution continues southward to Florida in the east and extreme eastern Wyoming, western Kansas, and Oklahoma southward to Mexico in the west. There is a disjunct population in Sonora, Mexico and formerly Arizona. Restoration efforts are underway in Arizona. In Mexico, this species is found in the northeast state of Tamaulipas southward to the Isthmus of Tehuantepec. In central Mexico it reaches further west and in the south reaches the Pacific Ocean. It is replaced by the black-throated bobwhite (*C. v. nigrogularis*) on the Yucatan Peninsula. Distribution just reaches Guatemala. Introduced populations are found on many Caribbean Islands, Washington, Oregon, Hawaii, British Columbia, New Zealand, Italy, and Germany.

HABITAT
Found in a variety of habitats as long as some type of early successional habitat is present. Most closely associated with fire maintained pine savannas and forest openings or clearcuts in typically forested areas. Now most often found in farmland in those regions. Highest populations are found in grass and brush rangelands except where these are intensively managed for livestock. In the southeastern United States, best populations are found on "plantations" comprised of pine savanna ecosystems in southern Georgia and northern Florida. These areas have actually been maintained and managed for quail.

BEHAVIOR
General behavior as in the family description. After the breeding season the birds go through an autumn "shuffle" to form larger coveys. This is time of greatest movement and some birds have been found to travel upwards of 60 mi (100 km). Home range sizes vary greatly. In better habitat might be 25–62 acres (10–25 ha) and much larger in poorer habitat.

FEEDING ECOLOGY AND DIET
Very well studied in this species. Shown to be primarily a seed eater with a wide range of seeds taken—a summary of studies identified 650 different types of seeds taken and 78 species that seemed to be more important. The types of seeds taken varies greatly with season and seems to be most related to seed abundance. Bobwhites will also take a variety of fruits and even

large items such as oak (*Quercus* spp.) acorns. Chicks are primarily insectivorous, but begin consuming mainly seeds by six to eight weeks of age.

REPRODUCTIVE BIOLOGY
The bobwhite quail was traditionally thought by lay people and scientists to form monogamous pairs with a great deal of parental care given by the males. More recent research is suggesting a much more complex social system where some individuals might be monogamous, but others are polygamous, including both polygyny and polyandry, and others appear to be promiscuous. Research is underway to try to clarify the bobwhite's mating system.

Unmated males give the "bob-white" call, which is a familiar spring call to anyone living within the distribution of this species. Pair formation is common, starting from January to March in the United States, with more northerly population beginning later. Like all the New World quails nests are built in a shallow depression on the ground, usually in dead grasses or other herbaceous vegetation. Clutch size averages 12–14 eggs, but ranges from seven to 28. Incubation is 23–24 days. The chicks are precocial and will leave the nest with the adults within hours of hatching. They can fly within two weeks. Mortality of nests, young, and adults is high. Hatching success ranges widely, but often from 20 to 40%. Hens are persistent renesters. Chick survival has been reported at 31% to one month of age. Annual survival is typically less than 30% and often less than 20%.

CONSERVATION STATUS
The species is widespread and common. However, there have been significant declines in populations throughout the eastern United States. This is thought to be due to a combination of farmland abandonment and reforestation, loss of fire maintained pine savannas, and intensification of remaining agriculture. In many parts of the East, populations have declined by 70–90%, since the 1960s. Populations are less well know in Mexico. One subspecies, the masked bobwhite, which inhabits scrubland in Sonora, Mexico and formerly Arizona, is considered Endangered in the United States. Because of its importance as a harvested gamebird, there are management programs in several parts of its distribution to try to restore numbers.

SIGNIFICANCE TO HUMANS
This species is one of the most important Galliformes. It is widely hunted and reared for human consumption. Large amounts of money are spent for its conservation for hunting. In some areas of the Southern Plains management of rangeland has shifted to emphasizing bobwhite by using livestock because hunting has become more lucrative than raising livestock. ◆

Resources

Books
Bonaccorso, Elisa. "Densidad y uso de habitat de perdíz *Odontophorus columbianus* en el Parque Nacional Henri Pittier." Trabajo Especial de Grado. Universidad Simón Bolívar Sartenejas.

Carroll, John P. "New World Quail." In *Handbook of Birds of the World.* Vol. 2. *New World Vultures to Guineafowl*, edited by Josep del Hoyo, Andrew Elliott, and Jordi Sargatal. Barcelona: BirdLife International and Lynx Edicions, 1994.

Johnsgard, Paul A. *Quails, Partridges, and Francolins of the World.* London: Oxford University Press, 1988.

Brennan, Leonard A. "Northern Bobwhite (*Colinus virginanus*)." In *The Birds of North America*, edited by Alan Poole and Frank Gill. No. 397. Philadelphia: The Birds of North America, Inc., 1999.

Fuller, Richard A., John P. Carroll, and Philip J.K. McGowan, eds. " Partridges, Quails, Francolins, Snowcocks, Guineafowl, and Turkeys. Status Survey and Conservation Action Plan

Resources

2000–2004.” Gland, Switzerland; Cambridge, United Kingdom; and Reading, United Kingdom: WPA/BirdLife/SSC Partridge, Quail and Francolin Specialist Group, IUCN, and World Pheasant Association, 2000.

McGowan, Philip J.K., Simon D. Dowell, John P. Carroll, and Nicholas J. Aebischer, eds. “Partridges, Quails, Francolins, Snowcocks, and Guineafowl. Status Survey and Conservation Action Plan 1995–1999.” Gland, Switzerland; Cambridge, United Kingdom; and Reading, United Kingdom: WPA/BirdLife/SSC Partridge, Quail and

Francolin Specialist Group, IUCN, and World Pheasant Association, 1995.

Organizations

WPA/BirdLife/SSC Partridge, Quail, and Francolin Specialist Group. c/o World Pheasant Association, PO Box 5, Lower Basildon, Reading, RG8 9PF United Kingdom. Phone: +44 1 189 845 140. Fax: +118 9843369. E-mail: wpa@gn.apc.org Web site: <http:/www.pheasant.org.uk/>. PQF: <http://www .gct.org.uk/pqf/>

John Patrick Carroll, PhD

Opisthocomiformes
Hoatzins
(Opisthocomidae)

Class Aves
Order Opisthocomiformes
Family Opisthocomidae
Number of families 1

Thumbnail description
Medium-sized, crested birds with bare face and long tail; obligate herbivore with foregut microbial fermentation

Size
24.5–27.5 in (62–70 cm); 1.4–1.9 lb (650–850 g)

Number of genera, species
1 genus; 1 species

Habitat
Riparian arboreal vegetation along rivers and streams, mangrove swamps

Conservation status
Not threatened

Distribution
East of the Andes in the Orinoco and Amazon river basins and in the Guianas

Evolution and systematics

The Opisthocomidae is a monotypic family that is restricted to tropical South America. Its sole representative, the hoatzin (*Opisthocomus hoazin*), is unique in appearance, behavior, morphological specializations, and physiological adaptations, and it is considered one of the most primitive of existing birds.

The hoatzin's phylogenetic relationship to other birds is uncertain and has been strongly debated since it was first described in 1776 (as *Phasianus hoazin* by P.L.S. Müller). Hoatzins were originally considered allied to the galliforms (fowl-like birds), and because of their chachalaca-like appearance hoatzins were placed near the Cracidae (chachalacas, guans). On the basis of osteology, mitochondrial and nuclear gene sequences, protein electrophoresis, and DNA-DNA hybridization, hoatzins have also been considered a cuculiform or a sister group to the cuculiforms, allied with the Neotropical anis (Crotophagidae) and the Guira cuckoo (*Guira guira*). However, hoatzins have an anisodactyl foot (a single toe pointing backwards), unlike the zygodactyl foot (two toes pointing backwards) of all cuckoos.

A more comprehensive comparative study of mitochondrial and gene sequences strongly supports placing hoatzins in the basal portion of a clade with turacos (Musophagidae) and not among or basal to cuckoos. Comparative studies of pterylosis (feather tract patterns), osteology, and growth pattern of primary wing feathers further support a sister relationship between hoatzins and turacos. However, until the relationship is better defined it is prudent to place hoatzins in their own order.

The relationship between hoatzins and turacos (the latter are restricted to sub-Saharan Africa) poses some interesting zoogeographic problems because the geographical distributions of these two groups are widely separated. Nevertheless, a fossil (*Foro panarium*) from the lower Eocene Green River formation of Wyoming in the United States shares osteological similarities to hoatzins and turacos. The only fossil of a hoatzin is *Hoazinoides magdalenae* from the Miocene of Colombia.

Hoatzin (*Opisthocomus hoazin*). (Illustration by Marguette Dongvillo)

Physical characteristics

Hoatzins are medium-sized birds with a long tail, long neck, and small head. The bright blue face is bare, the iris is red, and the eyelashes are prominent. The head is adorned with a fan-shaped shaggy crest that gives hoatzins an unusual and unmistakable appearance. Upperparts are mostly bronzy-olive streaked with buff, the chest is buff, the underparts are mostly chestnut, and the tail tip is buff. Sexes are similar in appearance and juveniles resemble adults.

Young hoatzins have two well-developed wing claws on digits two and three that are used to clamber about the vegetation when in danger. Researchers originally thought this feature was retained from ancestral *Archaeopteryx* but now believe that it evolved secondarily.

The sternum, pectoral girdle, and associated flight musculature of hoatzins are uniquely modified to accommodate a very large crop that functions as a fermentation chamber.

Distribution

Hoatzins are endemic to South America and are patchily distributed east of the Andes from Venezuela and the Guianas, southwards to Colombia, Ecuador, Peru, Brazil, and Bolivia. Hoatzins live along rivers and streams in the Orinoco and Amazon basins and on the Atlantic coast of the Guianas.

Habitat

Hoatzins are never found far from water. They are restricted to riparian arboreal vegetation bordering rivers, streams, lagoons, swamps, and oxbow lakes of the Amazon and Orinoco river systems. Along the Atlantic coast of the Guianas hoatzins frequently inhabit mangrove swamps. Hoatzins are limited to warmer lowlands from sea level to about 1,640 ft (500 m).

Behavior

In the central plains (Llanos) of Venezuela, hoatzins are territorial, colonial breeders. A monogamous breeding pair of adults and up to five nonbreeding helpers (retained from the previous year's nesting attempt) form the social unit. Male and female helpers seem to be equally common and both sexes participate in all reproductive activities except egg laying.

Territories are established throughout the breeding season (the rainy season) along waterways, and all activities are carried out within these multipurpose territories. When not tending a nest, hoatzins spend most daylight hours (70–80%) sitting or resting on a thickened callous of skin on the posterior tip of the sternum. Most of the rest of their time is spent foraging or in territorial disputes.

All members of the social unit (breeders and nonbreeders) defend territories against neighboring groups. Display copulations are a common form of territorial behavior and occur in response to intruders or similar displays by neighboring birds. Other territorial defense behaviors include postures, territorial vocalizations, chasing of intruders, and aerial battles. Maintenance behavior includes bathing during rainstorms and preening (but not allopreening, or mutual preening).

During the nonbreeding season most birds abandon their territories and form temporary flocks of up to 100 individuals. However, territories where reproductive success is high are defended year-round. Hoatzin groups are noisy; their hoarse shrieks, hisses, grunts, and growling sounds can be heard from afar. Adults are agile climbers but rather weak flyers; nevertheless, they are capable of flying up to 380 yd (350 m) without rest.

Feeding ecology and diet

Hoatzins are obligate herbivores, feeding almost exclusively on tree leaves. They prefer young leaves and shoots but also eat flowers and buds. In Venezuela, the diet of hoatzins includes the leaves of trees and shrubs belonging to families such as Fabaceae, Polygonaceae, Sterculiaceae, Combretaceae, Rutaceae, Vitaceae, and Lecythidaceae. Hoatzins have regularly spaced foraging bouts throughout the day; major bouts occur near sunrise and sunset, they also forage on moonlit nights.

Hoatzins are unique among birds in having a well-developed microbial fermentation system in the crop and caudal esophagus. In all other birds known to ferment plant material, microbial activity takes place in the hindgut. In hoatzins, the crop is greatly enlarged, thick walled, and muscular and is also the main site for leaf grinding. The microflora of the crop is composed of mixed bacterial populations and ciliate protozoa. Hoatzins retain the digesta for long periods of time, exposing leaves to bacterial enzyme attack. The end products of bacterial fermentation are volatile fatty acids that appear to be absorbed through the crop wall and used by the bird as energy substrates. Crop bacteria are likely to become a source of protein, carbon, and other nutrients because hoatzins have enzymes in the stomach that display bacteriolytic activity. Partially digested plant material, bacteria, and protozoa flow

Hoatzin (*Opisthocomus hoazin*) adult protecting its chicks. (Photo by François Gohier. Photo Researchers, Inc. Reproduced by permission.)

through the stomach and their constituents can then be absorbed through the intestinal wall.

The crop may also function as a detoxification chamber. Leaves of the plants eaten by hoatzins contain a broad array of phytotoxins, and it is likely that microorganisms in the crop degrade dietary toxins. Crop bacteria are implicated in the detoxification of saponins, a toxic compound present in several of the plants eaten by hoatzins in Venezuela. Hoatzins avoid highly tanniferous leaves and cannot handle tannin-rich diets.

Reproductive biology

Hoatzin parents and nonbreeders build the nest, which is an unlined platform made of twigs on branches 6.5–16.5 ft (2–5 m) high that are situated directly over the water. Courtship behavior is not very conspicuous, but reproductive copulation by breeders is common within several days of egg laying. Reproductive copulations are longer in duration than are display copulations. The clutch has one to six eggs, but two eggs are the most common; clutches of four to six eggs seem to be the result of two females laying in the same nest. Eggs are laid 1.5–2 days apart and hatch after 30–31 days of incubation. Incubation begins with the second egg when more than one egg is laid.

The young weigh 0.6–0.7 oz (17–21 g) at hatching and are almost naked. The crop of hoatzin chicks is sterile; the crop is inoculated during a chick's first two weeks of life through feeding by adults. At day 20 chicks are covered with

down and adult-like feathers begin to grow. Nestlings are brooded almost continuously for up to three weeks by all members of the social unit. If the nest is not disturbed, the young leave the nest at two to three weeks of age (3.9–5.6 oz; 110–160 g). However, four- to six-day-old chicks will plunge into the water when approached by a predator. At this age, they are capable of swimming underwater and can climb through the vegetation using their claws, wings, bill, and feet. The young will not return to the nest once disturbed. Young are brooded and fed off the nest for one or two months after nest departure. They are able to fly sometime between days 55 and 65 (12–16 oz; 350–450 g). Wing claws are shed at 70–100 days.

Hoatzins are seasonal breeders and reproductive activity coincides with the rainy season. In the Venezuelan Llanos about 47% of social units raise at least one young to independence per season. Nest failure is almost entirely attributable to predation, mainly by the wedge-capped Capuchin monkey (*Cebus olivaceus*).

Conservation status

Hoatzins are locally common but patchily distributed. They are not threatened or endangered but their use of restricted habitats and their specialized diet make them potentially vulnerable. Some watersheds that hoatzins inhabit are being converted to agricultural land and are heavily polluted with runoff pesticides and fertilizers from rice fields. The consequences of this threat are not yet known.

Significance to humans

Hoatzins are not regularly consumed for food because they have a strong musky odor that resembles the smell of cow manure. This strong smell is produced by volatile fatty acids in the crop. The local name given to hoatzins in Guyana (where the it is the national bird) is stinking pheasant, which aptly describes this feature of the bird. However, in Brazil, where hoatzins are called Cigana, locals consume hoatzin eggs and occasionally the birds themselves. Hoatzins have proven difficult to keep in captivity, but the Bronx Zoo in New York has been able to keep a group of birds in a public exhibit for more than ten years.

Resources

Books

del Hoyo, J., A. Elliot, and J. Sargatal, eds. *Handbook of the Birds of the World.* Vol. 3, *Hoatzin to Auks.* Barcelona: Lynx Edicions, 1996.

Periodicals

Blair Hedges, S., M.D. Simmons, M.A.M. van Dijk, G-J. Caspers, W.W. de Jong, and C.G. Sibley. "Phylogenetic Relationships of the Hoatzin, an Enigmatic South American Bird." *Proceedings of the National Academy of Science* 92 (1995): 11662–11665.

Domínguez-Bello, M.G., F. Michelangeli, M.C. Ruiz, A. Garcia, and E. Rodriguez. "Ecology of the Folivorous Hoatzin (*Opisthocomus hoazin*) on the Venezuelan Plains." *Auk* 111 (1994): 643–651.

Domínguez-Bello, M.G., M.C. Ruiz, and F. Michelangeli. "Evolutionary Significance of Foregut Fermentation in the Hoatzin (*Opisthocomus hoazin*, Aves: Opisthocomidae)." *Journal of Comparative Physiology B* 163 (1993): 594–601.

Grajal, A. "Structure and Function of the Digestive Tract of the Hoatzin *Opisthocomus hoazin*: a Folivorous Bird with Foregut Fermentation." *Auk* 97 (1995): 20–28.

Grajal, A., S.D. Strahl, R. Parra, M.G. Dominguez, and A. Neher. "Foregut Fermentation in the Hoatzin, a Neotropical Leaf-Eating Bird." *Science* 245 (1989): 1236–1238.

Hughes, J.M., and A.J. Baker. "Phylogenetic Relationships of the Enigmatic Hoatzin (*Opisthocomus hoazin*) Resolved using Mitochondrial and Nuclear Gene Sequences." *Molecular Biology and Evolution* 16 (1999): 1300–1307

Sibley, C.G., and J.E. Ahlquist. "The Relationships of the Hoatzin." *Auk* 90 (1973): 1–13.

Veron, G., and B.J. Winney. "Phylogenetic Relationships within the Turacos (Musophagidae)." *Ibis* 142 (2000): 446–456.

Carlos Bosque, PhD

For further reading

Ali, S. and Ripley, S. D. *Handbook of the Birds of India and Pakistan*. 2nd edition. 10 Vols. New York: Oxford University Press, 1978-1999.

American Ornithologists' Union. *Check-list of North American Birds: the Species of Birds of North America from the Arctic through Panama, including the West Indies and Hawaiian Islands*. 7th ed. Washington, DC: The American Ornithologists' Union, 1998.

Bennett, Peter M. and I. P. F. Owens. *Evolutionary Ecology of Birds: Life Histories, Mating Systems, and Extinction*. Oxford Series in Ecology and Evolution. Oxford: Oxford University Press, 2002.

Berthold, P. *Bird Migration: A General Survey*. Translated by H.-G. Bauer and V. Westhead. 2nd edition. Oxford Ornithology Series, no. 12. Oxford: Oxford University Press, 2001.

Boles, W. E. *The Robins and Flycatchers of Australia*. Sydney: Angus & Robertson, 1989.

Borrow, Nik, and Ron Demey. *Birds of Western Africa: An Identification Guide*. London: Christopher Helm, 2001.

Brewer, David, and B. K. Mackay. *Wrens, Dippers, and Thrashers*. New Haven: Yale University Press. 2001.

Brown, L. H., E. K. Urban and K. Newman, eds. *The Birds of Africa*. Vol. 1, *Ostriches to Falcons*. London: Academic Press, 1982.

Bruggers, R. L., and C. C. H. Elliott. *Quelea quelea: Africa's Bird Pest*. Oxford: Oxford University Press, 1990.

Burger, J., and Olla, B. I., eds. *Shorebirds: Breeding Behavior and Populations*. New York: Plenum Press, 1984.

Burton, J.A., Ed. *Owls of the World: Their Evolution, Structure and Ecology*. 2nd edition. London: Peter Lowe, 1992.

Byers, Clive, Jon Curson, and Urban Olsson. *Sparrows and Buntings: A Guide to the Sparrows and Buntings of North America and the World*. Boston: Houghton Mifflin Company, 1995.

Castro, I., and A. A. Phillips. *A Guide to the Birds of the Galápagos Islands*. Princeton: Princeton University Press, 1997.

Chantler, P., and G. Driessens. *Swifts: A Guide to the Swifts and Treeswifts of the World*. Sussex: Pica Press, 1995.

Cheke, R. A., and C. Mann. *Sunbirds: A Guide to the Sunbirds, Flowerpeckers, Spiderhunters and Sugarbirds of the World*. Christopher Helm, 2001.

Cleere, N., and D. Nurney. *Nightjars: A Guide to the Nightjars and Related Nightbirds*. Sussex: Pica Press, 1998.

Clement, P. *Finches and Sparrows: An Identification Guide*. Princeton: Princeton University Press, 1993.

Clement, P., et al. *Thrushes*. London: Christopher Helm, 2000.

Clements, J. F., and N. Shany. *A Field Guide to the Birds of Peru*. Temecula, California: Ibis Pub. 2001.

Coates, B. J. *The Birds of Papua New Guinea: Including the Bismarck Archipelago and Bougainville*. 2 vols. Alderley, Queensland: Dove Publications, 1985, 1990.

Coates, B. J., and K. D. Bishop. *A Guide to the Birds of Wallacea: Sulawesi, The Moluccas and Lesser Sunda Islands, Indonesia*. Alderley, Queensland: Dove Publications, 1997.

Cooke, Fred, Robert F. Rockwell, and David B. Lank. *The Snow Geese of La Pérouse Bay: Natural Selection in the Wild*. Oxford Ornithology Series, no. 4. Oxford: Oxford University Press, 1995.

Cooper, W. T., and J. M. Forshaw. *The Birds of Paradise and Bowerbirds*. Sydney: Collins, 1977.

Cramp, S., ed. *Handbook of the Birds of Europe the Middle East and North Africa. The Birds of the Western Palearctic*. Vol. 1, *Ostrich to Ducks*. Oxford: Oxford University Press, 1977.

Cramp, S., ed. *Handbook of the Birds of Europe the Middle East and North Africa. The Birds of the Western Palearctic*. Vol. 2, *Hawks to Bustards*. Oxford: Oxford University Press, .

Cramp, S., ed. *Handbook of the Birds of Europe the Middle East and North Africa. The Birds of the Western Palearctic*. Vol. 3, *Waders to Gulls*. Oxford: Oxford University Press, 1983.

Cramp, S., ed. *Handbook of the Birds of Europe the Middle East and North Africa. The Birds of the Western Palearctic*. Vol. 4, *Terns to Woodpeckers*. Oxford: Oxford University Press, 1985.

Cramp, S., ed. *Handbook of the Birds of Europe the Middle East and North Africa. The Birds of the Western Palearctic.* Vol. 6, *Warblers.* Oxford: Oxford University Press, 1992.

Cramp, S., ed. *Handbook of the Birds of Europe the Middle East and North Africa. The Birds of the Western Palearctic.* Vol. 7, *Flycatchers to Shrikes.* Oxford: Oxford University Press, 1993.

Cramp, S., ed. *Handbook of the Birds of Europe the Middle East and North Africa. The Birds of the Western Palearctic.* Vol. 8, *Crows to Finches.* Oxford: Oxford University Press, 1994.

Cramp, S., ed. *Handbook of the Birds of Europe the Middle East and North Africa. The Birds of the Western Palearctic.* Vol. 9, *Buntings and New World Warblers.* Oxford: Oxford University Press, 1994.

Davies, N. B. *Dunnock Behaviour and Social Evolution.* Oxford: Oxford University Press, 1992.

Davies, N. B. *Cuckoos, Cowbirds and Other Cheats.* London: T. & A. D. Poyser, 2000.

Davis, L. S., and J. Darby, eds. *Penguin Biology.* New York: Academic Press, 1990.

Deeming, D. C., ed. *Avian Incubation: Behaviour, Environment, and Evolution.* Oxford Ornithology Series, no. 13. Oxford: Oxford University Press, 2002.

del Hoyo, J., A. Elliott and J. Sargatal, eds. *Handbook of the Birds of the World.* Vol. 1, *Ostrich to Ducks.* Barcelona: Lynx Edicions, 1992.

del Hoyo, J., A. Elliott and J. Sargatal, eds. *Handbook of the Birds of the World.* Vol. 2, *New World Vultures to Guineafowl.* Barcelona: Lynx Edicions, 1994.

del Hoyo, J., A. Elliott and J. Sargatal, eds. *Handbook of the Birds of the World.* Vol. 3, *Hoatzin to Auks.* Barcelona: Lynx Edicions, 1996.

del Hoyo, J., A. Elliott and J. Sargatal, eds. *Handbook of the Birds of the World.* Vol. 4, *Sandgrouse to Cuckoos.* Barcelona: Lynx Edicions, 1997.

del Hoyo, J., A. Elliott and J. Sargatal, eds. *Handbook of the Birds of the World.* Vol. 5, *Barn-owls to Hummingbirds.* Barcelona: Lynx Edicions, 1999.

del Hoyo, J., A. Elliott and J. Sargatal, eds. *Handbook of the Birds of the World.* Vol. 6, *Mousebirds to Hornbills.* Barcelona: Lynx Edicions, 2001.

del Hoyo, J., A. Elliott and J. Sargatal, eds. *Handbook of the Birds of the World.* Vol. 7, *Jacamars to Woodpeckers.* Barcelona: Lynx Edicions, 2002.

Delacour, J., and D. Amadon. *Currasows and Related Birds.* New York: American Museum of Natural History, 1973.

Diamond, J., and A. B. Bond. *Kea, Bird of Paradox: The Evolution and Behavior of a New Zealand Parrot.* Berkeley: University of California Press, 1999.

Erritzoe, J. *Pittas of the World: a Monograph on the Pitta Family.* Cambridge: Lutterworth, 1998.

Feare, C., and A. Craig. *Starlings and Mynahs.* Princeton: Princeton University Press, 1999.

Feduccia, A. *The Origin and Evolution of Birds.* 2nd edition. New Haven: Yale University Press, 2001.

Ferguson-Lees, J., and D. A. Christie. *Raptors of the World.* Boston: Houghton Mifflin, 2001.

Fjeldsa, J., and N. Krabbe. *Birds of the High Andes.* Svendborg, Denmark: Apollo Books, 1990.

Forshaw, J. M. *Parrots of the World.* 3rd rev. edition. Melbourne: Lansdowne Editions, 1989.

Forshaw, J., ed. *Encyclopedia of Birds.* 2nd edition. McMahons Point, N.S.W.: Weldon Owen, 1998.

Frith, C. B., and B. M. Beehler. *The Birds of Paradise: Paradisaeidae.* Bird Families of the World, no. 6. Oxford University Press, 1998.

Fry, C. H., and K. Fry. *Kingfishers, Bee-eaters and Rollers: A Handbook.* Princeton: Princeton University Press, 1992.

Fry, C. H., S. Keith, and E. K. Urban, eds. *The Birds of Africa.* Vol. 3, *Parrots to Woodpeckers.* London: Academic Press, 1988.

Fry, C H., and S. Keith, eds. *The Birds of Africa.* Vol. 6. *Picathartes to Oxpeckers.* London: Academic Press, 2000.

Fuller, Errol. *Extinct Birds.* Rev. ed. Ithaca, N.Y.: Comstock Pub., 2001.

Gehlbach, Frederick R. *The Eastern Screech Owl: Life History, Ecology, and Behavior in the Suburbs and Countryside.* College Station: Texas A & M University Press, 1994.

Gibbs, D., E. Barnes, and J. Cox. *Pigeons and Doves: A Guide to the Pigeons and Doves of the World.* Robertsbridge: Pica Press, 2001.

Gill, F. B. *Ornithology.* 2nd edition. New York: W. H. Freeman, 1995.

Grant, P. R. *Ecology and Evolution of Darwin's Finches.* Princeton: Princeton University Press, 1986.

Hagemeijer, Ward J. M., and Michael J. Blair, eds. *The EBCC Atlas of European Breeding Birds: Their Distribution and Abundance.* London: T. & A. D. Poyser, 1997.

Hancock, J. A., J. A. Kushlan, and M. P. Kahl. *Storks, Ibises, and Spoonbills of the World.* New York: Academic Press, 1992.

Hancock, J. A., and J. A. Kushlan. *The Herons Handbook.* New York: Harper and Row, 1984.

Harris, Tony, and Kim Franklin. *Shrikes and Bush-shrikes: Including Wood-shrikes, Helmet-shrikes, Flycatcher-shrikes, Philentomas, Batises and Wattle-eyes.* London: Christopher Helm, 2000.

Harrap, S., and D. Quinn. *Chickadees, Tits, Nuthatches, and Treecreepers.* Princeton: Princeton University Press, 1995.

Heinrich, B. *Ravens in Winter.* New York: Summit Books, 1989.

Higgins, P. J., and S. J. J. F. Davies, eds. *The Handbook of Australian, New Zealand and Antarctic Birds.* Vol. 3, *Snipe to Pigeons.* Melbourne: Oxford University Press, 1996.

Higgins, P. J., ed. *The Handbook of Australian, New Zealand and Antarctic Birds.* Vol. 4, *Parrots to Dollarbirds.* Melbourne: Oxford University Press, 1999.

Hilty, S. L., and W. L. Brown. *A Guide to the Birds of Colombia.* Princeton, N. J.: Princeton University Press, 1986.

Holyoak, D.T. *Nightjars and Their Allies: The Caprimulgiformes. Bird Families of the World*, no. 7. Oxford: Oxford University Press, 2001.

Howard, R., and A. Moore. *A Complete Checklist of the Birds of the World.* 2nd edition. London: Macmillan, 1991.

Howell, S. N. G., and S. Webb. *A Guide to the Birds of Mexico and Northern Central America.* Oxford: Oxford University Press, 1995.

Howell, S. N. G. *Hummingbirds of North America: The Photographic Guide.* San Diego: Academic Press, 2002.

Isler, M. L., and P. R. Isler. *The Tanagers: Natural History, Distribution, and Identification.* Washington, D.C.: Smithsonian Institution Press, 1987.

Jaramillo, A., and P. Burke. *New World Blackbirds: The Icterids.* Princeton: Princeton University Press, 1999.

Jehl, Joseph R., Jr. *Biology of the Eared Grebe and Wilson's Phalarope in the Nonbreeding Season: A study of Adaptations to Saline Lakes. Studies in Avian Biology*, no. 12. San Diego: Cooper Ornithological Society, 1988.

Johnsgard, Paul A. *Bustards, Hemipodes, and Sandgrouses, Birds of Dry Places.* Oxford: Oxford University Press, 1991.

Johnsgard, Paul A. *Cormorants, Darters, and Pelicans of the World.* Washington, D.C.: Smithsonian Institution Press, 1993.

Johnsgard, Paul A. *Cranes of the World.* Bloomington: Indiana University Press, 1983.

Johnsgard, Paul A. *Diving Birds of North America.* Lincoln: University of Nebraska Press, 1987.

Johnsgard, Paul A. *The Hummingbirds of North America.* 2nd ed. Washington, D.C.: Smithsonian Institution Press, 1997.

Johnsgard, P. A. *The Plovers, Sandpipers, and Snipes of the World.* Lincoln: University of Nebraska Press, 1981.

Johnsgard, Paul A. *Trogons and Quetzals of the World.* Washington, D.C.: Smithsonian Institution Press, 2000.

Johnsgard, Paul A., and Montserrat Carbonell. *Ruddy Ducks and other Stifftails: Their Behavior and Biology.* Norman: University of Oklahoma Press, 1996.

Jones, D. N., R. W. R. J. Dekker, and C. S. Roselaar. *The Megapodes. Bird Families of the World*, no. 3. Oxford: Oxford University Press, 1995.

Juniper, Tony, and Mike Parr. *Parrots: A Guide to the Parrots of the World.* Sussex: Pica Press, 1998.

Kear, J., and N. Düplaix-Hall, eds. *Flamingos.* Berkhamsted: T. & A. D. Poyser, 1975.

Keith, S., E. K. Urban, and C. H. Fry, eds. *The Birds of Africa.* Vol. 4, *Broadbills to Chats.* London: Academic Press, 1992.

Kemp, A. *The Hornbills. Bird Families of the World*, no. 1. Oxford: Oxford University Press, 1995.

Kennedy, Robert S., et al. *A Guide to the Birds of the Philippines.* Oxford: Oxford University Press, 2000.

Lambert, F., and M. Woodcock. *Pittas, Broadbills and Asities.* Sussex: Pica Press, 1996.

Lefranc, Norbert, and Tim Worfolk. *Shrikes: A Guide to the Shrikes of the World.* Robertsbridge: Pica Press, 1997.

Lenz, Norbert. *Evolutionary Ecology of the Regent Bowerbird Sericulus chrysocephalus.* Special Issue of *Ecology of Birds*, Vol. 22, Supplement. Ludwigsburg, 1999.

MacKinnon, John R., and Karen Phillipps. *A Field Guide to the Birds of China.* Oxford Ornithology Series. Oxford; New York: Oxford University Press, 2000.

Madge, S. *Crows and Jays: A Guide to the Crows, Jays, and Magpies of the World.* New York: Houghton Mifflin, 1994.

Madge, Steve, and Phil McGowan. *Pheasants, Partridges and Grouse: A Guide to the Pheasants, Partridges, Quails, Grouse, Guineafowl, Buttonquails and Sandgrouse of the World.* London: Christopher Helm, 2002.

Marchant, S., and P. Higgins, eds. *The Handbook of Australian, New Zealand and Antarctic Birds.* Vol. 1, parts A , B, *Ratites to Ducks.* Melbourne: Oxford University Press, 1990.

Marchant, S., and P. Higgins, eds. *The Handbook of Australian, New Zealand and Antarctic Birds.* Vol. 2, *Raptors to Lapwings.* Melbourne: Oxford University Press, 1993

Higgins, P. J., J. M. Peter, and W. K. Steele. *The Handbook of Australian, New Zealand and Antarctic Birds.* Vol. 5, *Tyrant-Flycatcher to Chats.* Melbourne: Oxford University Press, 2001.

Marzluff, J. M., R. Bowman, and R. Donnelly, eds. *Avian Ecology and Conservation in an Urbanizing World.* Boston: Kluwer Academic Publishers, 2001.

Matthysen, E. The *Nuthatches.* London: T. A. & D. Poyser, 1998.

Mayfield, H. *The Kirtland's Warbler.* Bloomfield Hills, Michigan: Cranbrook Institute of Science, 1960.

Mayr, E. *The Birds of Northern Melanesia: Speciation, Ecology, and Biogeography.* Oxford: Oxford University Press, 2001.

McCabe, Robert A. *The Little Green Bird: Ecology of the Willow Flycatcher.* Madison, Wis.: Rusty Rock Press, 1991.

Mindell, D. P., ed. *Avian Molecular Evolution and Systematics.* New York: Academic Press, 1997.

Morse, D. H. *American Warblers, An Ecological and Behavioral Perspective.* Cambridge: Harvard University Press, 1989.

Mundy, P., D., et al. *The Vultures of Africa.* London: Academic Press, 1992.

Nelson, J. B. *The Sulidae: Gannets and Boobies.* Oxford: Oxford University Press, 1978.

Nelson, Bryan. *The Atlantic Gannet.* 2nd ed. Great Yarmouth: Fenix, 2002.

Olsen, Klaus Malling, and Hans Larsson. *Skuas and Jaegers: A Guide to the Skuas and Jaegers of the World.* New Haven: Yale University Press, 1997.

Olsen, Klaus Malling, and Hans Larsson. *Gulls of Europe, Asia and North America.* London: Christopher Helm, 2001.

Ortega, Catherine P. *Cowbirds and Other Brood Parasites.* Tucson: University of Arizona Press, 1998.

Padian, K., ed. *The Origin of Birds and the Evolution of Flight.* Memoirs of the California Academy of Sciences, no. 8. San Francisco: California Academy of Sciences, 1986.

Paul, Gregory S. *Dinosaurs of the Air: the Evolution and Loss of Flight in Dinosaurs and Birds.* Baltimore: Johns Hopkins University Press, 2001.

Poole, Alan F., P. Stettenheim, and Frank B. Gill, eds. *The Birds of North America.* 15 vols. to date. Philadelphia: The Academy of Natural Sciences; Washington, D.C.: American Ornithologists' Union, 1992-.

Raffaele, H., et al. *Birds of the West Indies.* London: Christopher Helm, A. & C. Black, 1998.

Ratcliffe, D. *The Peregrine Falcon.* 2nd ed. London: T. & A. D. Poyser, 1993.

Restall, R. *Munias and Mannikins.* Sussex: Pica Press, 1996.

Ridgely, R. S., and G. A. Gwynne, Jr. *A Guide to the Birds of Panama: with Costa Rica, Nicaragua, and Honduras.* 2nd ed. Princeton: Princeton University Press. 1989.

Ridgely, R. S., and G. Tudor. *The Birds of South America.* Vol. 1, *The Oscine Passerines.* Austin: University of Texas Press, 1989.

Ridgely, R. S., and G. Tudor. *The Birds of South America.* Vol. 2, *The Suboscine Passerines.* Austin: University of Texas Press, 1994.

Ridgely, Robert S., and Paul J. Greenfield. *The Birds of Ecuador.* Vol. 1, *Status, Distribution, and Taxonomy.* Ithaca, NY: Comstock Pub., 2001.

Ridgely, Robert S., and Paul J. Greenfield. *The Birds of Ecuador.* Vol. 2, *Field Guide.* Ithaca, NY: Comstock Pub., 2001.

Rising, J. D. *Guide to the Identification and Natural History of the Sparrows of the United States and Canada.* New York: Academic Press, 1996.

Rowley, I., and E. Russell. *Fairy-Wrens and Grasswrens. Bird Families of the World,* no. 4. Oxford: Oxford University Press, 1997.

Scott, J. M., S. Conant, and C. van Riper, III, eds. *Evolution, Ecology, Conservation, and Management of Hawaiian Birds: A Vanishing Avifauna. Studies in Avian Biology,* no. 22. Camarillo, CA : Cooper Ornithological Society, 2001.

Searcy, W. A., and K. Yasukawa. *Polygyny and Sexual Selection in Red-winged Blackbirds. Monographs in Behavior and Ecology.* Princeton: Princeton University Press, 1995.

Shirihai, H., G. Gargallo, and A. J. Helbig. *Sylvia Warblers.* Edited by G. M. Kirwan and L. Svensson. Princeton: Princeton University Press, 2001.

Short, L. L. *Woodpeckers of the World.* Greenville, Del.: Delaware Natural History Museum, 1984.

Short, L. L., and J. F.M. Horne. *Toucans, Barbets and Honeyguides: Ramphastidae, Capitonidae, and Indicatoridae. Bird Families of the World,* no. 8. Oxford: New York: Oxford University Press, 2001.

Sibley, C. G., and J. E. Ahlquist. *Phylogeny and Classification of Birds: A Study in Molecular Evolution.* New Haven: Yale University Press, 1990.

Sibley, C. G., and B. L. Monroe, Jr. *Distribution and Taxonomy of Birds of the World.* New Haven: Yale University Press, 1990.

Sibley, C. G., and B. L. Monroe, Jr. *A Supplement to Distribution and Taxonomy of Birds of the World.* New Haven: Yale University Press, 1993.

Sibley, David Allen. *The Sibley Guide to Birds.* New York: Knopf, 2000.

Sick, H. *Birds in Brazil, a Natural History.* Princeton University Press, 1993.

Simmons, Robert Edward. *Harriers of the World: Their Behaviour and Ecology. Oxford Ornithology Series,* no. 11. Oxford: Oxford University Press, 2000.

Sinclair, I., and O. Langrand. *Birds of the Indian Ocean Islands: Madagascar, Mauritius, Reunion, Rodrigues, Seychelles and the Comoros.* New Holland. 1998.

Skutch, A. F. *Antbirds and Ovenbirds: Their Lives and Homes.* Austin: University of Texas Press, 1996.

Skutch, A. F. *Orioles, Blackbirds, and Their Kin: A Natural History.* Tucson: University of Arizona Press, 1996.

Smith, S. M. *The Black-capped Chickadee: Behavioral Ecology and Natural History.* Ithaca: Cornell University Press, 1991.

Snow, D. W. *The Cotingas, Bellbirds, Umbrella-birds and Their Allies in Tropical America.* London: Brit. Museum (Nat. Hist.), 1982.

Snyder, Noel F. R., and Helen Snyder. *The California Condor: A Saga of Natural History and Conservation.* San Diego, Calif.: Academic Press, 2000.

Stacey, P. B., and W. D. Koenig, eds. *Cooperative Breeding in Birds.* Cambridge: Cambridge University Press, 1989.

Stattersfield, Alison J., et al. eds. *Threatened Birds of the World: The Official Source for Birds on the IUCN Red List.* Cambridge: BirdLife International, 2000.

Stiles, F. G., and A. F. Skutch. *A Guide to the Birds of Costa Rica.* Ithaca: Comstock, 1989.

Stokes, Donald W., and Lillian Q. Stokes. *A Guide to Bird Behavior.* 3 vols. Boston: Little, Brown, 1979-1989.

Stolz, D.F., J. W. Fitzpatrick, T. A. Parker III, and D. K. Moskovits. *Neotropical Birds, Ecology and Conservation.* Chicago: University of Chicago Press, 1996.

Summers-Smith, J. D. *The Sparrows: A Study of the Genus Passer.* Calton: T. & A. D. Poyser, 1988.

Taylor, B., and B. van Perlo. *Rails: A Guide to the Rails, Crakes, Gallinules and Coots of the World.* Sussex: Pica Press, 1998.

Terres, John K. *The Audubon Society Encyclopedia of North American Birds.* New York: Knopf, 1980.

Tickell, W. L. N. *Albatrosses.* Sussex: Pica Press, 2000.

Todd, F. S. *Natural History of the Waterfowl.* Ibis Publishing Co., San Diego Natural History Museum, 1996.

Turner, A., and C. Rose. *A Handbook to the Swallows and Martins of the World.* London: Christopher Helm, 1989.

Tyler, Stephanie J., and Stephen J. *The Dippers.* London: T. & A. D. Poyser, 1994.

Urban, E. K., C. H. Fry, and S. Keith, eds. *The Birds of Africa.* Vol. 2, *Gamebirds to Pigeons.* London: Academic Press, 1986.

Urban, E. K., C. H. Fry, and S. Keith, eds. *The Birds of Africa.* Vol. 5, *Thrushes to Puffback Flycatchers.* London: Academic Press, 1997.

van Rhijn, J. G. *The Ruff: Individuality in a Gregarious Wading Bird.* London: T. & A. D. Poyser, 1991.

Voous, Karel H. *Owls of the Northern Hemisphere.* Cambridge, Mass.: MIT Press, 1988.

Warham, J. *Behaviour and Population Ecology of the Petrels.* New York: Academic Press,1996.

Williams, T. D. *The Penguins. Bird Families of the World,* no. 2. Oxford: Oxford University Press, 1995.

Winkler, H., D. A. Christie, and D. Nurney. *Woodpeckers: A Guide to the Woodpeckers, Piculets and Wrynecks of the World.* Sussex: Pica Press, 1995.

Woolfenden, G. E., and J. W. Fitzpatrick. *The Florida Scrub Jay: Demography of a Cooperative-breeding Bird. Monographs in Population Biology,* no. 20. Princeton: Princeton University Press, 1984.

Zimmerman, D. A., D. A. Turner, and D. J. Pearson. *Birds of Kenya and Northern Tanzania.* London: Christopher Helm, 1996.

Compiled by Janet Hinshaw, Bird Division Collection Manager, University of Michigan Museum of Zoology

Organizations

African Bird Club
Wellbrook Court, Girton Road
Cambridge, Cambridgeshire CB3 0NA
United Kingdom
Phone: +44 1 223 277 318
Fax: +44-1-223-277-200
<http://www.africanbirdclub.org>

African Gamebird Research, Education and Development
(AGRED)
P.O. Box 1191
Hilton, KwaZulu-Natal 3245
South Africa
Phone: +27-33-343-3784

African-Eurasian Migratory Waterbird Agreement (AEWA)
UN Premises in Bonn, Martin Luther-King St.
Bonn D-53175 Germany
<http://www.wcmc.org.uk/AEWA>

American Ornithologists' Union
Suite 402, 1313 Dolley Madison Blvd
McLean, VA 22101
USA
<http://www.aou.org>

American Zoo and Aquarium Association
8403 Colesville Road
Suite 710
Silver Spring, Maryland 20910
<http://www.aza.org>

Association for BioDiversity Information
1101 Wilson Blvd., 15th Floor
Arlington, VA 22209
USA
<http://www.infonatura.org/>

Association for Parrot Conservation
Centro de Calidad Ambiental ITESM Sucursal de Correos
J., C.P. 64849
Monterrey, N.L.
Mexico

Australasian Raptor Association
415 Riversdale Road
Hawthorn East, Victoria 3123
Australia
Phone: +61 3 9882 2622
Fax: +61 3 9882 2677
<http://www.tasweb.com.au/ara/index.htm>

Australian National Wildlife Collection
GPO Box 284
Canberra, ACT 2601
Australia
Phone: +61 2 6242 1600
Fax: +61-2-6242-1688

The Bird Conservation Society of Thailand
69/12 Rarm Intra 24
Jarakhebua Lat Phrao, Bangkok 10230
Thailand
Phone: 943-5965
<http://www.geocities.com/TheTropics/Harbor/7503/
ruang_nok/princess_bird.html>

BirdLife International
Wellbrook Court, Girton Road
Cambridge, Cambridgeshire CB3 0NA
United Kingdom
Phone: +44 1 223 277 318
Fax: +44-1-223-277-200
<http://www.birdlife.net>

BirdLife International Indonesia Programme
P. O. Box 310/Boo
Bogor
Indonesia
Phone: +62 251 357222
Fax: +62 251 357961
<http://www.birdlife-indonesia.org>

BirdLife International, Panamerican Office
Casilla 17-17-717
Quito
Ecuador
Phone: +593 2 244 3261
Fax: +593 2 244 3261
<http://www.latinsynergy.org/birdlife.html>

BirdLife South Africa
P. O. Box 515
Randburg 2125
South Africa
Phone: +27-11-7895188
<http://www.birdlife.org.za>

Birds Australia
415 Riversdale Road
Hawthorn East, Victoria 3123
Australia
Phone: +61 3 9882 2622
Fax: +61 3 9882 2677
<http://www.birdsaustralia.com.au>

Birds Australia Parrot Association, Birds Australia
415 Riversdale Road
Hawthorn East, Victoria 3123
Australia
Phone: +61 3 9882 2622
Fax: +61 3 9882 2677
<http://www.birdsaustralia.com.au>

The Bishop Museum
1525 Bernice Street
Honolulu, HI 96817-0916
Phone: (808) 847-3511
<http://www.bishopmuseum.org>

British Trust for Ornithology
The Nunnery
Thetford, Norfolk IP24 2PU
United Kingdom
Phone: +44 (0) 1842 750050
Fax: +44 (0) 1842 750030
<http://www.bto.org>

Center for Biological Diversity
P.O. Box 710
Tucson, AZ 85702-0701
USA
Phone: (520) 623-5252
Fax: (520) 623-9797
<http://www.biologicaldiversity.org/swcbd/index.html>

Coraciiformes Taxon Advisory Group
<http://www.coraciiformestag.com>

The Cracid Specialist Group
PO Box 132038
Houston, TX 77219-2038
USA
Phone: (713) 639-4776
<http://www.angelfire.com/ca6/cracid>

Department of Ecology and Environmental Biology, Cornell University
E145 Corson Hall
Ithaca, NY 14853-2701
USA
Phone: (607) 254-4201
<http://www.es.cornell.edu/winkler/botw/fringillidae.html>

Department of Ecology and Evolutionary Biology, Tulane University
310 Dinwiddie Hall
New Orleans, LA 70118-5698
USA
Phone: (504) 865-5191
<http://www.tulane.edu/
eeob/Courses/Heins/Evolution/lecture17.html>

Department of Zoology, University of Toronto
25 Harbord Street
Toronto, Ontario M5S 3G5
Canada
Phone: (416) 978-3482
Fax: (416) 978-8532
<http://www.zoo.utoronto.ca>

Ducks Unlimited, Inc
One Waterfowl Way
Memphis, TN 38120
USA

Phone: (800) 453-8257
Fax: (901) 758-3850
<http://www.ducks.org>

Emu Farmers Federation of Australia
P.O Box 57
Wagin, Western Australia 6315
Australia
Phone: +61 8 9861 1136

Game Conservancy Trust
Fordingbridge, Hampshire SP6 1EF
United Kingdom
Phone: +44 1425 652381
Fax: +44 1425 651026
<http://www.gct.org.uk>

Gamebird Research Programme, Percy FitzPatrick Institute, University of Cape Town
Private Bag
Rondebosch, Western Cape 7701
South Africa
Phone: +27 21 6503290
Fax: +27 21 6503295
<http://www.uct.ac.za/depts/fitzpatrick>

Haribon Foundation for the Conservation of Natural Resources
9A Malingap Cot, Malumanay Streets, Teachers' Village, 1101 Diliman
Quezon City
Philippines
Phone: +63 2 9253332
<http://www.haribon.org.ph>

The Hawk and Owl Trust
11 St Marys Close
Newton Abbot, Abbotskerswell, Devon TQ12 5QF
United Kingdom
Phone: +44 (0)1626 334864
Fax: +44 (0)1626 334864
<http://www.hawkandowltrust.org>

Herons Specialist Group
Station Biologique de la Tour du Valat
Le Sambuc, Arles 13200
France
Phone: +33-4-90-97-20-13
Fax: 33-4-90-97-29-19
<http://www.tour-du-valat.com>

Hornbill Research Foundation
c/o Department of Microbiology, Faculty of Science, Mahidol University, Rama 6 Rd
Bangkok 10400
Thailand
Phone: +66 22 460 063, ext. 4006

International Crane Foundation
P.O. Box 447
Baraboo, WI 53913-0447
USA
Phone: (608) 356-9462
Fax: (608) 356-9465
<http://www.savingcranes.org>

International Shrike Working Group
"Het Speihuis," Speistraat, 17
Sint-Lievens-Esse (Herzele), B-9550

Belgium
Phone: +32 54 503 789

International Species Inventory System
<http://www.isis.org>

International Touraco Society
Brackenhurst, Grange Wood
Netherseal, Nr Swadlincote, Derbyshire DE12 8BE
United Kingdom
Phone: +44 (0)1283 760541

International Waterbird Census
<http://www.wetlands.org>

IUCN Species Survival Commission
219c Huntingdon Road
Cambridge, Cambridgeshire CB3 0DL
United Kingdom
<http://www.iucn.org/themes/ssc>

IUCN–The World Conservation Union
Rue Mauverney 28
Gland 1196
Switzerland
Phone: +41-22-999-0001
Fax: +41-22-999-0025
<http://www.iucn.org>

IUCN–World Conservation Union, USA Multilateral Office
1630 Connecticut Avenue
Washington, DC 20009
USA
Phone: (202) 387-4826
<http://www.iucn.org/places/usa/inter.html>

IUCN/SSC Grebes Specialist Group
Copenhagen DK 2100
Denmark
Phone: +45 3 532 1323
Fax: +45-35321010
<http://www.iucn.org>

Japanese Association for Wild Geese Protection
Minamimachi 16
Wakayangi 989-5502
Japan
Phone: +81 228 32 2004
Fax: +81 228 32 2004
<http://www.japwgp.org>

Ligue pour la Protection des Oiseaux
La Corderie Royale, B.P. 263
17305 Rochefort cedex
France
Phone: +33 546 821 234
Fax: +33 546 839 586
<http://www.lpo-birdlife.asso.fr>

Loro Parque Fundación
Loro Parque S.A. 38400 Puerto de la Cruz
Tenerife, Canary Islands
Spain

National Audubon Society
700 Broadway
New York, NY 10003
USA

Phone: (212) 979-3000
Fax: (212) 978-3188
<http://www.Audubon.org>

National Audubon Society Population & Habitat Program
1901 Pennsylvania Ave. NW, Suite 1100
Washington, DC 20006
USA
Phone: (202) 861-2242
<http://www.audubonpopulation.org>

Neotropical Bird Club
c/o The Lodge
Sandy, Bedfordshire SG19 2DL
United Kingdom

Oriental Bird Club, American Office
4 Vestal Street
Nantucket, MA 02554
USA
Phone: (508) 228-1782
<http://www.orientalbirdclub.org>

Ornithological Society of New Zealand
P.O. Box 12397
Wellington, North Island
New Zealand
<http://osnz.org.nz>

Pacific Island Ecosystems Research Center
3190 Maile Way, St. John Hall, Room 408
Honolulu, HI 96822
USA
Phone: (808) 956-5691
Fax: (808) 956-5687
<http://biology.usgs.gov/pierc/piercwebsite.htm>

ProAves Peru
P.O. Box 07
Piura
Peru

Raptor Conservation Group, Endangered Wildlife Trust
Private Bag X11
Parkview, Gauteng 2122
South Africa
Phone: +27-11-486-1102
Fax: +27-11-486-1506
<http://www.ewt.org.za>

Raptor Research Foundation
P.O. Box 1897, 810 E. 10th Street
Lawrence, KS 66044-88973
USA
<http://biology.biosestate.edu/raptor>

Research Centre for African Parrot Conservation Zoology and Entomology Department
Private Bag X01
Scottsville 3201
Natal Republic of South Africa

Roberts VII Project, Percy FitzPatrick Institute of African Ornithology, University of Cape Town
Rondebosch 7701
South Africa
Fax: (021) 650 3295
<http://www.uct.ac.za/depts/fitzpatrick/docs/r549.html>

Royal Society for the Protection of Birds
Admail 975 Freepost ANG 6335, The Lodge
Sandy, Bedfordshire SG19 2TN
United Kingdom
<http://www.rspb.org.uk>

Ruffed Grouse Society
451 McCormick Rd
Coraopolis, PA 15108
Phone: (888) 564-6747
Fax: (412) 262-9207
<http://www.ruffedgrousesociety.org>

Smithsonian Migratory Bird Center, Smithsonian National
Zoological Park
3001 Connecticut Avenue, NW
Washington, DC 20008
USA
Phone: (202) 673-4800
<http://www.natzoo.si.edu>

The Songbird Foundation
2367 Eastlake Ave. East
Seattle, WA 98102
USA
Phone: (206) 374-3674
Fax: (206) 374-3674
<http://www.songbird.org>

University of Michigan
3019 Museum of Zoology, 1109 Geddes Ave
Ann Arbor, MI 48109-1079
USA
Phone: (734) 647-2208
Fax: (734) 763-4080
<http://www.ummz.lsa.umich.edu/birds/index.html>

Wader Specialist Group, Mr. David Stroud
Monkstone House, City Road
Peterborough PE1 1JY
United Kingdom
Phone: +44 1733 866/810
Fax: +44 1733 555/448

Wader Study Group, The National Centre for Ornithology
The Nunnery
Thetford, Norfolk JP24 2PU
United Kingdom

Waterbird Society
National Museum of Natural History, Smithsonian
Institution
Washington, DC 20560
USA
<http://www.nmnh.si.edu/BIRDNET/cws>

Western Hemisphere Shorebird Reserve Network (WHSRN)
Manomet Center for Conservation Science, P O Box 1770
Manomet, MA 02345
USA
Phone: (508) 224-6521
Fax: (508) 224-9220
<http://www.manomet.org/WHSRN/index.html>

Wetlands International
Droevendaalsesteeg 3A

Wageningen 6700 CA
The Netherlands
Phone: +31 317 478884
Fax: +31 317 478885
<http://www.wetlands.agro.nl>

Wetlands International (the Americas)
7 Hinton Avenue North, Suite 200
Ottawa, Ontario K1Y 4P1
Canada
Phone: (613) 722-2090
<http://www.wetlands.org>

Wetlands International/Survival Service Commission Flamingo
Specialist Group
c/o Station Biologique de la Tour du Valat
Le Sambuc, Arles 13200
France

Wildfowl and Wetlands Trust
Slimbridge, Glos GL2 7BT
United Kingdom
Phone: +44 01453 891900

Woodcock and Snipe Specialist Group
Director, European Wildlife Research Institute
Bonndorf, Glashuette D-79848
Germany
Phone: 949 7653 1891
Fax: 949 7653 9269

Woodhoopoe Research Project, FitzPatrick Institute of African
Ornithology, University of Cape Town
P.O. Rondebosch
Cape Town, Western Cape 7700
South Africa
Phone: +27 (0)21 650-3290
Fax: +27-21-650-3295
<http://www.fitztitute.uct.ac.za>

Working Group on Birds in Madagascar and the Indian Ocean
Islands. World Wide Fund for Nature
Antananarivo 101 BP 738
Madagascar
Phone: +261 3207 80806

Working Group on International Wader and Waterfowl
Research (WIWO)
Stichting WIWO, c/o P O Box 925
Zeist 3700 AX
The Netherlands
<http://www.wiwo-international.org>

World Center for Birds of Prey, The Peregrine Fund
566 West Flying Hawk Lane
Boise, ID 83709
USA
Phone: (208) 362-3716
Fax: (208) 362-2376
<http://www.peregrinefund.org>

World Parrot Trust
Glanmor House
Hayle, Cornwall TR27 4HB
United Kingdom
<http://www.worldparrottrust.org>

Organizations

World Pheasant Association
    P. O. Box 5, Lower Basildon St
    Reading RG8 9PF
    United Kingdom
    Phone: +44 1 189 845 140
    Fax: +44 118 984 3369
    <http://www.pheasant.org.uk>

World Working Group on Birds of Prey and Owl
    P.O. Box 52
    Towcester NN12 7ZW
    United Kingdom
    Phone: +44 1 604 862 331
    Fax: +44 1 604 862 331
    <http://www.Raptors-International.de>

WPA/BirdLife/SSC Megapode Specialist Group
    c/o Department of Ornithology, National Museum of
    Natural History, P.O. Box 9517
    Leiden 2300 RA
    The Netherlands

WPA/BirdLife/SSC Partridge, Quail, and Francolin Specialist
Group
    c/o World Pheasant Association, PO Box 5
    Lower Basildon, Reading RG8 9PF
    United Kingdom
    Phone: +44 1 189 845 140
    Fax: +118 9843369
    <http:/www.pheasant.org.uk>
    PQF: <http://www.gct.org.uk/pqf>

• • • • •

Contributors to the first edition

The following individuals contributed chapters to the original edition of Grzimek's Animal Life Encyclopedia, *which was edited by Dr. Bernhard Grzimek, Professor, Justus Liebig University of Giessen, Germany; Director, Frankfurt Zoological Garden, Germany; and Trustee, Tanzanian National Parks, Tanzania.*

Dr. Michael Abs
Curator, Ruhr University
Bochum, Germany

Dr. Salim Ali
Bombay Natural History Society
Bombay, India

Dr. Rudolph Altevogt
Professor, Zoological Institute,
University of Münster
Münster, Germany

Dr. Renate Angermann
Curator, Institute of Zoology,
Humboldt University
Berlin, Germany

Edward A. Armstrong
Cambridge University
Cambridge, England

Dr. Peter Ax
Professor, Second Zoological Institute
and Museum, University of Göttingen
Göttingen, Germany

Dr. Franz Bachmaier
Zoological Collection of the State of
Bavaria
Munich, Germany

Dr. Pedru Banarescu
Academy of the Roumanian Socialist
Republic, Trajan Savulescu Institute of
Biology
Bucharest, Romania

Dr. A. G. Bannikow
Professor, Institute of Veterinary
Medicine
Moscow, Russia

Dr. Hilde Baumgärtner
Zoological Collection of the State of
Bavaria
Munich, Germany

C. W. Benson
Department of Zoology, Cambridge
University
Cambridge, England

Dr. Andrew Berger
Chairman, Department of Zoology,
University of Hawaii
Honolulu, Hawaii, U.S.A.

Dr. J. Berlioz
National Museum of Natural History
Paris, France

Dr. Rudolf Berndt
Director, Institute for Population
Ecology, Hiligoland Ornithological
Station
Braunschweig, Germany

Dieter Blume
Instructor of Biology, Freiherr-vom-
Stein School
Gladenbach, Germany

Dr. Maximilian Boecker
Zoological Research Institute and A.
Koenig Museum
Bonn, Germany

Dr. Carl-Heinz Brandes
Curator and Director, The Aquarium,
Overseas Museum
Bremen, Germany

Dr. Donald G. Broadley
Curator, Umtali Museum
Mutare, Zimbabwe

Dr. Heinz Brüll
Director; Game, Forest, and Fields
Research Station
Hartenholm, Germany

Dr. Herbert Bruns
Director, Institute of Zoology and the
Protection of Life
Schlangenbad, Germany

Hans Bub
Heligoland Ornithological Station
Wilhelmshaven, Germany

A. H. Chrisholm
Sydney, Australia

Herbert Thomas Condon
Curator of Birds, South Australian
Museum
Adelaide, Australia

Dr. Eberhard Curio
Director, Laboratory of Ethology,
Ruhr University
Bochum, Germany

Dr. Serge Daan
Laboratory of Animal Physiology,
University of Amsterdam
Amsterdam, The Netherlands

Dr. Heinrich Dathe
Professor and Director, Animal Park
and Zoological Research Station,
German Academy of Sciences
Berlin, Germany

Dr. Wolfgang Dierl
Zoological Collection of the State of
Bavaria
Munich, Germany

Dr. Fritz Dieterlen
Zoological Research Institute, A.
Koenig Museum
Bonn, Germany

Dr. Rolf Dircksen
Professor, Pedagogical Institute
Bielefeld, Germany

Josef Donner
Instructor of Biology
Katzelsdorf, Austria

Dr. Jean Dorst
Professor, National Museum of
Natural History
Paris, France

Dr. Gerti Dücker
Professor and Chief Curator,
Zoological Institute, University of
Münster
Münster, Germany

Dr. Michael Dzwillo
Zoological Institute and Museum,
University of Hamburg
Hamburg, Germany

Dr. Irenäus Eibl-Eibesfeldt
Professor and Director, Institute of
Human Ethology, Max Planck
Institute for Behavioral Physiology
Percha/Starnberg, Germany

Dr. Martin Eisentraut
Professor and Director, Zoological
Research Institute and A. Koenig
Museum
Bonn, Germany

Dr. Eberhard Ernst
Swiss Tropical Institute
Basel, Switzerland

R. D. Etchecopar
Director, National Museum of
Natural History
Paris, France

Dr. R. A. Falla
Director, Dominion Museum
Wellington, New Zealand

Dr. Hubert Fechter
Curator, Lower Animals, Zoological
Collection of the State of Bavaria
Munich, Germany

Dr. Walter Fiedler
Docent, University of Vienna, and
Director, Schönbrunn Zoo
Vienna, Austria

Wolfgang Fischer
Inspector of Animals, Animal Park
Berlin, Germany

Dr. C. A. Fleming
Geological Survey Department of
Scientific and Industrial Research
Lower Hutt, New Zealand

Dr. Hans Frädrich
Zoological Garden
Berlin, Germany

Dr. Hans-Albrecht Freye
Professor and Director, Biological
Institute of the Medical School
Halle a.d.S., Germany

Günther E. Freytag
Former Director, Reptile and
Amphibian Collection, Museum of
Cultural History in Magdeburg
Berlin, Germany

Dr. Herbert Friedmann
Director, Los Angeles County
Museum of Natural History
Los Angeles, California, U.S.A.

Dr. H. Friedrich
Professor, Overseas Museum
Bremen, Germany

Dr. Jan Frijlink
Zoological Laboratory, University of
Amsterdam
Amsterdam, The Netherlands

Dr. H.C. Karl Von Frisch
Professor Emeritus and former
Director, Zoological Institute,
University of Munich
Munich, Germany

Dr. H. J. Frith
C.S.I.R.O. Research Institute
Canberra, Australia

Dr. Ion E. Fuhn
Academy of the Roumanian Socialist
Republic, Trajan Savulescu Institute of
Biology
Bucharest, Romania

Dr. Carl Gans
Professor, Department of Biology,
State University of New York at
Buffalo
Buffalo, New York, U.S.A.

Dr. Rudolf Geigy
Professor and Director, Swiss Tropical
Institute
Basel, Switzerland

Dr. Jacques Gery
St. Genies, France

Dr. Wolfgang Gewalt
Director, Animal Park
Duisburg, Germany

Dr. H.C. Dr. H.C. Viktor Goerttler
Professor Emeritus, University of Jena
Jena, Germany

Dr. Friedrich Goethe
Director, Institute of Ornithology,
Heligoland Ornithological Station
Wilhelmshaven, Germany

Dr. Ulrich F. Gruber
Herpetological Section, Zoological
Research Institute and A. Koenig
Museum
Bonn, Germany

Dr. H. R. Haefelfinger
Museum of Natural History
Basel, Switzerland

Dr. Theodor Haltenorth
Director, Mammalology, Zoological
Collection of the State of Bavaria
Munich, Germany

Barbara Harrisson
Sarawak Museum, Kuching, Borneo
Ithaca, New York, U.S.A.

Dr. Francois Haverschmidt
President, High Court (retired)
Paramaribo, Suriname

Dr. Heinz Heck
Director, Catskill Game Farm
Catskill, New York, U.S.A.

Dr. Lutz Heck
Professor (retired), and Director,
Zoological Garden, Berlin
Wiesbaden, Germany

Dr. H.C. Heini Hediger
Director, Zoological Garder
Zurich, Switzerland

Dr. Dietrich Heinemann
Director, Zoological Garden, Münster
Dörnigheim, Germany

Dr. Helmut Hemmer
Institute for Physiological Zoology,
University of Mainz
Mainz, Germany

Dr. W. G. Heptner
Professor, Zoological Museum,
University of Moscow
Moscow, Russia

Dr. Konrad Herter
Professor Emeritus and Director
(retired), Zoological Institute, Free
University of Berlin
Berlin, Germany

Dr. Hans Rudolf Heusser
Zoological Museum, University of
Zurich
Zurich, Switzerland

Dr. Emil Otto Höhn
Associate Professor of Physiology,
University of Alberta
Edmonton, Canada

Dr. W. Hohorst
Professor and Director,
Parasitological Institute, Farbwerke
Hoechst A.G.
Frankfurt-Höchst, Germany

Dr. Folkhart Hückinghaus
Director, Senckenbergische Anatomy,
University of Frankfurt a.M.
Frankfurt a.M., Germany

Francois Hüe
National Museum of Natural
History
Paris, France

Dr. K. Immelmann
Professor, Zoological Institute,
Technical University of Braunschweig
Braunschweig, Germany

Dr. Junichiro Itani
Kyoto University
Kyoto, Japan

Dr. Richard F. Johnston
Professor of Zoology, University of
Kansas
Lawrence, Kansas, U.S.A.

Otto Jost
Oberstudienrat, Freiherr-vom-Stein
Gymnasium
Fulda, Germany

Dr. Paul Kähsbauer
Curator, Fishes, Museum of Natural
History
Vienna, Austria

Dr. Ludwig Karbe
Zoological State Institute and
Museum
Hamburg, Germany

Dr. N. N. Kartaschew
Docent, Department of Biology,
Lomonossow State University
Moscow, Russia

Dr. Werner Kästle
Oberstudienrat, Gisela Gymnasium
Munich, Germany

Dr. Reinhard Kaufmann
Field Station of the Tropical Institute,
Justus Liebig University, Giessen,
Germany
Santa Marta, Colombia

Dr. Masao Kawai
Primate Research Institute, Kyoto
University
Kyoto, Japan

Dr. Ernst F. Kilian
Professor, Giessen University and
Catedratico Universidad Austral,
Valdivia-Chile
Giessen, Germany

Dr. Ragnar Kinzelbach
Institute for General Zoology,
University of Mainz
Mainz, Germany

Dr. Heinrich Kirchner
Landwirtschaftsrat (retired)
Bad Oldesloe, Germany

Dr. Rosl Kirchshofer
Zoological Garden, University of
Frankfort a.M.
Frankfurt a.M., Germany

Dr. Wolfgang Klausewitz
Curator, Senckenberg Nature
Museum and Research Institute
Frankfurt a.M., Germany

Dr. Konrad Klemmer
Curator, Senckenberg Nature
Museum and Research Institute
Frankfurt a.M., Germany

Dr. Erich Klinghammer
Laboratory of Ethology, Purdue
University
Lafayette, Indiana, U.S.A.

Dr. Heinz-Georg Klös
Professor and Director, Zoological
Garden
Berlin, Germany

Ursula Klös
Zoological Garden
Berlin, Germany

Dr. Otto Koehler
Professor Emeritus, Zoological
Institute, University of Freiburg
Freiburg i. BR., Germany

Dr. Kurt Kolar
Institute of Ethology, Austrian
Academy of Sciences
Vienna, Austria

Dr. Claus König
State Ornithological Station of Baden-
Württemberg
Ludwigsburg, Germany

Dr. Adriaan Kortlandt
Zoological Laboratory, University of
Amsterdam
Amsterdam, The Netherlands

Dr. Helmut Kraft
Professor and Scientific Councillor,
Medical Animal Clinic, University of
Munich
Munich, Germany

Dr. Helmut Kramer
Zoological Research Institute and A.
Koenig Museum
Bonn, Germany

Dr. Franz Krapp
Zoological Institute, University of
Freiburg
Freiburg, Switzerland

Dr. Otto Kraus
Professor, University of Hamburg,
and Director, Zoological Institute and
Museum
Hamburg, Germany

Dr. Hans Krieg
Professor and First Director (retired),
Scientific Collections of the State of
Bavaria
Munich, Germany

Dr. Heinrich Kühl
Federal Research Institute for
Fisheries, Cuxhaven Laboratory
Cuxhaven, Germany

Dr. Oskar Kuhn
Professor, formerly University
Halle/Saale
Munich, Germany

Dr. Hans Kumerloeve
First Director (retired), State
Scientific Museum, Vienna
Munich, Germany

Dr. Nagamichi Kuroda
Yamashina Ornithological Institute,
Shibuya-Ku
Tokyo, Japan

Dr. Fred Kurt
Zoological Museum of Zurich
University, Smithsonian Elephant
Survey
Colombo, Ceylon

Dr. Werner Ladiges
Professor and Chief Curator,
Zoological Institute and Museum,
University of Hamburg
Hamburg, Germany

Leslie Laidlaw
Department of Animal Sciences,
Purdue University
Lafayette, Indiana, U.S.A.

Dr. Ernst M. Lang
Director, Zoological Garden
Basel, Switzerland

Dr. Alfredo Langguth
Department of Zoology, Faculty of
Humanities and Sciences, University
of the Republic
Montevideo, Uruguay

Leo Lehtonen
Science Writer
Helsinki, Finland

Bernd Leisler
Second Zoological Institute, University
of Vienna
Vienna, Austria

Dr. Kurt Lillelund
Professor and Director, Institute for
Hydrobiology and Fishery Sciences,
University of Hamburg
Hamburg, Germany

R. Liversidge
Alexander MacGregor Memorial
Museum
Kimberley, South Africa

Dr. Konrad Lorenz
Professor and Director, Max Planck
Institute for Behavioral Physiology
Seewiesen/Obb., Germany

Dr. Martin Lühmann
Federal Research Institute for the
Breeding of Small Animals
Celle, Germany

Dr. Johannes Lüttschwager
Oberstudienrat (retired)
Heidelberg, Germany

Dr. Wolfgang Makatsch
Bautzen, Germany

Dr. Hubert Markl
Professor and Director, Zoological
Institute, Technical University of
Darmstadt
Darmstadt, Germany

Basil J. Marlow, B.SC. (Hons)
Curator, Australian Museum
Sydney, Australia

Dr. Theodor Mebs
Instructor of Biology
Weissenhaus/Ostsee, Germany

Dr. Gerlof Fokko Mees
Curator of Birds, Rijks Museum of
Natural History
Leiden, The Netherlands

Hermann Meinken
Director, Fish Identification Institute,
V.D.A.
Bremen, Germany

Dr. Wilhelm Meise
Chief Curator, Zoological Institute
and Museum, University of Hamburg
Hamburg, Germany

Dr. Joachim Messtorff
Field Station of the Federal Fisheries
Research Institute
Bremerhaven, Germany

Dr. Marian Mlynarski
Professor, Polish Academy of
Sciences, Institute for Systematic and
Experimental Zoology
Cracow, Poland

Dr. Walburga Moeller
Nature Museum
Hamburg, Germany

Dr. H.C. Erna Mohr
Curator (retired), Zoological State
Institute and Museum
Hamburg, Germany

Dr. Karl-Heinz Moll
Waren/Müritz, Germany

Dr. Detlev Müller-Using
Professor, Institute for Game
Management, University of Göttingen
Hannoversch-Münden, Germany

Werner Münster
Instructor of Biology
Ebersbach, Germany

Dr. Joachim Münzing
Altona Museum
Hamburg, Germany

Dr. Wilbert Neugebauer
Wilhelma Zoo
Stuttgart-Bad Cannstatt, Germany

Dr. Ian Newton
Senior Scientific Officer, The Nature
Conservancy
Edinburgh, Scotland

Dr. Jürgen Nicolai
Max Planck Institute for Behavioral
Physiology
Seewiesen/Obb., Germany

Dr. Günther Niethammer
Professor, Zoological Research
Institute and A. Koenig Museum
Bonn, Germany

Dr. Bernhard Nievergelt
Zoological Museum, University of
Zurich
Zurich, Switzerland

Dr. C. C. Olrog
Institut Miguel Lillo San Miguel de
Tucuman
Tucuman, Argentina

Alwin Pedersen
Mammal Research and aRctic Explorer
Holte, Denmark

Dr. Dieter Stefan Peters
Nature Museum and Senckenberg
Research Institute
Frankfurt a.M., Germany

Dr. Nicolaus Peters
Scientific Councillor and Docent,
Institute of Hydrobiology and
Fisheries, University of Hamburg
Hamburg, Germany

Dr. Hans-Günter Petzold
Assistant Director, Zoological Garden
Berlin, Germany

Dr. Rudolf Piechocki
Docent, Zoological Institute,
University of Halle
Halle a.d.S., Germany

Dr. Ivo Poglayen-Neuwall
Director, Zoological Garden
Louisville, Kentucky, U.S.A.

Dr. Egon Popp
Zoological Collection of the State of
Bavaria
Munich, Germany

Dr. H.C. Adolf Portmann
Professor Emeritus, Zoological
Institute, University of Basel
Basel, Switzerland

Hans Psenner
Professor and Director, Alpine Zoo
Innsbruck, Austria

Dr. Heinz-Siburd Raethel
Oberveterinärrat
Berlin, Germany

Dr. Urs H. Rahm
Professor, Museum of Natural History
Basel, Switzerland

Dr. Werner Rathmayer
Biology Institute, University of
Konstanz
Konstanz, Germany

Walter Reinhard
Biologist
Baden-Baden, Germany

Dr. H. H. Reinsch
Federal Fisheries Research Institute
Bremerhaven, Germany

Dr. Bernhard Rensch
Professor Emeritus, Zoological
Institute, University of Münster
Münster, Germany

Dr. Vernon Reynolds
Docent, Department of Sociology,
University of Bristol
Bristol, England

Dr. Rupert Riedl
Professor, Department of Zoology,
University of North Carolina
Chapel Hill, North Carolina, U.S.A.

Dr. Peter Rietschel
Professor (retired), Zoological
Institute, University of Frankfurt a.M.
Frankfurt a.M., Germany

Dr. Siegfried Rietschel
Docent, University of Frankfurt;
Curator, Nature Museum and
Research Institute Senckenberg
Frankfurt a.M., Germany

Herbert Ringleben
Institute of Ornithology, Heligoland
Ornithological Station
Wilhelmshaven, Germany

Dr. K. Rohde
Institute for General Zoology, Ruhr
University
Bochum, Germany

Dr. Peter Röben
Academic Councillor, Zoological
Institute, Heidelberg University
Heidelberg, Germany

Dr. Anton E. M. De Roo
Royal Museum of Central Africa
Tervuren, South Africa

Dr. Hubert Saint Girons
Research Director, Center for
National Scientific Research
Brunoy (Essonne), France

Dr. Luitfried Von Salvini-Plawen
First Zoological Institute, University
of Vienna
Vienna, Austria

Dr. Kurt Sanft
Oberstudienrat, Diesterweg-Gymnasium
Berlin, Germany

Dr. E. G. Franz Sauer
Professor, Zoological Research
Institute and A. Koenig Museum,
University of Bonn
Bonn, Germany

Dr. Eleonore M. Sauer
Zoological Research Institute and A.
Koenig Museum, University of Bonn
Bonn, Germany

Dr. Ernst Schäfer
Curator, State Museum of Lower
Saxony
Hannover, Germany

Dr. Friedrich Schaller
Professor and Chairman, First
Zoological Institute, University of
Vienna
Vienna, Austria

Dr. George B. Schaller
Serengeti Research Institute, Michael
Grzimek Laboratory
Seronera, Tanzania

Dr. Georg Scheer
Chief Curator and Director,
Zoological Institute, State Museum of
Hesse
Darmstadt, Germany

Dr. Christoph Scherpner
Zoological Garden
Frankfurt a.M., Germany

Dr. Herbert Schifter
Bird Collection, Museum of Natural
History
Vienna, Austria

Dr. Marco Schnitter
Zoological Museum, Zurich
University
Zurich, Switzerland

Dr. Kurt Schubert
Federal Fisheries Research Institute
Hamburg, Germany

Eugen Schuhmacher
Director, Animals Films, I.U.C.N.
Munich, Germany

Dr. Thomas Schultze-Westrum
Zoological Institute, University of
Munich
Munich, Germany

Dr. Ernst Schüt
Professor and Director (retired), State
Museum of Natural History
Stuttgart, Germany

Dr. Lester L. Short Jr.
Associate Curator, American Museum
of Natural History
New York, New York, U.S.A.

Dr. Helmut Sick
National Museum
Rio de Janeiro, Brazil

Dr. Alexander F. Skutch
Professor of Ornithology, University
of Costa Rica
San Isidro del General, Costa Rica

Dr. Everhard J. Slijper
Professor, Zoological Laboratory,
University of Amsterdam
Amsterdam, The Netherlands

Bertram E. Smythies
Curator (retired), Division of Forestry
Management, Sarawak-Malaysia
Estepona, Spain

Dr. Kenneth E. Stager
Chief Curator, Los Angeles County
Museum of Natural History
Los Angeles, California, U.S.A.

Dr. H.C. Georg H.W. Stein
Professor, Curator of Mammals,
Institute of Zoology and Zoological
Museum, Humboldt University
Berlin, Germany

Dr. Joachim Steinbacher
Curator, Nature Museum and
Senckenberg Research Institute
Frankfurt a.M., Germany

Dr. Bernard Stonehouse
Canterbury University
Christchurch, New Zealand

Dr. Richard Zur Strassen
Curator, Nature Museum and
Senckenberg Research Institute
Frankfurt a.M., Germany

Dr. Adelheid Studer-Thiersch
Zoological Garden
Basel, Switzerland

Dr. Ernst Sutter
Museum of Natural History
Basel, Switzerland

Dr. Fritz Terofal
Director, Fish Collection, Zoological
Collection of the State of Bavaria
Munich, Germany

Dr. G. F. Van Tets
Wildlife Research
Canberra, Australia

Ellen Thaler-Kottek
Institute of Zoology, University of
Innsbruck
Innsbruck, Austria

Dr. Erich Thenius
Professor and Director, Institute of
Paleontolgy, University of Vienna
Vienna, Austria

Dr. Niko Tinbergen
Professor of Animal Behavior,
Department of Zoology, Oxford
University
Oxford, England

Alexander Tsurikov
Lecturer, University of Munich
Munich, Germany

Dr. Wolfgang Villwock
Zoological Institute and Museum,
University of Hamburg
Hamburg, Germany

Zdenek Vogel
Director, Suchdol Herpetological
Station
Prague, Czechoslovakia

Dieter Vogt
Schorndorf, Germany

Dr. Jiri Volf
Zoological Garden
Prague, Czechoslovakia
Otto Wadewitz
Leipzig, Germany

Dr. Helmut O. Wagner
Director (retired), Overseas Museum,
Bremen
Mexico City, Mexico

Dr. Fritz Walther
Professor, Texas A & M University
College Station, Texas, U.S.A.

John Warham
Zoology Department, Canterbury
University
Christchurch, New Zealand

Dr. Sherwood L. Washburn
University of California at Berkeley
Berkeley, California, U.S.A.

Eberhard Wawra
First Zoological Institute, University
of Vienna
Vienna, Austria

Dr. Ingrid Weigel
Zoological Collection of the State of
Bavaria
Munich, Germany

Dr. B. Weischer
Institute of Nematode Research,
Federal Biological Institute
Münster/Westfalen, Germany

Herbert Wendt
Author, Natural History
Baden-Baden, Germany

Dr. Heinz Wermuth
Chief Curator, State Nature Museum,
Stuttgart
Ludwigsburg, Germany

Dr. Wolfgang Von Westernhagen
Preetz/Holstein, Germany

Dr. Alexander Wetmore
United States National Museum,
Smithsonian Institution
Washington, D.C., U.S.A.

Dr. Dietrich E. Wilcke
Röttgen, Germany

Dr. Helmut Wilkens
Professor and Director, Institute of
Anatomy, School of Veterinary
Medicine
Hannover, Germany

Dr. Michael L. Wolfe
Utah, U.S.A.

Hans Edmund Wolters
Zoological Research Institute and A.
Koenig Museum
Bonn, Germany

Dr. Arnfrid Wünschmann
Research Associate, Zoological Garden
Berlin, Germany

Dr. Walter Wüst
Instructor, Wilhelms Gymnasium
Munich, Germany

Dr. Heinz Wundt
Zoological Collection of the State of
Bavaria
Munich, Germany

Dr. Claus-Dieter Zander
Zoological Institute and Museum,
University of Hamburg
Hamburg, Germany

Dr. Fritz Zumpt
Director, Entomology and
Parasitology, South African Institute
for Medical Research
Johannesburg, South Africa

Dr. Richard L. Zusi
Curator of Birds, United States
National Museum, Smithsonian
Institution
Washington, D.C., U.S.A.

CONTRIBUTORS TO THE FIRST EDITION

Glossary

The following glossary is not intended to be exhaustive, but rather includes primarily terms that (1) have some specific importance to our understanding of birds, (2) have been used in these volumes, (3) might have varying definitions relative to birds as opposed to common usage, or (4) are often misunderstood.

Accipiter—This is the genus name for a group of bird-eating hawks (Accipitridae; e.g., sharp-shinned hawk, Cooper's hawk). These birds show similar behavior and appearance and extreme sexual dimorphism. Females are much larger than males and the female of the sharp-shinned hawk often seems as large as the male of the Cooper's hawk, leading to some confusion on the part of birders. In the face of uncertainty, these birds are often just referred to as "Accipiters" and the name is now firmly ensconced in "birding" terminology.

Adaptive radiation—Diversification of a species or single ancestral type into several forms that are each adaptively specialized to a specific niche.

Aftershaft—A second rachis (= shaft) arising near the base of a contour feather, creating a feather that "branches." Aftershafts can be found in many birds (e.g., pheasants) but in most the aftershaft is much smaller than the main shaft of the feather. In ratites (ostrich-like birds), the aftershaft is about the same size as the main shaft. Sometimes the term "aftershaft" is restricted to the rachis that extends from the main rachis and the whole secondary structure is referred to as the "afterfeather."

Agonistic—Behavioral patterns that are aggressive in context. Most aggressive behavior in birds is expressed as song (in songbirds) or other vocal or mechanical sound (e.g. see Drumming). The next level of intensity is display, and only in extreme circumstances do birds resort to physical aggression.

Air sac—Thin-walled, extensions of the lungs, lying in the abdomen and thorax, and extending even into some bones of birds. Air sacs allow an increased respiratory capacity of birds and the removal of oxygen both as air passes in through the lungs and also as it passes back through the lungs as the bird exhales. The flow of air through the air sacs also helps dissipate the heat produced through muscle activity and increases a bird's volume while only minimally increasing weight—thus effectively making birds lighter relative to their size and more efficient in flying. Air sacs are best developed in the strongest flying birds and least developed in some groups that are flightless.

Alcid—Referring to a member of the family Alcidae; including puffins, auks, auklets, murres, razorbills, and guillemots.

Allopatric—Occurring in separate, nonoverlapping geographic areas.

Allopreening—Mutual preening; preening of the feathers of one bird by another; often a part of courtship or pair bond maintenance.

Alpha breeder—The reproductively dominant member of a social unit.

Alternate plumage—The breeding plumage of passerines, ducks, and many other groups; typically acquired through a partial molt prior to the beginning of courtship.

Altricial—An adjective referring to a bird that hatches with little, if any, down, is unable to feed itself, and initially has poor sensory and thermoregulatory abilities.

Alula—Small feathers at the leading edge of the wing and attached to the thumb; also called bastard wing; functions in controlling air flow over the surface of the wing, thus allowing a bird to land at a relatively slow speed.

Anatid—A collective term referring to members of the family Anatidae; ducks, geese, and swans.

Anisodactyl— An adjective that describes a bird's foot in which three toes point forward and one points backwards, a characteristic of songbirds.

Anserine—Goose-like.

Anting—A behavior of birds that involves rubbing live ants on the feathers, presumably to kill skin parasites.

Antiphonal duet—Vocalizations by two birds delivered alternately in response to one another; also known as responsive singing.

AOU—American Ornithologists' Union; the premier professional ornithological organization in North America; the organizational arbiter of scientific and standardized common names of North American birds as given in the periodically revised Check-list of North American Birds.

Arena—See Lek.

Aspect ratio—Length of a wing divided by width of the wing; High aspect wings are long and narrow. These are characteristic of dynamic soaring seabirds such as albatrosses. These birds have tremendous abilities to soar over the open ocean, but poor ability to maneuver in a small area. In contrast, low aspect ratio wings are short and broad, characteristic of many forest birds, and provide great ability to quickly maneuver in a small space.

Asynchronous—Not simultaneous; in ornithology often used with respect to the hatching of eggs in a clutch in which hatching occurs over two or more days, typically a result of initiation of incubation prior to laying of the last egg.

Auricular—An adjective referring to the region of the ear in birds, often to a particular plumage pattern over the ear.

Austral—May refer to "southern regions," typically meaning Southern Hemisphere. May also refer to the geographical region included within the Transition, Upper Austral, and Lower Austral Life Zones as defined by C. Hart Merriam in 1892–1898. These zones are often characterized by specific plant and animal communities and were originally defined by temperature gradients especially in the mountains of southwestern North America.

Autochthonous—An adjective that indicates that a species originated in the region where it now resides.

Barb—One of the hair-like extensions from the rachis of a feather. Barbs with barbules and other microstructures can adhere to one another, forming the strong, yet flexible vane needed for flight and protection and streamlining of body surfaces.

Barbules—A structural component of the barbs of many feathers; minute often interlocking filaments in a row at each side of a barb. As a result of their microstructure, barbules adhere to one another much like "Velcro®" thus assuring that feathers provide a stiff, yet flexible vane.

Basic plumage—The plumage an adult bird acquires as a result of its complete (or near complete) annual molt.

Bergmann's rule—Within a species or among closely related species of mammals and birds, those individuals in colder environments often are larger in body size. Bergmann's rule is a generalization that reflects the ability of warm-blooded animals to more easily retain body heat (in cold climates) if they have a high body surface to body volume ratio, and to more easily dissipate excess body heat (in hot environments) if they have a low body surface to body volume ratio.

Bioacoustics—The study of biological sounds such as the sounds produced by birds.

Biogeographic region—One of several major divisions of the earth defined by a distinctive assemblage of animals and plants. Sometimes referred to as "zoogeographic regions or realms" (for animals) or "phytogeographic regions or realms" (for plants). Such terminology dates from the late nineteenth century and varies considerably. Major biogeographic regions each have a somewhat distinctive flora and fauna. Those generally recognized include Nearctic, Neotropical, Palearctic, Ethiopian, Oriental, and Australian.

Biomagnification—Sometimes referred to as "bioaccumulation." Some toxic elements and chemical compounds are not readily excreted by animals and instead are stored in fatty tissues, removing them from active metabolic pathways. Birds that are low in a food chain (e.g., sparrows that eat seeds) accumulate these chemicals in their fatty tissues. When a bird that is higher in the food chain (e.g., a predator like a falcon) eats its prey (e.g., sparrows), it accumulates these chemicals from the fatty tissue of each prey individual, thus magnifying the level of the chemical in its own tissues. When the predator then comes under stress and all of these chemicals are released from its fat into its system, the effect can be lethal. Chemicals capable of such biomagnification include heavy metals such as lead and mercury, and such manmade compounds as organochlorine pesticides and polychlorinated biphenyls (PCBs).

Booming ground—See Lek.

Booted—An adjective describing a bird tarsus (leg) that has a smooth, generally undivided, rather than scaly (= scutellate) appearance. The extent of the smooth or scaly appearance of a bird tarsus varies among taxonomic groups and there are many different, more specific, patterns of tarsal appearance that are recognized.

Boreal—Often used as an adjective meaning "northern"; also may refer to the northern climatic zone immediately south of the Arctic; may also include the Arctic, Hudsonian, and Canadian Life Zones described by C. Hart Merriam.

Bristle—In ornithology, a feather with a thick, tapered rachis and no vane except for a remnant sometimes found near the bristle base.

Brood—As a noun: the young produced by a pair of birds during one reproductive effort. As a verb: to provide warmth and shelter to chicks by gathering them under the protection of breast and/or wings.

Brood parasitism—Reproductive strategy where one species of bird (the parasite) lays its eggs in the nests of another species (the host). An acceptable host will incubate the eggs and rear the chicks of the brood parasite, often to the detriment or loss of the host's own offspring.

Brood patch—A bare area of skin on the belly of a bird, the brood patch is enlarged beyond the normal apterium (bare area) as a result of loss of feathers. It becomes highly vascularized (many blood vessels just under the surface). The brood patch is very warm to the touch and the bird uses it to cover and warm its chicks. In terms of structure, the brood patch is the same as the incubation patch and the two terms are often used synonymously. Technically the brood patch and incubation patch differ in function: the incubation patch is used in incubating eggs, the brood patch is used to brood the young after the eggs hatch.

Brood reduction—Reduction in the number of young in the nest. Viewed from an evolutionary perspective, mechanisms that allow for brood reduction may assure that at least some offspring survive during stressful times and that during times of abundant resources all young may survive. Asynchronous hatching results in young of different ages and sizes in a nest and is a mechanism that facilitates brood reduction: the smallest chick often dies if there is a shortage of food. The barn owl (*Tyto alba*; Tytonidae) depends on food resources that vary greatly in availability from year to year and it often experiences brood reduction.

Buteo—This is the genus name for a group of hawks that have broad wings and soar. These hawks are often seen at a distance and are easily recognized as "Buteos" although they may not be identifiable as species. Hence the genus name has come into common English usage.

Caecum (pl. caeca)—Blindly-ending branch extending from the junction of the small and large intestine. Most birds have two caeca, but the number and their development in birds is highly variable. Caeca seem to be most highly developed and functional in facilitating microbial digestion of food in those birds that eat primarily plant materials.

Caruncle—An exposed, often brightly colored, fleshy protuberance or wrinkled facial skin of some birds.

Casque—An enlargement at the front of the head (e.g., on cassowaries, Casuaridae) or sometimes of the bill (e.g., on hornbills, Bucerotidae) of a bird. A casque may be bony, cartilaginous, or composed of feathers (e.g., Pri-

onopidae). A casque is often sexual ornamentation, but may protect the head of a cassowary crashing through underbrush, may be used for vocal amplification, or may serve a physiological function.

Cavity nester—A species that nests in some sort of a cavity. Primary cavity nesters (e.g., woodpeckers, Picidae; kingfishers, Alcedinidae; some swallows, Hirundinidae) are capable of excavating their own cavities; secondary cavity nesters (e.g., starlings, Sturnidae; House Sparrows, Passeridae; bluebirds, Turdidae) are not capable of excavating their own cavities.

Cere—The soft, sometimes enlarged, and often differently colored basal covering of the upper bill (maxilla) of many hawks (Falconiformes), parrots (Psitaciformes), and owls (Strigiformes). The nostrils are often within or at the edge of the cere. In parrots the cere is sometimes feathered.

Cladistic—Evolutionary relationships suggested as "tree" branches to indicate lines of common ancestry.

Cleidoic eggs—Cleidoic eggs are simply ones that are contained, hence protected, inside of a somewhat impervious shell—such as the eggs of birds. The presence of a shell around an egg freed the amphibian ancestors of reptiles from the need to return to the water to lay eggs and provided greater protection from dying.

Cline—A gradient in a measurable character, such as size and color, showing geographic differentiation. Various patterns of geographic variation are reflected as clines or clinal variation, and have been described as "ecogeographic rules."

Clutch—The set of eggs laid by a female bird during one reproductive effort. In most species, a female will lay one egg per day until the clutch is complete; in some species, particularly larger ones (e.g., New World vultures, Cathartidae), the interval between eggs may be more than one day.

Colony—A group of birds nesting in close proximity, interacting, and usually aiding in early warning of the presence of predators and in group defense.

Commensal—A relationship between species in which one benefits and the other is neither benefited nor harmed.

Congeneric—Descriptive of two or more species that belong to the same genus.

Conspecific—Descriptive of two or more individuals or populations that belong to the same species.

Conspecific colony—A colony of birds that includes only members of one species.

Contact call—Simple vocalization used to maintain communication or physical proximity among members of a social unit.

Contour feather—One of those feathers covering the body, head, neck, and limbs of a bird and giving rise to the shape (contours) of the bird.

Convergent evolution—When two evolutionarily unrelated groups of organisms develop similar characteristics due to adaptation to similar aspects of their environment or niche. The sharply pointed and curved talons of hawks and owls are convergent adaptations for their predatory lifestyle.

Cooperative breeding—A breeding system in which birds other than the genetic parents share in the care of eggs and young. There are many variants of cooperative breeding. The birds that assist with the care are usually referred to as "helpers" and these are often offspring of the same breeding pair, thus genetically related to the chicks they are tending. Cooperative breeding is most common among tropical birds and seems most common in situations where nest sites or breeding territories are very limited. Several studies have demonstrated that "helping" increases reproductive success. By helping a helper is often assuring survival of genes shared with the related offspring. The helper also may gain important experience and ultimately gain access to a breeding site.

Coracoid—A bone in birds and some other vertebrates extending from the scapula and clavicle to the sternum; the coracoid serves as a strut supporting the chest of the bird during powerful muscle movements associated with flapping flight.

Cosmopolitan—Adjective describing the distribution pattern of a bird found around the world in suitable habitats.

Countershading—A color pattern in which a bird or other animal is darker above and lighter below. The adaptive value of the pattern is its ability to help conceal the animal: a predator looking down from above sees the darker back against the dark ground; a predator looking up from below sees the lighter breast against the light sky; a predator looking from the side sees the dark back made lighter by the light from above and the light breast made darker by shading.

Covert—A feather that covers the gap at the base between flight feathers of the wing and tail; coverts help create smooth wing and tail contours that make flight more efficient.

Covey—A group of birds, often comprised of family members that remain together for periods of time; usually applied to game birds such as quail (Odontophorinae).

Crepuscular—Active at dawn and at dusk.

Crèche—An aggregation of young of many colonially-nesting birds (e.g., penguins, Spheniscidae; terns, Laridae). There is greater safety from predators in a crèche.

Crissum—The undertail coverts of a bird; often distinctively colored.

Critically Endangered—A technical category used by IUCN for a species that is at an extremely high risk of extinction in the wild in the immediate future.

Cryptic—Hidden or concealed; i.e., well-camouflaged patterning.

Dichromic—Occurring in two distinct color patterns (e.g., the bright red of male and dull red-brown of female northern cardinals, *Cardinalis cardinalis*)

Diurnal—Active during the day.

Dimorphic—Occurring in two distinct forms (e.g., in reference to the differences in tail length of male and female boat-tailed grackles, *Cassidix major*).

Disjunct—A distribution pattern characterized by populations that are geographically separated from one another.

Dispersal—Broadly defined: movement from an area; narrowly defined: movement from place of hatching to place of first breeding.

Dispersion—The pattern of spatial arrangement of individuals, populations, or other groups; no movement is implied.

Disruptive color—A color pattern such as the breast bands on a killdeer (*Charadrius vociferus*) that breaks up the outline of the bird, making it less visible to a potential predator, when viewed from a distance

DNA-DNA hybridization—A technique whereby the genetic similarity of different bird groups is determined based on the extent to which short stretches of their DNA, when mixed together in solution in the laboratory, are able to join with each other.

Dominance hierarchy—"Peck order"; the social status of individuals in a group; each animal can usually dominate those animals below it in a hierarchy.

Dummy nest—Sometimes called a "cock nest." An "extra" nest, often incomplete, sometimes used for roosting, built by aggressive males of polygynous birds. Dummy nests may aid in the attraction of additional mates, help define a male's territory, or confuse potential predators.

Dump nest—A nest in which more than one female lays eggs. Dump nesting is a phenomenon often linked to young, inexperienced females or habitats in which nest sites are scarce. The eggs in dump nest are usually not incubated. Dump nesting may occur within a species or between species.

Dynamic soaring—A type of soaring characteristic of oceanic birds such as albatrosses (Diomedeidae) in which the bird takes advantage of adjacent wind currents that are of different speeds in order to gain altitude and effortlessly stay aloft.

Echolocation—A method of navigation used by some swifts (Apodidae) and oilbirds (Steatornithidae) to move in darkness, such as through caves to nesting sites. The birds emit audible "clicks" and determine pathways by using the echo of the sound from structures in the area.

Eclipse plumage—A dull, female-like plumage of males of Northern Hemisphere ducks (Anatidae) and other birds such as house sparrows (*Passer domesticus*) typically attained in late summer prior to the annual fall molt. Ducks are flightless at this time and the eclipse plumage aids in their concealment at a time when they would be especially vulnerable to predators.

Ecotourism—Travel for the primary purpose of viewing nature. Ecotourism is now "big business" and is used as a non-consumptive but financially rewarding way to protect important areas for conservation.

Ectoparasites—Relative to birds, these are parasites such as feather lice and ticks that typically make their home on the skin or feathers.

Emarginate—Adjective referring to the tail of a bird that it notched or forked or otherwise has an irregular margin as a result of tail feathers (rectrices) being of different lengths. Sometimes refers to individual flight feather that is particularly narrowed at the tip.

Endangered—A term used by IUCN and also under the Endangered Species Act of 1973 in the United States in reference to a species that is threatened with imminent extinction or extirpation over all or a significant portion of its range.

Endemic—Native to only one specific area.

Eocene—Geological time period; subdivision of the Tertiary, from about 55.5 to 33.7 million years ago.

Erythrocytes—Red blood cells; in birds, unlike mammals, these retain a nucleus and are longer lived. Songbirds tend to have smaller, more numerous (per volume) erythrocytes that are richer in hemoglobin than are the erythrocytes of more primitive birds.

Ethology—The study of animal behavior.

Exotic—Not native.

Extant—Still in existence; not destroyed, lost, or extinct.

Extinct—Refers to a species that no longer survives anywhere.

Extirpated—Referring to a local extinction of a species that can still be found elsewhere.

Extra-pair copulation—In a monogamous species, refers to any mating that occurs between unpaired males and females.

Facial disc—Concave arrangement of feathers on the face of an owl. The facial discs on an owl serve as sound parabolas, focusing sound into the ears around which the facial discs are centered, thus enhancing their hearing.

Fecal sac—Nestling songbirds (Passeriformes) and closely related groups void their excrement in "packages"—enclosed in thin membranes—allowing parents to remove the material from the nest. Removal of fecal material likely reduces the potential for attraction of predators.

Feminization—A process, often resulting from exposure to environmental contaminants, in which males produce a higher levels of female hormones (or lower male hormone levels), and exhibit female behavioral or physiological traits.

Feral—Gone wild; i.e., human-aided establishment of non-native species.

Fledge—The act of a juvenile making its first flight; sometimes generally used to refer to a juvenile becoming independent.

Fledgling—A juvenile that has recently fledged. An emphasis should be placed on "recently." A fledgling generally lacks in motor skills and knowledge of its habitat and fledglings are very vulnerable, hence under considerable parental care. Within a matter of a few days, however, they gain skills and knowledge and less parental care is needed.

Flight feathers—The major feathers of the wing and tail that are crucial to flight. (See Primary, Secondary, Tertial, Alula, Remex, Rectrix)

Flyway—A major pathway used by a group of birds during migration. The flyway concept was developed primarily with regard to North American waterfowl (Anatidae) and has been used by government agencies in waterfowl management. Major flyways described include the Atlantic, Mississippi, Central, and Pacific flyways. While the flyway concept is often used in discussions of other groups of birds, even for waterfowl the concept is an oversimplification. The patterns of movements of migrant waterfowl and other birds vary greatly among species.

Frugivorous—Feeds on fruit.

Galliform—Chicken-like, a member of the Galliformes.

Gape—The opening of the mouth of a bird; the act of opening the mouth, as in begging.

Gizzard—The conspicuous, muscular portion of the stomach of a bird. Birds may swallow grit or retain bits of bone or hard parts of arthropods in the gizzard and these function in a manner analogous to teeth as the strong muscles of the gizzard contract, thus breaking food into smaller particles. The gizzard is best developed in birds that eat seeds and other plant parts; in some fruit-eating birds the gizzard is very poorly developed.

Glareolid—A member of the family Glareolidae.

Gloger's rule—Gloger's rule is an ecogeographic generalization that suggests that within a species or closely related group of birds there is more melanin (a dark pigment) in feathers in warm humid parts of the species' or groups' range, and less melanin in feathers in dry or cooler parts of the range.

Gorget—Colorful throat patch or bib (e.g., of many hummingbirds, Trochilidae).

Graduated—An adjective used to describe the tail of a bird in which the central rectrices are longest and those to the outside are increasingly shorter.

Granivorous—Feeding on seeds.

Gregarious—Occuring in large groups.

Gular—The throat region.

Hallux—The innermost digit of a hind or lower limb.

Hawk—Noun: a member of the family Accipitridae. Verb: catching insects by flying around with the mouth open (e.g. swallows, Hirundinidae; nightjars, Caprimulgidae).

Heterospecific colony—A colony of birds with two or more species.

Heterothermy—In birds, the ability to go into a state of torpor or even hibernation, lowering body temperature through reduced metabolic activity and thus conserving energy resources during periods of inclement weather or low food.

Hibernation—A deep state of reduced metabolic activity and lowered body temperature that may last for weeks; attained by few birds, resulting from reduced food supplies and cool or cold weather.

Holarctic—The Palearctic and Nearctic bigeographic regions combined.

Homeothermy—In birds the metabolic ability to maintain a constant body temperature. The lack of development of homeothermy in new-hatched chicks is the underlying need for brooding behavior.

Hover-dip—A method of foraging involving hovering low over the water, and then dipping forward to pick up prey from the surface (e.g., many herons, Ardeidae).

Hybrid—The offspring resulting from a cross between two different species (or sometimes between distinctive subspecies).

Imprinting—A process that begins with an innate response of a chick to its parent or some other animal (or object!) that displays the appropriate stimulus to elicit the chick's response. The process continues with the chick rapidly learning to recognize its parents. Imprinting typically occurs within a few hours (often 13–16 hours) after hatching. Imprinting then leads to learning behavioral characteristics that facilitate its survival, including such things as choice of foraging sites and foods, shelter, recognition of danger, and identification of a potential mate. The most elaborate (and best studied) imprinting is associated with precocial chicks such as waterfowl (Anatidae).

Incubation patch—See Brood patch.

Indigenous—See Endemic.

Innate—An inherited characteristic; e.g., see Imprinting.

Insectivorous—In ornithology technically refers to a bird that eats insects; generally refers to in birds that feed primarily on insects and other arthropods.

Introduced species—An animal or plant that has been introduced to an area where it normally does not occur.

Iridescent—Showing a rainbow-like play of color caused by differential refraction of light waves that change as the angle of view changes. The iridescence of bird feathers is a result of a thinly laminated structure in the barbules of those feathers. Iridescent feathers are made more brilliant by pigments that underlie this structure, but the pigments do not cause the iridescence.

Irruptive—A species of bird that is characterized by irregular long-distance movements, often in response to a fluctuating food supply (e.g., red crossbill, *Loxia curvirostra*, Fringillidae; snowy owl, *Nyctea scandiaca*, Strigidae).

IUCN—The World Conservation Union; formerly the International Union for the Conservation of Nature, hence IUCN. It is the largest consortium of governmental and nongovernmental organizations focused on conservation issues.

Juvenal—In ornithology (contrary to most dictionaries), restricted to use as an adjective referring to a characteristic (usually the plumage) of a juvenile bird.

Juvenile—A young bird, typically one that has left the nest.

Kleptoparasitism—Behavior in which one individual takes ("steals") food, nest materials, or a nest site from another.

Lachrymal—Part of the skull cranium, near the orbit; lachrymal and Harderian glands in this region lubricate and protect the surface of the eye.

Lamellae—Transverse tooth-like or comb-like ridges inside the cutting edge of the bill of birds such as ducks (Anatidae) and flamingos (Phoenicopteridae). Lamellae serve as a sieve during feeding: the bird takes material into its mouth, then uses its tongue to force water out through the lamellae, while retaining food particles.

Lek—A loose to tight association of several males vying for females through elaborate display; lek also refers to the specific site where these males gather to display. Lek species include such birds as prairie chickens (Phasianidae) and manakins (Pipridae).

Lobed feet—Feet that have toes with stiff scale-covered flaps that extend to provide a surface analogous to webbing on a duck as an aid in swimming.

Lore—The space between the eye and bill in a bird. The loral region often differs in color from adjacent areas of a bird's face. In some species the area is darker, thus helping to reduce glare, serving the same function as the dark pigment some football players apply beneath each eye. In predatory birds, a dark line may extend from the eye to the bill, perhaps decreasing glare, but also serving as a sight to better aim its bill. The color and pattern of plumage and skin in the loral region is species-specific and often of use in helping birders identify a bird.

Malar—Referring to the region of the face extending from near the bill to below the eye; markings in the region are often referred to as "moustache" stripes.

Mandible—Technically the lower half of a bird's bill. The plural, mandibles, is used to refer to both the upper and lower bill. The upper half of a bird's bill is technically the maxilla, but often called the "upper mandible."

Mantle—Noun: The plumage of the back of the bird, including wing coverts evident in the back region on top of the folded wing (especially used in describing hawks (Accipitridae) and gulls (Laridae). Verb: The behavior in which a raptor (typically on the ground) shields its acquired prey to protect it from other predators.

Mesoptile—On chicks, the second down feathers; these grow attached to the initial down, or protoptile.

Metabolic rate—The rate of chemical processes in living organisms, resulting in energy expenditure and growth. Hummingbirds (Trochilidae), for example, have a very high metabolic rate. Metabolic rate decreases when a bird is resting and increases during activity.

Miocene—The geological time period that lasted from about 23.8 to 5.6 million years ago.

Migration—A two-way movement in birds, often dramatically seasonal. Typically latitudinal, though in some species is altitudinal or longitudinal. May be short-distance or long-distance. (See Dispersal)

Mitochondrial DNA—Genetic material located in the mitochondria (a cellular organelle outside of the nucleus). During fertilization of an egg, only the DNA from the nucleus of a sperm combines with the DNA from the nucleus of an egg. The mitochondrial DNA of each offspring is inherited only from its mother. Changes in mitochondrial DNA occur quickly through mutation and studying differences in mitochondrial DNA helps scientists better understand relationships among groups.

Mobbing—A defensive behavior in which one or more birds of the same or different species fly toward a potential predator, such as a hawk, owl, snake, or a mammal, swooping toward it repeatedly in a threatening manner, usually without actually striking the predator. Most predators depend on the element of surprise in capturing their prey and avoid the expenditure of energy associated with a chase. Mobbing alerts all in the neighborhood that a potential predator is at hand and the predator often moves on. Rarely, a predator will capture a bird that is mobbing it.

Molecular phylogenetics—The use of molecular (usually genetic) techniques to study evolutionary relationships between or among different groups of organisms.

Molt—The systematic and periodic loss and replacement of feathers. Once grown, feathers are dead structures that continually wear. Birds typically undergo a complete or near-complete molt each year and during this molt feathers are usually lost and replaced with synchrony between right and left sides of the body, and gradually, so that the bird retains the ability to fly. Some species, such as northern hemisphere ducks, molt all of their flight feathers at once, thus become flightless for a short time. Partial molts, typically involving only contour feathers, may occur prior to the breeding season.

Monophyletic—A group (or clade) that shares a common ancestor.

Monotypic—A taxonomic category that includes only one form (e.g., a genus that includes only one species; a species that includes no subspecies).

Montane—Of or inhabiting the biogeographic zone of relatively moist, cool upland slopes below timberline dominated by large coniferous trees.

Morphology—The form and structure of animals and plants.

Mutualism—Ecological relationship between two species in which both gain benefit.

Nail—The horny tip on the leathery bill of ducks, geese, and swans (Anatidae).

Nectarivore—A nectar-eater (e.g., hummingbirds, Trochilidae; Hawaiian honeycreepers, Drepaniidae).

Near Threatened—A category defined by the IUCN suggesting possible risk of extinction in the medium term future.

Nearctic—The biogeographic region that includes temperate North America faunal region.

Neotropical—The biogeographic region that includes South and Central America, the West Indies, and tropical Mexico.

Nestling—A young bird that stays in the nest and needs care from parents.

New World—A general descriptive term encompassing the Nearctic and Neotropical biogeographic regions.

Niche—The role of an organism in its environment; multidimensional, with habitat and behavioral components.

Nictitating membrane—The third eyelid of birds; may be transparent or opaque; lies under the upper and lower eyelids. When not in use, the nictitating membrane is held at the corner of the eye closest to the bill; in use it moves horizontally or diagonally across the eye. In flight it keeps the bird's eyes from drying out; some aquatic birds have a lens-like window in the nictitating membrane, facilitating vision underwater.

Nidicolous—An adjective describing young that remain in the nest after hatching until grown or nearly grown.

Nidifugous—An adjective describing young birds that leave the nest soon after hatching.

Nocturnal—Active at night.

Nominate subspecies—The subspecies described to represent its species, the first described, bearing the specific name.

Nuclear DNA—Genetic material from the nucleus of a cell from any part of a bird's body other than its reproductive cells (eggs or sperm).

Nuptial displays—Behavioral displays associated with courtship.

Oligocene—The geologic time period occurring from about 33.7 to 23.8 million years ago.

Old World—A general term that usually describes a species or group as being from Eurasia or Africa.

Omnivorous—Feeding on a broad range of foods, both plant and animal matter.

Oscine—A songbird that is in the suborder Passeri, order Passeriformes; their several distinct pairs of muscles within the syrinx allow these birds to produce the diversity of sounds that give meaning to the term "songbird."

Osteological—Pertaining to the bony skeleton.

Palearctic—A biogeographic region that includes temperate Eurasia and Africa north of the Sahara.

Paleocene—Geological period, subdivision of the Tertiary, from 65 to 55.5 million years ago.

Pamprodactyl—The arrangement of toes on a bird's foot in which all four toes are pointed forward; characteristic of swifts (Apodidae).

Parallaxis—Comparing the difference in timing and intensity of sounds reaching each ear (in owls).

Passerine—A songbird; a member of the order Passeriformes.

Pecten—A comb-like structure in the eye of birds and reptiles, consisting of a network of blood vessels projecting inwards from the retina. The main function of the pecten seems to be to provide oxygen to the tissues of the eye.

Pectinate—Having a toothed edge like that of a comb. A pectinate claw on the middle toe is a characteristic of nightjars, herons, and barn owls. Also known as a "feather comb" since the pectinate claw is used in preening.

Pelagic—An adjective used to indicate a relationship to the open sea.

Phalloid organ—Penis-like structure on the belly of buffalo weavers; a solid rod, not connected to reproductive or excretory system.

GLOSSARY

Philopatry—Literally "love of homeland"; a bird that is philopatric is one that typically returns to nest in the same area in which it was hatched. Strongly philopatric species (e.g., hairy woodpecker, *Picoides borealis*) tend to accumulate genetic characteristics that adapt them to local conditions, hence come to show considerable geographic variation; those species that show little philopatry tend to show little geographic variation.

Phylogenetics—The study of racial evolution.

Phylogeny—A grouping of taxa based on evolutionary history.

Picid—A member of the family Picide (woodpeckers, wrynecks, piculets).

Piscivorous—Fish-eating.

Pleistocene—In general, the time of the great ice ages; geological period variously considered to include the last 1 to 1.8 million years.

Pliocene—The geological period preceding the Pleistocence; the last subdivision of what is known as the Tertiary; lasted from 5.5 to 1.8 million years ago.

Plumage—The complete set of feathers that a bird has.

Plunge-diving—A method of foraging whereby the bird plunges from at least several feet up, head-first into the water, seizes its prey, and quickly takes to the wing (e.g., terns, Laridae; gannets, Sulidae).

Polygamy—A breeding system in which either or both male and female may have two or more mates.

Polyandry—A breeding system in which one female bird mates with two or more males. Polyandry is relatively rare among birds.

Polygyny—A breeding system in which one male bird mates with two or more females.

Polyphyletic—A taxonomic group that is believed to have originated from more than one group of ancestors.

Powder down—Specialized feathers that grow continuously and break down into a fine powder. In some groups (e.g., herons, Ardeidae) powder downs occur in discrete patches (on the breast and flanks); in others (e.g., parrots, Psitacidae) they are scattered throughout the plumage. Usually used to waterproof the other feathers (especially in birds with few or no oil glands).

Precocial—An adjective used to describe chicks that hatch in an advanced state of development such that they generally can leave the nest quickly and obtain their own food, although they are often led to food, guarded, and brooded by a parent (e.g., plovers, Charadriidae; chicken-like birds, Galliformes).

Preen—A verb used to describe the behavior of a bird when it cleans and straightens its feathers, generally with the bill.

Primaries—Unusually strong feathers, usually numbering nine or ten, attached to the fused bones of the hand at the tip of a bird's wing.

Protoptile—The initial down on chicks.

Pterylosis—The arrangement of feathers on a bird.

Quaternary—The geological period, from 1.8 million years ago to the present, usually including two subdivisions: the Pleistocene, and the Holocene.

Quill—An old term that generally refers to a primary feather.

Rachis—The shaft of a feather.

Radiation—The diversification of an ancestral species into many distinct species as they adapt to different environments.

Ratite—Any of the ostrich-like birds; characteristically lack a keel on the sternum (breastbone).

Rectrix (pl. rectrices)—A tail feather of a bird; the rectrices are attached to the fused vertebrae that form a bird's bony tail.

Remex (pl. remiges)—A flight feather of the wing; remiges include the primaries, secondaries, tertials, and alula).

Reproductive longevity—The length of a bird's life over which it is capable of reproduction.

Resident—Nonmigratory.

Rhampotheca—The horny covering of a bird's bill.

Rictal bristle— A specialized tactile, stiff, hairlike feather with elongated, tapering shaft, sometimes with short barbs at the base. Rictal bristles prominently surround the mouth of birds such as many nightjars (Caprimulgidae), New World flycatchers (Tyrannidae), swallows (Hirundinidae), hawks (Accipitridae) and owls (Strigidae). They are occasionally, but less precisely referred to as "vibrissae," a term more appropriate to the "whiskers" on a mammal.

Rookery—Originally a place where rooks nest; now a term often used to refer to a breeding colony of gregarious birds.

Sally—A feeding technique that involves a short flight from a perch or from the ground to catch a prey item before returning to a perch.

Salt gland—Also nasal gland because of their association with the nostrils; a gland capable of concentrating and excreting salt, thus allowing birds to drink saltwater. These glands are best developed in marine birds.

Scapulars—Feathers at sides of shoulders.

Schemochrome—A structural color such as blue or iridescence; such colors result from the structure of the feather rather than from the presence of a pigment.

Scutellation—An arrangement or a covering of scales, as that on a bird's leg.

Secondaries—Major flight feathers of the wing that are attached to the ulna.

Sexual dichromatism—Male and female differ in color pattern (e.g., male hairy woodpecker [*Picoides villosus*, Picidae] has a red band on the back of the head, female has no red).

Sexual dimorphism—Male and female differ in morphology, such as size, feather size or shape, or bill size or shape.

Sibling species—Two or more species that are very closely related, presumably having differentiated from a common ancestor in the recent past; often difficult to distinguish, often interspecifically territorial.

Skimming—A method of foraging whereby the skimmers (Rhynchopidae) fly low over the water with the bottom bill slicing through the water and the tip of the bill above. When the bird hits a fish, the top bill snaps shut.

Slotting—Abrupt narrowing of the inner vane at the tip of some outer primaries on birds that soar; slotting breaks up wing-tip turbulence, thus facilitating soaring.

Sonagram—A graphic representation of sound.

Speciation—The evolution of new species.

Speculum—Colored patch on the wing, typically the secondaries, of many ducks (Anatidae).

Spur—A horny projection with a bony core found on the tarsometatarsus.

Sternum—Breastbone.

Structural color—See Schemochrome.

Suboscine—A songbird in the suborder Passeri, order Passeriformes, whose songs are thought to be innate, rather than learned.

Sympatric—Inhabiting the same range.

Syndactyl—Describes a condition of the foot of birds in which two toes are fused near the base for part of their length (e.g., kingfishers, Alcedinidae; hornbills, Bucerotidae).

Synsacrum—The expanded and elongated pelvis of birds that is fused with the lower vertebrae.

Syrinx (pl. syringes)—The "voice box" of a bird; a structure of cartilage and muscle located at the junction of the trachea and bronchi, lower on the trachea than the larynx of mammals. The number and complexity of muscles in the syrinx vary among groups of birds and have been of value in determining relationships among groups.

Systematist—A specialist in the classification of organisms; systematists strive to classify organisms on the basis of their evolutionary relationships.

Tarsus—In ornithology also sometimes called Tarsometatarsus or Metatarsus; the straight part of a bird's foot immediately above its toes. To the non-biologist, this seems to be the "leg" bone—leading to the notion that a bird's "knee" bends backwards. It does not. The joint at the top of the Tarsometatarsus is the "heel" joint, where the Tarsometatarsus meets the Tibiotarsus. The "knee" joint is between the Tibiotarsus and Femur.

Taxon (pl. taxa)—Any unit of scientific classification (e.g., species, genus, family, order).

Taxonomist—A specialist in the naming and classification of organisms. (See also Systematist. Taxonomy is the older science of naming things; identification of evolutionary relationships has not always been the goal of taxonomists. The modern science of Systematics generally incorporates taxonomy with the search for evolutionary relationships.)

Taxonomy—The science of identifying, naming, and classifying organisms into groups.

Teleoptiles—Juvenal feathers.

Territory—Any defended area. Typically birds defend a territory with sound such as song or drumming. Territorial defense is typically male against male, female against female, and within a species or between sibling species. Area defended varies greatly among taxa, seasons, and habitats. A territory may include the entire home range, only the area immediately around a nest, or only a feeding or roosting area.

Tertiary—The geological period including most of the Cenozoic; from about 65 to 1.8 million years ago.

Tertial—A flight feather of the wing that is loosely associated with the humerus; tertials fill the gap between the secondary feathers and the body.

Thermoregulation—The ability to regulate body temperature; can be either behavioral or physiological. Birds can regulate body temperature by sunning or moving to shade or water, but also generally regulate their body temperature through metabolic processes. Baby birds initially have poor thermoregulatory abilities and thus must be brooded.

Threatened—A category defined by IUCN and by the Endangered Species Act of 1973 in the United States to refer to a species that is at risk of becoming endangered.

Tomium (pl. tomia)—The cutting edges of a bird's bill.

Torpor—A period of reduced metabolic activity and lowered body temperature; often results from reduced availability of food or inclement weather; generally lasts for only a few hours (e.g., hummingbirds, Trochilidae; swifts, Apodidae).

Totipalmate—All toes joined by webs, a characteristic that identifies members of the order Pelecaniformes.

Tribe—A unit of classification below the subfamily and above the genus.

Tubercle—A knob- or wart-like projection.

Urohydrosis—A behavior characteristic of storks and New World vultures (Ciconiiformes) wherein these birds excrete on their legs and make use of the evaporation of the water from the excrement as an evaporative cooling mechanism.

Uropygial gland—A large gland resting atop the last fused vertebrae of birds at the base of a bird's tail; also known as oil gland or preen gland; secretes an oil used in preening.

Vane—The combined barbs that form a strong, yet flexible surface extending from the rachis of a feather.

Vaned feather—Any feather with vanes.

Viable population—A population that is capable of maintaining itself over a period of time. One of the major conservation issues of the twenty-first century is determining what is a minimum viable population size. Population geneticists have generally come up with estimates of about 500 breeding pairs.

Vibrissae—See Rictal bristle.

Vulnerable—A category defined by IUCN as a species that is not Critically Endangered or Endangered, but is still facing a threat of extinction.

Wallacea—The area of Indonesia transition between the Oriental and Australian biogeographical realms, named after Alfred Russell Wallace, who intensively studied this area.

Wattles—Sexual ornamentation that usually consists of flaps of skin on or near the base of the bill.

Zoogeographic region—See Biogeographic region.

Zygodactyl—Adjective referring to the arrangement of toes on a bird in which two toes project forward and two to the back.

Compiled by Jerome A. Jackson, PhD

Aves species list

Struthioniformes [Order]
Struthionidae [Family]
Struthio [Genus]
S. camelus [Species]

Rheidae [Family]
Rhea [Genus]
R. Americana [Species]
Pterocnemia [Genus]
P. pennata [Species]

Casuaridae [Family]
Casuarius [Genus]
C. bennetti [Species]
C. casuarius
C. unappendiculatus

Dromaiidae [Family]
Dromaius [Genus]
D. novaehollandiae [Species]
D. diemenianus

Apterygidae [Family]
Apteryx [Genus]
A. australis [Species]
A. owenii
A. haastii

Tinamiiformes [Order]
Tinamidae [Family]
Tinamus [Genus]
T. tao [Species]
T. solitarius
T. osgoodi
T. major
T. guttatus
Nothocercus [Genus]
N. bonapartei [Species]
N. julius
N. nigrocapillus
Crypturellus [Genus]
C. berlepschi [Species]
C. cinereus
C. soui
C. ptaritepui
C. obsoletus
C. undulatus
C. transfasciatus

C. strigulosus
C. duidae
C. erythropus
C. noctivagus
C. atrocapillus
C. cinnamomeus
C. boucardi
C. kerriae
C. variegatus
C. brevirostris
C. bartletti
C. parvirostris
C. casiquiare
C. tataupa
Rhynchotus [Genus]
R. rufescens [Species]
Nothoprocta [Genus]
N. taczanowski [Species]
N. kalinowskii
N. omata
N. perdicaria
N. cinerascens
N. pentlandii
N. curvirostris
Nothura [Genus]
N. boraquira [Species]
N. minor
N. darwinii
N. maculosa
Taoniscus [Genus]
T. nanus [Species]
Eudromia [Genus]
E. elegans [Species]
E. formosa
Tinamotis [Genus]
T. pentlandii [Species]
T. ingoufi

Procellariiformes [Order]
Diomedidae [Family]
Diomedea [Genus]
D. exulans [Species]
D. epomophora
D. irrorata
D. albatrus
D. nigripes
D. immutabilis

D. melanophrys
D. cauta
D. chrysostoma
D. chlororhynchos
D. bulleri
Phoebetria [Genus]
P. fusca [Species]
P. palpebrata
Macronectes [Genus]
M. giganteus [Species]
M. halli
Fulmarus [Genus]
F. glacialoides [Species]
F. glacialis
Thalassoica [Genus]
T. antarctica [Species]
Daption [Genus]
D. capense [Species]
Pagodroma [Genus]
P. nivea [Species]
Pterodroma [Genus]
P. macroptera [Species]
P. lessonii
P. incerta
P. solandri
P. magentae
P. rostrata
P. macgillivrayi
P. neglecta
P. arminjoniana
P. alba
P. ultima
P. brevirostris
P. mollis
P. inexpectata
P. cahow
P. hasitata
P. externa
P. baraui
P. phaeopygia
P. hypoleuca
P. nigripennis
P. axillaris
P. cookii
P. defilippiana
P. longirostris
P. leucoptera

Halobaena [Genus]
  H. caerulea
Pachyptila [Genus]
  P. vittata [Species]
  P. desolata
  P. belcheri
  P. turtur
  P. crassirostris
Bulweria [Genus]
  B. bulwerii [Species]
  B. fallax
Procellaria [Genus]
  P. aequinoctialis [Species]
  P. westlandica
  P. parkinsoni
  P. cinerea
Calonectris [Genus]
  C. diomedea [Species]
  C. leucomelas
Puffinus [Genus]
  P. pacificus [Species]
  P. bulleri
  P. carneipes
  P. creatopus
  P. gravis
  P. griseus
  P. tenuirostris
  P. nativitatis
  P. puffinus
  P. gavia
  P. huttoni
  P. lherminieri
  P. assimilis
Oceanites [Genus]
  O. oceanicus [Species]
  O. gracilis
Garrodia [Genus]
  G. nereis [Species]
Pelagodroma [Genus]
  P. marina [Species]
Fregetta [Genus]
  F. tropica [Species]
  F. grallaria
Nesofregetta [Genus]
  N. fuliginosa [Species]
Hydrobates [Genus]
  H. pelagicus [Species]
Halocyptena [Genus]
  H. microsoma [Species]
Oceanodroma [Genus]
  O. tethys [Species]
  O. castro
  O. monorhis
  O. leucorhoa
  O. macrodactyla
  O. markhami
  O. tristami
  O. melania
  O. matsudairae
  O. homochroa

O. hornbyi
O. furcata
Pelecanoides [Genus]
  P. garnotii [Species]
  P. magellani
  P. georgicus
  P. urinator

Sphenisciformes [Order]
Spheniscidae [Family]
  Aptenodytes [Genus]
    A. patagonicus [Species]
    A. forsteri
  Pygoscelis [Genus]
    P. papua [Species]
    P. adeliae
    P. antarctica
  Eudyptes [Genus]
    E. chrysocome [Species]
    E. pachyrhynchus
    E. robustus
    E. sclateri
    E. chryoslophus
  Megadyptes [Genus]
    M. antipodes [Species]
  Eudyptula [Genus]
    E. minor [Species]
  Spheniscus [Genus]
    S. demersus [Species]
    S. humboldti
    S. magellanicus
    S. mendiculus

Gaviiformes [Order]

Gaviidae [Family]
  Gavia [Genus]
    G. stellata [Species]
    G. arctica
    G. immer
    G. adamsii

Podicipediformes [Order]

Podicipedidae [Family]
  Rollandia [Genus]
    R. rolland [Species]
    R. microptera
  Tachybaptus [Genus]
    T. novaehollandiae [Species]
    T. ruficollis
    T. rufolavatus
    T. pelzelnii
    T. dominicus
  Podilymbus [Genus]
    P. podiceps [Species]
    P. gigas
  Poliocephalus [Genus]
    P. poliocephalus [Species]
    P. rufopectus
  Podiceps [Genus]
    P. major [Species]

P. auritus
P. grisegena
P. cristatus
P. nigricollis
P. occipitalis
P. taczanowskii
P. gallardoi
Aechmophorus [Genus]
  A. occidentalis [Species]

Pelecaniformes [Order]
Phaethontidae [Family]
  Phaethon [Genus]
    P. aethereus [Species]
    P. rubricauda
    P. lepturus

Fregatidae [Family]
  Fregata [Genus]
    F. magnificens [Species]
    F. minor
    F. ariel
    F. andrewsi

Phalacrocoracidae [Family]
  Phalacrocorax [Genus]
    P. carbo [Species]
    P. capillatus
    P. nigrogularis
    P. varius
    P. harrisi
    P. auritus
    P. olivaceous
    P. fuscicollis
    P. sulcirostris
    P. penicillatus
    P. capensis
    P. neglectus
    P. punctatus
    P. aristotelis
    P. perspicillatus
    P. urile
    P. pelagicus
    P. gaimardi
    P. magellanicus
    P. bouganvillii
    P. atriceps
    P. albiventer
    P. carunculatus
    P. campbelli
    P. fuscescens
    P. melanoleucos
    P. niger
    P. pygmaeus
    P. africanus
  Anhinga [Genus]
    A. anhinga [Species]
    A. melanogaster

Sulidae [Family]
  Sula [Genus]
    S. bassana [Species]

S. capensis
S. serrator
S. nebouxii
S. variegata
S. dactylatra
S. sula
S. leucogaster
S. abbotti

Pelecanidae [Family]
Pelecanus [Genus]
P. onocrotalus [Species]
P. rufescens
P. philippensis
P. conspicillatus
P. erythrorhynchos
P. occidentalis

Ciconiiformes [Order]
Ardeidae [Family]
Syrigma [Genus]
S. sibilatrix [Species]
Pilherodius [Genus]
P. pileatus [Species]
Ardea [Genus]
A. cinerea [Species]
A. herodias
A. cocoi
A. pacifica
A. melanocephala
A. hombloti
A. imperialis
A. sumatrana
A. goliath
A. purpurea
A. alba
Egretta [Genus]
E. rufescens [Species]
E. picata
E. vinaceigula
E. ardesiaca
E. tricolor
E. intermedia
E. ibis
E. novaehollandiae
E. caerulea
E. thula
E. garzetta
E. gularis
E. dimorpha
E. eulophotes
E. sacra
Ardeola [Genus]
A. ralloides [Species]
A. grayii
A. bacchus
A. speciosa
A. idae
A. rufiventris
A. striata

Agamia [Genus]
A. agami [Species]
Nyctanassa [Genus]
N. violacea [Species]
Nycticorax [Genus]
N. nycticorax [Species]
N. caledonicus
N. leuconotus
N. magnificus
N. goisagi
N. melanolophus
Cochlearius [Genus]
C. cochlearius [Species]
Tigrisoma [Genus]
T. mexicanum [Species]
T. fasciatum
T. lineatum
Zonerdius [Genus]
Z. heliosylus [Species]
Tigriornis [Genus]
T. leucolophus [Species]
Zebrilus [Genus]
Z. undulatus [Species]
Ixobrychus [Genus]
I. involucris [Species]
I. exilis
I. minutus
I. sinensis
I. eurhythmus
I. cinnamomeus
I. sturmii
I. flavicollis
Botaurus [Genus]
B. pinnatus [Species]
B. lentiginosus
B. stellaris
B. poiciloptilus

Scopidae [Family]
Scopus [Genus]
S. umbretta [Species]

Ciconiidae [Family]
Mycteria [Genus]
M. americana [Species]
M. cinerea
M. ibis
M. leucocephala
Anastomus [Genus]
A. oscitans [Species]
A. lamelligerus
Ciconia [Genus]
C. nigra [Species]
C. abdimii
C. episcopus
C. maguari
C. ciconia
Ephippiorhynchus [Genus]
E. asiaticus [Species]
E. senegalensis
Jabiru [Genus]
J. mycteria [Species]

Leptoptilos [Genus]
L. javanicus [Species]
L. dubius
L. crumeniferus

Balaenicipitidae [Family]
Balaeniceps [Genus]
B. rex [Species]

Threskiornithidae [Family]
Eudocimus [Genus]
E. albus [Species]
E. ruber
Phimosus [Genus]
P. infuscatus [Species]
Plegadis [Genus]
P. falcinellus [Species]
P. chihi
P. ridgwayi
Cercibis [Genus]
C. oxycerca [Species]
Theristicus [Genus]
T. caerulescens [Species]
T. caudatus
T. melanopsis
Mesembrinibis [Genus]
M. cayennensis [Species]
Bostrychia [Genus]
B. hagedash [Species]
B. carunculata
B. olivacea
B. rara
Lophotibis [Genus]
L. cristata [Species]
Threskiornis [Genus]
T. aethiopicus [Species]
T. spinicollis
Geronticus [Genus]
G. eremita [Species]
G. calvus
Pseudibis [Genus]
P. papillosa [Species]
P. gigantea
Nipponia [Genus]
N. nippon [Species]
Platalea [Genus]
P. leucocorodia [Species]
P. minor
P. alba
P. flavipes
P. ajaja

Phoenicopteriformes [Order]
Phoenicopteridae [Family]
Phoenicopterus [Genus]
P. ruber [Species]
P. chilensis
Phoeniconaias [Genus]
P. minor [Species]
Phoenicoparrus [Genus]
P. andinus [Species]
P. jamesii

Falconiformes [Order]
Cathartidae [Family]
  Coragyps [Genus]
    C. atratus [Species]
  Cathartes [Genus]
    C. burrovianus [Species]
    C. melambrotus
  Gymnogyps [Genus]
    G. californianus [Species]
  Vultur [Genus]
    V. gryphus [Species]
  Sarcoramphus [Genus]
    S. papa [Species]

Accipitridae [Family]
  Pandion [Genus]
    P. haliaetus [Species]
  Aviceda [Genus]
    A. cuculoides [Species]
    A. madagascariensis
    A. jerdoni
    A. subcristata
    A. leuphotes
  Leptodon [Genus]
    L. cayanensis [Species]
  Chondrohierax [Genus]
    C. uncinatus [Species]
  Henicopernis [Genus]
    H. longicauda [Species]
    H. infuscata
  Pernis [Genus]
    P. apivorus [Species]
    P. ptilorhynchus
    P. celebensis
  Elanoides [Genus]
    E. forficatus [Species]
  Macheiramphus [Genus]
    M. alcinus [Species]
  Gampsonyx [Genus]
    G. swainsonii [Species]
  Elanus [Genus]
    E. leucurus [Species]
    E. caeruleus
    E. notatus
    E. scriptus
  Chelictinia [Genus]
    C. riocourii [Species]
  Rostrhamus [Genus]
    R. sociabilis [Species]
    R. hamatus
  Harpagus [Genus]
    H. bidentatus [Species]
    H. diodon
  Ictinia [Genus]
    I. plumbea [Species]
    I. misisippiensis
  Lophoictinia [Genus]
    L. isura [Species]
  Hamirostra [Genus]
    H. melanosternon [Species]

Milvus [Genus]
  M. milvus [Species]
  M. migrans
Haliastur [Genus]
  H. sphenurus [Species]
  H. indus
Haliaeetus [Genus]
  H. leucogaster [Species]
  H. sanfordi
  H. vocifer
  H. vociferoides
  H. leucoryphus
  H. albicilla
  H. leucocephalus
  H. pelagicus
Ichthyophaga [Genus]
  I. humilis [Species]
  I. ichthyaetus
Gypohierax [Genus]
  G. angolensis [Species]
Gypaetus [Genus]
  G. barbatus [Species]
Neophron [Genus]
  N. percnopterus [Species]
Necrosyrtes [Genus]
  N. monachus [Species]
Gyps [Genus]
  G. bengalensis [Species]
  G. africanus
  G. indicus
  G. rueppellii
  G. himalayensis
  G. fulvus
Aegypius [Genus]
  A. monachus [Species]
  A. tracheliotus
  A. occipitalis
  A. calvus
Circaetus [Genus]
  C. gallicus [Species]
  C. cinereus
  C. fasciolatus
  C. cinerascens
Terathopius [Genus]
  T. ecaudatus [Species]
Spilornis [Genus]
  S. cheela [Species]
  S. elgini
Dryotriorchis [Genus]
  D. spectabilis [Species]
Eutriorchis [Genus]
  E. astur [Species]
Polyboroides [Genus]
  P. typus [Species]
  P. radiatus
Circus [Genus]
  C. assimilis [Species]
  C. maurus
  C. cyaneus
  C. cinereus

C. macrourus
C. melanoleucos
C. pygargus
C. ranivorus
C. aeruginosus
C. spilonotus
C. approximans
C. maillardi
C. buffoni
Melierax [Genus]
  M. gabar [Species]
  M. metabates
  M. canorus
Accipiter [Genus]
  A. poliogaster [Species]
  A. trivirgatus
  A. griseiceps
  A. tachiro
  A. castanilius
  A. badius
  A. brevipes
  A. butleri
  A. soloensis
  A. francesii
  A. trinotatus
  A. fasciatus
  A. novaehollandiae
  A. melanochlamys
  A. albogularis
  A. rufitorques
  A. haplochrous
  A. henicogrammus
  A. luteoschistaceus
  A. imitator
  A. poliocephalus
  A. princeps
  A. superciliosus
  A. collaris
  A. erythropus
  A. minullus
  A. gularis
  A. virgatus
  A. nanus
  A. cirrhocephalus
  A. brachyurus
  A. erythrauchen
  A. rhodogaster
  A. ovampensis
  A. madagascariensis
  A. nisus
  A. rufiventris
  A. striatus
  A. bicolor
  A. cooperii
  A. gundlachi
  A. melanoleucus
  A. henstii
  A. gentilis
  A. meyerianus
  A. buergersi

A. radiatus
A. doriae
Urotriorchis [Genus]
    U. macrourus [Species]
Butastur [Genus]
    B. rufipennis [Species]
    B. liventer
    B. teesa
    B. indicus
Kaupifalco [Genus]
    K. monogrammicus [Species]
Geranospiza [Genus]
    G. caerulescens [Species]
Leucopternis [Genus]
    L. schistacea [Species]
    L. plumbea
    L. princeps
    L. melanops
    L. kuhli
    L. lacernulata
    L. semiplumbea
    L. albicollis
    L. polionota
Asturina [Genus]
    A. nitida [Species]
Buteogallus [Genus]
    B. aequinoctialis [Species]
    B. subtilis
    B. anthracinus
    B. urubitinga
    B. meridionalis
Parabuteo [Genus]
    P. unicinctus [Species]
Busarellus [Genus]
    B. nigricollis [Species]
Geranoaetus [Genus]
    G. melanoleucus [Species]
Harpyhaliaetus [Genus]
    H. solitarius [Species]
    H. coronatus
Buteo [Genus]
    B. magnirostris [Species]
    B. leucorrhous
    B. ridgwayi
    B. lineatus
    B. platypterus
    B. brachyurus
    B. swainsoni
    B. galapagoensis
    B. albicaudatus
    B. polyosoma
    B. poecilochrous
    B. albonotatus
    B. solitarius
    B. ventralis
    B. jamaicensis
    B. buteo
    B. oreophilus
    B. brachypterus
    B. rufinus

B. hemilasius
B. regalis
B. lagopus
B. auguralis
B. rufofuscus
Morphnus [Genus]
    M. guianensis [Species]
Harpia [Genus]
    H. harpyja [Species]
Pithecophaga [Genus]
    P. jeffreyi [Species]
Ictinaetus [Genus]
    I. malayensis [Species]
Aquila [Genus]
    A. pomarina [Species]
    A. clanga
    A. rapax
    A. heliaca
    A. wahlbergi
    A. gurneyi
    A. chrysaetos
    A. audax
    A. verreauxii
Hieraaetus [Genus]
    H. fasciatus [Species]
    H. spilogaster
    H. pennatus
    H. morphnoides
    H. dubius
    H. kienerii
Spizastur [Genus]
    S. melanoleucus [Species]
Lophaetus [Genus]
    L. occipitalis [Species]
Spizaetus [Genus]
    S. africanus [Species]
    S. cirrhatus
    S. nipalensis
    S. bertelsi
    S. lanceolatus
    S. philippensis
    S. alboniger
    S. nanus
    S. tyrannus
    S. ornatus
Stephanoaetus [Genus]
    S. coronatus [Species]
Oroaetus [Genus]
    O. isidori [Species]
Polemaetus [Genus]
    P. bellicosus [Species]

Sagittariidae [Family]
    Sagittarius [Genus]
        S. serpentarius [Species]

Falconidae [Family]
    Daptrius [Genus]
        D. ater [Species]
        D. americanus

Phalcoboenus [Genus]
    P. megalopterus [Species]
    P. australis
Polyborus [Genus]
    P. plancus [Species]
Milvago [Genus]
    M. chimachima [Species]
    M. chimango
Herpetotheres [Genus]
    H. cachinnans [Species]
Micrastur [Genus]
    M. ruficollis [Species]
    M. gilvicollis
    M. mirandollei
    M. semitorquatus
    M. buckleyi
Spiziapteryx [Genus]
    S. circumcinctus [Species]
Polihierax [Genus]
    P. semitorquatus [Species]
    P. insignis
Microhierax [Genus]
    M. caerulescens [Species]
    M fringillarius
    M. latifrons
    M. erythrogerys
    M. melanoleucus
Falco [Genus]
    F. berigora [Species]
    F. naumanni
    F. sparverius
    F. tinnunculus
    F. newtoni
    F. punctatus
    F. araea
    F. moluccensis
    F. cenchroides
    F. rupicoloides
    F. alopex
    F. ardosiaceus
    F. dickinsoni
    F. zoniventris
    F. chicquera
    F. vespertinus
    F. amurensis
    F. eleonorae
    F. concolor
    F. femoralis
    F. columbarius
    F. rufigularis
    F. subbuteo
    F. cuvieri
    F. severus
    F. longipennis
    F. novaeseelandiae
    F. hypoleucos
    F. subniger
    F. mexicanus
    F. jugger
    F. biarmicus

AVES SPECIES LIST

F. cherrug
F. rusticolus
F. kreyenborgi
F. peregrinus
F. deiroleucus
F. fasciinucha

Anseriformes [Order]
Anatidae [Family]
Anseranas [Genus]
  A. semipalmata [Species]
Dendrocygna [Genus]
  D. guttata [Species]
  D. eytoni
  D. bicolor
  D. arcuata
  D. javanica
  D. viduata
  D. arborea
  D. autumnalis
Thalassornis [Genus]
  T. leuconotus [Species]
Cygnus [Genus]
  C. olor [Species]
  C. atratus
  C. melanocoryphus
  C. buccinator
  C. cygnus
  C. bewickii
  C. columbianus
Coscoroba [Genus]
  C. coscoroba [Species]
Anser [Genus]
  A. cygnoides [Species]
  A. fabalis
  A. albifrons
  A. erythropus
  A. anser
  A. indicus
  A. caerulescens
  A. rossii
  A. canagicus
Branta [Genus]
  B. sandvicensis [Species]
  B. canadensis
  B. leucopsis
  B. bernicla
  B. ruficollis
Cereopsis [Genus]
  C. novaehollandiae [Species]
Stictonetta [Genus]
  S. naevosa [Species]
Cyanochen [Genus]
  C. cyanopterus [Species]
Chloephaga [Genus]
  C. melanoptera [Species]
  C. picta
  C. hybrida
  C. poliocephala
  C. rubidiceps

Neochen [Genus]
  N. jubata [Species]
Alopochen [Genus]
  A. aegyptiaca [Species]
Tadorna [Genus]
  T. ferruginea [Species]
  T. cana
  T. variegata
  T. cristata
  T. tadornoides
  T. tadorna
  T. radjah
Tachyeres [Genus]
  T. pteneres [Species]
  T. brachypterus
  T. patachonicus
Plectropterus [Genus]
  P. gambensis [Species]
Cairina [Genus]
  C. moschata [Species]
  C. scutulata
Pteronetta [Genus]
  P. hartlaubii
Sarkidiornis [Genus]
  S. melanotos [Species]
Nettapus [Genus]
  N. pulchellus [Species]
  N. coromandelianus
  N. auritus
Callonetta [Genus]
  C. leucophrys [Species]
Aix [Genus]
  A. sponsa [Species]
  A. galericulata
Chenonetta [Genus]
  C. jubata [Species]
Amazonetta [Genus]
  A. brasiliensis [Species]
Merganetta [Genus]
  M. armata [Species]
Hymenolaimus [Genus]
  H. malacorhynchos [Species]
Anas [Genus]
  A. waigiuensis [Species]
  A. penelope
  A. americana
  A. sibilatrix
  A. falcata
  A. strepera
  A. formosa
  A. crecca
  A. flavirostris
  A. capensis
  A. gibberifrons
  A. bernieri
  A. castanea
  A. aucklandica
  A. platyrhynchos
  A. rubripes
  A. undulata

A. melleri
A. poecilorhyncha
A. superciliosa
A. luzonica
A. sparsa
A. specularioides
A. specularis
A. acuta
A. georgica
A. bahamensis
A. erythrorhyncha
A. versicolor
A. hottentota
A. querquedula
A. discors
A. cyanoptera
A. platalea
A. smithii
A. rhynchotis
A. clypeata
Malacorhynchus [Genus]
  M. membranaceus [Species]
Marmaronetta [Genus]
  M. angustirostris [Species]
Rhodonessa [Genus]
  R. caryophyllacea [Species]
Netta [Genus]
  N. rufina [Species]
  N. peposaca
  N. erythrophthalma
Aythya [Genus]
  A. valisineria [Species]
  A. ferina
  A. americana
  A. collaris
  A. australis
  A. baeri
  A. nyroca
  A. innotata
  A. novaeseelandiae
  A. fuligula
  A. marila
  A. affinis
Somateria [Genus]
  S. mollissima [Species]
  S. spectabilis
  S. fischeri
Polysticta [Genus]
  P. stelleri [Species]
Camptorhynchus [Genus]
  C. labradorius [Species]
Histrionicus [Genus]
  H. histrionicus [Species]
Clangula [Genus]
  C. hyemalis [Species]
Melanitta [Genus]
  M. nigra [Species]
  M. perspicillata
  M. fusca

Bucephala [Genus]
  B. clangula [Species]
  B. islandica
  B. albeola
Mergus [Genus]
  M. albellus [Species]
  M. cucullatus
  M. octosetaceous
  M. serrator
  M. squamatus
  M. merganser
  M. australis
Heteronetta [Genus]
  H. atricapilla [Species]
Oxyura [Genus]
  O. dominica [Species]
  O. jamaicensis
  O. leucocephala
  O. maccoa
  O. vittata
  O. australis
Biziura [Genus]
  B. lobata [Species]

Anhimidae [Family]
Anhima [Genus]
  A. cornuta [Species]
Chauna [Genus]
  C. chavaria [Species]
  C. torquata

Galliformes [Order]
Megapodiidae [Family]
Megapodius [Genus]
  M. nicobariens [Species]
  M. tenimberens
  M. reinwardt
  M. affinis
  M. eremita
  M. freycinet
  M. laperouse
  M. layardi
  M. pritchardii
Eulipoa [Genus]
  E. wallacei [Species]
Leipoa [Genus]
  L. ocellata [Species]
Alectura [Genus]
  A. lathami [Species]
Talegalla [Genus]
  T. cuvieri [Species]
  T. fuscirostris
  T. jobiensis
Aepypodius [Genus]
  A. arfakianus [Species]
  A. bruijnii
Macrocephalon [Genus]
  M. maleo [Species]

Cracidae [Family]
Nothocrax [Genus]
  N. urumutum [Species]

Mitu [Genus]
  M. tomentosa [Species]
  M. salvini
  M. mitu
Pauxi [Genus]
  P. pauxi [Species]
Crax [Genus]
  C. nigra [Species]
  C. alberti
  C. fasciolata
  C. pinima
  C. globulosa
  C. blumenbachii
  C. rubra
Penelope [Genus]
  P. purpurascens [Species]
  P. ortoni
  P. albipennis
  P. marail
  P. montagnii
  P. obscura
  P. superciliaris
  P. jacu-caca
  P. ochrogaster
  P. pileata
  P. argyrotis
Ortalis [Genus]
  O. motmot [Species]
  O. spixi
  O. araucuan
  O. superciliaris
  O. guttata
  O. columbiana
  O. wagleri
  O. vetula
  O. ruficrissa
  O. ruficauda
  O. garrula
  O. canicollis
  O. erythroptera
Penelopina [Genus]
  P. nigra [Species]
Chamaepetes [Genus]
  C. goudotii [Species]
  C. unicolor
Pipile [Genus]
  P. pipile [Species]
  P. cumanensis
  P. jacutinga
Aburria [Genus]
  A. aburri [Species]
Oreophasis [Genus]
  O. derbianus [Species]

Tetraonidae [Family]
Tetrao [Genus]
  T. urogallus [Species]
  T. parvirostris
Lyrurus [Genus]
  L. tetrix [Species]
  L. mlokosiewiczi

Dendragapus [Genus]
  D. obscurus [Species]
Lagopus [Genus]
  L. scoticus [Species]
  L. lagopus
  L. mutus
  L. leucurus
Canachites [Genus]
  C. canadensis [Species]
  C. franklinii
Falcipennis [Genus]
  F. falcipennis [Species]
Tetrastes [Genus]
  T. bonasia [Species]
  T. sewerzowi
Bonasa [Genus]
  B. umbellus [Species]
Pedioecetes [Genus]
  P. phasianellus [Species]
Tympanuchus [Genus]
  T. cupido [Species]
  T. palladicinctus
Centrocercus [Genus]
  C. urophasianus [Species]

Phasianidae [Family]
Dendrortyx [Genus]
  D. barbatus [Species]
  D. macroura
  D. leucophrys
  D. hypospodius
Oreortyx [Genus]
  O. picta [Species]
Callipepla [Genus]
  C. squamota [Species]
Lophortyx [Genus]
  L. californica [Species]
  L. gambelli
  L. leucoprosopon
  L. douglasii
Philortyx [Genus]
  P. fasciatus [Species]
Colinus [Genus]
  C. virginianus [Species]
  C. nigrogularis
  C. leucopogon
  C. cristatus
Odontophorus [Genus]
  O. gujanensis [Species]
  O. capueira
  O. erythrops
  O. hyperythrus
  O. melanonotus
  O. speciosus
  O. loricatus
  O. parambae
  O. strophium
  O. atrifrons
  O. leucolaemus
  O. columbianus
  O. soderstromii

O. balliviani
O. stellatus
O. guttatus
Dactylortyx [Genus]
   D. thoracicus [Species]
Cyrtonyx [Genus]
   C. montezumae [Species]
   C. sallei
   C. ocellatus
Rhynchortyx [Genus]
   R. cinctus [Species]
Lerwa [Genus]
   L. lerwa [Species]
Ammoperdix [Genus]
   A. griseogularis [Species]
   A. heyi
Tetraogallus [Genus]
   T. caucasicus [Species]
   T. caspius
   T. tibetanus
   T. altaicus
   T. himalayensis
Tetraophasis [Genus]
   T. obscurus [Species]
   T. szechenyii
Alectoris [Genus]
   A. graeca [Species]
   A. rufa
   A. barbara
   A. melanocephala
Anurophasis [Genus]
   A. monorthonyx [Species]
Francolinus [Genus]
   F. francolinus [Species]
   F. pictus
   F. pintadeanus
   F. pondicerianus
   F. gularis
   F. lathami
   F. nahani
   F. streptophorus
   F. coqui
   F. albogularis
   F. sephaena
   F. africanus
   F. shelleyi
   F. levaillantii
   F. finschi
   F. gariepensis
   F. adspersus
   F. capensis
   F. natalensis
   F. harwoodi
   F. bicalcaratus
   F. icterorhynchus
   F. clappertoni
   F. hartlaubi
   F. swierstrai
   F. hildebrandti
   F. squamatus

F. ahantensis
F. griseostriatus
F. camerunensis
F. nobilis
F. jacksoni
F. castaneicollis
F. atrifrons
F. erckelii
Pternistis [Genus]
   P. rufopictus [Species]
   P. afer
   P. swainsonii
   P. leucoscepus
Perdix [Genus]
   P. perdix [Species]
   P. barbata
   P. hodgsoniae
Rhizothera [Genus]
   R. longirostris [Species]
Margaroperdix [Genus]
   M. madagarensis [Species]
Melanoperdix [Genus]
   M. nigra [Species]
Coternix [Genus]
   C. coturnix [Species]
   C. coromandelica
   C. delegorguei
   C. pectoralis
   C. novaezelandiae
Synoicus [Genus]
   S. ypsilophorus [Species]
Excalfactoria [Genus]
   E. adansonii [Species]
   E. chinensis
Perdicula [Genus]
   P. asiatica [Species]
Cryptoplectron [Genus]
   C. erythrorhynchum [Species]
   C. manipurensis
Arborophila [Genus]
   A. torqueola [Species]
   A. rufogularis
   A. atrogularis
   A. crudigularis
   A. mandellii
   A. brunneopectus
   A. rufipectus
   A. gingica
   A. davidi
   A. cambodiana
   A. orientalis
   A. javanica
   A. rubrirostris
   A. hyperythra
   A. ardens
Tropicoperdix [Genus]
   T. charltonii [Species]
   T. chloropus
   T. merlini
Caloperdix [Genus]
   C. oculea [Species]

Haematortyx [Genus]
   H. sanguiniceps [Species]
Rollulus [Genus]
   R. roulroul [Species]
Ptilopachus [Genus]
   P. petrosus [Species]
Bambusicola [Genus]
   B. fytchii [Species]
   B. thoracica
Galloperdix [Genus]
   G. spadicea [Species]
   G. lunulata
   G. bicalcarata
Ophrysia [Genus]
   O. superciliosa [Species]
Ithaginis [Genus]
   I. cruentus [Species]
Tragopan [Genus]
   T. melanocephalus [Species]
   T. satyra
   T. blythii
   T. temminckii
   T. caboti
Lophophorus [Genus]
   L. impejanus [Species]
   L. sclateri
   L. lhuysii
Crossoptilon [Genus]
   C. mantchuricum [Species]
   C. auritum
   C. crossoptilon
Gennaeus [Genus]
   G. leucomelanos [Species]
   G. horsfieldii
   G. lineatus
   G. nycthemerus
Hierophasis [Genus]
   H. swinhoii [Species]
   H. imperialis
   H. edwardsi
Houppifer [Genus]
   H. erythrophthalmus [Species]
   H. inornatus
Lophura [Genus]
   L. rufa [Species]
   L. ignita
Diardigallus [Genus]
   D. diardi [Species]
Lobiophasis [Genus]
   L. bulweri [Species]
Gallus [Genus]
   G. gallus [Species]
   G. lafayetii
   G. sonneratii
   G. varius
Pucrasia [Genus]
   P. macrolopha [Species]
Catreus [Genus]
   C. wallichii [Species]
Phasianus [Genus]
   P. colchicus [Species]

Syrmaticus [Genus]
  S. reevesii [Species]
  S. soemmerringii
  S. humiae
  S. ellioti
  S. mikado
Chrysolophus [Genus]
  C. pictus [Species]
  C. amherstiae
Chalcurus [Genus]
  C. inopinatus [Species]
  C. chalcurus
Polyplecton [Genus]
  P. bicalcaratum [Species]
  P. germaini
  P. malacensis
  P. schleiermacheri
  P. emphanum
Rheinardia [Genus]
  R. ocellata [Species]
Argusianus [Genus]
  A. argus [Species]
Pavo [Genus]
  P. cristatus [Species]
  P. muticus

Numididae [Family]
  Phasidus [Genus]
    P. niger [Species]
  Agelastes [Genus]
    A. meleagrides [Species]
  Numida [Genus]
    N. meleagris [Species]
  Guttera [Genus]
    G. plumifera [Species]
    G. edouardi
    G. pucherani
  Acryllium [Genus]
    A. vulturinum [Species]

Meleagridae [Family]
  Meleagris [Genus]
    M. gallopavo [Species]
  Agriocharis [Genus]
    A. ocellata [Species]

Opisthocomidae [Family]
  Opisthocomus [Genus]
    O. hoazin [Species]

Gruiformes [Order]
Mesoenatidae [Family]
  Mesoenas [Genus]
    M. variegata [Species]
    M. unicolor
  Monias [Genus]
    M. benschi [Species]

Turnicidae [Family]
  Turnix [Genus]
    T. sylvatica [Species]
    T. worcesteri

T. nana
T. hottentotta
T. tanki
T. suscitator
T. nigricollis
T. ocellata
T. melanogaster
T. varia
T. castanota
T. pyrrhothorax
T. velox
Ortyxelos [Genus]
  O. meiffrenii [Species]
Pedionomus [Genus]
  P. torquatus [Species]

Gruidae [Family]
  Grus [Genus]
    G. grus [Species]
    G. nigricollis
    G. monacha
    G. canadensis
    G. japonensis
    G. americana
    G. vipio
    G. antigone
    G. rubicunda
    G. leucogeranus
  Bugeranus [Genus]
    B. carunculatus [Species]
  Anthropoides [Genus]
    A. virgo [Species]
    A. paradisea
  Balearica [Genus]
    B. pavonina [Species]

Aramidae [Family]
  Aramus [Genus]
    A. scolopaceus [Species]

Psophiidae [Family]
  Psophia [Genus]
    P. crepitans [Species]
    P. leucoptera
    P. viridis

Rallidae [Family]
  Rallus [Genus]
    R. longirostris [Species]
    R. elegans
    R. limicola
    R. semiplumbeus
    R. aquaticus
    R. caerulescens
    R. madagascariensis
    R. pectoralis
    R. muelleri
    R. striatus
    R. philippensis
    R. ecaudata
    R. torquatus
    R. owstoni
    R. wakensis

Nesolimnas [Genus]
  N. dieffenbachii [Species]
Cabalus [Genus]
  C. modestus [Species]
Atlantisia [Genus]
  A. rogersi [Species]
Tricholimnas [Genus]
  T. conditicius [Species]
  T. lafresnayanus
  T. sylvestris
Ortygonax [Genus]
  O. rytirhynchos [Species]
  O. nigricans
Pardirallus [Genus]
  P. maculatus [Species]
Dryolimnas [Genus]
  D. cuvieri [Species]
Rougetius [Genus]
  R. rougetii [Species]
Amaurolimnas [Genus]
  A. concolor [Species]
Rallina [Genus]
  R. fasciata [Species]
  R. eurizonoides
  R. canningi
  R. tricolor
Rallicula [Genus]
  R. rubra [Species]
  R. leucospila
Cyanolimnas [Genus]
  C. cerverai [Species]
Aramides [Genus]
  A. mangle [Species]
  A. cajanea
  A. wolfi
  A. gutturalis
  A. ypecaha
  A. axillaris
  A. calopterus
  A. saracura
Aramidopsis [Genus]
  A. plateni [Species]
Nesoclopeus [Genus]
  N. poeciloptera [Species]
  N. woodfordi
Gymnocrex [Genus]
  G. rosenbergii [Species]
  G. plumbeiventris
Gallirallus [Genus]
  G. australis [Species]
  G. troglodytes
Habropteryx [Genus]
  H. insignis [Species]
Habroptila [Genus]
  H. wallacii [Species]
Megacrex [Genus]
  M. inepta [Species]
Eulabeornis [Genus]
  E. castaneoventris [Species]
Himantornis [Genus]
  H. haematopus [Species]

AVES SPECIES LIST

Aves species list

Canirallus [Genus]
  C. oculeus [Species]
Mentocrex [Genus]
  M. kioloides [Species]
Crecopsis [Genus]
  C. egregria [Species]
Crex [Genus]
  C. crex [Species]
Anurolimnas [Genus]
  A. castaneiceps [Species]
Limnocorax [Genus]
  L. flavirostra [Species]
Porzana [Genus]
  P. parva [Species]
  P. pusilla
  P. porzana
  P. fluminea
  P. carolina
  P. spiloptera
  P. flaviventer
  P. albicollis
  P. fusca
  P. paykullii
  P. olivieri
  P. bicolor
  P. tabuensis
Porzanula [Genus]
  P. palmeri [Species]
Pennula [Genus]
  P. millsi [Species]
  P. sandwichensis
Nesophylax [Genus]
  N. ater [Species]
Aphanolimnas [Genus]
  A. monasa [Species]
Laterallus [Genus]
  L. jamaicensis [Species]
  L. spilonotus
  L. exilis
  L. albigularis
  L. melanophaius
  L. ruber
  L. levraudi
  L. viridis
  L. hauxwelli
  L. leucopyrrhus
Micropygia [Genus]
  M. schomburgkii [Species]
Coturnicops [Genus]
  C. exquisita [Species]
  C. noveboracensis
  C. notata
  C. ayresi
Neocrex [Genus]
  N. erythrops [Species]
Sarothura [Genus]
  S. rufa [Species]
  S. lugeus
  S. pulchra
  S. elegans

S. bohmi
S. antonii
S. lineata
S. insularis
S. watersi
Aenigmatolimnas [Genus]
  A. marginalis [Species]
Poliolimnas [Genus]
  P. cinereus [Species]
Porphyriops [Genus]
  P. melanops [Species]
Tribonyx [Genus]
  T. ventralis [Species]
  T. mortierii
Amaurornis [Genus]
  A. akool [Species]
  A. olivacea
  A. isabellina
  A. phoenicurus
Gallicrex [Genus]
  G. cinerea [Species]
Gallinula [Genus]
  G. tenebrosa [Species]
  G. chloropus
  G. angulata
Porphyriornis [Genus]
  P. nesiotis [Species]
  P. comeri
Pareudiastes [Genus]
  P. pacificus [Species]
Porphyrula [Genus]
  P. alleni [Species]
  P. martinica
  P. parva
Porphyrio [Genus]
  P. porphyrio [Species]
  P. madagascariensis
  P. poliocephalus
  P. albus
  P. pulverulentus
Notornis [Genus]
  N. mantelli [Species]
Fulica [Genus]
  F. atra [Species]
  F. cristata
  F. americana
  F. ardesiaca
  F. armillata
  F. caribaea
  F. leucoptera
  F. rufifrons
  F. gigantea
  F. cornuta

Heliornithidae [Family]
  Podica [Genus]
    P. senegalensis [Species]
  Heliopais [Genus]
    H. personata [Species]
  Heliornis [Genus]
    H. fulica [Species]

Rhynochetidae [Family]
  Rhynochetos [Genus]
    R. jubatus [Species]
Eurypygidae [Family]
  Eurypyga [Genus]
    E. helias [Species]
Cariamidae [Family]
  Cariama [Genus]
    C. cristata [Species]
  Chunga [Genus]
    C. burmeisteri [Species]
Otidae [Family]
  Tetrax [Genus]
    T. tetrax [Species]
  Otis [Genus]
    O. tarda [Species]
  Neotis [Genus]
    N. cafra [Species]
    N. ludwigii
    N. burchellii
    N. Nuba
    N. heuglinii
  Choriotius [Genus]
    C. arabs [Species]
    C. kori
    C. nigriceps
    C. australis
  Chlamydotis [Genus]
    C. undulata [Species]
  Lophotis [Genus]
    L. savilei [Species]
    L. ruficrista
  Afrotis [Genus]
    A. atra [Species]
  Eupodotis [Genus]
    E. vigorsii [Species]
    E. ruppellii
    E. humilis
    E. senegalensis
    E. caerulescens
  Lissotis [Genus]
    L. melanogaster [Species]
    L. hartlaubii
  Houbaropsis [Genus]
    H. bengalensis [Species]
  Sypheotides [Genus]
    S. indica [Species]

Charadriiformes [Order]
Jacanidae [Family]
  Microparra [Genus]
    M. capensis [Species]
  Actophilornis [Genus]
    A. africana [Species]
    A. albinucha
  Irediparra [Genus]
    I. gallinacea [Species]
  Hydrophasianus [Genus]
    H. chirurgus [Species]

Metopidius [Genus]
  M. indicus [Species]
Jacana [Genus]
  J. spinosa [Species]

Rostratulidae [Family]
  Rostratula [Genus]
    R. benghalensis [Species]
  Nycticryphes [Genus]
    N. semicollaris [Species]

Haematopodidae [Family]
  Haematopus [Genus]
    H. ostralegus [Species]
    H. leucopodus
    H. fuliginosus
    H. ater
  Chettusia [Genus]
    C. leucura [Species]
    C. gregaria

Charadriidae [Family]
  Vanellus [Genus]
    V. vanellus [Species]
  Belonopterus [Genus]
    B. chilensis [Species]
  Hemiparra [Genus]
    H. crassirostris [Species]
  Tylibyx [Genus]
    T. melanocephalus [Species]
  Microsarcops [Genus]
    M. cinereus [Species]
  Lobivanellus [Genus]
    L. indicus [Species]
  Xiphidiopterus [Genus]
    X. albiceps [Species]
  Rogibyx [Genus]
    R. tricolor [Species]
  Lobibyx [Genus]
    L. novaehollandiae [Species]
    L. miles
  Afribyx [Genus]
    A. senegallus [Species]
  Stephanibyx [Genus]
    S. lugubris [Species]
    S. melanopterus
    S. coronatus
  Hoplopterus [Genus]
    H. spinosus [Species]
    H. armatus
    H. duvaucelii
  Hoploxypterus [Genus]
    H. cayanus [Species]
  Ptilocelys [Genus]
    P. resplendens [Species]
  Zonifer [Genus]
    Z. tricolor [Species]
  Anomalophrys [Genus]
    A. superciliosus [Species]
  Lobipluvia [Genus]
    L. malabarica [Species]

Sarciophorus [Genus]
  S. tectus [Species]
Squatarola [Genus]
  S. squatarola [Species]
Pluvialis [Genus]
  P. apricaria [Species]
  P. dominica
Pluviorhynchus [Genus]
  P. obscurus [Species]
Charadrius [Genus]
  C. rubricollis [Species]
  C. hiaticula
  C. melodus
  C. dubius
  C. alexandrinus
  C. venustus
  C. falklandicus
  C. alticola
  C. bicinctus
  C. peronii
  C. collaris
  C. pecuarius
  C. sanctaehelenae
  C. thoracicus
  C. placidus
  C. vociferus
  C. tricollaris
  C. mongolus
  C. wilsonia
  C. leschenaultii
Elseyornis [Genus]
  E. melanops [Species]
Eupoda [Genus]
  E. asiatica [Species]
  E. veredus
  E. montana
Oreopholus [Genus]
  O. ruficollis [Species]
Erythrogonys [Genus]
  E. cinctus [Species]
Eudromias [Genus]
  E. morinellus [Species]
Zonibyx [Genus]
  Z. modestus [Species]
Thinornis [Genus]
  T. novaeseelandiae [Species]
Anarhynchus [Genus]
  A. frontalis [Species]
Pluvianellus [Genus]
  P. socialis [Species]
Phegornis [Genus]
  P. mitchellii [Species]

Scopacidae [Family]
  Aechmorhynchus [Genus]
    A. cancellatus [Species]
    A. parvirostris
  Prosobonia [Genus]
    P. leucoptera [Species]
  Bartramia [Genus]
    B. longicauda [Species]

Numenius [Genus]
  N. minutus [Species]
  N. borealis
  N. phaeopus
  N. tahitiensis
  N. tenuirostris
  N. arquata
  N. madagascariensis
  N. americanus
Limosa [Genus]
  L. limosa [Species]
  L. haemastica
  L. lapponica
  L. fedoa
Tringa [Genus]
  T. erythropus [Species]
  T. totanus
  T. flavipes
  T. stagnatilis
  T. nebularia
  T. melanoleuca
  T. ocrophus
  T. solitaria
  T. glareola
Pseudototanus [Genus]
  P. guttifer [Species]
Xenus [Genus]
  X. cinereus [Species]
Actitis [Genus]
  A. hypoleucos [Species]
  A. macularia
Catoptrophorus [Genus]
  C. semipalmatus [Species]
Heteroscelus [Genus]
  H. brevipes [Species]
  H. incanus
Aphriza [Genus]
  A. virgata [Species]
Arenaria [Genus]
  A. interpres [Species]
  A. melanocephala
Limnodromus [Genus]
  L. griseus [Species]
  L. semipalmatus
Coenocorypha [Genus]
  C. aucklandica [Species]
Capella [Genus]
  C. solitaria [Species]
  C. hardwickii
  C. nemoricola
  C. stenura
  C. megala
  C. nigripennis
  C. macrodactyla
  C. media
  C. gallinago
  C. delicata
  C. paraguaiae
  C. nobilis
  C. undulata

Chubbia [Genus]
  C. imperialis [Species]
  C. jamesoni
  C. stricklandii
Scolopax [Genus]
  S. rusticola [Species]
  S. saturata
  S. celebensis
  S. rochussenii
Philohela [Genus]
  P. minor [Species]
Lymnocryptes [Genus]
  L. minima [Species]
Calidris [Genus]
  C. canutus [Species]
  C. tenuirostris
Crocethia [Genus]
  C. alba [Species]
Ereunetes [Genus]
  E. pusillus [Species]
  E. mauri
Eurynorhynchus [Genus]
  E. pygmeus [Species]
Erolia [Genus]
  E. ruficollis [Species]
  E. minuta
  E. temminckii
  E. subminuta
  E. minutilla
  E. fuscicollis
  E. bairdii
  E. melanotos
  E. acuminata
  E. maritima
  E. ptilocnemis
  E. alpina
  E. testacea
Limicola [Genus]
  L. falcinellus [Species]
Micropalama [Genus]
  M. himantopus [Species]
Tryngites [Genus]
  T. subruficollis [Species]
Philomachus [Genus]
  P. pugnax [Species]

Recurvostridae [Family]
  Ibidorhyncha [Genus]
    I. struthersii [Species]
  Himantopus [Genus]
    H. himantopus [Species]
  Cladorhynchus [Genus]
    C. leucocephala [Species]
  Recurvirostra [Genus]
    R. avosetta [Species]
    R. americana
    R. novaehollandiae
    R. andina

Phalaropodidae [Family]
  Phalaropus [Genus]
    P. fulicarius [Species]

Steganopus [Genus]
  S. tricolor [Species]
Lobipes [Genus]
  L. lobatus [Species]

Dromadidae [Family]
  Dromas [Genus]
    D. ardeola [Species]

Burhinidae [Family]
  Burhinus [Genus]
    B. oedicnemus [Species]
    B. senegalensis
    B. vermiculatus
    B. capensis
    B. bistriatus
    B. superciliaris
    B. magnirostris
  Esacus [Genus]
    E. recurvirostris [Species]
  Orthoramphus [Genus]
    O. magnirostris [Species]

Glareolidae [Family]
  Pluvianus [Genus]
    P. aegyptius [Species]
  Cursorius [Genus]
    C. cursor [Species]
    C. temminckii
    C. coromandelicus
  Rhinoptilus [Genus]
    R. africanus [Species]
    R. cinctus
    R. chalcopterus
    R. bitorquatus
  Peltohyas [Genus]
    P. australis [Species]
  Stiltia [Genus]
    S. isabella [Species]
  Glareola [Genus]
    G. pratincola [Species]
    G. maldivarum
    G. nordmanni
    G. ocularis
    G. nuchalis
    G. cinerea
    G. lactea
  Attagis [Genus]
    A. gayi [Species]
    A. malouinus
  Thinocorus [Genus]
    T. orbignyianus [Species]
    T. rumicivorus

Chionididae [Family]
  Chionis [Genus]
    C. alba [Species]
    C. minor

Stercorariidae [Family]
  Catharacta [Genus]
    C. skua [Species]

Stercorarius [Genus]
  S. pomarinus [Species]
  S. parasiticus
  S. longicaudus

Laridae [Family]
  Gabianus [Genus]
    G. pacificus [Species]
    G. scoresbii
  Pagophila [Genus]
    P. eburnea [Species]
  Larus [Genus]
    L. fuliginosus [Species]
    L. modestus
    L. heermanni
    L. leucophthalmus
    L. hemprichii
    L. belcheri
    L. crassirostris
    L. audouinii
    L. delawarensis
    L. canus
    L. argentatus
    L. fuscus
    L. californicus
    L. occidentalis
    L. dominicanus
    L. schistisagus
    L. marinus
    L. glaucescens
    L. hyperboreus
    L. leucopterus
    L. ichthyaetus
    L. atricilla
    L. brunnicephalus
    L. cirrocephalus
    L. serranus
    L. pipixcan
    L. novaehollandiae
    L. melanocephalus
    L. bulleri
    L. maculipennis
    L. ridibundus
    L. genei
    L. philadelphia
    L. minutus
    L. saundersi
  Rhodostethia [Genus]
    R. rosea [Species]
  Rissa [Genus]
    R. tridactyla [Species]
    R. brevirostris
  Creagrus [Genus]
    C. furcatus [Species]
  Xema [Genus]
    X. sabini [Species]
  Chlidonias [Genus]
    C. hybrida [Species]
    C. leucoptera
    C. nigra

Phaetusa [Genus]
    P. simplex [Species]
Gelochelidon [Genus]
    G. nilotica [Species]
Hydroprogne [Genus]
    H. tschegrava [Species]
Sterna [Genus]
    S. aurantia [Species]
    S. hirundinacea
    S. hirundo
    S. paradisaea
    S. vittata
    S. virgata
    S. forsteri
    S. trudeaui
    S. dougallii
    S. striata
    S. repressa
    S. sumatrana
    S. melanogaster
    S. aleutica
    S. lunata
    S. anaethetus
    S. fuscata
    S. nereis
    S. albistriata
    S. superciliaris
    S. balaenarum
    S. iorata
    S. albifrons
Thalasseus [Genus]
    T. bergii [Species]
    T. maximus
    T. bengalensis
    T. zimmermanni
    T. eurygnatha
    T. elegans
    T. sandvicensis
Larosterna [Genus]
    L. inca [Species]
Procelsterna [Genus]
    P. cerulea [Species]
Anous [Genus]
    A. stolidus [Species]
    A. tenuirostris
    A. minutus
Gygis [Genus]
    G. alba [Species]

Rynchopidae [Family]
Rynchops [Genus]
    R. nigra [Species]
    R. flavirostris
    R. albicollis

Alcidae [Family]
Plautus [Genus]
    P. alle [Species]
Pinguinis [Genus]
    P. impennis [Species]
Alca [Genus]
    A. torda [Species]

Uria [Genus]
    U. lomvia [Species]
    U. aalge
Cepphus [Genus]
    C. grylle [Species]
    C. columba
    C. carbo
Brachyramphus [Genus]
    B. marmoratus [Species]
    B. brevirostris
    B. hypoleucus
    B. craveri
Synthliboramphus [Genus]
    S. antiquus [Species]
    S. wumizusume
Ptychoramphus [Genus]
    P. aleuticus [Species]
Cyclorrhynchus [Genus]
    C. psittacula [Species]
Aethia [Genus]
    A. cristatella [Species]
    A. pusilla
    A. pygmaea
Cercorhinca [Genus]
    C. monocerata [Species]
Fratercula [Genus]
    F. arctica [Species]
    F. corniculata
Lunda [Genus]
    L. cirrhata [Species]

Columbiformes [Order]
Pteroclididae [Family]
Syrrhaptes [Genus]
    S. tibetanus [Species]
    S. paradoxus
Pterocles [Genus]
    P. alchata [Species]
    P. namaqua
    P. exustus
    P. senegallus
    P. orientalis
    P. coronatus
    P. gutturalis
    P. burchelli
    P. personatus
    P. decoratus
    P. lichtensteinii
    P. bicinctus
    P. indicus
    P. quadricinctus

Raphidae [Family]
Raphus [Genus]
    R. cucullatus [Species]
    R. solitarius
Pezophaps [Genus]
    P. solitaria [Species]

Columbidae [Family]
Sphenurus [Genus]
    S. apicauda [Species]

    S. seimundi
    S. oxyura
    S. sphenurus
    S. korthalsi
    S. sieboldii
    S. farmosae
Butreron [Genus]
    B. capellei [Species]
Treron [Genus]
    T. curvirostra [Species]
    T. pompadora
    T. fulvicollis
    T. olax
    T. vernans
    T. bicincta
    T. s. thomae
    T. australis
    T. calva
    T. delalandii
    T. waalia
    T. phoenicoptera
Phapitreron [Genus]
    P. leucotis [Species]
    P. amethystina
Leucotreron [Genus]
    L. occipitalis [Species]
    L. fischeri
    L. merrilli
    L. marchei
    L. subgularis
    L. leclancheri
    L. cincta
    L. dohertyi
    L. porphyrea
Ptilinopus [Genus]
    P. dupetithouarsii [Species]
    P. regina
    P. mercierii
    P. purpuratus
    P. coralensis
    P. insularis
    P. rarotongensis
    P. huttoni
    P. porphyraceus
    P. greyii
    P. richardsii
    P. ponapensis
    P. pelewensis
    P. roseicapilla
    P. perousii
    P. superbus
    P. pulchellus
    P. coronulatus
    P. monacha
    P. iozonus
    P. insolitus
    P. rivoli
    P. miquelli
    P. bellus
    P. solomonensis

P. viridis
P. eugeniae
P. geelvinkiana
P. pectoralis
P. naina
P. byogastra
P. granulifrons
P. melanospila
P. jambu
P. wallacii
P. aurantiifrons
P. ornatus
P. perlatus
P. tannensis
Chrysoena [Genus]
C. victor [Species]
C. viridis
C. luteovirens
Alectroenas [Genus]
A. pulcherrima [Species]
A. sganzini
A. madagascariensis
A. nitidissima
Drepanoptila [Genus]
D. holosericea [Species]
Megaloprepia [Genus]
M. magnifica [Species]
M. formosa
Ducula [Genus]
D. galeata [Species]
D. aurorae
D. oceanica
D. pacifica
D. rubricera
D. myristicivora
D. concinna
D. aenea
D. oenothorax
D. pistrinaria
D. whartoni
D. rosacea
D. perspicillata
D. pickeringii
D. latrans
D. bakeri
D. brenchleyi
D. goliath
D. bicolor
D. luctuosa
D. melanura
D. spilorrhoa
D. cineracea
D. lacernulata
D. badia
D. mullerii
D. pinon
D. melanochroa
D. poliocephala
D. forsteni
D. mindorensis

D. radiata
D. rufigaster
D. finschii
D. chalconota
D. zoeae
D. carola
Cryptophaps [Genus]
C. poecilorrhoa [Species]
Hemiphaga [Genus]
H. novaeseelandiae [Species]
Lopholaimus [Genus]
L. antarcticus [Species]
Gymnophaps [Genus]
G. albertisii [Species]
G. solomonensis
G. mada
Columba [Genus]
C. leuconota [Species]
C. rupestris
C. livia
C. oenas
C. eversmanni
C. oliviae
C. albitorques
C. palumbus
C. trocaz
C. junoniae
C. leucocephala
C. picazuro
C. gymnophtalmos
C. squamosa
C. maculosa
C. unicincta
C. guinea
C. hodgsonii
C. arquatrix
C. thomensis
C. albinucha
C. flavirostris
C. oenops
C. inornata
C. caribaea
C. rufina
C. fasciata
C. albilinea
C. araucana
C. elphinstonii
C. torringtoni
C. pulchricollis
C. punicea
C. palumboides
C. janthina
C. versicolor
C. jouyi
C. vitiensis
C. pallidiceps
C. norfolciensis
C. argentina
C. pollenii
C. speciosa

C. nigriristris
C. goodsoni
C. subvinacea
C. plumbea
C. chiriquensis
C. purpureotincta
C. delegorguei
C. iriditorques
C. malherbii
Nesoenas [Genus]
. mayeri [Species]
Turacoena [Genus]
T. manadensis [Species]
T. modesta
Macropygia [Genus]
M. unchall [Species]
M. amboinensis
M. ruficeps
M. magna
M. phasianella
M. rufipennis
M. nigrirostris
M. mackinlayi
Reinwardtoena [Genus]
R. reinwardtsi [Species]
R. browni
Coryphoenas [Genus]
C. crassirostris [Species]
Ectopistes [Genus]
E. migratoria [Species]
Zenaidura [Genus]
Z. macroura [Species]
Z. graysoni
Z. auriculata
Zenaida [Genus]
Z. aurita [Species]
Z. asiatica
Nesopelia [Genus]
N. galapagoensis [Species]
Streptopelia [Genus]
S. turtur [Species]
S. orientalis
S. lugens
S. picturata
S. decaocto
S. roseogrisea
S. semitorquata
S. decipiens
S. capicola
S. vinacea
S. reichenowi
S. fulvopectoralis
S. bitorquata
S. tranquebarica
S. chinensis
S. senegalensis
Geopelia [Genus]
G. bumeralis [Species]
G. striata
G. cuneata

Metriopelia [Genus]
    M. ceciliae [Species]
    M. morenoi
    M. melanoptera
    M. aymara
Scardafella [Genus]
    S. inca [Species]
    S. squammata
Uropelia [Genus]
    U. campestris [Species]
Columbina [Genus]
    C. picui [Species]
Columbigallina [Genus]
    C. passerina [Species]
    C. talpacoti
    C. minuta
    C. buckleyi
    C. cruziana
Oxypelia [Genus]
    O. cyanopis [Species]
Claravis [Genus]
    C. pretiosa [Species]
    C. mondetoura
    C. godefrida
Oena [Genus]
    O. capensis [Species]
Tympanistria [Genus]
    T. tympanistria [Species]
Turtur [Genus]
    T. afer [Species]
    T. abyssinicus
    T. chalcospilos
    T. brehmeri
Chalcophaps [Genus]
    C. indica [Species]
    C. stephani
Henicophaps [Genus]
    H. albifrons [Species]
    H. foersteri
Petrophassa [Genus]
    P. albipennis [Species]
    P. rufipennis
Phaps [Genus]
    P. chalcoptera [Species]
    P. elegans
Ocyphaps [Genus]
    O. lophotes [Species]
Lophophaps [Genus]
    L. plumifera [Species]
    L. ferruginea
Geophaps [Genus]
    G. scripta [Species]
    G. smithii
Histriophaps [Genus]
    H. histrionica [Species]
Aplopelia [Genus]
    A. larvata [Species]
    A. simplex
Leptotila [Genus]
    L. verreauxi [Species]

L. megalura
L. jamaicensis
L. plumbeiceps
L. rufaxilla
L. wellsi
L. cassini
L. ochraceiventris
Osculatia [Genus]
    O. saphirina [Species]
Oreopeleia [Genus]
    O. veraguensis [Species]
    O. lawrencii
    O. goldmani
    O. costaricensis
    O. chrysia
    O. mystacea
    O. martinica
    O. violacea
    O. montana
    O. caniceps
    O. albifacies
    O. chiriquensis
    O. linearis
    O. bourcieri
    O. erythropareia
Geotrygon [Genus]
    G. versicolor [Species]
Gallicolumba [Genus]
    G. luzonica [Species]
    G. platenae
    G. keayi
    G. criniger
    G. menagei
    G. rufigula
    G. tristigmata
    G. beccarii
    G. salamonis
    G. sanctaecrucis
    G. stairi
    G. canifrons
    G. xanthonura
    G. kubaryi
    G. jobiensis
    G. erythroptera
    G. rubescens
    G. hoedtii
Leucosarcia [Genus]
    L. melanoleuca [Species]
Trugon [Genus]
    T. terrestris [Species]
Microgoura [Genus]
    M. meeki [Species]
Starnoenas [Genus]
    S. cyanocephala [Species]
Otidiphaps [Genus]
    O. nobilis [Species]
Caloenas [Genus]
    C. nicobarica [Species]
Goura [Genus]
    G. cristata [Species]

G. scheepmakeri
G. victoria
Didunculus [Genus]
    D. strigirostris [Species]

Psittaciformes [Order]
Psittacidae [Family]
Strigops [Genus]
    S. habroptilus [Species]
Nestor [Genus]
    N. meridionalis [Species]
    N. notabilis
    N. productus
Chalcopsitta [Genus]
    C. atra [Species]
    C. insignis
    C. sintillata
    C. duivenbodei
    C. cardinalis
Eos [Genus]
    E. cyanogenia [Species]
    E. reticulata
    E. squamata
    E. histrio
    E. bornea
    E. semilarvata
    E. goodfellowi
Trichoglossus [Genus]
    T. ornatus [Species]
    T. haematod
    T. rubiginosus
    T. chlorolepidotus
    T. euteles
Psitteuteles [Genus]
    P. flavoviridis [Species]
    P. johnstoniae
    P. goldiei
    P. versicolor
    P. iris
Pseudeos [Genus]
    P. fuscata [Species]
Domicella [Genus]
    D. hypoinochroa [Species]
    D. amabilis
    D. lory
    D. domicella
    D. tibialis
    D. chlorocercus
    D. albidinucha
    D. garrula
Phigys [Genus]
    P. solitarius [Species]
Vini [Genus]
    V. australis [Species]
    V. kuhlii
    V. stepheni
    V. peruviana
    V. ultramarina
Glossopsitta [Genus]
    G. concinna [Species]
    G. porphyrocephala
    G. pusilla

Charmosyna [Genus]
   C. palmarum [Species]
   C. meeki
   C. rubrigularis
   C. aureicincta
   C. diadema
   C. toxopei
   C. placentis
   C. rubronotata
   C. multistriata
   C. wilhelminae
   C. pulchella
   C. margarethae
   C. josefinae
   C. papou
Oreopsittacus [Genus]
   O. arfaki [Species]
Neopsittacus [Genus]
   N. musschenbroekii [Species]
   N. pullicauda
Psittaculirostris [Genus]
   P. desmaresti [Species]
   P. salvadorii
Opopsitta [Genus]
   P. gulielmitertii [Species]
   P. diophthalma
Lathamus [Genus]
   L. discolor [Species]
Micropsitta [Genus]
   M. bruijnii [Species]
   M. keiensis
   M. geelvinkiana
   M. pusio
   M. meeki
   M. finschii
Probosciger [Genus]
   P. aterrimus [Species]
Calyptorhynchus [Genus]
   C. baudinii [Species]
   C. funereus
   C. magnificus
   C. lathami
Callocephalon [Genus]
   C. fimbriatum [Species]
Kakatoe [Genus]
   K. galerita [Species]
   K. sulphurea
   K. alba
   K. moluccensis
   K. Haematuropygia
   K. leadbeateri
   K. ducrops
   K. sanguinea
   K. tenuirostris
   K. roseicapilla
Nymphicus [Genus]
   N. hollandicus [Species]
Anodorhynchus [Genus]
   A. hyacinthinus [Species]
   A. glaucus
   A. leari

Ara [Genus]
   A. ararauna [Species]
   A. caninde
   A. militaris
   A. ambigua
   A. macao
   A. chloroptera
   A. tricolor
   A. rubrogenys
   A. auricollis
   A. severa
   A. spixii
   A. manilata
   A. maracana
   A. couloni
   A. nobilis
Aratinga [Genus]
   A. acuticaudata [Species]
   A. guarouba
   A. holochlora
   A. strenua
   A. finschi
   A. wagleri
   A. mitrata
   A. erythrogenys
   A. leucophthalmus
   A. chloroptera
   A. euops
   A. auricapillus
   A. jandaya
   A. solstitialis
   A. weddellii
   A. astec
   A. nana
   A. canicularis
   A. pertinax
   A. cactorum
   A. aurea
Nandayus [Genus]
   N. nenday [Species]
Leptosittaca [Genus]
   L. branickii [Species]
Conuropsis [Genus]
   C. carolinensis [Species]
Rhynchopsitta [Genus]
   R. pachyrhyncha [Species]
Cyanoliseus [Genus]
   C. patagonus [Species]
   C. whitleyi
Ognorhynchus [Genus]
   O. icterotis [Species]
Pyrrhura [Genus]
   P. cruentata [Species]
   P. devillei
   P. frontalis
   P. perlata
   P. rhodogaster
   P. molinae
   P. hypoxantha
   P. hoematotis

   P. leucotis
   P. picta
   P. viridicata
   P. egregria
   P. melanura
   P. berlepschi
   P. rupicola
   P. albipectus
   P. calliptera
   P. rhodocephala
   P. hoffmanni
Microsittace [Genus]
   M. ferruginea [Species]
Enicognathus [Genus]
   E. leptorhynchus [Species]
Myiopsitta [Genus]
   M. monachus [Species]
Amoropsittaca [Genus]
   A. aymara [Species]
Psilopsaigon [Genus]
   P. aurifrons [Species]
Bolborhynchus [Genus]
   B. lineola [Species]
   B. ferrugineifrons
   B. andicolus
Forpus [Genus]
   F. cyanopygius [Species]
   F. passerinus
   F. conspicillatus
   F. sclateri
   F. coelestis
Brotogeris [Genus]
   B. tirica [Species]
   B. versicolurus
   B. pyrrhopterus
   B. jugularis
   B. gustavi
   B. chrysopterus
   B. sanctithomae
Nannopsittaca [Genus]
   N. panychlora [Species]
Touit [Genus]
   T. batavica [Species]
   T. purpurata
   T. melanonotus
   T. huetii
   T. dilectissima
   T. surda
   T. stictoptera
   T. emmae
Pionites [Genus]
   P. melanocephala [Species]
   P. leucogaster
Pionopsitta [Genus]
   P. pileata [Species]
   P. haematotis
   P. caica
   P. barrabandi
   P. pyrilia
Hapalopsittaca [Genus]
   H. melanotis [Species]

H. fuertesi
H. amazonina
H. pyrrhops
Gypopsitta [Genus]
  G. vulturina [Species]
Graydidascalus [Genus]
  G. brachyurus [Species]
Pionus [Genus]
  P. menstruus [Species]
  P. sordidus
  P. maximiliani
  P. tumultuosus
  P. seniloides
  P. senilis
  P. chalcopterus
  P. fuscus
Amazona [Genus]
  A. collaria [Species]
  A. leucocephala
  A. ventralis
  A. xantholora
  A. albifrons
  A. agilis
  A. vittata
  A. pretrei
  A. viridigenalis
  A. finschi
  A. autumnalis
  A. dufresniana
  A. brasiliensis
  A. arausiaca
  A. festiva
  A. xanthops
  A. barbadensis
  A. aestiva
  A. ochrocephala
  A. amazonica
  A. mercenaria
  A. farinosa
  A. vinacea
  A. guildingii
  A. versicolor
  A. imperialis
Deroptyus [Genus]
  D. accipitrinus [Species]
Triclaria [Genus]
  T. malachitacea [Species]
Poicephalus [Genus]
  P. robustus [Species]
  P. gulielmi
  P. flavifrons
  P. cryptoxanthus
  P. senegalus
  P. meyeri
  P. rufiventris
  P. ruppellii
Psittacus [Genus]
  P. erithacus [Species]
Coracopsis [Genus]
  C. vasa [Species]
  C. nigra

Psittrichas [Genus]
  P. fulgidus [Species]
Lorius [Genus]
  L. roratus [Species]
Geoffroyus [Genus]
  G. geoffroyi [Species]
  G. simplex
  G. heteroclitus
Prioniturus [Genus]
  P. luconensis [Species]
  P. discurus
  P. flavicans
  P. platurus
  P. mada
Tanygnathus [Genus]
  T. lucionensis [Species]
  T. mulleri
  T. gramineus
  T. heterurus
  T. megalorynchos
Mascarinus [Genus]
  M. mascarin [Species]
Psittacula [Genus]
  P. eupatria [Species]
  P. krameri
  P. alexandri
  P. caniceps
  P. exsul
  P. derbyana
  P. longicauda
  P. cyanocephala
  P. intermedia
  P. himalayana
  P. calthorpae
  P. columboides
Polytelis [Genus]
  P. swainsonii [Species]
  P. anthopeplus
  P. alexandrae
Aprosmictus [Genus]
  A. jonquillaceus [Species]
  A. erythropterus
Alisterus [Genus]
  A. amboinensis [Species]
  A. chloropterus
  A. scapularis
Prosopeia [Genus]
  P. tabuensis [Species]
  P. personata
Psittacella [Genus]
  P. brehmii [Species]
  P. picta
  P. modesta
Bolbopsittacus [Genus]
  B. lunulatus [Species]
Psittinus [Genus]
  P. cyanurus [Species]
Agapornis [Genus]
  A. cana [Species]
  A. pullaria
  A. roseicollis

A. taranta
A. swinderniana
A. fischeri
A. personata
A. lilianae
A. nigrigenis
Loriculus [Genus]
  L. vernalis [Species]
  L. beryllinus
  L. pusillus
  L. philippensis
  L. amabilis
  L. stigmatus
  L. galgulus
  L. exilis
  L. flosculus
  L. aurantiifrons
Platycercus [Genus]
  P. elegans [Species]
  P. caledonicus
  P. eximius
  P. icterotis
  P. adscitus
  P. venustus
  P. zonarius
Purpureicephalus [Genus]
  P. spurius [Species]
Northiella [Genus]
  N. haematogaster [Species]
Psephotus [Genus]
  P. haematonotus [Species]
  P. varius
  P. pulcherrimus
  P. chrysopterygius
Neophema [Genus]
  N. elegans [Species]
  N. chrysostomus
  N. chrysogaster
  N. petrophila
  N. pulchella
  N. splendida
  N. bourkii
Eunymphicus [Genus]
  E. cornutus [Species]
Cyanoramphus [Genus]
  C. unicolor [Species]
  C. novaezelandiae
  C. zealandicus
  C. auriceps
  C. malherbi
  C. ulietanus
Melopsittacus [Genus]
  M. undulatus [Species]
Pezoporus [Genus]
  P. wallicus [Species]
Geopsittacus [Genus]
  G. occidentalis [Species]

Cuculiformes [Order]
Musophagidae [Family]
  Tauraco [Genus]
    T. persa [Species]

T. livingstonii
T. corythaix
T. schuttii
T. fischeri
T. erythrolophus
T. bannermani
T. ruspolii
T. leucotis
T. macrorhynchus
T. hartlaubi
T. leucolophus
Gallirex [Genus]
  G. porphyreolophus [Species]
Ruwenzorornis [Genus]
  R. johnstoni [Species]
Musophaga [Genus]
  M. violacea [Species]
Corythaeola [Genus]
  C. cristata [Species]
Crinifer [Genus]
  C. leucogaster [Species]
  C. africanus
  C. concolor
  C. personata

Cuculidae [Family]
  Clamator [Genus]
    C. glandarius [Species]
    C. coromandus
    C. serratus
    C. jacobinus
    C. cafer
  Pachycoccyx [Genus]
    P. audeberti [Species]
  Cuculus [Genus]
    C. crassirostris [Species]
    C. sparverioides
    C. varius
    C. vagans
    C. fugax
    C. solitarius
    C. clamosus
    C. micropterus
    C. canorus
    C. saturatus
    C. poliocephalus
    C. pallidus
  Cercococcyx [Genus]
    C. mechowi [Species]
    C. olivinus
    C. montanus
  Penthoceryx [Genus]
    P. sonneratii [Species]
  Cacomantis [Genus]
    C. merulinus [Species]
    C. variolosus
    C. castaneiventris
    C. heinrichi
    C. pyrrophanus
  Rhamphomantis [Genus]
    R. megarhynchus [Species]

Misocalius [Genus]
  M. osculans [Species]
Chrysococcyx [Genus]
  C. cupreus [Species]
  C. flavigularis
  C. klaas
  C. caprius
Chalcites [Genus]
  C. maculatus [Species]
  C. xanthorhynchus
  C. basalis
  C. lucidus
  C. malayanus
  C. crassirostris
  C. ruficollis
  C. meyeri
Caliechthrus [Genus]
  C. leucolophus [Species]
Surniculus [Genus]
  S. lugubris [Species]
Microdynamis [Genus]
  M. parva [Species]
Eudynamys [Genus]
  E. scolopacea [Species]
Urodynamis [Genus]
  U. taitensis [Species]
Scythrops [Genus]
  S. novaehollandiae [Species]
Coccyzus [Genus]
  C. pumilus [Species]
  C. cinereus
  C. erythropthalmus
  C. americanus
  C. euleri
  C. minor
  C. melacoryphus
  C. lansbergi
Piaya [Genus]
  P. rufigularis [Species]
  P. pluvialis
  P. cayana
  P. melanogaster
  P. minuta
Saurothera [Genus]
  S. merlini [Species]
  S. vetula
Ceuthmochares [Genus]
  C. aereus [Species]
Rhopodytes [Genus]
  R. diardi [Species]
  R. sumatranus
  R. tristis
  R. viridirostris
Taccocua [Genus]
  T. leschenaulti [Species]
Rhinortha [Genus]
  R. chlorophaea [Species]
Zanclostomus [Genus]
  Z. javanicus [Species]
Rhamphococcyx [Genus]

R. calyorhynchus [Species]
R. curvirostris
Phaenicophaeus [Genus]
  P. pyrrhocephalus [Species]
Dasylophus [Genus]
  D. superciliosus [Species]
Lepidogrammus [Genus]
  L. cumingi [Species]
Crotophaga [Genus]
  C. major [Species]
  C. ani
  C. sulcirostris
Guira [Genus]
  G. guira [Species]
Tapera [Genus]
  T. naevia [Species]
Morococcyx [Genus]
  M. erythropygus [Species]
Dromococcyx [Genus]
  D. phasianellus [Species]
  D. pavoninus
Geococcyx [Genus]
  G. californiana [Species]
  G. velox
Neomorphus [Genus]
  N. geoffroyi [Species]
  N. squaminger
  N. radiolosus
  N. rufipennis
  N. pucheranii
Carpococcyx [Genus]
  C. radiceus [Species]
  C. renauldi
Coua [Genus]
  C. delalandei [Species]
  C. gigas
  C. coquereli
  C. serriana
  C. reynaudii
  C. cursor
  C. ruficeps
  C. cristata
  C. verreauxi
  C. caerulea
Centropus [Genus]
  C. milo [Species]
  C. goliath
  C. violaceus
  C. menbeki
  C. ateralbus
  C. chalybeus
  C. phasianinus
  C. spilopterus
  C. bernsteini
  C. chlororhynchus
  C. rectunguis
  C. steerii
  C. sinensis
  C. andamanensis
  C. nigrorufus
  C. viridis

C. toulou
C. bengalensis
C. grillii
C. epomidis
C. leucogaster
C. anselli
C. monachus
C. senegalensis
C. superciliosus
C. melanops
C. celebensis
C. unirufus

Strigiformes [Order]
Tytonidae [Family]
Tyto [Genus]
T. soumagnei [Species]
T. alba
T. rosenbergii
T. inexpectata
T. novaehollandiae
T. aurantia
T. tenebricosa
T. capensis
T. longimembris
Phodilus [Genus]
P. badius [Species]

Strigidae [Family]
Otus [Genus]
O. sagittatus [Species]
O. rufescens
O. icterorhynchus
O. spilocephalus
O. vandewateri
O. balli
O. alfredi
O. brucei
O. scops
O. umbra
O. senegalensis
O. flammeolus
O. brookii
O. rutilus
O. manadensis
O. beccarii
O. silvicola
O. whiteheadi
O. insularis
O. bakkamoena
O. asio
O. trichopsis
O. barbarus
O. guatemalae
O. roboratus
O. cooperi
O. choliba
O. atricapillus
O. ingens
O. watsonii
O. nudipes

O. clarkii
O. albogularis
O. minimus
O. leucotis
O. hartlaubi
Pyrroglaux [Genus]
P. podargina [Species]
Mimizuku [Genus]
M. gurneyi [Species]
Jubula [Genus]
J. lettii [Species]
Lophostrix [Genus]
L. cristata [Species]
Bubo [Genus]
B. virginianus [Species]
B. bubo
B. capensis
B. africanus
B. poensis
B. nipalensis
B. sumatrana
B. shelleyi
B. lacteus
B. coromandus
B. leucostictus
Pseudoptynx [Genus]
P. philippensis [Species]
Ketupa [Genus]
K. blakstoni [Species]
K. zeylonensis
K. flavipes
K. ketupu
Scotopelia [Genus]
S. peli [Species]
S. ussheri
S. bouvieri
Pulsatrix [Genus]
P. perspicillata [Species]
P. koeniswaldiana
P. melanota
Nyctea [Genus]
N. scandiaca [Species]
Surnia [Genus]
S. ulula [Species]
Glaucidium [Genus]
G. passerinum [Species]
G. gnoma
G. siju
G. minutissimum
G. jardinii
G. brasilianum
G. perlatum
G. tephronotum
G. capense
G. brodiei
G. radiatum
G. cuculoides
G. sjostedti
Micrathene [Genus]
M. whitneyi [Species]

Uroglaux [Genus]
U. dimorpha [Species]
Ninox [Genus]
N. rufa [Species]
N. strenua
N. connivens
N. novaeseelandiae
N. scutulata
N. affinis
N. superciliaris
N. philippensis
N. spilonota
N. spilocephala
N. perversa
N. squamipila
N. theomacha
N. punctulata
N. meeki
N. solomonis
N. odiosa
N. jacquinoti
Gymnoglaux [Genus]
G. lawrencii [Species]
Sceloglaux [Genus]
S. albifacies [Species]
Athene [Genus]
A. noctua [Species]
A. brama
A. blewitti
Speotyto [Genus]
S. cunicularia [Species]
Ciccaba [Genus]
C. virgata [Species]
C. nigrolineata
C. huhula
C. albitarsus
C. woodfordii
Strix [Genus]
S. butleri [Species]
S. seloputo
S. ocellata
S. leptogrammica
S. aluco
S. occidentalis
S. varia
S. hylophila
S. rufipes
S. uralensis
S. davidi
S. nebulosa
Rhinoptynx [Genus]
R. clamator [Species]
Asio [Genus]
A. otus [Species]
A. stygius
A. abyssinicus
A. madagascariensis
A. flammeus
A. capensis
Pseudoscops [Genus]
P. grammicus [Species]

Nesasio [Genus]
   N. solomonensis [Species]
Aegolius [Genus]
   A. funereus [Species]
   A. acadicus
   A. ridgwayi
   A. harrisii

Caprimulgiformes [Order]
Steatornithidae [Family]
   Steatornis [Genus]
      S. caripensis [Species]

Podargidae [Family]
   Podargus [Genus]
      P. strigoides [Species]
      P. papuensis
      P. ocellatus
   Batrachostomus [Genus]
      B. auritus [Species]
      B. harteri
      B. septimus
      B. stellatus
      B. moniliger
      B. hodgsoni
      B. poliolophus
      B. javensis
      B. affinis

Nyctibiidae [Family]
   Nyctibius [Genus]
      N. grandis [Species]
      N. aethereus
      N. griseus
      N. leucopterus
      N. bracteatus

Aegothelidae [Family]
   Aegotheles [Genus]
      A. crinifrons [Species]
      A. insignis
      A. cristatus
      A. savesi
      A. bennettii
      A. wallacii
      A. albertisi

Caprimulgidae [Family]
   Lurocalis [Genus]
      L. semitorquatus [Species]
   Chordeiles [Genus]
      C. pusillus [Species]
      C. rupestris
      C. acutipennis
      C. minor
   Nyctiprogne [Genus]
      N. leucopyga [Species]
   Podager [Genus]
      P. nacunda [Species]
   Eurostopodus [Genus]
      E. guttatus [Species]
      E. albogularis

E. diabolicus
E. papuensis
E. archboldi
E. temminckii
E. macrotis
Veles [Genus]
   V. binotatus [Species]
Nyctidromus [Genus]
   N. albicollis [Species]
Phalaenoptilus [Genus]
   P. nuttallii [Species]
Siphonorhis [Genus]
   S. americanus [Species]
Otophanes [Genus]
   O. mcleodii [Species]
   O. yucatanicus
Nyctiphrynus [Genus]
   N. ocellatus [Species]
Caprimulgus [Genus]
   C. carolinensis [Species]
   C. rufus
   C. cubanensis
   C. sericocaudatus
   C. ridgwayi
   C. vociferus
   C. saturatus
   C. longirostris
   C. cayennensis
   C. maculicaudus
   C. parvulus
   C. maculosus
   C. nigrescens
   C. hirundinaceus
   C. ruficollis
   C. indicus
   C. europaeus
   C. aegyptius
   C. mahrattensis
   C. nubicus
   C. eximius
   C. madagascariensis
   C. macrurus
   C. pectoralis
   C. rufigena
   C. donaldsoni
   C. poliocephalus
   C. asiaticus
   C. natalensis
   C. inornatus
   C. stellatus
   C. ludovicianus
   C. monticolus
   C. affinis
   C. tristigma
   C. concretus
   C. pulchellus
   C. enarratus
   C. batesi
Scotornis [Genus]
   S. fossii [Species]
   S. climacurus

Macrodipteryx [Genus]
   M. longipennis [Species]
Semeiophorus [Genus]
   S. vexillarius [Species]
Hydropsalis [Genus]
   H. climacocerca [Species]
   H. brasiliana
Uropsalis [Genus]
   U. segmentata [Species]
   U. lyra
Macropsalis [Genus]
   M. creagra [Species]
Eleothreptus [Genus]
   E. anomalus [Species]

Apodiformes [Order]

Apodidae [Family]
   Collocalia [Genus]
      C. gigas [Species]
      C. whiteheadi
      C. lowi
      C. fuciphaga
      C. brevirostris
      C. francica
      C. inexpectata
      C. inquieta
      C. vanikorensis
      C. leucophaea
      C. vestita
      C. spodiopygia
      C. hirundinacea
      C. troglodytes
      C. marginata
      C. esculenta
   Hirundapus [Genus]
      H. caudacutus [Species]
      H. giganteus
      H. ernsti
   Streptoprocne [Genus]
      S. zonaris [Species]
      S. biscutata
   Aerornis [Genus]
      A. senex [Species]
      A. semicollaris
   Chaetura [Genus]
      C. chapmani [Species]
      C. pelagica
      C. vauxi
      C. richmondi
      C. gaumeri
      C. leucopygialis
      C. sabini
      C. thomensis
      C. sylvatica
      C. nubicola
      C. cinereiventris
      C. spinicauda
      C. martinica
      C. rutila
      C. ussheri

C. andrei
C. melanopygia
C. brachyura
Zoonavena [Genus]
Z. grandidieri [Species]
Mearnsia [Genus]
M. picina [Species]
M. novaeguineae
M. cassini
M. bohmi
Cypseloides [Genus]
C. cherriei [Species]
C. fumigatus
C. major
Nephoecetes [Genus]
N. niger [Species]
Apus [Genus]
A. melba [Species]
A. aequatorialis
A. reichenowi
A. apus
A. sladeniae
A. toulsoni
A. pallidus
A. acuticaudus
A. pacificus
A. unicolor
A. myoptilus
A. batesi
A. caffer
A. horus
A. affinis
A. andecolus
Aeronautes [Genus]
A. saxatalis [Species]
A. montivagus
Panyptila [Genus]
P. sanctihieronymi [Species]
P. cayennensis
Tachornis [Genus]
T. phoenicobia [Species]
Micropanyptila [Genus]
M. furcata [Species]
Reinarda [Genus]
R. squamata [Species]
Cypsiurus [Genus]
C. parvus [Species]

Hemiprocnidae [Family]
Hemiprocne [Genus]
H. longipennis [Species]
H. mystacea
H. comata

Trochilidae [Family]
Doryfera [Genus]
D. johannae [Species]
D. ludovicae
Androdon [Genus]
A. aequatorialis [Species]
Ramphodon [Genus]

R. naevius [Species]
R. dohrnii
Glaucis [Genus]
G. hirsuta [Species]
Threnetes [Genus]
T. niger [Species]
T. leucurus
T. ruckeri
Phaethornis [Genus]
P. yaruqui [Species]
P. guy
P. syrmatophorus
P. superciliosus
P. malaris
P. eurynome
P. hispidus
P. anthophilus
P. bourcieri
P. philippii
P. squalidus
P. augusti
P. pretrei
P. subochraceus
P. nattereri
P. gounellei
P. rupurumii
P. porcullae
P. ruber
P. griseogularis
P. longuemareus
P. zonura
Eutoxeres [Genus]
E. aquila [Species]
E. condamini
Phaeochroa [Genus]
P. cuvierii [Species]
Campylopterus [Genus]
C. curvipennis [Species]
C. largipennis
C. rufus
C. hyperythrus
C. hemileucurus
C. ensipennis
C. falcatus
C. phainopeplus
C. villaviscensio
Eupetomana [Genus]
E. macroura [Species]
Florisuga [Genus]
F. mellivora [Species]
Melanotrochilus [Genus]
M. fuscus [Species]
Colibri [Genus]
C. delphinae [Species]
C. thalassinus
C. coruscans
C. serrirostris
Anthracothorax [Genus]
A. viridigula [Species]
A. prevostii

A. nigricollis
A. veraguensis
A. dominicus
A. viridis
A. mango
Avocettula [Genus]
A. recurvirostris [Species]
Eulampis [Genus]
E. jugularis [Species]
Sericotes [Genus]
S. holosericeus [Species]
Chrysolampis [Genus]
C. mosquitus [Species]
Orthorhyncus [Genus]
O. cristatus [Species]
Klais [Genus]
K. guimeti [Species]
Abeillia [Genus]
A. albeillei [Species]
Stephanoxis [Genus]
S. lalandi [Species]
Lophornis [Genus]
L. ornata [Species]
L. gouldii
L. magnifica
L. delattrei
L. stictolopha
L. melaniae
Polemistria [Genus]
P. chalybea [Species]
P. pavonina
Lithiophanes [Genus]
L. insignibarbis [Species]
Paphosia [Genus]
P. helenae [Species]
P. adorabilis
Popelairia [Genus]
P. popelairii [Species]
P. langsdorffi
P. letitiae
P. conversii
Discosura [Genus]
D. longicauda [Species]
Chlorestes [Genus]
C. notatus [Species]
Chlorostilbon [Genus]
C. prasinus [Species]
C. vitticeps
C. aureoventris
C. canivetti
C. ricordii
C. swainsonii
C. maugaeus
C. russatus
C. gibsoni
C. inexpectatus
C. stenura
C. alice
C. poortmani
C. euchloris
C. auratus

Cynanthus [Genus]
 C. sordidus [Species]
 C. latirostris
Ptochoptera [Genus]
 P. iolaima [Species]
Cyanophaia [Genus]
 C. bicolor [Species]
Thalurania [Genus]
 T. furcata [Species]
 T. watertonii
 T. glaucopis
 T. lerchi
Neolesbia [Genus]
 N. nehrkorni [Species]
Panterpe [Genus]
 P. insignis [Species]
Damophila [Genus]
 D. julie [Species]
Lepidopyga [Genus]
 L. coeruleogularis [Species]
 L. goudoti
 L. luminosa
Hylocharis [Genus]
 H. xantusii [Species]
 H. leucotis
 H. eliciae
 H. sapphirina
 H. cyanus
 H. chrysura
 H. grayi
Chrysuronia [Genus]
 C. oenone [Species]
Goldmania [Genus]
 G. violiceps [Species]
Goethalsia [Genus]
 G. bella [Species]
Trochilus [Genus]
 T. polytmus [Species]
Leucochloris [Genus]
 L. albicollis [Species]
Polytmus [Genus]
 P. guainumbi [Species]
Waldronia [Genus]
 W. milleri [Species]
Smaragdites [Genus]
 S. theresiae [Species]
Leucippus [Genus]
 L. fallax [Species]
 L. baeri
 L. chionogaster
 L. viridicauda
Talaphorus [Genus]
 T. hypostictus [Species]
 T. taczanowskii
 T. chlorocercus
Amazilia [Genus]
 A. candida [Species]
 A. chionopectus
 A. versicolor
 A. hollandi

A. luciae
A. fimbriata
A. lactea
A. amabilis
A. cyaneotincta
A. rosenbergi
A. boucardi
A. franciae
A. veneta
A. leucogaster
A. cyanocephala
A. microrhyncha
A. cyanifrons
A. beryllina
A. cyanura
A. saucerrottei
A. tobaci
A. viridigaster
A. edward
A. rutila
A. yucatanensis
A. tzacatl
A. castaneiventris
A. amazilia
A. violiceps
Eupherusa [Genus]
 E. eximia [Species]
 E. nigriventris
Elvira [Genus]
 E. chionura [Species]
 E. cupreiceps
Microchera [Genus]
 M. albocoronata [Species]
Chalybura [Genus]
 C. buffonii [Species]
 C. urochrysia
Aphantochroa [Genus]
 A. cirrochloris [Species]
Lampornis [Genus]
 L. clemenciae [Species]
 L. amethystinus
 L. viridipallens
 L. hemileucus
 L. castaneoventris
 L. cinereicauda
Lamprolaima [Genus]
 L. rhami [Species]
Adelomyia [Genus]
 A. melanogenys [Species]
Anthocephala [Genus]
 A. floriceps [Species]
Urosticte [Genus]
 U. ruficrissa [Species]
 U. benjamini
Phlogophilus [Genus]
 P. hemileucurus [Species]
 P. harterti
Clytolaema [Genus]
 C. rubricauda [Species]
Polyplancta [Genus]
 P. aurescens [Species]

Heliodoxa [Genus]
 H. rubinoides [Species]
 H. leadbeateri
 H. jacula
 H. xanthogonys
Ionolaima [Genus]
 I. schreibersii [Species]
Agapeta [Genus]
 A. gularis [Species]
Lampraster [Genus]
 L. branickii [Species]
Eugenia [Genus]
 E. imperatrix [Species]
Eugenes [Genus]
 E. fulgens [Species]
Hylonympha [Genus]
 H. macrocerca [Species]
Sternoclyta [Genus]
 S. cyanopectus [Species]
Topaza [Genus]
 T. pella [Species]
 T. pyra
Oreotrochilus [Genus]
 O. chimborazo [Species]
 O. stolzmanni
 O. melanogaster
 O. estella
 O. bolivianus
 O. leucopleurus
 O. adela
Urochroa [Genus]
 U. bougueri [Species]
Patagona [Genus]
 P. gigas [Species]
Aglaeactis [Genus]
 A. cupripennis [Species]
 A. aliciae
 A. castelnaudii
 A. pamela
Lafresnaya [Genus]
 L. lafresnayi [Species]
Pterophanes [Genus]
 P. cyanopterus [Species]
Coeligena [Genus]
 C. coeligena [Species]
 C. wilsoni
 C. prunellei
 C. torquata
 C. phalerata
 C. eos
 C. bonapartei
 C. helianthea
 C. lutetiae
 C. violifer
 C. iris
Ensifera [Genus]
 E. ensifera [Species]
Sephanoides [Genus]
 S. sephanoides [Species]
 S. fernandensis

Boissoneaua [Genus]
    B. flavescens [Species]
    B. matthewsii
    B. jardini
Heliangelus [Genus]
    H. mavors [Species]
    H. clarisse
    H. amethysticollis
    H. strophianus
    H. exortis
    H. viola
    H. micraster
    H. squamigularis
    H. speciosa
    H. rothschildi
    H. luminosus
Eriocnemis [Genus]
    E. nigrivestis [Species]
    E. soderstromi
    E. vestitus
    E. godini
    E. cupreoventris
    E. luciani
    E. isaacsonii
    E. mosquera
    E. glaucopoides
    E. alinae
    E. derbyi
Haplophaedia [Genus]
    H. aureliae [Species]
    H. lugens
Ocreatus [Genus]
    O. underwoodii [Species]
Lesbia [Genus]
    L. victoriae [Species]
    L. nuna
Sappho [Genus]
    S. sparganura [Species]
Polyonymus [Genus]
    P. caroli [Species]
Zodalia [Genus]
    Z. glyceria [Species]
Ramphomicron [Genus]
    R. microrhynchum [Species]
    R. dorsale
Metallura [Genus]
    M. phoebe [Species]
    M. theresiae
    M. purpureicauda
    M. aeneocauda
    M. melagae
    M. eupogon
    M. williami
    M. tyrianthina
    M. ruficeps
Chalcostigma [Genus]
    C. olivaceum [Species]
    C. stanleyi
    C. heteropogon
    C. herrani

Oxypogon [Genus]
    O. guerinii [Species]
Opisthoprora [Genus]
    O. euryptera [Species]
Taphrolesbia [Genus]
    T. griseiventris [Species]
Aglaiocercus [Genus]
    A. kingi [Species]
    A. emmae
    A. coelestis
Oreonympha [Genus]
    O. nobilis [Species]
Augastes [Genus]
    A. scutatus [Species]
    A. lumachellus
Schistes [Genus]
    S. geoffroyi [Species]
Heliothryx [Genus]
    H. barroti [Species]
    H. aurita
Heliactin [Genus]
    H. cornuta [Species]
Loddigesia [Genus]
    L. mirabilis [Species]
Heliomaster [Genus]
    H. constantii [Species]
    H. longirostris
    H. squamosus
    H. furcifer
Rhodopis [Genus]
    R. vesper [Species]
Thaumastura [Genus]
    T. cora [Species]
Philodice [Genus]
    P. evelynae [Species]
    P. bryantae
    P. mitchellii
Doricha [Genus]
    D. enicura [Species]
    D. eliza
Tilmatura [Genus]
    T. dupontii [Species]
Microstilbon [Genus]
    M. burmeisteri [Species]
Calothorax [Genus]
    C. lucifer [Species]
    C. pulcher
Archilochus [Genus]
    A. colubris [Species]
    A. alexandri
Calliphlox [Genus]
    C. amethystina [Species]
Mellisuga [Genus]
    M. minima [Species]
Calypte [Genus]
    C. anna [Species]
    C. costae
    C. helenae
Stellula [Genus]
    S. calliope [Species]

Atthis [Genus]
    A. heloisa [Species]
Myrtis [Genus]
    M. fanny [Species]
Eulidia [Genus]
    E. yarrellii [Species]
Myrmia [Genus]
    M. micrura [Species]
Acestrura [Genus]
    A. mulsanti [Species]
    A. decorata
    A. bombus
    A. heliodor
    A. berlepschi
    A. harteri
Chaetocercus [Genus]
    C. jourdanii [Species]
Selasphorus [Genus]
    S. platycercus [Species]
    S. rufus
    S. sasin
    S. flammula
    S. torridus
    S. simoni
    S. ardens
    S. scintilla

Coliiformes [Order]
Coliidae [Family]
    Colius [Genus]
        C. striatus [Species]
        C. castanotus
        C. colius
        C. leucocephalus
        C. indicus
        C. macrourus

Trogoniformes [Order]
Trogonidae [Family]
    Pharomachrus [Genus]
        P. mocinno [Species]
        P. fulgidus
        P. pavoninus
    Euptilotis [Genus]
        E. neoxenus [Species]
    Priotelus [Genus]
        P. temnurus [Species]
    Temnotrogon [Genus]
        T. roseigaster [Species]
    Trogon [Genus]
        T. massena [Species]
        T. clathratus
        T. melanurus
        T. strigilatus
        T. citreolus
        T. mexicanus
        T. elegans
        T. collaris
        T. aurantiiventris
        T. personatus
        T. rufus

T. surrucura
T. curucui
T. violaceus
Apaloderma [Genus]
  A. narina [Species]
  A. aequatoriale
Heterotrogon [Genus]
  H. vittatus [Species]
Harpactes [Genus]
  H. reinwardtii [Species]
  H. fasciatus
  H. kasumba
  H. diardii
  H. ardens
  H. whiteheadi
  H. orrhophaeus
  H. duvaucelii
  H. oreskios
  H. erythrocephalus
  H. wardi

Coraciiformes [Order]
Alcedinidae [Family]
  Ceryle [Genus]
    C. lugubris [Species]
    C. maxima
    C. torquata
    C. alcyon
    C. rudis
  Chloroceryle [Genus]
    C. amazona [Species]
    C. americana
    C. inda
    C. aenea
  Alcedo [Genus]
    A. hercules [Species]
    A. atthis
    A. semitorquata
    A. meninting
    A. quadribrachys
    A. euryzona
    A. coerulescens
    A. cristata
    A. leucogaster
  Myioceyx [Genus]
    M. lecontei [Species]
  Ispidina [Genus]
    I. picta [Species]
    I. madagascariensis
  Ceyx [Genus]
    C. cyanopectus [Species]
    C. argentatus
    C. goodfellowi
    C. lepidus
    C. azureus
    C. websteri
    C. pusillus
    C. erithacus
    C. rufidorsum
    C. melanurus
    C. fallax

Pelargopsis [Genus]
  P. amauroptera [Species]
  P. capensis
  P. melanorhyncha
Lacedo [Genus]
  L. pulchella [Species]
Dacelo [Genus]
  D. novaeguineae [Species]
  D. leachii
  D. tyro
  D. gaudichaud
Clytoceyx [Genus]
  C. rex [Species]
Melidora [Genus]
  M. macrorrhina [Species]
Cittura [Genus]
  C. cyanotis [Species]
Halcyon [Genus]
  H. coromanda [Species]
  H. badia
  H. smyrnensis
  H. pileata
  H. cyanoventris
  H. leucocephala
  H. senegalensis
  H. senegaloides
  H. malimbica
  H. albiventris
  H. chelicuti
  H. nigrocyanea
  H. winchelli
  H. diops
  H. macleayii
  H. albonotata
  H. leucopygia
  H. farquhari
  H. pyrrhopygia
  H. torotoro
  H. megarhyncha
  H. australasia
  H. sancta
  H. cinnamomina
  H. funebris
  H. chloris
  H. saurophaga
  H. recurvirostris
  H. venerata
  H. tuta
  H. gambieri
  H. godeffroyi
  H. miyakoensis
  H. bougainvillei
  H. concreta
  H. lindsayi
  H. fulgida
  H. monacha
  H. princeps
Tanysiptera [Genus]
  T. hydrocharis [Species]
  T. galatea

T. riedelii
T. carolinae
T. ellioti
T. nympha
T. danae
T. sylvia

Todidae [Family]
  Todus [Genus]
    T. multicolor [Species]
    T. angustirostris
    T. todus
    T. mexicanus
    T. subulatus

Momotidae [Family]
  Hylomanes [Genus]
    H. momotula [Species]
  Aspatha [Genus]
    A. gularis [Species]
  Electron [Genus]
    E. platyrhynchum [Species]
    E. carinatum
  Eumomota [Genus]
    E. superciliosa [Species]
  Baryphthengus [Genus]
    B. ruficapillus [Species]
  Momotus [Genus]
    M. mexicanus [Species]
    M. momota

Meropidae [Family]
  Dicrocercus [Genus]
    D. hirundineus [Species]
  Melittophagus [Genus]
    M. revoilii [Species]
    M. pusillus
    M. variegatus
    M. lafresnayii
    M. bullockoides
    M. bulocki
    M. gularis
    M. mulleri
  Aerops [Genus]
    A. albicollis [Species]
    A. boehmi
  Merops [Genus]
    M. leschenaulti [Species]
    M. apiaster
    M. superciliosus
    M. ornatus
    M. orientalis
    M. viridis
    M. malimbicus
    M. nubicus
    M. nubicoides
  Bombylonax [Genus]
    B. breweri [Species]
  Nyctyornis [Genus]
    N. amicta [Species]
    N. athertoni

Meropogon [Genus]
  M. forsteni [Species]

Leptosomatidae [Family]
  Leptosomus [Genus]
    L. discolor [Species]

Coraciidae [Family]
  Brachypteracias [Genus]
    B. leptosomus [Species]
    B. squamigera
  Atelornis [Genus]
    A. pittoides [Species]
    A. crossleyi
  Uratelornis [Genus]
    U. chimaera [Species]
  Coracias [Genus]
    C. garrulus [Species]
    C. abyssinica
    C. caudata
    C. spatulata
    C. noevia
    C. benghalensis
    C. temminckii
    C. cyanogaster
  Eurystomus [Genus]
    E. glaucurus [Species]
    E. gularis
    E. orientalis

Upupidae [Family]
  Upupa [Genus]
    U. epops [Species]

Phoeniculidae [Family]
  Phoeniculus [Genus]
    P. purpureus [Species]
    P. bollei
    P. castaneiceps
    P. aterrimus
  Rhinopomastus [Genus]
    R. minor [Species]
    R. cyanomelas
Bucerotidae [Family]
  Tockus [Genus]
    T. birostris [Species]
    T. fasciatus
    T. alboterminatus
    T. bradfieldi
    T. pallidirostris
    T. nasutus
    T. hemprichii
    T. monteiri
    T. griseus
    T. hartlaubi
    T. camurus
    T. erythrorhynchus
    T. flavirostris
    T. deckeni
    T. jacksoni
  Berenicornis [Genus]
    B. comatus [Species]
    B. albocristatus

Ptiloaemus [Genus]
  P. tickelli [Species]
Anorrhinus [Genus]
  A. galeritus [Species]
Penelopides [Genus]
  P. panini [Species]
  P. exarhatus
Aceros [Genus]
  A. nipalensis [Species]
  A. corrugatus
  A. leucocephalus
  A. cassidix
  A. undulatus
  A. plicatus
  A. everetti
  A. narcondami
Anthracoceros [Genus]
  A. malayanus [Species]
  A. malabaricus
  A. coronatus
  A. montani
  A. marchei
Bycanistes [Genus]
  B. bucinator [Species]
  B. cylindricus
  B. subcylindricus
  B. brevis
Ceratogymna [Genus]
  C. atrata [Species]
  C. elata
Buceros [Genus]
  B. rhinoceros [Species]
  B. bicornis
  B. hydrocorax
Rhinoplax [Genus]
  R. vigil [Species]
Bucorvus [Genus]
  B. abyssinicus [Species]
  B. leadbeateri

Piciformes [Order]
Galbulidae [Family]
  Galbalcyrhynchus [Genus]
    G. leucotis [Species]
  Brachygalba [Genus]
    B. lugubris [Species]
    B. phaeonota
    B. goeringi
    B. salmoni
    B. albogularis
  Jacamaralcyon [Genus]
    J. tridactyla [Species]
  Galbula [Genus]
    G. albirostris [Species]
    G. galbula
    G. tombacea
    G. cyanescens
    G. pastazae
    G. ruficauda
    G. leucogastra
    G. dea

Jacamerops [Genus]
  J. aurea [Species]

Bucconidae [Family]
  Notharchus [Genus]
    N. macrorhynchos [Species]
    N. pectoralis
    N. ordii
    N. tectus
  Bucco [Genus]
    B. macrodactylus [Species]
    B. tamatia
    B. noanamae
    B. capensis
  Nystalus [Genus]
    N. radiatus [Species]
    N. chacuru
    N. striolatus
    N. maculatus
  Hypnelus [Genus]
    H. ruficollis [Species]
    H. bicinctus
  Malacoptila [Genus]
    M. striata [Species]
    M. fusca
    M. fulvogularis
    M. rufa
    M. panamensis
    M. mystacalis
  Micromonacha [Genus]
    M. lanceolata [Species]
  Nonnula [Genus]
    N. rubecula [Species]
    N. sclateri
    N. brunnea
    N. frontalis
    N. ruficapilla
    N. amaurocephala
  Hapaloptila [Genus]
    H. castanea [Species]
  Monasa [Genus]
    M. atra [Species]
    M. nigrifrons
    M. morphoeus
    M. flavirostris
  Chelidoptera [Genus]
    C. tenebrosa [Species]

Capitonidae [Family]
  Capito [Genus]
    C. aurovirens [Species]
    C. maculicoronatus
    C. squamatus
    C. hypoleucus
    C. dayi
    C. quinticolor
    C. niger
  Eubucco [Genus]
    E. richardsoni [Species]
    E. bourcierii
    E. versicolor

Semnornis [Genus]
  S. frantzii [Species]
  S. ramphastinus
Psilopogon [Genus]
  P. pyrolophus [Species]
Megalaima [Genus]
  M. virens [Species]
  M. lagrandieri
  M. zeylanica
  M. viridis
  M. faiostricta
  M. corvina
  M. chrysopogon
  M. rafflesii
  M. mystacophanos
  M. javensis
  M. flavifrons
  M. franklinii
  M. oorti
  M. asiatica
  M. incognita
  M. henricii
  M. armillaris
  M. pulcherrima
  M. robustirostris
  M. australis
  M. eximia
  M. rubricapilla
  M. haemacephala
Calorhamphus [Genus]
  C. fuliginosus [Species]
Gymnobucco [Genus]
  G. calvus [Species]
  G. peli
  G. sladeni
  G. bonapartei
Smilorhis [Genus]
  S. leucotis [Species]
Stactolaema [Genus]
  S. olivacea [Species]
  S. anchietae
  S. whytii
Pogoniulus [Genus]
  P. duchaillui [Species]
  P. scolopaceus
  P. leucomystax
  P. simplex
  P. coryphaeus
  P. pusillus
  P. chrysoconus
  P. bilineatus
  P. subsulphureus
  P. atroflavus
Tricholaema [Genus]
  T. lacrymosum [Species]
  T. leucomelan
  T. diadematum
  T. melanocephalum
  T. flavibuccale
  T. hirsutum

Lybius [Genus]
  L. undatus [Species]
  L. vieilloti
  L. torquatus
  L. guifsobalito
  L. rubrifacies
  L. chaplini
  L. leucocephalus
  L. minor
  L. melanopterus
  L. bidentatus
  L. dubius
  L. rolleti
Trachyphonus [Genus]
  T. purpuratus [Species]
  T. vaillantii
  T. erythrocephalus
  T. darnaudii
  T. margaritatus

Indicatoridae [Family]
  Prodotiscus [Genus]
    P. insignis [Species]
    P. regulus
  Melignomon [Genus]
    M. zenkeri [Species]
  indicator [Genus]
    I. exilis [Species]
    I. propinquus
    I. minor
    I. conirostris
    I. variegatus
    I. maculatus
    I. archipelagicus
    I. indicator
    I. xanthonotus
  Melichneutes [Genus]
    M. robustus [Species]

Ramphastidae [Family]
  Aulacorhynchus [Genus]
    A. sulcatus [Species]
    A. calorhynchus
    A. derbianus
    A. prasinus
    A. haematopygus
    A. coeruleicinctis
    A. huallagae
  Pteroglossus [Genus]
    P. torquatus [Species]
    P. sanguineus
    P. erythropygius
    P. castanotis
    P. aracari
    P. pluricinctus
    P. viridis
    P. bitorquatus
    P. olallae
    P. flavirostris
    P. mariae
    P. beauharnaesii

Selenidera [Genus]
  S. spectabilis [Species]
  S. culik
  S. reinwardtii
  S. langsdorffi
  S. nattereri
  S. maculirostris
Andigena [Genus]
  A. bailloni [Species]
  A. laminirostris
  A. hypoglauca
  A. cucullata
  A. nigrirostris
Ramphastos [Genus]
  R. vitellinus [Species]
  R. dicolorus
  R. citreolaemus
  R. sulfuratus
  R. swainsonii
  R. ambiguus
  R. aurantiirostris
  R. tucanus
  R. cuvieri
  R. inca
  R. toco

Picidae [Family]
  Jynx [Genus]
    J. torquilla [Species]
    J. ruficollis
  Picumnus [Genus]
    P. cinnamomeus [Species]
    P. rufiventris
    P. fuscus
    P. castelnau
    P. leucogaster
    P. limae
    P. olivaceus
    P. granadensis
    P. nebulosus
    P. exilis
    P. borbae
    P. aurifrons
    P. temminckii
    P. cirratus
    P. sclateri
    P. steindachneri
    P. squamulatus
    P. minutissimus
    P. pallidus
    P. albosquamatus
    P. guttifer
    P. varzeae
    P. pygmaeus
    P. asterias
    P. pumilus
    P. innominatus
  Nesoctites [Genus]
    N. micromegas [Species]
  Verreauxia [Genus]
    V. africana [Species]

Sasia [Genus]
  S. ochracea [Species]
  S. abnormis
Geocolaptes [Genus]
  G. olivaceus [Species]
Colaptes [Genus]
  C. cafer [Species]
  C. auratus
  C. chrysoides
  C. rupicola
  C. pitius
  C. campestris
Nesoceleus [Genus]
  N. fernandinae [Species]
Chrysoptilus [Genus]
  C. melanochloros [Species]
  C. punctigula
  C. atricollis
Piculus [Genus]
  P. rivolii [Species]
  P. auricularis
  P. aeruginosus
  P. rubiginosus
  P. simplex
  P. flavigula
  P. leucolaemus
  P. aurulentus
  P. chrysochloros
Campethera [Genus]
  C. punctuligera [Species]
  C. nubica
  C. bennettii
  C. cailliautii
  C. notata
  C. abingoni
  C. taeniolaema
  C. tullbergi
  C. maculosa
  C. permista
  C. caroli
  C. nivosa
Celeus [Genus]
  C. flavescens [Species]
  C. spectabilis
  C. castaneus
  C. immaculatus
  C. elegans
  C. jumana
  C. grammicus
  C. loricatus
  C. undatus
  C. flavus
  C. torquatus
Micropternus [Genus]
  M. brachyurus [Species]
Picus [Genus]
  P. viridis [Species]
  P. vaillantii
  P. awokera
  P. squamatus

P. viridanus
P. vittatus
P. xanthopygaeus
P. canus
P. rabieri
P. erythropygius
P. flavinucha
P. puniceus
P. chlorolophus
P. mentalis
P. mineaceus
Dinopium [Genus]
  D. benghalense [Species]
  D. shorii
  D. javanense
  D. rafflesii
Gecinulus [Genus]
  G. grantia [Species]
  G. viridis
Meiglyptes [Genus]
  M. tristis [Species]
  M. jugularis
  M. tukki
Mulleripicus [Genus]
  M. pulverulentus [Species]
  M. funebris
  M. fuliginosus
  M. fulvus
Dryocopus [Genus]
  D. martius [Species]
  D. javensis
  D. pileatus
  D. lineatus
  D. erythrops
  D. schulzi
  D. galeatus
Asyndesmus [Genus]
  A. lewis [Species]
Melanerpes [Genus]
  M. erythrocephalus [Species]
  M. portoricensis
  M. herminieri
  M. formicivorus
  M. hypopolius
  M. carolinus
  M. aurifrons
  M. chrysogenys
  M. superciliaris
  M. caymanensis
  M. radiolatus
  M. striatus
  M. rubricapillus
  M. pucherani
  M. chrysauchen
  M. flavifrons
  M. cruentatus
  M. rubrifrons
Leuconerpes [Genus]
  L. candidus [Species]
Sphyrapicus [Genus]

S. varius [Species]
S. thyroideus
Trichopicus [Genus]
  T. cactorum [Species]
Veniliornis [Genus]
  V. fumigatus [Species]
  V. spilogaster
  V. passerinus
  V. frontalis
  V. maculifrons
  V. cassini
  V. affinis
  V. kirkii
  V. callonotus
  V. sanguineus
  V. dignus
  V. nigriceps
Dendropicos [Genus]
  D. fuscescens [Species]
  D. stierlingi
  D. elachus
  D. abyssinicus
  D. poecilolaemus
  D. gabonensis
  D. lugubris
Dendrocopos [Genus]
  D. major [Species]
  D. leucopterus
  D. syriacus
  D. assimilis
  D. himalayensis
  D. darjellensis
  D. medius
  D. leucotos
  D. cathpharius
  D. hyperythrus
  D. auriceps
  D. atratus
  D. macei
  D. mahrattensis
  D. minor
  D. canicapillus
  D. wattersi
  D. kizuki
  D. moluccensis
  D. maculatus
  D. temminckii
  D. obsoletus
  D. dorae
  D. albolarvatus
  D. villosus
  D. pubescens
  D. borealis
  D. nuttallii
  D. scalaris
  D. arizonae
  D. stricklandi
  D. mixtus
  D. lignarius
Picoides [Genus]

P. tridactylus [Species]
P. arcticus
Sapheopipo [Genus]
  S. noguchii [Species]
Xiphidiopicus [Genus]
  X. percussus [Species]
Polipicus [Genus]
  P. johnstoni [Species]
  P. elliotii
Mesopicos [Genus]
  M. goertae [Species]
  M. griseocephalus
Thripias [Genus]
  T. namaquus [Species]
  T. xantholophus
  T. pyrrhogaster
Hemicircus [Genus]
  H. concretus [Species]
  H. canente
Blythipicus [Genus]
  B. pyrrhotis [Species]
  B. rubiginosus
Chrysocolaptes [Genus]
  C. validus [Species]
  C. festivus
  C. lucidus
Phloeoceastes [Genus]
  P. guatemalensis [Species]
  P. melanoleucos
  P. leucopogon
  P. rubricollis
  P. robustus
  P. pollens
  P. haematogaster
Campephilus [Genus]
  C. principalis [Species]
  C. imperialis
  C. magellanicus

Passeriformes [Order]
Eurylaimidae [Family]
Smithornis [Genus]
  S. capensis [Species]
  S. rufolateralis
  S. sharpei
Pseudocalyptomena [Genus]
  P. graueri [Species]
Corydon [Genus]
  C. sumatranus [Species]
Cymbirhynchus [Genus]
  C. macrorhynchos [Species]
Eurylaimus [Genus]
  E. javanicus [Species]
  E. ochromalus
  E. steerii
Serilophus [Genus]
  S. lunatus [Species]
Psarisomus [Genus]
  P. dalhousiae [Species]
Calyptomena [Genus]

C. viridis [Species]
C. hosii
C. whiteheadi

Dendrocolaptidae [Family]
Dendrocincla [Genus]
  D. tyrannina [Species]
  D. macrorhyncha
  D. fuliginosa
  D. anabatina
  D. merula
  D. homochroa
Deconychura [Genus]
  D. longicauda [Species]
  D. stictolaema
Sittasomus [Genus]
  S. griseicapillus [Species]
Glyphorynchus [Genus]
  G. spirurus [Species]
Drymornis [Genus]
  D. bridgesii [Species]
Nasica [Genus]
  N. longirostris [Species]
Dendrexetastes [Genus]
  D. rufigula [Species]
Hylexetastes [Genus]
  H. perrotii [Species]
  H. stresemanni
Xiphocolaptes [Genus]
  X. promeropirhynchus [Species]
  X. albicollis
  X. falcirostris
  X. franciscanus
  X. major
Dendrocolaptes [Genus]
  D. certhia [Species]
  D. concolor
  D. hoffmannsi
  D. picumnus
  D. platyrostris
Xiphorhynchus [Genus]
  X. picus [Species]
  X. necopinus
  X. obsoletus
  X. ocellatus
  X. spixii
  X. elegans
  X. pardalotus
  X. guttatus
  X. flavigaster
  X. striatigularis
  X. lachrymosus
  X. erythropygius
  X. triangularis
Lepidocolaptes [Genus]
  L. leucogaster [Species]
  L. souleyetii
  L. angustirostris
  L. affinis
  L. squamatus
  L. fuscus
  L. albolineatus

Campylorhamphus [Genus]
  C. pucherani [Species]
  C. trochilirostris
  C. pusillus
  C. procurvoides

Furnariidae [Family]
Geobates [Genus]
  G. poecilopterus [Species]
Geositta [Genus]
  G. maritima [Species]
  G. peruviana
  G. saxicolina
  G. isabellina
  G. rufipennis
  G. punensis
  G. cunicularia
  G. antarctica
  G. tenuirostris
  G. crassirostris
Upucerthia [Genus]
  U. dumetaria [Species]
  U. albigula
  U. validirostris
  U. serrana
  U. andaecola
Ochetorhynchus [Genus]
  O. ruficaudus [Species]
  O. certhioides
  O. harteri
Eremobius [Genus]
  E. phoenicurus [Species]
Chilia [Genus]
  C. melanura [Species]
Cinclodes [Genus]
  C. antarcticus [Species]
  C. patagonicus
  C. oustaleti
  C. fuscus
  C. comechingonus
  C. atacamensis
  C. palliatus
  C. taczanowskii
  C. nigrofumosus
  C. excelsior
Clibanornis [Genus]
  C. dendrocolaptoides [Species]
Furnarius [Genus]
  F. rufus [Species]
  F. leucopus
  F. torridus
  F. minor
  F. figulus
  F. cristatus
Limnornis [Genus]
  L. curvirostris [Species]
Sylviorthorhynchus [Genus]
  S. desmursii [Species]
Aphrastura [Genus]
  A. spinicauda [Species]
  A. masafuerae

Phleocryptes [Genus]
  P. melanops [Species]
Leptasthenura [Genus]
  L. andicola [Species]
  L. striata
  L. pileata
  L. xenothorax
  L. striolata
  L. aegithaloides
  L. platensis
  L. fuliginiceps
  L. yanacensis
  L. setaria
Spartonoica [Genus]
  S. maluroides [Species]
Schizoeaca [Genus]
  S. coryi [Species]
  S. fuliginosa
  S. griseomurina
  S. palpebralis
  S. helleri
  S. harterti
Schoeniophylax [Genus]
  S. phryganophila [Species]
Oreophylax [Genus]
  O. moreirae [Species]
Synallaxis [Genus]
  S. ruficapilla [Species]
  S. superciliosa
  S. poliophrys
  S. azarae
  S. frontalis
  S. moesta
  S. cabanisi
  S. spixi
  S. hypospodia
  S. subpudica
  S. albescens
  S. brachyura
  S. albigularis
  S. gujanensis
  S. propinqua
  S. cinerascens
  S. tithys
  S. cinnamomea
  S. fuscorufa
  S. unirufa
  S. rutilans
  S. erythrothorax
  S. cherriei
  S. stictothorax
Hellmayrea [Genus]
  H. gularis [Species]
Gyalophylax [Genus]
  G. hellmayri [Species]
Certhiaxis [Genus]
  C. cinnamomea [Species]
  C. mustelina
Limnoctites [Genus]
  L. rectirostris [Species]

Poecilurus [Genus]
  P. candei [Species]
  P. kollari
  P. scutatus
Cranioleuca [Genus]
  C. sulphurifera [Species]
  C. semicinerea
  C. obsoleta
  C. pyrrhophia
  C. subcristata
  C. hellmayri
  C. curtata
  C. furcata
  C. demissa
  C. erythrops
  C. vulpina
  C. pallida
  C. antisiensis
  C. marcapatae
  C. albiceps
  C. baroni
  C. albicapilla
  C. mulleri
  C. gutturata
Siptornopsis [Genus]
  S. hypochondriacus [Species]
Asthenes [Genus]
  A. pyrrholeuca [Species]
  A. dorbignyi
  A. berlepschi
  A. baeri
  A. patagonica
  A. steinbachi
  A. humicola
  A. modesta
  A. pudibunda
  A. ottonis
  A. heterura
  A. wyatti
  A. humilis
  A. anthoides
  A. sclateri
  A. hudsoni
  A. virgata
  A. maculicauda
  A. flammulata
  A. urubambensis
Thripophaga [Genus]
  T. macroura [Species]
  T. cherriei
  T. fusciceps
  T. berlepschi
Phacellodomus [Genus]
  P. sibilatrix [Species]
  P. rufifrons
  P. striaticeps
  P. erythrophthalmus
  P. ruber
  P. striaticollis
  P. dorsalis

Coryphistera [Genus]
  C. alaudina [Species]
Anumbius [Genus]
  A. annumbi [Species]
Siptornis [Genus]
  S. striaticollis [Species]
Xenerpestes [Genus]
  X. minlosi [Species]
  X. singularis
Metopothrix [Genus]
  M. aurantiacus [Species]
Roraimia [Genus]
  R. adusta [Species]
Margarornis [Genus]
  M. squamiger [Species]
  M. bellulus
  M. rubiginosus
  M. stellatus
Premnornis [Genus]
  P. guttuligera [Species]
Premnoplex [Genus]
  P. brunnescens [Species]
Pseudocolaptes [Genus]
  P. lawrencii [Species]
  P. boissonneautii
Berlepschia [Genus]
  B. rikeri [Species]
Pseudoseisura [Genus]
  P. cristata [Species]
  P. lophotes
  P. gutturalis
Hyloctistes [Genus]
  H. subulatus [Species]
Ancistrops [Genus]
  A. strigilatus [Species]
Anabazenops [Genus]
  A. fuscus [Species]
Syndactyla [Genus]
  S. rufosuperciliata [Species]
  S. subalaris
  S. guttulata
  S. mirandae
Simoxenops [Genus]
  S. ucayalae [Species]
  S. striatus
Anabacerthia [Genus]
  A. striaticollis [Species]
  A. temporalis
  A. amaurotis
Philydor [Genus]
  P. atricapillus [Species]
  P. erythrocercus
  P. pyrrhodes
  P. dimidiatus
  P. baeri
  P. lichtensteini
  P. rufus
  P. erythropterus
  P. ruficaudatus
Automolus [Genus]

A. leucophthalmus [Species]
A. infuscatus
A. dorsalis
A. rubiginosus
A. albigularis
A. ochrolaemus
A. rufipileatus
A. ruficollis
A. melanopezus
Hylocryptus [Genus]
H. erythrocephalus [Species]
H. rectirostris
Cichlocolaptes [Genus]
C. leucophrus [Species]
Heliobletus [Genus]
H. contaminatus [Species]
Thripadectes [Genus]
T. flammulatus [Species]
T. holostictus
T. melanorhynchus
T. rufobrunneus
T. virgaticeps
T. scrutator
T. ignobilis
Xenops [Genus]
X. milleri [Species]
X. tenuirostris
X. rutilans
X. minutus
Megaxenops [Genus]
M. parnaguae [Species]
Pygarrhichas [Genus]
P. albogularis [Species]
Sclerurus [Genus]
S. scansor [Species]
S. albigularis
S. mexicanus
S. rufigularis
S. caudacutus
S. guatemalensis
Lochmias [Genus]
L. nematura [Species]

Formicariidae [Family]
Cymbilaimus [Genus]
C. lineatus [Species]
Hypoedaleus [Genus]
H. guttatus [Species]
Batara [Genus]
B. cinerea [Species]
Mackenziaena [Genus]
M. leachii [Species]
M. severa
Frederickena [Genus]
F. viridis [Species]
U. unduligera
Taraba [Genus]
T. major [Species]
Sakesphorus [Genus]
S. canadensis [Species]
S. cristatus

S. bernardi
S. melanonotus
S. melanothorax
S. luctuosus
Biatas [Genus]
B. nigropectus [Species]
Thamnophilus [Genus]
T. doliatus [Species]
T. multistriatus
T. palliatus
T. bridgesi
T. nigriceps
T. praecox
T. nigrocinereus
T. aethiops
T. unicolor
T. schistaceus
T. murinus
T. aroyae
T. punctatus
T. amazonicus
T. insignis
T. caerulescens
T. torquatus
T. ruficapillus
Pygiptila [Genus]
P. stellaris [Species]
Megastictus [Genus]
M. margaritatus [Species]
Neoctantes [Genus]
N. niger [Species]
Clytoctantes [Genus]
C. alixii [Species]
Xenornis [Genus]
X. setifrons [Species]
Thamnistes [Genus]
T. anabatinus [Species]
Dysithamnus [Genus]
D. stictothorax [Species]
D. mentalis
D. striaticeps
D. puncticeps
D. xanthopterus
D. ardesiacus
D. saturninus
D. occidentalis
D. plumbeus
Thamnomanes [Genus]
T. caesius [Species]
Myrmotherula [Genus]
M. brachyura [Species]
M. obscura
M. sclateri
M. klagesi
M. surinamensis
M. ambigua
M. cherriei
M. guttata
M. longicauda
M. hauxwelli

M. gularis
M. gutturalis
M. fulviventris
M. leucophthalma
M. haematonota
M. ornata
M. erythrura
M. erythronotos
M. axillaris
M. schisticolor
M. sunensis
M. longipennis
M. minor
M. iheringi
M. grisea
M. unicolor
M. behni
M. urosticta
M. menetriesii
M. assimilis
Dichrozona [Genus]
D. cincta [Species]
Myrmorchilus [Genus]
M. strigilatus [Species]
Herpsilochmus [Genus]
H. pileatus [Species]
H. sticturus
H. stictocephalus
H. dorsimaculatus
H. roraimae
H. pectoralis
H. longirostris
H. axillaris
H. rufimarginatus
Microrhopias [Genus]
M. quixensis [Species]
Formicivora [Genus]
F. iheringi [Species]
F. grisea
F. serrana
F. melanogaster
F. rufa
Drymophila [Genus]
D. ferruginea [Species]
D. genei
D. ochropyga
D. devillei
D. caudata
D. malura
D. squamata
Terenura [Genus]
T. maculata [Species]
T. callinota
T. humeralis
T. sharpei
T. spodioptila
Cercomacra [Genus]
C. cinerascens [Species]
C. brasiliana
C. tyrannina

C. nigriscens
C. serva
C. nigricans
C. carbonaria
C. melanaria
C. ferdinandi
Sipia [Genus]
  S. berlepschi [Species]
  S. rosenbergi
Pyriglena [Genus]
  P. leuconota [Species]
  P. atra
  P. leucoptera
Rhopornis [Genus]
  R. ardesiaca [Species]
Myrmoborus [Genus]
  M. leucophrys [Species]
  M. lugubris
  M. myotherinus
  M. melanurus
Hypocnemis [Genus]
  H. cantator [Species]
  H. hypoxantha
Hypocnemoides [Genus]
  H. melanopogon [Species]
  H. maculicauda
Myrmochanes [Genus]
  M. hemileucus [Species]
Gymnocichla [Genus]
  G. nudiceps [Species]
Sclateria [Genus]
  S. naevia [Species]
Percnostola [Genus]
  P. rufifrons [Species]
  P. schistacea
  P. leucostigma
  P. caurensis
  P. lophotes
Myrmeciza [Genus]
  M. longipes [Species]
  M. exsul
  M. ferruginea
  M. ruficauda
  M. laemosticta
  M. disjuncta
  M. pelzelni
  M. hemimelaena
  M. hyperythra
  M. goeldii
  M. melanoceps
  M. fortis
  M. immaculata
  M. griseiceps
Myrmoderus [Genus]
  M. loricatus [Species]
  M. squamosus
Myrmophylax [Genus]
  M. atrothorax [Species]
  M. stictothorax
Formicarius [Genus]

F. colma [Species]
F. analis
F. nigricapillus
F. rufipectus
Chamaeza [Genus]
  C. campanisona [Species]
  C. nobilis
  C. ruficauda
  C. mollissima
Pithys [Genus]
  P. albifrons [Species]
  P. castanea
Gymnopithys [Genus]
  G. rufigula [Species]
  G. salvini
  G. lunulata
  G. leucaspis
Rhegmatorhina [Genus]
  R. gymnops [Species]
  R. berlepschi
  R. cristata
  R. hoffmannsi
  R. melanosticta
Hylophylax [Genus]
  H. naevioides [Species]
  H. naevia
  H. punctulata
  H. poecilonota
Phlegopsis [Genus]
  P. nigromaculata [Species]
  P. erythroptera
  P. borbae
Phaenostictus [Genus]
  P. mcleannani [Species]
Myrmornis [Genus]
  M. torquata [Species]
Pittasoma [Genus]
  P. michleri [Species]
  P. rufopileatum
Grallaricula [Genus]
  G. flavirostris [Species]
  G. ferrugineipectus
  G. nana
  G. loricata
  G. peruviana
  G. lineifrons
  G. cucullata
Myrmothera [Genus]
  M. campanisona [Species]
  M. simplex
Thamnocharis [Genus]
  T. dignissima [Species]
Grallaria [Genus]
  G. squamigera [Species]
  G. excelsa
  G. gigantea
  G. guatimalensis
  G. varia
  G. alleni
  G. haplonota

G. milleri
G. bangsi
G. quitensis
G. erythrotis
G. hypoleuca
G. przewalskii
G. capitalis
G. nuchalis
G. albigula
G. ruficapilla
G. erythroleuca
G. rufocinerea
G. griseonucha
G. rufula
G. andicola
G. macularia
G. fulviventris
G. berlepschi
G. perspicillata
G. ochroleuca

Conopophagidae [Family]
Conopophaga [Genus]
  C. lineata [Species]
  C. cearae
  C. aurita
  C. roberti
  C. peruviana
  C. ardesiaca
  C. castaneiceps
  C. melanops
  C. melanogaster
Corythopis [Genus]
  C. delalandi [Species]
  C. torquata

Rhinocryptidae [Family]
Pteroptochos [Genus]
  P. castaneus [Species]
  P. tarnii
  P. megapodius
Scelorchilus [Genus]
  S. albicollis [Species]
  S. rubecula
Rhinocrypta [Genus]
  R. lanceolata [Species]
Teledromas [Genus]
  T. fuscus [Species]
Liosceles [Genus]
  L. thoracicus [Species]
Merulaxis [Genus]
  M. ater [Species]
Melanopareia [Genus]
  M. torquata [Species]
  M. maximiliani
  M. maranonicus
  M. elegans
Scytalopus [Genus]
  S. unicolor [Species]
  S. speluncae
  S. macropus

AVES SPECIES LIST

S. femoralis
S. argentifrons
S. chiriquensis
S. panamensis
S. latebricola
S. indigoticus
S. magellanicus
Psilorhamphus [Genus]
P. guttatus [Species]
Myornis [Genus]
M. senilis [Species]
Eugralla [Genus]
E. paradoxa [Species]
Acropternis [Genus]
A. orthonyx [Species]

Tyrannidae [Family]
Phyllomyias [Genus]
P. fasciatus [Species]
P. burmeisteri
P. virescens
P. sclateri
P. griseocapilla
P. griseiceps
P. plumbeiceps
P. nigrocapillus
P. cinereiceps
P. uropygialis
Zimmerius [Genus]
Z. vilissimus [Species]
Z. bolivianus
Z. cinereicapillus
Z. gracilipes
Z. viridiflavus
Ornithion [Genus]
O. inerme [Species]
O. semiflavum
O. brunneicapillum
Camptostoma [Genus]
C. imberbe [Species]
C. obsoletum
Phaeomyias [Genus]
P. murina [Species]
Sublegatus [Genus]
S. modestus [Species]
S. obscurior
Suiriri [Genus]
S. suiriri [Species]
Tyrannulus [Genus]
T. elatus [Species]
Myiopagis [Genus]
M. gaimardii [Species]
M. caniceps
M. subplacens
M. flavivertex
M. cotta
M. viridicata
M. leucospodia
Elaenia [Genus]
E. martinica [Species]
E. flavogaster

E. spectabilis
E. albiceps
E. parvirostris
E. mesoleuca
E. strepera
E. gigas
E. pelzelni
E. cristata
E. ruficeps
E. chiriquensis
E. frantzii
E. obscura
E. dayi
E. pallatangae
E. fallax
Mecocerculus [Genus]
M. leucophrys [Species]
M. poecilocercus
M. hellmayri
M. calopterus
M. minor
M. stictopterus
Serpophaga [Genus]
S. cinerea [Species]
S. hypoleuca
S. nigricans
S. araguayae
S. subcristata
Inezia [Genus]
I. inornata [Species]
I. tenuirostris
I. subflava
Stigmatura [Genus]
S. napensis [Species]
S. budytoides
Anairetes [Genus]
A. alpinus [Species]
A. agraphia
A. agilis
A. reguloides
A. flavirostris
A. fernandezianus
A. parulus
Tachuris [Genus]
T. rubrigastra [Species]
Culicivora [Genus]
C. caudacuta [Species]
Polystictus [Genus]
P. pectoralis [Species]
P. superciliaris
Pseudocolopteryx [Genus]
P. sclateri [Species]
P. dinellianus
P. acutipennis
P. flaviventris
Euscarthmus [Genus]
E. meloryphus [Species]
E. rufomarginatus
Mionectes [Genus]
M. striaticollis [Species]

M. oliveceus
M. oleagineus
M. macconnelli
M. rufiventris
Leptopogon [Genus]
L. rufipectus [Species]
L. taczanowskii
L. amaurocephalus
L. superciliaris
Phylloscartes [Genus]
P. nigrifrons [Species]
P. poecilotis
P. chapmani
P. ophthalmicus
P. eximius
P. gualaquizae
P. flaviventris
P. venezuelanus
P. orbitalis
P. flaveolus
P. roquettei
P. ventralis
P. paulistus
P. oustaleti
P. difficilis
P. flavovirens
P. virescens
P. superciliaris
P. sylviolus
Pseudotriccus [Genus]
P. pelzelni [Species]
P. simplex
P. ruficeps
Myiornis [Genus]
M. auricularis [Species]
M. albiventris
M. ecaudatus
Lophotriccus [Genus]
L. pileatus [Species]
L. eulophotes
L. vitiosus
L. galeatus
Atalotriccus [Genus]
A. pilaris [Species]
Poecilotriccus [Genus]
P. ruficeps [Species]
P. capitale
P. tricolor
P. andrei
Oncostoma [Genus]
O. cinereigulare [Species]
O. olivaceum
Hemitriccus [Genus]
H. minor [Species]
H. josephinae
H. diops
H. obsoletus
H. flammulatus
H. zosterops
H. aenigma

H. orbitatus
H. iohannis
H. striaticollis
H. nidipendulus
H. spodiops
H. margaritaceiventer
H. inoratus
H. granadensis
H. mirandae
H. kaempferi
H. rufigularis
H. furcatus
Todirostrum [Genus]
T. senex [Species]
T. russatum
T. plumbeiceps
T. fumifrons
T. latirostre
T. sylvia
T. maculatum
T. poliocephalum
T. cinereum
T. pictum
T. chrysocrotaphum
T. nigriceps
T. calopterum
Cnipodectes [Genus]
C. subbrunneus [Species]
Ramphotrigon [Genus]
R. megacephala [Species]
R. fuscicauda
R. ruficauda
Rhynchocyclus [Genus]
R. brevirostris [Species]
R. olivaceus
R. fulvipectus
Tolmomyias [Genus]
T. sulphurescens [Species]
T. assimilis
T. poliocephalus
T. flaviventris
Platyrinchus [Genus]
P. saturatus [Species]
P. cancrominus
P. mystaceus
P. coronatus
P. flavigularis
P. platyrhynchos
P. leucoryphus
Onychorhynchus [Genus]
O. coronatus [Species]
Myiotriccus [Genus]
M. ornatus [Species]
Terenotriccus [Genus]
T. erythrurus [Species]
Myiobius [Genus]
M. villosus [Species]
M. barbatus
M. atricaudus
Myiophobus [Genus]

M. flavicans [Species]
M. phoenicomitra
M. inornatus
M. roraimae
M. lintoni
M. pulcher
M. ochraceiventris
M. cryptoxanthus
M. fasciatus
Aphanotriccus [Genus]
A. capitalis [Species]
A. audax
Xenotriccus [Genus]
X. callizonus [Species]
X. mexicanus
Pyrrhomyias [Genus]
P. cinnamomea [Species]
Mitrephanes [Genus]
M. phaeocercus [Species]
M. olivaceus
Contopus [Genus]
C. borealis [Species]
C. fumigatus
C. ochraceus
C. sordidulus
C. virens
C. cinereus
C. nigrescens
C. albogularis
C. caribaeus
C. latirostris
Empidonax [Genus]
E. flaviventris [Species]
E. virescens
E. alnorum
E. traillii
E. albigularis
E. euleri
E. griseipectus
E. minimus
E. hammondii
E. wrightii
E. oberholseri
E. affinis
E. difficilis
E. flavescens
E. fulvifrons
E. atriceps
Nesotriccus [Genus]
N. ridgwayi [Species]
Cnemotriccus [Genus]
C. fuscatus [Species]
Sayornis [Genus]
S. phoebe [Species]
S. saya
S. nigricans
Pyrocephalus [Genus]
P. rubinus [Species]
Ochthoeca [Genus]
O. cinnamomeiventris [Species]
O. diadema

O. frontalis
O. pulchella
O. rufipectoralis
O. fumicolor
O. oenanthoides
O. parvirostris
O. leucophrys
O. piurae
O. littoralis
Myiotheretes [Genus]
M. striaticollis [Species]
M. erythropygius
M. rufipennis
M. pernix
M. fumigatus
M. fuscorufus
Xolmis [Genus]
X. pyrope [Species]
X. cinerea
X. coronata
X. velata
X. dominicana
X. irupero
Neoxolmis [Genus]
N. rubetra [Species]
N. ruficentris
Agriornis [Genus]
A. montana [Species]
A. andicola
A. livida
A. microptera
A. murina
Muscisaxicola [Genus]
M. maculirostris [Species]
M. fluviatilis
M. macloviana
M. capistrata
M. rufivertex
M. juninensis
M. albilora
M. alpina
M. cinerea
M. albifrons
M. flavinucha
M. frontalis
Lessonia [Genus]
L. oreas [Species]
L. rufa
Knipolegus [Genus]
K. striaticeps [Species]
K. hudsoni
K. poecilocercus
K. signatus
K. cyanirostris
K. poecilurus
K. orenocensis
K. aterrimus
K. nigerrimus
K. lophotes
Hymenops [Genus]
H. perspicillata [Species]

Fluvicola [Genus]
   F. pica [Species]
   F. nengeta
   F. leucocephala
Colonia [Genus]
   C. colonus [Species]
Alectrurus [Genus]
   A. tricolor [Species]
   A. risora
Gubernetes [Genus]
   G. yetapa [Species]
Satrapa [Genus]
   S. icterophrys [Species]
Tumbezia [Genus]
   T. salvini [Species]
Muscigralla [Genus]
   M. brevicauda [Species]
Hirundinea [Genus]
   H. ferruginea [Species]
Machetornis [Genus]
   M. rixosus [Species]
Muscipipra [Genus]
   M. vetula [Species]
Attila [Genus]
   A. phoenicurus [Species]
   A. cinnamomeus
   A. torridus
   A. citriniventris
   A. bolivianus
   A. rufus
   A. spadiceus
Casiornis [Genus]
   C. rufa [Species]
   C. fusca
Rhytipterna [Genus]
   R. simplex [Species]
   R. holerythra
   R. immunda
Laniocera [Genus]
   L. hypopyrrha [Species]
   L. rufescens
Sirystes [Genus]
   S. sibilator [Species]
Myiarchus [Genus]
   M. semirufus [Species]
   M. yucatanensis
   M. barbirostris
   M. tuberculifer
   M. swainsoni
   M. venezuelensis
   M. panamensis
   M. ferox
   M. cephalotes
   M. phaeocephalus
   M. apicalis
   M. cinerascens
   M. nuttingi
   M. crinitus
   M. tyrannulus
   M. magnirostris

   M. nugator
   M. validus
   M. sagrae
   M. stolidus
   M. antillarum
   M. oberi
Deltarhynchus [Genus]
   D. flammulatus [Species]
Pitangus [Genus]
   P. lictor [Species]
   P. sulphuratus
Megarhynchus [Genus]
   M. pitangua [Species]
Myiozetetes [Genus]
   M. cayanensis [Species]
   M. similis
   M. granadensis
   M. luteiventris
Conopias [Genus]
   C. inornatus [Species]
   C. parva
   C. trivirgata
   C. cinchoneti
Myiodynastes [Genus]
   M. hemichrysus [Species]
   M. chrysocephalus
   M. bairdii
   M. maculatus
   M. luteiventris
Legatus [Genus]
   L. leucophaius [Species]
Empidonomus [Genus]
   E. varius [Species]
   E. aurantioatrocristatus
Tyrannopsis [Genus]
   T. sulphurea [Species]
Tyrannus [Genus]
   T. niveigularis [Species]
   T. albogularis
   T. melancholicus
   T. couchii
   T. vociferans
   T. crassirostris
   T. verticalis
   T. forficata
   T. savana
   T. tyrannus
   T. dominicensis
   T. caudifasciatus
   T. cubensis
Xenopsaris [Genus]
   X. albinucha [Species]
Pachyramphus [Genus]
   P. viridis [Species]
   P. versicolor
   P. spodiurus
   P. rufus
   P. castaneus
   P. cinnamomeus
   P. polychopterus

   P. marginatus
   P. albogriseus
   P. major
   P. surinamus
   P. aglaiae
   P. homochrous
   P. minor
   P. validus
   P. niger
Tityra [Genus]
   T. cayana [Species]
   T. semifasciata
   T. inquisitor
   T. leucura

Pipridae [Family]
Schiffornis [Genus]
   S. major [Species]
   S. turdinus
   S. virescens
Sapayoa [Genus]
   S. aenigma [Species]
Piprites [Genus]
   P. griseiceps [Species]
   P. chloris
   P. pileatus
Neopipo [Genus]
   N. cinnamomea [Species]
Chloropipo [Genus]
   C. flavicapilla [Species]
   C. holochlora
   C. uniformis
   C. unicolor
Xenopipo [Genus]
   X. atronitens [Species]
Antilophia [Genus]
   A. galeata [Species]
Tyranneutes [Genus]
   T. stolzmanni [Species]
   T. virescens
Neopelma [Genus]
   N. chrysocephalum [Species]
   N. pallescens
   N. aurifrons
   N. sulphureiventer
Heterocercus [Genus]
   H. flavivertex [Species]
   H. aurantiivertex
   H. lineatus
Machaeropterus [Genus]
   M. regulus [Species]
   M. pyrocephalus
   M. deliciosus
Manacus [Genus]
   M. manacus [Species]
Corapipo [Genus]
   C. leucorrhoa [Species]
   C. gutturalis
Ilicura [Genus]
   I. militaris [Species]
Masius [Genus]

M. chrysopterus [Species]
Chiroxiphia [Genus]
    C. linearis [Species]
    C. lanceolata
    C. pareola
    C. caudata
Pipra [Genus]
    P. pipra [Species]
    P. coronata
    P. isidorei
    P. coeruleocapilla
    P. nattereri
    P. vilasboasi
    P. iris
    P. serena
    P. aureola
    P. fasciicauda
    P. filicauda
    P. mentalis
    P. erythrocephala
    P. rubrocapilla
    P. chloromeros
    P. cornuta

Cotingidae [Family]
Phoenicircus [Genus]
    P. carnifex [Species]
    P. nigricollis
Laniisoma [Genus]
    L. elegans [Species]
Phibalura [Genus]
    P. flavirostris [Species]
Tijuca [Genus]
    T. atra [Species]
Carpornis [Genus]
    C. cucullatus [Species]
    C. melanocephalus
Ampelion [Genus]
    A. rubrocristatus [Species]
    A. rufaxilla
    A. sclateri
    A. stresemanni
Pipreola [Genus]
    P. riefferii [Species]
    P. intermedia
    P. arcuata
    P. auroeopectus
    P. frontalis
    P. chlorolepidota
    P. formosa
    P. whitelyi
Ampelioides [Genus]
    A. tschudii [Species]
Iodopleura [Genus]
    I. pipra [Species]
    I. fusca
    I. isabellae
Calyptura [Genus]
    C. cristata [Species]
Lipaugus [Genus]
    L. subalaris [Species]

    L. cryptolophus
    L. fuscocinereus
    L. vociferans
    L. unirufus
    L. lanioides
    L. streptophorus
Chirocylla [Genus]
    C. uropygialis [Species]
Porphyrolaema [Genus]
    P. porphyrolaema [Species]
Cotinga [Genus]
    C. amabilis [Species]
    C. ridgwayi
    C. nattererii
    C. maynana
    C. cotinga
    C. maculata
    C. cayana
Xipholena [Genus]
    X. punicea [Species]
    X. lamellipennis
    X. atropurpurea
Carpodectes [Genus]
    C. nitidus [Species]
    C. antoniae
    C. hopkei
Conioptilon [Genus]
    C. mcilhennyi [Species]
Gymnoderus [Genus]
    G. foetidus [Species]
Haematoderus [Genus]
    H. militaris [Species]
Querula [Genus]
    Q. purpurata [Species]
Pyroderus [Genus]
    P. scutatus [Species]
Cephalopterus [Genus]
    C. glabricollis [Species]
    C. penduliger
    C. ornatus
Perissocephalus [Genus]
    P. tricolor [Species]
Procnias [Genus]
    P. tricarunculata [Species]
    P. alba
    P. averano
    P. nudicollis
Rupicola [Genus]
    R. rupicola [Species]
    R. peruviana

Oxyruncidae [Family]
Oxyruncus [Genus]
    O. cristatus [Species]

Phytotomidae [Family]
Phytotoma [Genus]
    P. raimondii [Species]
    P. rara
    P. rutila

Pittidae [Family]
Pitta [Genus]
    P. phayrei [Species]
    P. nipalensis
    P. soror
    P. oatesi
    P. schneideri
    P. caerulea
    P. cyanea
    P. elliotii
    P. guajana
    P. gurneyi
    P. kochi
    P. erythrogaster
    P. arcuata
    P. granatina
    P. venusta
    P. baudii
    P. sordida
    P. brachyura
    P. nympha
    P. angolensis
    P. superba
    P. maxima
    P. steerii
    P. moluccensis
    P. versicolor
    P. anerythra

Philepittidae [Family]
Philepitta [Genus]
    P. castanea [Species]
    P. schlegeli
Neodrepanis [Genus]
    N. coruscans [Species]
    N. hypoxantha

Acanthisittidae [Family]
Acanthisitta [Genus]
    A. chloris [Species]
Xenicus [Genus]
    X. longipes [Species]
    X. gilviventris
    X. lyalli

Menuridae [Family]
Menura [Genus]
    M. novaehollandiae [Species]
    M. alberti

Atrichornithidae [Family]
Atrichornis [Genus]
    A. clamosus [Species]
    A. rufescens

Alaudidae [Family]
Mirafra [Genus]
    M. javanica [Species]
    M. hova
    M. cordofanica
    M. williamsi
    M. cheniana

AVES SPECIES LIST

M. albicauda
M. passerina
M. candida
M. pulpa
M. hypermetra
M. somalica
M. africana
M. chuana
M. angolensis
M. rufocinnamomea
M. apiata
M. africanoides
M. collaris
M. assamica
M. rufa
M. gilleti
M. poecilosterna
M. sabota
M. erythroptera
M. nigricans
Heteromirafra [Genus]
H. ruddi [Species]
Certhilauda [Genus]
C. curvirostris [Species]
C. albescens
C. albofasciata
Eremopterix [Genus]
E. australis [Species]
E. leucotis
E. signata
E. verticalis
E. nigriceps
E. grisea
E. leucopareia
Ammomanes [Genus]
A. cincturus [Species]
A. phoenicurus
A. deserti
A. dunni
A. grayi
A. burrus
Alaemon [Genus]
A. alaudipes [Species]
A. hamertoni
Ramphocoris [Genus]
R. clotbey [Species]
Melanocorypha [Genus]
M. calandra [Species]
M. bimaculata
M. maxima
M. mongolica
M. leucoptera
M. yeltoniensis
Calandrella [Genus]
C. cinerea [Species]
C. blanfordi
C. acutirostris
C. raytal
C. rufescens
C. razae

C. conirostris
C. starki
C. sclateri
C. fringillaris
C. obbiensis
C. personata
Chersophilus [Genus]
C. duponti [Species]
Pseudalaemon [Genus]
P. fremantlii [Species]
Galerida [Genus]
G. cristata [Species]
G. theklae
G. malabarica
G. deva
G. modesta
G. magnirostris
Lullula [Genus]
L. arborea [Species]
Alauda [Genus]
A. arvensis [Species]
A. gulgula
Eremophila [Genus]
E. alpestris [Species]
E. bilopha

Hirundinidae [Family]
Pseudochelidon [Genus]
P. eurystomina [Species]
Tachycineta [Genus]
T. bicolor [Species]
T. albilinea
T. albiventer
T. leucorrhoa
T. leucopyga
T. thalassina
Callichelidon [Genus]
C. cyaneoviridis [Species]
Kalochelidon [Genus]
K. euchrysea [Species]
Progne [Genus]
P. tapera [Species]
P. subis
P. dominicensis
P. chalybea
P. modesta
Notiochelidon [Genus]
N. murina [Species]
N. cyanoleuca
N. flavipes
N. pileata
Atticora [Genus]
A. fasciata [Species]
A. melanoleuca
Neochelidon [Genus]
N. tibialis [Species]
Alopochelidon [Genus]
A. fucata [Species]
Stelgidopteryx [Genus]
S. ruficollis [Species]
Cheramoeca [Genus]

C. leucosternum [Species]
Pseudhirundo [Genus]
P. griseopyga [Species]
Riparia [Genus]
R. paludicola [Species]
R. congica
R. riparia
R. cincta
Phedina [Genus]
P. borbonica [Species]
P. brazzae
Ptyonoprogne [Genus]
P. rupestris [Species]
P. obsoleta
P. fuligula
P. concolor
Hirundo [Genus]
H. rustica [Species]
H. lucida
H. angolensis
H. tahitica
H. albigularis
H. aethiopica
H. smithii
H. atrocaerulea
H. nigrita
H. leucosoma
H. megaensis
H. nigrorufa
H. dimidiata
Cecropis [Genus]
C. cucullata [Species]
C. abyssinica
C. semirufa
C. senegalensis
C. daurica
C. striolata
Petrochelidon [Genus]
P. rufigula [Species]
P. preussi
P. andecola
P. nigricans
P. spilodera
P. pyrrhonota
P. fulva
P. fluvicola
P. ariel
P. fuliginosa
Delichon [Genus]
D. urbica [Species]
D. dasypus
D. nipalensis
Psalidoprocne [Genus]
P. nitens [Species]
P. fuliginosa
P. albiceps
P. pristoptera
P. oleaginea
P. antinorii
P. petiti

P. holomelaena
P. orientalis
P. mangebettorum
P. chalybea
P. obscura

Motacillidae [Family]
  Dendronanthus [Genus]
    D. indicus [Species]
  Motacilla [Genus]
    M. flava [Species]
    M. citreola
    M. cinerea
    M. alba
    M. grandis
    M. madaraspatensis
    M. aguimp
    M. clara
    M. capensis
    M. flaviventris
  Tmetothylacus [Genus]
    T. tenellus [Species]
  Macronyx [Genus]
    M. capensis [Species]
    M. croceus
    M. fullebornii
    M. sharpei
    M. flavicollis
    M. aurantiigula
    M. ameliae
    M. grimwoodi
  Anthus [Genus]
    A. novaeseelandiae [Species]
    A. leucophrys
    A. vaalensis
    A. pallidiventris
    A. melindae
    A. campestris
    A. godlewskii
    A. berthelotii
    A. similis
    A. brachyurus
    A. caffer
    A. trivialis
    A. nilghiriensis
    A. hodgsoni
    A. gustavi
    A. pratensis
    A. cervinus
    A. roseatus
    A. spinoletta
    A. sylvanus
    A. spragueii
    A. furcatus
    A. hellmayri
    A. chacoensis
    A. lutescens
    A. correndera
    A. nattereri
    A. bogotensis
    A. antarcticus

A. gutturalis
A. sokokensis
A. crenatus
A. lineiventris
A. chloris

Campephagidae [Family]
  Pteropodocys [Genus]
    P. maxima [Species]
  Coracina [Genus]
    C. novaehollandiae [Species]
    C. fortis
    C. atriceps
    C. pollens
    C. schistacea
    C. caledonica
    C. caeruleogrisea
    C. temminckii
    C. larvata
    C. striata
    C. bicolor
    C. lineata
    C. boyeri
    C. leucopygia
    C. papuensis
    C. robusta
    C. longicauda
    C. parvula
    C. abbotti
    C. analis
    C. caesia
    C. pectoralis
    C. graueri
    C. cinerea
    C. azurea
    C. typica
    C. newtoni
    C. coerulescens
    C. dohertyi
    C. tenuirostris
    C. morio
    C. schisticeps
    C. melaena
    C. montana
    C. holopolia
    C. mcgregori
    C. panayensis
    C. polioptera
    C. melaschistos
    C. fimbriata
    C. melanoptera
  Campochaera [Genus]
    C. sloetii [Species]
  Chlamydochaera [Genus]
    C. jefferyi [Species]
  Lalage [Genus]
    L. melanoleuca [Species]
    L. nigra
    L. sueurii
    L. aurea
    L. atrovirens

L. leucomela
L. maculosa
L. sharpei
L. leucopygia
  Campephaga [Genus]
    C. phoenicea [Species]
    C. quiscalina
    C. lobata
  Pericrocotus [Genus]
    P. roseus [Species]
    P. divaricus
    P. cinnamomeus
    P. lansbergei
    P. erythropygius
    P. solaris
    P. ethologus
    P. brevirostris
    P. miniatus
    P. flammeus
  Hemipus [Genus]
    H. picatus [Species]
    H. hirundinaceus
  Tephrodornis [Genus]
    T. gularis [Species]
    T. pondicerianus

Pycnonotidae [Family]
  Spizixos [Genus]
    S. canifrons [Species]
    S. semitorques
  Pycnonotus [Genus]
    P. zeylanicus [Species]
    P. striatus
    P. leucogrammicus
    P. tympanistrigus
    P. melanoleucos
    P. priocephalus
    P. atriceps
    P. melanicterus
    P. squamatus
    P. cyaniventris
    P. jocosus
    P. xanthorrhous
    P. sinensis
    P. taivanus
    P. leucogenys
    P. cafer
    P. aurigaster
    P. xanthopygos
    P. nigricans
    P. capensis
    P. barbatus
    P. eutilotus
    P. nieuwenhuisii
    P. urostictus
    P. bimaculatus
    P. finlaysoni
    P. xantholaemus
    P. penicillatus
    P. flavescens
    P. goiavier

AVES SPECIES LIST

P. luteolus
P. plumosus
P. blanfordi
P. simplex
P. brunneus
P. erythrophthalmos
P. masukuensis
P. montanus
P. virens
P. gracilis
P. ansorgei
P. curvirostris
P. importunus
P. latirostris
P. gracilirostris
P. tephrolaemus
P. milanjensis
Calyptocichla [Genus]
C. serina [Species]
Baeopogon [Genus]
B. indicator [Species]
B. clamans
Ixonotus [Genus]
I. guttatus [Species]
Chlorocichla [Genus]
C. falkensteini [Species]
C. simplex
C. flavicollis
C. flaviventris
C. laetissima
Thescelocichla [Genus]
T. leucopleura [Species]
Phyllastrephus [Genus]
P. scandens [Species]
P. terrestris
P. strepitans
P. cerviniventris
P. fulviventris
P. poensis
P. hypochloris
P. baumanni
P. poliocephalus
P. flavostriatus
P. debilis
P. lorenzi
P. albigularis
P. fischeri
P. orostruthus
P. icterinus
P. xavieri
P. madagascariensis
P. zosterops
P. tenebrosus
P. xanthophrys
P. cinereiceps
Bleda [Genus]
B. syndactyla [Species]
B. eximia
B. canicapilla
Nicator [Genus]

N. chloris [Species]
N. gularis
N. vireo
Criniger [Genus]
C. barbatus [Species]
C. calurus
C. ndussumensis
C. olivaceus
C. finschii
C. flaveolus
C. pallidus
C. ochraceus
C. bres
C. phaeocephalus
Setornis [Genus]
S. criniger [Species]
Hypsipetes [Genus]
H. viridescens [Species]
H. propinquus
H. charlottae
H. palawanensis
H. criniger
H. philippinus
H. siquijorensis
H. everetti
H. affinis
H. indicus
H. mcclellandii
H. malaccensis
H. virescens
H. flavala
H. amaurotis
H. crassirostris
H. borbonicus
H. madagascariensis
H. nicobariensis
H. thompsoni
Neolestes [Genus]
N. torquatus [Species]
Tylas [Genus]
T. eduardi [Species]

Irenidae [Family]
Aegithina [Genus]
A. tiphia [Species]
A. nigrolutea
A. viridissima
A. lafresnayei
Chloropsis [Genus]
C. flavipennis [Species]
C. palawanensis
C. sonnerati
C. cyanopogon
C. cochinchinensis
C. aurifrons
C. hardwickei
C. venusta
Irena [Genus]
I. puella [Species]
I. cyanogaster
Eurocephalus [Genus]

E. ruppelli [Species]
E. anguitimens

Laniidae [Family]
Prionops [Genus]
P. plumata [Species]
P. poliolopha
P. caniceps
P. alberti
P. retzii
P. gabela
P. scopifrons
Lanioturdus [Genus]
L. torquatus [Species]
Nilaus [Genus]
N. afer [Species]
Dryoscopus [Genus]
D. pringlii [Species]
D. gambensis
D. cubla
D. senegalensis
D. angolensis
D. sabini
Tchagra [Genus]
T. minuta [Species]
T. senegala
T. tchagra
T. australis
T. jamesi
T. cruenta
Laniarius [Genus]
L. ruficeps [Species]
L. luhderi
L. ferrugineus
L. barbarus
L. mufumbiri
L. atrococcineus
L. atroflavus
L. fulleborni
L. funebris
L. leucorhynchus
Telophorus [Genus]
T. bocagei [Species]
T. sulfureopectus
T. olivaceus
T. nigrifrons
T. multicolor
T. kupeensis
T. zeylonus
T. viridis
T. quadricolor
T. dohertyi
Malaconotus [Genus]
M. cruentus [Species]
M. lagdeni
M. gladiator
M. blanchoti
M. alius
Corvinella [Genus]
C. corvina [Species]
C. melanoleuca

Lanius [Genus]
    L. tigrinus [Species]
    L. souzae
    L. bucephalus
    L. cristatus
    L. collurio
    L. collueioides
    L. gubernator
    L. vittatus
    L. schach
    L. validirostris
    L. mackinnoni
    L. minor
    L. ludovicianus
    L. excubitor
    L. excubitoroides
    L. sphenocercus
    L. cabanisi
    L. dorsalis
    L. somalicus
    L. collaris
    L. newtoni
    L. senator
    L. nubicus
Pityriasis [Genus]
    P. gymnocephala [Species]

Vangidae [Family]
  Calicalicus [Genus]
    C. madagascariensis [Species]
  Schetba [Genus]
    S. rufa [Species]
  Vanga [Genus]
    V. curvirostris [Species]
  Xenopirostris [Genus]
    X. xenopirostris [Species]
    X. damii
    X. polleni
  Falculea [Genus]
    F. palliata [Species]
  Leptopterus [Genus]
    L. viridis [Species]
    L. chabert
    L. madagascarinus
  Oriolia [Genus]
    O. bernieri [Species]
  Euryceros [Genus]
    E. prevostrii [Species]

Bombycillidae [Family]
  Bombycilla [Genus]
    B. garrulus [Species]
    B. japonica
    B. cedrorum
  Ptilogonys [Genus]
    P. cinereus [Species]
    P. caudatus
  Phainopepla [Genus]
    P. nitens [Species]
  Phainoptila [Genus]
    P. melanoxantha [Species]

Hypocolius [Genus]
    H. ampelinus [Species]

Dulidae [Family]
  Dulus [Genus]
    D. dominicus [Species]

Cinclidae [Family]
  Cinclus [Genus]
    C. cinclus [Species]
    C. pallasii
    C. mexicanus
    C. leucocephalus

Troglodytidae [Family]
  Campylorhynchus [Genus]
    C. jocosus [Species]
    C. gularis
    C. yucatanicus
    C. brunneicapillus
    C. griseus
    C. rufinucha
    C. turdinus
    C. nuchalis
    C. fasciatus
    C. zonatus
    C. megalopterus
  Odontorchilus [Genus]
    O. cinereus [Species]
    O. branickii
  Salpinctes [Genus]
    S. obsoletus [Species]
    S. mexicanus
  Hylorchilus [Genus]
    H. sumichrasti [Species]
  Cinnycerthia [Genus]
    C. unirufa [Species]
    C. peruana
  Cistothorus [Genus]
    C. platensis [Species]
    C. meridae
    C. apolinari
    C. palustris
  Thryomanes [Genus]
    T. bewickii [Species]
    T. sissonii
  Ferminia [Genus]
    F. cerverai [Species]
  Thryothorus [Genus]
    T. atrogularis [Species]
    T. fasciatoventris
    T. euophrys
    T. genibarbis
    T. coraya
    T. felix
    T. maculipectus
    T. rutilus
    T. nigricapillus
    T. thoracicus
    T. pleurostictus
    T. ludovicianus

    T. rufalbus
    T. nicefori
    T. sinaloa
    T. modestus
    T. leucotis
    T. superciliaris
    T. guarayanus
    T. longirostris
    T. griseus
  Troglodytes [Genus]
    T. troglodytes [Species]
    T. aedon
    T. solstitialis
    T. rufulus
    T. browni
  Uropsila [Genus]
    U. leucogastra [Species]
  Henicorhina [Genus]
    H. leucosticta [Species]
    H. leucophrys
  Microcerculus [Genus]
    M. marginatus [Species]
    M. ustulatus
    M. bambla
  Cyphorhinus [Genus]
    C. thoracicus [Species]
    C. aradus

Mimidae [Family]
  Dumetalla [Genus]
    D. carolinensis [Species]
  Melanoptila [Genus]
    M. glabrirostris [Species]
  Melanotis [Genus]
    M. caerulescens [Species]
    M. hypoleucus
  Mimus [Genus]
    M. polyglottos [Species]
    M. gilvus
    M. gundlachii
    M. thenca
    M. longicaudatus
    M. saturninus
    M. patagonicus
    M. triurus
    M. dorsalis
  Nesomimus [Genus]
    N. trifasciatus [Species]
  Mimodes [Genus]
    M. graysoni [Species]
  Oreoscoptes [Genus]
    O. montanus [Species]
  Toxostoma [Genus]
    T. rufum [Species]
    T. longirostre
    T. guttatum
    T. cinereum
    T. bendirei
    T. ocellatum
    T. curvirostre
    T. lecontei

T. redivivum
T. dorsale
Cinclocerthia [Genus]
   C. ruficauda [Species]
Ramphocinclus [Genus]
   R. brachyurus [Species]
Donacobius [Genus]
   D. atricapillus [Species]
Allenia [Genus]
   A. fusca [Species]
Margarops [Genus]
   M. fuscatus [Species]

Prunellidae [Family]
  Prunella [Genus]
   P. collaris [Species]
   P. himalayana
   P. rubeculoides
   P. strophiata
   P. montanella
   P. fulvescens
   P. ocularis
   P. atrogularis
   P. koslowi
   P. modularis
   P. rubida
   P. immaculata

Turdidae [Family]
  Brachypteryx [Genus]
   B. stellata [Species]
   B. hyperythra
   B. major
   B. calligyna
   B. leucophrys
   B. montana
  Zeledonia [Genus]
   Z. coronata [Species]
  Erythropygia [Genus]
   E. coryphaeus [Species]
   E. leucophrys
   E. hartlaubi
   E. galactotes
   E. paena
   E. leucosticta
   E. quadrivirgata
   E. barbata
   E. signata
  Namibornis [Genus]
   N. herero [Species]
  Cercotrichas [Genus]
   C. podobe [Species]
  Pinarornis [Genus]
   P. plumosus [Species]
  Chaetops [Genus]
   C. frenatus [Species]
  Drymodes [Genus]
   D. brunneopygia [Species]
   D. superciliaris
  Pogonocichla [Genus]
   P. stellata [Species]

P. swynnertoni
Erithacus [Genus]
   E. gabela [Species]
   E. cyornithopsis
   E. aequatorialis
   E. erythrothorax
   E. sharpei
   E. gunningi
   E. rubecula
   E. akahige
   E. komadori
   E. sibilans
   E. luscinia
   E. megarhynchos
   E. calliope
   E. svecicus
   E. pectoralis
   E. ruficeps
   E. obscurus
   E. pectardens
   E. brunneus
   E. cyane
   E. cyanurus
   E. chrysaeus
   E. indicus
   E. hyperythrus
   E. johnstoniae
Cossypha [Genus]
   C. roberti [Species]
   C. bocagei
   C. polioptera
   C. archeri
   C. isabellae
   C. natalensis
   C. dichroa
   C. semirufa
   C. heuglini
   C. cyanocampter
   C. caffra
   C. anomala
   C. humeralis
   C. ansorgei
   C. niveicapilla
   C. heinrichi
   C. albicapilla
Modulatrix [Genus]
   M. stictigula [Species]
Cichladusa [Genus]
   C. guttata [Species]
   C. arquata
   C. ruficauda
Alethe [Genus]
   A. diademata [Species]
   A. poliophrys
   A. fuelleborni
   A. montana
   A. lowei
   A. poliocephala
   A. choloensis
Copsychus [Genus]

C. saularis [Species]
C. sechellarum
C. albospecularis
C. malabaricus
C. stricklandii
C. luzoniensis
C. niger
C. pyrropygus
Irania [Genus]
   I. gutturalis [Species]
Phoenicurus [Genus]
   P. alaschanicus [Species]
   P. erythronotus
   P. caeruleocephalus
   P. ochruros
   P. phoenicurus
   P. hodgsoni
   P. frontalis
   P. schisticeps
   P. auroreus
   P. moussieri
   P. erythrogaster
Rhyacornis [Genus]
   R. bicolor [Species]
   R. fuliginosus
Hodgsonius [Genus]
   H. phaenicuroides [Species]
Cinclidium [Genus]
   C. leucurum [Species]
   C. diana
   C. frontale
Grandala [Genus]
   G. coelicolor [Species]
Sialia [Genus]
   S. sialis [Species]
   S. mexicana
   S. currucoides
Enicurus [Genus]
   E. scouleri [Species]
   E. velatus
   E. ruficapillus
   E. immaculatus
   E. schistaceus
   E. leschenaulti
   E. maculatus
Cochoa [Genus]
   C. purpurea [Species]
   C. viridis
   C. azurea
Myadestes [Genus]
   M. townsendi [Species]
   M. obscurus
   M. elisabeth
   M. genibarbis
   M. ralloides
   M. unicolor
   M. leucogenys
Entomodestes [Genus]
   E. leucotis [Species]
   E. coracinus

Stizorhina [Genus]
   S. fraseri [Species]
   S. finschii
Neocossyphus [Genus]
   N. rufus [Species]
   N. poensis
Cercomela [Genus]
   C. sinuata [Species]
   C. familiaris
   C. tractrac
   C. schlegelii
   C. fusca
   C. dubia
   C. melanura
   C. scotocerca
   C. sordida
Saxicola [Genus]
   S. rubetra [Species]
   S. macrorhyncha
   S. insignis
   S. dacotiae
   S. torquata
   S. leucura
   S. caprata
   S. jerdoni
   S. ferrea
   S. gutturalis
Myrmecocichla [Genus]
   M. tholloni [Species]
   M. aethiops
   M. formicivora
   M. nigra
   M. arnotti
   M. albifrons
   M. melaena
Thamnolaea [Genus]
   T. cinnamomeiventris [Species]
   T. coronata
   T. semirufa
Oenanthe [Genus]
   O. bifasciata [Species]
   O. isabellina
   O. bottae
   O. xanthoprymna
   O. oenanthe
   O. deserti
   O. hispanica
   O. finschii
   O. picata
   O. lugens
   O. monacha
   O. alboniger
   O. pleschanka
   O. leucopyga
   O. leucura
   O. monticola
   O. moesta
   O. pileata
Chaimarrornis [Genus]
   C. leucocephalus [Species]

Saxicoloides [Genus]
   S. fulicata [Species]
Pseudocossyphus [Genus]
   P. imerinus [Species]
Monticola [Genus]
   M. rupestris [Species]
   M. explorator
   M. brevipes
   M. rufocinereus
   M. angolensis
   M. saxatilis
   M. cinclorhynchus
   M. rufiventris
   M. solitarius
Myophonus [Genus]
   M. blighi [Species]
   M. melanurus
   M. glaucinus
   M. robinsoni
   M. horsfieldii
   M. insularis
   M. caeruleus
Geomalia [Genus]
   G. heinrichi [Species]
Zoothera [Genus]
   Z. schistacea [Species]
   Z. dumasi
   Z. interpres
   Z. erythronota
   Z. wardii
   Z. cinerea
   Z. peronii
   Z. citrina
   Z. everetti
   Z. sibirica
   Z. naevia
   Z. pinicola
   Z. piaggiae
   Z. oberlaenderi
   Z. gurneyi
   Z. cameronensis
   Z. princei
   Z. crossleyi
   Z. guttata
   Z. spiloptera
   Z. andromedae
   Z. mollissima
   Z. dixoni
   Z. dauma
   Z. talaseae
   Z. margaretae
   Z. monticola
   Z. marginata
   Z. terrestris
Amalocichla [Genus]
   A. sclateriana [Species]
   A. incerta
Cataponera [Genus]
   C. turdoides [Species]
Nesocichla [Genus]

   N. eremita [Species]
Cichlherminia [Genus]
   C. lherminieri [Species]
Phaeornis [Genus]
   P. obscurus [Species]
   P. palmeri
Catharus [Genus]
   C. gracilirostris [Species]
   C. aurantiirostris
   C. fuscater
   C. occidentalis
   C. mexicanus
   C. dryas
   C. fuscescens
   C. minimus
   C. ustulatus
   C. guttatus
Hylocichla [Genus]
   H. mustelina [Species]
Platycichla [Genus]
   P. flavipes [Species]
   P. leucops
Turdus [Genus]
   T. bewsheri [Species]
   T. olivaceofuscus
   T. olivaceus
   T. abyssinicus
   T. helleri
   T. libonyanus
   T. tephronotus
   T. menachensis
   T. ludoviciae
   T. litsipsirupa
   T. dissimilis
   T. unicolor
   T. cardis
   T. albocinctus
   T. torquatus
   T. boulboul
   T. merula
   T. poliocephalus
   T. chrysolaus
   T. celaenops
   T. rubrocanus
   T. kessleri
   T. feae
   T. pallidus
   T. obscurus
   T. ruficollis
   T. naumanni
   T. pilaris
   T. iliacus
   T. philomelos
   T. mupinensis
   T. viscivorus
   T. aurantius
   T. ravidus
   T. plumbeus
   T. chiguanco
   T. nigriscens

T. fuscater
T. serranus
T. nigriceps
T. reevei
T. olivater
T. maranonicus
T. fulviventris
T. rufiventris
T. falcklandii
T. leucomelas
T. amaurochalinus
T. plebejus
T. ignobilis
T. lawrencii
T. fumigatus
T. hauxwelli
T. haplochrous
T. grayi
T. nudigenis
T. jamaicensis
T. albicollis
T. rufopalliatus
T. swalesi
T. rufitorques
T. migratorius

Orthonychidae [Family]
  Orthonyx [Genus]
    O. temminckii [Species]
    O. spaldingii
  Androphobus [Genus]
    A. viridis [Species]
  Psophodes [Genus]
    P. olivaceus [Species]
    P. nigrogularis
  Sphenostoma [Genus]
    S. cristatum [Species]
  Cinclostoma [Genus]
    C. punctatum [Species]
    C. castanotum
    C. cinnamomeum
    C. ajax
  Ptilorrhoa [Genus]
    P. leucosticta [Species]
    P. caerulescens
    P. castanonota
  Eupetes [Genus]
    E. macrocercus [Species]
  Melampitta [Genus]
    M. lugubris [Species]
    M. gigantea
  Ifrita [Genus]
    I. kowaldi [Species]

Timaliidae [Family]
  Pellorneum [Genus]
    P. ruficeps [Species]
    P. palustre
    P. fuscocapillum
    P. capistratum
    P. albiventre

Trichastoma [Genus]
  T. tickelli [Species]
  T. pyrrogenys
  T. malaccense
  T. cinereiceps
  T. rostratum
  T. bicolor
  T. separium
  T. celebense
  T. abbotti
  T. perspicillatum
  T. vanderbilti
  T. pyrrhopterum
  T. cleaveri
  T. albipectus
  T. rufescens
  T. rufipenne
  T. fulvescens
  T. puveli
  T. poliothorax
Leonardina [Genus]
  L. woodi [Species]
Ptyrticus [Genus]
  P. turdinus [Species]
Malacopteron [Genus]
  M. magnirostre [Species]
  M. affine
  M. cinereum
  M. magnum
  M. palawanense
  M. albogulare
Pomatorhinus [Genus]
  P. hypoleucos [Species]
  P. erythrogenys
  P. horsfieldii
  P. schisticeps
  P. montanus
  P. ruficollis
  P. ochraceiceps
  P. ferruginosus
Garritornis [Genus]
  G. isidorei [Species]
Pomatostomus [Genus]
  P. temporalis [Species]
  P. superciliosus
  P. ruficeps
Xiphirhynchus [Genus]
  X. superciliaris [Species]
Jabouilleia [Genus]
  J. danjoui [Species]
Rimator [Genus]
  R. malacoptilus [Species]
Ptilocichla [Genus]
  P. leucogrammica [Species]
  P. mindanensis
  P. falcata
Kenopia [Genus]
  K. striata [Species]
Napothera [Genus]
  N. rufipectus [Species]

N. atrigularis
N. macrodactyla
N. marmorata
N. crispifrons
N. brevicaudata
N. crassa
N. rabori
N. epilepidota
Pnoepyga [Genus]
  P. albiventer [Species]
  P. pusilla
Spelaeornis [Genus]
  S. caudatus [Species]
  S. troglodytoides
  S. formosus
  S. chocolatinus
  S. longicaudatus
Sphenocichla [Genus]
  S. humei [Species]
Neomixis [Genus]
  N. tenella [Species]
  N. viridis
  N. striatigula
  N. flavoviridis
Stachyris [Genus]
  S. rodolphei [Species]
  S. rufifrons
  S. ambigua
  S. ruficeps
  S. pyrrhops
  S. chrysaea
  S. plateni
  S. capitalis
  S. speciosa
  S. whiteheadi
  S. striata
  S. nigrorum
  S. hypogrammica
  S. grammiceps
  S. herberti
  S. nigriceps
  S. poliocephala
  S. striolata
  S. oglei
  S. maculata
  S. leucotis
  S. nigricollis
  S. thoracica
  S. erythroptera
  S. melanothorax
Dumetia [Genus]
  D. hyperythra [Species]
Rhopocichla [Genus]
  R. atriceps [Species]
Macronous [Genus]
  M. flavicollis [Species]
  M. gularis
  M. kelleyi
  M. striaticeps
  M. ptilosus

Micromacronus [Genus]
 M. leytensis [Species]
Timalia [Genus]
 T. pileata [Species]
Chrysomma [Genus]
 C. sinense [Species]
Moupinia [Genus]
 M. altirostris [Species]
 M. poecilotis
Chamaea [Genus]
 C. fasciata [Species]
Turdoides [Genus]
 T. nipalensis [Species]
 T. altirostris
 T. caudatus
 T. earlei
 T. gularis
 T. longirostris
 T. malcolmi
 T. squamiceps
 T. fulvus
 T. aylmeri
 T. rubiginosus
 T. subrufus
 T. striatus
 T. affinis
 T. melanops
 T. tenebrosus
 T. reinwardtii
 T. plebejus
 T. jardineii
 T. squamulatus
 T. leucopygius
 T. hindei
 T. hypoleucus
 T. bicolor
 T. gymnogenys
Babax [Genus]
 B. lanceolatus [Species]
 B. waddelli
 B. koslowi
Garrulax [Genus]
 G. cinereifrons [Species]
 G. palliatus
 G. rufifrons
 G. perspicillatus
 G. albogularis
 G. leucolophus
 G. monileger
 G. pectoralis
 G. lugubris
 G. striatus
 G. strepitans
 G. milleti
 G. maesi
 G. chinensis
 G. vassali
 G. galbanus
 G. delesserti
 G. variegatus

G. davidi
G. sukatschewi
G. cineraceus
G. rufogularis
G. lunulatus
G. maximus
G. ocellatus
G. caerulatus
G. mitratus
G. ruficollis
G. merulinus
G. canorus
G. sannio
G. cachinnans
G. lineatus
G. virgatus
G. austeni
G. squamatus
G. subunicolor
G. elliotii
G. henrici
G. affinis
G. erythrocephalus
G. yersini
G. formosus
G. milnei
Liocichla [Genus]
 L. phoenicea [Species]
 L. steerii
Leiothrix [Genus]
 L. argentauris [Species]
 L. lutea
Cutia [Genus]
 C. nipalensis [Species]
Pteruthius [Genus]
 P. rufiventer [Species]
 P. flaviscapis
 P. xanthochlorus
 P. melanotis
 P. aenobarbus
Gampsorhynchus [Genus]
 G. rufulus [Species]
Actinodura [Genus]
 A. egertoni [Species]
 A. ramsayi
 A. nipalensis
 A. waldeni
 A. souliei
 A. morrisoniana
Minla [Genus]
 M. cyanouroptera [Species]
 M. strigula
 M. ignotincta
Alcippe [Genus]
 A. chrysotis [Species]
 A. variegaticeps
 A. cinerea
 A. castaneceps
 A. vinipectus
 A. striaticollis

A. ruficapilla
A. cinereiceps
A. rufogularis
A. brunnea
A. brunneicauda
A. poioicephala
A. pyrrhoptera
A. peracensis
A. morrisonia
A. nipalensis
A. abyssinica
A. atriceps
Lioptilus [Genus]
 L. nigricapillus [Species]
 L. gilberti
 L. rufocinctus
 L. chapini
Parophasma [Genus]
 P. galinieri [Species]
Phyllanthus [Genus]
 P. atripennis [Species]
Crocias [Genus]
 C. langbianis [Species]
 C. albonotatus
Heterophasia [Genus]
 H. annectens [Species]
 H. capistrata
 H. gracilis
 H. melanoleuca
 H. auricularis
 H. pulchella
 H. picaoides
Yuhina [Genus]
 Y. castaniceps [Species]
 Y. bakeri
 Y. flavicollis
 Y. gularis
 Y. diademata
 Y. occipitalis
 Y. brunneiceps
 Y. nigrimenta
 Y. zantholeuca
Malia [Genus]
 M. grata [Species]
Myzornis [Genus]
 M. pyrrhoura [Species]
Horizorhinus [Genus]
 H. dohrni [Species]
Oxylabes [Genus]
 O. madagascariensis [Species]
Mystacornis [Genus]
 M. crossleyi [Species]

Panuridae [Family]
Panurus [Genus]
 P. biarmicus [Species]
Conostoma [Genus]
 C. oemodium [Species]
Paradoxornis [Genus]
 P. paradoxus [Species]
 P. unicolor

P. flavirostris
P. guttaticollis
P. conspicillatus
P. ricketti
P. webbianus
P. alphonsianus
P. zappeyi
P. przewalskii
P. fulvifrons
P. nipalensis
P. davidianus
P. atrosuperciliaris
P. ruficeps
P. gularis
P. heudei

Picathartidae [Family]
   Picathartes [Genus]
      P. gymnocephalus [Species]
      P. oreas

Polioptilidae [Family]
   Microbates [Genus]
      M. collaris [Species]
      M. cinereiventris
   Ramphocaenus [Genus]
      R. melanurus [Species]
   Polioptila [Genus]
      P. caerulea [Species]
      P. melanura
      P. lembeyei
      P. albiloris
      P. plumbea
      P. lactea
      P. guianensis
      P. schistaceigula
      P. dumicola

Sylviidae [Family]
   Oligura [Genus]
      O. castaneocoronata [Species]
   Tesia [Genus]
      T. superciliaris [Species]
      T. olivea
      T. cyaniventer
   Urosphena [Genus]
      U. subulata [Species]
      U. whiteheadi
      U. squameiceps
      U. pallidipes
   Cettia [Genus]
      C. diphone [Species]
      C. annae
      C. parens
      C. ruficapilla
      C. fortipes
      C. vulcania
      C. major
      C. flavolivacea
      C. robustipes
      C. brunnifrons
      C. cetti

Bradypterus [Genus]
   B. baboecala [Species]
   B. graueri
   B. grandis
   B. carpalis
   B. alfredi
   B. sylvaticus
   B. barratti
   B. victorini
   B. cinnamomeus
   B. thoracicus
   B. major
   B. tacsanowskius
   B. luteoventris
   B. palliseri
   B. seebohmi
   B. caudatus
   B. accentor
   B. castaneus
Bathmocercus [Genus]
   B. cerviniventris [Species]
   B. rufus
   B. winifredae
Dromaeocercus [Genus]
   D. brunneus [Species]
   D. seeboehmi
Nesillas [Genus]
   N. typica [Species]
   N. aldabranus
   N. mariae
Thamnornis [Genus]
   T. chloropetoides [Species]
Melocichla [Genus]
   M. mentalis [Species]
Achaetops [Genus]
   A. pycnopygius [Species]
Sphenoeacus [Genus]
   S. afer [Species]
Megalurus [Genus]
   M. pryeri [Species]
   M. timoriensis
   M. palustris
   M. albolimbatus
   M. gramineus
   M. punctatus
Cincloramphus [Genus]
   C. cruralis [Species]
   C. mathewsi
Eremiornis [Genus]
   E. carteri [Species]
Megalurulus [Genus]
   M. bivittata [Species]
   M. mariei
Cichlornis [Genus]
   C. whitneyi [Species]
   C. llaneae
   C. grosvenori
Ortygocichla [Genus]
   O. rubiginosa [Species]
   O. rufa

Chaetornis [Genus]
   C. striatus [Species]
Graminicola [Genus]
   G. bengalensis [Species]
Schoenicola [Genus]
   S. platyura [Species]
Locustella [Genus]
   L. lanceolata [Species]
   L. naevia
   L. certhiola
   L. ochotensis
   L. pleskei
   L. fluvialtilis
   L. luscinioides
   L. fasciolata
   L. amnicola
Acrocephalus [Genus]
   A. melanopogon [Species]
   A. paludicola
   A. schoenobaenus
   A. sorghophilus
   A. bistrigiceps
   A. agricola
   A. concinens
   A. scirpaceus
   A. cinnamomeus
   A. baeticatus
   A. palustris
   A. dumetorum
   A. arundinaceus
   A. stentoreus
   A. orinus
   A. orientalis
   A. luscinia
   A. familiaris
   A. aequinoctialis
   A. caffer
   A. atyphus
   A. vaughani
   A. rufescens
   A. brevipennis
   A. gracilirostris
   A. newtoni
   A. aedon
Bebrornis [Genus]
   B. rodericanus [Species]
   B. sechellensis
Hippolais [Genus]
   H. caligata [Species]
   H. pallida
   H. languida
   H. olivetorum
   H. polyglotta
   H. icterina
Chloropeta [Genus]
   C. natalensis [Species]
   C. similis
   C. gracilirostris
Cisticola [Genus]
   C. erythrops [Species]
   C. lepe

C. cantans
C. lateralis
C. woosnami
C. anonyma
C. bulliens
C. chubbi
C. hunteri
C. nigriloris
C. aberrans
C. bodessa
C. chiniana
C. cinereola
C. ruficeps
C. rufilata
C. subruficapilla
C. lais
C. restricta
C. njombe
C. galactotes
C. pipiens
C. carruthersi
C. tinniens
C. robusta
C. aberdare
C. natalensis
C. fulvicapilla
C. angusticauda
C. melanura
C. brachyptera
C. rufa
C. troglodytes
C. nana
C. incana
C. juncidis
C. cherina
C. haesitata
C. aridula
C. textrix
C. eximia
C. dambo
C. brunnescens
C. ayresii
C. exilis
Scotocerca [Genus]
S. inquieta [Species]
Rhopophilus [Genus]
R. pekinensis [Species]
Prinia [Genus]
P. burnesi [Species]
P. criniger
P. polychroa
P. atrogularis
P. cinereocapilla
P. buchanani
P. rufescens
P. hodgsoni
P. gracilis
P. sylvatica
P. familiaris
P. flaviventris

P. socialis
P. subflava
P. somalica
P. fluviatilis
P. maculosa
P. flavicans
P. substriata
P. molleri
P. robertsi
P. leucopogon
P. leontica
P. bairdii
P. erythroptera
P. pectoralis
Drymocichla [Genus]
D. incana [Species]
Urolais [Genus]
U. epichlora [Species]
Spiloptila [Genus]
S. clamans [Species]
Apalis [Genus]
A. thoracica [Species]
A. pulchra
A. ruwenzori
A. nigriceps
A. jacksoni
A. chariessa
A. binotata
A. flavida
A. ruddi
A. rufogularis
A. sharpii
A. goslingi
A. bamendae
A. porphyrolaema
A. melanocephala
A. chirindensis
A. cinerea
A. alticola
A. karamojae
A. rufifrons
Stenostira [Genus]
S. scita [Species]
Phyllolais [Genus]
P. pulchella [Species]
Orthotomus [Genus]
O. metopias [Species]
O. moreaui
O. cucullatus
O. sutorius
O. atrogularis
O. derbianus
O. sericeus
O. ruficeps
O. sepium
O. cinereiceps
O. nigriceps
O. samaransis
Camaroptera [Genus]
C. brachyura [Species]

C. brevicauda
C. harterti
C. superciliaris
C. chloronota
Calamonastes [Genus]
C. simplex [Species]
C. stierlingi
C. fasciolatus
Euryptila [Genus]
E. subcinnamomea [Species]
Poliolais [Genus]
P. lopesi [Species]
Graueria [Genus]
G. vittata [Species]
Eremomela [Genus]
E. icteropygialis [Species]
E. flavocrissalis
E. scotops
E. pusilla
E. canescens
E. gregalis
E. badiceps
E. turneri
E. atricollis
E. usticollis
Randia [Genus]
R. pseudozosterops [Species]
Newtonia [Genus]
N. brunneicauda [Species]
N. amphichroa
N. archboldi
N. fanovanae
Sylvietta [Genus]
S. virens [Species]
S. denti
S. leucophrys
S. brachyura
S. philippae
S. whytii
S. ruficapilla
S. rufescens
S. isabellina
Hemitesia [Genus]
H. neumanni [Species]
Macrosphenus [Genus]
M. kempi [Species]
M. flavicans
M. concolor
M. pulitzeri
M. kretschmeri
Amaurocichla [Genus]
A. bocagei [Species]
Hypergerus [Genus]
H. atriceps [Species]
H. lepidus
Hyliota [Genus]
H. flavigaster [Species]
H. australis
H. violacea
Hylia [Genus]

H. prasina [Species]
Phylloscopus [Genus]
  P. ruficapilla [Species]
  P. laurae
  P. laetus
  P. herberti
  P. budongoensis
  P. umbrovirens
  P. trochilus
  P. collybita
  P. sindianus
  P. neglectus
  P. bonelli
  P. sibilatrix
  P. fuscatus
  P. fuligiventer
  P. affinis
  P. griseolus
  P. armandii
  P. schwarzi
  P. pulcher
  P. maculipennis
  P. proregulus
  P. subviridis
  P. inornatus
  P. borealis
  P. trochiloides
  P. nitidus
  P. plumbeitarsus
  P. tenellipes
  P. magnirostris
  P. tytleri
  P. occipitalis
  P. coronatus
  P. ijimae
  P. reguloides
  P. davisoni
  P. cantator
  P. ricketti
  P. olivaceus
  P. cebuensis
  P. trivirgatus
  P. sarasinorum
  P. presbytes
  P. poliocephalus
  P. makirensis
  P. amoenus
Seicercus [Genus]
  S. burkii [Species]
  S. xanthoschistos
  S. affinis
  S. poliogenys
  S. castaniceps
  S. montis
  S. grammiceps
Tickellia [Genus]
  T. hodgsoni [Species]
Abroscopus [Genus]
  A. albogularis [Species]
  A. schisticeps
  A. superciliaris

Parisoma [Genus]
  P. buryi [Species]
  P. lugens
  P. boehmi
  P. layardi
  P. subcaeruleum
Sylvia [Genus]
  S. atricapilla [Species]
  S. borin
  S. communis
  S. curruca
  S. nana
  S. nisoria
  S. hortensis
  S. leucomelaena
  S. rueppelli
  S. melanocephala
  S. melanothorax
  S. mystacea
  S. cantillans
  S. conspicillata
  S. deserticola
  S. undata
  S. sarda
Regulus [Genus]
  R. ignicapillus [Species]
  R. regulus
  R. goodfellowi
  R. satrapa
  R. calendula
Leptopoecile [Genus]
  L. sophiae [Species]
  L. elegans

Muscicapidae [Family]
Melaenornis [Genus]
  M. semipartitus [Species]
  M. pallidus
  M. infuscatus
  M. mariquensis
  M. microrhynchus
  M. chocolatinus
  M. fischeri
  M. brunneus
  M. edolioides
  M. pammelaina
  M. ardesiacus
  M. annamarulae
  M. ocreatus
  M. cinerascens
  M. silens
Rhinomyias [Genus]
  R. addita [Species]
  R. oscillans
  R. brunneata
  R. olivacea
  R. umbratilis
  R. ruficauda
  R. colonus
  R. gularis
  R. insignis
  R. goodfellowi

Muscicapa [Genus]
  M. striata [Species]
  M. gambagae
  M. griseisticta
  M. sibirica
  M. dauurica
  M. ruficauda
  M. muttui
  M. ferruginea
  M. sordida
  M. thalassina
  M. panayensis
  M. albicaudata
  M. indigo
  M. infuscata
  M. ussheri
  M. boehmi
  M. aquatica
  M. olivascens
  M. lendu
  M. adusta
  M. epulata
  M. sethsmithii
  M. comitata
  M. tessmanni
  M. cassini
  M. caerulescens
  M. griseigularis
Myioparus [Genus]
  M. plumbeus [Species]
Humblotia [Genus]
  H. flavirostris [Species]
Ficedula [Genus]
  F. hypoleuca [Species]
  F. albicollis
  F. zanthopygia
  F. narcissina
  F. mugimaki
  F. hodgsonii
  F. dumetoria
  F. strophiata
  F. parva
  F. subruba
  F. monileger
  F. solitaris
  F. hyperythra
  F. basilanica
  F. rufigula
  F. buruensis
  F. henrici
  F. harterti
  F. platenae
  F. bonthaina
  F. westermanni
  F. superciliaris
  F. tricolor
  F. sapphira
  F. nigrorufa
  F. timorensis
  F. cyanomelana

Niltava [Genus]
  N. grandis [Species]
  N. macgrigoriae
  N. davidi
  N. sundara
  N. sumatrana
  N. vivida
  N. hyacinthina
  N. hoevelli
  N. sanfordi
  N. concreta
  N. ruecki
  N. herioti
  N. hainana
  N. pallipes
  N. poliogenys
  N. unicolor
  N. rubeculoides
  N. banyumas
  N. superba
  N. caerulata
  N. turcosa
  N. tickelliae
  N. rufigastra
  N. hodgsoni
Culicicapa [Genus]
  C. ceylonensis [Species]
  C. helianthea

Platysteiridae [Family]
Bias [Genus]
  B. flammulatus [Species]
  B. musicus
Pseudobias [Genus]
  P. wardi [Species]
Batis [Genus]
  B. diops [Species]
  B. margaritae
  B. mixta
  B. dimorpha
  B. capensis
  B. fratrum
  B. molitor
  B. soror
  B. pririt
  B. senegalensis
  B. orientalis
  B. minor
  B. perkeo
  B. minulla
  B. minima
  B. ituriensis
  B. poensis
Platysteira [Genus]
  P. cyanea [Species]
  P. albifrons
  P. peltata
  P. laticincta
  P. castanea
  P. tonsa
  P. blissetti

  P. chalybea
  P. jamesoni
  P. concreta

Maluridae [Family]
Clytomyias [Genus]
  C. insignis [Species]
malurus [Genus]
  M. wallacii [Species]
  M. grayi
  M. alboscapulatus
  M. melanocephalus
  M. leucopterus
  M. cyaneus
  M. splendens
  M. lamberti
  M. amabilis
  M. pulcherrimus
  M. elegans
  M. coronatus
  M. cyanocephalus
Stipiturus [Genus]
  S. malachurus [Species]
  S. mallee
  M. ruficeps
Amytornis [Genus]
  A. textilis [Species]
  A. purnelli
  A. housei
  A. woodwardi
  A. dorotheae
  A. striatus
  A. barbatus
  A. goyderi

Acanthizidae [Family]
Dasyornis [Genus]
  D. brachypterus [Species]
  D. broadbenti
Pycnoptilus [Genus]
  P. floccosus [Species]
Origma [Genus]
  O. solitaria [Species]
Crateroscelis [Genus]
  C. gutturalis [Species]
  C. murina
  C. nigrorufa
  C. robusta
Sericornis [Genus]
  S. citreogularis [Species]
  S. maculatus
  S. humilis
  S. frontalis
  S. beccarii
  S. nouhuysi
  S. magnirostris
  S. keri
  S. spilodera
  S. perspicillatus
  S. rufescens
  S. papuensis

  S. arfakianus
  S. magnus
Pyrrholaemus [Genus]
  P. brunneus [Species]
Chthonicola [Genus]
  C. sagittatus [Species]
Calamanthus [Genus]
  C. fuliginosus [Species]
  C. campestris
Hylacola [Genus]
  H. pyrrhopygius [Species]
  H. cautus
Acanthiza [Genus]
  A. murina [Species]
  A. inornata
  A. reguloides
  A. iredalei
  A. katherina
  A. pusilla
  A. apicalis
  A. ewingii
  A. chrysorrhoa
  A. uropygialis
  A. robustirostris
  A. nana
  A. lineata
Smicrornis [Genus]
  S. brevirostris [Species]
Gerygone [Genus]
  G. cinerea [Species]
  G. chloronota
  G. palpebrosa
  G. olivacea
  G. dorsalis
  G. chrysogaster
  G. ruficauda
  G. magnirostris
  G. sulphurea
  G. inornata
  G. ruficollis
  G. fusca
  G. tenebrosa
  G. laevigaster
  G. flavolateralis
  G. insularis
  G. mouki
  G. modesta
  G. igata
  G. albofrontata
Aphelocephala [Genus]
  A. leucopsis [Species]
  A. pectoralis
  A. nigricincta
Mohoua [Genus]
  M. ochrocephala [Species]
Finschia [Genus]
  F. novaeseelandiae [Species]
Epthianura [Genus]
  E. albifrons [Species]
  E. tricolor

E. aurifrons
E. crocea
Ashbyia [Genus]
  A. lovensis [Species]

Monarchidae [Family]
Erythrocercus [Genus]
  E. mccallii [Species]
  E. holochlorus
  E. livingstonei
Elminia [Genus]
  E. longicauda [Species]
  E. albicauda
Trochocercus [Genus]
  T. nigromitratus [Species]
  T. albiventris
  T. albonotatus
  T. cyanomelas
  T. nitens
Philentoma [Genus]
  P. pyrhopterum [Species]
  P. velatum
Hypothymis [Genus]
  H. azurea [Species]
  H. helenae
  H. coelestris
Eutrichomyias [Genus]
  E. rowleyi [Species]
Terpsiphone [Genus]
  T. rufiventer [Species]
  T. bedfordi
  T. rufocinerea
  T. viridis
  T. paradisi
  T. atrocaudata
  T. cyanescens
  T. cinnamomea
  T. atrochalybeia
  T. mutata
  T. corvina
  T. bourbonnensis
Chasiempis [Genus]
  C. sandwichensis [Species]
Pomarea [Genus]
  P. dimidiata [Species]
  P. nigra
  P. mendozae
  P. iphis
  P. whitneyi
Mayrornis [Genus]
  M. versicolor [Species]
  M. lessoni
  M. schistaceus
Neolalage [Genus]
  N. banksiana [Species]
Clytorhynchus [Genus]
  C. pachycephaloides [Species]
  C. vitiensis
  C. nigrogularis
  C. hamlini
Metabolus [Genus]

M. rugensis [Species]
Monarcha [Genus]
  M. axillaris [Species]
  M. rubiensis
  M. cinerascens
  M. melanopsis
  M. frater
  M. erythrostictus
  M. castaneiventris
  M. richardsii
  M. leucotis
  M. guttulus
  M. mundus
  M. sacerdotum
  M. trivirgatus
  M. leucurus
  M. julianae
  M. manadensis
  M. brehmii
  M. infelix
  M. menckei
  M. verticalis
  M. barbatus
  M. browni
  M. viduus
  M. godeffroyi
  M. takatsukasae
  M. chrysomela
Arses [Genus]
  A. insularis [Species]
  A. telescophthalmus
  A. kaupi
Myiagra [Genus]
  M. oceanica [Species]
  M. galeata
  M. atra
  M. rubecula
  M. ferrocyanea
  M. cervinicauda
  M. caledonica
  M. vanikorensis
  M. albiventris
  M. azureocapilla
  M. ruficollis
  M. cyanoleuca
  M. alecto
  M. hebetior
  M. inquieta
Lamprolia [Genus]
  L. victoriae [Species]
Machaerirhynchus [Genus]
  M. flaviventer [Species]
  M. nigripectus
Peltops [Genus]
  P. blainvillii [Species]
  P. montanus
Rhipidura [Genus]
  R. hypoxantha [Species]
  R. superciliaris
  R. cyaniceps

R. phoenicura
R. nigrocinnamomea
R. albicollis
R. euryura
R. aureola
R. javanica
R. perlata
R. leucophrys
R. rufiventris
R. cockerelli
R. albolimbata
R. hyperythra
R. threnothorax
R. maculipectus
R. leucothorax
R. atra
R. fuliginosa
R. drownei
R. tenebrosa
R. rennelliana
R. spilodera
R. nebulosa
R. brachyrhyncha
R. personata
R. dedemi
R. superflua
R. teysmanni
R. lepida
R. opistherythra
R. rufidorsa
R. dahli
R. matthiae
R. malaitae
R. rufifrons

Eopsaltriidae [Family]
Monachella [Genus]
  M. muelleriana [Species]
Microeca [Genus]
  M. leucophaea [Species]
  M. flavigaster
  M. hemixantha
  M. griseoceps
  M. flavovirescens
  M. papuana
Eugerygone [Genus]
  E. rubra [Species]
Petroica [Genus]
  P. bivittata [Species]
  P. archboldi
  P. multicolor
  P. goodenovii
  P. phoenicea
  P. rosea
  P. rodinogaster
  P. cucullata
  P. vittata
  P. macrocephala
  P. australis
  P. traversi
Tregellasia [Genus]

T. capito [Species]
T. leucops
Eopsaltria [Genus]
    E. australis [Species]
    E. flaviventris
    E. georgiana
Peneoenanthe [Genus]
    P. pulverulenta [Species]
Peocilodryas [Genus]
    P. brachyura [Species]
    P. hypoleuca
    P. placens
    P. albonotata
    P. superciliosa
Peneothello [Genus]
    P. sigillatus [Species]
    P. cryptoleucus
    P. cyanus
    P. bimaculatus
Heteromyias [Genus]
    H. cinereifrons [Species]
    H. albispecularis
Pachycephalopsis [Genus]
    P. hattamensis [Species]
    P. poliosoma

Pachycephalidae [Family]
Eulacestoma [Genus]
    E. nigropectus [Species]
Falcunculus [Genus]
    F. frontatus [Species]
Oreoica [Genus]
    O. gutturalis [Species]
Pachycare [Genus]
    P. flavogrisea [Species]
Rhagologus [Genus]
    R. leucostigma [Species]
Hylocitrea [Genus]
    H. bonensis [Species]
Pachycephala [Genus]
    P. raveni [Species]
    P. rufinucha
    P. tenebrosa
    P. olivacea
    P. rufogularis
    P. inornata
    P. hypoxantha
    P. cinerea
    P. phaionota
    P. hyperythra
    P. modesta
    P. philippensis
    P. sulfuriventer
    P. meyeri
    P. soror
    P. simplex
    P. orpheus
    P. pectoralis
    P. flavifrons
    P. caledonica
    P. implicata

P. nudigula
P. lorentzi
P. schlegelii
P. aurea
P. rufiventris
P. lanioides
Colluricincla [Genus]
    C. megarhyncha [Species]
    C. parvula
    C. boweri
    C. harmonica
    C. woodwardi
Pitohui [Genus]
    P. kirhocephalus [Species]
    P. dichrous
    P. incertus
    P. ferrugineus
    P. cristatus
    P. nigrescens
    P. tenebrosus
Turnagra [Genus]
    T. capensis [Species]

Aegithalidae [Family]
Aegithalos [Genus]
    A. caudatus [Species]
    A. leucogenys
    A. concinnus
    A. iouschistos
    A. fuliginosus
Psaltria [Genus]
    P. exilis [Species]
Psaltriparus [Genus]
    P. minimus [Species]
    P. melanotis

Remizidae [Family]
Remiz [Genus]
    R. pendulinus [Species]
Anthoscopus [Genus]
    A. punctifrons [Species]
    A. parvulus
    A. musculus
    A. flavifrons
    A. caroli
    A. sylviella
    A. minutus
Auriparus [Genus]
    A. flaviceps [Species]
Cephalopyrus [Genus]
    C. flammiceps [Species]

Paridae [Family]
Parus [Genus]
    P. palustris [Species]
    P. lugubris
    P. montanus
    P. atricapillus
    P. carolinensis
    P. sclateri
    P. gambeli

P. superciliosus
P. davidi
P. cinctus
P. hudsonicus
P. rufescens
P. wollweberi
P. rubidiventris
P. melanolophus
P. ater
P. venustulus
P. elegans
P. amabilis
P. cristatus
P. dichrous
P. afer
P. griseiventris
P. niger
P. leucomelas
P. albiventris
P. leuconotus
P. funereus
P. fasciiventer
P. fringillinus
P. rufiventris
P. major
P. bokharensis
P. monticolus
P. nuchalis
P. xanthogenys
P. spilonotus
P. holsti
P. caeruleus
P. cyanus
P. varius
P. semilarvatus
P. inornatus
P. bicolor
Melanochlora [Genus]
    M. sultanea [Species]
Sylviparus [Genus]
    S. modestus [Species]
Hypositta [Genus]
    H. corallirostris [Species]

Sittidae [Family]
Sitta [Genus]
    S. europaea [Species]
    S. nagaensis
    S. castanea
    S. himalayensis
    S. victoriae
    S. pygmaea
    S. pusilla
    S. whiteheadi
    S. yunnanensis
    S. canadensis
    S. villosa
    S. leucopsis
    S. carolinensis
    S. krueperi
    S. neumayer

S. tephronota
S. frontalis
S. solangiae
S. azurea
S. magna
S. formosa
Neositta [Genus]
   N. chrysoptera [Species]
   N. papuensis
Daphoenositta [Genus]
   D. miranda [Species]
Tichodroma [Genus]
   T. muraria [Species]

Certhiidae [Family]
  Certhia [Genus]
   F. familiaris [Species]
   F. brachydactyla
   F. himalayana
   F. nipalensis
   F. discolor
  Salpornis [Genus]
   S. spilonotus [Species]

Rhabdornithidae [Family]
  Rhabdornis [Genus]
   R. mysticalis [Species]
   R. inornatus

Climacteridae [Family]
  Climacteris [Genus]
   C. erythrops [Species]
   C. affinis
   C. picumnus
   C. rufa
   C. melanura
   C. leucophaea

Dicaeidae [Family]
  Melanocharis [Genus]
   M. arfakiana [Species]
   M. nigra
   M. longicauda
   M. versteri
   M. striativentris
  Rhamphocharis [Genus]
   R. crassirostris [Species]
  Prionochilus [Genus]
   P. olivaceus [Species]
   P. maculatus
   P. percussus
   P. plateni
   P. xanthopygius
   P. thoracicus
  Dicaeum [Genus]
   D. annae [Species]
   D. agile
   D. everetti
   D. aeruginosum
   D. proprium
   D. chrysorrheum
   D. melanoxanthum

D. vincens
D. aureolimbatum
D. nigrilore
D. anthonyi
D. bicolor
D. quadricolor
D. australe
D. retrocinctum
D. trigonostigma
D. hypoleucum
D. erythrorhynchos
D. concolor
D. pygmaeum
D. nehrkorni
D. vulneratum
D. erythrothorax
D. pectorale
D. eximium
D. aeneum
D. tristrami
D. igniferum
D. maugei
D. sanguinolentum
D. hirundinaceum
D. celebicum
D. monticolum
D. ignipectus
D. cruentatum
D. trochileum
Oreocharis [Genus]
   O. arfaki [Species]
Paramythia [Genus]
   P. montium [Species]
Pardalotus [Genus]
   P. quadragintus [Species]
   P. punctatus
   P. xanthopygus
   P. rubricatus
   P. striatus
   P. ornatus
   P. substriatus
   P. melanocephalus

Nectariniidae [Family]
  Anthreptes [Genus]
   A. gabonicus [Species]
   A. fraseri
   A. reichenowi
   A. anchietae
   A. simplex
   A. malacensis
   A. rhodolaema
   A. singalensis
   A. longuemarei
   A. orientalis
   A. neglectus
   A. aurantium
   A. pallidogaster
   A. pujoli
   A. rectirostris
   A. collaris
   A. platurus

Hypogramma [Genus]
   H. hypogrammicum [Species]
Nectarina [Genus]
   N. seimundi [Species]
   N. batesi
   N. olivacea
   N. ursulae
   N. veroxii
   N. balfouri
   N. reichenbachii
   N. hartlaubii
   N. newtonii
   N. thomensis
   N. oritis
   N. alinae
   N. bannermani
   N. verticalis
   N. cyanolaema
   N. fuliginosa
   N. rubescens
   N. amethystina
   N. senegalensis
   N. adelberti
   N. zeylonica
   N. minima
   N. sperata
   N. sericea
   N. calcostetha
   N. dussumeiri
   N. lotenia
   N. jugularis
   N. buettikoferi
   N. solaris
   N. asiatica
   N. souimanga
   N. humbloti
   N. comorensis
   N. coquerellii
   N. venusta
   N. talatala
   N. oustaleti
   N. fusca
   N. chalybea
   N. afra
   N. mediocris
   N. preussi
   N. neergaardi
   N. chloropygia
   N. minulla
   N. regia
   N. loveridgei
   N. rockefelleri
   N. violacea
   N. habessinica
   N. bouvieri
   N. osea
   N. cuprea
   N. tacazze
   N. bocagii
   N. purpureiventris

N. shelleyi
N. mariquensis
N. bifasciata
N. pembae
N. chalcomelas
N. coccinigastra
N. erythrocerca
N. congensis
N. pulchella
N. nectarinioides
N. famosa
N. johnstoni
N. notata
N. johannae
N. superba
N. kilimensis
N. reichenowi
Aethopyga [Genus]
A. primigenius [Species]
A. boltoni
A. flagrans
A. pulcherrima
A. duyvenbodei
A. shelleyi
A. gouldiae
A. nipalensis
A. eximia
A. christinae
A. saturata
A. siparaja
A. mysticalis
A. ignicauda
Arachnothera [Genus]
A. longirostra [Species]
A. crassirostris
A. robusta
A. flavigaster
A. chrysogenys
A. clarae
A. affinis
A. magna
A. everetti
A. juliae

Zosteropidae [Family]
Zosterops [Genus]
Z. erythropleura [Species]
Z. japonica
Z. palpebrosa
Z. ceylonensis
Z. conspicillata
Z. salvadorii
Z. atricapilla
Z. everetti
Z. nigrorum
Z. montana
Z. wallacei
Z. flava
Z. chloris
Z. consibrinorum
Z. grayi

Z. uropygialis
Z. anomala
Z. atriceps
Z. atrifrons
Z. mysorensis
Z. fuscicapilla
Z. buruensis
Z. kuehni
Z. novaeguineae
Z. metcalfi
Z. natalis
Z. lutea
Z. griseotincta
Z. rennelliana
Z. vellalavella
Z. luteirostris
Z. rendovae
Z. murphyi
Z. ugiensis
Z. stresemanni
Z. sanctaecrucis
Z. samoensis
Z. explorator
Z. flavifrons
Z. minuta
Z. xanthochroa
Z. lateralis
Z. strenua
Z. tenuirostris
Z. albogularis
Z. inornata
Z. cinerea
Z. abyssinica
Z. pallida
Z. senegalensis
Z. virens
Z. borbonica
Z. ficedulina
Z. griseovirescens
Z. maderaspatana
Z. mayottensis
Z. modesta
Z. mouroniensis
Z. olivacea
Z. vaughani
Woodfordia [Genus]
W. superciliosa [Species]
W. lacertosa
Rukia [Genus]
R. palauensis [Species]
R. oleaginea
R. ruki
R. longirostra
Tephrozosterops [Genus]
T. stalkeri [Species]
Madanga [Genus]
M. ruficollis [Species]
Lophozosterops [Genus]
L. pinaiae [Species]
L. goodfellowi

L. squamiceps
L. javanica
L. superciliaris
L. dohertyi
Oculocincta [Genus]
O. squamifrons [Species]
Heleia [Genus]
H. muelleri [Species]
H. crassirostris
Chlorocharis [Genus]
C. emiliae [Species]
Hypocryptadius [Genus]
H. cinnamomeus [Species]
Speirops [Genus]
S. brunnea [Species]
S. leucophoeca
S. lugubris

Meliphagidae [Family]
Timeliopsis [Genus]
T. fulvigula [Species]
T. griseigula
Melilestes [Genus]
M. megarhynchus [Species]
M. bouganvillei
Toxorhamphus [Genus]
T. novaeguineae [Species]
T. poliopterus
Oedistoma [Genus]
O. iliolophum [Species]
O. pygmaeum
Glycichaera [Genus]
G. fallax [Species]
Lichmera [Genus]
L. lombokia [Species]
L. argentauris
L. indistincta
L. incana
L. alboauricularis
L. squamata
L. deningeri
L. monticola
L. flavicans
L. notabilis
L. cockerelli
Myzomela [Genus]
M. blasii [Species]
M. albigula
M. cineracea
M. eques
M. obscura
M. cruentata
M. nigrita
M. pulchella
M. kuehni
M. erythrocephala
M. adolphinae
M. sanguinolenta
M. cardinalis
M. chermesina
M. sclateri

M. lafargei
M. melanocephala
M. eichhorni
M. malaitae
M. tristrami
M. jugularis
M. erythromelas
M. vulnerata
M. rosenbergii
Certhionyx [Genus]
   C. niger [Species]
   C. variegatus
Meliphaga [Genus]
   M. mimikae [Species]
   M. montana
   M. orientalis
   M. albonotata
   M. aruensis
   M. analoga
   M. vicina
   M. gracilis
   M. notata
   M. flavirictus
   M. lewinii
   M. flava
   M. albilineata
   M. virescens
   M. versicolor
   M. fasciogularis
   M. inexpectata
   M. fusca
   M. plumula
   M. chrysops
   M. cratitia
   M. keartlandi
   M. penicillata
   M. ornata
   M. reticulata
   M. leucotis
   M. flavicollis
   M. melanops
   M. cassidix
   M. unicolor
   M. flaviventer
   M. polygramma
   M. macleayana
   M. frenata
   M. subfrenata
   M. obscura
Oreornis [Genus]
   O. chrysogenys [Species]
Foulehaio [Genus]
   F. carunculata [Species]
   F. provocator
Cleptornis [Genus]
   C. marchei [Species]
Apalopteron [Genus]
   A. familiare [Species]
Melithreptus [Genus]
   M. brevirostris [Species]

M. lunatus
M. albogularis
M. affinis
M. gularis
M. laetior
M. validirostris
Entomyzon [Genus]
   E. cyanotis [Species]
Notiomystis [Genus]
   N. cincta [Species]
Pycnopygius [Genus]
   P. ixoides [Species]
   P. cinereus
   P. stictocephalus
Philemon [Genus]
   P. meyeri [Species]
   P. brassi
   P. citreogularis
   P. inornatus
   P. gilolensis
   P. fuscicapillus
   P. subcorniculatus
   P. moluccensis
   P. buceroides
   P. novaeguineae
   P. cockerelli
   P. eichhorni
   P. albitorques
   P. argenticeps
   P. corniculatus
   P. diemenensis
Ptiloprora [Genus]
   P. plumbea [Species]
   P. meekiana
   P. erythropleura
   P. guisei
   P. perstriata
Melidectes [Genus]
   M. fuscus [Species]
   M. princeps
   M. nouhuysi
   M. ochromelas
   M. leucostephes
   M. belfordi
   M. torquatus
Melipotes [Genus]
   M. gymnops [Species]
   M. fumigatus
   M. ater
Vosea [Genus]
   V. whitemanensis [Species]
Myza [Genus]
   M. celebensis [Species]
   M. sarasinorum
Meliarchus [Genus]
   M. sclateri [Species]
Gymnomyza [Genus]
   G. viridis [Species]
   G. samoensis
   G. aubryana

Moho [Genus]
   M. braccatus [Species]
   M. bishopi
   M. apicalis
   M. nobilis
Chaetoptila [Genus]
   C. angustipluma [Species]
Phylidonyris [Genus]
   P. pyrrhoptera [Species]
   P. novaehollandiae
   P. nigra
   P. albifrons
   P. melanops
   P. undulata
   P. notabilis
Ramsayornis [Genus]
   R. fasciatus [Species]
   R. modestus
Plectorhyncha [Genus]
   P. lanceolata [Species]
Conopophila [Genus]
   C. whitei [Species]
   C. albogularis
   C. rufogularis
   C. picta
Xanthomyza [Genus]
   X. phrygia [Species]
Cissomela [Genus]
   C. pectoralis [Species]
Acanthorhynchus [Genus]
   A. tenuirostris [Species]
   A. superciliosus
Manorina [Genus]
   M. melanophrys [Species]
   M. melanocephala
   M. flavigula
   M. melanotis
Anthornis [Genus]
   A. melanura [Species]
Anthochaera [Genus]
   A. rufogularis [Species]
   A. chrysoptera
   A. carunculata
   A. paradoxa
Prosthemadera [Genus]
   P. novaeseelandiae [Species]
Promerops [Genus]
   P. cafer [Species]
   P. gurneyi

Emberizidae [Family]
Melophus [Genus]
   M. lathami [Species]
Latoucheornis [Genus]
   L. siemsseni [Species]
Emberiza [Genus]
   E. calandra [Species]
   E. citrinella
   E. leucocephala
   E. cia
   E. cioides

E. jankowskii
E. buchanani
E. stewarti
E. cineracea
E. hortulana
E. caesia
E. cirlus
E. striolata
E. impetuani
E. tahapisi
E. socotrana
E. capensis
E. yessoensis
E. tristami
E. fucata
E. pusilla
E. chrysophrys
E. rustica
E. elegans
E. aureola
E. poliopleura
E. flaviventris
E. affinis
E. cabanisi
E. rutila
E. koslowi
E. melanocephala
E. bruniceps
E. sulphurata
E. spodocephala
E. variabilis
E. pallasi
E. schoeniclus
Calcarius [Genus]
  C. mccownii [Species]
  C. lapponicus
  C. pictus
  C. ornatus
Plectrophenax [Genus]
  P. nivalis [Species]
Calamospiza [Genus]
  C. melanocorys [Species]
Zonotrichia [Genus]
  Z. iliaca [Species]
  Z. melodia
  Z. lincolnii
  Z. georgiana
  Z. capensis
  Z. querula
  Z. leucophrys
  Z. albicollis
  Z. atricapilla
Junco [Genus]
  J. vulcani [Species]
  J. hyemalis
  J. phaeonotus
Ammodramus [Genus]
  A. sandwichensis [Species]
  A. maritimus
  A. caudacutus

A. leconteii
A. bairdii
A. baileyi
A. henslowii
A. savannarum
A. humeralis
A. aurifrons
Spizella [Genus]
  S. arborea [Species]
  S. passerina
  S. pusilla
  S. atrogularis
  S. pallida
  S. breweri
Pooecetes [Genus]
  P. gramineus [Species]
Chondestes [Genus]
  C. grammacus [Species]
Amphispiza [Genus]
  A. bilineata [Species]
  A. belli
Aimophila [Genus]
  A. mystacalis [Species]
  A. humeralis
  A. ruficauda
  A. sumichrasti
  A. stolzmanni
  A. strigiceps
  A. aestivalis
  A. botterii
  A. cassinii
  A. quinquestriata
  A. carpalis
  A. ruficeps
  A. notosticta
  A. rufescens
Torreornis [Genus]
  T. inexpectata [Species]
Oriturus [Genus]
  O. superciliosus [Species]
Phrygilus [Genus]
  P. atriceps [Species]
  P. gayi
  P. patagonicus
  P. fruticeti
  P. unicolor
  P. dorsalis
  P. erythronotus
  P. plebejus
  P. carbonarius
  P. alaudinus
Melanodera [Genus]
  M. melanodera [Species]
  M. xanthogramma
Haplospiza [Genus]
  H. rustica [Species]
  H. unicolor
Acanthidops [Genus]
  A. bairdii [Species]
Lophospingus [Genus]

L. pusillus [Species]
L. griseocristatus
Donacospiza [Genus]
  D. albifrons [Species]
Rowettia [Genus]
  R. goughensis [Species]
Nesospiza [Genus]
  N. acunhae [Species]
  N. wilkinsi
Diuca [Genus]
  D. speculifera [Species]
  D. diuca
Idiopsar [Genus]
  I. brachyurus [Species]
Piezorhina [Genus]
  P. cinerea [Species]
Xenospingus [Genus]
  X. concolor [Species]
Incaspiza [Genus]
  I. pulchra [Species]
  I. ortizi
  I. laeta
  I. watkinsi
Poospiza [Genus]
  P. thoracica [Species]
  P. boliviana
  P. alticola
  P. hypochondria
  P. erythrophrys
  P. ornata
  P. nigrorufa
  P. lateralis
  P. rubecula
  P. garleppi
  P. baeri
  P. caesar
  P. hispaniolensis
  P. torquata
  P. cinerea
Sicalis [Genus]
  S. citrina [Species]
  S. lutea
  S. uropygialis
  S. luteocephala
  S. auriventris
  S. olivascens
  S. columbiana
  S. flaveola
  S. luteola
  S. raimondii
  S. taczanowskii
Emberizoides [Genus]
  E. herbicola [Species]
Embernagra [Genus]
  E. platensis [Species]
  E. longicauda
Volatinia [Genus]
  V. jacarina [Species]
Sporophila [Genus]
  S. frontalis [Species]
  S. falcirostris

S. schistacea
S. intermedia
S. plumbea
S. americana
S. torqueola
S. collaris
S. lineola
S. luctuosa
S. nigricollis
S. ardesiaca
S. melanops
S. obscura
S. caerulescens
S. albogularis
S. leucoptera
S. peruviana
S. simplex
S. nigrorufa
S. bouvreuil
S. insulata
S. minuta
S. hypoxantha
S. hypochroma
S. ruficollis
S. palustris
S. castaneiventris
S. cinnamomea
S. melanogaster
S. telasco
Oryzoborus [Genus]
O. crassirostris [Species]
O. angolensis
Amaurospiza [Genus]
A. concolor [Species]
A. moesta
Melopyrrha [Genus]
M. nigra [Species]
Dolospingus [Genus]
D. fringilloides [Species]
Catamenia [Genus]
C. analis [Species]
C. inornata
C. homochroa
C. oreophila
Tiaris [Genus]
T. canora [Species]
T. olivacea
T. bicolor
T. fuliginosa
Loxipasser [Genus]
L. anoxanthus [Species]
Loxigilla [Genus]
L. portorocensis [Species]
L. violacea
L. noctis
Melanospiza [Genus]
M. richardsoni [Species]
Geospiza [Genus]
G. magnirostris [Species]
G. fortis

G. fuliginosa
G. difficilis
G. scandens
G. conirostris
Camarhynchus [Genus]
C. crassirostris [Species]
C. psittacula
C. pauper
C. parvulus
C. pallidus
C. heliobates
Certhidea [Genus]
C. olivacea [Species]
Pinaroloxias [Genus]
P. inornata [Species]
Pipilo [Genus]
P. chlorurus [Species]
P. ocai
P. erythrophthalmus
P. socorroensis
P. fuscus
P. aberti
P. albicollis
Melozone [Genus]
M. kieneri [Species]
M. biarcuatum
M. leucotis
Arremon [Genus]
A. taciturnus [Species]
A. flavirostris
A. aurantiirostris
A. schlegeli
A. abeillei
Arremonops [Genus]
A. rufivirgatus [Species]
A. tocuyensis
A. chlorinotus
A. conirostris
Atlapetes [Genus]
A. albinucha [Species]
A. pallidinucha
A. rufinucha
A. leucopis
A. pileatus
A. melanocephalus
A. flaviceps
A. fuscoolivaceus
A. tricolor
A. albofrenatus
A. schistaceus
A. nationi
A. leucopterus
A. albiceps
A. pallidiceps
A. rufigenis
A. semirufus
A. personatus
A. fulviceps
A. citrinellus
A. brunneinucha
A. torquatus

Pezopetes [Genus]
P. capitalis [Species]
Oreothraupis [Genus]
O. arremonops [Species]
Pselliophorus [Genus]
P. tibialis [Species]
P. luteoviridis
Lysurus [Genus]
L. castaneiceps [Species]
Urothraupis [Genus]
U. stolzmanni [Species]
Charitospiza [Genus]
C. eucosma [Species]
Coryphaspiza [Genus]
C. melanotis [Species]
Saltatricula [Genus]
S. multicolor [Species]
Gubernatrix [Genus]
G. cristata [Species]
Coryphospingus [Genus]
C. pileatus [Species]
C. cucullatus
Rhodospingus [Genus]
R. cruentus [Species]
Paroaria [Genus]
P. coronata [Species]
P. dominicana
P. gularis
P. baeri
P. capitata
Catamblyrhynchus [Genus]
C. diadema [Species]
Spiza [Genus]
S. americana [Species]
Pheucticus [Genus]
P. chrysopeplus [Species]
P. aureoventris
P. ludovicianus
P. melanocephalus
Cardinalis [Genus]
C. cardinalis [Species]
C. phoeniceus
C. sinuatus
Caryothraustes [Genus]
C. canadensis [Species]
C. humeralis
Rhodothraupis [Genus]
R. celaeno [Species]
Periporphyrus [Genus]
P. erythromelas [Species]
Pitylus [Genus]
P. grossus [Species]
Saltator [Genus]
S. atriceps [Species]
S. maximus
S. atripennis
S. similis
S. coerulescens
S. orenocensis
S. maxillosus

S. aurantiirostris
S. cinctus
S. atricollis
S. rufiventris
S. albicollis
Passerina [Genus]
P. glaucocaerulea [Species]
P. cyanoides
P. brissonii
P. parellina
P. caerulea
P. cyanea
P. amoena
P. versicolor
P. ciris
P. rositae
P. leclancherii
P. caerulescens
Orchesticus [Genus]
O. albeillei [Species]
Schistochlamys [Genus]
S. ruficapillus [Species]
S. melanopis
Neothraupis [Genus]
N. fasciata [Species]
Cypsnagra [Genus]
C. hirundinacea [Species]
Conothraupis [Genus]
C. speculigera [Species]
C. mesoleuca
Lamprospiza [Genus]
L. melanoleuca [Species]
Cissopis [Genus]
C. leveriana [Species]
Chlorornis [Genus]
C. reifferii [Species]
Compsothraupis [Genus]
C. loricata [Species]
Sericossypha [Genus]
S. albocristata [Species]
Nesospingus [Genus]
N. speculiferus [Species]
Chlorospingus [Genus]
C. ophthalmicus [Species]
C. tacarcunae
C. inornatus
C. punctulatus
C. semifuscus
C. zeledoni
C. pileatus
C. parvirostris
C. flavigularis
C. flavovirens
C. canigularis
Cnemoscopus [Genus]
C. rubrirostris [Species]
Hemispingus [Genus]
H. atropileus [Species]
H. superciliaris
H. reyi

H. frontalis
H. melanotis
H. goeringi
H. verticalis
H. xanthophthalmus
H. trifasciatus
Pyrrhocoma [Genus]
P. ruficeps [Species]
Thlypopsis [Genus]
T. fulviceps [Species]
T. ornata
T. pectoralis
T. sordida
T. inornata
T. ruficeps
Hemithraupis [Genus]
H. guira [Species]
H. ruficapilla
H. flavicollis
Chrysothlypis [Genus]
C. chrysomelas [Species]
C. salmoni
Nemosia [Genus]
N. pileata [Species]
N. rourei
Phaenicophilus [Genus]
P. palmarum [Species]
P. poliocephalus
Calyptophilus [Genus]
C. frugivorus [Species]
Rhodinocichla [Genus]
R. rosea [Species]
Mitrospingus [Genus]
M. cassinii [Species]
M. oleagineus
Chlorothraupis [Genus]
C. carmioli [Species]
C. olivacea
C. stolzmanni
Orthogonys [Genus]
O. chloricterus [Species]
Eucometis [Genus]
E. penicillata [Species]
Lanio [Genus]
L. fulvus [Species]
L. versicolor
L. aurantius
L. leucothorax
Creurgops [Genus]
C. verticalis [Species]
C. dentata
Heterospingus [Genus]
H. xanthopygius [Species]
Tachyphonus [Genus]
T. cristatus [Species]
T. rufiventer
T. surinamus
T. luctuosus
T. delatrii
T. coronatus

T. rufus
T. phoenicius
Trichothraupis [Genus]
T. melanops [Species]
Habia [Genus]
H. rubica [Species]
H. fuscicauda
H. atrimaxillaris
H. gutturalis
H. cristata
Piranga [Genus]
P. bidentata [Species]
P. flava
P. rubra
P. roseogularis
P. olivacea
P. ludoviciana
P. leucoptera
P. erythrocephala
P. rubriceps
Calochaetes [Genus]
C. coccineus [Species]
Ramphocelus [Genus]
R. sanguinolentus [Species]
R. nigrogularis
R. dimidiatus
R. melanogaster
R. carbo
R. bresilius
R. passerinii
R. flammigerus
Spindalis [Genus]
S. zena [Species]
Thraupis [Genus]
T. episcopus [Species]
T. sayaca
T. cyanoptera
T. ornata
T. abbas
T. palmarum
T. cyanocephala
T. bonariensis
Cyanicterus [Genus]
C. cyanicterus [Species]
Buthraupis [Genus]
B. arcaei [Species]
B. melanochlamys
B. rothschildi
B. edwardsi
B. aureocincta
B. montana
B. eximia
B. wetmorei
Wetmorethraupis [Genus]
W. sterrhopteron [Species]
Anisognathus [Genus]
A. lacrymosus [Species]
A. igniventris
A. flavinuchus
A. notabilis

Stephanophorus [Genus]
    S. diadematus [Species]
Iridosornis [Genus]
    I. porphyrocephala [Species]
    I. analis
    I. jelskii
    I. rufivertex
Dubusia [Genus]
    D. taeniata [Species]
Delothraupis [Genus]
    D. castaneoventris [Species]
Pipraeidea [Genus]
    P. melanonota [Species]
Euphonia [Genus]
    E. jamaica [Species]
    E. plumbea
    E. affinis
    E. luteicapilla
    E. chlorotica
    E. trinitatis
    E. concinna
    E. saturata
    E. finschi
    E. violacea
    E. laniirostris
    E. hirundinacea
    E. chalybea
    E. musica
    E. fulvicrissa
    E. imitans
    E. gouldi
    E. chrysopasta
    E. mesochrysa
    E. minuta
    E. anneae
    E. xanthogaster
    E. rufiventris
    E. pectoralis
    E. cayennensis
Chlorophonia [Genus]
    C. flavirostris [Species]
    C. cyanea
    C. pyrrhophrys
    C. occipitalis
Chlorochrysa [Genus]
    C. phoenicotis [Species]
    C. calliparaea
    C. nitidissima
Tangara [Genus]
    T. inornata [Species]
    T. cabanisi
    T. palmeri
    T. mexicana
    T. chilensis
    T. fastuosa
    T. seledon
    T. cyanocephala
    T. desmaresti
    T. cyanoventris
    T. johannae

T. schrankii
T. florida
T. arthus
T. icterocephala
T. xanthocephala
T. chrysotis
T. parzudakii
T. xanthogastra
T. punctata
T. guttata
T. varia
T. rufigula
T. gyrola
T. lavinia
T. cayana
T. cucullata
T. peruviana
T. preciosa
T. vitriolina
T. rufigenis
T. ruficervix
T. labradorides
T. cyanotis
T. cyanicollis
T. larvata
T. nigrocincta
T. dowii
T. nigroviridis
T. vassorii
T. heinei
T. viridicollis
T. argyrofenges
T. cyanoptera
T. pulcherrima
T. velia
T. callophrys
Dacnis [Genus]
    D. albiventris [Species]
    D. lineata
    D. flaviventer
    D. hartlaubi
    D. nigripes
    D. venusta
    D. cayana
    D. viguieri
    D. berlepschi
Chlorophanes [Genus]
    C. spiza [Species]
Cyanerpes [Genus]
    C. nitidus [Species]
    C. lucidus
    C. caeruleus
    C. cyaneus
Xenodacnis [Genus]
    X. parina [Species]
Oreomanes [Genus]
    O. fraseri [Species]
Diglossa [Genus]
    D. baritula [Species]
    D. lafresnayii

    D. carbonaria
    D. venezuelensis
    D. albilatera
    D. duidae
    D. major
    D. indigotica
    D. glauca
    D. caerulescens
    D. cyanea
Euneornis [Genus]
    E. campestris [Species]
Tersina [Genus]
    T. viridis [Species]

Parulidae [Family]
    Mniotilta [Genus]
        M. varia [Species]
    Vermivora [Genus]
        V. bachmanii [Species]
        V. chrysoptera
        V. pinus
        V. peregrina
        V. celata
        V. ruficapilla
        V. virginiae
        V. crissalis
        V. luciae
        V. gutturalis
        V. superciliosa
    Parula [Genus]
        P. americana [Species]
        P. pitiayumi
    Dendroica [Genus]
        D. petechia [Species]
        D. pensylvanica
        D. cerulea
        D. caerulescens
        D. plumbea
        D. pharetra
        D. pinus
        D. graciae
        D. adelaidae
        D. pityophila
        D. dominica
        D. nigrescens
        D. townsendi
        D. occidentalis
        D. chrysoparia
        D. virens
        D. discolor
        D. vitellina
        D. tigrina
        D. fusca
        D. magnolia
        D. coronata
        D. palmarum
        D. kirtlandii
        D. striata
        D. castanea
    Catharopeza [Genus]
        C. bishopi [Species]

Setophaga [Genus]
   S. ruticilla [Species]
Seiurus [Genus]
   S. aurocapillus [Species]
   S. noveboracensis
   S. motacilla
Limnothlypis [Genus]
   L. swainsonii [Species]
Helmitheros [Genus]
   H. vermivorus [Species]
Protonotaria [Genus]
   P. citrea [Species]
Geothlypis [Genus]
   G. trichas [Species]
   G. beldingi
   G. flavovelata
   G. rostrata
   G. semiflava
   G. speciosa
   G. nelsoni
   G. chiriquensis
   G. aequinoctialis
   G. poliocephala
   G. formosa
   G. agilis
   G. philadelphia
   G. tolmiei
Microligea [Genus]
   M. palustris [Species]
Teretistris [Genus]
   T. fernandinae [Species]
   T. fornsi
Leucopeza [Genus]
   L. semperi [Species]
Wilsonia [Genus]
   W. citrina [Species]
   W. pusilla
   W. canadensis
Cardellina [Genus]
   C. rubrifrons [Species]
Ergaticus [Genus]
   E. ruber [Species]
   E. versicolor
Myioborus [Genus]
   M. pictus [Species]
   M. miniatus
   M. brunniceps
   M. pariae
   M. cardonai
   M. torquatus
   M. ornatus
   M. melanocephalus
   M. albifrons
   M. flavivertex
   M. albifacies
Euthlypis [Genus]
   E. lachrymosa [Species]
Basileuterus [Genus]
   B. fraseri [Species]
   B. bivittatus

   B. chrysogaster
   B. flaveolus
   B. luteoviridis
   B. signatus
   B. nigrocristatus
   B. griseiceps
   B. basilicus
   B. cinereicollis
   B. conspicillatus
   B. coronatus
   B. culicivorus
   B. rufifrons
   B. belli
   B. melanogenys
   B. tristriatus
   B. trifasciatus
   B. hypoleucus
   B. leucoblepharus
   B. leucophrys
Phaeothlypis [Genus]
   P. fulvicauda [Species]
   P. rivularis
Peucedramus [Genus]
   P. taeniatus [Species]
Xenoligea [Genus]
   X. montana [Species]
Granatellus [Genus]
   G. venustus [Species]
   G. sallaei
   G. pelzelni
Icteria [Genus]
   I. virens [Species]
Conirostrum [Genus]
   C. speciosum [Species]
   C. leucogenys
   C. bicolor
   C. margaritae
   C. cinereum
   C. ferrugineiventre
   C. rufum
   C. sitticolor
   C. albifrons
Coereba [Genus]
   C. flaveola [Species]

Drepanididae [Family]
  Himatione [Genus]
   H. sanguinea [Species]
  Palmeria [Genus]
   P. dolei [Species]
  Vestiaria [Genus]
   V. coccinea [Species]
  Drepanis [Genus]
   D. funerea [Species]
   D. pacifica
  Ciridops [Genus]
   C. anna [Species]
  Viridonia [Genus]
   V. virens [Species]
   V. parva
   V. sagittirostris

Hemignathus [Genus]
   H. obscurus [Species]
   H. lucidus
   H. wilsoni
Loxops [Genus]
   L. coccinea [Species]
Paroreomyza [Genus]
   P. maculata [Species]
Pseudonester [Genus]
   P. xanthophrys [Species]
Psittirostra [Genus]
   P. psittacea [Species]
Loxioides [Genus]
   L. cantans [Species]
   L. palmeri
   L. flaviceps
   L. bailleui
   L. kona

Vireonidae [Family]
  Cyclarhis [Genus]
   C. gujanensis [Species]
   C. nigrirostris
  Vireolanius [Genus]
   V. melitophrys [Species]
   V. pulchellus
   V. leucotis
  Vireo [Genus]
   V. brevipennis [Species]
   V. huttoni
   V. atricapillus
   V. griseus
   V. pallens
   V. caribaeus
   V. bairdi
   V. gundlachii
   V. crassirostris
   V. bellii
   V. vicinior
   V. nelsoni
   V. hypochryseus
   V. modestus
   V. nanus
   V. latimeri
   V. osburni
   V. carmioli
   V. solitarius
   V. flavifrons
   V. philadelphicus
   V. olivaceus
   V. magister
   V. altiloquus
   V. gilvus
Hylophilus [Genus]
   H. poicilotis [Species]
   H. thoracicus
   H. semicinereus
   H. pectoralis
   H. sclateri
   H. muscicapinus
   H. brunneiceps

H. semibrunneus
H. aurantifrons
H. hypoxanthus
H. flavipes
H. ochraceiceps
H. decurtatus

Icteridae [Family]
Psarocolius [Genus]
P. oseryi [Species]
P. latirostris
P. decumanus
P. viridis
P. atrovirens
P. angustifrons
P. wagleri
P. montezuma
P. cassini
P. bifasciatus
P. guatimozinus
P. yuracares
Cacicus [Genus]
C. cela [Species]
C. uropygialis
C. chrysopterus
C. koepckeae
C. leucoramphus
C. chrysonotus
C. sclateri
C. solitarius
C. melanicterus
C. holosericeus
Icterus [Genus]
I. cayanensis [Species]
I. chrysater
I. nigrogularis
I. leucopteryx
I. auratus
I. mesomelas
I. auricapillus
I. graceannae
I. xantholemus
I. pectoralis
I. gularis
I. pustulatus
I. cucullatus
I. icterus
I. galbula
I. spurius
I. dominicensis
I. wagleri
I. laudabilis
I. bonana
I. oberi
I. graduacauda
I. maculialatus
I. parisorum
Nesopsar [Genus]
N. nigerrimus [Species]
Xanthopsar [Genus]
X. flavus [Species]

Gymnomystax [Genus]
G. mexicanus [Species]
Xanthocephalus [Genus]
X. xanthocephalus [Species]
Agelaius [Genus]
A. thilius [Species]
A. phoeniceus
A. tricolor
A. icterocephalus
A. humeralis
A. xanthomus
A. cyanopus
A. ruficapillus
Leistes [Genus]
L. militaris [Species]
Pezites [Genus]
P. militaris [Species]
Sturnella [Genus]
S. magna [Species]
S. neglecta
Pseudoleistes [Genus]
P. guirahuro [Species]
P. virescens
Amblyramphus [Genus]
A. holosericeus [Species]
Hypopyrrhus [Genus]
H. pyrohypogaster [Species]
Curaeus [Genus]
C. curaeus [Species]
C. forbesi
Gnorimopsar [Genus]
G. chopi [Species]
Oreopsar [Genus]
O. bolivianus [Species]
Lampropsar [Genus]
L. tanagrinus [Species]
Macroagelaius [Genus]
M. subalaris [Species]
Dives [Genus]
D. atroviolacea [Species]
D. dives
Quiscalus [Genus]
Q. mexicanus [Species]
Q. major
Q. palustris
Q. nicaraguensis
Q. quiscula
Q. niger
Q. lugubris
Euphagus [Genus]
E. carolinus [Species]
E. cyanocephalus
Molothrus [Genus]
M. badius [Species]
M. rufoaxillaris
M. bonariensis
M. aeneus
M. ater
Scaphidura [Genus]
S. oryzivorus [Species]

Fringillidae [Family]
Fringilla [Genus]
F. coelebs [Species]
F. teydea
F. montifringilla
Serinus [Genus]
S. pusillus [Species]
S. serinus
S. syriacus
S. canaria
S. citrinella
S. thibetanus
S. canicollis
S. nigriceps
S. citrinelloides
S. frontalis
S. capistratus
S. koliensis
S. scotops
S. leucopygius
S. atrogularis
S. citrinipectus
S. mozambicus
S. donaldsoni
S. flaviventris
S. sulphuratus
S. albogularis
S. gularis
S. mennelli
S. tristriatus
S. menschensis
S. striolatus
S. burtoni
S. rufobrunneus
S. leucopterus
S. totta
S. alario
S. estherae
Neospiza [Genus]
N. concolor [Species]
Linurgus [Genus]
L. olivaceus [Species]
Rhynchostruthus [Genus]
R. socotranus [Species]
Carduelis [Genus]
C. chloris [Species]
C. sinica
C. spinoides
C. ambigua
C. spinus
C. pinus
C. atriceps
C. spinescens
C. yarrellii
C. cucullata
C. crassirostris
C. magellanica
C. dominicensis
C. siemiradzkii
C. olivacea

C. notata
C. xanthogastra
C. atrata
C. uropygialis
C. barbata
C. tristis
C. psaltria
C. lawrencei
C. carduelis
Acanthis [Genus]
    A. flammea [Species]
    A. hornemanni
    A. flavirostris
    A. cannabina
    A. yemenensis
    A. johannis
Leucosticte [Genus]
    L. nemoricola [Species]
    L. brandti
    L. arctoa
Callacanthis [Genus]
    C. burtoni [Species]
Rhodopechys [Genus]
    R. sanguinea [Species]
    R. githaginea
    R. mongolica
    R. obsoleta
Uragus [Genus]
    U. sibiricus [Species]
Urocynchramus [Genus]
    U. pylzowi [Species]
Carpodacus [Genus]
    C. rubescens [Species]
    C. nipalensis
    C. erythrinus
    C. purpureus
    C. cassinii
    C. mexicanus
    C. pulcherrimus
    C. eos
    C. rhodochrous
    C. vinaceus
    C. edwardsii
    C. synoicus
    C. roseus
    C. trifasciatus
    C. rhodopeplus
    C. thura
    C. rhodochlamys
    C. rubicilloides
    C. rubicilla
    C. puniceus
    C. roborowskii
Chaunoproctus [Genus]
    C. ferreorostris [Species]
Pinicola [Genus]
    P. enucleator [Species]
    P. subhimachalus
Haematospiza [Genus]
    H. sipahi [Species]

Loxia [Genus]
    L. pytyopsittacus [Species]
    L. curvirostra
    L. leucoptera
Pyrrhula [Genus]
    P. nipalensis [Species]
    P. leucogenys
    P. aurantiaca
    P. erythrocephala
    P. erythaca
    P. pyrrhula
Coccothraustes [Genus]
    C. coccothraustes [Species]
    C. migratorius
    C. personatus
    C. icterioides
    C. affinis
    C. melanozanthos
    C. carnipes
    C. vespertinus
    C. abeillei
Pyrrhoplectes [Genus]
    P. epauletta [Species]

Estrildidae [Family]
Parmoptila [Genus]
    P. woodhousei [Species]
Nigrita [Genus]
    N. fusconota [Species]
    N. bicolor
    N. luteifrons
    N. canicapilla
Nesocharis [Genus]
    N. shelleyi [Species]
    N. ansorgei
    N. capistrata
Pytilia [Genus]
    P. phoenicoptera [Species]
    P. hypogrammica
    P. afra
    P. melba
Mandingoa [Genus]
    M. nitidula [Species]
Cryptospiza [Genus]
    C. reichenovii [Species]
    C. salvadorii
    C. jacksoni
    C. shelleyi
Pyrenestes [Genus]
    P. sanguineus [Species]
    P. ostrinus
    P. minor
Spermophaga [Genus]
    P. poliogenys [Species]
    P. haematina
    P. ruficapilla
Clytospiza [Genus]
    C. monteiri [Species]
Hypargos [Genus]
    H. margaritatus [Species]
    H. niveoguttatus

Euschistospiza [Genus]
    E. dybowskii [Species]
    E. cinereovinacea
Lagonosticta [Genus]
    L. rara [Species]
    L. rufopicta
    L. nitidula
    L. senegala
    L. rubricata
    L. landanae
    L. rhodopareia
    L. larvata
Uraeginthus [Genus]
    U. angolensis [Species]
    U. bengalus
    U. cyanocephala
    U. granatina
    U. ianthinogaster
Estrilda [Genus]
    E. caerulescens [Species]
    E. perreini
    E. thomensis
    E. melanotis
    E. paludicola
    E. melpoda
    E. rhodopyga
    E. rufibarba
    E. troglodytes
    E. astrild
    E. nigriloris
    E. nonnula
    E. atricapilla
    E. erythronotos
    E. charmosyna
Amandava [Genus]
    A. amandava [Species]
    A. formosa
    A. subflava
Ortygospiza [Genus]
    O. atricollis [Species]
    O. gabonensis
    O. locustella
Aegintha [Genus]
    A. temporalis [Species]
Emblema [Genus]
    E. picta [Species]
    E. bella
    E. oculata
    E. guttata
Oreostruthus [Genus]
    O. fuliginosus [Species]
Neochmia [Genus]
    N. phaeton [Species]
    N. ruficauda
Poephila [Genus]
    P. guttata [Species]
    P. bichenovii
    P. personata
    P. acuticauda
    P. cincta

Erythrura [Genus]
   E. hyperythra [Species]
   E. prasina
   E. viridifacies
   E. tricolor
   E. coloria
   E. trichroa
   E. papuana
   E. psittacea
   E. cyaneovirens
   E. kleinschmidti
Chloebia [Genus]
   C. gouldiae [Species]
Aidemosyne [Genus]
   A. modesta [Species]
Lonchura [Genus]
   L. malabarica [Species]
   L. griseicapilla
   L. nana
   L. cucullata
   L. bicolor
   L. fringilloides
   L. striata
   L. leucogastroides
   L. fuscans
   L. molucca
   L. punctulata
   L. kelaarti
   L. leucogastra
   L. tristissima
   L. leucosticta
   L. quinticolor
   L. malacca
   L. maja
   L. pallida
   L. grandis
   L. vana
   L. caniceps
   L. nevermanni
   L. spectabilis
   L. forbesi
   L. hunsteini
   L. flaviprymna
   L. castaneothorax
   L. stygia
   L. teerinki
   L. monticola
   L. montana
   L. melaena
   L. pectoralis
Padda [Genus]
   P. fuscata [Species]
   P. oryzivora
Amadina [Genus]
   A. erythrocephala [Species]
   A. fasciata
Pholidornis [Genus]
   P. rushiae [Species]

Ploceidae [Family]
  Vidua [Genus]

   V. chalybeata [Species]
   V. funerea
   V. wilsoni
   V. hypocherina
   V. fischeri
   V. regia
   V. macroura
   V. paradisaea
   V. orientalis
Bubalornis [Genus]
   B. albirostris [Species]
Dinemellia [Genus]
   D. dinemelli [Species]
Plocepasser [Genus]
   P. mahali [Species]
   P. superciliosus
   P. donaldsoni
   P. rufoscapulatus
Histurgops [Genus]
   H. ruficauda [Species]
Pseudonigrita [Genus]
   P. arnaudi [Species]
   P. cabanisi
Philetairus [Genus]
   P. socius [Species]
Passer [Genus]
   P. ammodendri [Species]
   P. domesticus
   P. hispaniolensis
   P. pyrrhonotus
   P. castanopterus
   P. rutilans
   P. flaveolus
   P. moabiticus
   P. iagoensis
   P. melanurus
   P. griseus
   P. simplex
   P. montanus
   P. luteus
   P. eminibey
Petronia [Genus]
   P. brachydactyla [Species]
   P. xanthocollis
   P. petronia
   P. superciliaris
   P. dentata
Montifringilla [Genus]
   M. nivalis [Species]
   M. adamsi
   M. taczanowskii
   M. davidiana
   M. ruficollis
   M. blanfordi
   M. theresae
Sporopipes [Genus]
   S. squamifrons [Species]
   S. frontalis
Amblyospiza [Genus]
   A. albifrons [Species]

Ploceus [Genus]
   P. baglafecht [Species]
   P. bannermani
   P. batesi
   P. nigrimentum
   P. bertrandi
   P. pelzelni
   P. subpersonatus
   P. luteolus
   P. ocularis
   P. nigricollis
   P. alienus
   P. melanogaster
   P. capensis
   P. subaureus
   P. xanthops
   P. aurantius
   P. heuglini
   P. bojeri
   P. castaneiceps
   P. princeps
   P. xanthopterus
   P. castanops
   P. galbula
   P. taeniopterus
   P. intermedius
   P. velatus
   P. spekei
   P. spekeoides
   P. cucullatus
   P. grandis
   P. nigerrimus
   P. weynsi
   P. golandi
   P. dicrocephalus
   P. melanocephalus
   P. jacksoni
   P. badius
   P. rubiginosus
   P. aureonucha
   P. tricolor
   P. albinucha
   P. nelicourvi
   P. sakalava
   P. hypoxanthus
   P. superciliosus
   P. benghalensis
   P. manyar
   P. philippinus
   P. megarhynchus
   P. bicolor
   P. flavipes
   P. preussi
   P. dorsomaculatus
   P. olivaceiceps
   P. insignis
   P. angolensis
   P. sanctithomae
Malimbus [Genus]
   M. coronatus [Species]
   M. cassini

M. scutatus
M. racheliae
M. ibadanensis
M. nitens
M. rubricollis
M. erythrogaster
M. malimbicus
M. rubriceps
Quelea [Genus]
  Q. cardinalis [Species]
  Q. erythrops
  Q. quelea
Foudia [Genus]
  F. madagascariensis [Species]
  F. eminentissima
  F. rubra
  F. bruante
  F. sechellarum
  F. flavicans
Euplectes [Genus]
  E. anomalus [Species]
  E. afer
  E. diadematus
  E. gierowii
  E. nigroventris
  E. hordeaceus
  E. orix
  E. aureus
  E. capensis
  E. axillaris
  E. macrourus
  E. hartlaubi
  E. albonotatus
  E. ardens
  E. progne
  E. jacksoni
Anomalospiza [Genus]
  A. imberbis [Species]

Sturnidae [Family]
Aplonis [Genus]
  A. zelandica [Species]
  A. santovestris
  A. pelzelni
  A. atrifusca
  A. corvina
  A. mavornata
  A. cinerascens
  A. tabuensis
  A. striata
  A. fusca
  A. opaca
  A. cantoroides
  A. crassa
  A. feadensis
  A. insularis
  A. dichroa
  A. mysolensis
  A. magna
  A. minor
  A. panayensis

A. metallica
A. mystacea
A. brunneicapilla
Poeoptera [Genus]
  P. kenricki [Species]
  P. stuhlmanni
  P. lugubris
Grafisia [Genus]
  G. torquata [Species]
Onychognathus [Genus]
  O. walleri [Species]
  O. nabouroup
  O. morio
  O. blythii
  O. frater
  O. tristramii
  O. fulgidus
  O. tenuirostris
  O. albirostris
  O. salvadorii
Lamprotornis [Genus]
  L. iris [Species]
  L. cupreocauda
  L. purpureiceps
  L. curruscus
  L. purpureus
  L. nitens
  L. chalcurus
  L. chalybaeus
  L. chloropterus
  L. acuticaudus
  L. splendidus
  L. ornatus
  L. australis
  L. mevesii
  L. purpuropterus
  L. caudatus
Cinnyricinclus [Genus]
  C. femoralis [Species]
  C. sharpii
  C. leucogaster
Speculipastor [Genus]
  S. bicolor [Species]
Neocichla [Genus]
  N. gutturalis [Species]
Spreo [Genus]
  S. fischeri [Species]
  S. bicolor
  S. albicapillus
  S. superbus
  S. pulcher
  S. hildebrandti
Cosmopsarus [Genus]
  C. regius [Species]
  C. unicolor
Saroglossa [Genus]
  S. aurata [Species]
  S. spiloptera
Creatophora [Genus]
  C. cinerea [Species]

Necropsar [Genus]
  N. leguati [Species]
Fregilupus [Genus]
  F. varius [Species]
Sturnus [Genus]
  S. senex [Species]
  S. malabaricus
  S. erythropygius
  S. pagodarum
  S. sericeus
  S. philippensis
  S. sturninus
  S. roseus
  S. vulgaris
  S. unicolor
  S. cinerascens
  S. contra
  S. nigricollis
  S. burmannicus
  S. melanopterus
  S. sinensis
Leucopsar [Genus]
  L. rothschildi [Species]
Acridotheres [Genus]
  A. tristis [Species]
  A. ginginianus
  A. fuscus
  A. grandis
  A. albocinctus
  A. cristatellus
Ampeliceps [Genus]
  A. coronatus [Species]
Mino [Genus]
  M. anais [Species]
  M. dumontii
Basilornis [Genus]
  B. celebensis [Species]
  B. galeatus
  B. corythaix
  B. miranda
Streptocitta [Genus]
  S. albicollis [Species]
  S. albertinae
Sarcops [Genus]
  S. calvus [Species]
Gracula [Genus]
  G. ptilogenys [Species]
  G. religiosa
Enodes [Genus]
  E. erythrophris [Species]
Scissirostrum [Genus]
  S. dubium [Species]
Buphagus [Genus]
  B. africanus [Species]
  B. erythrorhynchus

Oriolidae [Family]
Oriolus [Genus]
  O. szalayi [Species]
  O. phaeochromus
  O. forsteni

O. bouroensis
O. viridifuscus
O. sagittatus
O. flavocinctus
O. xanthonotus
O. albiloris
O. isabellae
O. oriolus
O. auratus
O. chinensis
O. chlorocephalus
O. crassirostris
O. brachyrhynchus
O. monacha
O. larvatus
O. nigripennis
O. xanthornus
O. hosii
O. crentus
O. traillii
O. mellianus
Sphecotheres [Genus]
S. vieilloti [Species]
S. flaviventris
S. viridis
S. hypoleucus

Dicruridae [Family]
Chaetorhynchus [Genus]
C. papuensis [Species]
Dicrurus [Genus]
D. ludwigii [Species]
D. atripennis
D. adsimilis
D. fuscipennis
D. aldabranus
D. forficatus
D. waldenii
D. macrocercus
D. leucophaeus
D. caerulescens
D. annectans
D. aeneus
D. remifer
D. balicassius
D. hottentottus
D. megarhynchus
D. montanus
D. andamanensis
D. paradiseus

Callaeidae [Family]
Callaeas [Genus]
C. cinerea [Species]
Creadion [Genus]
C. carunculatus [Species]
Heterolocha [Genus]
H. acutirostris [Species]

Grallinidae [Family]
Grallina [Genus]

G. cyanoleuca [Species]
G. brujini
Corcorax [Genus]
C. melanorhamphos [Species]
Struthidea [Genus]
S. cinerea [Species]

Artamidae [Family]
Artamus [Genus]
A. fuscus [Species]
A. leucorhynchus
A. monachus
A. maximus
A. insignis
A. personatus
A. superciliosus
A. cinereus
A. cyanopterus
A. minor

Cracticidae [Family]
Cracticus [Genus]
C. mentalis [Species]
C. torquatus
C. cassicus
C. louisiadensis
C. nigrogularis
C. quoyi
Gymnorhina [Genus]
G. tibicen [Species]
Strepera [Genus]
S. graculina [Species]
S. fuliginosa
S. versicolor

Ptilonorhynchidae [Family]
Ailuroedus [Genus]
A. buccoides [Species]
A. crassirostris
Scenopoeetes [Genus]
S. dentirostris [Species]
Archboldia [Genus]
A. papuensis [Species]
Amblyornis [Genus]
A. inornatus [Species]
A. macgregoriae
A. subalaris
A. flavifrons
Prionodura [Genus]
P. newtoniana [Species]
Sericulus [Genus]
S. aureus [Species]
S. bakeri
S. chrysocephalus
Ptilonorhynchus [Genus]
P. violaceus [Species]
Chlamydera [Genus]
C. maculata [Species]
C. nuchalis
C. lauterbachi
C. cerviniventris

Paradisaeidae [Family]
Loria [Genus]
L. loriae [Species]
Loboparadisea [Genus]
L. sericea [Species]
Cnemophilus [Genus]
C. macgregorii [Species]
Macgregoria [Genus]
M. pulchra [Species]
Lycocorax [Genus]
L. pyrrhopterus [Species]
Manucodia [Genus]
M. ater [Species]
M. jobiensis
M. chalybatus
M. comrii
Phonygammus [Genus]
P. keraudrenii [Species]
Ptiloris [Genus]
P. paradiseus [Species]
P. victoriae
P. magnificus
Semioptera [Genus]
S. wallacei [Species]
Seleucidis [Genus]
S. melanuleuca [Species]
Paradigalla [Genus]
P. carunculata [Species]
Drepanornis [Genus]
D. albertisi [Species]
D. brujini
Epimachus [Genus]
E. fastuosus [Species]
E. meyeri
Astrapia [Genus]
A. nigra [Species]
A. splendidissima
A. mayeri
A. stephaniae
A. rothschildi
Lophorina [Genus]
L. superba [Species]
Parotia [Genus]
P. sefilata [Species]
P. carolae
P. lawesii
P. wahnesi
Pteridophora [Genus]
P. alberti [Species]
Cicinnurus [Genus]
C. regius [Species]
Diphyllodes [Genus]
D. magnificus [Species]
D. respublica
Paradisaea [Genus]
P. apoda [Species]
P. minor
P. decora
P. rubra
P. guilielmi
P. rudolphi

Corvidae [Family]
  Platylophus [Genus]
    P. galericulatus [Species]
  Platysmurus [Genus]
    P. leucopterus [Species]
  Gymnorhinus [Genus]
    G. cyanocephala [Species]
  Cyanocitta [Genus]
    C. cristata [Species]
    C. stelleri
  Aphelocoma [Genus]
    A. coerulescens [Species]
    A. ultramarina
    A. unicolor
  Cyanolyca [Genus]
    C. viridicyana [Species]
    C. pulchra
    C. cucullata
    C. pumilo
    C. nana
    C. mirabilis
    C. argentigula
  Cissilopha [Genus]
    C. melanocyanea [Species]
    C. sanblasiana
    C. beecheii
  Cyanocorax [Genus]
    C. caeruleus [Species]
    C. cyanomelas
    C. violaceus
    C. cristatellus
    C. heilprini
    C. cayanus
    C. affinis
    C. chrysops
    C. mysticalis
    C. dickeyi
    C. yncas
  Psilorhinus [Genus]
    P. morio [Species]
  Calocitta [Genus]
    C. formosa [Species]
  Garrulus [Genus]

    G. glandarius [Species]
    G. lanceolatus
    G. lidthi
  Perisoreus [Genus]
    P. canadensis [Species]
    P. infaustus
    P. internigrans
  Urocissa [Genus]
    U. ornata [Species]
    U. caerulea
    U. flavirostris
    U. erythrorhyncha
    U. whiteheadi
  Cissa [Genus]
    C. chinensis [Species]
    C. thalassina
  Cyanopica [Genus]
    C. cyana [Species]
  Dendrocitta [Genus]
    D. vagabunda [Species]
    D. occipitalis
    D. formosae
    D. leucogastra
    D. frontalis
    D. baileyi
  Crypsirina [Genus]
    C. temia [Species]
    C. cucullata
  Temnurus [Genus]
    T. temnurus [Species]
  Pica [Genus]
    P. pica [Species]
    P. nuttali
  Zavattariornis [Genus]
    Z. stresemanni [Species]
  Podoces [Genus]
    P. hendersoni [Species]
    P. biddulphi
    P. panderi
    P. pleskei
  Pseudopodoces [Genus]
    P. humilis [Species]
  Nucifraga [Genus]

    N. columbiana [Species]
    N. caryocatactes
  Pyrrhocorax [Genus]
    P. pyrrhocorax [Species]
    P. graculus
  Ptilostomus [Genus]
    P. afer [Species]
  Corvus [Genus]
    C. monedula [Species]
    C. dauuricus
    C. splendens
    C. moneduloides
    C. enca
    C. typicus
    C. florensis
    C. kubaryi
    C. validus
    C. woodfordi
    C. fuscicapillus
    C. tristis
    C. capensis
    C. frugilegus
    C. brachyrhynchos
    C. caurinus
    C. imparatus
    C. ossifragus
    C. palmarum
    C. jamaicensis
    C. nasicus
    C. leucognaphalus
    C. corone
    C. macrorhynchos
    C. orru
    C. bennetti
    C. coronoides
    C. torquatus
    C. albus
    C. tropicus
    C. cryptoleucus
    C. ruficollis
    C. corax
    C. rhipidurus
    C. albicollis
    C. crassirostris

A brief geologic history of animal life

A note about geologic time scales: A cursory look will reveal that the timing of various geological periods differs among textbooks. Is one right and the others wrong? Not necessarily. Scientists use different methods to estimate geological time—methods with a precision sometimes measured in tens of millions of years. There is, however, a general agreement on the magnitude and relative timing associated with modern time scales. The closer in geological time one comes to the present, the more accurate science can be—and sometimes the more disagreement there seems to be. The following account was compiled using the more widely accepted boundaries from a diverse selection of reputable scientific resources.

Geologic time scale

Era	Period	Epoch	Dates	Life forms
Proterozoic			2,500-544 mya*	First single-celled organisms, simple plants, and invertebrates (such as algae, amoebas, and jellyfish)
Paleozoic	Cambrian		544-490 mya	First crustaceans, mollusks, sponges, nautiloids, and annelids (worms)
	Ordovician		490-438 mya	Trilobites dominant. Also first fungi, jawless vertebrates, starfish, sea scorpions, and urchins
	Silurian		438-408 mya	First terrestrial plants, sharks, and bony fish
	Devonian		408-360 mya	First insects, arachnids (scorpions), and tetrapods
	Carboniferous	Mississippian	360-325 mya	Amphibians abundant. Also first spiders, land snails
		Pennsylvanian	325-286 mya	First reptiles and synapsids
	Permian		286-248 mya	Reptiles abundant. Extinction of trilobytes
Mesozoic	Triassic		248-205 mya	Diversification of reptiles: turtles, crocodiles, therapsids (mammal-like reptiles), first dinosaurs
	Jurassic		205-145 mya	Insects abundant, dinosaurs dominant in later stage. First mammals, lizards, frogs, and birds
	Cretaceous		145-65 mya	First snakes and modern fish. Extinction of dinosaurs, rise and fall of toothed birds
Cenozoic	Tertiary	Paleocene	65-55.5 mya	Diversification of mammals
		Eocene	55.5-33.7 mya	First horses, whales, and monkeys
		Oligocene	33.7-23.8 mya	Diversification of birds. First anthropoids (higher primates)
		Miocene	23.8-5.6 mya	First hominids
		Pliocene	5.6-1.8 mya	First australopithecines
	Quaternary	Pleistocene	1.8 mya-8,000 ya	Mammoths, mastodons, and Neanderthals
		Holocene	8,000 ya-present	First modern humans

*Millions of years ago (mya)

Index

Bold page numbers indicate the primary discussion of a topic; page numbers in italics indicate illustrations.

Amami thrushes, 10:485
Amami woodcocks, 9:180
Amandava amandava. See Red avadavats
Amandava formosa. See Green avadavats
Amandava subflava. See Zebra waxbills
Amani sunbirds, 11:210
Amaurornis phoenicurus. See White-breasted
 waterhens
Amazilia spp., 9:443, 9:448
Amazilia castaneiventris. See Chestnut-bellied
 hummingbirds
Amazilia distans. See Táchira emeralds
Amazilia tzacatl. See Rufous-tailed
 hummingbirds
Amazon kingfishers, 10:*11*, 10:*21–22*
Amazona autumnalis. See Red-lored Amazons
Amazona guildingii. See St. Vincent Amazons
Amazona ochrocephala. See Yellow-crowned
 Amazons
Amazona vittata. See Puerto Rican Amazon
 parrots
Amazonetta brasiliensis. See Brazilian teals
Amazonian umbrellabirds, 10:305, 10:306,
 10:308, 10:*309*, 10:*316–317*
Amblycercus holosericeus, 11:301
Amblynura kleinschmidti. See Pink-billed
 parrotfinches
Amblyornis spp., 11:477, 11:478, 11:480,
 11:481
Amblyornis inornatus. See Vogelkops
Amblyornis macgregoriae. See Macgregor's
 bowerbirds
Amblyornis subalaris. See Streaked bowerbirds
Amblyospiza albifrons. See Grosbeak weavers
Amblyramphus holosericeus. See Scarlet-headed
 blackbirds
American anhingas, 8:*10*, 8:*204*, 8:209
American avocets, 9:133, 9:*134*, 9:*136*,
 9:*139–140*
American black ducks, 8:371
American black oystercatchers, 9:126
American black rails, 9:1
American black swifts. *See* Black swifts
American black vultures, 8:234, 8:276, 8:278,
 8:*279*, 8:*281–282*
American cliff swallows, 10:358, 10:*359*,
 10:*362*, 10:*364–365*
American coots, 9:*46*, 9:50, 9:51
American crows, 11:505, 11:506, 11:*508*,
 11:*511*, 11:*520–521*
American cuckoos, 9:311
American dabchicks. *See* Least grebes
American darters. *See* American anhingas
American dippers, 10:475, 10:476, 10:477,
 10:*478*, 10:480
American finfoot. *See* Sungrebes
American golden-plovers, 8:29, 9:162, 9:*165*,
 9:*166–167*
American goldfinches, 11:*323–324*, 11:*326*,
 11:*331–332*
American jabirus. *See* Jabirus
American jacanas. *See* Northern jacanas
American lesser golden-plovers. *See* American
 golden-plovers
American mourning doves, 9:*244*, 9:*248*,
 9:251, 9:*254*, 9:*259–260*
American oystercatchers, 9:*126*, 9:*127*, 9:*129*,
 9:*130*
American painted-snipes. *See* South American
 painted-snipes

American pied oystercatchers. *See* American
 oystercatchers
American pipits, 10:372
American redstarts, 11:*287*, 11:288, 11:289
American robins, 10:484–*485*, 10:*491*,
 10:*502–503*
American striped cuckoos, 9:312, 9:313,
 9:*316*, 9:*327–328*
American treecreepers. *See* Brown creepers
American tufted titmice, 11:156, 11:157
American white pelicans, 8:186, 8:225, 8:226,
 8:229, 8:230–*231*
American wigeons, 8:*375*, 8:*383–384*
American woodcocks, 9:177, 9:180
Amethyst woodstars, 9:439
Ammodramus leconteii. See Le Conte's sparrows
Ammodramus nelsoni. See Nelson's sharp-tailed
 sparrows
Ammomanes spp. *See* Larks
Ammomanes cincturus. See Bar-tailed larks
Ammomanes grayi. See Gray's larks
Ampeliceps spp., 11:408
Ampeliceps coronatus. See Golden-crested
 mynas
Ampelion stresemanni. See White-cheeked
 cotingas
Amphispiza bilineata. See Black-throated
 sparrows
Amsterdam albatrosses, 8:113
Amur eastern red-footed falcons. *See* Amur
 falcons
Amur falcons, 8:353, 8:*358*
Amytornis housei. See Black grasswrens
Amytornis striatus. See Striated grasswrens
Amytornithinae. *See* Grasswrens
Anabathmis spp. *See* Sunbirds
Anabathmis reichenbachii. See Reichenbach's
 sunbirds
Andalusian hemipodes. *See* Barred
 buttonquails
Anairetes alpinus. See Ash-breasted tit tyrants
Anambra waxbills, 11:357
Anaplectes rubriceps. See Red-headed finches
Anarhynchus frontalis. See Wrybills
Anas acuta. See Northern pintails
Anas albogularis. See Andaman teals
Anas americana. See American wigeons
Anas aucklandica. See Brown teals
Anas castanea. See Chestnut teals
Anas chlorotis. See Brown teals
Anas clypeata. See Northern shovelers
Anas eatoni. See Eaton's pintails
Anas falcata. See Falcated teals
Anas formosa. See Baikal teals
Anas gibberifrons. See Sunda teals
Anas gracilis. See Gray teals
Anas laysanensis. See Laysan ducks
Anas melleri. See Meller's duck
Anas nesiotis. See Campbell Island teals
Anas platyrhynchos. See Mallards
Anas waigiuensis. See Salvadori's teals
Anas wyvilliana. See Hawaiian ducks
Anastomus spp. *See* Storks
Anastomus oscitans. See Asian openbills
Anatidae. *See* Ducks; Geese; Swans
Anatinae. *See* Dabbling ducks
Anchieta's barbets, 10:115
Andaman barn owls, 9:338, 9:341
Andaman dark serpent eagles. *See* Andaman
 serpent-eagles

Andaman red-whiskered bulbuls, 10:398
Andaman serpent-eagles, 8:*325*, 8:*336*
Andaman teals, 8:364
Andean avocets, 9:134
Andean cocks-of-the-rock, 10:305, 10:*310*,
 10:*318*
Andean condors, 8:233, 8:236, 8:275, 8:276,
 8:277, 8:278, 8:279, 8:*284*
Andean flamingos, 8:303, 8:304, 8:305, 8:306,
 8:*307*, 8:*310*
Andean hillstars, 9:442, 9:444
Anderson, Atholl, 8:98
Andigena spp. *See* Mountain toucans
Andigena hypoglauca. See Gray-breasted
 mountain toucans
Andigena laminirostris. See Plate-billed
 mountain toucans
Andrews' frigatebirds. *See* Christmas
 frigatebirds
Andropadus spp. *See* Hummingbirds
Andropadus spp., 10:396
Androphobus viridis. See Papuan whipbirds
Angolan pittas. *See* African pittas
Anhima cornuta. See Horned screamers
Anhimidae. *See* Screamers
Anhinga anhinga. See American anhingas
Anhingas, 8:183–186, **8:201–210**
Anhingidae. *See* Anhingas
Anianiaus, 11:*345*, 11:*347–348*
Anis, 9:311–313, 9:*314*, 9:325–326
Ankober serins, 11:324
Annam pheasants. *See* Edwards' pheasants
Anna's hummingbirds, 8:*10*, 9:443, 9:*444*,
 9:*447*, 9:*453*, 9:*467*
Annobón paradise-flycatchers, 11:97
Annobón white-eyes, 11:227
Anodorhynchus hyacinthinus. See Hyacinth
 macaws
Anomalopteryx didiformis. See Moas
Anomalospiza imberbis. See Cuckoo finches
Anoplolepis gracilipes. See Yellow crazy ants
Anorrhinus tickelli. See Tickell's brown hornbills
Anous spp. *See* Noddies
Anser anser. See Greylag geese
Anser caerulescens. See Snow geese
Anser canagica. See Emperor geese
Anser indicus. See Bar-headed geese
Anseranas semipalmata. See Magpie geese
Anseriformes, **8:363–368**
Anserinae. *See* Geese; Swans
Ant thrushes, **10:239–256**, 10:*243–244*,
 10:487
 behavior, 10:240–241
 conservation status, 10:241–242
 distribution, 10:*239*, 10:240
 evolution, 10:239
 feeding ecology, 10:241
 habitats, 10:240
 humans and, 10:242
 physical characteristics, 10:239–240
 reproduction, 10:241
 taxonomy, 10:239
Antarctic giant petrels. *See* Southern giant
 petrels
Antarctic petrels, 8:108, 8:*126*
Antbirds, 10:170
Anteater chats, 10:141, 10:*490*, 10:*498*
Anthobaphes spp. *See* Sunbirds
Anthobaphes violacea. See Orange-breasted
 sunbirds

Anthocephala spp. *See* Hummingbirds
Anthochaera carunculata. See Red wattlebirds
Anthoscopus spp., 11:148, 11:149
Anthracoceros marchei. See Palawan hornbills
Anthracoceros montani. See Sulu hornbills
Anthracothorax spp. *See* Hummingbirds
Anthreptes spp. *See* Sunbirds
Anthreptes fraseri. See Scarlet-tufted sunbirds
Anthreptes malacensis. See Plain-throated sunbirds
Anthreptes rectirostris. See Green sunbirds
Anthreptes reichenowi. See Plain-backed sunbirds
Anthreptes rhodolaema. See Red-throated sunbirds
Anthreptes rubritorques. See Banded sunbirds
Anthropoides spp., 9:23, 9:25
Anthropoides virgo. See Demoiselle cranes
Anthroscopus spp. *See* African penduline tits
Anthus spp. *See* Pipits
Anthus antarcticus. See South Georgia pipits
Anthus berthelotii. See Berthelot's pipits
Anthus caffer. See Bush pipits
Anthus campestris. See Tawny pipits
Anthus cervinus. See Red-throated pipits
Anthus chacoensis. See Chaco pipits
Anthus chloris. See Yellow-breasted pipits
Anthus correndera. See Correndera pipits
Anthus crenatus. See Yellow-tufted pipits
Anthus gustavi. See Pechora pipits
Anthus gutturalis. See Alpine pipits
Anthus hodgsoni. See Olive-backed pipits
Anthus lineiventris. See Striped pipits
Anthus longicaudatus. See Long-tailed pipits
Anthus lutescens. See Yellowish pipits
Anthus melindae. See Malindi pipits
Anthus nattereri. See Ochre-breasted pipits
Anthus nilghiriensis. See Nilgiri pipits
Anthus novaeseelandiae. See Australasian pipits
Anthus nyassae. See Woodland pipits
Anthus petrosus. See Rock pipits
Anthus pratensis. See Meadow pipits
Anthus richardi. See Richard's pipits
Anthus roseatus. See Rosy pipits
Anthus rubescens. See American pipits
Anthus sokokensis. See Sokoke pipits
Anthus spragueii. See Sprague's pipits
Anthus trivialis. See Tree pipits
Antibyx spp. *See* Blacksmith plovers
Antillean cloud swifts. *See* White-collared swifts
Antillean emeralds. *See* Puerto Rican emeralds
Antillean grackles, 11:302–303
Antilophia bokermanni. See Araripe manakins
Antioquia bristle tyrants, 10:275
Antipodean albatrosses, 8:113
Apalharpactes narina. See Narina trogons
Apalharpactes reinwardtii. See Javan trogons
Apalis spp., 11:7
Apalis flavida. See Yellow-breasted apalis
Apaloderma vittatum. See Collared trogons
Apalopteron familiare. See Bonin white-eyes
Apapanes, 11:342, 11:345, 11:348
Aphelocephala spp. *See* Whitefaces
Aphelocephala leucopsis. See Southern whitefaces
Aphelocephala pectoralis. See Tasmanian thornbills
Aphelocoma californica. See Western scrub-jays
Aphelocoma coerulescens. See Florida scrub-jays

Aphelocoma ultramarina. See Gray-breasted jays
Aphrastura spinicauda. See Thorn-tailed rayaditos
Aphriza virgata. See Surfbirds
Apis dorsata. See Rock bees
Aplonis spp., 11:408, 11:410
Aplonis brunneicapilla. See White-eyed starlings
Aplonis cinerascens. See Rarotonga starlings
Aplonis corvina. See Kosrae Mountain starlings
Aplonis crassa. See Tanimbar starlings
Aplonis feadensis. See Atoll starlings
Aplonis fusca bulliana. See Lord Howe Island starlings
Aplonis fusca fusca. See Norfolk Island starlings
Aplonis grandis. See Brown-winged starlings
Aplonis mystacea. See Yellow-eyed starlings
Aplonis pelzelni. See Pohnpei Mountain starlings
Aplonis santovestris. See Mountain starlings
Aplonis zelandica. See Tanimbar starlings
Apo mynas, 11:410
Apo sunbirds, 11:210
Apodi, 9:415
Apodidae. *See* Swifts
Apodiformes, **9:415–419**, 9:417
Apodinae. *See* Swifts
Apolinar's wrens, 10:529
Apostlebirds, 11:453, 11:454, 11:455, 11:457–458
Aptenodytes spp. *See* Penguins
Aptenodytes forsteri. See Emperor penguins
Apterygidae. *See* Kiwis
Apteryx australis. See Brown kiwis
Apteryx haastii. See Great spotted kiwis
Apteryx owenii. See Little spotted kiwis
Aptoenodytes patavonicus. See King penguins
Apus spp. *See* Swifts
Apus apus. See Common swifts
Apus batesi. See Bates's swifts
Apus caffer. See White-rumped swiftlets
Apus horus. See Horus swifts
Apus melba. See Alpine swifts
Apus pacificus. See Pacific swifts
Apus pallidus. See Pallid swifts
Aquatic warblers, 11:6, 11:7
Aquila adalberti. See Spanish imperial eagles
Aquila audax. See Australian wedge-tailed eagles
Aquila chrysaetos. See Golden eagles
Aquila gurneyi. See Gurney's eagles
Aquila heliaca. See Imperial eagles
Aquila nipalensis. See Steppe eagles
Aquila verreauxii. See Verraux's eagles
Ara ararauna. See Blue and yellow macaws
Ara chloroptera. See Green-winged macaws
Ara macao. See Scarlet macaws
Arabian babblers, 10:505, 10:507, 10:510, 10:514
Arabian woodpeckers, 10:150
Aracaris, 10:125–128
Arachnothera spp. *See* Sunbirds
Arachnothera chrysogenys. See Yellow-eared spiderhunters
Arachnothera juliae. See Whitehead's spiderhunters
Aramidae. *See* Limpkins
Aramides axillaris. See Rufous-necked wood-rails
Aramides ypecaha. See Giant wood-rails

Araripe manakins, 10:297, 10:298, 10:299–300
Araripe's soldiers. *See* Araripe manakins
Aratinga pertinax. See Brown-throated parakeets
Aratinga wagleri. See Scarlet-fronted parrots
Arborescent puyas, 9:442
Arborophila spp. *See* Hill-partridges
Arborophila ardens. See Hainan hill-partridges
Arborophila cambodiana. See Chestnut-headed hill-partridges
Arborophila davidi. See Orange-necked hill-partridges
Arborophila rubirostris. See Red-billed hill-partridges
Arborophila rufipectus. See Sichuan hill-partridges
Arborophila torqueola. See Asian hill-partridges
Archaeoganga spp., 9:231
Archaeopsittacus verreauxi, 9:275
Archaeopteryx, 8:3, 8:8, 8:14
Archboldia spp., 11:477, 11:478
Archboldia papuensis. See Archbold's bowerbirds
Archbold's bowerbirds, 11:478, 11:481, 11:482, 11:485
Archbold's owlet-nightjars, 9:388
Archer's larks, 10:346
Archilochus colubris. See Ruby-throated hummingbirds
Arctic fulmars. *See* Northern fulmars
Arctic jaegers. *See* Arctic skuas
Arctic loons, 8:159, 8:162, 8:164–165
Arctic peregrine falcons, 8:349
Arctic skuas, 9:204, 9:210, 9:211
Arctic terns, 8:29, 8:31, 9:102
Arctic warblers, 11:7, 11:8, 11:20–21
Ardea alba. See Great white egrets
Ardea americana. See Whooping cranes
Ardea antigone. See Sarus cranes
Ardea carunculata. See Wattled cranes
Ardea cinerea. See Gray herons
Ardea goliath. See Goliath herons
Ardea grus. See Eurasian cranes
Ardea herodias. See Great blue herons
Ardea humbloti, 8:245
Ardea idea, 8:245
Ardea insignis, 8:245
Ardea melanocephala. See Black-headed herons
Ardea purpurea. See Purple herons
Ardea virgo. See Demoiselle cranes
Ardeola spp. *See* Herons
Ardeola ralloides. See Squacco herons
Ardeotis spp. *See* Bustards
Ardeotis kori. See Kori bustards
Ardeotis nigriceps. See Great Indian bustards
Arenaria interpres. See Ruddy turnstones
Arenariinae. *See* Turnstones
Arfak berrypeckers, 11:189, 11:190
Argusianus argus. See Great argus pheasants
Aristotle, hoopoes and, 10:63
Arkansas kingbirds. *See* Western kingbirds
Armitage, K. B., 9:444
Army ants, 10:241
Art, birds and, 8:26
Artamidae. *See* Woodswallows
Artamus cinereus. See Black-faced woodswallows
Artamus cyanopterus. See Dusky woodswallows
Artamus fuscus. See Ashy woodswallows
Artamus insignis. See Bismark woodswallows

INDEX

Blue-necked jacamars. *See* Yellow-billed
jacamars
Blue plantain-eaters. *See* Great blue turacos
Blue swallows, 10:360
Blue-tailed emeralds, 9:438
Blue-tailed trogons. *See* Javan trogons
Blue-throated hummingbirds, 9:416
Blue-throated motmots, 10:32, 10:*34*, 10:*35*
Blue-throated rollers, 10:51–52
Blue titmice, 11:*156*, 11:158
Blue vangas, 10:439, 10:440, 10:441, 10:442
Blue-winged leafbirds, 10:417, 10:*418*, 10:*421*
Blue-winged parrots, 9:277
Blue-winged pittas, 10:194, 10:*195*
Blue-winged warblers, 11:285, 11:286, 11:*292*,
11:*294*
Blue-winged yellow warblers. *See* Blue-
winged warblers
Bluebills, 11:353, 11:354
Bluebirds, 10:485, 10:*487*, 10:488, 10:489,
10:*494*
Bluethroats, 10:483, 10:484, 10:486
Bluish-slate antshrikes. *See* Cinereous
antshrikes
Blyth's reed warblers, 9:314
Boat-billed herons, 8:239, 8:241, 8:*246*,
8:*258–259*
Bobolinks, 11:301–303, 11:*307*, 11:*321–322*
Bobwhites. *See* Northern bobwhite quails
Bock, W. J., 10:329
Body temperature, 8:9–10
Bogotá sunangels, 9:449, 9:450–451
Bohemian waxwings, 10:447, 10:448, 10:*450*,
10:451–*452*
Boiga irregularis. *See* Brown tree snakes
Boissonneaua spp. *See* Hummingbirds
Boissonneaua jardini. *See* Velvet-purple coronets
Boland, Chris, 11:457
Boles, W. E., 10:385, 11:459
Bolivian blackbirds, 11:293, 11:304
Bolivian earthcreepers, 10:*212*, 10:*216*
Bolivian spinetails, 10:211
Bombycilla cedrorum. *See* Cedar waxwings
Bombycilla garrulus. *See* Bohemian waxwings
Bombycilla japonica. *See* Japanese waxwings
Bombycillidae, **10:447–454**, 10:*450*
Bombycillinae. *See* Waxwings
Bonaparte's tinamous. *See* Highland tinamous
Bonaparte's gulls, 9:205
Bonaparte's nightjars, 9:405
Bonasa bonasia. *See* Hazel grouse
Bonasa sewerzowi. *See* Chinese grouse
Bonasa umbellus. *See* Ruffed grouse
Bonin Island honeyeaters. *See* Bonin white-eyes
Bonin siskins, 11:324
Bonin white-eyes, 11:227, 11:235
Boobies, 8:183, 8:184, **8:211–223**
behavior, 8:212–213
conservation status, 8:214–215
distribution, 8:*211*, 8:212
evolution, 8:211
feeding ecology, 8:213
habitats, 8:212
humans and, 8:215
physical characteristics, 8:211
reproduction, 8:213–214
species of, 8:*217–222*
taxonomy, 8:211
Boobook owls. *See* Southern boobook owls
Booted eagles, 8:321

Boreal chickadees, 11:156
Boreal owls, 9:350
Boreal peewees. *See* Olive-sided flycatchers
Boreal tits, 11:157
Bornean bristleheads, 10:425, 10:429, 11:469,
11:*470*, 11:*471*
Bornean frogmouths, 9:379
Bornean peacock-pheasants, 8:437
Bostrychia bocagei. *See* Dwarf olive ibises
Bostrychia hagedash. *See* Hadada ibises
Botaurinae. *See* Bitterns
Botaurus poiciloptilus. *See* Australasian bitterns
Botaurus stellaris. *See* Eurasian bitterns
Botha's larks, 10:346
Bothas. *See* Bush-shrikes
Boucard, Adolphe, 9:452
Boucard tinamus. *See* Slaty-breasted tinamous
Boulton's hill-partridges. *See* Sichuan hill-
partridges
Bowerbirds, 10:174, **11:477–488**, 11:*482*
behavior, 11:479
conservation status, 11:481
distribution, 11:*477*, 11:478
evolution, 11:477
feeding ecology, 11:479
habitats, 11:478–479
humans and, 11:481
physical characteristics, 11:477–478
reproduction, 11:479–480
species of, 11:*483–488*
taxonomy, 11:477
Brace's emeralds, 9:450
Brachygalba goeringi. *See* Pale-headed jacamars
Brachygalba lugubris. *See* Brown jacamars
Brachypteracias leptosomus. *See* Short-legged
ground-rollers
Brachypteracias squamigera. *See* Scaly ground-
rollers
Brachypteraciidae. *See* Ground-rollers
Brachyramphus spp. *See* Murrelets
Brachyramphus marmoratus. *See* Marbled
murrelets
Bradypterus spp., 11:3, 11:4, 11:7
Bradypterus baboecala. *See* Little rush-warblers
Bradypterus sylvaticus, 11:5
Brahminy ducks. *See* Ruddy shelducks
Brahminy kites, 8:320, 8:323
Brahminy starlings, 11:*408*
Bramblings, 11:*326*, 11:*329–330*
Brandt's cormorants, 8:*203*, 8:*204*, 8:*206*
Branta canadensis. *See* Canada geese
Branta sandvicensis. *See* Hawaiian geese
Brazilian mergansers, 8:365, 8:367, 8:*375*,
8:*389–390*
Brazilian tapirs, 9:78
Brazilian teals, 8:371
Brazza's swallows, 10:360
Breeding. *See* Reproduction
Breeding Bird Survey, 8:23, 10:448, 10:453
Breeding seasons. *See* Reproduction
Brenowitz, E. A., 8:40
Brewer's blackbirds, 11:302–303, 11:304
Brewster's warblers, 11:285–286, 11:294
Bridled titmice, 11:*158*, 11:*159*, 11:162
Bright-headed cisticolas. *See* Golden-headed
cisticolas
Bristle-nosed barbets, 10:115
Bristle-thighed curlews, 9:180
Bristlebirds, 11:55–57, 11:57, 11:*58*, 11:59,
11:*59*

Bristleheads, 11:467–469, 11:*470*, 11:471
Broad-billed dove-petrels. *See* Broad-billed
prions
Broad-billed hummingbirds, 9:*439*
Broad-billed motmots, 10:31–32, 10:*34*,
10:*35–36*
Broad-billed prions, 8:*127*, 8:130–131
Broad-billed todies, 10:26
Broad-tailed hummingbirds, 9:443, 9:445
Broad-winged hawks, 8:321
Broadbills, 10:170, **10:177–186**, 10:*180*
Broken-wing display, 9:74
Brolga cranes, 9:25
Brontë, Emily, 10:462
Bronze-cuckoos, 11:57
Bronze sunbirds, 11:209
Bronze-tailed sicklebills. *See* White-tipped
sicklebills
Bronzed drongos, 11:438
Bronzed shags. *See* New Zealand king shags
Brood parasitism
cowbirds, 11:*256*, 11:291, 11:321
cuckoos, 9:313–314, 11:445
honeyguides as, 10:139
indigobirds, 11:356, 11:378
Viduinae, 11:378
weaver-finches, 11:356
whydahs, 11:356, 11:378
See also Reproduction
Brooks, D., 10:127
Brotogeris jugularis. *See* Orange-chinned
parakeets
Brown, William L., 10:130
Brown-and-yellow marshbirds, 11:*307*,
11:*316–317*
Brown-backed flowerpeckers, 11:191
Brown-backed honeyeaters, 11:*236*
Brown birds. *See* Brown tremblers
Brown boobies, 8:*216*, 8:222
Brown-capped tits, 11:157
Brown-capped vireos. *See* Warbling vireos
Brown-cheeked hornbills, 10:75
Brown-chested flycatchers. *See* Brown-chested
jungle-flycatchers
Brown-chested jungle-flycatchers, 11:29,
11:*33–34*
Brown creepers, 11:118, 11:177, 11:*179*,
11:*180–181*
Brown dippers, 10:475, 10:*478*, 10:*479–480*
Brown ducks. *See* Brown teals
Brown-eared bulbuls. *See* Ashy bulbuls
Brown eared-pheasants, 8:*438*, 8:*451*
Brown falcons, 8:353, 8:*358–359*
Brown flycatchers. *See* Jacky winters
Brown hawk-owls, 9:347, 9:349
Brown hawks. *See* Brown falcons
Brown-headed cowbirds, 11:291, 11:293,
11:304, 11:*307*, 11:*320–321*
Brown-headed nuthatches, 11:*169*, 11:*170*,
11:*172*
Brown-headed stork-billed kingfishers. *See*
Stork-billed kingfishers
Brown honeyeaters, 11:*240*, 11:246–247
Brown jacamars, 10:91
Brown jays, 11:504
Brown kingfishers. *See* Laughing kookaburras
Brown kiwis, 8:*10*, 8:89, 8:90, 8:*90*, 8:91,
8:92, 8:*92–93*
Brown mesites, 9:5, 9:6, 9:7, 9:*8–9*
Brown nightjars, 9:368, 9:401, 9:404

Brown owls. *See* Tawny owls
Brown pelicans, 8:*183*, 8:185, 8:186, 8:225,
  8:*226*, 8:*227*, 8:*229*, 8:*231–232*
Brown roatelos. *See* Brown mesites
Brown shrike-thrushes. *See* Gray shrike-
  thrushes
Brown shrikes, 10:430
Brown skuas, 9:163, 9:206
Brown-tailed apalis. *See* Yellow-breasted
  apalis
Brown-tailed sicklebills. *See* White-tipped
  sicklebills
Brown teals, 8:363, 8:364, 8:367, 8:*374*, 8:*384*
Brown thornbills, 11:56, 11:57
Brown thrashers, 8:39, 10:*465*, 10:*466*,
  10:*468*, 10:*472–473*
Brown-throated conures. *See* Brown-throated
  parakeets
Brown-throated parakeets, 9:*281*, 9:*290*–291
Brown-throated sunbirds. *See* Plain-throated
  sunbirds
Brown-throated treecreepers, 11:178
Brown tree snakes, 11:98, 11:230, 11:508
Brown treecreepers, 11:*134*
Brown tremblers, 10:*469*, 10:*473–474*
Brown waxbills. *See* Common waxbills
Brown-winged starlings, 11:408
Brown woodpeckers. *See* Smoky-brown
  woodpeckers
Brown wren-babblers. *See* Pygmy wren-
  babblers
Brown's cormorants. *See* Brandt's cormorants
Brubrus, 10:425–430
Bruce, Murray, 9:338
Bruce's green pigeons, 9:*254*, 9:*263–264*
Bruijn's brush turkeys, 8:401
Brunnich's guillemots. *See* Thick-billed
  murres
Brush-tailed possums, 10:331
Brush-turkeys. *See* Moundbuilders
Brushland tinamous, 8:58, 8:59, 8:*60*, 8:*65*
Bryant's sparrows. *See* Savannah sparrows
Bubalornis niger. See Red-billed buffalo weavers
Bubalornithinae. *See* Buffalo weavers
Bubo spp. *See* Eagle-owls
Bubo ascalaphus. See Savigny's eagle-owls
Bubo blakistoni. See Blakiston's eagle-owls
Bubo bubo. See Eurasian eagle-owls
Bubo lacteus. See Verreaux's eagle owls
Bubo poirrieri, 9:331, 9:345
Bubo sumatranus. See Barred eagle-owls
Bubo virginianus. See Great horned owls
Bucco capensis. See Collared puffbirds
Bucconidae. *See* Puffbirds
Bucephala spp. *See* Golden-eye ducks
Buceros bicornis. See Great hornbills
Buceros hydrocorax. See Rufous hornbills
Buceros rhinoceros. See Rhinoceros hornbills
Bucerotes, 10:1–4
Bucerotidae. *See* Hornbills
Bucerotinae. *See* Hornbills
Bucorvinae. *See* Hornbills
Bucorvus abyssinicus. See Northern ground-
  hornbills
Bucorvus leadbeateri. See Southern ground-
  hornbills
Budgerigars, 8:24, 8:40, 9:276, 9:278–280,
  9:*282*, 9:*285–286*
Budgies. *See* Budgerigars
Buff-backed herons. *See* Cattle egrets

Buff-banded rails, 9:*49*, 9:52
Buff-breasted buttonquails, 9:14, 9:19
Buff-breasted flowerpeckers. *See* Fire-breasted
  flowerpeckers
Buff-breasted sandpipers, 9:177–178, 9:179
Buff-breasted scrubwrens. *See* White-browed
  scrubwrens
Buff-breasted warblers. *See* Mangrove
  gerygones
Buff-rumped thornbills, 11:56
Buff-spotted flufftails, 9:48, 9:52, 9:*55*, 9:57
Buff-spotted woodpeckers, 10:144
Buffalo weavers, 11:375, 11:376, 11:377–378,
  11:*380*, 11:382
Bugeranus spp., 9:23
Bugeranus carunculatus. See Wattled cranes
Bulbuls, **10:395–413**, 10:*400–401*
  behavior, 10:397
  conservation status, 10:398–399
  distribution, 10:*395*, 10:396
  evolution, 10:395
  feeding ecology, 10:398
  habitats, 10:396–397
  humans and, 10:399
  physical characteristics, 10:395–396
  reproduction, 10:398
  species of, 10:*402–413*
  taxonomy, 10:395
Bulky kakapos, 9:275
Bullbirds. *See* Umbrellabirds
Buller, Lawry, 10:207
Buller's mollymawks, 8:114, 8:115
Bullfinches, 11:323, 11:324, 11:*327*, 11:335
Bullock's orioles. *See* Baltimore orioles
Bulo Berti boubous, 10:427, 10:429
Bulweria bulwerii. See Bulwer's petrels
Bulweria fallax. See Joaquin's petrels
Bulwer's petrels, 8:*132*
Bulwer's pheasants, 8:434
Buntings. *See* New World finches
Buphaginae. *See* Oxpeckers
Buphagus erythrorhynchus. See Red-billed
  oxpeckers
Burchell's coursers, 9:*154*, 9:*158*
Burchell's sandgrouse, 9:232
Burhinidae. *See* Thick-knees
Burhinus spp. *See* Thick-knees
Burhinus capensis. See Spotted dikkops
Burhinus grallarius. See Bush thick-knees
Burhinus oedicnemus. See Stone-curlews
Burhinus senegalensis. See Senegal thick-knees
Burhinus vermiculatus. See Water dikkops
Burmeister's seriemas. *See* Black-legged
  seriemas
Burmese hornbills. *See* Plain-pouched
  hornbills
Burrowing owls, 9:333, 9:346, 9:349, 9:350,
  9:*353*, 9:*362–363*
Burrowing parakeets, 9:278
Buru Island monarchs. *See* White-tipped
  monarchs
Bush canaries. *See* Yellowheads
Bush pipits, 10:373, 10:*377*, 10:*382–383*
Bush-quails, 8:434
Bush-shrikes, 10:425–430, 10:*431*,
  10:*432–433*, 10:*433*, 10:*434*
Bush stone-curlews, 9:*144*
Bush thick-knees, 9:144, 9:145–146
Bush turkeys. *See* Australian brush-turkeys
Bush wrens, 10:203

Bushcreepers, 10:209
Bushtits, 10:169, 11:141, 11:142, 11:*146*
Bushveld pipits. *See* Bush pipits
Bustard quail. *See* Barred buttonquails
Bustards, 9:1–4, **9:91–100**, 9:95
Bustards, Hemipodes and Sandgrouse
  (Johnsgard), 9:1, 9:11
Butcherbirds, 11:467–468, 11:*470*,
  11:472–473
Buteo jamaicensis. See Red-tailed hawks
Buteo lagopus. See Rough-legged buzzards
Buteo magnirostris. See Roadside hawks
Buteo platypterus. See Broad-winged hawks
Buteo swainsoni. See Swainson's hawks
Buteogallus aequinoctialis. See Crab hawks
Buttonquails, **9:11–22**, 9:*12*, 9:15
  behavior, 9:12–13
  conservation status, 9:14
  distribution, 9:*11*, 9:12
  evolution, 9:11
  feeding ecology, 9:13
  habitats, 9:12–13
  humans and, 9:14
  physical characteristics, 9:12
  reproduction, 9:13–14
  species of, 9:*16–21*
  taxonomy, 9:1, 9:11–12
Buzzards, 8:318
Bycanistes brevis. See Silvery-cheeked hornbills
Bycanistes cylindricus. See Brown-cheeked
  hornbills

C

Cabanis's greenbuls, 10:397
Cabbage birds. *See* Spotted bowerbirds
Cacatua spp. *See* Cockatoos
Cacatua banksii. See Black cockatoos
Cacatua galerita. See Sulphur-crested
  cockatoos
Cacatua pastinator. See Western corella parrots
Cacatua tenuirostris. See Slender-billed parrots
Caches. *See* Food caches
Cacicus spp. *See* Caciques
Cacicus cela. See Yellow-rumped caciques
Cacicus koepckeae. See Selva caciques
Cacicus melanicterus. See Yellow-winged
  caciques
Cacicus montezuma. See Montezuma's
  oropendolas
Cacicus sclateri. See Ecuadorian caciques
Caciques, 11:301–304, 11:306, 11:*308*,
  11:309–310
Cacklers. *See* Gray-crowned babblers
Cactospiza pallida. See Woodpecker finches
Cactus ground-finches, 11:*265*
Cactus wrens, 10:174, 10:526, 10:527, 10:528,
  10:*530*, 10:*531*
Caerulean paradise-flycatchers, 11:97
Cagebirds, 8:22, 8:24
  bulbuls as, 10:399
  canaries as, 11:325, 11:339
  parrots as, 9:280
  red-billed leiothrix as, 10:517
  turacos as, 9:302
  weaverfinches as, 11:356, 11:357
  See also Humans
Cahows. *See* Bermuda petrels

Dendrocopos dorae. See Arabian woodpeckers
Dendrocopos leucotos. See White-backed woodpeckers
Dendrocopos major. See Great spotted woodpeckers
Dendrocygna eytoni. See Plumed whistling ducks
Dendrocygna javanica. See Lesser whistling-ducks
Dendrocygna viduata. See White-faced whistling ducks
Dendroica spp., 11:287
Dendroica angelae. See Elfin-wood warblers
Dendroica caerulescens. See Black-throated blue warblers
Dendroica castanea. See Bay-breasted warblers
Dendroica cerulea. See Cerulean warblers
Dendroica coronata. See Yellow-rumped warblers
Dendroica discolor. See Prairie warblers
Dendroica graciae. See Grace's warblers
Dendroica kirtlandii. See Kirtland's warblers
Dendroica occidentalis. See Hermit warblers
Dendroica pensylvanica. See Chestnut-sided warblers
Dendroica petichia. See Yellow warblers
Dendroica striata. See Blackpoll warblers
Dendroica townsendi. See Townsend's warblers
Dendronanthus indicus. See Forest wagtails
Dendropicos goertae. See Gray woodpeckers
Dendrortyx barbatus. See Bearded wood-partridges
Dendrortyx macroura. See Long-tailed wood-partridges
Dendroscansor decurvirostris, 10:203
Denham's bustards, 9:*92,* 9:93
Deroptyus accipitrinus. See Red-fan parrots
Des Murs's wiretails, 10:*212,* 10:*218–219*
Desert birds. *See* Gibberbirds
Desert chats. *See* Gibberbirds
Desert larks, 10:345
Diamond birds. *See* Pardalotes
Diamond doves, 9:*254,* 9:*259*
Diamond firetails, 11:356, 11:*359,* 11:*366–367*
Diamond Java sparrows. *See* Diamond firetails
Diamond sparrows. *See* Diamond firetails
Diard's trogons, 9:479
Dicaeidae. *See* Flowerpeckers
Dicaeum spp. *See* Flowerpeckers
Dicaeum aeneum. See Midget flowerpeckers
Dicaeum agile. See Thick-billed flowerpeckers
Dicaeum anthonyi. See Flame-crowned flowerpeckers
Dicaeum celebicum. See Gray-sided flowerpeckers
Dicaeum chrysorrheum. See Yellow-vented flowerpeckers
Dicaeum concolor. See Plain flowerpeckers
Dicaeum cruentatum. See Scarlet-backed flowerpeckers
Dicaeum everetti. See Brown-backed flowerpeckers
Dicaeum geelvinkianum. See Red-capped flowerpeckers
Dicaeum haematostictum. See Black-belted flowerpeckers
Dicaeum hirundinaceum. See Mistletoebirds
Dicaeum ignipectus. See Fire-breasted flowerpeckers
Dicaeum proprium. See Whiskered flowerpeckers

Dicaeum quadricolor. See Cebu flowerpeckers
Dicaeum vincens. See White-throated flowerpeckers
Dicruridae. *See* Drongos
Dicrurus spp. *See* Drongos
Dicrurus adsimilis. See Fork-tailed drongos
Dicrurus aldabranus, 11:439
Dicrurus andamanensis, 11:439
Dicrurus bracteatus. See Spangled drongos
Dicrurus fuscipennis. See Comoro drongos
Dicrurus leucophaeus. See Ashy drongos
Dicrurus ludwigii. See Square-tailed drongos
Dicrurus macrocercus. See Black drongos
Dicrurus megarhynchus. See Ribbon-tailed drongos
Dicrurus modestus. See Velvet-mantled drongos
Dicrurus paradiseus. See Greater racket-tailed drongos
Dicrurus sumatranus. See Sumatran drongos
Dicrurus waldenii. See Mayotte drongos
The Dictionary of Birds, 11:505
Dideric cuckoos, 9:*317,* 9:*320,* 11:385
Didric cuckoos. *See* Dideric cuckoos
Didunculines. *See* Tooth-billed pigeons
Didunculus strigirostris. See Tooth-billed pigeons
Didus solitaria. See Rodrigues solitaires
Diederick cuckoos. *See* Dideric cuckoos
Diederik cuckoos. *See* Dideric cuckoos
Dieldrin. *See* Pesticides
Diet. *See* Feeding ecology
Digestion, avian, 8:10–11
See also Feeding ecology
Dimorphic fantails, 11:84, 11:*88,* 11:*91–92*
Dimorphic rufous fantails. *See* Dimorphic fantails
Dimorphism, 8:54
Dinopium benghalense. See Lesser flame-backed woodpeckers
Dinornis giganteus, 8:96, 8:97
Dinornis novaezealandiae, 8:96, 8:97
Dinornis struthoides, 8:96
Dinornithidae. *See* Moas
Diomedea spp. *See* Great albatrosses
Diomedea albatrus. See Short-tailed albatrosses
Diomedea amsterdamensis. See Amsterdam albatrosses
Diomedea antipodensis. See Antipodean albatrosses
Diomedea bulleri. See Buller's mollymawks
Diomedea cauta. See Shy albatrosses
Diomedea cauta cauta. See White-capped albatrosses
Diomedea cauta eremita. See Chatham mollymawks
Diomedea chlororhynchos. See Yellow-nosed mollymawks
Diomedea chrysostoma. See Gray-headed mollymawks
Diomedea dabbenena. See Tristan albatrosses
Diomedea epomophora. See Southern royal albatrosses
Diomedea epomophora sanfordi. See Northern royal albatrosses
Diomedea exulans. See Wandering albatrosses; Wandering royal albatrosses
Diomedea gibsoni. See Gibson's albatrosses
Diomedea immutabilis. See Laysan albatrosses
Diomedea impavida. See Campbell black-browed mollymawks

Diomedea irrorata. See Waved albatrosses
Diomedea melanophris. See Black-browed mollymawks
Diomedea nigripes. See Black-footed albatrosses
Diomedea platei. See Pacific mollymawks
Diomedea salvini. See Salvin's mollymawks
Diomedeidae. *See* Albatrosses
Diomedes (Greek hero), 8:116
Dippers, 10:173, **10:475–482,** 10:*478*
Directional hearing, owls, 9:337
The Directory of Australian Birds (Schodde and Mason), 10:329
Discosura spp. *See* Hummingbirds
Discosura longicauda, 9:440
Diseases
bird, 8:16
bird carriers, 8:25–26
Distribution
albatrosses, 8:*113,* 8:114, 8:*118–121*
Alcidae, 9:219, 9:221, 9:*224–228*
anhingas, 8:201, 8:202, 8:*207–209*
Anseriformes, 8:363, 8:364
ant thrushes, 10:239, 10:240, 10:*245–256*
Apodiformes, 9:417
asities, 10:187, 10:188, 10:*190*
Australian chats, 11:65, 11:*67–68*
Australian creepers, 11:*133,* 11:134, 11:*137–139*
Australian fairy-wrens, 11:45, 11:*48–53*
Australian honeyeaters, 11:235, 11:236, 11:*241–253*
Australian robins, 11:*105,* 11:*108–112*
Australian warblers, 11:55, 11:*59–63*
babblers, 10:505, 10:506, 10:*511–523*
barbets, 10:113, 10:114, 10:*119–122*
barn owls, 9:335–356, 9:*340–343*
bee-eaters, 10:39
birds of paradise, 11:489, 11:490, 11:*495–501*
Bombycillidae, 10:447–448, 10:*451–454*
boobies, 8:211, 8:212
bowerbirds, 11:477, 11:478, 11:*483–488*
broadbills, 10:177, 10:178, 10:*181–186*
bulbuls, 10:395, 10:396, 10:*402–412*
bustards, 9:*91–92,* 9:*96–99*
buttonquails, 9:11, 9:12, 9:*16–21*
caracaras, 8:*347,* 8:348
cassowaries, 8:75, 8:76
Charadriidae, 9:*161,* 9:162, 9:*166–172*
chats, 10:*483,* 10:485, 10:*492–499*
chickadees, 11:*155,* 11:156, 11:*160–165*
chowchillas, 11:*69,* 11:70
Ciconiiformes, 8:234
Columbidae, 9:247–248, 9:*255–266*
Columbiformes, 9:243
condors, 8:275, 8:276–277
Coraciiformes, 10:*1,* 10:3–4
cormorants, 8:201, 8:202, 8:205–206, 8:*207–209*
Corvidae, 11:*503,* 11:504–505, 11:*512–523*
cotingas, 10:305, 10:306, 10:*312–322*
crab plovers, 9:121
Cracidae, 8:*413,* 8:414
cranes, 9:23, 9:24, 9:*31–35*
cuckoo-shrikes, 10:385–386, 10:*389–393*
cuckoos, 9:311, 9:312, 9:*318–329*
dippers, 10:475–476, 10:*479–481*
diving-petrels, 8:*143,* 8:*145–146*
doves, 9:243, 9:247–248, 9:*255–266*
drongos, 11:437, 11:438, 11:*441–445*

Doves *(continued)*
  habitats, 9:248–249
  humans and, 9:252
  physical characteristics, 9:241–243, 9:247
  reproduction, 9:244, 9:250–251
  species of, 9:255–266
  taxonomy, 9:241, 9:247
Dowitchers, 9:177
Drepanididae. *See* Hawaiian honeycreepers
Drepanidinae, 11:342
Drepanis spp. *See* Sicklebills
Drepanis coccinea. See Iiwis
Drepanoplectes jacksoni. See Jackson's widow-birds
Drepanorhynchus spp. *See* Sunbirds
Drepanorhynchus reichenowi. See Golden-winged sunbirds
Drepanornis spp. *See* Birds of paradise
Drepanornis bruijnii. See Pale-billed sicklebills
Dreptes spp. *See* Sunbirds
Dreptes thomensis. See São Tomé sunbirds
Dromadidae. *See* Crab plovers
Dromaius ater. See King Island emus
Dromaius baudinianus. See Kangaroo Island emus
Dromaius novaehollandiae. See Emus
Dromas ardeola. See Crab plovers
Dromococcyz phasianellus. See Pheasant cuckoos
Drongo cuckoos. *See* Asian drongo-cuckoos
Drongos, **11:437–445**, 11:*440*
Drymocichla spp., 11:4
Drymodes brunneopygia. See Southern scrub robins
Drymophila squamata. See Scaled antbirds
Drymornis bridgesi. See Scimitar-billed woodcreepers
Dryocopus galeatus. See Helmeted woodpeckers
Dryocopus lineatus. See Lineated woodpeckers
Dryocopus martius. See Black woodpeckers
Dryocopus pileatus. See Pileated woodpeckers
Dryolimnas cuvieri. See White-throated rails
Dryoscopus spp. *See* Bush-shrikes
Dryoscopus gambensis. See Northern puffbacks
DuBois, Sieur, 9:269
Duck hawks. *See* Peregrine falcons
Ducks, 8:20, 8:21, **8:369–392**
  behavior, 8:371–372
  conservation status, 8:372–373
  distribution, 8:*369*, 8:370–371
  evolution, 8:369
  feeding ecology, 8:372
  habitats, 8:371
  humans and, 8:373
  physical characteristics, 8:369–370
  reproduction, 8:372
  species of, 8:*376*–392
  taxonomy, 8:369
Ducula spp. *See* Fruit pigeons
Ducula luctuosa. See White imperial pigeons
Ducula spilorrhoa. See Torresian imperial pigeons
Duets. *See* Songs, bird
Dulidae. *See* Palmchats
Dulit frogmouths, 9:378, 9:379
Dull-blue flycatchers, 11:*29*, 11:*35*–36
Dulus dominicus. See Palmchats
Dumbacher, Jack, 11:76, 11:118
Dumetella carolinensis. See Gray catbirds
Dunlins, 9:175

Dunnocks, 10:*460*, 10:*461*, 10:*462*
Dupont's larks, 10:342, 10:343, 10:*345*, 10:346
Dürer, Albrecht, 9:451
d'Urville, Dumont, 8:147
Dusky Asian orioles, 11:427–428
Dusky barn owls. *See* Sooty owls
Dusky-blue flycatchers. *See* Dull-blue flycatchers
Dusky broadbills, 10:177, 10:179, 10:*180*, 10:*185*
Dusky brown orioles, 11:*428*
Dusky buttonquails. *See* Barred buttonquails
Dusky fantails, 11:87
Dusky flycatchers, 10:271
Dusky friarbirds, 11:*428*
Dusky indigobirds, 11:*381*, 11:*393*
Dusky moorhens, 9:47, 9:50
Dusky redshanks. *See* Spotted redshanks
Dusky São Tomé sunbirds. *See* São Tomé sunbirds
Dusky sunbirds, 11:208
Dusky thrushes, 10:485
Dusky woodswallows, 11:*460*, 11:*461*, 11:*463*
Dvořák, Antonín, 8:19
Dwarf bitterns, 8:233
Dwarf buttonquails. *See* Black-rumped buttonquails
Dwarf cassowaries. *See* Bennett's cassowaries
Dwarf geese. *See* African pygmy geese
Dwarf honeyguides, 10:139
Dwarf olive ibises, 8:293
Dwarf tinamous, 8:59
Dyaphorophyia castanea. See Chestnut wattle-eyes
Dyck, Jan, 9:276
Dysithamnus mentalis. See Plain antvireos
Dysithamnus puncticeps. See Spot-crowned antvireos
Dysmorodrepanis, 11:343
Dysmorodrepanis munroi. See Lanai hookbills

E

Eagle-owls, 9:331–333, 9:345–347, 9:349, 9:351–*352*, 9:357–359
Eagles, **8:317–341**, 8:318
  behavior, 8:320–321
  conservation status, 8:323
  distribution, 8:*317*, 8:319–320
  evolution, 8:317–318
  feeding ecology, 8:321–322
  habitats, 8:320
  humans and, 8:323–324
  pesticides, 8:26
  physical characteristics, 8:318–319
  reproduction, 8:321, 8:322–323
  species of, 8:*329*–341
  symbolism, 8:20
  taxonomy, 8:317–318
Ear tufts, 9:331–332
Eared doves, 9:243
Eared grebes, 8:169, 8:172, 8:*173*, 8:178–*179*
Eared pheasants, 8:434
Eared quetzals, 9:479, 9:*480*, 9:*484*
Eared trogons. *See* Eared quetzals
Ears, avian, 8:6, 8:39–40
  See also Physical characteristics

Earthcreepers, 10:*212*, 10:*215*, 10:*216*–217
Eastern bearded greenbuls, 10:*400*, 10:*408*–409
Eastern black-headed orioles, 11:*430*, 11:*433*–434
Eastern bluebirds, 10:485, 10:*487*, 10:488, 10:*490*, 10:*494*
Eastern bristlebirds. *See* Bristlebirds
Eastern broad-billed rollers. *See* Dollarbirds
Eastern grass owls, 9:336, 9:337, 9:338, 9:*339*, 9:*342*–343
Eastern kingbirds, 10:272, 10:273, 10:275
Eastern meadowlarks, 11:*304*
Eastern phoebes, 10:273, 10:276, 10:284–285
Eastern red-footed falcons. *See* Amur falcons
Eastern rosellas, 9:282, 9:*285*
Eastern screech owls, 9:*331*, 9:*352*, 9:355
Eastern scrub-birds. *See* Rufous scrub-birds
Eastern shrike-tits, 11:*119*, 11:*120*
Eastern towhees, 11:*267*, 11:*279*
Eastern warbling vireos. *See* Warbling vireos
Eastern whipbirds, 10:331, 11:70, 11:75–76, 11:77, 11:79–*80*
Eastern white pelicans. *See* Great pelicans
Eastern whitefaces. *See* Southern whitefaces
Eastern wood-pewees, 10:275
Eastern yellow robins, 11:*106*, 11:*107*, 11:*109*
Eaton's pintails, 8:364
Echolocation, 8:6
  oilbirds, 9:374
  swiftlets, 9:422–423
Eclairs sur L'Au-Dela (Messiaen), 10:332
Eclectus parrots, 9:276, 9:282, 9:287–288
Eclectus roratus. See Eclectus parrots
Ecology, birds and, 8:15–17
  See also Feeding ecology
Ecotourism. *See* Humans
Ectopistes migratorius. See Passenger pigeons
Ecuadorian caciques, 11:304
Ecuadorian hillstars, 9:444
Edible-nest swiftlets, 9:424, 9:*425*, 9:*428*
Edolius ludwigii. See Square-tailed drongos
Edolius megarhynchus. See Ribbon-tailed drongos
Edward lyrebirds. *See* Superb lyrebirds
Edwards' pheasants, 8:433, 8:437, 8:*438*, 8:*450*–451
Egg laying. *See* Reproduction
Egg-manipulation programs, Chatham Islands black robins, 11:110–111
Egg mimicry, 9:314
Egg predation
  wrens, 10:527–528
Eggs, avian, 8:7
  elephant birds, 8:104
  incubation, 8:12–14
  kiwis, 8:90
  ostriches, 8:22, 8:26, 8:101
  See also Reproduction
Egrets, 8:26, 8:233–236, 8:241, 8:244, 8:245
Egretta spp. *See* Egrets
Egretta ardesiaca. See Black herons
Egretta eulophotes, 8:245
Egretta garzetta. See Little egrets
Egretta ibis. See Cattle egrets
Egretta thula. See Snowy egrets
Egretta tricolor. See Tricolor herons
Egretta vinaceigula, 8:245
Egyptian geese, 8:262
Egyptian plovers, 9:151, 9:152, 9:*154*, 9:*157*–*158*

Euptilotis neoxenus. See Eared quetzals
Eurasian avocets. See Pied avocets
Eurasian bitterns, 8:246, 8:256–257
Eurasian blackbirds. See Blackbirds
Eurasian bullfinches, 11:327, 11:335
Eurasian chiffchaffs. See Chiffchaffs
Eurasian collared doves, 8:24
Eurasian coots, 9:52
Eurasian crag martins. See Crag martins
Eurasian cranes, 9:24–27, 9:29, 9:30, 9:34
Eurasian crows. See Carrion crows
Eurasian dippers, 10:475, 10:476, 10:477,
    10:478, 10:479
Eurasian dotterels, 9:161, 9:163
Eurasian eagle-owls, 9:331, 9:345, 9:347,
    9:348, 9:352, 9:357
Eurasian golden orioles, 11:301–302, 11:429,
    11:430, 11:434
Eurasian green woodpeckers, 10:148
Eurasian griffons, 8:323
Eurasian hobbies, 8:315, 8:349
Eurasian jays, 11:504, 11:506–507, 11:510,
    11:513–514
Eurasian kingfishers. See Common kingfishers
Eurasian magpies, 11:506, 11:510, 11:516–517
Eurasian nutcrackers. See Spotted nutcrackers
Eurasian nuthatches. See Nuthatches
Eurasian oystercatchers, 9:126, 9:127
Eurasian penduline tits. See European
    penduline tits
Eurasian plovers. See Northern lapwings
Eurasian pratincoles, 9:152
Eurasian pygmy-owls, 9:347
Eurasian rooks, 11:505
Eurasian scops-owls, 9:352, 9:354
Eurasian siskins, 11:327, 11:333
Eurasian sparrowhawks, 8:319, 8:321, 8:323
Eurasian spoonbills. See Spoonbills
Eurasian tawny owls. See Tawny owls
Eurasian thick-knees, 9:144
Eurasian tree sparrows. See Tree sparrows
Eurasian treecreepers, 11:177, 11:178, 11:179,
    11:180
Eurasian turtledoves, 9:253
Eurasian wood-owls. See Tawny owls
Eurasian woodcocks, 9:177, 9:179–180, 9:181,
    9:182
Eurasian wrynecks. See Northern wrynecks
Eurocephalus spp. See Helmet-shrikes
European bee-eaters, 10:42, 10:43, 10:47–48
European common quail, 8:25
European goldfinches, 11:327, 11:332–333
European goshawks. See Northern goshawks
European greenfinches. See Greenfinches
European kingfishers. See Common kingfishers
European nightjars, 9:368, 9:370, 9:402–404,
    9:407, 9:411–412
European penduline tits, 11:147, 11:148,
    11:149, 11:150, 11:151
European pied flycatchers. See Pied flycatchers
European rock-thrushes. See Rock thrushes
European rollers, 10:52, 10:53, 10:54, 10:55
European serins, 11:323, 11:327, 11:334
European starlings, 8:19, 8:24, 11:407–409,
    11:410, 11:413, 11:421–422
European storm-petrels, 8:136, 8:137, 8:138
European swallows. See Barn swallows
European turtledoves, 9:253, 9:256–257
European Union Birds Directive. See EU
    Birds Directive

European white storks, 8:20, 8:236, 8:265,
    8:266, 8:267, 8:269, 8:272
European wrynecks. See Northern wrynecks
Eurostopodus spp., 9:401
Eurostopodus diabolicus. See Heinrich's
    nightjars
Eurostopodus guttatus. See Spotted nightjars
Eurostopodus mystacalis. See White-throated
    nightjars
Euryapteryx curtus, 8:96, 8:97
Euryapteryx geranoides, 8:96, 8:97
Euryceros prevostii. See Helmet vangas
Eurychelidon sirintarae. See White-eyed river
    martins
Eurylaimidae. See Broadbills
Eurylaimus blainvillii. See Clicking peltops
Eurylaimus ochromalus. See Black-and-yellow
    broadbills
Eurylaimus samarensis. See Visayan wattled
    broadbills
Eurylaimus steerii. See Mindanao broadbills
Eurynorhynchus pygmeus. See Spoon-billed
    sandpipers
Eurypyga helias. See Sunbitterns
Eurypyga helias meridionalis, 9:73–74
Eurypygidae. See Sunbitterns
Eurystomus azureus. See Azure rollers
Eurystomus gularis. See Blue-throated rollers
Eurystomus orientalis. See Dollarbirds
Eutoxeres spp., 9:443
Eutoxeres aquila, 9:446–447, 9:456
Eutriorchis astur. See Serpent-eagles
Evening grosbeaks, 8:24, 11:323, 11:325,
    11:327, 11:336
Everett's hornbills. See Sumba hornbills
Everett's monarchs. See White-tipped
    monarchs
Evolution
    adaptive radiation and, 11:342, 11:350
    albatrosses, 8:107, 8:113
    Alcidae, 9:219
    anhingas, 8:201, 8:207–209
    Anseriformes, 8:363
    ant thrushes, 10:239
    Apodiformes, 9:415
    asities, 10:187
    Australian chats, 11:65
    Australian creepers, 11:133
    Australian fairy-wrens, 11:45
    Australian honeyeaters, 11:235
    Australian robins, 11:105
    Australian warblers, 11:55
    babblers, 10:505
    barbets, 10:113
    barn owls, 9:335
    bee-eaters, 10:39
    bird songs and, 8:42
    birds of paradise, 11:489
    bitterns, 8:239
    Bombycillidae, 10:447
    boobies, 8:211
    bowerbirds, 11:477
    broadbills, 10:177
    bulbuls, 10:395
    bustards, 9:23, 9:91
    buttonquails, 9:11
    Caprimulgiformes, 9:367
    caracaras, 8:347
    cassowaries, 8:75
    Charadriidae, 9:161

Charadriiformes, 9:101, 9:107
chats, 10:483
chickadees, 11:155
chowchillas, 11:69
Ciconiiformes, 8:233
Columbidae, 9:247
Columbiformes, 9:241
condors, 8:275
convergent, 8:3
Coraciiformes, 10:3–4
cormorants, 8:183, 8:201
Corvidae, 11:503
cotingas, 10:305
crab plovers, 9:121
Cracidae, 8:413
cranes, 9:23
cuckoo-shrikes, 10:385
cuckoos, 9:311
dippers, 10:475
diving-petrels, 8:143
doves, 9:247
drongos, 11:437
ducks, 8:369
eagles, 8:317–318
elephant birds, 8:103
emus, 8:83
Eupetidae, 11:75
fairy bluebirds, 10:415
Falconiformes, 8:313
falcons, 8:347
false sunbirds, 10:187
fantails, 11:83
finches, 11:323
flamingos, 8:303
flowerpeckers, 11:189
frigatebirds, 8:183, 8:193
frogmouths, 9:377
Galliformes, 8:399
gannets, 8:211
geese, 8:369
Glareolidae, 9:151
grebes, 8:169
Gruiformes, 9:1, 9:107
hammerheads, 8:261
Hawaiian honeycreepers, 11:341
hawks, 8:317–318
hedge sparrows, 10:459
herons, 8:233, 8:239
hoatzins, 8:465
honeyguides, 10:137
hoopoes, 10:61
hornbills, 10:71
hummingbirds, 9:437
ibises, 8:291
Icteridae, 11:301
ioras, 10:415
jacamars, 10:91
jacanas, 9:107
kagus, 9:41
kingfishers, 10:5–6
kiwis, 8:89
Laridae, 9:203–204
larks, 10:341
leafbirds, 10:415
logrunners, 11:69
long-tailed titmice, 11:141
lyrebirds, 10:329
magpie-shrikes, 11:467
manakins, 10:295
mesites, 9:5

G

INDEX

Index

(Habitats (continued))

Habitats (continued)
trogons, 9:478, 9:481–485
tropicbirds, 8:184, 8:187, 8:190–191
trumpeters, 9:78, 9:79, 9:81–82
turacos, 9:300, 9:304–310
typical owls, 9:347–348, 9:354–365
tyrant flycatchers, 10:271, 10:278–288
vanga shrikes, 10:440, 10:443–444
Vireonidae, 11:255–256, 11:258–262
wagtails, 10:373, 10:378–383
weaverfinches, 11:354, 11:360–372
weavers, 11:376–377, 11:382–394
whistlers, 11:116, 11:120–125
white-eyes, 11:228, 11:232–233
woodcreepers, 10:229, 10:232–236
woodhoopoes, 10:66, 10:68–69
woodswallows, 11:459, 11:462–464
wrens, 10:527, 10:531–538
Hachisuka, Masauji, 9:269
Hadada ibises, 8:295, 8:298
Hadedahs. See Hadada ibises
Hadedas. See Hadada ibises
Haematopidae. See Oystercatchers
Haematops validirostris. See Strong-billed
honeyeaters
Haematopus ater. See Blackish oystercatchers
Haematopus bachmani. See American black
oystercatchers
Haematopus chathamensis. See Chatham Islands
oystercatchers
Haematopus fuliginosus. See Sooty
oystercatchers
Haematopus leucopodis. See Magellanic
oystercatchers
Haematopus longirostris. See Australian pied
oystercatchers
Haematopus meadewaldoi. See Canary Islands
oystercatchers
Haematopus moquini. See African black
oystercatchers
Haematopus ostralegus. See Eurasian
oystercatchers
Haematopus palliatus. See American
oystercatchers
Haematopus sulcatus, 9:125
Haematopus unicolor. See Variable oystercatchers
Haemotoderus spp. See Cotingas
Haffer, J., 10:91, 10:125
Hainan hill-partridges, 8:434
Hair-crested drongos, 11:438
Hairy hermits, 9:454, 9:456–457
Halcyon chelicuti. See Striped kingfishers
Halcyon chloris. See Collared kingfishers
Halcyon godeffroyi. See Marquesas kingfishers
Halcyon leucocephala. See Gray-headed
kingfishers
Halcyon sancta. See Sacred kingfishers
Halcyon tuta. See Chattering kingfishers
Halcyoninae, 10:5–7
Half-collared flycatchers. See Collared
flycatchers
Haliaeetus albicilla. See White-tailed eagles
Haliaeetus leucocephalus. See Bald eagles
Haliaeetus leucogaster. See White-bellied sea-
eagles
Haliaeetus leucoryphus. See Pallas's sea-eagles
Haliaeetus pelagicus. See Steller's sea-eagles
Haliaeetus sanfordi. See Sanford's sea-eagles
Haliaeetus vociferoides. See Madagascar fish-
eagles

Haliastur indus. See Brahminy kites
Haliastur sphenurus. See Whistling kites
Hall, Michelle, 11:456
Hall's babblers, 11:127, 11:128
Halocyptena microsoma. See Least storm-petrels
Hamerkops. See Hammerheads
Hamirostra melanosternon. See Australian
black-breasted buzzards; Black-breasted
buzzards
Hammerheads, 8:183, 8:225, 8:233, 8:234,
8:239, 8:261–263, 8:262, 8:272
Hammond's flycatchers, 10:271, 10:276,
10:285–286
Hanging parrots, 9:279
Hapaloptila castanea. See White-faced nunbirds
Haplophaedia spp. See Hummingbirds
Happy jacks. See Gray-crowned babblers
Harlequin ducks, 8:364, 8:375, 8:388, 8:389
Harlequins. See Harlequin ducks
Harpactes diardii. See Diard's trogons
Harpactes duvaucelii. See Scarlet-rumped trogons
Harpactes kasumba. See Red-naped trogons
Harpactes oreskios. See Orange-breasted
trogons
Harpactes orrhophaeus. See Cinnamon-rumped
trogons
Harpactes wardi. See Ward's trogons
Harpactes whiteheadi. See Whitehead's trogons
Harpagornis moorei. See Haast's eagles
Harpia harpyja. See Harpy eagles
Harpy eagles, 8:319, 8:326, 8:340–341
Harrier hawks, 8:318, 8:319, 8:320
Harris, T., 10:425
Harris' hawks, 8:322, 8:325, 8:338–339
Hartert, Ernst, 10:505
Hartlaub's turacos, 9:300, 9:302, 9:303, 9:307
Hatching. See Reproduction
Hawaii amakihis, 11:343
Hawaiian crows, 8:27, 11:508
Hawaiian ducks, 8:364, 8:367
Hawaiian geese, 8:366, 8:367, 8:370, 8:371
Hawaiian honeycreepers, 10:172,
11:341–352, 11:345
behavior, 11:342
conservation status, 11:343–344
distribution, 11:341–342
evolution, 11:341
feeding ecology, 11:342–343
habitats, 11:342
humans and, 11:344
physical characteristics, 11:341
reproduction, 11:343
species of, 11:346–351
taxonomy, 11:341
Hawfinches, 11:324, 11:327, 11:334–335
Hawk-eagles, 8:321
Hawk owls. See Northern hawk-owls
Hawkmoths, 9:445
Hawks, 8:317–341
behavior, 8:320–321
buzzard-like, 8:318
conservation status, 8:323
distribution, 8:317, 8:319–320
evolution, 8:317–318
feeding ecology, 8:321–322
habitats, 8:320
humans and, 8:323–324
pesticides, 8:26
physical characteristics, 8:318–319
reproduction, 8:321, 8:322–323

species of, 8:329–341
taxonomy, 8:317–318
Hazel grouse, 8:434
Hearing, avian, 8:6, 8:39–40, 9:337
See also Physical characteristics
Heathwrens, 11:56
Hedge accentors. See Dunnocks
Hedge sparrows, 10:459–464, 10:461
Hedydipna spp. See Sunbirds
Hedydipna collaris. See Collared sunbirds
Hedydipna metallica. See Nile Valley sunbirds
Hedydipna pallidigaster. See Amani sunbirds
Hedydipna platura. See Pygmy sunbirds
Hegner, Robert, 10:40
Heinrich's nightjars, 9:370, 9:405
Heinroth, Katharina, 9:255
Heinroth, Oskar, 9:255
Helena's parotias. See Lawes's parotias
Heliangelus spp. See Hummingbirds
Heliangelus regalis. See Royal sunangels
Heliangelus zusii. See Bogotá sunangels
Heliconia psittacorum. See Costa Rican
heliconias
Heliodoxa spp. See Hummingbirds
Heliolais erythroptera. See Red-winged warblers
Heliomaster spp. See Hummingbirds
Heliomaster longirostris. See Long-billed
starthroats
Heliopais personata. See Masked finfoot
Heliornis senegalensis. See African finfoot
Heliornithidae, 9:1–4, 9:45
Heliothryx spp., 9:443, 9:446
Heliothryx barroti. See Purple-crowned fairies
Helmet-shrikes, 10:425–430, 10:432
Helmet vangas, 10:439, 10:440, 10:441
Helmetcrests, 9:442
Helmeted curassows. See Northern helmeted
curassows
Helmeted guineafowl, 8:426, 8:426–428,
8:429, 8:430
Helmeted honeyeaters, 11:244
Helmeted hornbills, 10:72, 10:75–76, 10:77,
10:80–81
Helmeted mynas, 11:410, 11:412, 11:415
Helmeted woodpeckers, 10:150
Helmitheros vermivorus. See Worm-eating
warblers
Heliornis fulica. See Sungrebes
Hemerodromus cinctus. See Three-banded
coursers
Hemignathus spp. See Amakihis
Hemignathus lucidus. See Maui nukupuus
Hemignathus virens. See Hawaii amakihis
Hemiphaga spp., 9:248
Hemipodius melanogaster. See Black-breasted
buttonquails
Hemipodius pyrrhothorax. See Red-chested
buttonquails
Hemiprocne comata. See Whiskered tree swifts
Hemiprocne coronata. See Crested tree swifts
Hemiprocne longipennis. See Gray-rumped tree
swifts
Hemiprocne mystacea. See Moustached tree
swifts
Hemiprocnidae. See Tree swifts
Hemipus spp. See Flycatcher-shrikes
Hemipus hirundinaceus. See Black-winged
flycatcher-shrikes
Hemitriccus furcatus. See Fork-tailed pygmy
tyrants

Hemitriccus kaempferi. See Kaempfer's tody
tyrants
Hemprich's hornbills, 10:73
Hen harriers, 8:318, 8:326, 8:336–337
Henicopernis longicauda. See Long-tailed
buzzards
Henicorhina spp., 10:527
Henicorhina leucophrys. See Gray-breasted
wood wrens
Henicorhina levcoptera. See Bar-winged wood
wrens
Henst's goshawks, 8:323
Herbert, Thomas, 9:270
Hermit hummingbirds, 9:446
Hermit ibises, 8:293, 8:295, 8:297–298
Hermit thrushes, 10:484, 10:485, 10:491,
10:501
Hermit warblers, 11:288
Hermits, 9:415, 9:438–439, 9:441–442,
9:446–449, 9:454–455, 11:286
Herons, 8:233–236, 8:234, **8:239–260**
behavior, 8:234–235, 8:242–243
conservation status, 8:236, 8:244–245
distribution, 8:239, 8:241
evolution, 8:239
feeding ecology, 8:235, 8:242, 8:243–244
flight, 8:234
habitats, 8:241–242
humans and, 8:245
pesticides, 8:26
physical characteristics, 8:233–234,
8:239–241
reproduction, 8:234, 8:236, 8:244
species of, 8:248–259
taxonomy, 8:239
Herpetotheres cachinnans. See Laughing falcons
Herpsilochmus atricapillus. See Black-capped
antwrens
Herpsilochmus parkeri. See Ash-throated
antwrens
Herring gulls, 9:206, 9:209, 9:210, 9:211–212
Heteralocha acutirostris. See Huias
Heteromirafra spp. See Larks
Heteromirafra archeri. See Archer's larks
Heteromirafra ruddi. See Rudd's larks
Heteromirafra sidamoensis. See Sidamo bushlarks
Heteromyias albispecularis. See Gray-headed
robins
Heteronetta atricapilla. See Black-headed ducks
Heuglin's coursers. See Three-banded
coursers
Heuglin's robins. See White-browed robin
chats
Hibernation. See Energy conservation
Hieraaetus pennatus. See Booted eagles
Higher vocal center, 8:39
Highland guans, 8:413
Highland motmots, 10:31, 10:32
Highland tinamous, 8:58, 8:59, 8:59, 8:60,
8:62
Highland wood wrens. See Gray-breasted
wood wrens
Hihis. See Stitchbirds
Hill mynas, 11:412, 11:417–418
Hill-partridges, 8:399, 8:434
Hill swallows. See House swallows
Hillstars, 9:438, 9:442, 9:444, 9:447, 9:450
Hilty, Steven L., 10:130
Himalayan brown owls. See Asian brown
wood-owls

Himalayan griffon vultures, 8:313, 8:316,
8:319
Himalayan monals, 8:435
Himalayan quails, 8:437
Himalayan wood-owls. See Asian brown
wood-owls
Himantopus spp. See Stilts
Himantopus himantopus. See Black-winged stilts
Himantopus mexicanus. See Black-necked stilts
Himantopus novaezelandiae. See Black stilts
Himantornis haematopus. See Nkulengu rails
Himantornithinae, 9:45
Himatione sanguinea. See Apapanes
Hippolais spp., 11:3, 11:4
Hippolais icterina, 11:5
Hippolais languida, 11:4
Hippolais polyglotta. See Melodious warblers
Hippopotamus, jacanas and, 9:109
Hirundapus spp., 9:418, 9:424
Hirundapus caudacutus. See White-throated
needletails
Hirundapus celebensis. See Purple needletails
Hirundinidae. See Martins; Swallows
Hirundo apus. See Common swifts
Hirundo atrocaerulea. See Blue swallows
Hirundo caudacuta. See White-throated
needletails
Hirundo coronata. See Crested tree swifts
Hirundo fuciphaga. See Edible-nest swiftlets
Hirundo megaensis. See White-tailed swallows
Hirundo melba. See Alpine swifts
Hirundo nigra. See Black swifts
Hirundo pelagica. See Chimney swifts
Hirundo perdita. See Red Sea cliff-swallows
Hirundo pratinocola. See Collared pratincoles
Hirundo pyrrhonota. See American cliff
swallows
Hirundo rustica. See Barn swallows
Hirundo tahitica. See House swallows
Hirundo zonaris. See White-collared swifts
Hispaniolan crossbills, 11:324
Hispaniolan palm crows, 11:508
Hispaniolan trogons, 9:479, 9:480, 9:483–484
Histoplasma capsulatum, 8:25
Histrionicus histrionicus. See Harlequin ducks
Hitchcock, Alfred, 8:22
Hoary-headed dabchicks. See Hoary-headed
grebes
Hoary-headed grebes, 8:170, 8:171, 8:173,
8:176–177
Hoary redpolls, 11:331
Hoary-throated spinetails, 10:211
Hoatzins, **8:465–468**, 8:466, 8:467, 10:326
Hodgson's broadbills. See Silver-breasted
broadbills
Hodgson's frogmouths, 9:380, 9:384
Holarctic rails, 9:47
Holarctic wrens. See Winter wrens
Holdaway, Richard, 11:83–84
Homing, 8:32–35
See also Behavior
Homing pigeons, 8:22, 9:439
Honey badgers, 10:139
Honey buzzards, 8:314
Honeycreepers, Hawaiian. See Hawaiian
honeycreepers
Honeyeaters, Australian. See Australian
honeyeaters
Honeyguides, 10:85–88, **10:137–145**, 10:140,
11:219

Hood mockingbirds, 10:467, 10:469, 10:472
Hooded cranes, 9:3, 9:24–27
Hooded crows, 11:506
Hooded grebes, 8:171, 8:173, 8:179–180
Hooded mergansers, 8:364
Hooded orioles, 11:306
Hooded pittas, 10:194, 10:195, 10:196,
10:197–198
Hooded plovers, 9:164
Hooded robins, 11:106
Hooded vultures, 8:320
Hooded warblers, 11:292, 11:297
Hook-billed bulbuls, 10:399
Hook-billed hermits, 9:443, 9:449
Hook-billed kingfishers, 10:6, 10:12, 10:14
Hook-billed kites, 8:322, 8:327, 8:330
Hook-billed vangas, 10:439, 10:440, 10:441,
10:444
Hoopoe larks. See Greater hoopoe-larks
Hoopoes, 8:26, 10:1–2, **10:61–63**, 10:62
Hormones, bird songs and, 8:39
Hornbill Specialist Group (IUCN), 10:75
Hornbills, 10:1–4, **10:71–84**, 10:77
behavior, 10:73
conservation status, 10:75
evolution, 10:71
feeding ecology, 10:73–74
habitats, 10:73
humans and, 10:75–76
physical characteristics, 10:71–72
reproduction, 10:74–75
species of, 10:78–84
taxonomy, 10:71
The Hornbills (Kemp), 10:71
Hornby's storm-petrels, 8:138
Horned coots, 9:50
Horned frogmouths. See Sunda frogmouths
Horned grebes, 8:170, 8:171
Horned guans, 8:401, 8:413, 8:417, 8:420–421
Horned larks, 10:342, 10:343, 10:345, 10:347,
10:354, 11:272, 11:273
Horned puffins, 9:102
Horned screamers, 8:393, 8:394, 8:395, 8:396
Horneros. See Ovenbirds
Horsfield's bronze-cuckoos, 9:317, 9:321,
11:66
Horus (Egyptian god), 8:22
Horus swifts, 9:424
Horwich, Robert, 9:28
Hose's broadbills, 10:178, 10:179, 10:180,
10:181–182
Host species. See Brood parasitism
Hottentott buttonquails. See Black-rumped
buttonquails
Houbara bustards, 9:92, 9:93, 9:95, 9:97
Houbaropsis spp. See Bustards
Houbaropsis bengalensis. See Bengal floricans
Houde, P., 9:41
House crows, 11:505, 11:509, 11:511,
11:519–520
House finches, 8:31, 11:325
House martins, 10:362, 10:365–366
House mynas. See Common mynas
House sparrows, 8:24, 9:440, 10:361, 11:397,
11:398, 11:400, 11:401
House swallows, 10:362, 10:365
House wrens, 10:174, 10:525–529, 10:527,
10:530, 10:536–537
Huebner, S., 10:71
Huet huets. See Black-throated huet-huets

INDEX

INDEX

Leucophaeus scoresbii. See Dolphin gulls
Leucopsar spp., 11:409
Leucopsar rothschildi. See Bali mynas
Leucopternis lacernulata. See White-necked hawks
Leucosarcia melanoleuca. See Wonga pigeons
Leucosticte arctoa. See Asian rosy-finches
Leucosticte sillemi. See Sillem's mountain-finches
Leucosticte tephrocotis. See Gray-crowned rosy finches
Lichenostomus spp. *See* Australian honeyeaters
Lichenostomus melanops. See Yellow-tufted honeyeaters
Lichmera spp. *See* Australian honeyeaters
Lichmera indistincta. See Brown honeyeaters
Lichtenstein's sandgrouse, 9:231, *9:234,* 9:237–238
The Life of Birds (Attenborough), 11:507
The Life of Lyrebird (Smith), 10:330
Light-mantled albatrosses, 8:114, 8:*117,* 8:*119*–120
Light-mantled sooties. *See* Light-mantled albatrosses
Lilac-cheeked kingfishers, 10:6, 10:*12,* 10:*16*
Lilac-crowned wrens. *See* Purple-crowned fairy-wrens
Lilac kingfishers. *See* Lilac-cheeked kingfishers
Lily trotters. *See* Jacanas
Limnodromus semipalmatus. See Asian dowitchers
Limnofregata azygosternon, 8:186
Limosa spp. *See* Godwits
Limosa lapponica. See Bar-tailed godwits
Limosa limosa. See Black-tailed godwits
Limpkins, 9:1–4, 9:23, **9:37–39,** 9:*38,* 9:*39,* 9:69
Lina's sunbirds, 11:210
Lineated woodpeckers, 10:*148*
Linnaeus
  albatrosses and, 8:116
  on dippers, 10:475
  on flickers, 10:155
  on herons, 8:239
  on mimids, 10:465
  on owls, 9:345
  on palmchats, 10:455
  on Piciformes, 10:85
  on todies, 10:25
Linnets, 11:323
Liocichla spp. *See* Liocichlas
Liocichla omeiensis. See Omei Shan liocichlas
Liocichlas, 10:505
Liosceles spp. *See* Tapaculos
Liosceles thoracicus. See Rusty-belted tapaculos
Lipaugus spp. *See* Cotingas
Lipaugus lanioides, 10:308
Lipaugus uropygialis, 10:308
Lissotis spp. *See* Bustards
Lissotis melanogaster. See Black-bellied bustards
Literature
  albatrosses, 8:116
  birds and, 8:20, 8:26
  dodos, 9:270
  hedge sparrows, 10:462
  larks, 10:346
  See also Mythology
Little auks. *See* Dovekies
Little bitterns, 8:*243*

Little blue penguins. *See* Little penguins
Little brown bustards, 9:93
Little brown cranes. *See* Sandhill cranes
Little bustards, 9:92, 9:93, 9:*95,* 9:*99*
Little cassowaries. *See* Bennett's cassowaries
Little corellas, 9:276
Little egrets, 8:*247,* 8:253–*254*
Little grassbirds, 11:*9,* 11:*21*–22
Little gray kiwis. *See* Little spotted kiwis
Little grebes, 8:169, 8:170, 8:171, 8:172, 8:*173,* 8:174–*175*
Little green bee-eaters, 10:*40*
Little ground-jays. *See* Hume's ground-jays
Little hermits, 9:*441*
Little king birds of paradise. *See* King birds of paradise
Little marshbirds. *See* Little grassbirds
Little owls, 9:*333,* 9:*348*
Little Papuan frogmouths. *See* Marbled frogmouths
Little penguins, 8:148, 8:150, 8:*152,* 8:*157*
Little reedbirds. *See* Little grassbirds
Little-ringed plovers, 9:163
Little rush-warblers, 11:4, 11:*9,* 11:*15*–16
Little shearwaters, 8:108
Little shrike-thrushes, 11:*119,* 11:*123*
Little slaty flycatchers, 11:*29,* 11:*33*
Little sparrowhawks. *See* African little sparrowhawks
Little spotted kiwis, 8:90, 8:*91,* 8:92–*93*
Little stints, 9:177
Little treecreepers. *See* White-throated treecreepers
Little turtledoves. *See* Diamond doves
Little woodstars, 9:450
Little woodswallows, 11:*461,* 11:*462*–463
Live-bird trade. *See* Cagebirds
Livezey, Bradley, 8:369, 9:45
Livingstone's turacos, 9:300
Loango weavers, 11:379
Lobed ducks. *See* Musk ducks
Lobopardisea sericea. See Yellow-breasted birds of paradise
Locust birds. *See* Collared pratincoles
Locust starlings. *See* Common mynas
Locustella spp., 11:3, 11:4, 11:7
Locustella naevia. See Grasshopper warblers
Loggerhead shrikes, 10:425, 10:427, 10:428, 10:*429,* 10:*431,* 10:*437*–438
Logrunners, **11:69–74,** 11:*72–73,* 11:*75*
Lonchura spp., 11:353
Lonchura cantans. See African silverbills
Lonchura maja. See White-headed munias
Lonchura punctulata. See Spotted munias
Lonchura striata. See White-backed munias
Long-billed prions. *See* Broad-billed prions
Long-billed crombecs, 11:*2,* 11:4, 11:*8,* 11:*19*
Long-billed curlews, 9:*181,* 9:*183*
Long-billed gnatwrens, 11:*8,* 11:*10*
Long-billed larks. *See* Cape long-billed larks
Long-billed marsh wrens. *See* Marsh wrens
Long-billed murrelets. *See* Marbled murrelets
Long-billed plovers, 9:162
Long-billed rhabdornis. *See* Greater rhabdornis
Long-billed starthroats, 9:445
Long-billed sunbirds, 11:207
Long-billed vultures, 8:323
Long-billed woodcreepers, 10:*231,* 10:*233*
Long-crested eagles, 8:321

Long-eared owls, 9:333, 9:348
Long-legged pratincoles. *See* Australian pratincoles
Long-tailed broadbills, 10:179, 10:*180,* 10:*182–183*
Long-tailed buzzards, 8:*327,* 8:331, 8:*331*
Long-tailed cuckoos doves. *See* Barred cuckoo-doves
Long-tailed ducks. *See* Oldsquaws
Long-tailed emerald sunbirds. *See* Malachite sunbirds
Long-tailed fantails, 11:87
Long-tailed fiscals, 10:428
Long-tailed frogmouths. *See* Sunda frogmouths
Long-tailed ground-rollers, 10:52, 10:*54,* 10:*57*
Long-tailed hermits, 9:441
Long-tailed honey-buzzards. *See* Long-tailed buzzards
Long-tailed koels, 9:*316,* 9:*323*
Long-tailed manakins, 10:*297,* 10:*298,* 10:*299*
Long-tailed meadowlarks, 11:302, 11:*307,* 11:*315–316*
Long-tailed minivets, 10:386
Long-tailed munias. *See* Pin-tailed parrotfinches
Long-tailed paradigallas, 11:492
Long-tailed pipits, 10:375
Long-tailed potoos, 9:397
Long-tailed shrikes, 10:426, 10:427, 10:*431,* 10:*436*–437
Long-tailed silky flycatchers, 10:447
Long-tailed skuas, 9:204
Long-tailed tailorbirds. *See* Common tailorbirds
Long-tailed titmice, 10:*169,* 11:*141,* **11:141–146,** 11:*142,* 11:*144,* 11:*145*
Long-tailed widows, 11:379
Long-tailed wood-partridges, 8:456
Long-tailed woodcreepers, 10:*231,* 10:232–*233*
Long-trained nightjars, 9:402
Long-wattled umbrellabirds, 10:308, 10:*311,* 10:*317*–318
Long-whiskered owlets, 9:332
Longclaws, 10:371, 10:372–373, 10:373, 10:374, 10:375, 10:*377,* 10:*380*
Longspurs, 11:272
Loons, **8:159–167,** 8:*162*
Lophaetus occipitalis. See Long-crested eagles
Lopholaimus spp. *See* Fruit pigeons
Lophophanes spp. *See* Crested tits
Lophophorus spp. *See* Monals
Lophophorus impejanus. See Himalayan monals
Lophophorus lhuysii. See Chinese monals
Lophorina spp. *See* Birds of paradise
Lophornis spp. *See* Hummingbirds
Lophornis brachylophus. See Short-crested coquettes
Lophornis magnificus. See Frilled coquettes
Lophotibis cristata. See Madagascar crested ibises
Lophotis spp. *See* Bustards
Lophozosterops spp., 11:228
Lophozosterops dohertyi, 11:228
Lophura bulweri. See Bulwer's pheasants
Lophura edwardsi. See Edwards' pheasants
Lophura hatinhensis, Vietnamese pheasants
Lophura ignita. See Crested firebacks
Lophura imperialis. See Imperial pheasants
Lophura nycthemera berliozi. See Berlioz's silver pheasants

INDEX

Nicator vireo. See Yellow-throated nicators
Niceforo's wrens, 10:529
Nicobar bulbuls, 10:396, 10:398
Nicobar megapodes, 8:404
Nicobar scrubfowl, 8:401
Nicolai, J., 11:375
Nidicolous species, 8:15
Niethammer, Günther, 9:255
Night-adapted vision, owls, 9:332
Night herons, 8:233, 8:234, 8:242
    See also specific types of night herons
Night parrots, 9:277
Nighthawks, 9:367–370, 9:401–402,
    9:404–405, 9:409
Nightingale wrens, 10:527
Nightingales, 8:19, 10:484, 10:485–486,
    10:488, 10:489, 10:490, 10:492
Nightjars, 9:367–371, **9:401–414,** 9:407
    behavior, 9:403
    conservation status, 9:405
    distribution, 9:401, 9:402
    evolution, 9:401
    feeding ecology, 9:403–404
    habitats, 9:402–403
    humans and, 9:405–406
    physical characteristics, 9:401–402
    reproduction, 9:404–405
    species of, 9:408–414
    taxonomy, 9:335, 9:401
Nigrita fusconota. See White-breasted negro-
    finches
Nihoa finches, 11:342, 11:343
Nilaus spp. *See* Bush-shrikes
Nilaus afer. See Brubrus
Nile Valley sunbirds, 11:208
Nilgiri laughing thrushes, 10:507
Nilgiri pipits, 10:375
Niltava grandis. See Large niltavas
Nimba flycatchers, 11:27
Ninox spp., 9:332, 9:346, 9:347, 9:349
Ninox boobook. See Southern boobook owls
Ninox novaeseelandiae. See Owl moreporks
Ninox scutulata. See Brown hawk-owls
Ninox superciliaris. See White-browed hawk-
    owls
Ninoxini, 9:347
Nipponia nippon. See Japanese ibises
Nitlins. *See* Least bitterns
Nkulengu rails, 9:45
Noble snipes, 9:177
Nocturnal behavior. *See* Behavior
Nocturnal curassows, 8:413
Noddies, 9:203
Noisy friarbirds, 11:240, 11:242
Noisy miners, 11:238
Noisy scrub-birds, 10:337–338, 10:339–340
Nomadism, avian, 8:31
Non-native species introduction. *See*
    Introduced species
Nonnula ruficapilla. See Rufous-capped nunlets
Nordmann's greenshanks, 9:177, 9:179, 9:180
Norfolk Island parakeets, 9:280
Norfolk Island starlings, 11:410
Noronha vireos. *See* Red-eyed vireos
North Atlantic gannets. *See* Northern gannets
North Island rifleman. *See* Riflemen
Northern anteater chats. *See* Anteater chats
Northern bald ibises. *See* Hermit ibises
Northern beardless flycatchers. *See* Northern
    beardless-tyrannulets

Northern beardless-tyrannulets, 10:269,
    10:270, 10:276, 10:281–282
Northern bobwhite quails, 8:455, 8:456,
    8:457, 8:458, 8:459, 8:461–462
Northern bobwhites. *See* Northern bobwhite
    quails
Northern boobooks. *See* Southern boobook
    owls
Northern brownbuls, 10:397
Northern cassowaries. *See* One-wattled
    cassowaries
Northern eagle-owls. *See* Eurasian eagle-owls
Northern fairy flycatchers. *See* African blue-
    flycatchers
Northern fantails, 11:83, 11:88, 11:93
Northern flickers, 10:88, 10:149, 10:150,
    10:152, 10:155–156
Northern fulmars, 8:108, 8:124, 8:125, 8:127,
    8:129–130
Northern gannets, 8:214, 8:216, 8:217–218
Northern giant petrels, 8:109
Northern goshawks, 8:323, 8:326, 8:338
Northern ground-hornbills, 10:72
Northern harriers. *See* Hen harriers
Northern hawk-owls, 9:335, 9:346,
    9:349–350, 9:353, 9:361–362
Northern helmeted curassows, 8:413, 8:417,
    8:421–422
Northern horned screamers, 8:393
Northern house-martins. *See* House martins
Northern house wrens. *See* House wrens
Northern jacanas, 9:108, 9:111, 9:112
Northern lapwings, 9:165, 9:172–173
Northern logrunners. *See* Chowchillas
Northern lyrebirds. *See* Albert's lyrebirds
Northern mockingbirds, 10:465, 10:466,
    10:468, 10:471
Northern needletails. *See* White-throated
    needletails
Northern orioles. *See* Baltimore orioles
Northern oystercatchers. *See* Variable
    oystercatchers
Northern Pacific albatrosses, 8:113, 8:114
Northern parulas, 11:287
Northern pintails, 8:364
Northern potoos, 9:395
Northern puffbacks, 10:431, 10:432–433
Northern pygmy owls, 9:349, 9:350
Northern quetzals. *See* Resplendent quetzals
Northern ravens, 11:505–507, 11:511,
    11:522–523
Northern royal albatrosses, 8:113, 8:114,
    8:115, 8:117, 8:118
Northern saw-whet owls, 9:353, 9:363–364
Northern screamers, 8:395, 8:396–397
Northern shovelers, 8:375, 8:385–386
Northern shrikes, 10:425, 10:427, 10:428,
    10:429
Northern starlings. *See* European starlings
Northern three-toed woodpeckers. *See*
    Three-toed woodpeckers
Northern treecreepers. *See* Eurasian
    treecreepers
Northern waterthrush, 11:287, 11:288
Northern wheatears. *See* Wheatears
Northern wrens. *See* Winter wrens
Northern wrynecks, 10:149, 10:153, 10:154
Northwestern crows, 11:506
Notharchus macrorhynchos. See White-necked
    puffbirds

Nothocercus bonapartei. See Highland tinamous
Nothocrax urumutum. See Nocturnal curassows
Nothoprocta cinerascens. See Brushland
    tinamous
Nothoprocta kalinowskii. See Kalinowski's
    tinamous
Nothoprocta ornata. See Ornate tinamous
Nothoprocta taczanowskii. See Taczanowski's
    tinamous
Nothura maculosa. See Spotted nothuras
Notiomystis spp. *See* Stitchbirds
Notornis mantelli. See Takahe rails
Nowicki, S., 8:39
Nubian bustards, 9:93
Nubian vultures. *See* Lappet-faced vultures
Nucifraga spp. *See* Corvidae
Nucifraga caryocatactes. See Spotted nutcrackers
Nucifraga columbiana. See Clark's nutcrackers
Nukupu'us, 11:343
Numeniini. *See* Curlews; Godwits
Numenius spp. *See* Curlews
Numenius americanus. See Long-billed curlews
Numenius borealis. See Eskimo curlews
Numenius phaeopus. See Whimbrels
Numenius tahitiensis. See Bristle-thighed
    curlews
Numenius tenuirostris. See Slender-billed
    curlews
Numida meleagris. See Helmeted guineafowl
Numididae. *See* Guineafowl
Nunbirds, 10:101–103
Nutcrackers. *See* Spotted nutcrackers
Nuthatch vangas, 10:439, 10:440
Nuthatches, **11:167–175,** 11:170
    behavior, 11:168, 11:171–175
    conservation status, 11:169, 11:171–175
    distribution, 11:167, 11:168, 11:171–174
    evolution, 11:167
    feeding ecology, 11:168, 11:171–175
    habitats, 11:168, 11:171–174
    humans and, 11:169, 11:171–175
    physical characteristics, 11:167–168,
        11:171–174
    reproduction, 11:168–169, 11:171–175
    species of, 11:171–175
    taxonomy, 11:167, 11:171–174
Nutmeg finches. *See* Spotted munias
Nutmeg mannikins. *See* Spotted munias
Nutmeg pigeons. *See* White imperial pigeons
Nutting's flycatchers, 10:276, 10:283–284
Nyctanassa violacea. See Yellow-crowned night
    herons
Nyctea scandiaca. See Snowy owls
Nyctibiidae. *See* Potoos
Nyctibius aethereus. See Long-tailed potoos
Nyctibius bracteatus. See Rufous potoos
Nyctibius grandis. See Great potoos
Nyctibius griseus. See Gray potoos
Nyctibius jamaicensis. See Northern potoos
Nyctibius leucopterus. See White-winged potoos
Nycticorax magnificus. See White-eared night
    herons
Nycticorax nycticorax. See Black-crowned night
    herons
Nycticryphes semicollaris. See South American
    painted-snipes
Nyctiperdix spp., 9:231
Nyctiphrynus rosenbergi. See Chocó poorwills
Nyctiphrynus vielliardi. See Plain-tailed
    nighthawks

INDEX

INDEX

INDEX

INDEX

INDEX

INDEX